national electrical code®

electrical systems

based on the 1996 NEC®

AMERICAN TECHNICAL PUBLISHERS, INC.
HOMEWOOD, ILLINOIS 60430

Michael I. Callanan
Bill Wusinich

ACKNOWLEDGMENTS

The authors and publisher are grateful to the following companies and organizations for providing technical information and assistance.

ABB Power T&D Company, Inc.
Crouse-Hinds Division,
Cooper Industries, Inc.
Federal Signal Corporation
General Electric Co.
Square D Company

1 2 3 4 5 6 7 8 9 – 96 – 9 8 7 6 5 4 3 2

Printed in the United States of America

Callanan, Michael I.
Electrical systems based on the 1996 NEC® / Micahel I. Callanan, Bill Wusinich.
p. cm.
"Based on the 1996 NEC®."
Includes index.
ISBN 0-8269-1690-2 (paper)
1. Electric engineering--United States--Insurance requirements. 2. Electric engineering--United States--Examinations, questions, etc. 3. National Fire Protection Association. National Electrical Code® (1996) I. Wusinich, Bill. II. Title.
TK260.C34 1997
621.319'24'0218--dc20 96-42422
CIP

CONTENTS

1	THE NATIONAL ELECTRICAL CODE®	Text	1
		Review Questions 1	19
		Trade Test 1	21
2	WIRING METHODS	Text	23
		Review Questions 2	39
		Trade Test 2	43
3	WIRING MATERIALS – RACEWAYS AND BOXES	Text	45
		Review Questions 3	79
		Trade Test 3	81
4	WIRING MATERIALS – SWITCHES, SWITCHBOARDS, AND PANELBOARDS	Text	83
		Review Questions 4	97
		Trade Test 4	99
5	CONDUCTORS AND OVERCURRENT PROTECTION	Text	101
		Review Questions 5	131
		Trade Test 5	135
6	BRANCH CIRCUITS AND FEEDERS	Text	137
		Review Questions 6	161
		Trade Test 6	163
7	GROUNDING	Text	165
		Review Questions 7	199
		Trade Test 7	203
8	TRANSFORMERS	Text	205
		Review Questions 8	231
		Trade Test 8	235

9	**SERVICES**	Text	237
		Review Questions 9	263
		Trade Test 9	267
10	**MOTORS, GENERATORS, A/C & REFRIGERATION, AND FIRE PUMPS**	Text	269
		Review Questions 10	293
		Trade Test 10	295
11	**EQUIPMENT FOR GENERAL USE**	Text	299
		Review Questions 11	327
		Trade Test 11	329
12	**SPECIAL LOCATIONS**	Text	331
		Review Questions 12	361
		Trade Test 12	363
13	**CALCULATIONS**	Text	365
		Review Questions 13	387
		Trade Test 13	389
14	**FINAL EXAM**		391
	APPENDIX		397
	GLOSSARY		415
	INDEX		421

INTRODUCTION

PREFACE

Electrical Systems is designed for use by journeyman and master electricians, inspectors, contractors, electrical designers, and others who use the National Electrical Code®. The text makes extensive use of CAD-drawn art to show and tell how to apply the NEC®. A comprehensive Appendix provides useful tables and the Glossary defines terms used in the text.

The *Instructor's Guide* for this book contains answers for all Review questions. Answers and NEC® substantiation are given for all Trade Tests and the Final Exam. Solutions are given as appropriate. Copies of the *Instructor's Guide* may be purchased from American Technical Publishers, Inc.

Blank copies of calculation forms, designed specifically for this book, may be reproduced for instructional use only. These forms, located at the back of this book, shall not be reproduced and sold to others. Calculation forms that may be reproduced for instructional use only are clearly marked:

NATIONAL ELECTRICAL CODE®

The National Electrical Code® is sponsored and controlled by the National Fire Protection Association (NFPA). The primary function of the NEC® is to safeguard people and property against electrical hazards. It is mandatory that *Electrical Systems* be used only in conjunction with the current NEC®. Copies of the current NEC® (NFPA No. 70) may be ordered directly from its publisher

National Fire Protection Association, Inc.
Batterymarch Park
Quincy, MA 02269

WATTS AND VOLT-AMPERES

In general, within the NEC®, the term *watts* (W) has been superseded by the term *volt-amperes* (VA) for the computation of loads. However, references to nameplate ratings still reflect the term watts on certain loads.

CALCULATIONS

When total wattage or VA is to be divided by phase-to-phase (3ϕ) voltage times 1.732, the following values may be substituted:

for 208 V × 1.732, use 360
for 230 V × 1.732, use 398
for 240 V × 1.732, use 416
for 440 V × 1.732, use 762
for 460 V × 1.732, use 797
for 480 V × 1.732, use 831
for 2400 V × 1.732, use 4157
for 4160 V × 1.732, use 7205

MANDATORY USE OF SHALL

Section 110-1 states that mandatory rules use the word *shall.* Always refer to the NEC® for mandatory rules.

THW Cu

Unless otherwise specified, copper conductors are sized on THW per Table 310-16.

REVIEW QUESTIONS AND TRADE TESTS

Electrical Systems contains Review Questions and Trade Tests after Chapters 1 – 13. These tests should be completed after studying the corresponding chapter. Chapter 14 contains the Final Exam. This test includes questions that cover the content of the entire text. The Final Exam should be taken after completing the text, Review Questions, and Trade Tests.

Review question formats include True-False, Multiple Choice, and Completion. Always record the answer in the space provided or encircle either T or F as appropriate. Trade Tests and the Final Exam contain an additional blank for providing the NEC® reference, if required by your instructor.

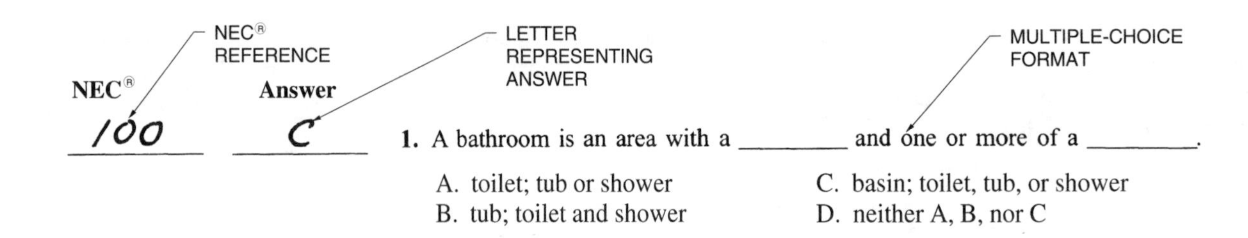

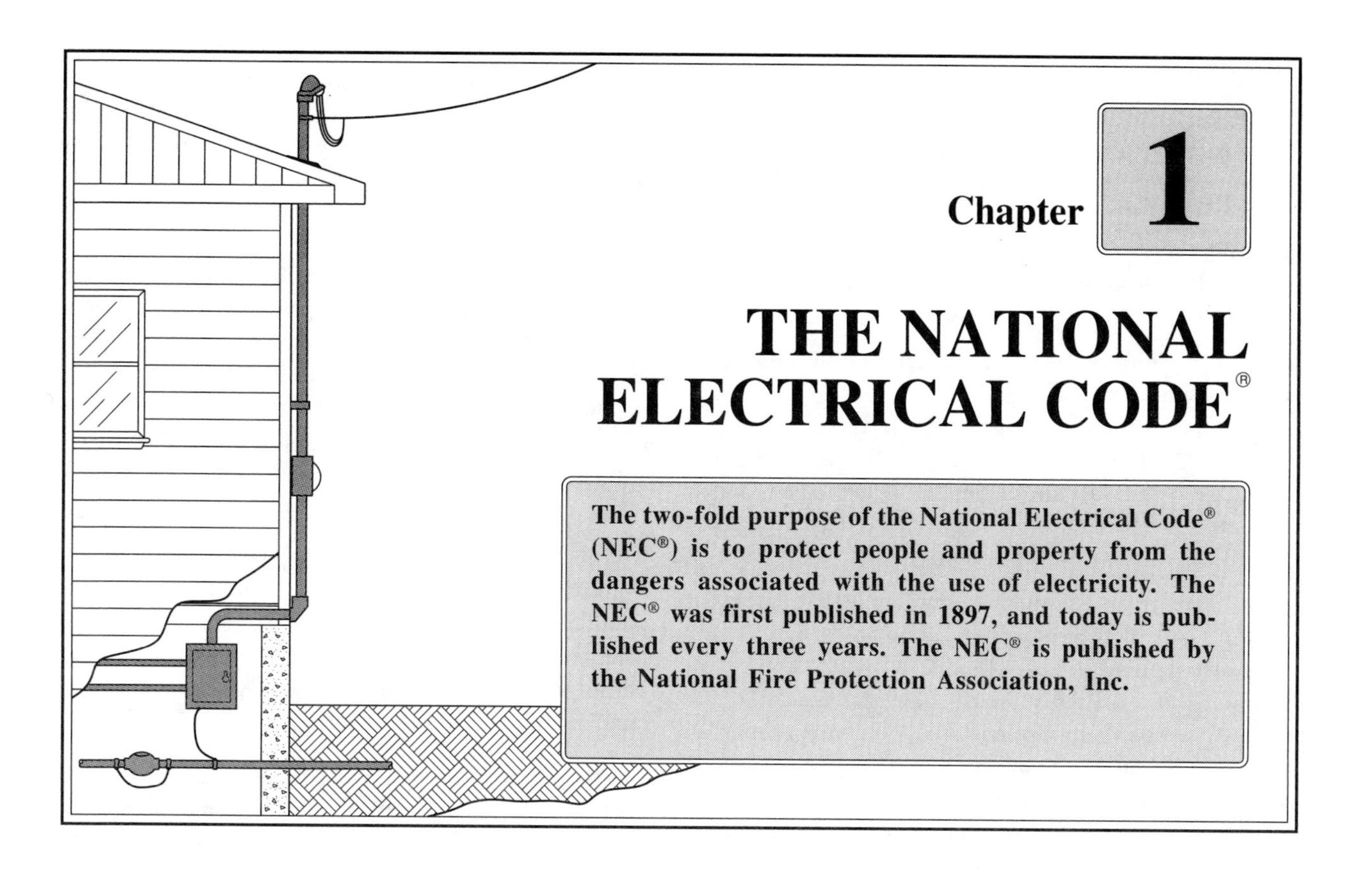

Chapter 1

THE NATIONAL ELECTRICAL CODE®

The two-fold purpose of the National Electrical Code® (NEC®) is to protect people and property from the dangers associated with the use of electricity. The NEC® was first published in 1897, and today is published every three years. The NEC® is published by the National Fire Protection Association, Inc.

THE NATIONAL ELECTRICAL CODE®

The National Electrical Code® (NEC®) is one of the most widely used and recognized consensus standards in the world today. It is a true consensus standard because members from throughout the electrical industry contribute to its development. The NEC® is revised and updated every three years to reflect current trends in the electrical industry. The purpose of the NEC® is to protect people and property from hazards that arise from the use of electricity.

NEC® Process

The National Fire Protection Association (NFPA) sponsors the development of the NEC®. The NFPA publishes guidelines in the Regulations Governing Committee Projects for the procedures for all of the standards it publishes. For the NEC®, the procedures call for essentially a four-step process:

1. Receipt of Proposals (ROP)
2. Receipt of Comments (ROC)
3. NFPA Annual Meeting
4. Standards Council Issuance

Report on Proposals. The first step in the code process is the receipt of proposals. Anyone can submit a proposal to change the NEC® provided it contains the required information. See Figure 1-1.

NEC® PROPOSALS

NEC® proposals shall contain:

1. Identification of the submitter.
   (name and organization or company affiliation)
2. Identification of the specific Code section and type of revision that is proposed.
3. A statement of the problem and substantiation for why the change is necessary.
4. The actual wording of the revised text or the wording to be deleted.

Figure 1-1. NEC® proposals shall contain the required information in order to be acted on.

The key to successful code proposals is proper substantiation for the proposed changes. The time window to submit proposals is very small. Usually, the closing date for receipt of proposals is in the fall of the same year a new Code is issued. For example, the last date to submit changes for the 1996 NEC® was November 5, 1993. After the last date to receive proposals, code panels meet to discuss and vote on each of the proposed changes. These votes are recorded and published in the Report on Proposals.

Receipt of Comments. Once the Report on Proposals is published, everyone has the opportunity to submit a comment on each of the proposed changes whether the Code-Making Panel voted to accept or reject them. A closing date for comments is published, and after the closing date the Code-Making Panel meets to vote on each of the comments.

The Report on Comments contains all comments and the votes of the respective code panels. If a proposal or comment is not accepted by the Code-Making Panel, the statement must state the reason the proposal or comment was not accepted. New proposals cannot be submitted during this stage of the Code cycle. The Code-Making Panel can only take action on proposals that have received adequate public review during the proposal stage.

NFPA Annual Meeting. Once the Code-Making Panels have reviewed the proposals and comments, the next step is to present the changes at the NFPA Annual Meeting. Floor action on individual proposals can occur during this meeting. Usually there are very few of these actions on the floor of the NFPA Annual Meeting, because most members choose to support the actions of the Code panels during the initial stages of the Code process.

Voting on floor actions, at the NFPA Annual Meeting, is limited to NFPA members only. A simple majority vote is required for a floor action to pass. Actions that occur on the floor of the NFPA Annual Meeting are still subject to review by the NFPA Standards Council.

Standards Council Issuance. The Standards Council has the responsibility for overseeing all of the codes and standards developed for the NFPA. The NEC® Correlating Committee works directly under the Standards Council. The Correlating Committee steers the Code panels through the process, ensuring that each proposal and comment received is handled according to an established operating procedure. Once the process is complete, the Standards Council reviews the entire process and actually issues the document for publication.

The NEC® is a legal document designed to be adopted by local and/or state governmental bodies. Local jurisdictions may choose to adopt the Code in its entirety, with specific additions or exceptions, or they may choose not to adopt the Code at all.

Code-Panel Membership. Each Code proposal or comment is reviewed by the representatives of various segments of the electrical industry. There are 20 Code-Making Panels to cover the articles in the NEC®. The members of each Code panel represent labor, manufacturing, electrical utilities, electrical inspectors, contractor associations, and testing laboratories.

NFPA Operating Procedures for the National Electrical Code® Committee state that no interest group shall comprise more than ⅓ of the total voting panel membership. Membership is designated as either principal or alternate. Alternate members assume the participation and voting rights only when the principal member is not present.

The official vote on panel proposals and comments occurs on a written ballot after the meetings conclude. Although each proposal is voted on during the Code-Making Panel discussions, the official vote on each proposal occurs when the Code-Making Panel member returns a written ballot which was mailed after the conclusion of the Code-Making Panel meetings. All members who vote against a panel action must state the reasons for doing so, and their comments are recirculated to each panel member.

Using the NEC®

The NEC® is available in soft cover, loose-leaf, and PC versions. While the NEC® is not difficult to use, it does have its own method of organization which must be understood for ease of use.

Revisions, Deletions, and Extracted Text. The NEC® specifies that any material in the current edition of the text that has changed from the previous edition is identified by a vertical line in the margin. Text that has been deleted from the previous edition of the NEC® is shown as a bullet (•) that also appears in the margin. Text that has been extracted from another NFPA document is designated by the use of the superscript letter (^{X}). These editorial markings are important for helping to trace the development of Code requirements. See Figure 1-2.

NEC® EDITORIAL SYMBOLS		
•	**BULLET**	A bullet is used to show where text has been deleted from the previous edition of the NEC®. For example, see 210-8(b) where the definition of bathroom has been relocated to 100.
|	**VERTICAL LINE**	Vertical lines appearing in the margins of the NEC® represent sections where the text has been revised from the previous edition of the NEC®. For example, see 100 where the definition of bathroom has been relocated.
x	**SUPERSCRIPT "X"**	The superscript letter "X" is used whenever text that appears in the NEC® has been extracted from another NFPA document. For example, see 517-25 where text that has been extracted from NFPA 99, Health Care Facilities, appears.

Figure 1-2. NEC® editorial symbols are useful tools for studying the latest NEC® changes.

Outline Format. The NEC® is arranged in a simple outline format. Section numbers are designated by a dash following the article number. For example, 210-8 indicates that the NEC® section is taken from Article 210, Section 8.

Section numbers often proceed in a normal numbering sequence, but sometimes sections are not used in numerical sequence. This occurs when sections have been deleted from the NEC® or perhaps if the Code panel wanted to leave room for future Code sections.

Subsections are part of the sections and are designated by lowercase letters set off in parentheses. For example, 210-8(a) is a subsection of 210-8. Subsections are further broken down by numbers to indicate parts of subsections. For example, 210-8(a)(4) is a part of subsection 210-8(a). In a few instances, the NEC® has further subsections of subsections. In most cases, the format remains in this number-letter-number sequence. The only time the NEC® deviates from this fashion is when sections do not contain any subsections, but do contain numbered items. For example, 230-43(4) is a numbered item of Section 230-43.

Because the NEC® is organized in an outline format, exceptions are applied only to the section they follow or as noted. For example, the exception for receptacles that are not readily accessible follows 210-8(a)(2) and cannot be applied to 210-8(a)(1).

Exceptions. Exceptions permit alternate methods to the general (main) rule. Exceptions are permitted in lieu of the main rule when all of the provisions of the exception are met. For example, 210-8(a)(5) requires that, in general, all receptacles in dwelling unit unfinished basements shall be GFCI-protected. Exception 2 allows receptacles to be installed without GFCI protection provided that a single receptacle is used. This ensures that the receptacles cannot be easily accessed for other uses. A duplex receptacle is also permitted to be used if two appliances, located in dedicated spaces and not easily moved, are cord- and plug-connected per 400. The purpose of this exception is to allow appliances, such as freezers or refrigerators which may be located in unfinished basements to be installed without GFCI protection provided there is no means for connecting portable cords or other appliances to the same receptacle.

FPNs. Fine Print Notes are used throughout the NEC® to explain information that is contained in the Code. For the purposes of enforcement, FPNs are not mandatory and do not contain any mandatory provisions. For example, 210-4(d) requires that each ungrounded conductor in a building containing more than one voltage system be identified as to its phase and voltage. The FPN clarifies that the methods used to identify the ungrounded conductor can be color coding, marking tape, tagging, or other effective means.

Layout. The arrangement of the NEC® is specified in 90-3. There is an Introduction and nine chapters in the NEC®. The first four chapters apply generally, unless they are modified by the latter chapters. For example, 300-14 requires that, in general, a minimum of 6″ of free conductor be provided at each outlet for splices or terminations. This applies throughout the NEC® unless modified.

Chapter 8 is independent of the other chapters. None of the NEC® provisions apply unless they are directly referenced. Chapter 8 is independent of the other chapters because it covers communication circuits. These circuits are part of the premises wiring, but are treated separately because they do not, in general, pose a threat of electrical shock to persons. None of the NEC® provisions therefore apply, unless they are directly referenced in Chapter 8.

Chapter 9 contains Tables and Examples which are referenced throughout the entire NEC®. The Appendix, which appears at the end of the Code, does not contain any mandatory provisions, but it does contain information that may be helpful in designing or installing electrical installations. For example, Appendix C contains many useful tables for determining the maximum number of conductors permitted in a raceway.

All applications of the NEC® are subject to the approval of the AHJ.

Applying the NEC®

The National Electrical Code® is adopted for use by local (village, town, city, etc.), county or parish, and state authorities. These authorities are generally represented by building officials who issue permits for jobs and make periodic inspections of work in progress. Additionally, these authorities certify that the completed building meets all applicable codes before occupancy occurs.

Purpose and Intent – 90-1. The two-fold purpose of the NEC® is to protect people and property from the dangers associated with the use of electricity. This two-fold purpose should be kept in mind whether applying the NEC® or submitting new proposals to modify the NEC®.

The NEC® is not intended to be used as an instruction manual per 90-1(c). The rules are stated without any reasons for their existence. For example, 210-8(a)(1) states that all 15 A and 20 A receptacles installed in dwelling bathrooms shall have GFCI protection. In order to understand the reason for the rule, the Report on Proposals would have to be read to determine the substantiation when the Code panel first accepted the change.

The NEC® is also not a design specification per 90-1(c). It contains the rules and necessary provisions for electrical systems, but leaves the design and layout of the electrical systems to others.

Scope – 90-2. Before applying the NEC®, the scope should be reviewed to determine if the electrical installation falls within the jurisdiction of the Code. Installations which are not covered are given in 90-2(b). Often, the distinction between what is covered and what is not covered is very fine and can lead to some difficult interpretations. For example, installations for electrical utilities, are not covered if they involve the metering, generation, transmission, or distribution of electric energy. Those installations, however, that are used by an electric utility, but are not an integral part of the electrical energy transmission or distribution are covered. Another example is watercraft. Installations on ships are not covered, but installations on floating buildings are.

Enforcement – 90-4. All applications of the NEC® are ultimately subject to approval by the Authority Having Jurisdiction (AHJ). The NEC® is designed to be used as a legal document. Any interpretations and approval of equipment rest with the Authority Having Jurisdiction. In many cases, the AHJ is a local or municipal building official.

NEW/EXISTING PRODUCTS

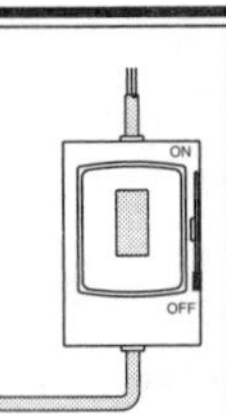

The NEC® requires that new products be developed or existing products be modified to meet new Code rules. For example, in 1990, the NEC® required that the main bonding jumper, where it is a screw, be identified with a green finish. The AHJ, per 90-4, was allowed to permit the use of products that complied with the 1987 NEC® until the green screws became available or until the next Code was published.

Also in 1990, the NEC® required that all motor-operated water pumps, including sump pumps, be grounded. Installers confronted with the two-wire sump pumps had the option of altering the new equipment or asking the AHJ to invoke their enforcement power.

Some localities permit the use of independent third party electrical inspection agencies. In other cases, the AHJ may be a governmental official, local fire marshal, or an insurance official with jurisdiction over the installation. In any case, whoever the AHJ is, they can permit alternate methods or even waive specific requirements if they believe the installation is equally effective at meeting the intended objective. See Figure 1-3.

AUTHORITY HAVING JURISDICTION

Duties and responsibilities:

1. Make interpretations of NEC® rules.
2. Approve equipment and materials.
3. Grant special permission.
4. Waive specific requirements.
5. Permit alternate methods.

Figure 1-3. The AHJ is responsible for enforcement of the NEC® per 90-4.

Future Use – 90-8. Sections 90-1(b) and 90-8, taken together, indicate that installations completed in accordance with the NEC® merely met a minimum established standard. The installations may not be adequate for future use, and while it is not required, allowance for future expansion is generally in the best interest of all concerned. Essentially, the NEC® is a minimum standard. The designer or installer can choose to provide for future use if desired. Installations in accordance with the NEC® are not necessarily efficient and may not meet the future needs and demands for electricity.

Definitions – 100

Applying and interpreting the NEC® incorrectly can often be traced to a failure to properly understand the terms used. To assist in the proper application of the NEC®, terms that are used in two or more articles are defined in 100. In addition, terms may be defined in the particular article to which they apply. Many articles contain a separate section with the specific definitions for that article only.

FUTURE CONSIDERATIONS

Although it is not required, electrical contractors and installers of electrical systems should take future use into consideration when planning electrical installations. The tendency to "low-bid" in order to keep the overall job cost down should be considered very carefully. Often, during the course of a project, additional items (extras) are added to the initial installation. Installers who allow for these considerations may find themselves in a better position to take on the extras at a reduced cost.

Some electrical inspectors may require that future consideration be given to all electrical installations. For example, although a ½" conduit may meet the NEC® requirements for a particular installation, some inspectors do not allow conduit in trade sizes smaller than ¾". Although this is not required by the NEC®, 90-4 grants the AHJ the power of enforcement for the Code requirements. Always consult with the AHJ before beginning an electrical installation if there are any concerns regarding future consideration.

Other definitions may be included within the particular section to which they apply. For example, 210-8(a)(5) defines an unfinished basement for the purposes of applying the general rule which requires all receptacles in unfinished basements of dwellings to be GFCI-protected. See Figure 1-4.

Accessible – 100. *Accessible* equipment admits close approach and is not guarded by locked doors, elevation, etc. Many NEC® references require that the wiring methods used for electrical installations be accessible. For example, 370-29 requires that conduit bodies, junction boxes, and outlet boxes be installed so that the wiring within is accessible. Such wiring is capable of being removed without damage to the building or the finish. Boxes with accessible wiring methods should not be installed in such a method that the wall would be taken down or damaged to gain access to the boxes. See Figure 1-5.

NEC® DEFINITIONS	
100	If a term is used in two or more NEC® articles, its definition appears in 100.
OTHER ARTICLES	Some articles contain a section that has definitions of terms used in that particular article only. For example, see 550-2, 551-2, 517-3, 680-4, and 690-2.
INDIVIDUAL SECTIONS	Sometimes the definition is included in the specific NEC® section to which the definition is applicable. For example, see the definition of unfinished basement in 210-8(a)(5).

Figure 1-4. Definitions in the NEC® appear in Article 100 or in the particular article or section in which they are used.

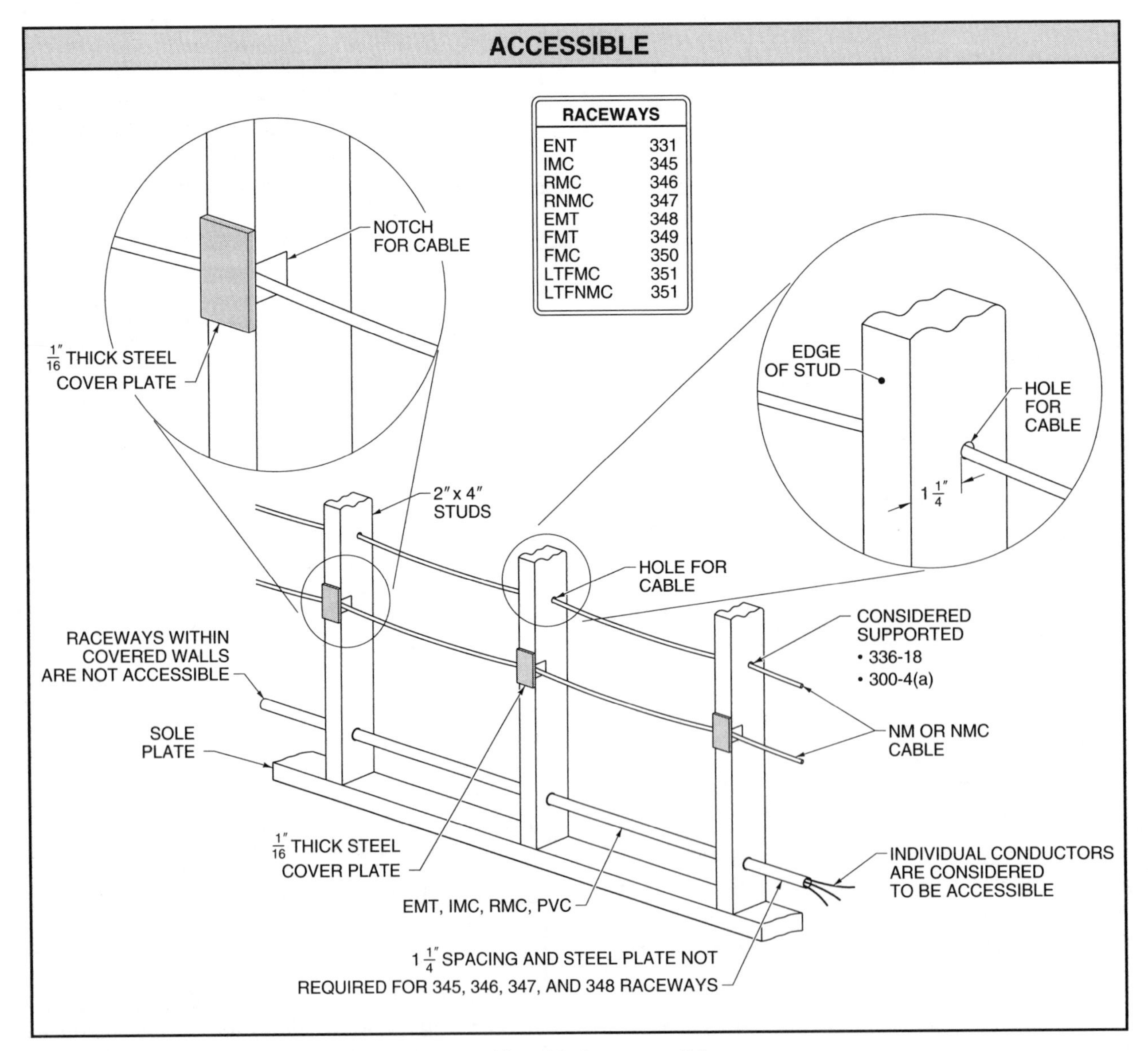

Figure 1-5. Raceways within walls are not considered to be accessible.

Readily Accessible – 100. *Readily accessible* equipment is capable of being reached quickly. Equipment is often required to be installed in a readily accessible location. Such equipment cannot be located behind locked doors. It cannot be located so that a ladder is necessary to reach it, and it cannot be located behind obstacles. See Figure 1-6.

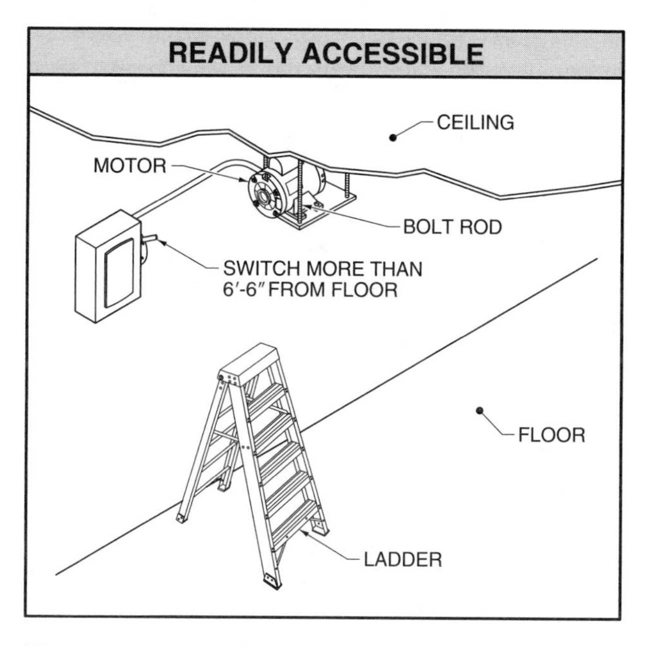

Figure 1-6. If portable ladders are necessary to reach equipment, it is not readily accessible.

Ampacity – 100. *Ampacity* is the current that a conductor can carry continuously, under the conditions of use. Ampacity is derived from the words ampere and capacity.

The type of insulation is a major factor in determining a conductor's ampacity. For example, Table 310-16, which lists the allowable ampacities or insulated conductors in a raceway, cable, or earth classifies conductor insulation according to the temperature rating of the conductor. There are three ratings for copper (Cu) and aluminum (Al). Notice how the allowable ampacity of a No. 8 Cu conductor changes from 40 A for 60°C, to 50 A for 75°C, and to 55 A for 90°C.

Other factors that affect a conductor's ampacity are the ambient temperature surrounding the conductor, and the number of conductors in a raceway or cable. *Ambient temperature* is the temperature of air around a piece of equipment. A FPN to 310-10 states that the temperature rating of any conductor is determined in part by the maximum temperature at any point along its length that the conductor can withstand over a long period of time without sustaining damage to the conductor's insulation. If a conductor passes through an area with a high ambient temperature, the conductor's ampacity is based on the highest ambient temperature to which the conductor is exposed.

Table 310-16 has correction factors at the bottom of the Table for making ambient temperature corrections. Likewise, the allowable ampacities listed in Table 310-16 are based on not more than three current-carrying conductors in a raceway or cable. If the number of conductors is increased beyond three, then the conductor's ampacity shall be derated. This is because as the number of conductors increases, the ability of the conductor to dissipate heat away from the raceway or cable decreases. Note 8 of Notes to Ampacity Tables of 0 to 2000 Volts lists the derating percentages for more than three conductors.

ONE SERVICE PER BUILDING

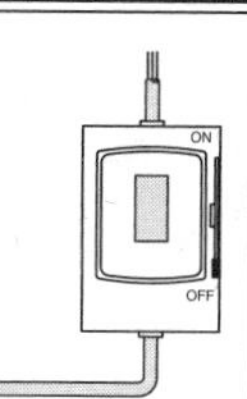

Each building or structure shall be provided with only one service per 230-2. While it is true that there are seven exceptions to this rule, confusion often results from an improper understanding of what constitutes a building. For example, attached garages are generally not considered to be a separate building and are not eligible for a separate service. Detached garages, however, are clearly buildings and 230-2 is applicable.

Many local electrical utilities provide electricians and contractors with books that contain the electrical utilities' service requirements. These requirements must be followed no matter what the NEC® requirements state, before the electrical utility supplies the power. Contractors interested in installing electrical services should contact their local utility to determine what their electrical service requirements are and their procedures for obtaining services and meter information.

Approved – 100. *Approved* is acceptable to the AHJ. Conductors and equipment shall be approved in order to be acceptable per 110-2. Per the definition, the AHJ has reviewed the conductors, equipment, or installation and found it to be acceptable. The AHJ, in determining whether to grant permission or approval, evaluates the equipment to see if it has been listed, labeled, or identified for the use.

Building – 100. A *building* is a stand-alone structure or is separated from adjoining structures by fire walls. One of two conditions must be present in order to meet the definition of a building for the purposes of the NEC®. First, the structure must stand alone. Additions to an existing building do not constitute a building. For example, a detached garage is a building while an attached garage may not be.

The second condition involves a structure that is attached to or part of an existing structure, but is separated by fire walls and rated fire openings. In this case, a single structure can be classified as two buildings. See Figure 1-7.

Conductors – 100. A *conductor* is a slender rod or wire that is used to control the flow of electrons in an electrical circuit. The NEC® recognizes three types of conductors: copper, aluminum, and copper-clad aluminum.

A *bare conductor* is a conductor with no insulation or covering of any type. Bare conductors are often permitted to be used for the equipment grounding conductor (EGC), bonding jumpers, and sometimes for the grounded conductor.

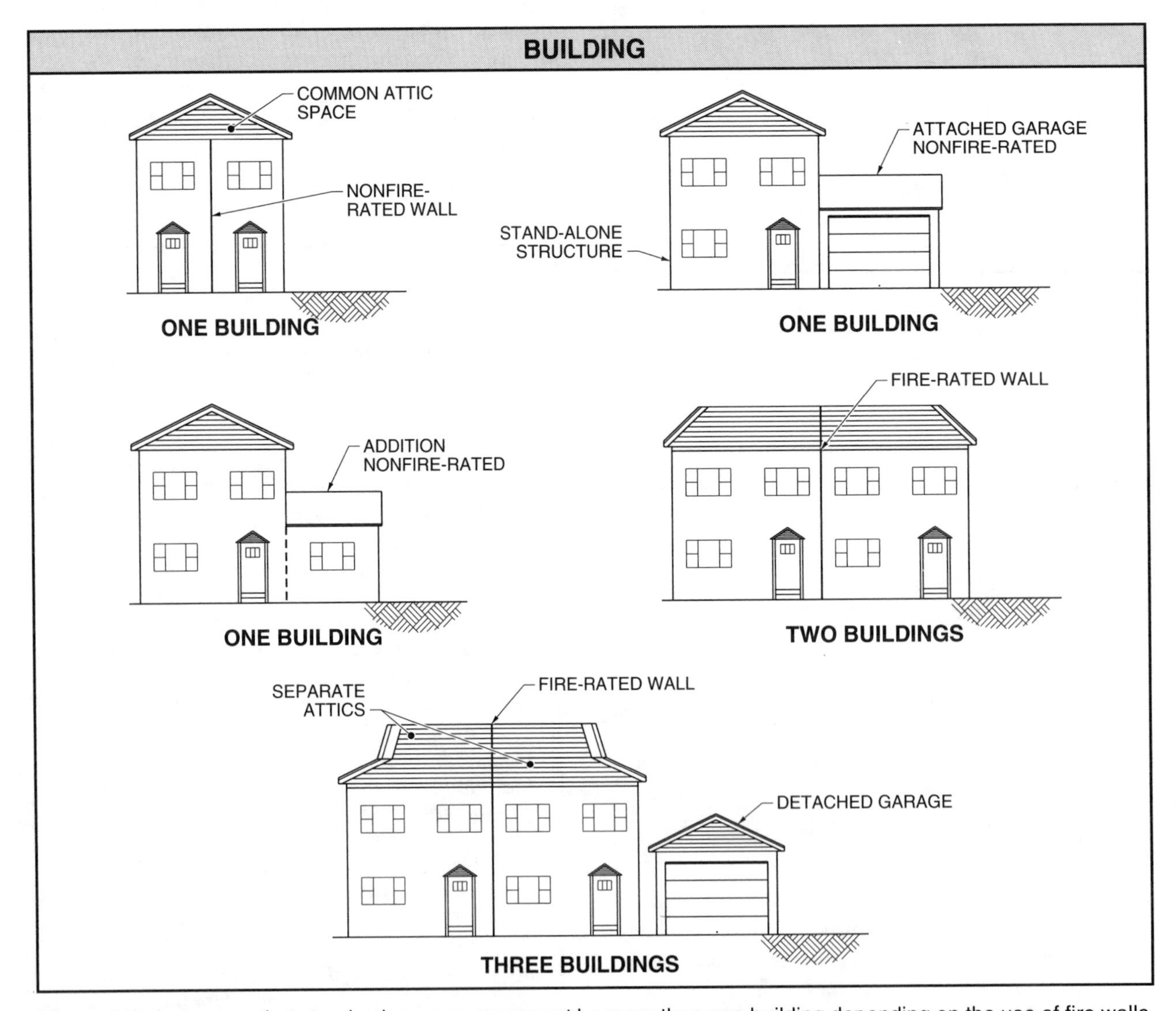

Figure 1-7. A structure that stands alone may or may not be more than one building depending on the use of fire walls.

A *covered conductor* is a conductor not encased in a material recognized by the NEC®. Conductors installed in free air may not require a specific insulation and can merely be covered with a suitable material. An example of a material that is not classified as an insulation is the rubber covering used on conductors installed in free air. The rubber covering protects the actual conductor, but the covering has not been evaluated for its insulating properties.

An *insulated conductor* is a conductor covered with a material classified as electrical insulation. Most conductors are insulated conductors. These conductors, whether part of a cable assembly or building wire installed in a raceway, are covered to a thickness with a material classified as electrical insulation. Table 310-13 lists many insulations recognized by the NEC®.

Continuous Load – 100. A *continuous load* is a load in which the maximum current may continue for three hours or more. Much of the equipment installed and serviced under the NEC® is intermittent in its use. An *intermittent load* is a load in which the maximum current does not continue for three hours. Equipment, and the conductors feeding it, has varying periods of load and no-load conditions. Other equipment stays on, under load conditions, for longer periods of time. Such equipment operating for more than 3 hours constitutes a continuous load.

Device – 100. A *device* is any unit of an electrical system that carries, but does not use electricity. Receptacles and switches are two types of devices. Both of them carry, but do not, in and of themselves, use electricity. See Figure 1-8.

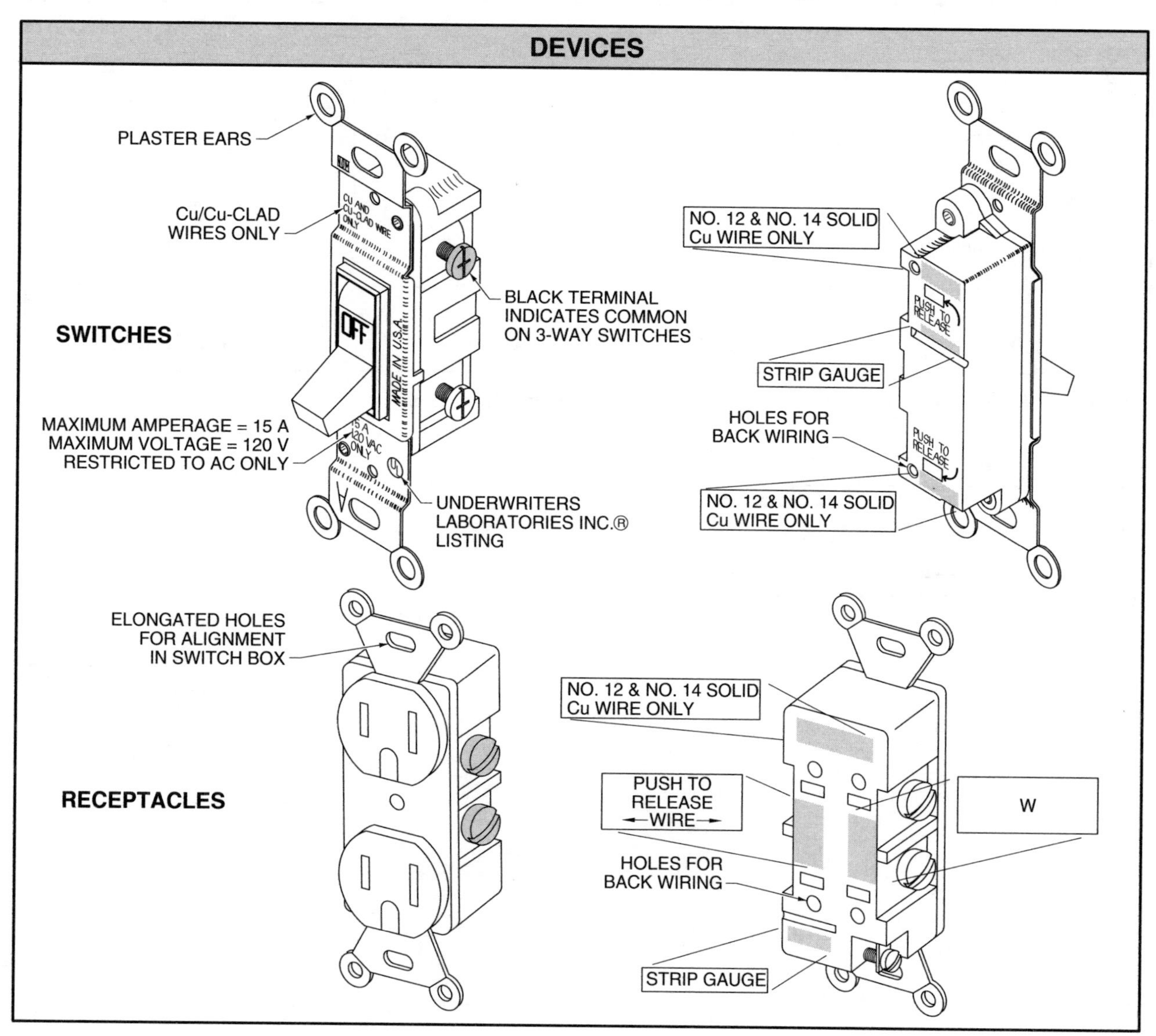

Figure 1-8. Switches and receptacles are devices because they carry electricity, but do not use it.

Dwelling – 100. Four types of occupancies are included under the definition of dwelling in 100. These are dwelling unit, multifamily dwelling, one-family dwelling, and two-family dwelling. A *dwelling* contains eating, living, sleeping space, and permanent provisions for cooking and sanitation. See Figure 1-9.

A *dwelling unit* is a dwelling with one or more rooms used by one or more people for housekeeping. No definition is given for housekeeping. A *multifamily dwelling* is a dwelling with three or more dwelling units. A *one-family dwelling* is a dwelling with one dwelling unit. A *two-family dwelling* is a dwelling with two dwelling units.

Section 210-8(b) refers to occupancies that are classified as other than dwelling units. The NEC® does not define commonly used classifications such as residential, commercial, or industrial even though the terms are used throughout the NEC®.

Exposed – 100. *Exposed* as applied to wiring methods is on a surface or behind panels which allow access. For example, wiring methods such as AC and NM have requirements which must be adhered to when installed in exposed locations. Exposed locations are not limited to those locations that are visible. Installations behind panels, such as drop ceiling tiles which are designed to be removed, are considered to be exposed. See Figure 1-10.

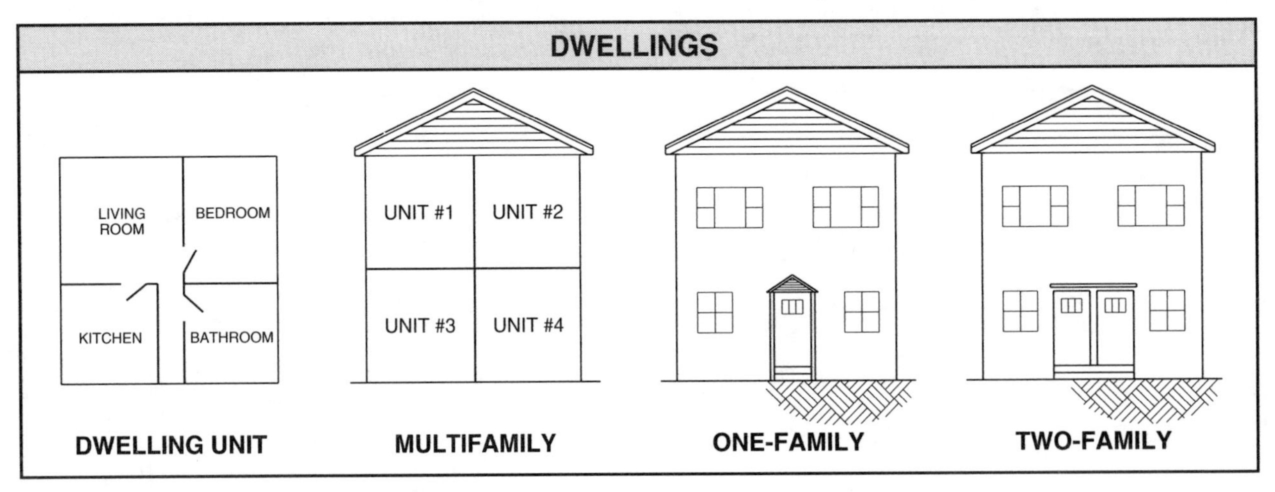

Figure 1-9. Four types of occupancies are included in the definition of dwelling.

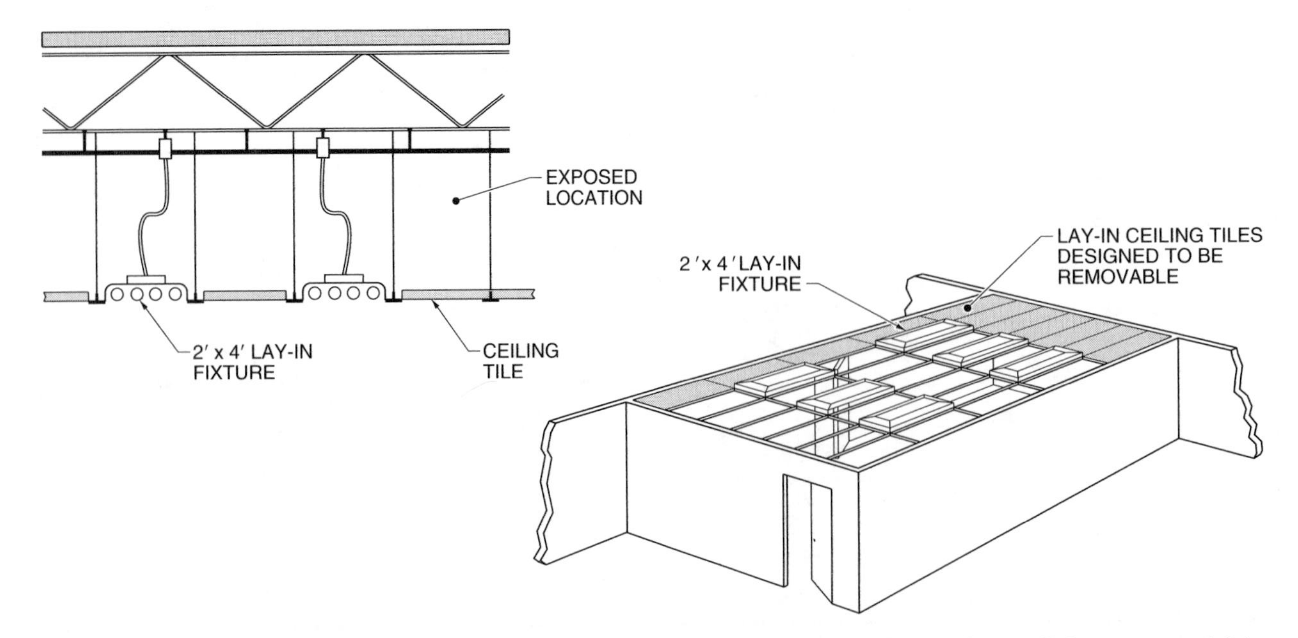

Figure 1-10. Wiring methods installed above lay-in ceilings are considered to be exposed, even if they are not visible.

Fitting – 100. A *fitting* is an electrical system accessory that performs a mechanical function. Connectors, couplings, locknuts, bushings, etc., are components of the electrical system that serve a mechanical function. Although not designed to serve an electrical function, they can be an important part of the equipment grounding path when the raceway serves as an equipment grounding conductor. See Figure 1-11.

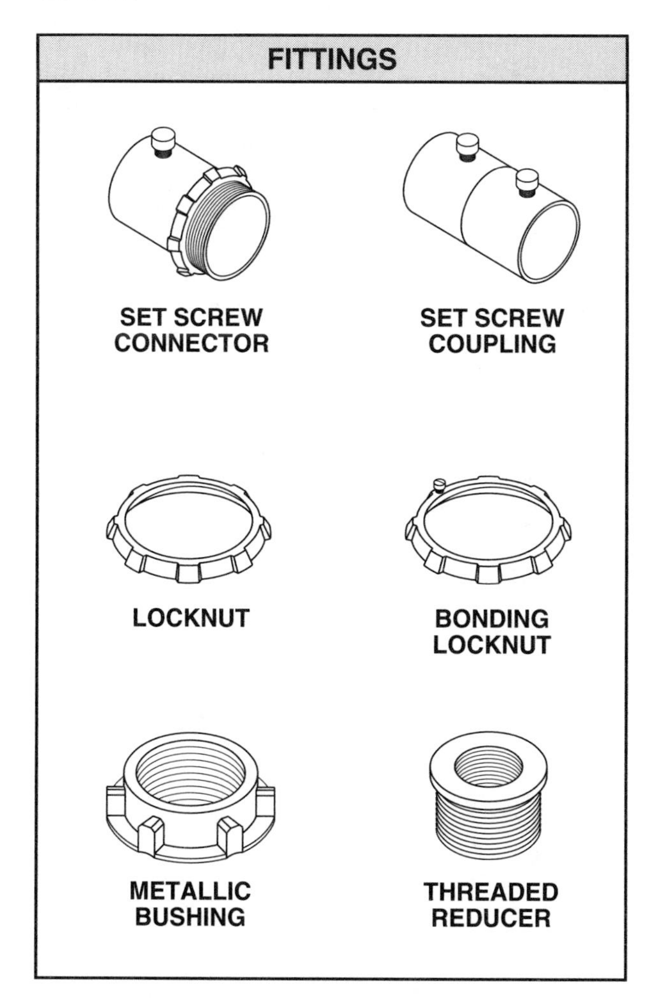

Figure 1-11. Fittings provide a mechanical function in constructing electrical systems.

Identified – 100. *Identified* is recognized as suitable for the use, purpose, etc. Misapplication of the NEC® often results when equipment which is not suitable for the purpose is used. For example, equipment suitable for damp locations cannot be installed in wet locations. Equipment that is required to be identified, is required to be suitable for its use or function. For example, boxes used to support ceiling fans are required to be identified and a label is attached to the box to indicate that it is identified.

In Sight From – 100. *In sight from* is visible and not more than 50′ away. Some NEC® installation provisions require certain equipment to be located in sight from, within sight, or within sight from other equipment. Anytime this provision is stated, the equipment shall be visible and not more than 50′ from the other equipment. Visible means an unobstructed view. See Figure 1-12.

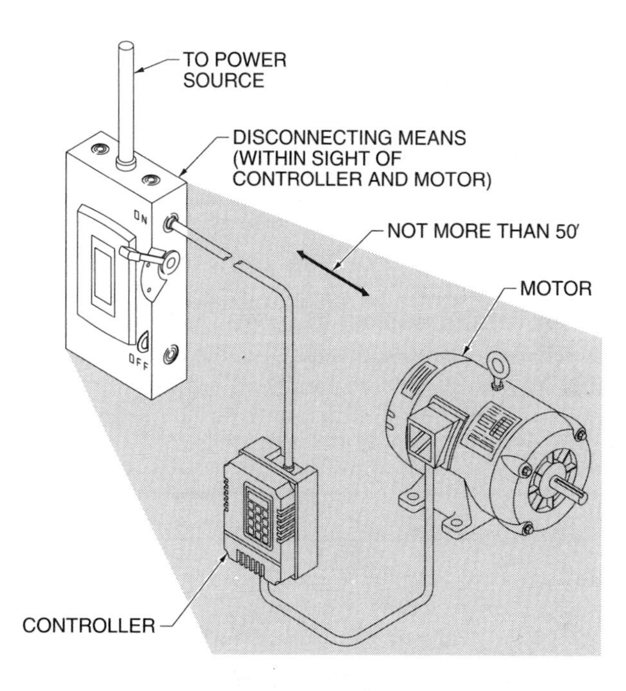

Figure 1-12. The motor disconnecting means shall be located within sight from the controller.

Labeled – 100. *Labeled* is equipment acceptable to the AHJ to which a label has been attached. Labeling identifies equipment or material which has been evaluated by a suitable testing laboratory and found to be acceptable and in compliance with established performance and/or construction standards. Some equipment, connectors, couplings, etc., may be difficult to label because of the physical makeup or size of the equipment. Often, this type of equipment contains labeling information on the carton or box in which it is packaged.

Listed – 100. *Listed* is equipment or material approved by the AHJ in a list. Like labeling, listing of a product certifies that the equipment has been evaluated by a testing laboratory suitable to the AHJ. Equipment which has been listed is published in a directory by the testing laboratory certifying that it has met established standards and is periodically reviewed to ensure standards compliance. Earlier Code editions required that the listing be done by a nationally recognized testing laboratory such as Underwriters' Laboratories, Inc. or Factory Mutual. This requirement no longer appears in the NEC®.

Location – 100. Electrical equipment can be installed in various types of locations subjecting it to all types of weather. Locations are classified as dry, damp, or wet to help determine suitability of equipment.

A *dry location* is a location which is not normally damp or wet. Most electrical equipment is installed in dry locations. Dry locations are out of the direct effects of moisture. If a location is normally dry, but temporarily subject to moisture, it may still be classified as a dry location.

A *damp location* is a partially protected area subject to some moisture. Damp locations are subject to moderate degrees of moisture, but are generally out of the weather. Outdoor locations which are partially covered and interior locations such as basements are classified as damp locations.

A *wet location* is subject to water or other liquids. Locations that are subject to more than moderate degrees of moisture are classified as wet locations. Equipment to be installed in wet locations should be designed so that it is suitable for locations exposed to weather without any protection. Underground installations and those in direct contact with the earth are classified as wet locations. See Figure 1-13.

Outlet – 100. An *outlet* is any point in the electrical system where current supplies utilization equipment. The purpose of electrical systems is to deliver electrical power at convenient points in order to operate electrical equipment. The points at which the electrical system is accessed are outlets. *Receptacle outlets* are outlets that provide power for cord- and plug-connected equipment. *Lighting outlets* are outlets that provide power for lighting fixtures. See Figure 1-14.

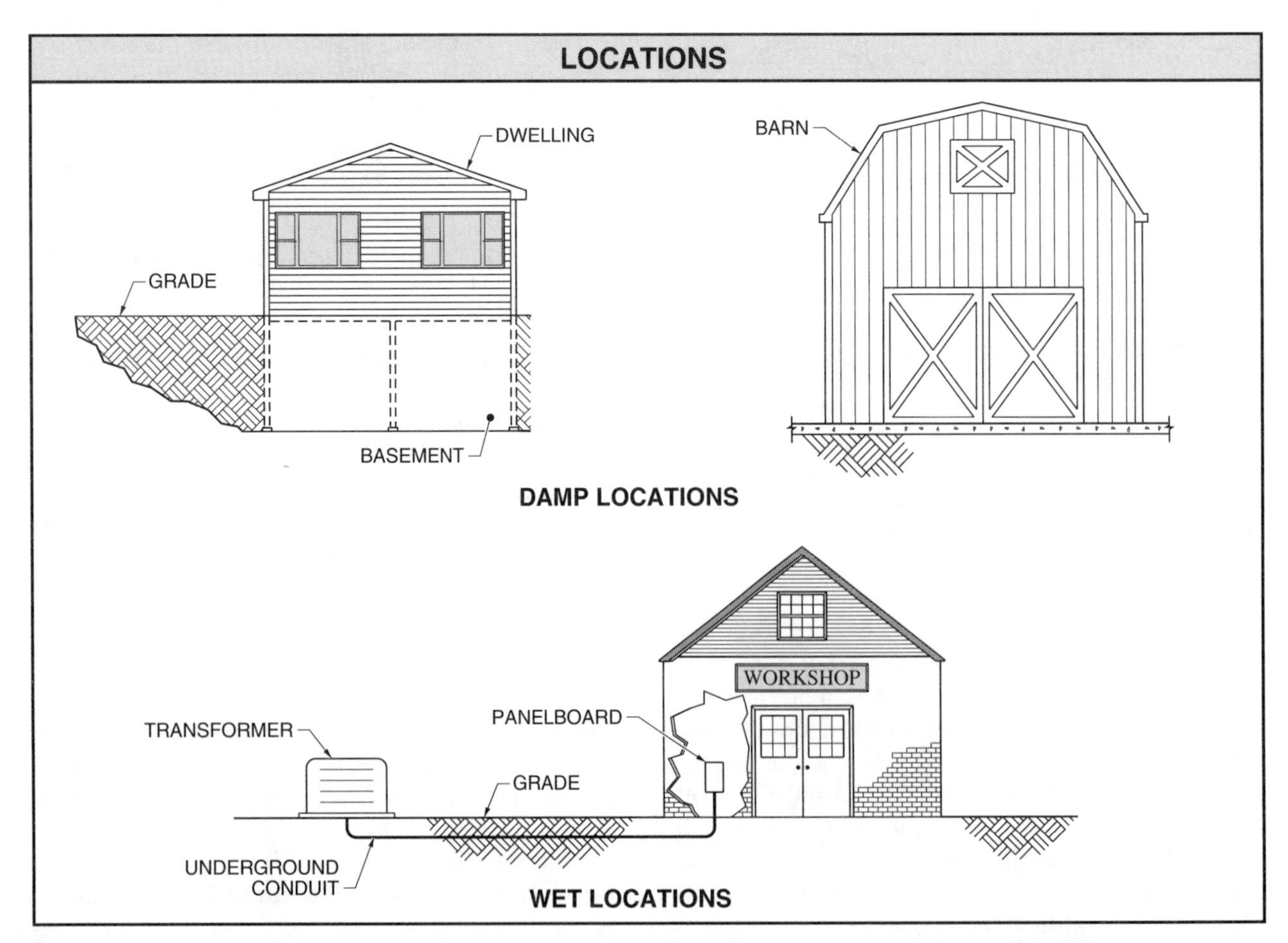

Figure 1-13. Equipment and materials shall be suitable for the location in which they are installed.

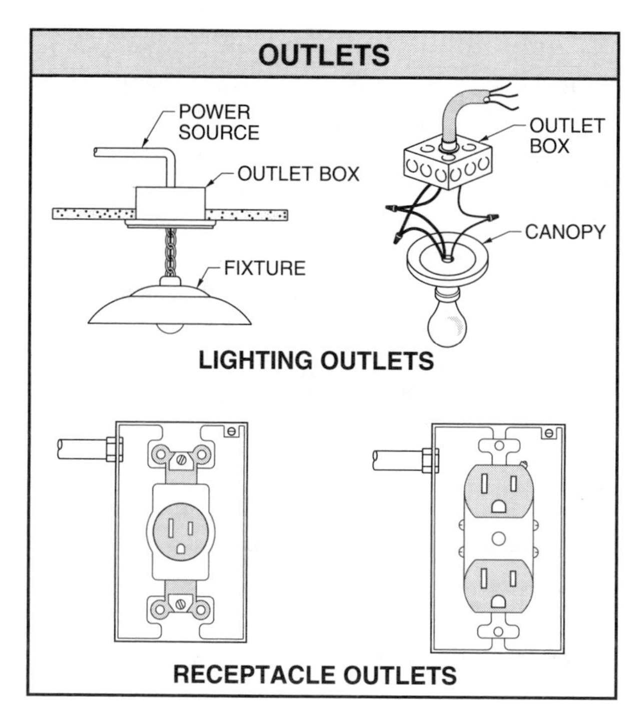

Figure 1-14. Outlets do not use electricity, but provide a means to access it.

Overcurrent – 100. *Overcurrent* is any current in excess of that for which the conductor or equipment is rated. All equipment and conductors are designed to operate at a specified current rating. When the rated current is exceeded, the result is an overcurrent. Overcurrents are caused by overloads, ground faults, or short circuits.

A *ground fault* is an unintentional connection between an ungrounded conductor and any grounded raceway, box, enclosure, fitting, etc. Ground faults can result in a large magnitude current flow in the ground path. The amount of current that flows depends upon the resistance or impedance of the ground path. For example, if a 120 V, ungrounded conductor in a metal raceway is nicked during installation, a ground fault occurs. The amount of current that flows depends upon the impedance of the metal conduit back to the source of power.

A *short circuit* is the unintentional connection of two ungrounded conductors that have a potential difference between them. A short circuit also occurs when an ungrounded conductor is connected to a grounded conductor. Approximately 75% of the available short circuit current can flow in a ground fault. The largest overcurrent results when a short circuit occurs. A short circuit occurs when two ungrounded conductors or an ungrounded and grounded conductor come in contact with each other. A short circuit is not an overload. In general, all equipment should be suitable for the available short-circuit current. See 110-9, 110-10, and 230-65.

Overload – 100. An *overload* is the lowest magnitude of overcurrent. A small-magnitude overcurrent, that over a period of time, leads to an overcurrent which may operate the overcurrent protection device (fuse or CB). Overloads can result from operating equipment on loads for which it was not designed or simply by overloading the electrical circuit.

Raceway – 100. A *raceway* is a metal or nonmetallic enclosed channel for conductors. The primary purpose of raceways is to support the electrical conductors and protect them from physical damage. Raceways may have other functions such as equipment grounding conductors. See 250-91(b).

Installation requirements for specific raceways are found in 300 and within the particular raceway article. For example, 300-11 requires that all raceways be securely fastened in place while 348-12 requires that electrical metallic tubing be securely fastened in place at least every 10′.

Voltage, Nominal – 100. *Nominal voltage* is any voltage within an acceptable range. Most voltages referenced in the NEC® are nominal voltages. That is, the values may not represent actual values but a range of voltages that are acceptable. For example, 210-8(a) lists the requirements for GFCI protection of 125 V receptacles in dwelling units. The 125 V is a nominal voltage. Receptacles of 110 V, 115 V, and 120 V are also covered.

Voltage-to-Ground – 100. *Voltage-to-ground* is the difference of potential between a given conductor and ground. However, for ungrounded systems, the voltage-to-ground is the maximum voltage between any two conductors of the circuit. In some instances, the NEC® specifies voltage between conductors. In these cases, voltage shall be considered to be either line-to-line or line-to-neutral voltage depending on the type of circuit.

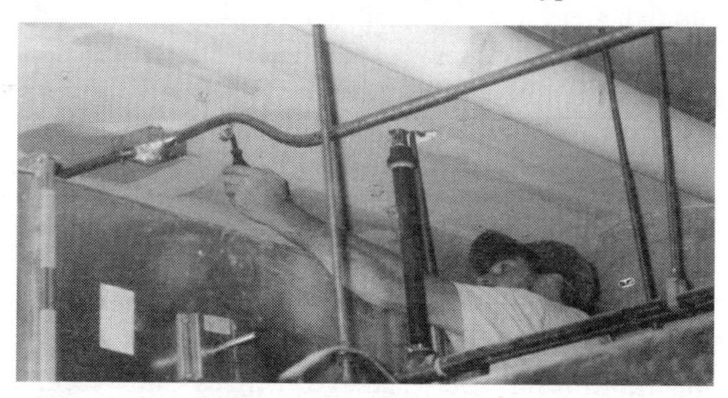

Raceways support electrical conductors and protect them from physical damage.

ELECTRICAL INSTALLATIONS – 110

The basic building block for electrical installations throughout the NEC® is 110. Electrical installation requirements concerning equipment, conductors, and terminations are included. Additionally, work space clearance requirements which are critical for protection of workers are also included.

Approval – 110-2

Section 110-2 requires that all electrical equipment and conductors installed under the requirements of the NEC® shall be approved to be deemed acceptable. The AHJ is charged with the responsibility for approving both equipment and installations. The AHJ, in deciding whether to grant approval, should determine if the equipment is identified, listed, or labeled for the intended use.

Usage of Equipment – 110-3

Equipment shall be suitable for its intended use. Factors such as mechanical strength, conductor termination space, conductor insulation, thermal effects, and arcing effects should be considered along with the primary factor, which is to ensure that equipment will not in any way place people at risk during operation.

Misapplication of equipment is one of the most frequent factors contributing to poor electrical installations. Section 110-3(b) requires that all listed or labeled equipment shall be installed, used, or both in a manner consistent with any instructions included with the listing.

Conductors – 110-5

In general, the NEC® recognizes three types of conductors: copper, aluminum, and copper-clad aluminum. Unless specified, all conductors in the NEC® are considered to be copper. There are better conductors than copper. For example, silver is a better conductor of electricity than copper, however, copper offers the most ampacity at the least cost per foot.

Conductor sizes are expressed in American Wire Gage (AWG) or circular mils (CM) per 110-6. Table 8, Chapter 9, contains some useful information on conductor properties and can also be used to show the relationship between AWG sizes and circular mils. Basically, the AWG is a system for comparing the relative area of a conductor. For the purposes of the NEC®, sizes start at No. 18 (small) and run through No. 1 (larger).

After No. 1, the AWG uses the aught sizes. Four sizes are listed using this method: No. 1/0, No. 2/0, No. 3/0, and No. 4/0. As the number value increases, the size of the conductor decreases. Thus, a No. 4 conductor is larger than a No. 8 conductor.

After the No. 4/0 size, the Table switches to the use of kcmils. A *mil* is .0001″. A *circular mil* is a measurement used to determine the cross-sectional area of a conductor. The prefix "k" stands for thousands, thus kcmil is equal to thousands of circular mils. As the number of kcmils increases, the size of the conductor increases.

LISTED EQUIPMENT

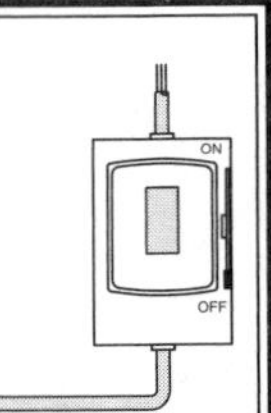

Contractors or installers frequently interpret 110-3(b) to mean that the use of nonlisted material and equipment is prohibited. This section does not require that only listed equipment be installed. Such confusion may result from job specifications that require only listed equipment or from Code rules that may require some particular material or equipment to be listed. For example, 110-14(b) requires that all splicing devices for direct earth burial be listed.

Section 110-3(b) requires that if listed or labeled equipment is used or installed, then it must be done so in accordance with any instructions included in the listing or labeling. For example, ceiling fans are listed with specific instructions detailing the minimum clearance from the floor for the blades. If the clearances are not met, the installation would be a Code violation despite the fact that there are no specific minimum height requirements for ceiling fans provided in the NEC®.

Equipment – 110-9

In addition to the factors listed in 110-3(a), other factors shall be considered before making a final determination as to the suitability of equipment. Section 110-9 requires that all equipment used to interrupt current at fault levels shall have an interrupting rating suitable for the available nominal voltage and circuit current. Proper selection of fuses and CBs requires that the installer know what the available fault current is at the line terminals of the equipment.

Many manufacturers of these devices publish information designed to assist the installer in calculating the available fault current at the line terminals of the equipment. Equipment that is not designed to open the circuit under fault conditions, such as snap switches and contactors etc. shall have an interrupting rating suitable for the current it is designed to interrupt.

Electrical equipment shall be installed in a neat and workmanlike manner.

Mechanical Installation – 110-12

All electrical equipment shall be installed in a neat and workmanlike manner. There is no definition of what constitutes neat and workmanlike, but equipment that is installed without any consideration given to it being level, plumb, or adequately supported would violate this section.

Unused openings in equipment shall be effectively closed per 110-12(a). For the purposes of this section, an effective seal is one that is equivalent to the material from which the equipment is constructed.

Section 110-12(c) requires that electrical equipment, especially internal components, shall be protected against foreign materials. Often, equipment may be left unprotected from materials like paint, plaster, abrasives, etc., that are used in construction of the job. Suitable protection shall be given to the equipment to ensure that nothing will adversely affect its safe operation.

Mounting and Cooling – 110-13

Aside from the requirement to install all equipment in a neat and workmanlike manner, all equipment shall be securely fastened on the surface to which it is mounted per 110-13(a). Proper installation requires that the correct fastener for the specific installation be used. For example, wood plugs are not suitable for use in holes in masonry, concrete, or plaster.

One of the primary considerations that must be taken into account when designing electrical installations is the effect of temperature on the equipment. Section 110-13(b) requires that proper ventilation be provided for electrical equipment which depends on the natural circulation of air.

Some equipment, such as transformers, are constructed with ventilation openings which shall not be obstructed by adjacent walls or equipment. The transformer nameplate specifies the required distance from the wall the transformer shall be installed to provide for proper air circulation and cooling. See 450-9.

TERMINALS

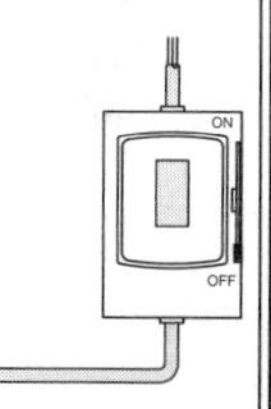

All terminals for more than one wire shall be identified per 110-14(a). Identified means suitable for the specific use. If the termination is to be used for more than one conductor, then it must have been evaluated for such use and identified as such. A common misapplication of this rule occurs in service equipment panelboards where contractors terminate the grounded conductor (neutral) and the EGC (ground) of a branch circuit on the same terminal strip and under the same screw. While these conductors are permitted at the service to be terminated on the same neutral block, they cannot be terminated under the same screw unless the terminal strip is so identified. As a rule of thumb, single screw time terminations commonly found on panelboard neutral blocks are not suitable for more than one conductor.

Electrical Connections – 110-14

At the core of any electrical system is the electrical connection and termination. Just like equipment, electrical connections shall be properly designed and installed to avoid problems associated with excessive heat. Improper electrical terminations lead to high resistance connections which in turn leads to electrical system failures.

The first consideration in making good electrical connections is the type of conductor material involved. The NEC® recognizes copper, aluminum, and copper-clad aluminum, each of which has different physical characteristics. The basic rule when selecting termination devices is to always use a termination device identified for the type of conductor material. For example, when using lugs with aluminum conductors, make sure the lug is identified for use with aluminum conductors. Similarly, if conductors of dissimilar materials are to be used, the termination device shall be identified for use with both types of conductor material.

When the termination is a terminal block, care must be taken to ensure a good electrical connection without damaging the conductor. Terminals should not be used for more than one conductor, unless they are identified. When it is necessary to splice conductors, identified splicing devices shall be used. If the splice is underground, the device shall be listed for direct burial use. See Figure 1-15. Although not common, splices are permitted to be made by brazing, welding, or soldering with a fusible metal or alloy provided the conductors are mechanically and electrically secured first.

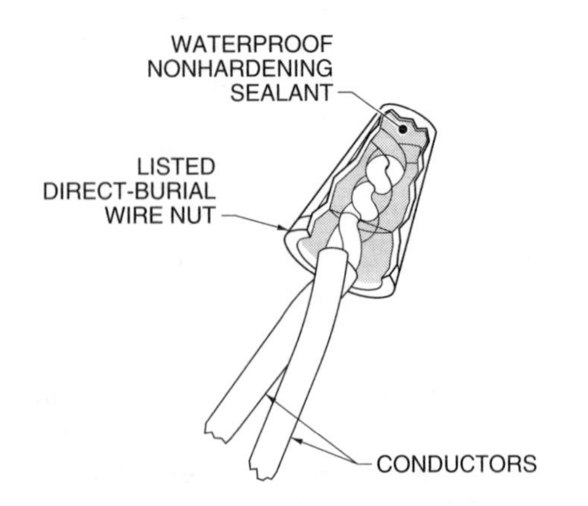

Figure 1-15. Listed direct-burial wire nuts contain a waterproof, nonhardening sealant.

Another consideration when making electrical terminations is the temperature rating of the termination device. Equipment, such as conductors, are rated according to how much heat rise can occur at the conductor termination. For example, if a panelboard is labeled with a 75°C temperature rating, the temperature at the conductor termination can rise up to 75°C. Conductors, in this case, would require insulation ratings of at least 75°C. If 60°C conductors were used, the insulation on the conductor would not be sufficient for the termination heat rise, and insulation failure could occur.

Section 110-14(c) requires that, in general, equipment terminations for circuits rated 100 A or less or designed for No. 14 through No. 1 conductors are rated for use on 60°C conductors. Equipment terminations for circuits rated over 100 A, or conductors larger than No. 1, are rated for use on 75°C conductors. Conductors with a higher rated insulation termination are allowed per 110-14(c), Ex. 1 provided the conductor's ampacity is selected from the lower rating. When listed equipment is provided with higher rated termination provisions than allowed in the general rules, 110-14(c), Ex. 2 applies. In these cases, the higher temperature rating is permitted to be used.

Both the conductor and equipment temperature ratings shall be considered when designing or installing electrical systems. The ampacity of the conductor shall be selected so that the lowest temperature rating of the conductor or any equipment in the circuit is never exceeded.

Work Space Clearance – 110-16

Section 110-16 includes installation and design considerations to provide for safe and efficient operation of electrical equipment. Requirements regarding work space clearance limitations, lighting, headroom, and means of entrance and exit are specified to help assure that workers can perform repairs and maintenance operations safely.

Working Clearances – 110-16(a). Equipment which may need to be repaired, serviced, maintained, etc., while energized, shall be provided with minimum working clearances. Table 110-16(a) lists the minimum working clearances.

Two factors affect the minimum working clearance dimensions. The first factor is the nominal voltage-to-ground at which the equipment operates. In most cases, higher voltage-to-ground equipment requires more work space clearance. The second factor is the condition of use. For voltages-to-ground above 150 V and up to 600 V, the minimum working clearances increase depending on the relationship of exposed live parts to ground and other exposed live parts. See Figure 1-16.

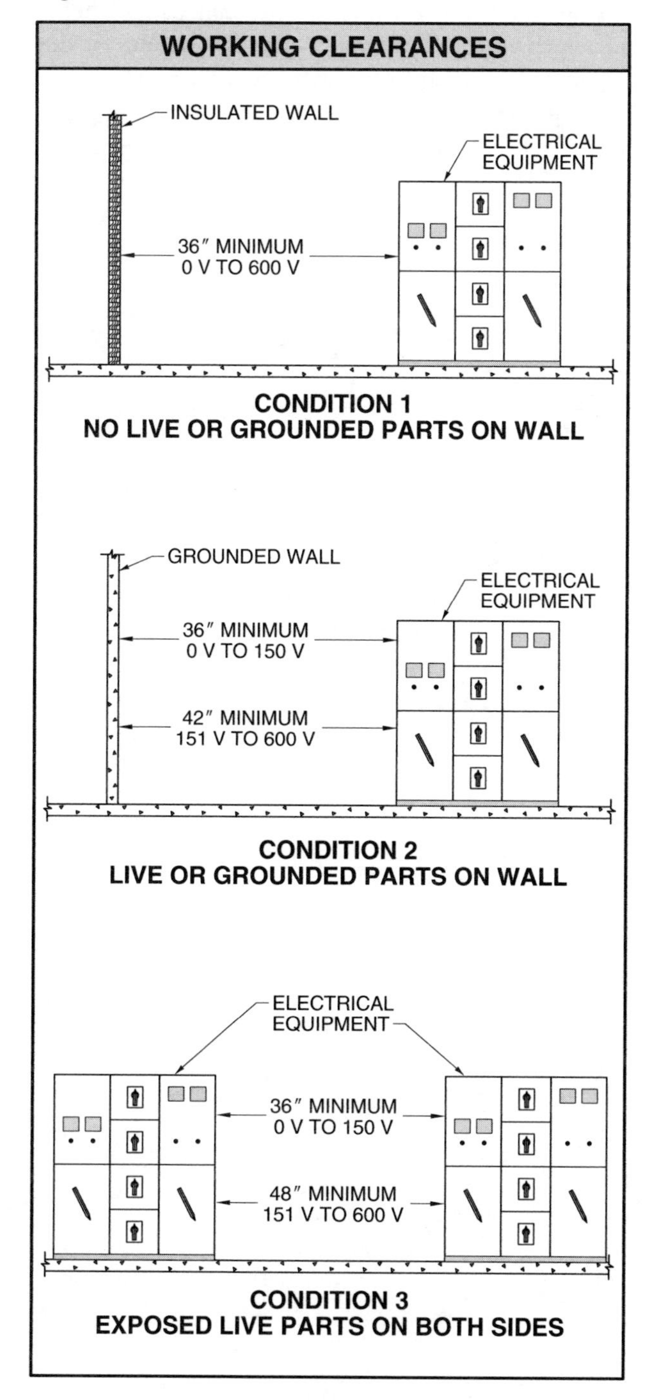

Figure 1-16. Three conditions determine working clearances of electrical equipment less than 600 V.

In general, these minimum work space clearances only apply to front working space, unless there are renewable components located in the rear of the equipment or all connections are not accessible from either the front or the sides. In these cases, the minimum clearances apply to the rear work space as well. See 110-16(a), Ex. 1.

Other conditions in which these work space clearances may be reduced are specialized conditions with strict limitations. See 110-16(a), Ex. 2 and 3. In any of these cases, the distances required shall be measured from the live parts, if exposed, or from the face of the enclosure, if one is provided. Concrete, brick, and tile walls are considered to be grounded when determining distances to ground.

Clear Spaces – 110-16(b). In addition to the front and rear minimum working clearances, 110-16 requires that the work space clearance width in front of the equipment shall be a minimum of 30″. The 30″ measurement is not required to be measured from the center of the equipment. See Figure 1-17. These work space clearances must extend from the floor or platform to a height of 6½′ or the height of the equipment, whichever is greater.

Equipment with doors or hinged panels shall be installed so doors or panels can be opened a minimum of 90°. Section 110-16(b) requires that the spaces required by 110-16 shall not be used for storage.

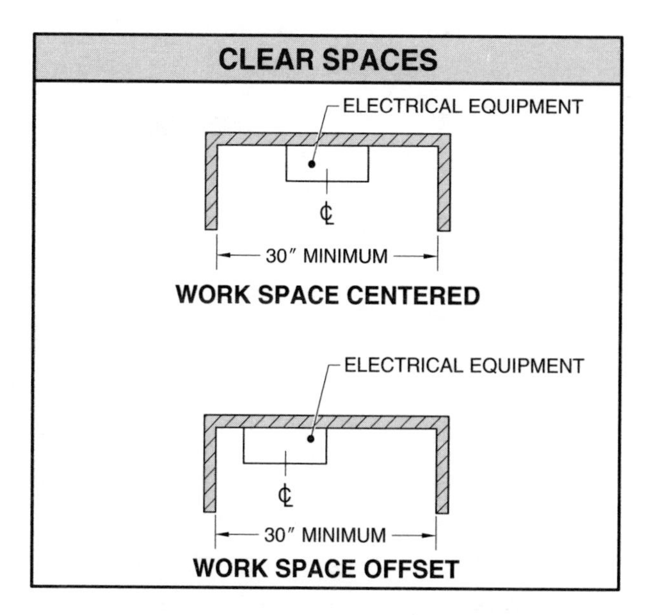

Figure 1-17. A 30″ clear work space is required for the safety of persons working on the live parts of equipment.

Access and Entrance – 110-16(c). Because of the potential hazards when servicing, maintaining, or repairing electric equipment, at least one entrance, of adequate size, shall be provided for entrance and exit purposes about electrical equipment. If the equipment is rated at 1200 A or more and over 6′ wide, and contains overcurrent, switching, or control devices, there shall be two entrances, one at each end. The entrances shall be a minimum of 2′ wide by 6½′ in height. See Figure 1-18.

The requirement for an additional entrance is not necessary if there is a clear and direct means of exit provided or if the working space clearance specified in Table 110-16(a) is doubled. See 110-16(c), Ex. 2.

Illumination – 110-16(d). Proper illumination is required about electric equipment. The work space around service equipment, panelboards, switchboards, main control centers (MCCs), etc. shall be sufficiently illuminated. It is not the intent of this section to require a dedicated light fixture to meet the illumination requirement.

Headroom – 110-16(e). This section establishes a general rule for minimum headroom about service equipment, panelboards, switchboards, MCCs, etc. at 6½′ or the height of the equipment whichever is greater. The headroom requirement is not necessary in dwelling units or for service equipment or panelboards rated at 200 A or less.

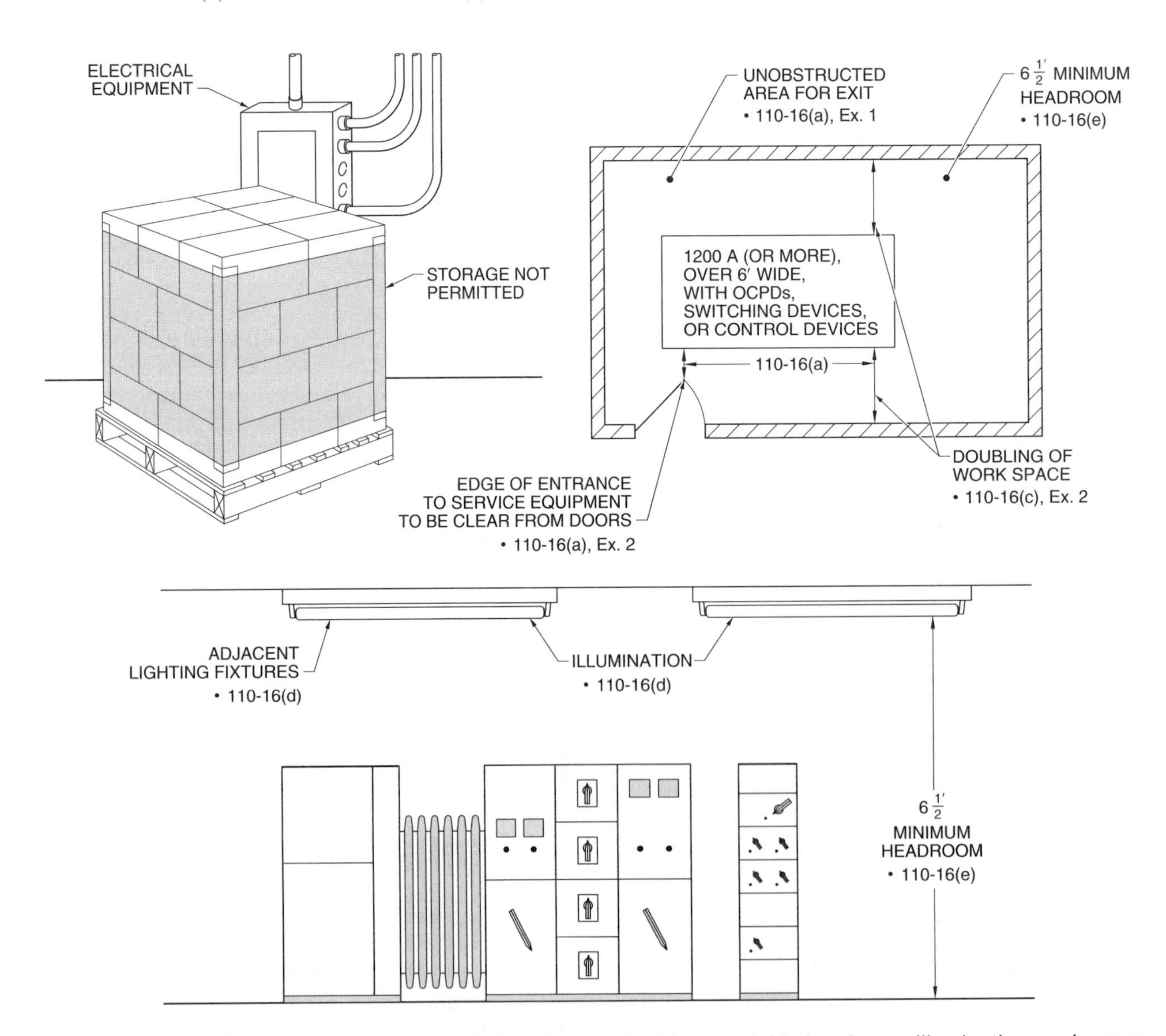

Figure 1-18. Persons working on live electrical equipment shall have suitable headroom, illumination, and means of access and entrance.

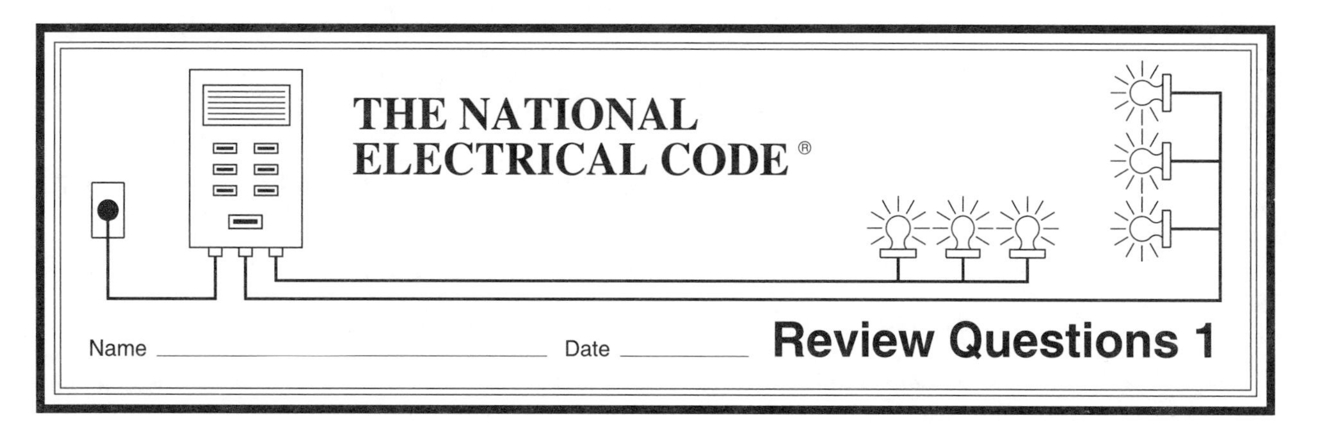

Review Questions

__________ **1.** The first four chapters of the NEC® apply generally unless they are ________ by the latter chapters.

__________ **2.** Chapter ________ of the NEC® is independent of the other chapters, and none of the provisions apply unless they are directly referenced.

T F **3.** Equipment designed to open circuits under fault conditions shall have an interrupting rating suitable for the available voltage and circuit current at the line side of the equipment.

T F **4.** Weatherproof equipment is constructed so that water will not enter it.

T F **5.** Installations underground in mines are not covered by the NEC®.

__________ **6.** Equipment that is required to be readily accessible shall be capable of being reached ________.

__________ **7.** A conductor's ________ is the amount of amperes it can carry continuously without exceeding its insulation rating.

__________ **8.** Boxes, installed within walls that are required to be taken down to gain access, are not considered ________.

__________ **9.** The word ________ indicates that an NEC® rule is mandatory.

T F **10.** Feeders are the circuit conductors between the final overcurrent device protecting the circuit and the outlets.

T F **11.** At least two entrances to electrical rooms shall be provided to give access to the working space about all electrical equipment.

T F **12.** All terminations and equipment are marked with tightening torques.

__________ **13.** The two-fold purpose of the NEC® is to protect ________ and ________ from the dangers associated with the use of electricity.

__________ **14.** The ________ of the NEC® contains a list of the electrical installations that are covered by the NEC®.

__________ **15.** Interpretations and approval of equipment are granted by the ________.

__________ **16.** The AHJ can permit alternate ________ for installation, if they believe the installation is equally effective at meeting the intended objective.

__________ **17.** Although done in conformance with the NEC®, all electrical installations may not be adequate for ________ use.

__________ **18.** The NEC® is not intended to be used as a(n) ________ specification.

T F **19.** Explanatory material appears in the NEC® in Fine Print Notes.

T F **20.** The minimum headroom of working spaces about electrical equipment in commercial buildings is $6\frac{1}{4}'$.

T F **21.** Electric equipment with ventilating openings shall be installed so that openings are not blocked.

T F **22.** The NEC® may be used as a design specification by qualified persons.

__________ **23.** If the AHJ finds the equipment or installation to be acceptable, it is ________.

__________ **24.** ________ conductors are conductors not encased in a material recognized by the NEC®.

__________ **25.** Connectors, couplings, locknuts, and bushings are electrical ________.

__________ **26.** Equipment that is ________ has been found to be suitable for a specific use or function.

__________ **27.** Equipment that is required to be located in sight from other equipment, shall be visible and not more than ________′ from the other equipment.

__________ **28.** Equipment that has been ________ is published in a directory by the testing lab certifying that it has met established standards.

__________ **29.** One of the primary purposes of raceways is to protect conductors from physical ________.

__________ **30.** The NEC® requires that all electric equipment and conductors be ________ to be deemed acceptable.

__________ **31.** When not specified, all conductors in the NEC® are considered to be ________.

__________ **32.** All unused openings in equipment are required to be effectively ________.

__________ **33.** Terminals for more than one conductor shall be ________.

__________ **34.** Wire connectors for use with conductors in direct burial applications shall be ________.

__________ **35.** Each ________ required by the NEC® shall be legibly marked to indicate its purpose.

See Electrical Equipment for problems 1 through 4.

NEC®	Answer		
________	T	F	**1.** The electrical equipment at A is readily accessible.
________	T	F	**2.** The electrical equipment at B is readily accessible.
________	T	F	**3.** The electrical equipment at C is readily accessible.
________	T	F	**4.** The electrical equipment at D is readily accessible.

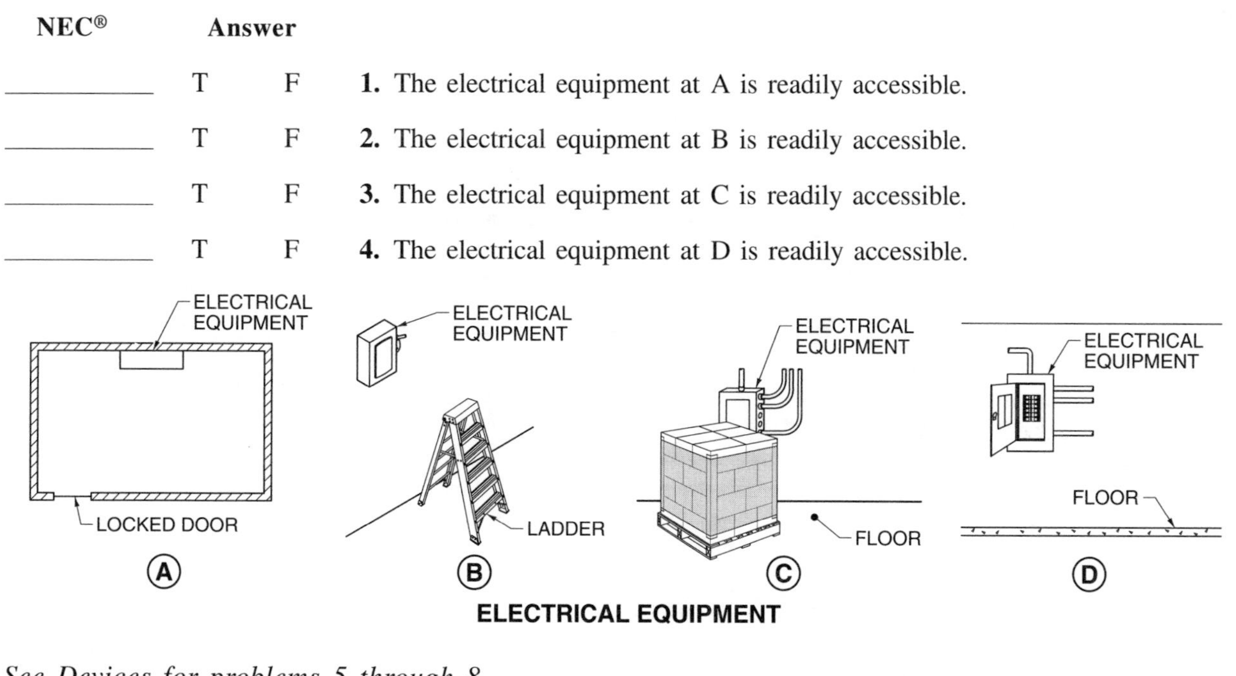

ELECTRICAL EQUIPMENT

See Devices for problems 5 through 8.

________	T	F	**5.** The electrical system unit at A is a device.
________	T	F	**6.** The electrical system unit at B is a device.
________	T	F	**7.** The electrical system unit at C is a device.
________	T	F	**8.** The electrical system unit at D is a device.

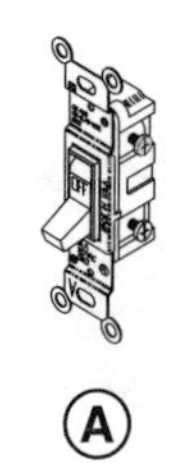

Ⓐ

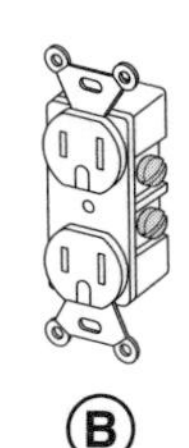

Ⓑ

Ⓒ

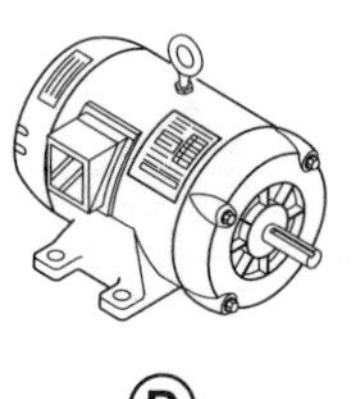

Ⓓ

DEVICES

See Loads for problems 9 through 11.

__________ __________ **9.** The load at A is a(n) ________ load.

__________ __________ **10.** The load at B is a(n) ________ load.

__________ __________ **11.** The load at C is a(n) ________ load.

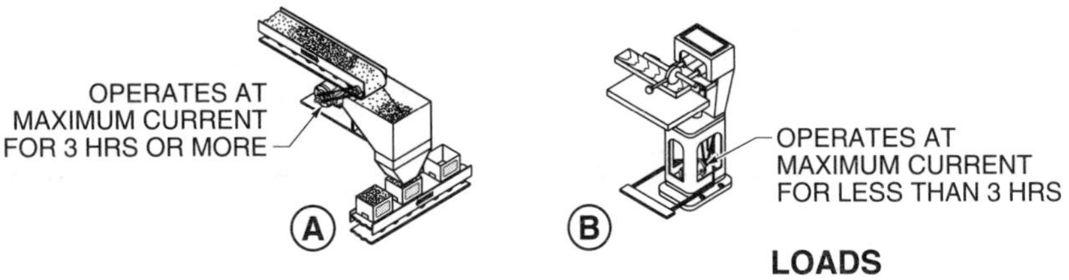

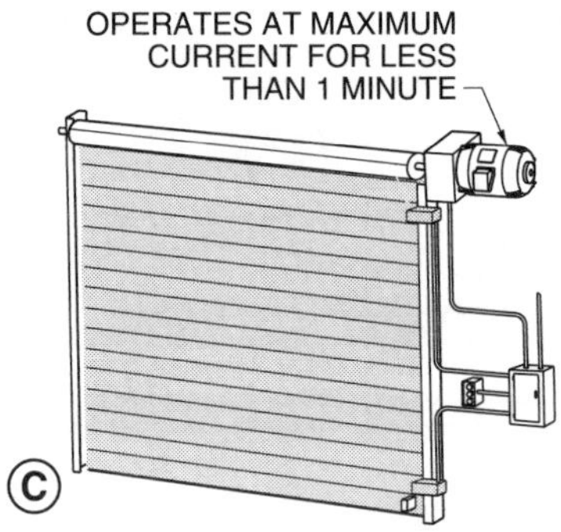

LOADS

See Within Sight for problems 12 through 15.

__________ T F **12.** The equipment at A is within sight.

__________ T F **13.** The equipment at B is within sight.

__________ T F **14.** The equipment at C is within sight.

__________ T F **15.** The equipment at D is within sight.

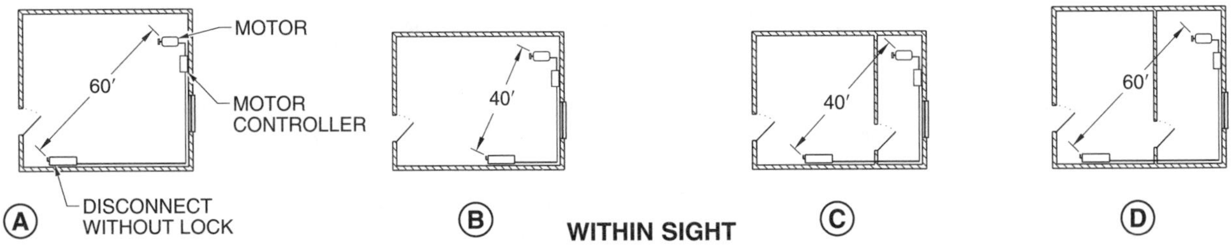

WITHIN SIGHT

See Working Clearances for problems 16 through 20.

__________ __________ **16.** The minimum working clearance at A is ________″.

__________ __________ **17.** The minimum working clearance at B is ________″.

__________ __________ **18.** The minimum working clearance at C is ________″.

__________ __________ **19.** The minimum working clearance at D is ________″.

__________ __________ **20.** The minimum working clearance at E is ________″.

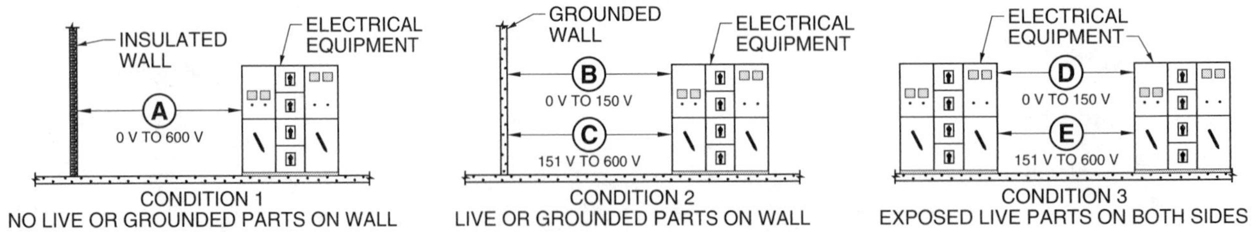

WORKING CLEARANCES

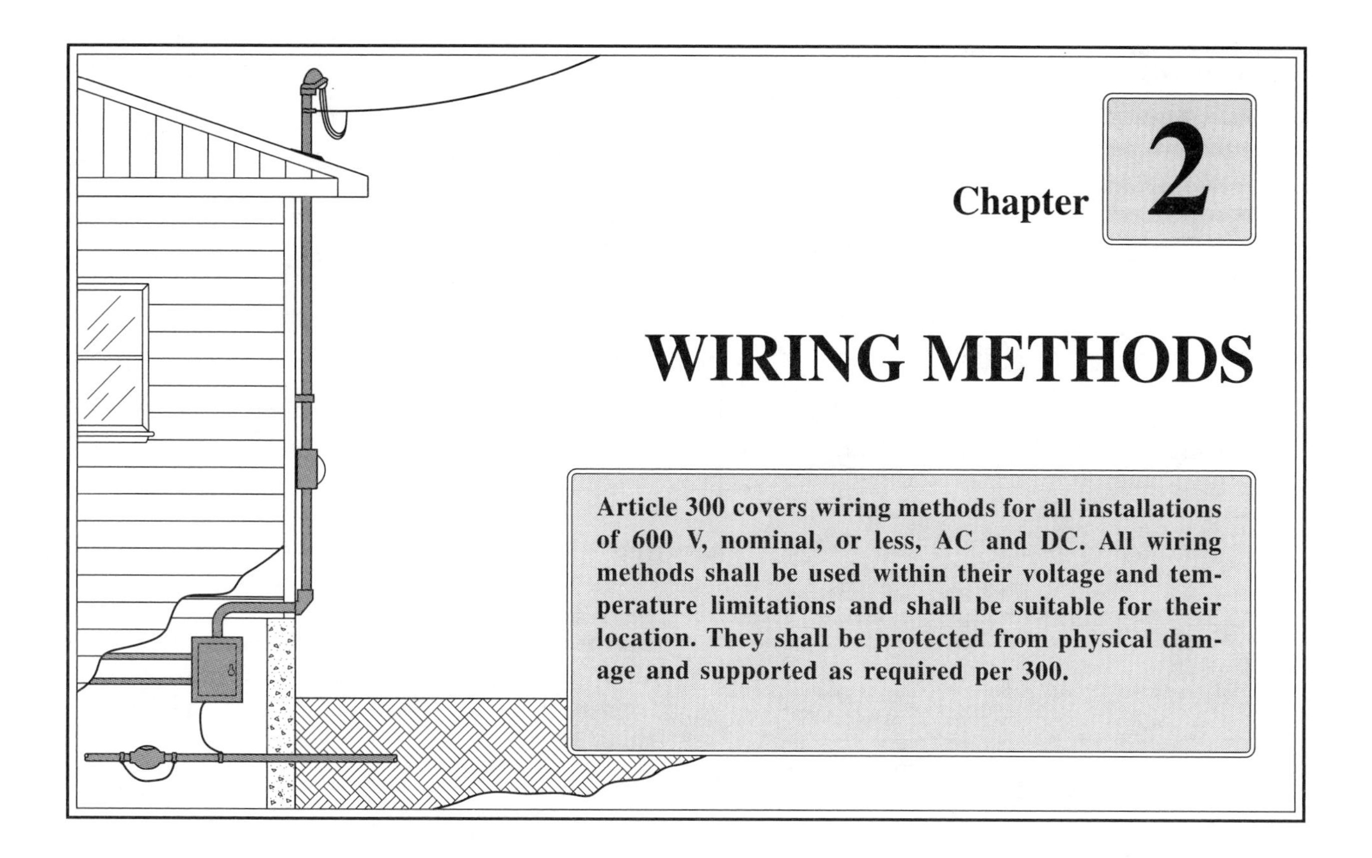

WIRING METHODS – 300

Requirements for wiring methods in all installations are covered in 300. Article 300 is concerned with the following aspects of the electrical system:

- Protecting the electrical conductors from physical damage, water damage, corrosion, and when they are installed in underground installations
- Securing and supporting raceways, cable assemblies, boxes, and cabinets
- Supporting conductors in vertical raceways
- Preventing the heating effect of inductive currents in metallic parts of an electrical system
- Securing the integrity of fire-resistant-rated walls
- Preventing the spread of toxic fumes in an air handling system in the event of a fire to the electrical system

Article 300 does not apply to wiring that is part of equipment, such as motors, motor control centers, or other factory-assembled control equipment. Article 300 does not apply to the following articles unless they are specifically referenced:

- 504 – Intrinsically Safe Systems
- 725 – Class 1, Class 2, and Class 3 Circuits
- 760 – Fire Protective Signaling Circuits
- 770 – Optical Fiber Cables
- 800 – Communications Systems
- 810 – Radio and Television Systems
- 820 – Community Antenna Television and Radio Distribution Systems

Conductors of Same Circuit – 300-3(b)

All conductors of the same circuit, including the grounded conductor and the EGCs shall be run together in the same raceway, cable tray, trench, cable, or cord. However, 300-3(b), Ex. 3 permits EGCs and EBJs to be run as a single conductor on the exterior of the raceway. See Figure 2-1.

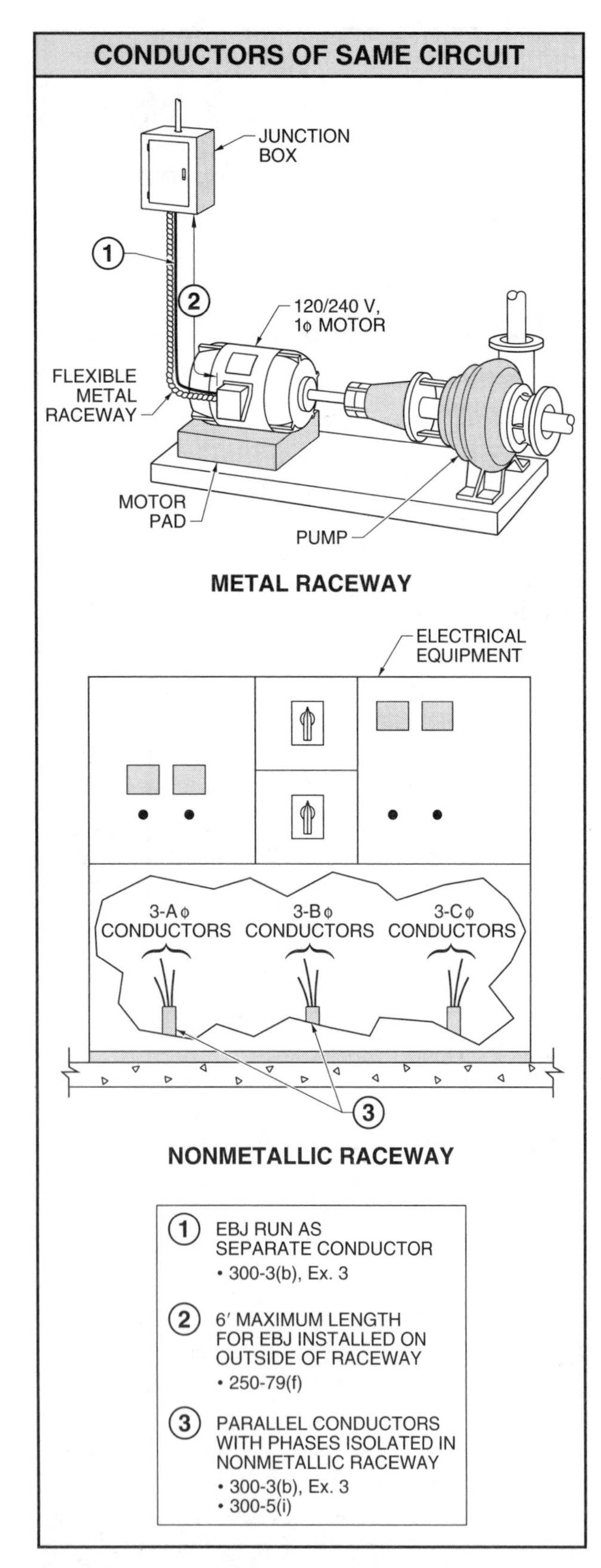

Figure 2-1. The EBJ may be run as a separate conductor on the exterior of the raceway.

Another example of 300-3(b), Ex. 3 is when the conductors are run in parallel and the phases are isolated in a nonmetallic raceway per 300-5(i). There is no inductive heating within these raceways because they are nonmetallic. The raceways shall not enter a metallic electrical enclosure individually. This causes inductive heating of the metal where the individual phase conductors penetrate the enclosures.

Conductors of Different Systems – 300-3(c)

Conductors of different systems (600 V or less, AC and DC) may occupy the same raceway, wiring enclosure, or equipment. All conductors shall have an insulation rating equal to the highest circuit voltage applied to any conductor within that raceway or wiring enclosure. For example, 300-3(c) permits the use of power and control wires from two different systems to be installed in a single raceway to supply a motor load. See Figure 2-2.

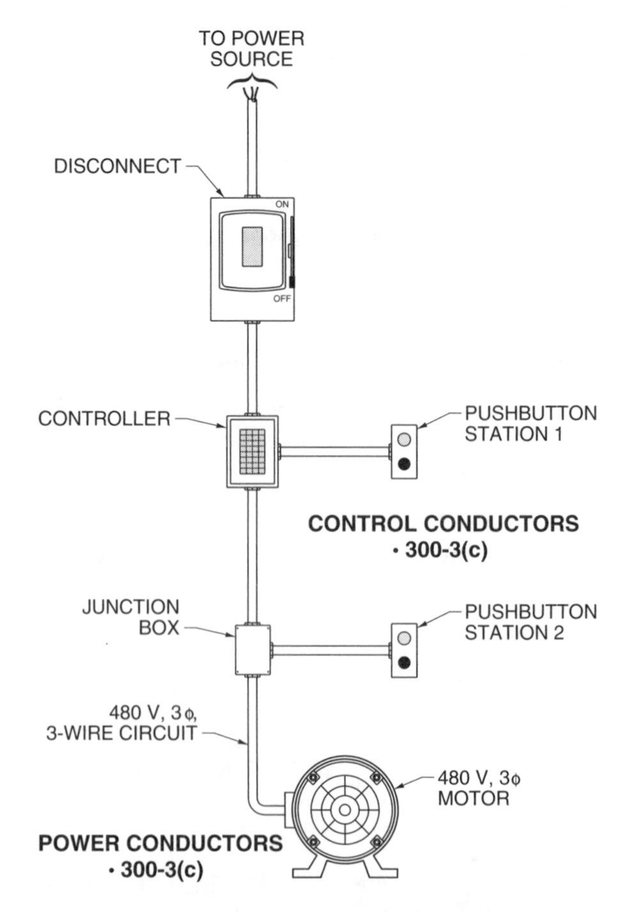

Figure 2-2. Conductors of different systems (600 V or less, AC and DC) may occupy the same raceway.

EQUIPMENT BONDING JUMPER

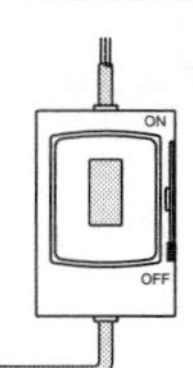

The closer the EGC or the EBJ is run to the current-carrying conductors, the lower its impedance. The EBJ may be run outside of a piece of flexible metal conduit. It shall never be wrapped around the flexible metal conduit. Wrapping the EBJ around the flexible metal conduit would cause the impedance to rise dramatically in the event of a ground fault. Per 250-79(f), an EBJ installed on the outside of the raceway or enclosure shall not exceed 6′ and shall be routed with the raceway or enclosure. It intentionally does not say that the EBJ shall be run around the raceway.

Protection from Physical Damage – 300-4

Where conductors are subject to physical damage, they shall be adequately protected. The conductors inside raceways and cable assemblies require protection against the penetration of nails and screws. Holes that are drilled into framing members to permit the installation of raceways or cables shall be at least 1¼″ from the nearest edge of the wood member. This assures that nails or screws do not penetrate installed raceways or cables. See Figure 2-3.

Bored Holes and Notches in Wood – 300-4(a). When a cable or raceway-type wiring method is installed through a bored hole of a wooden framing member, it shall be at least 1¼″ from the nearest edge of the joist or rafter per 300-4(a)(1). If the cable or raceway is less than 1¼″ from the edge, a steel plate or bushing at least ¹⁄₁₆″ thick shall be used to protect the cable or raceway from being damaged by nails or screws per 300-4(a)(1). The steel plate shall be wide enough and long enough to cover the wiring area.

Cables or raceways are permitted to be laid in notches in wooden studs, joists, rafters, or other wooden members provided the conductors are protected against nails or screws by a steel plate at least ¹⁄₁₆″ thick per 300-4(a)(2). The notch cannot weaken the building structure. The requirement for bored holes and notches does not apply to 345–IMC, 346–RMC, 347–PVC, or 348–EMT.

Cables and Nonmetallic Tubing Through Metal Framing Members – 300-4(b). Where nonmetallic-sheathed cables pass through cut or drilled slots or holes in metal members, the cable shall be protected by bushings or grommets securely fastened in the opening prior to installation of the cable per 300-4(b)(1). This applies to factory and field slots and holes.

A steel sleeve, steel plate, or steel clip shall be used to protect nonmetallic-sheathed cables or ENT cable or tubing where nails or screws are likely to penetrate per 300-4(b)(2). The steel plate shall be at least ¹⁄₁₆″ thick.

Cables Through Spaces Behind Panels Designed to Allow Access – 300-4(c). Cable or raceway-type wiring methods, installed above suspended ceiling panels, shall not be laid on the suspended ceiling. The wiring methods shall be supported per their applicable articles.

Cables and raceways shall not be laid on a suspended ceiling.

Cables and Raceways Parallel to Framing Members – 300-4(d). Where cables and raceways are installed parallel to framing members such as joists, rafters, or studs, there shall be 1¼″ minimum clearance between the outside surface of the raceway and the nearest edge of the framing member. If this distance falls below 1¼″, a ¹⁄₁₆″ steel plate shall be used to protect the raceway from nails, screws, etc. This applies to exposed and concealed locations.

Cables and Raceways in Shallow Grooves – 300-4(e). Cables and raceways installed in shallow grooves and covered by drywall, paneling, etc. shall be protected by 1¼″ minimum free space. If this distance falls below 1¼″, a ¹⁄₁₆″ steel plate shall be used to protect the raceway from nails, screws, etc.

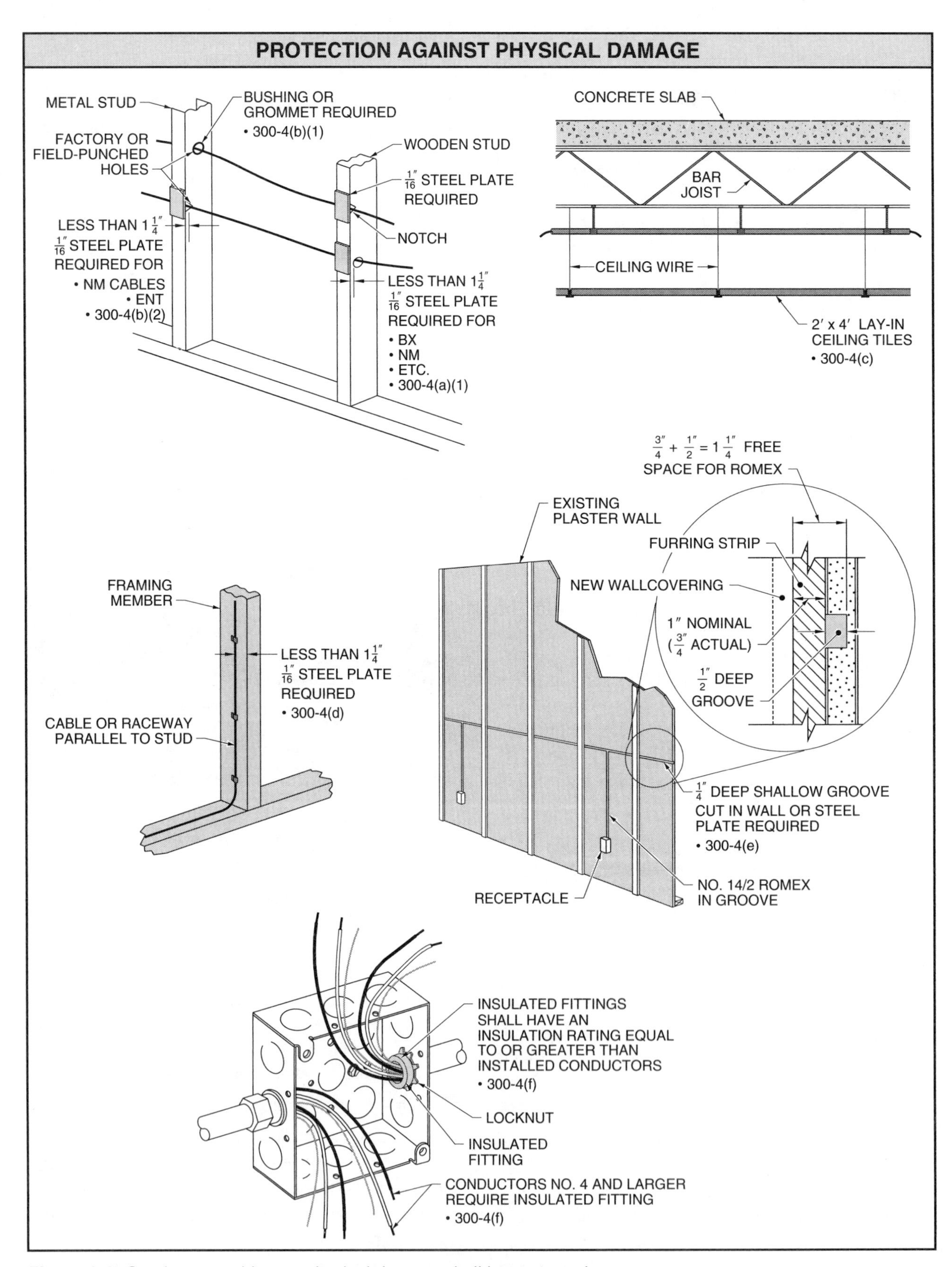

Figure 2-3. Conductors subject to physical damage shall be protected.

Underground Installations – 300-5

The general requirements for installing cables, conduits, and other raceway systems underground are covered by 300-5. The main concerns are for the protection of the conductors, splices, and taps and the prevention of moisture from coming in contact with energized parts of the electrical system. The burial depths depend on the location of the installation and the wiring method used.

Minimum Burial Depths of Cables and Raceways. Table 300-5 establishes the minimum cover requirements that cables and raceways shall have when installed in the ground. *Cover* is the shortest distance measured between a point on the top surface of any direct buried conductor, cable, conduit, or other raceway and the top surface of finished grade, concrete, or similar cover. For example, the minimum cover for any wiring method not listed in Table 300-5 is 24″ from the top surface of the wiring method to the top surface of the finished grade, concrete, or similar cover. The minimum cover for RMC under a one-family dwelling driveway is 18″ as listed in Table 300-5. See Figure 2-4.

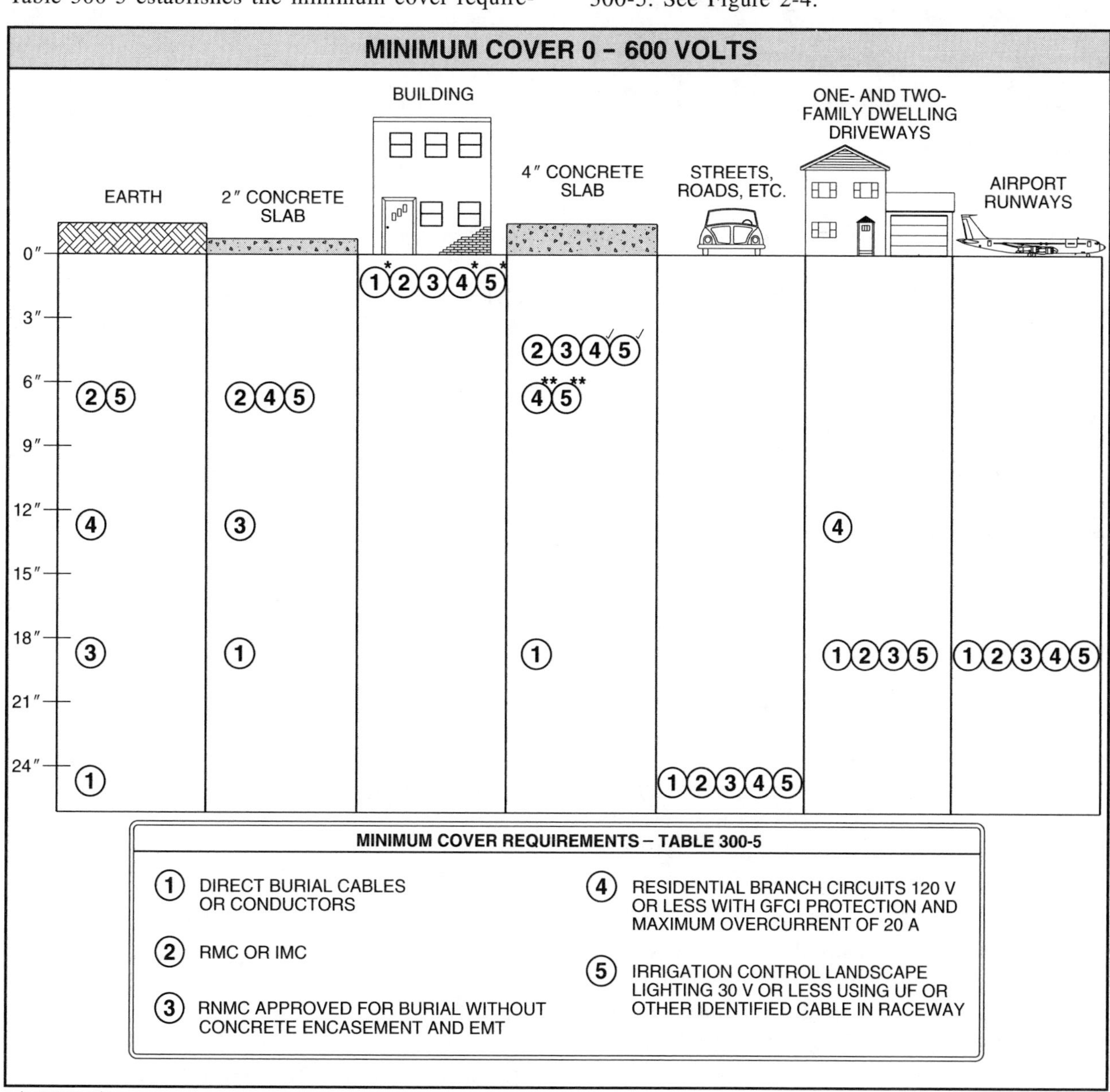

Figure 2-4. All cables and raceways approved for direct burial shall meet minimum cover requirements.

Raceways shall maintain the required burial depth of Table 300-5.

EMT is not shown in Table 300-5. However, this does not mean that EMT cannot be buried in earth or concrete. Section 348-1 permits EMT to be installed in concrete and in direct contact with the earth. It shall not be installed in cinder concrete or cinder fill, unless protected on all sides by a layer of noncinder concrete at least 2″ thick or unless the tubing is at least 18″ under the fill.

Cables or raceways buried in a trench below a 2″ concrete slab or equivalent shall maintain the required burial depths shown in Table 300-5. Cables or conductors that are installed under a building have no minimum burial depth provided the cables and conductors are installed in an approved raceway.

Cables or raceways installed under a 4″ minimum thick concrete exterior slab without vehicular traffic shall meet the requirements shown in Table 300-5. The concrete slab shall extend at least 6″ beyond the cables or raceways. All cables and raceways installed under streets, highways, roads, driveways, and parking lots shall be buried a minimum depth of 24″.

Cables and raceways installed under one- and two-family dwelling unit driveways and outdoor parking areas shall be buried a minimum of 18″. These driveways and parking areas are to be used for dwelling-related purposes only. Underground, residential branch circuits that are rated 20 A, 120 V, 1ϕ or less and are GFCI-protected have a minimum burial depth of 12″.

Cables and raceways installed in or under airport runways, including adjacent areas where trespassing is prohibited, shall be buried to a minimum depth of 18″. Running conduit on the surface of the ground is not prohibited as long as the conduit is securely fastened in place and not exposed to physical damage such as vehicular traffic. Depth requirements may be reduced when cables or conductors rise for termination or splices or when other access is required per Table 300-5, Note 3.

Protection from Damage – 300-5(d). All cables and conductors emerging from the ground shall be protected from physical damage to a maximum depth of 18″ below finished grade. Cables and conductors installed on the side of a pole or building shall be provided with physical protection up to a height of 8′ above finished grade. This protection may be provided by using RMC, IMC, Schedule 80 RNMC, or equivalent as determined by the AHJ. See Figure 2-5.

Buried Splices and Taps – 300-5(e). Splices and taps are permitted to be buried in a trench without the use of a splice box, provided approved methods are used, with listed materials per 110-14(b).

Backfill – 300-5(f). Care must be taken so as not to damage cables and raceways when backfilling trenches. Heavy rocks or any sharp or corrosive materials shall not be used in the backfill material. Where necessary, running boards, sleeves, or some form of granular material shall be used to prevent damage to the cable or conduit.

Underground Bushings – 300-5(h). A bushing or terminal fitting shall be used on the end of a conduit where direct burial cables leave the conduit. A seal may be used provided it offers the same protection as a bushing.

Ground Movement – 300-5(j). Direct-buried cables, raceways, and conductors that are subject to ground movement shall be arranged to prevent damage to the equipment to which these cables, raceways, and conductors are connected. The ground movement may be caused by settlement or frost heaves. Section 300-5(j), FPN recognizes "S" type loops in underground direct burial cable to raceway transitions and expansion joints in pipe risers to fixed equipment installations.

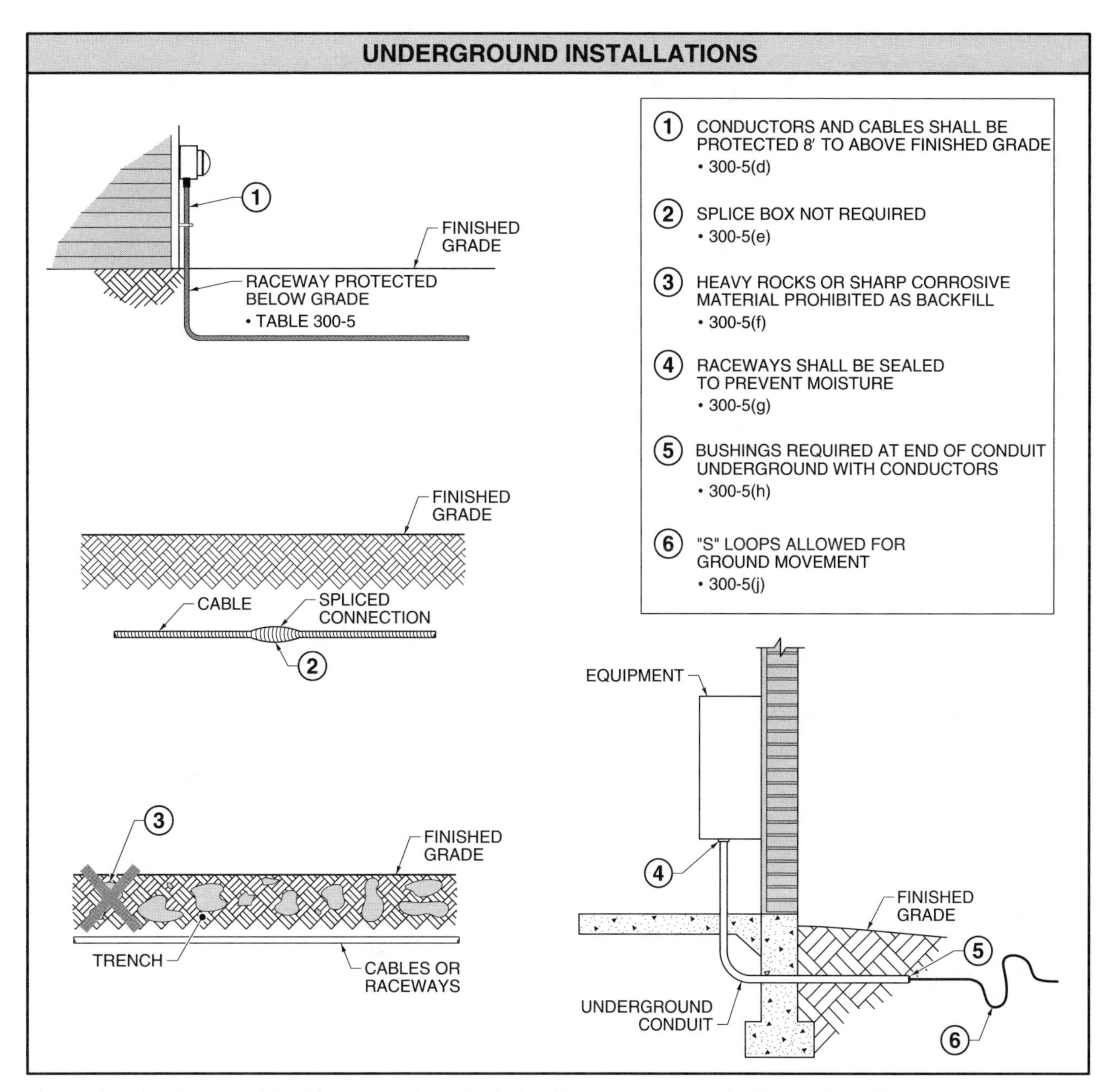

Figure 2-5. Underground installation of direct-buried cables or raceways shall be protected.

Protection Against Corrosion – 300-6

All of the metallic components that make up an electrical system shall be of materials suitable for the environment in which they are to be installed. This includes the metal raceways, cable armor, boxes, cable sheathing, cabinets, elbows, couplings, fittings, supports, and support hardware. Boxes or cabinets marked “Raintight,” “Rainproof,” or “Outdoor Type” can be installed out-of-doors.

Indoor Wet Locations – 300-6(c). In dairies, breweries, canneries, or wherever the walls are frequently washed down, the electrical system shall not entrap water or moisture between itself and the surface to which it is mounted. This is true for the entire electrical system including boxes, fittings, and conduits. There shall be at least 1/4″ airspace between the electrical system and the wall or the supporting surface. See Figure 2-6.

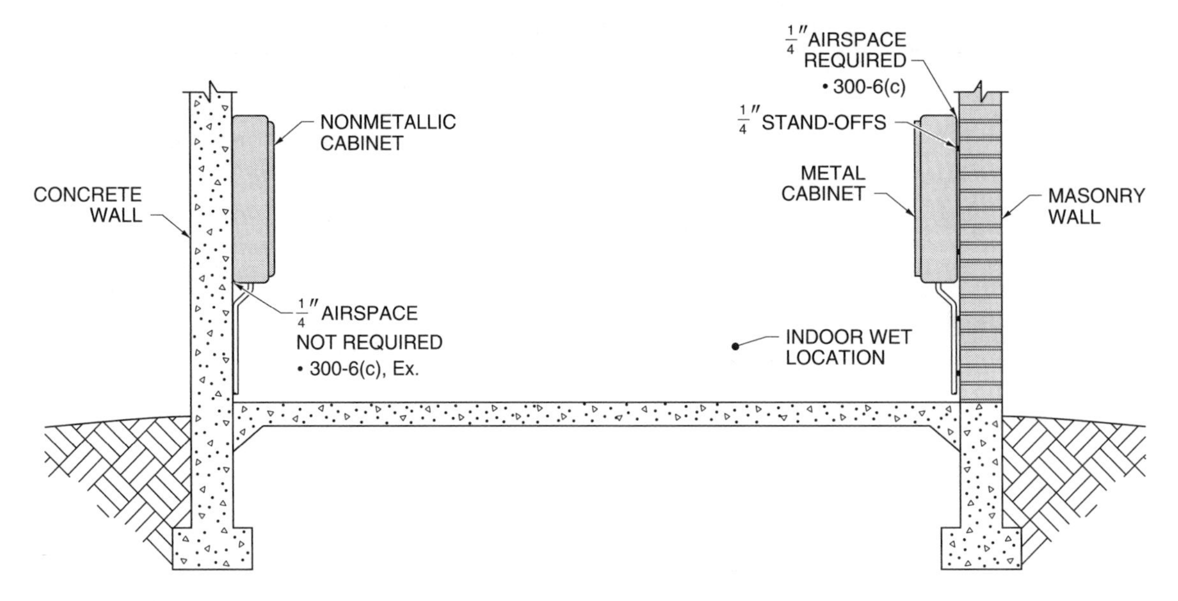

Figure 2-6. A ¼″ airspace is required in wet locations so that the electrical system cannot entrap water or moisture between itself and the surface.

Raceways Exposed to Different Temperatures – 300-7. Installed raceways that are exposed to different temperatures present two problems. See Figure 2-7. The first is that condensation may build up inside the raceway system. The second is the expansion and contraction of the raceway system.

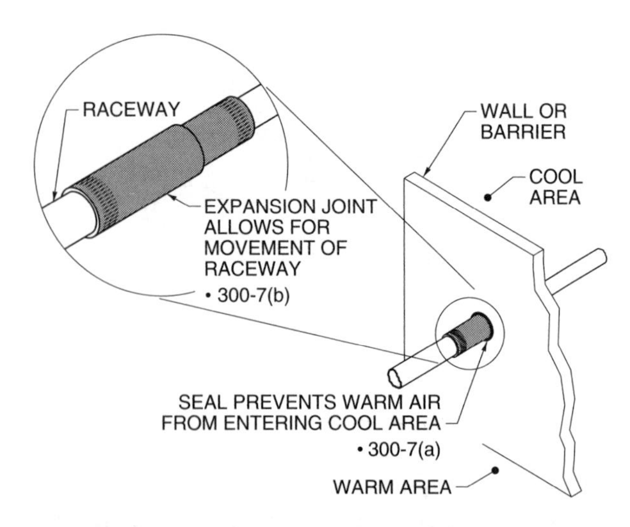

Figure 2-7. Installed raceways that are exposed to different temperatures shall avoid condensation and allow for expansion and contraction.

Section 300-7(a) requires seals to be installed in the raceway system whenever the raceway changes from one environmental temperature to another. Air, traveling through a raceway and going from a warm environment into a cooler environment, condensates in the raceway as it travels through the cooler environment. Accumulation of this condensed moisture could travel into equipment that contains live parts. The required seal shall be installed where the raceway changes from one environment to the other.

Section 300-7(b) requires expansion joints in the raceway system to compensate for thermal expansion and contraction. Expansion joints allow for the movement of the conduit within the expansion fitting. This relieves any strain on other connections within the conduit system.

Electrical Continuity of Metal Raceways and Enclosures – 300-10

All noncurrent-carrying metallic parts of an electrical system shall be bonded together with a permanent and continuous bonding method. The metal raceways, cable armor, and other metallic enclosures shall be electrically connected into a continuous low-impedance path. This low-impedance path is critical in allowing enough fault current to flow to facilitate the opening of the overcurrent device.

If the raceway system were nonmetallic, all noncurrent-carrying metal parts of that system would have to be bonded to the EGC. If a noncurrent-carrying metal part were to come in contact with a live conductor, this low-impedance path for fault current to flow through would have to be present in order to trip or open the overcurrent device. See Figure 2-8.

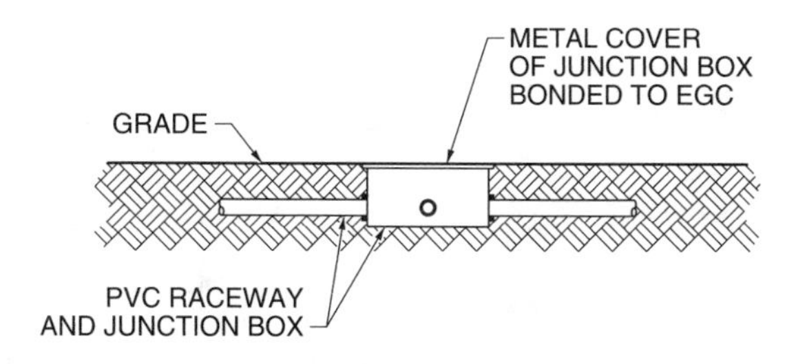

Figure 2-8. All noncurrent-carrying metallic parts of an electrical system shall be bonded together with a permanent and continuous bonding method.

Securing and Supporting Raceways, Cable Assemblies, Boxes, and Cabinets – 300-11

Wiring systems installed above a fire-rated floor, ceiling, or roof or ceiling assembly shall have an independent means of being secured and supported. See Figure 2-9. These wiring systems shall not be secured or supported by the ceiling assembly or the ceiling support wires. The ceiling assembly may support the wiring system if the ceiling assembly and support of the wiring system were both tested together as part of the fire-rated assembly.

Raceways Used as Means of Support – 300-11(b). Raceway systems shall not be used to support raceways, cables, or other nonelectric equipment. The intent is to prevent other cabling systems such as telephone and coaxial cables from being wrapped around existing electrical raceway systems. This could lead to heat dissipation problems for the conductors within the electrical raceway system. Another problem could be that the additional weight might cause undue stress on the electrical raceway system and cause it to fail.

Mechanical Continuity of Raceways and Cables – 300-12

Metal or nonmetallic raceways, cable armors, and cable sheaths shall be continuous between all cabinets, boxes, fittings, or other enclosures or outlets. However, per 300-12, Ex., a short section of raceway used to provide support or protection of cable assemblies from physical damage does not have to be continuous.

Mechanical and Electrical Continuity of Conductors – 300-13

Conductors in raceways shall be continuous between all outlets, boxes, devices, etc. There shall be no splice or tap within a raceway. Raceways that provide removable or hinged covers may have splices or taps contained within them.

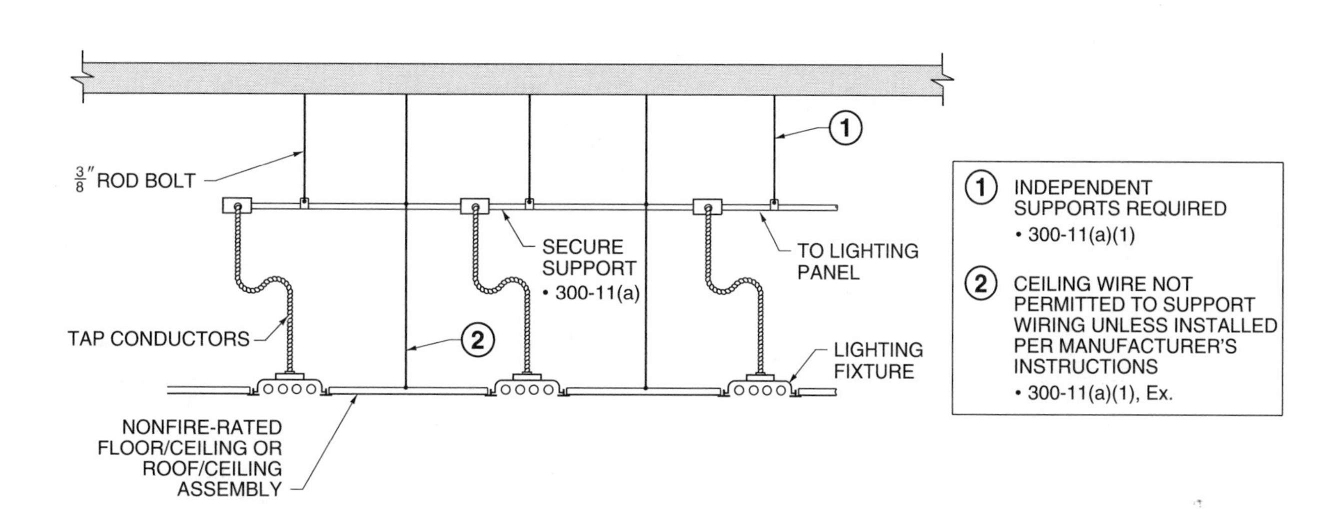

Figure 2-9. Wiring systems installed above a roof/ceiling assembly shall have an independent means of support.

Removing Devices in Multiwire Circuits – 300-13(b). The grounded conductor (neutral) of a multiwire branch circuit shall never be opened. If the grounded conductor were opened and the loads were not equal, then the voltage across these loads would be changed to something other than 120 V. The more the unbalance, the more dramatic the voltage change. See Figure 2-10.

When a multiwire branch circuit is run from outlet box to outlet box, the terminal screws of the device, such as a duplex receptacle, shall not be used to make the grounded conductor continuous. All of the grounded conductors of a multiwire branch circuit within an outlet box shall be spliced together with a tail provided for the terminal screw of the device. If this device were replaced while the circuit was energized, the grounded conductor would not become open. See Figure 2-11.

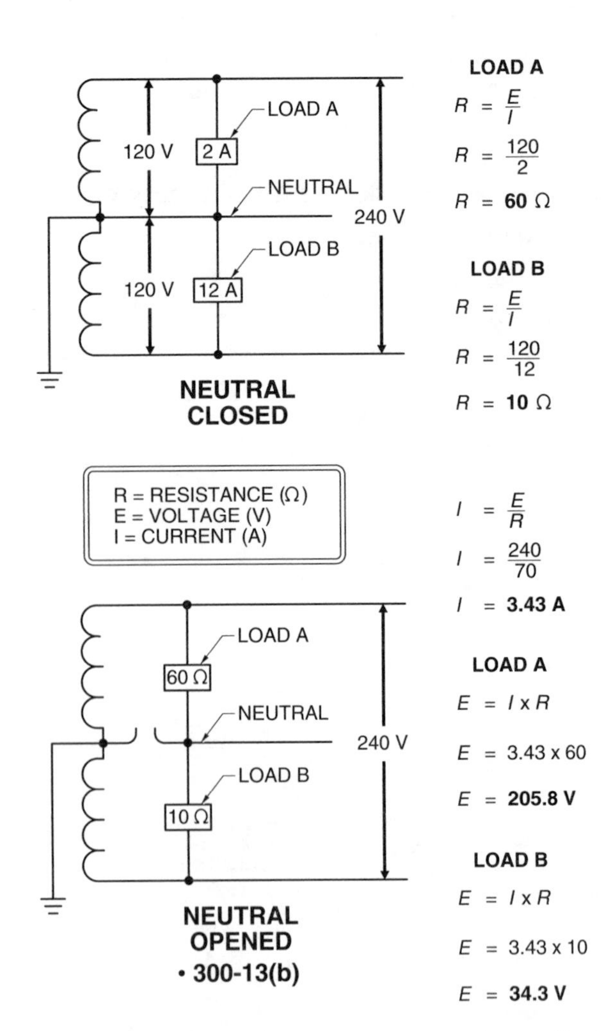

Figure 2-10. The grounded conductor (neutral) of a multiwire branch circuit shall never be opened.

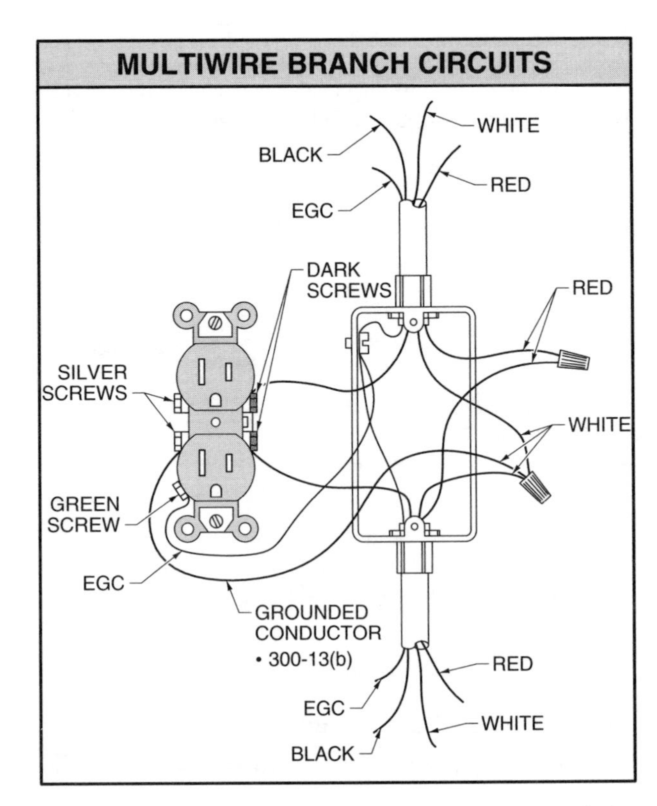

Figure 2-11. The grounded conductors shall be spliced together to prevent an open neutral.

Length of Conductors at Outlet Box – 300-14

At least 6″ of free conductor shall be left at each outlet, junction, or switch point. The free conductor is to be used for terminating receptacles, fixtures, or other devices. Section 300-14, Ex. does not require the 6″ of free conductor if the conductors are not spliced or terminated at this point.

Boxes, Conduit Bodies, or Fittings Required – 300-15

A box, conduit body, or fitting shall be installed at each conductor splice connection point, outlet, switch point, junction point, or pull point. See Figure 2-12. Section 300-15(a) requires a box or conduit body for the connection of conduit, electrical metallic tubing, surface raceway, or other raceways. There are two exceptions to this basic rule. Section 300-15(a), Ex. 1 permits conductor splices in surface raceways, wireways, header-ducts, multioutlet assemblies, auxiliary gutters, and cable trays. Section 300-15(a), Ex. 2 permits a fixture that is used as a raceway to have conductor splices within the fixture.

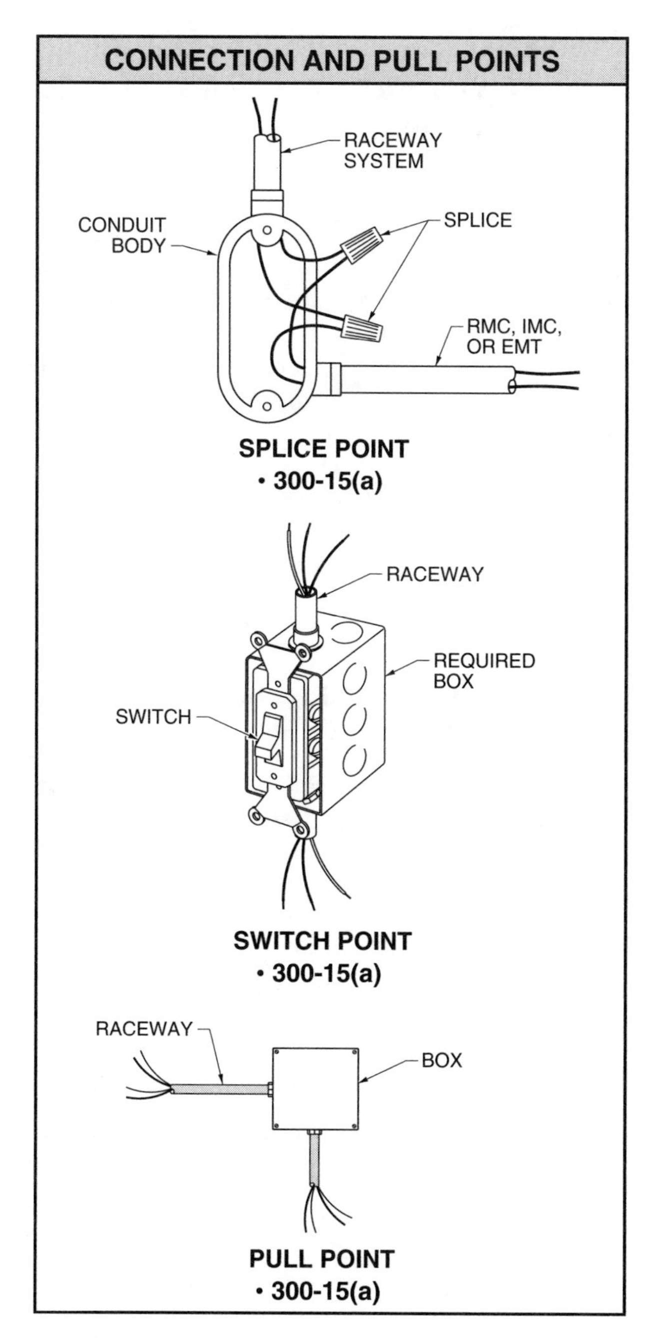

Figure 2-12. A box, conduit body, or fitting shall be installed at each connection or pull point.

Section 300-15(b) requires that only a box may be installed for the connection of AC cable, MC cable, MI cable, metal-sheathed cable, NM cable, or any other cable assembly. Per 300-15(b), Ex. 1, conduit is permitted to be used as protection for cables that are buried and exit the ground. A box is not required where cable enters or leaves this conduit.

Section 300-15(b), Ex. 8 permits the use of a fitting in lieu of a box when there is a transition from a cable to a raceway system. The fitting shall be identified for this use, and the conductors shall not be spliced at this point. See Figure 2-13.

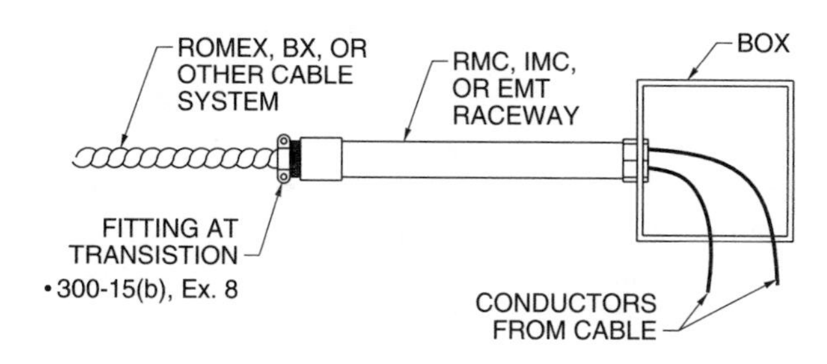

Figure 2-13. A box is not required at the transition from a cable to a raceway system.

Section 300-15(b), Ex. 9 permits splices and taps that are listed for direct burial to be installed without the use of a box. There is no requirement for a box based on the length of a raceway.

Fittings and connectors shall only be used for applications for which they are designed. For example, it is a violation of the NEC® to terminate MC cable in a connector listed for a BX cable only.

Number and Size of Conductors in a Raceway – 300-17. The number and size of conductors in any raceway shall not be more than permits the dissipation of heat given off by the conductors when they carry current. The installation and withdrawal of conductors from a raceway system should be possible without causing damage to the conductors or to their insulation.

The general rule for conduit fill as found in Chapter 9, Table 1 still applies. When a conduit contains three or more conductors, the cross-sectional area of the conductors shall not exceed 40% of the cross-sectional area of the conduit or tubing. Chapter 9, Appendix C lists the maximum number of conductors permitted in the various types of raceways according to the diameter size of the raceway. For example, IMC has a thinner wall thickness when compared to RMC. Therefore the ID is larger for IMC as opposed to RMC.

Raceway Installations – 300-18. Raceway systems, including all required boxes, shall be installed before the conductors can be pulled. The primary function of the raceway system is to provide protection for the conductors. There have been many reports of damage to conductors that were pulled into incomplete raceway and enclosure systems.

SUPPORTING CONDUCTORS IN A VERTICAL RACEWAY – 300-19

In long vertical runs, the weight of the conductors shall not rely on the termination points for their sole support. The conductors shall be supported at the top of the raceway or as close as possible to the top.

Spacing Conductor Supports – 300-19(a)

The intervals at which intermediate cable supports are required are determined by weight of the conductor. The larger the conductor, the more it weighs, thereby requiring more frequent supports. The conductor material is also a factor to consider. Copper is heavier than aluminum, therefore it requires supports at more frequent intervals. Table 300-19(a) shows the required spacing for conductor supports in vertical runs. For example, No. 250 kcmil Al conductors require supports every 135′, while No. 250 kcmil Cu conductors require supports every 60′.

Support Methods – 300-19(b)

Vertical conductors may be supported by insulating wedges installed in the end of the raceway. This helps prevent damage to the insulation due to the weight of the conductors.

Another recognized method of providing support for vertical conductors is to install boxes at required intervals per Table 300-19(a). Insulating supports or cleats are installed to provide the necessary support for each conductor.

The third method of providing support for vertical conductors is to deflect the conductors by at least 90°. The conductors would have to be run horizontally a distance not less than twice the diameter of the conductors. The conductors would have to be carried on two or more insulating supports. The conductors should be secured to these supports by tie wires.

PREVENTING HEATING EFFECT OF INDUCTIVE CURRENTS IN METALLIC PARTS – 300-20

AC conductors produce a fluctuating magnetic field around the conductor. If these conductors are in close proximity to ferrous metals, the fluxing magnetic field causes the molecules in the ferrous metal to move back and forth producing a heating effect in the ferrous metal. If conductors of opposite polarities are brought together, their fluxing magnetic fields cancel one another out. Therefore conductors of opposite polarities, which are run in close proximity to one another, do not produce a heating effect on ferrous materials.

Conductors Grouped Together – 300-20(a)

AC conductors installed in metal raceways or enclosures shall be arranged so as not to cause induction heating of the metal raceways or enclosures. To accomplish this, all phase conductors, grounded conductors (if used), and EGCs shall be grouped together. See Figure 2-14. Since all of the circuit conductors are grouped together, the magnetic fields around each individual conductor cancel out. This eliminates the inductive heating effect on the ferrous metal.

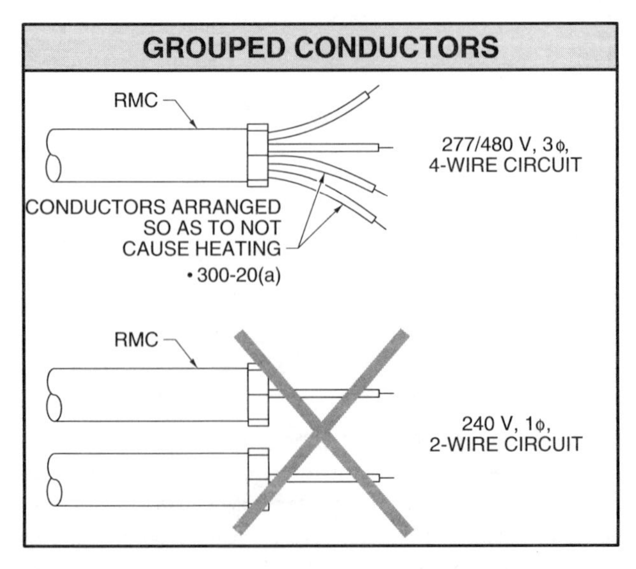

Figure 2-14. All phase conductors, grounded conductors, and EGCs shall be grouped together.

Single Conductors Passing Through Metal – 300-20(b)

Single conductors carrying AC current must not pass through metal with magnetic properties unless the inductive heating effect is minimized by cutting slots in the metal between the individual holes through which the individual conductors pass. See Figure 2-15.

If conductors carrying AC currents are run into a metal enclosure, they should all enter the enclosure through the same opening. See Figure 2-16.

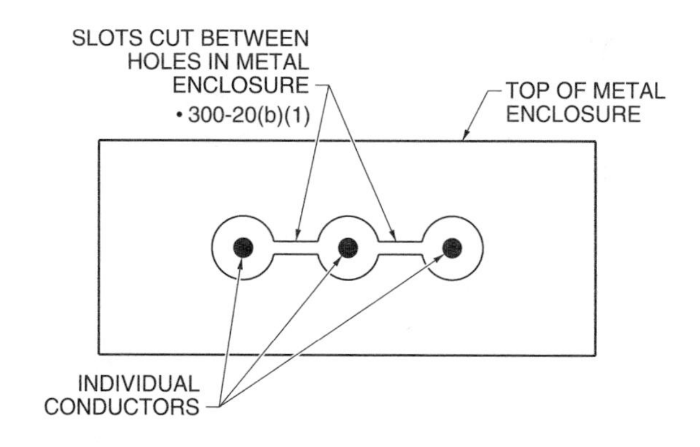

Figure 2-15. The inductive heating effect of metal is minimized by cutting slots between the holes.

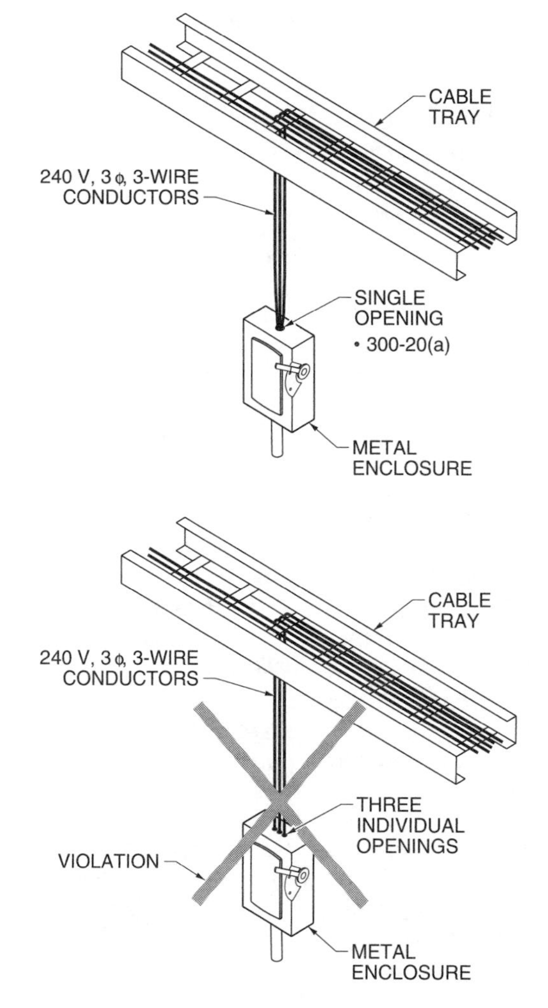

Figure 2-16. Conductors carrying AC currents should enter the enclosure through the same hole.

SECURING INTEGRITY OF FIRE-RESISTANT-RATED WALLS – 300-21

The installation of electrical systems in hollow spaces, vertical shafts, and air-handling ducts shall be made without decreasing the integrity of the fire resistance rating of the installation. The concern here is the spread of fire or products of combustion. Whenever an electrical system penetrates a fire-resistant-rated wall, floor, or ceiling, the opening around the penetration shall be firestopped using approved methods to restore the fire resistance rating. See Figure 2-17.

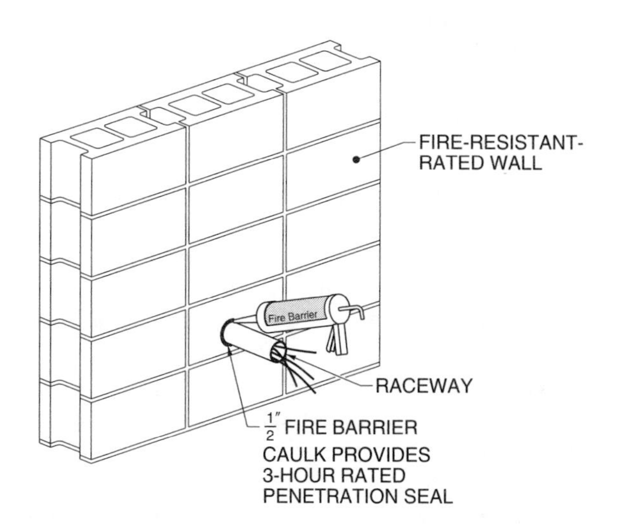

Figure 2-17. Openings in fire-resistant-rated walls shall be firestopped.

MAGNETIC FIELDS AND CONDUCTORS

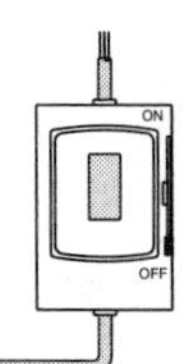

An AC conductor produces a fluxing magnetic field around the conductor. The more current that flows through the conductor, the stronger the fluxing magnetic field becomes. When a ferrous metal (iron or steel) surrounds this single current-carrying conductor, there is an interaction between the molecules of the ferrous material and the fluxing magnetic field. The molecules of the ferrous material move back and forth producing a heating effect on the ferrous material. The more current, the stronger the magnetic field. The stronger the fluxing magnetic field becomes, the more heat produced in the ferrous material. This heat may be hot enough to ignite combustible materials.

Two conductors with current flowing in opposite directions produce offsetting magnetic fields that cancel one another out. With the magnetic fields canceled out, there is no induction heating effect on the ferrous material. An example of two such conductors is the ungrounded conductors of a 1ϕ, 3-wire system. Another example of conductors whose magnetic fields cancel one another out is the three ungrounded conductors of a 3ϕ system.

PREVENTING SPREAD OF TOXIC FUMES IN AN AIR-HANDLING SYSTEM – 300-22

Electrical wiring and equipment may give off toxic fumes in the event of a fire. The requirements for the installation of electrical equipment and their associated wiring systems within ducts, plenums, and other air-handling spaces are covered in 300-22. The concern here is to limit the use of materials that could contribute smoke and products of combustion during a fire in an area that handles environmental air. The four categories of air-handling installations are ducts for dust, loose stock, or vapor removal; ducts or plenums used for environmental air; other space used for environmental air; and data processing systems.

Wiring in Ducts for Dust, Loose Stock, or Vapor Removal – 300-22(a)

Wiring systems are not permitted to be installed in ducts used to handle dust, loose stock, or flammable vapors. No wiring system shall be installed in any shaft containing ducts used for vapor removal or for ventilation of commercial-type cooking equipment.

Wiring in Ducts or Plenums Used for Environmental Air – 300-22(b)

The wiring systems permitted in ducts or plenums specifically made to handle the supply and return of conditioned air are:

- MI cable
- MC cable, the smooth or corrugated sheath type, not the MC cable with an interlocking outer jacket
- EMT
- FMT
- IMC
- RMC

Flexible metal conduit and liquidtight flexible metal conduit are permitted in these ducts and plenums provided their length does not exceed 4′ and they are used to connect physically-adjustable equipment and devices. The connectors used with this flexible conduit shall effectively close any openings in the connection. See Figure 2-18.

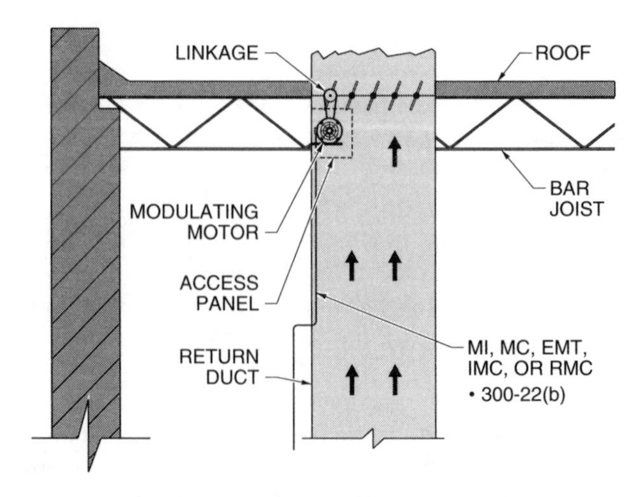

Figure 2-18. MI, MC, EMT, FMT, IMC, or RMC may be run in ducts or plenums used for environmental air.

The equipment and devices permitted in these ducts or plenums shall be associated with the movement or sensing of the contained air. Lighting fixtures installed to facilitate the maintenance and repair of this equipment shall be the enclosed gasketed type.

Wiring in Other Spaces Used for Environmental Air – 300-22(c)

"Other spaces" refers to space used for environmental air-handling purposes other than ducts and plenums. An example of "other spaces" is the space over a hung ceiling used for environmental air-handling purposes. See Figure 2-19.

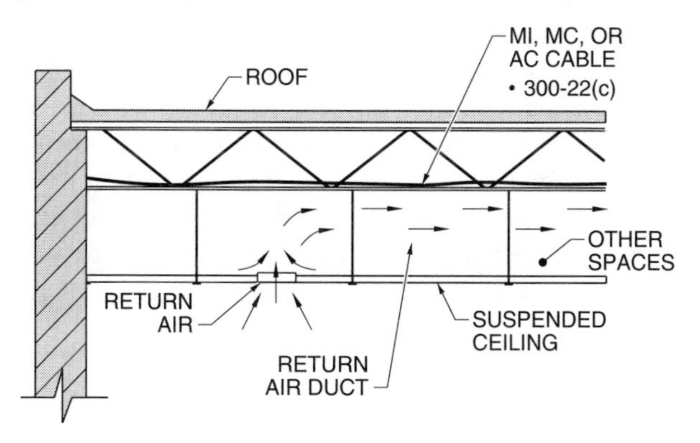

Figure 2-19. An "other space" is the space over a hung ceiling used for environmental air.

The wiring methods permitted in other spaces are:

- Totally enclosed nonventilated insulated busway without provisions for plug-in connections
- MI cable
- MC cable without an overall nonmetallic covering
- AC cable
- Other factory-assembled multiconductor control or power cable that is listed for the use

Other type cables and conductors permit the installation of factory-assembled cords and cables such as SJ, SO, and other types listed in 400. Also permitted are the standard recognized type building wires such as THW, THWN, and other types listed in 310.

The installation of other type cables and conductors shall be installed in one of the following: EMT, flexible metallic tubing, IMC, RMC, flexible metal conduit, and where accessible, surface metal raceway or metal wireway with metal covers or solid bottom metal cable tray with solid metal covers.

Other type cables and conductors shall be installed in one of the following:

- EMT
- FMT
- IMC
- RMC
- FMC
- Where accessible, surface metal raceway or metal wireway with metal covers or solid bottom metal cable tray with solid metal covers

Section 300-22, Ex. 5 permits NM cable to pass through a closed-in stud or joist space used as a return in a forced-air heating/air conditioning system. See Figure 2-20. This exception applies only to dwelling units.

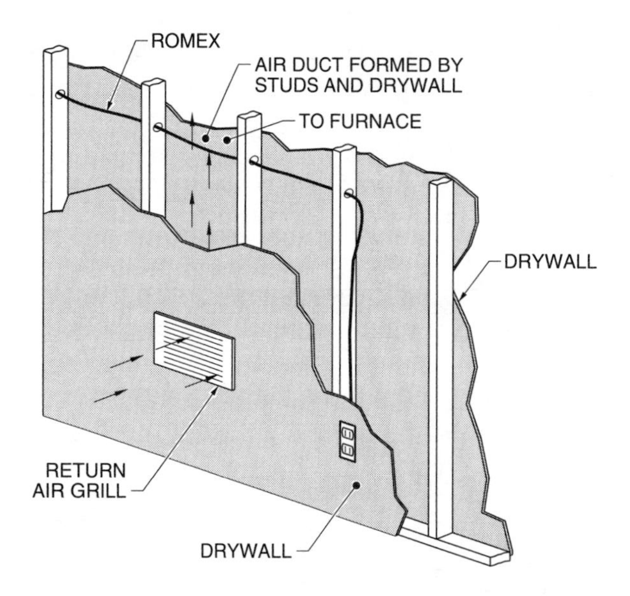

Figure 2-20. NM cable may pass through a closed-in stud or joist space used as a return in a forced-air heating/air conditioning system.

Wiring in Spaces Used for Data Processing Systems – 300-22(d)

Wiring in air-handling areas beneath raised floors of computer/data processing equipment is permitted per 645, formerly entitled "Electronic Computer/Data Processing Equipment," and now named "Information Technology Equipment." For a room to fall into this category, it shall comply with all six conditions required by 645-2. See Figure 2-21. Only when the six conditions are met is a room considered a dedicated Information Technology Equipment room and the provisions specified in 300-22(c) for Other Space Used For Environmental Air do not apply to the area under the raised floor of the room.

Supply circuits and interconnecting cables under raised floors in information technology equipment rooms are permitted provided five conditions are met. Per 645-5(d)(1), the raised floor is of suitable construction to support the computer equipment which will be placed on top of it. These floor assemblies are available with UL listings. They come with vertical floor stands that are mounted on a heavy-duty base plate which sit on top of the subfloor. On the top of this floor stand is a platform which accepts the 2′ × 2′ fire-resistance floor panels. These floor stands are adjustable in height enabling the installer to guarantee a level floor. For more information see NFPA 75-1995 (ANSI).

In addition to the raised floor meeting structural standards, the area under the floor shall be accessible. Most of these floors are constructed by using 2′ × 2′ panel sections which lay in a grid structure. These panels may be removed with a plunger suction device for access to the wiring system. Per 645-5(d)(2), these wiring methods may be used to supply the branch-circuit supply conductors: RMC, RNMC, IMC, EMT, metal wireway, surface metal raceway with metal cover, FMC, LTFMC, LTFNMC, MI cable, MC cable, or AC cable.

Per 645-5(d)(3), ventilation in the underfloor area is used for the data processing equipment and data processing area only. The HVAC system can be either dedicated to the information technology equipment use and separated from other areas of occupancy, or at the point of penetration to the information technology equipment room, be provided with dampers that operate in unison with the smoke detectors or by the disconnecting means in 645-2(a). Per 645-5(d)(4), the openings in the raised floor that permit the cables to enter and leave shall offer protection to the cables from abrasions. This protection could be offered by providing a bushing fitting at the opening. Per 645-5(d)(5), cables other than those covered in (2), shall be listed as Type DP (data processing).

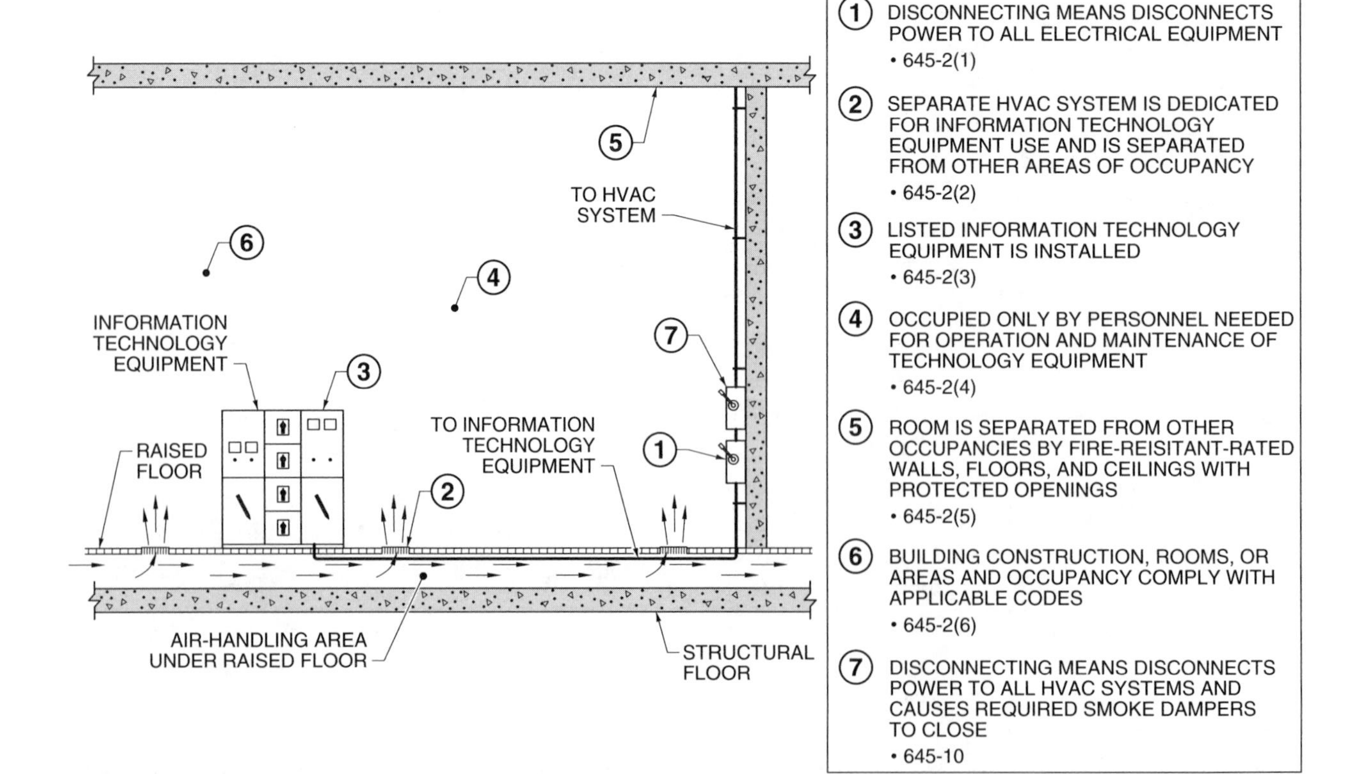

Figure 2-21. Wiring is permitted in air-handling areas beneath raised floors of computer/data processing equipment.

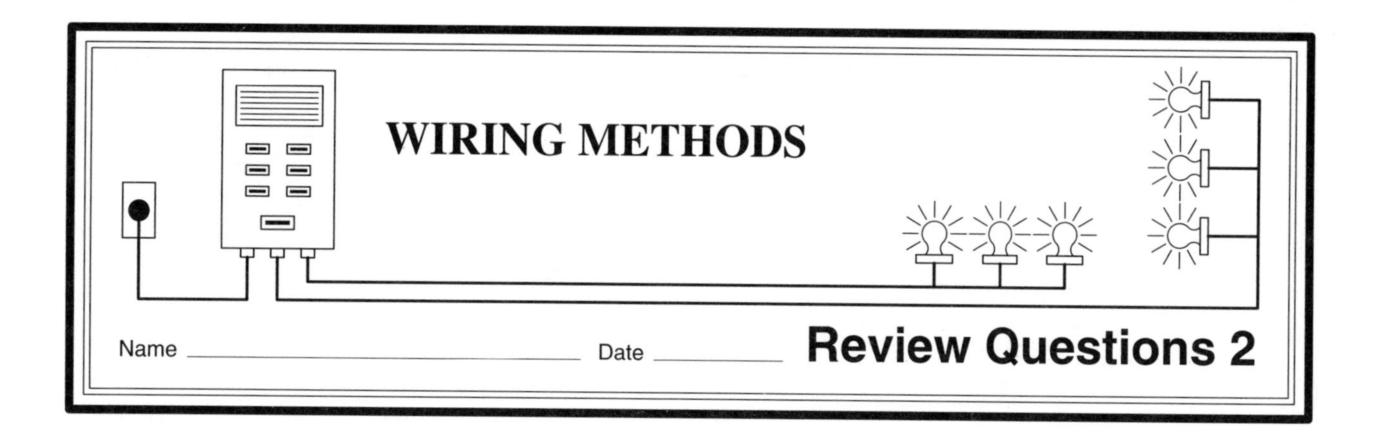

Review Questions

T F **1.** Article 300 applies to wiring that is part of equipment.

T F **2.** Conductors of different systems (600 V or less, AC and DC) may occupy the same raceway.

_______ **3.** Holes that are drilled into framing members to permit the installation of raceways or cables shall be at least _______″ from the nearest edge of the wood member.

A. $\frac{1}{4}$
B. 1
C. $1\frac{1}{4}$
D. neither A, B, nor C

_______ **4.** Cables or raceways in notches of framing members shall be protected from nails or screws by a steel plate at least _______″ thick.

A. $\frac{1}{32}$
B. $\frac{1}{16}$
C. $\frac{1}{8}$
D. neither A, B, nor C

_______ **5.** The minimum cover for any wiring method not listed in Table 300-5 is _______″.

T F **6.** Per Table 300-5, RMC shall be buried under 12″ of earth beneath a highway.

T F **7.** Seals shall be installed in the raceway system whenever the raceway changes from one environmental temperature to another.

_______ **8.** All cables and conductors emerging from the ground shall be protected from physical damage to a maximum depth of _______″.

A. 12
B. 18
C. 24
D. 36

_______ **9.** In an indoor wet location, the electrical panelboard metal enclosure shall be at least _______″ from the masonry wall.

A. $\frac{1}{8}$
B. $\frac{1}{4}$
C. $\frac{3}{8}$
D. $\frac{1}{2}$

T F **10.** There shall be no splice or tap within a raceway.

T F **11.** Raceway systems shall be used to support raceways, cables, or other nonelectric equipment.

T F 12. The grounded conductor (neutral) of a multiwire branch circuit shall never be opened.

_______ 13. At least _______″ of free conductor shall be left at each outlet, junction, or switch point.

A. 6
B. 12
C. 18
D. 24

_______ 14. A(n) _______ space is the space over a hung ceiling used for environmental air.

A. roof
B. duct
C. other
D. neither A, B, nor C

T F 15. Conductors of opposite polarities, which are run in close proximity to one another, produce a heating effect on ferrous metal.

T F 16. AC conductors installed in metal raceways or enclosures shall be arranged to provide induction heating of the metal raceways or enclosures.

T F 17. Copper conductors in a vertical raceway require more supports than an equivalent length of aluminum conductors.

_______ 18. When a conduit contains three or more conductors, the cross-sectional area of the conductors shall not exceed _______% of the cross-sectional area of the conduit.

_______ 19. The inductive heating effect of metal is minimized by cutting _______ between the holes for conductors.

_______ 20. Underground, residential branch circuits that are rated 20 A, 120 V, 1φ or less and are GFCI-protected have a minimum burial depth of _______″.

_______ 21. Cables and raceways in or under airport runways shall be buried to a minimum depth of _______″.

_______ 22. Cables and raceways installed under streets, highways, roads, driveways, and parking lots shall be buried to a minimum depth of _______″.

A. 12
B. 18
C. 24
D. neither A, B, nor C

_______ 23. _______ is the shortest distance measured between a point on the top surface of any direct buried conductor, cable, conduit, or other raceway and the top surface of finished grade, concrete, or similar cover.

_______ 24. Cables and conductors installed on the side of a pole or building shall be protected up to _______′ above finished grade.

A. 5
B. 7½
C. 10
D. neither A, B, nor C

_______ 25. Boxes or cabinets marked "_______" can be installed out-of-doors.

A. Raintight
B. Rainproof
C. Outdoor Type
D. A, B, and C

Minimum Cover 0 – 600 Volts

1 DIRECT BURIAL CABLES OR CONDUCTORS

2 RMC OR IMC

3 RNMC APPROVED FOR BURIAL WITHOUT CONCRETE ENCASEMENT AND EMT

4 RESIDENTIAL BRANCH CIRCUITS 120 V OR LESS WITH GFCI PROTECTION AND MAXIMUM OVERCURRENT OF 20 A

5 IRRIGATION CONTROL LANDSCAPE LIGHTING 30 V OR LESS USING UF OR OTHER IDENTIFIED CABLE IN RACEWAY

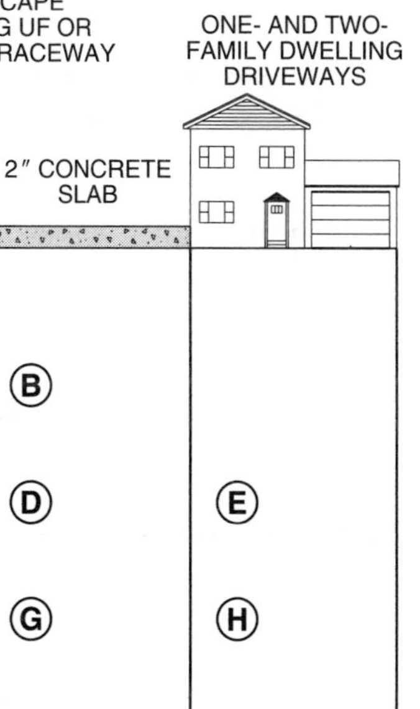

__________ **1.** A = ________ and ________

__________ **2.** B = ________, ________, and ________

__________ **3.** C = ________

__________ **4.** D = ________

__________ **5.** E = ________

__________ **6.** F = ________

__________ **7.** G = ________

__________ **8.** H = ________, ________, ________, and ________

__________ **9.** I = ________

Neutral Closed

__________ **1.** The load at A is ________ Ω.

__________ **2.** The load at B is ________ Ω.

__________ **3.** The load at C is ________ Ω.

__________ **4.** The load at D is ________ Ω.

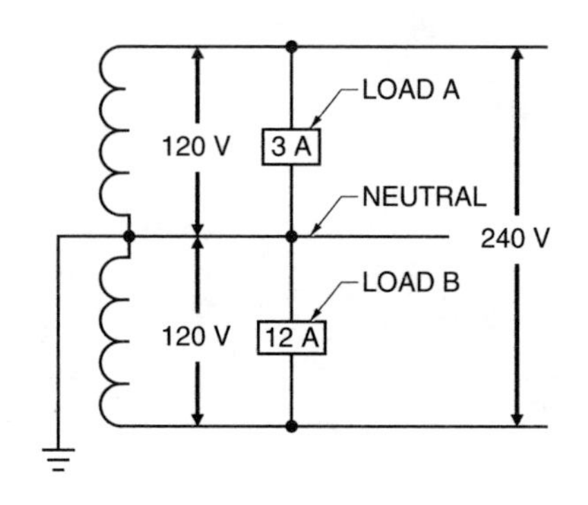

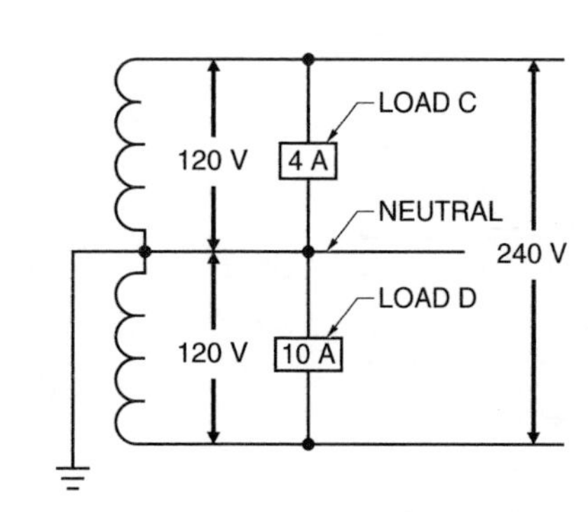

Neutral Opened

__________ **1.** The load at A is ________ V.

__________ **2.** The load at B is ________ V.

__________ **3.** The load at C is ________ V.

__________ **4.** The load at D is ________ V.

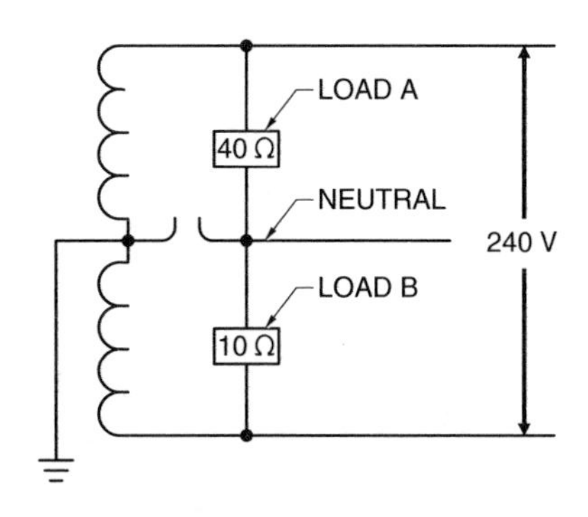

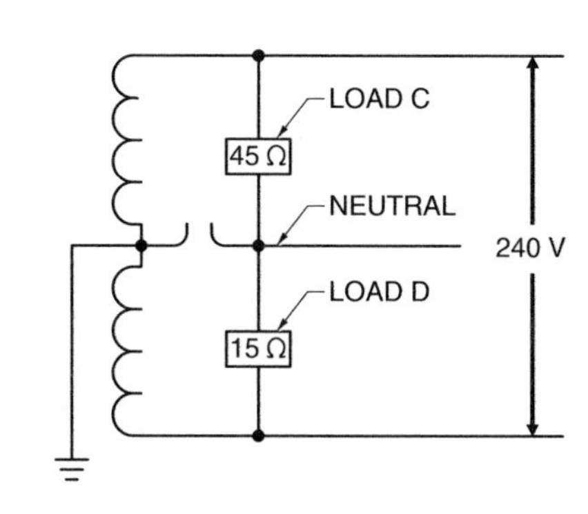

Underground Installations

__________ **1.** Conductors and cables shall be protected 8′ to above finished grade.

__________ **2.** Splice box not required.

__________ **3.** Heavy rocks or sharp corrosive material prohibited as backfill.

__________ **4.** Raceways shall be sealed to prevent moisture.

__________ **5.** Bushings required at end of conduit underground with conductors.

__________ **6.** "S" loops allowed for ground movement.

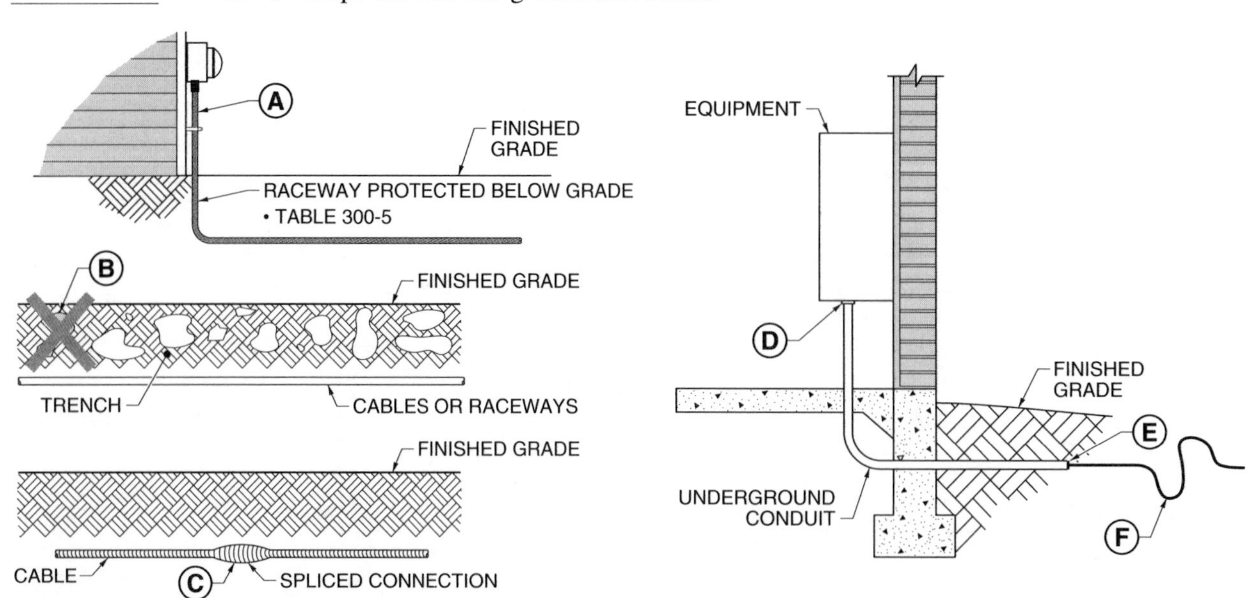

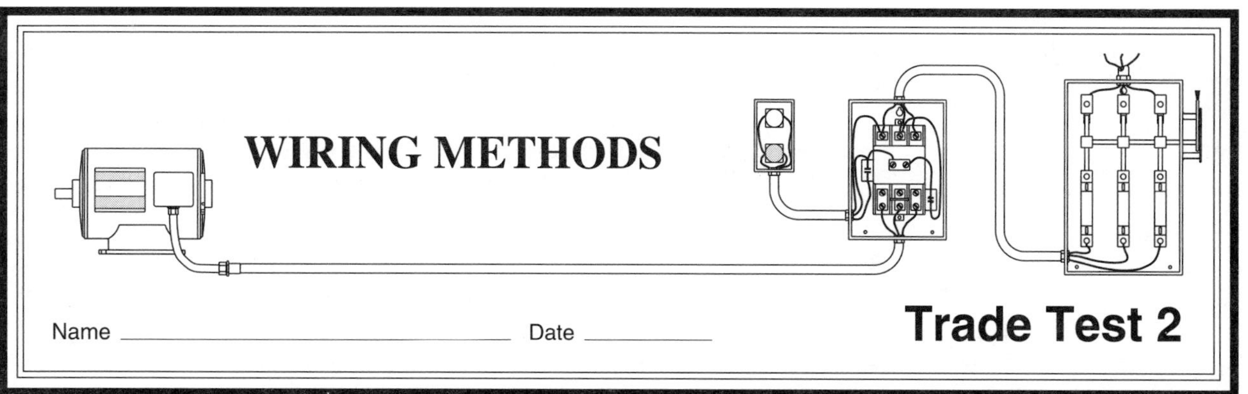

WIRING METHODS

Name ______________________ Date __________

Trade Test 2

NEC®	Answer	
________	________	**1.** Determine the minimum burial depth for a Type UF cable installed to supply an outdoor receptacle in a one-family dwelling. The branch circuit is rated at 20 A, 120 V and provided with GFCI protection.
________	________	**2.** Determine the minimum burial depth for Schedule 80 PVC conduit installed under the driveway of a two-family dwelling unit used only for dwelling unit purposes.
________	________	**3.** A direct burial cable emerges from the ground and is installed on the service of a wooden utility pole. Determine the minimum burial depth and maximum height to which the cable shall be protected from physical damage.
________	________	**4.** *See Figure 1.* An electrical service is installed in an area with frost heaves and a good degree of ground movement. The electrical contractor installs the service using Schedule 80 PVC conduit per 230-43(11) and includes expansion joints where the conduit emerges from the ground. Does this installation violate Article 300 of the NEC®?

METER SOCKET
EXPANSION FITTING
PVC STRAP
2″ PVC RACEWAY
SERVICE LATERAL

FIGURE 1

NEC®	Answer	
________	________	**5.** An electrical contractor installs branch-circuit conductors in RNMC in a one-family dwelling. The contractor supports the horizontal run of RNMC in notches cut in the 2″ × 4″ framing members which are spaced on 24″ centers. The conduit is supported within 3′ of the termination point. Steel plates are not used to cover the notches in the framing members. Does this installation violate Article 300 of the NEC®?
________	________	**6.** When installing a 5′ length of Type MC cable, the electrician runs out of cable connectors. Instead of a connector for Type MC cable, the electrician uses a fitting listed for use with Type NM cable only. Does this installation violate Article 300 of the NEC®?

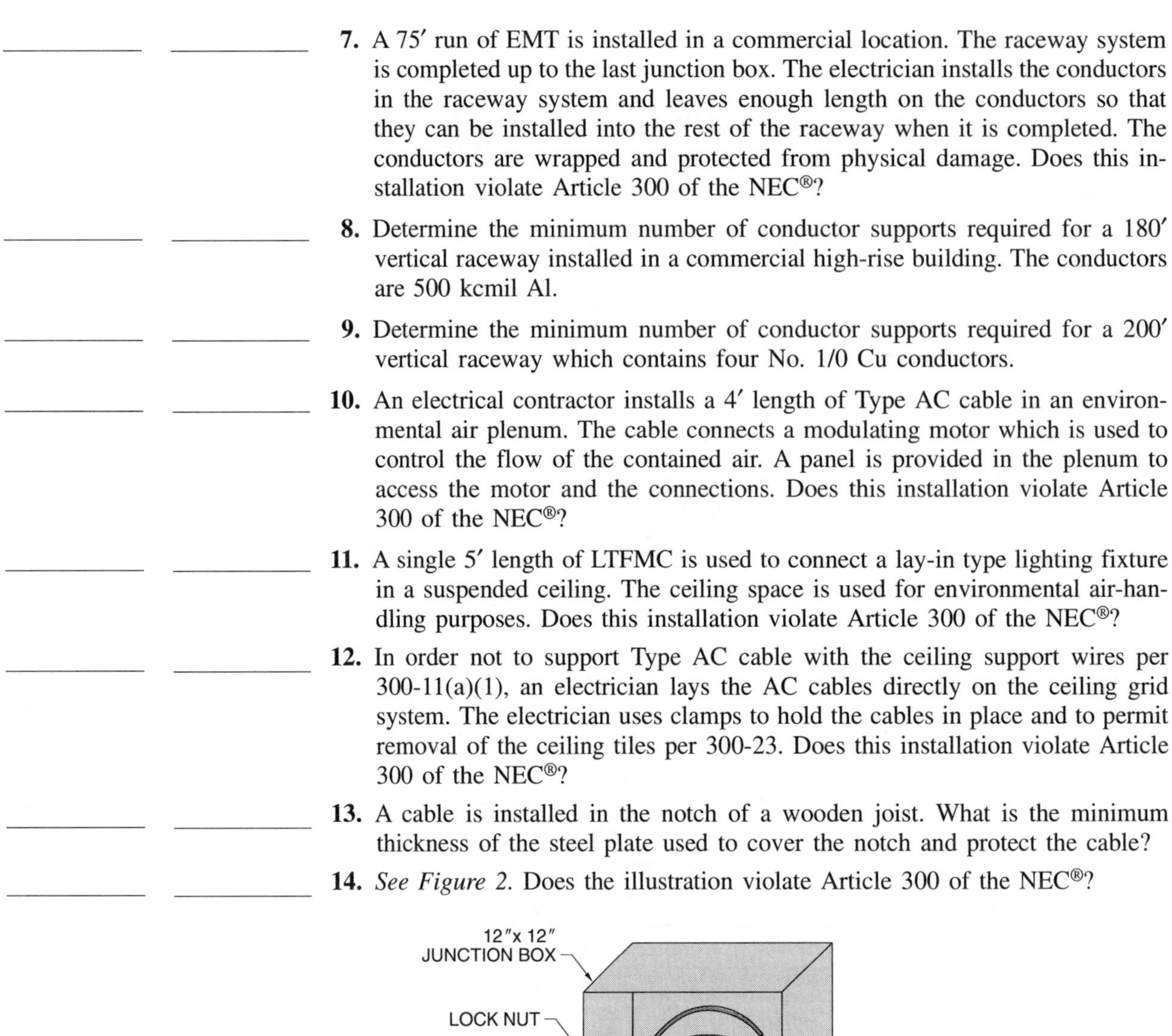

7. A 75′ run of EMT is installed in a commercial location. The raceway system is completed up to the last junction box. The electrician installs the conductors in the raceway system and leaves enough length on the conductors so that they can be installed into the rest of the raceway when it is completed. The conductors are wrapped and protected from physical damage. Does this installation violate Article 300 of the NEC®?

8. Determine the minimum number of conductor supports required for a 180′ vertical raceway installed in a commercial high-rise building. The conductors are 500 kcmil Al.

9. Determine the minimum number of conductor supports required for a 200′ vertical raceway which contains four No. 1/0 Cu conductors.

10. An electrical contractor installs a 4′ length of Type AC cable in an environmental air plenum. The cable connects a modulating motor which is used to control the flow of the contained air. A panel is provided in the plenum to access the motor and the connections. Does this installation violate Article 300 of the NEC®?

11. A single 5′ length of LTFMC is used to connect a lay-in type lighting fixture in a suspended ceiling. The ceiling space is used for environmental air-handling purposes. Does this installation violate Article 300 of the NEC®?

12. In order not to support Type AC cable with the ceiling support wires per 300-11(a)(1), an electrician lays the AC cables directly on the ceiling grid system. The electrician uses clamps to hold the cables in place and to permit removal of the ceiling tiles per 300-23. Does this installation violate Article 300 of the NEC®?

13. A cable is installed in the notch of a wooden joist. What is the minimum thickness of the steel plate used to cover the notch and protect the cable?

14. *See Figure 2.* Does the illustration violate Article 300 of the NEC®?

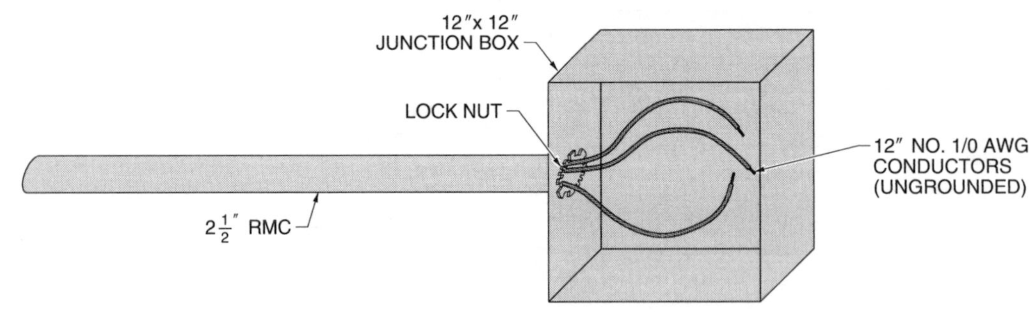

FIGURE 2

15. *See Figure 3.* Does the illustration violate Article 300 of the NEC®?

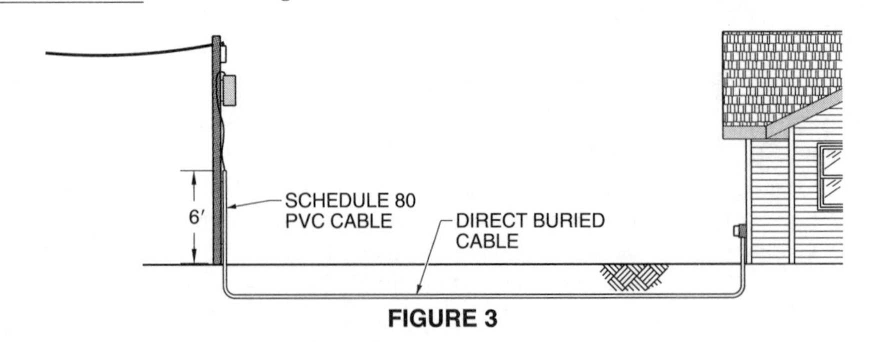

FIGURE 3

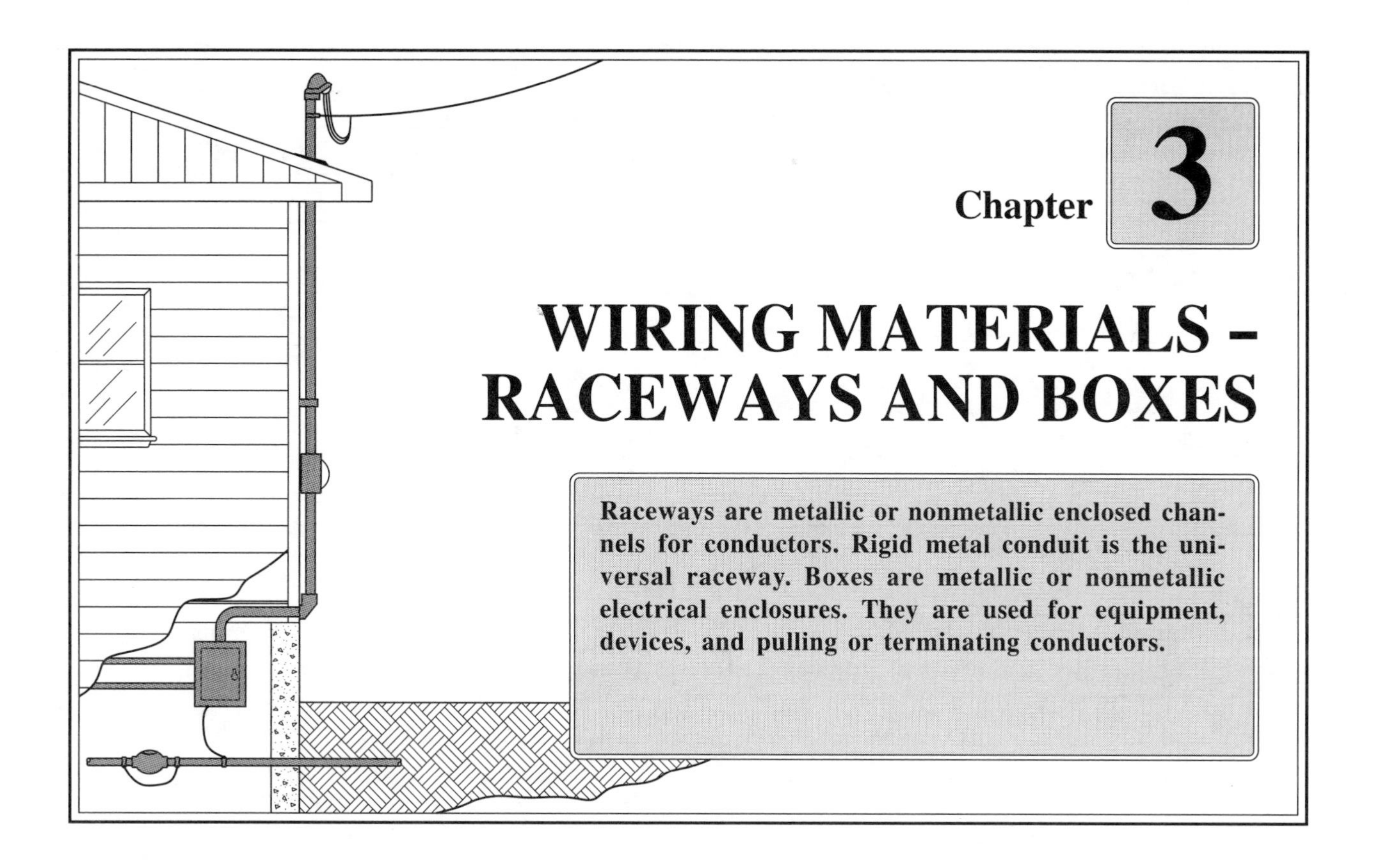

RACEWAY SYSTEMS

A *raceway* is an enclosed channel for conductors. A raceway includes all of the enclosures used for running conductors among different components of an electrical system per 100. This includes switches, panels, controllers, and cabinets. Examples of raceways include RMC, RNMC, IMC, LTFMC, FMT, FMC, ENMT, EMT, wireways, and busways. *Note:* When the NEC® uses the word conduit, it means only those raceways that contain the word "conduit" in its title.

A *raceway system* is an enclosed channel of metal or nonmetallic materials used to contain the wires or cables of an electrical system. Raceway systems are composed of raceways. Raceways are commercially available in a variety of lengths. They provide protection to the conductors and give the system flexibility in that existing conductors can be removed and new conductors pulled in. New conductors may be added to an existing system provided there is enough room. Some examples of raceways are rigid metal conduit, rigid nonmetallic conduit, intermediate metal conduit, liquidtight flexible metallic conduit, flexible metal conduit, electrical nonmetallic tubing, electrical metallic tubing, underfloor raceways, cellular concrete floor raceways, cellular metal floor raceways, surface raceways, wireways, and busways.

Each of these raceways has its own article in the NEC®. The sections entitled "Uses Permitted" and "Uses Not Permitted" should be reviewed for the particular raceway being considered. Some of these articles require the raceway to be listed. If that is the case, the raceway shall be installed in accordance with any instructions included in the listing or labeling per 110-3(b).

Rigid Metal Conduit (RMC) – 346

Rigid metal conduit (RMC) is a conduit made of metal. It is the universal raceway. See Figure 3-1. It is permitted under all atmospheric conditions and in all types of occupancies per 346-1. When installing RMC, dissimilar metals that could cause galvanic action shall be avoided. *Note:* Aluminum fittings and enclosures are permitted to be used with RMC and vice-versa.

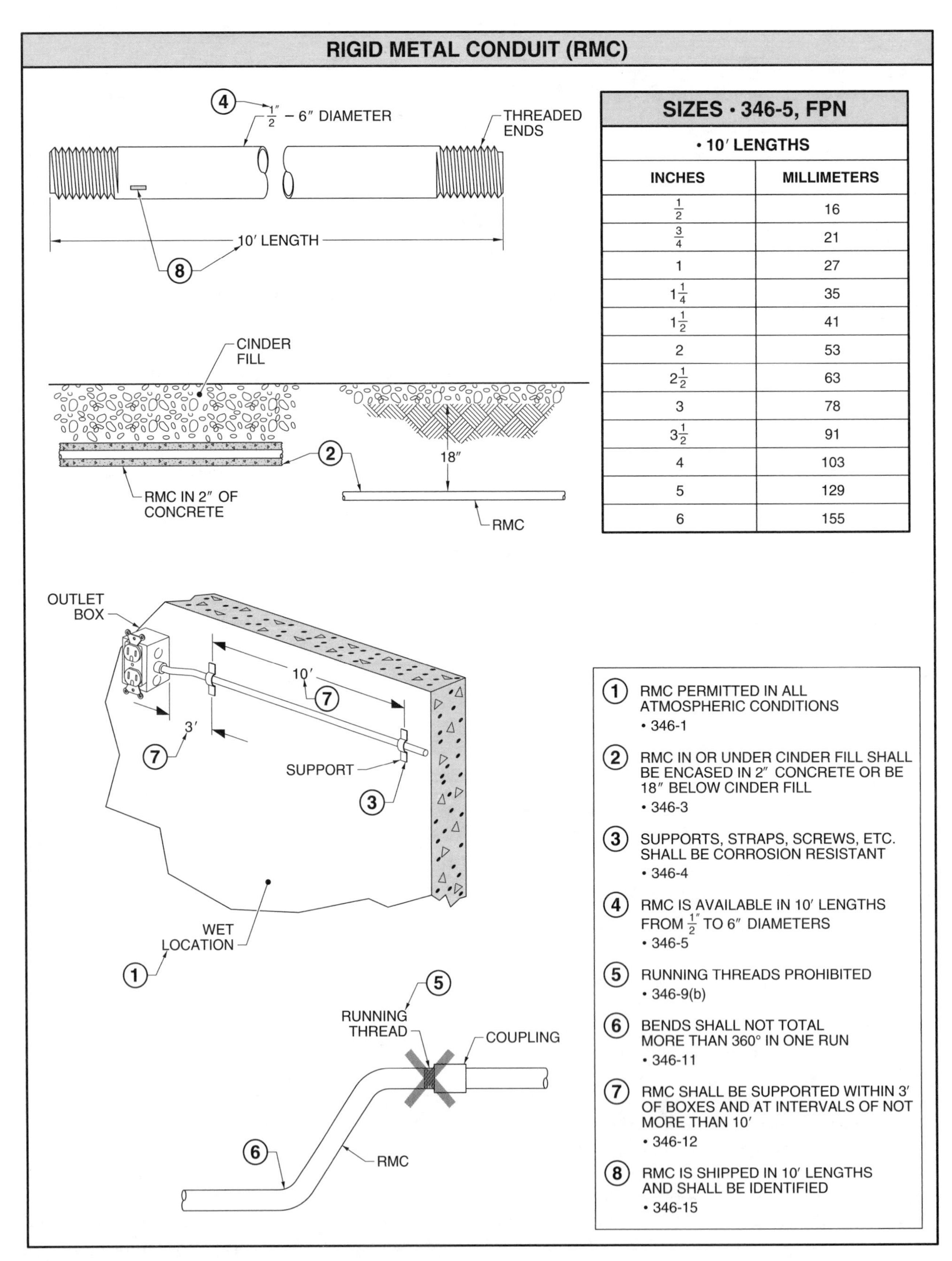

INCHES	MILLIMETERS
$\frac{1}{2}$	16
$\frac{3}{4}$	21
1	27
$1\frac{1}{4}$	35
$1\frac{1}{2}$	41
2	53
$2\frac{1}{2}$	63
3	78
$3\frac{1}{2}$	91
4	103
5	129
6	155

Figure 3-1. Rigid metal conduit (RMC) is a conduit made of metal.

RMC shall not be run in or under cinder fill unless it is contained in an envelope of at least 2″ of noncinder concrete or is buried at least 18″ below the cinder fill per 346-3. All supports, bolts, straps, etc. shall be corrosion-resistant per 346-4.

RMC is available in 10′ lengths with a coupling on one end. Per 346-15(a), lengths shorter or longer than 10′ may be shipped. This is considered a special order by the manufacturer and a large quantity should be ordered or the price is prohibitive. It is rare that the manufacturer provides lengths other than the standard 10′ length. RMC is available in sizes from ½″ in diameter up to 6″ in diameter per 346-5. Sizes may be given in inches or millimeters.

Section 430-145(b) permits ⅜″ RMC to be used between a motor and its junction box when the junction box is not part of the motor housing. Section 346-9(b) prohibits the use of running threads when installing RMC.

RUNNING THREADS PROHIBITED

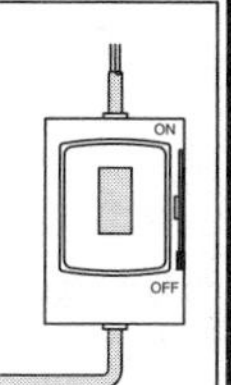

An attempt is made to screw a length of RMC with an offset onto a straight length of RMC installed in a narrow ditch. The offset prevents the installer from turning the RMC onto the straight length of conduit. An approved split coupling or Erickson coupling is needed to join these two lengths of conduit as the NEC® does not permit the use of running threads.

A running thread is twice as long as a standard thread. The coupling is run all the way up this extended thread. The two lengths of conduit are then aligned and the coupling on the running thread is backed off onto the length with the standard thread. This exposes half of the running threads which would eventually lead to a corrosion problem.

Nipples. A *nipple* is a short piece of conduit or tubing that does not exceed 24″ in length. See Figure 3-2. Nipples are used to join enclosures that are mounted close together. The wiring passes through the nipple from one enclosure to the other. The requirements for installing nipples are found in Chapter 9, Note 4.

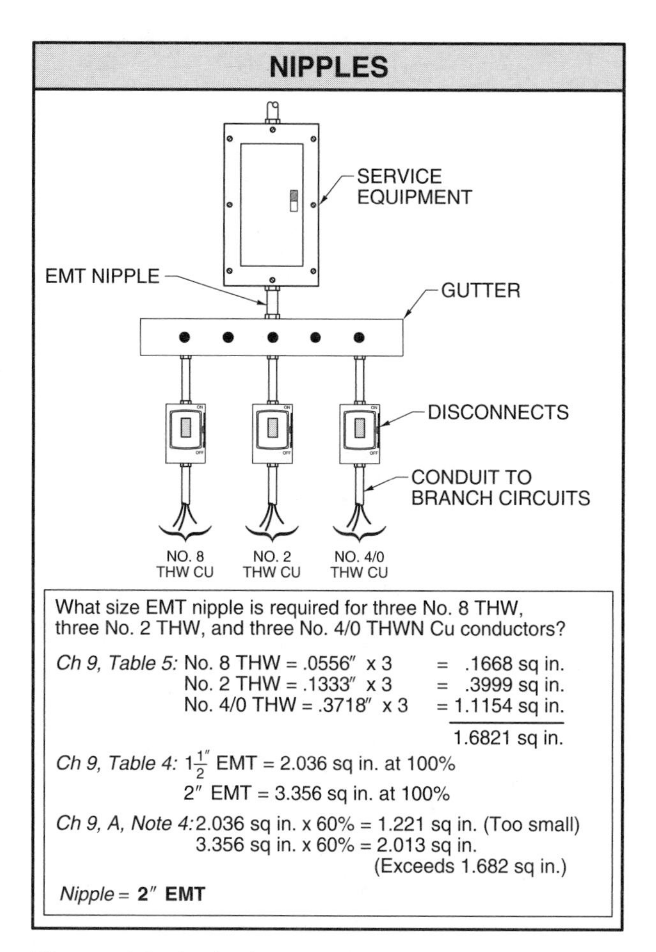

Figure 3-2. A nipple is a short piece of conduit or tubing that does not exceed 24″ in length.

Nipples are permitted to be installed between boxes, cabinets, and similar enclosures. The conductor fill for a nipple may be as high as 60% of its total cross-sectional area. The requirement for derating conductor ampacity in Note 8 of Notes to Ampacity Tables of 0 to 2000 Volts does not apply to nipples.

Rigid Nonmetallic Conduit (RNMC) – 347

Rigid nonmetallic conduit (RNMC) is a conduit made of materials other than metal. See Figure 3-3. There are several products that come under the regulation of 347. These materials are recognized as having suitable physical characteristics for direct burial in the earth. Some of the materials used in the manufacture of these products include fiber, asbestos cement, soapstone, rigid polyvinyl chloride, fiberglass epoxy, high-density polyethylene (for underground use), and rigid polyvinyl chloride (for aboveground use).

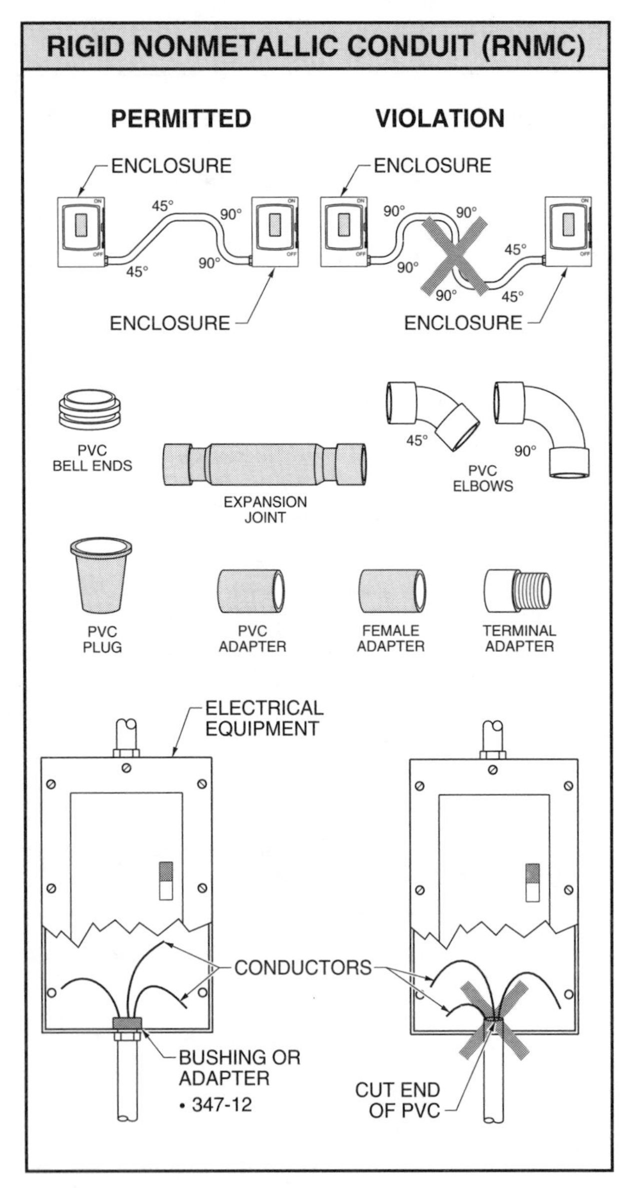

Figure 3-3. Rigid nonmetallic conduit (RNMC) is a conduit made of materials other than metal.

Rigid polyvinyl chloride (PVC) conduit is available in thin wall (Schedule 40) and thick wall (Schedule 80). In the early 1900s, PVC water pipes were given a number designation depending of the wall thickness. The thinnest wall thickness was given the designation number 10. The next thicker size was 20, then 30, 40, and up to 100. When PVC was presented to the electrical industry, the two sizes adopted were 40 and 80.

Schedule 40 is manufactured from fibers of polyvinyl chloride. It is waterproof, rustproof, and does not corrode in most locations. Because Schedule 40 does not have the strength of metal conduits, it shall not be used where subject to physical damage. Schedule 80 has a heavier wall thickness and is suitable for most locations where RMC is used. PVC must be supported as required in Table 347-8.

If the length change in a run of PVC exceeds $\frac{1}{4}$″ due to thermal expansion, an expansion fitting shall be provided as required by 347-9. Chapter 9, Table 10 gives the expansion characteristics of PVC conduit.

EXPANSION JOINTS

A 20′ run of PVC is installed on the outside of a building. In the winter months, the temperature averages 30°F. In the summer months, the temperature averages 85°F. The change in average temperature from winter to summer is 55°F (85°F – 30°F = 55°F). Per Chapter 9, Table 10, this change in temperature causes a length change of 2.2″ per 100′ of PVC or .22″ in a 10′ length ($\frac{10}{100}$ × 2.2″ = .22″). Section 347-9 requires an expansion fitting when there is more than .25″ of expansion and contraction length change. This 20′ run of PVC requires an expansion joint, since the total expansion of the 20′ of PVC is .44″ (.22″ × 2 = .44″).

A total run of PVC shall not exceed 360° of bends (including kicks and offsets). A *bend* is any change in direction of a raceway. See Figure 3-4. There shall not be more than the equivalent of four 90° bends between pull points. A *kick* is a single bend in a raceway. Kicks are often used for a raceway entrance and exit in an enclosure. An *offset* is a double bend in a raceway, each containing the same number of degrees. Offsets are commonly used when raceways are routed around obstacles, other raceways, etc. Bends can be any angle necessary to meet job requirements.

Bushings – 347-12. A *bushing* is a fitting placed on the end of a conduit to protect the conductor's insulation from abrasion. A bushing or adapter shall be provided on the end of RNMC when it enters a box, fitting, or other enclosure.

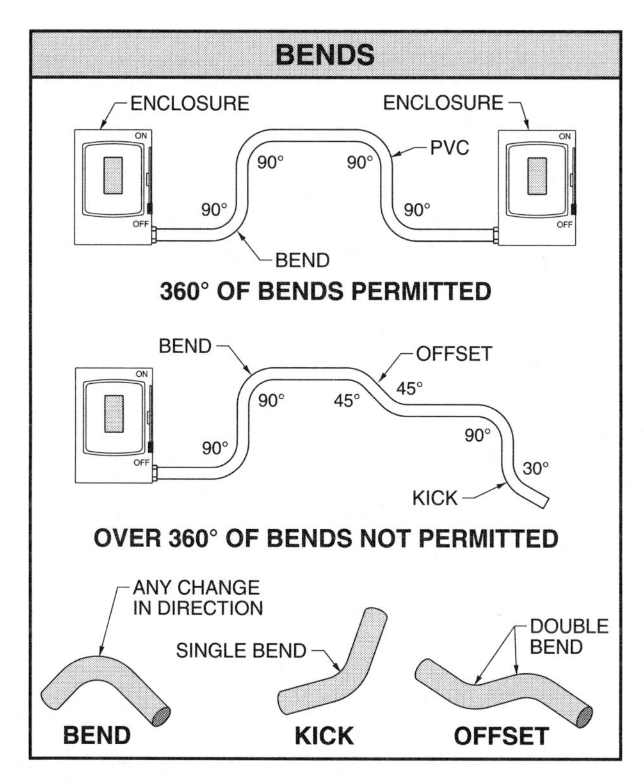

Figure 3-4. A total run of PVC shall not exceed 360° of bends.

Electrical Metallic Tubing (EMT) – 348

Electrical metallic tubing (EMT) is a lightweight tubular steel raceway without threads on the ends. It is the lightest raceway manufactured. The lengths are joined together by using set screws or compression fittings. See Figure 3-5.

The wall thickness of EMT is much thinner than RMC. A piece of ½″ EMT has an ID of .620″ and an OD of .706″ while a piece of ½″ RMC has an ID of .632″ and an OD of .840″. The EMT is about 40% thinner in wall thickness. There are five restrictions to the use of EMT:

(1) Where subject to physical damage.

(2) In corrosive atmospheres.

(3) Where buried in cinder fill.

(4) In hazardous locations, except as permitted in 502-4(b) Class II, Division 2 locations; 503-3 Class III, Division 1 and 2 locations; and 504-20 intrinsically safe apparatus.

(5) For the support of fixtures or other equipment except conduit bodies no larger than the largest trade size of the EMT.

The total number of bends in a run of EMT shall not exceed 360° per 348-10. EMT shall be supported every 10′ and within 3′ of every outlet box per 348-12. There are two exceptions to support this requirement. First, 348-12, Ex. 1, where the building structure does not permit fastening within 3′ of the outlet box, it can be extended to 5′ of the outlet box. Second, 348-12, Ex. 2, when concealed in finished buildings or prefinished wall panels and supporting is impracticable, EMT may be fished. The fittings used with EMT are either set screws or compression type. EMT is shipped in 10′ lengths. Couplings are not provided on one end as is the case with RMC.

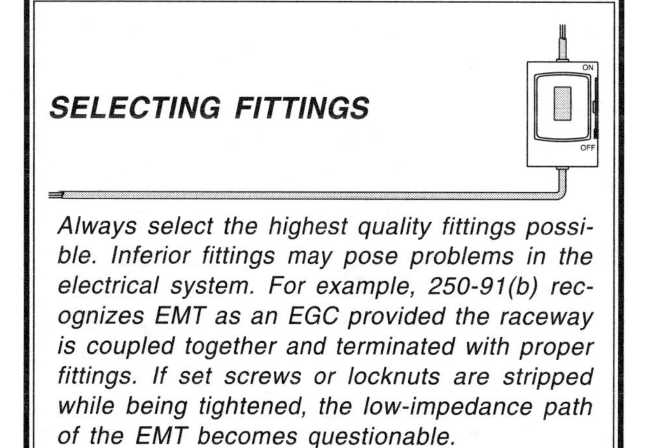

Always select the highest quality fittings possible. Inferior fittings may pose problems in the electrical system. For example, 250-91(b) recognizes EMT as an EGC provided the raceway is coupled together and terminated with proper fittings. If set screws or locknuts are stripped while being tightened, the low-impedance path of the EMT becomes questionable.

Flexible Metal Conduit (FMC) – 350

Flexible metal conduit (FMC) is a raceway of metal strips which are formed into a circular cross-sectional raceway. See Figure 3-6. The metal strips are helically wound, formed, and interlocked. FMC shall be listed as required by 350-4 and it is permitted to be used in exposed and concealed locations. FMC is known in the field as Greenfield. FMC is not permitted in seven locations:

(1) In wet locations unless the conductors are approved for that condition and it is installed in such a manner that liquid will not enter the raceway or enclosure to which it is connected.

(2) In hoistways.

(3) In storage battery rooms.

(4) In hazardous locations except as permitted in 501-4(b) Class I, Division 2 (limited flexibility at motor terminals) and 504-20 intrinsically safe apparatus.

(5) Where exposed to materials that could deteriorate the conductors such as gasoline and oil products.

(6) Underground or in poured concrete.

(7) Where subject to physical damage.

FMC is manufactured in diameters from ⅜″ through 4″. It is shipped in coils of 100′, 50′, and 25′ depending on its diameter. FMC in diameters of ⅜″, ½″, and ¾″ is shipped in 100′ coils. FMC in diameters of 1″ and 1¼″ is shipped in 50′ coils. FMC in diameters of 1½″ and larger is shipped in 25′ coils. Nationally recognized listing agencies do not accept FMC as an EGC. They do, however, recognize FMC per 350-14, Ex. If the total length in any ground path is 6′ or less, the conduit is terminated in fittings listed for grounding, and the overcurrent protection devices are rated at 20 A or less.

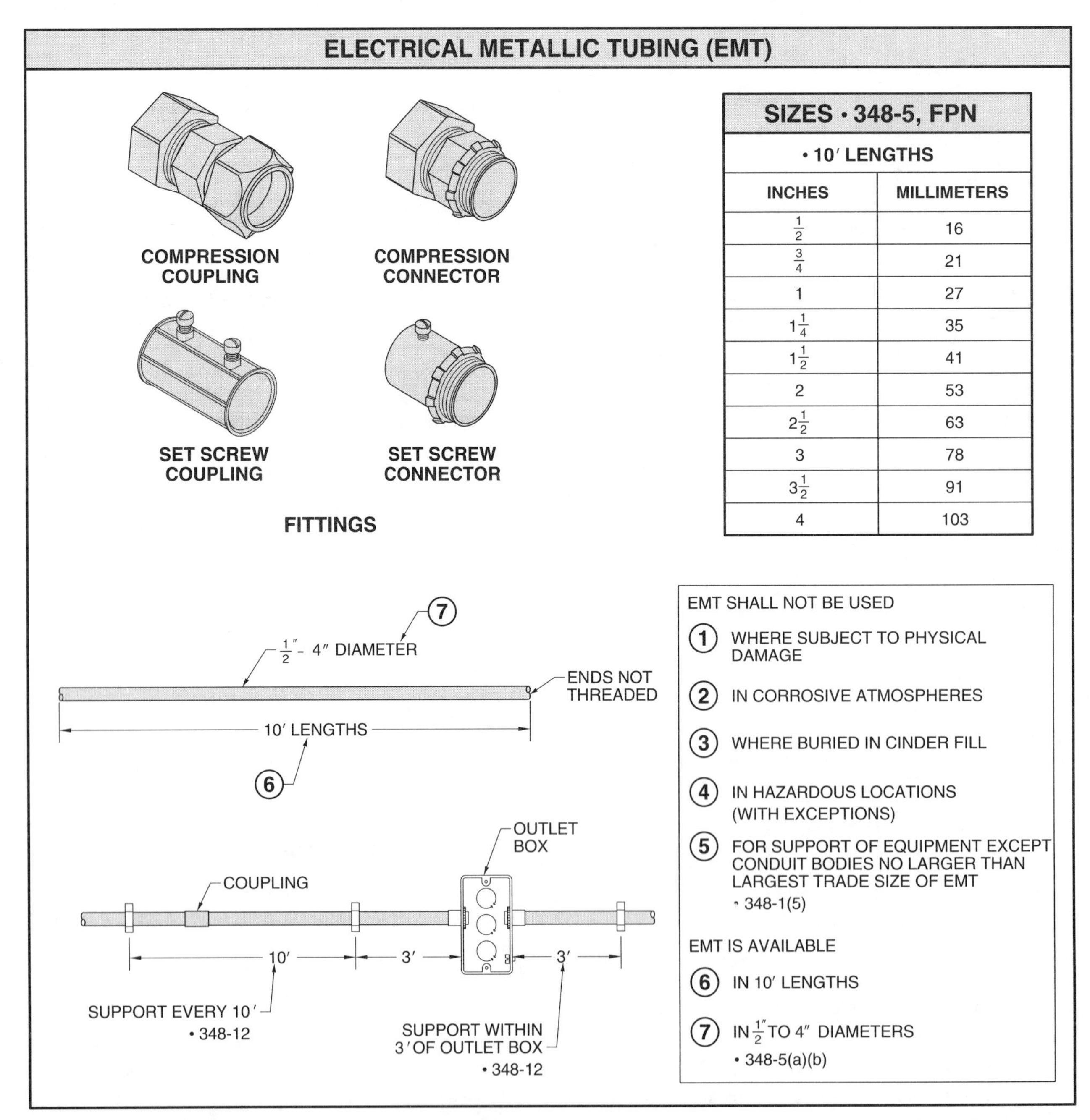

Figure 3-5. Electrical metallic tubing (EMT) is a light-weight tubular steel raceway without threads on the ends.

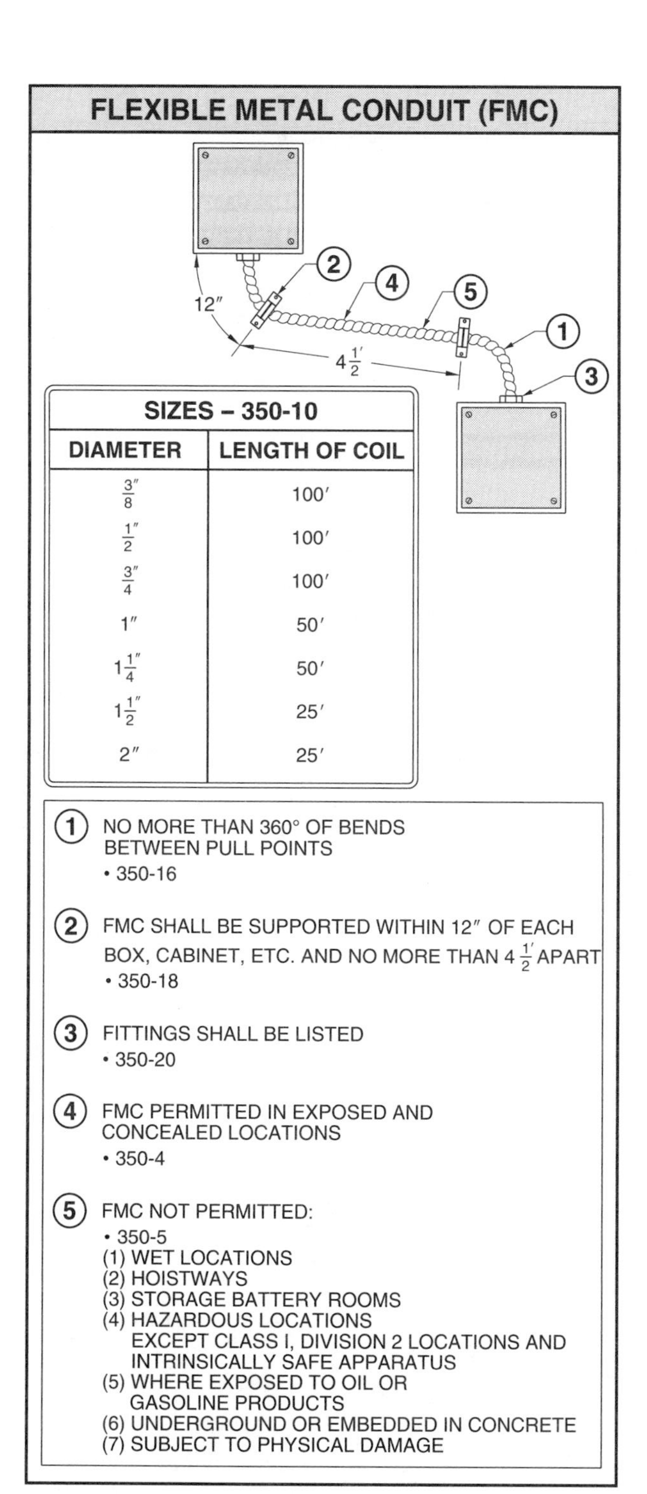

Figure 3-6. Flexible metal conduit (FMC) is a raceway of metal strips which are formed into a circular cross-sectional raceway.

FMC shall be supported every 4½′ and be securely fastened-in-place within 12″ of each box, cabinet, or conduit body per 350-18. There are three exceptions to this rule:

- Ex. 1 – Where FMC is fished.
- Ex. 2 – Lengths of FMC not exceeding 3′ may be used at terminals where flexibility is required.
- Ex. 3 – Lengths of FMC not exceeding 6′ may be used from a fixture terminal connection for tap connections to light fixtures per 410-67(c).

LISTED

During the 1996 NEC® cycle, CMP 8 included the word "listed" for several raceway systems. This listing requirement now ensures that these raceway systems are manufactured to a published standard which states the required physical and electrical characteristics upon which the requirements are based. It also ensures that the follow-up service of the listing agency helps maintain this consistency.

Section 350-20 prohibits the use of angle connectors for concealed raceway installations. An angle connector for FMC makes a 90° bend in a very short radius. Angle connectors have a backplate which is removed when installing the conductors. Once the conductors are installed through the angle connector, the backplate is reattached. It is very difficult, if not impossible, to fish conductors through an angle connector with the backplate attached. FMC angle connectors shall be accessible after installation.

RECOGNIZING FMC AS AN EGC

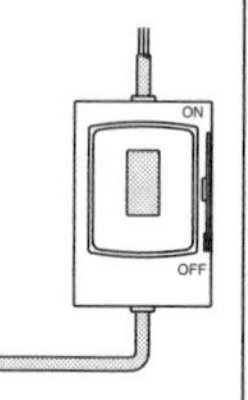

Although the national listing agencies do not recognize FMC as an EGC, some states and local municipalities recognize FMC as an EGC. For example, the city of Los Angeles recognizes FMC as an EGC.

Liquidtight Flexible Metal Conduit (LTFMC) and Liquidtight Flexible Nonmetallic Conduit (LTFNMC) – 351

Liquidtight flexible metal conduit (LTFMC) and liquidtight flexible nonmetallic conduit (LTFNMC) are raceways of circular cross section. See Figure 3-7. *Liquidtight flexible metal conduit (LTFMC)* is a raceway of circular cross section with an outer liquidtight, nonmetallic, sunlight-resistant jacket over an inner helically-wound metal strip. The most common name for this product is "Sealtite" which is a registered name for the product manufactured by the Anaconda Co., Metal Hose Division. There are associated couplings, connectors, and fittings that shall be used when running this raceway. Sizes smaller than ½″ are prohibited per 351-5(a) except for ⅜″ size for flexible connections to motors and ⅜″ size as part of an approved assembly or for fixture whips not over 6′.

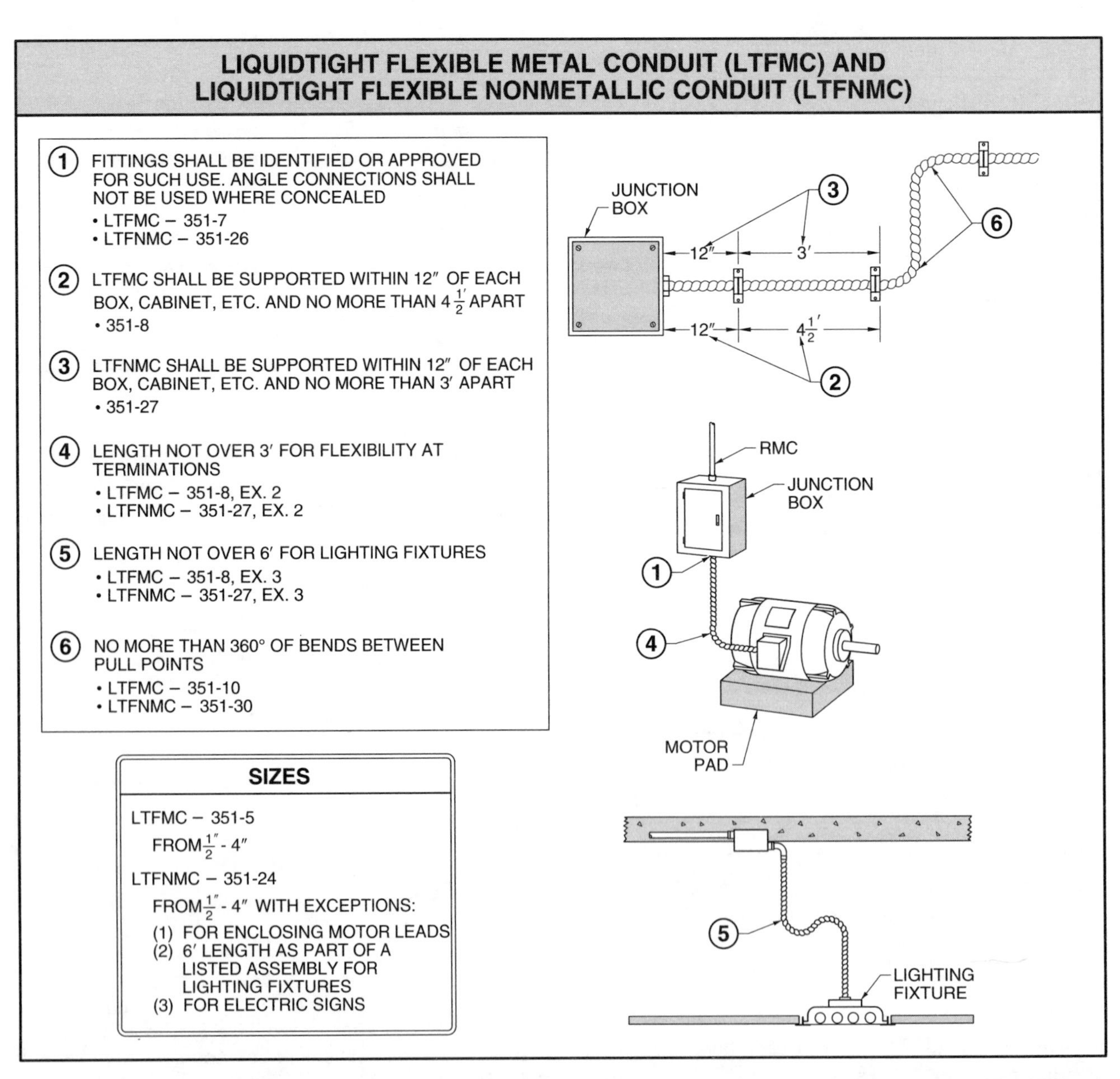

Figure 3-7. Liquidtight flexible metal conduit (LTFMC) and liquidtight flexible nonmetallic conduit (LTFNMC) are raceways of circular cross section.

LTFMC shall be supported every 4½′ and within 12″ of every outlet per 351-8. There are three exceptions to this rule:

- Ex. 1 – Where LTFMC is fished.
- Ex. 2 – Lengths of LTFMC not exceeding 3′ at terminals where flexibility is necessary.
- Ex. 3 – Lengths of LTFMC not exceeding 6′ for lighting fixture whips per 410-67(c).

The outer jacket of LTFMC permits use in wet locations or where exposed to mineral oils, grease, and other chemicals. LTFMC is not intended to be installed in locations where it is exposed to petroleum products unless identified for such use.

LTFMC shall not be used in areas where subject to physical damage per 351-4(b)(1). Per 351-4(b)(2), it shall not be used where any combination of ambient and conductor temperatures produce an operating temperature in excess of that for which the material is approved. LTFMC is permitted for direct burial in the earth per 351-4(a)(3) if it is listed and marked for such use.

Per 351-9, Ex., LTFMC is permitted to be used as an EGC in sizes of 1¼″ or less, provided the following three conditions are met:

(1) There is no more than 6′ of LTFMC in the return ground path.

(2) The LTFMC is terminated in fittings listed for grounding.

(3) For trade sizes ¾″ through 1¼″, the conductors within the LTFMC are protected by an overcurrent device rated 60 A or less. For trade sizes ⅜″ and ½″, the conductors contained within the LTFMC are protected by an overcurrent device rated 20 A or less.

Liquidtight flexible nonmetallic conduit (LTFNMC) is a raceway of circular cross section in one of three types. The first type has a smooth, seamless inner core and cover bonded together, making it look somewhat like a heavy-duty garden hose. The second type has a smooth inner surface. On the outside, it looks like LTFMC. The third type has a corrugated internal and external surface and is similar in appearance to ENT. None of these three types has a helically-wound metal strip on the inside. One of the main applications for LTFNMC is on industrial machinery. For example, the wiring associated with a lathe, milling machine, or punch press may be run in LTFNMC.

Lengths of FMC not exceeding 6′ may be used to connect lighting fixtures.

Surface Metal and Nonmetallic Raceways – 352

A *surface raceway* is an enclosed channel for conductors which is attached to a surface. See Figure 3-8. Surface raceways are available in several sizes. Each size has a complete line of boxes and accessories needed to install that size surface raceway as a system. There are a number of products that fall under the requirements 352. These systems are very convenient to use when adding to or extending an existing system in an installation where fishing cables is difficult.

Section 352-5 permits surface metal raceways to pass through dry walls, dry partitions, and dry floors provided the length passing through is unbroken. Access to the conductors shall be maintained on both sides of the wall, partition, or floor.

Section 352-7 permits splices and taps in surface metal raceways that have removable covers to give access to the conductors after the installation. Conductors, including splices and taps, shall not fill the raceway to more than 75% of its cross-sectional area.

Section 352-9 requires a means to connect an EGC to the surface metal raceway enclosures when a transition is made to another wiring method. For example, a Romex cable terminated in the back of a metallic wiremold outlet enclosure requires a means to connect the EGC of the cable to the metal enclosure.

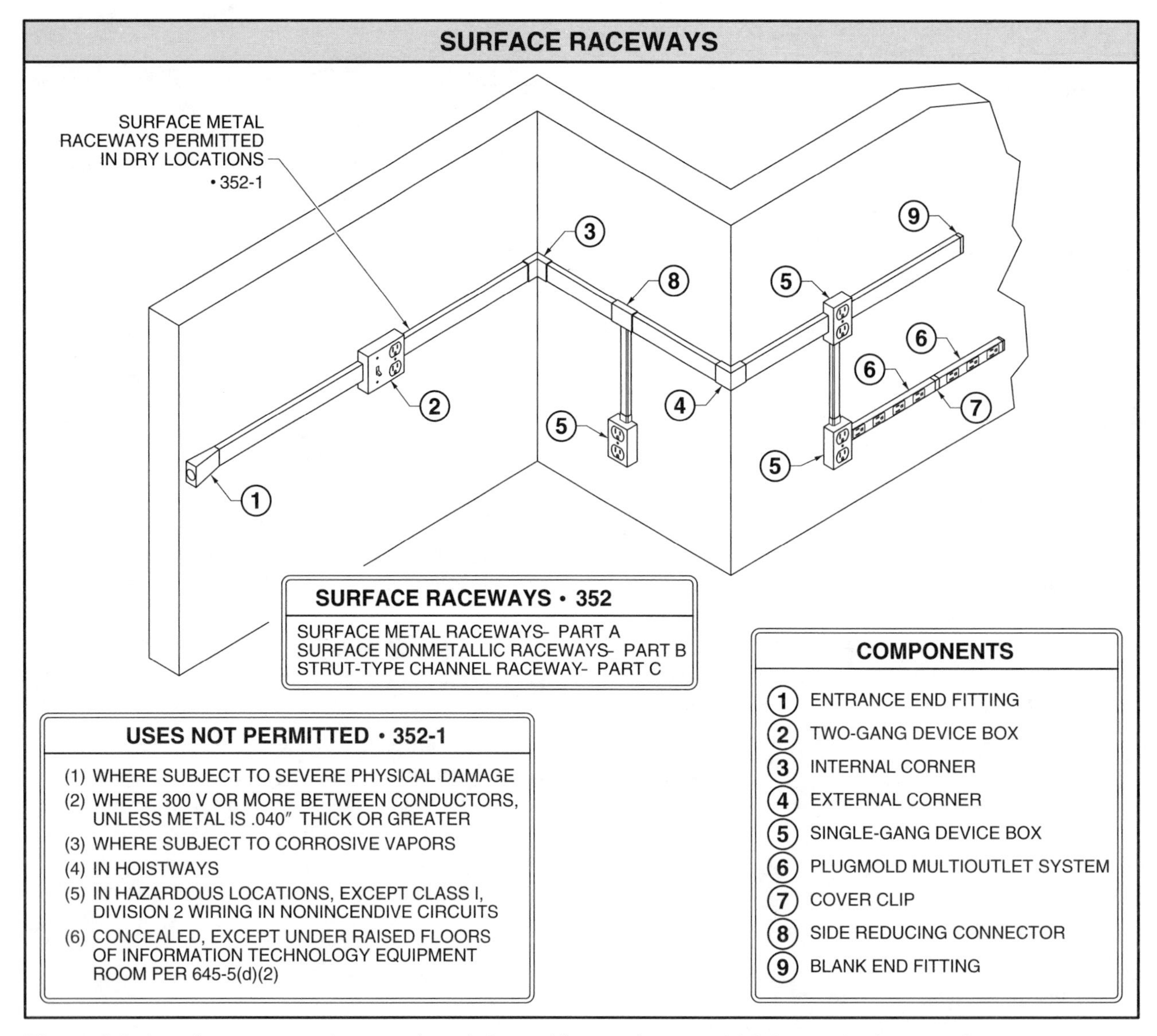

Figure 3-8. A surface raceway is an enclosed channel for conductors which is attached to a surface.

Strut-Type Channel Raceway. A *strut-type channel raceway* is a surface raceway formed of moisture-resistant and corrosion-resistant metal. These channel raceways may be galvanized, stainless, enameled, PVC-coated, or aluminum. Covers shall be either metallic or nonmetallic. See Figure 3-9. Section 352-41 permits these raceways to be used:

(1) Where exposed.

(2) In damp locations.

(3) In locations subject to corrosive vapors where protected by finishes judged suitable for the condition.

(4) Where voltage does not exceed 600 V.

(5) As power poles.

Section 352-42 prohibits the use of strut-type channel raceways where the raceway is concealed and in hazardous locations. Section 352-45 specifies that the number of conductors permitted in strut-type raceways shall not exceed the percentage fill shown in Table 352-45. Derating factors of Note 8(a) of Notes to Ampacity Tables of 0 to 2000 Volts does not apply to conductors installed in strut-type raceways when all of the following conditions are met:

(1) The cross-sectional area of the raceway exceeds 4 sq in.

(2) There are no more than 30 current-carrying conductors.

STRUT-TYPE CHANNEL RACEWAYS

CHANNEL HANGER
CLOSURE STRIP
ENDCAP WITH KNOCKOUT
CONDUCTORS INSTALLED IN FIELD
CHANNEL
OUTLET BOX
FIXTURE HANGER
① 10′
3′

USES • 352-41

- WHERE EXPOSED
- IN DAMP LOCATIONS
- IN CORROSIVE LOCATIONS WHERE PROTECTED BY FINISH
- WHERE NOT OVER 600 V
- AS POWER POLES

① STRUT-TYPE CHANNEL RACEWAY SHALL BE SUPPORTED WITHIN 3′ OF OUTLET BOXES AND NO MORE THAN 10′ APART
• 352-47

Figure 3-9. A strut-type channel raceway is a surface raceway formed of moisture-resistant and corrosion-resistant metal.

(3) The sum of the cross-sectional areas of all the conductors does not exceed 20% of the interior cross-sectional area of the raceway.

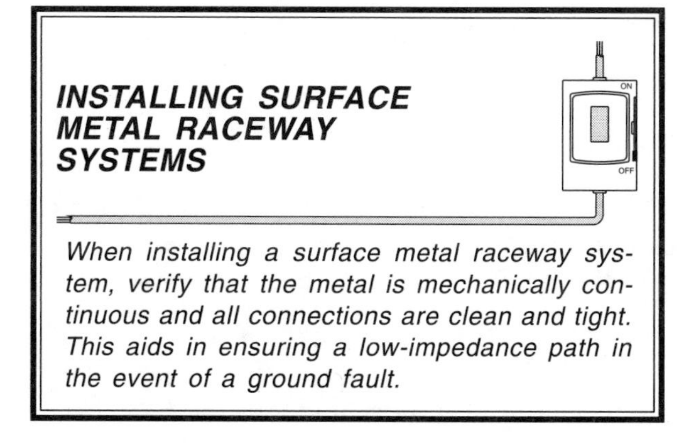

Section 352-47 requires that strut-type channel raceways be secured at least every 10′ and within 3′ of each outlet box, junction box, or other channel raceway termination. Section 352-50 requires that strut-type channel raceway enclosures which provide a transition to or from other wiring methods shall have a means for connecting an EGC.

Multioutlet Assembly – 353. A *multioutlet assembly* is a metal raceway with factory-installed conductors and attachment plug receptacles. See Figure 3-10. There is no provision for field installation of additional conductors except where the product is marked to indicate the number, type, and size of additional conductors which may be field installed. Multioutlet assemblies are similar to surface raceways. The difference is that they contain conductors and receptacles which are assembled in the field or at the factory. Multioutlet assemblies are popular for work benches, laboratories, countertop areas, and for upgrading older systems to meet today's requirements.

Section 353-3 permits the extension of metal multioutlet assemblies through (not within) dry partitions. Provisions shall be made so that the cap or cover is removable on all exposed portions and no outlet is located within the partitions.

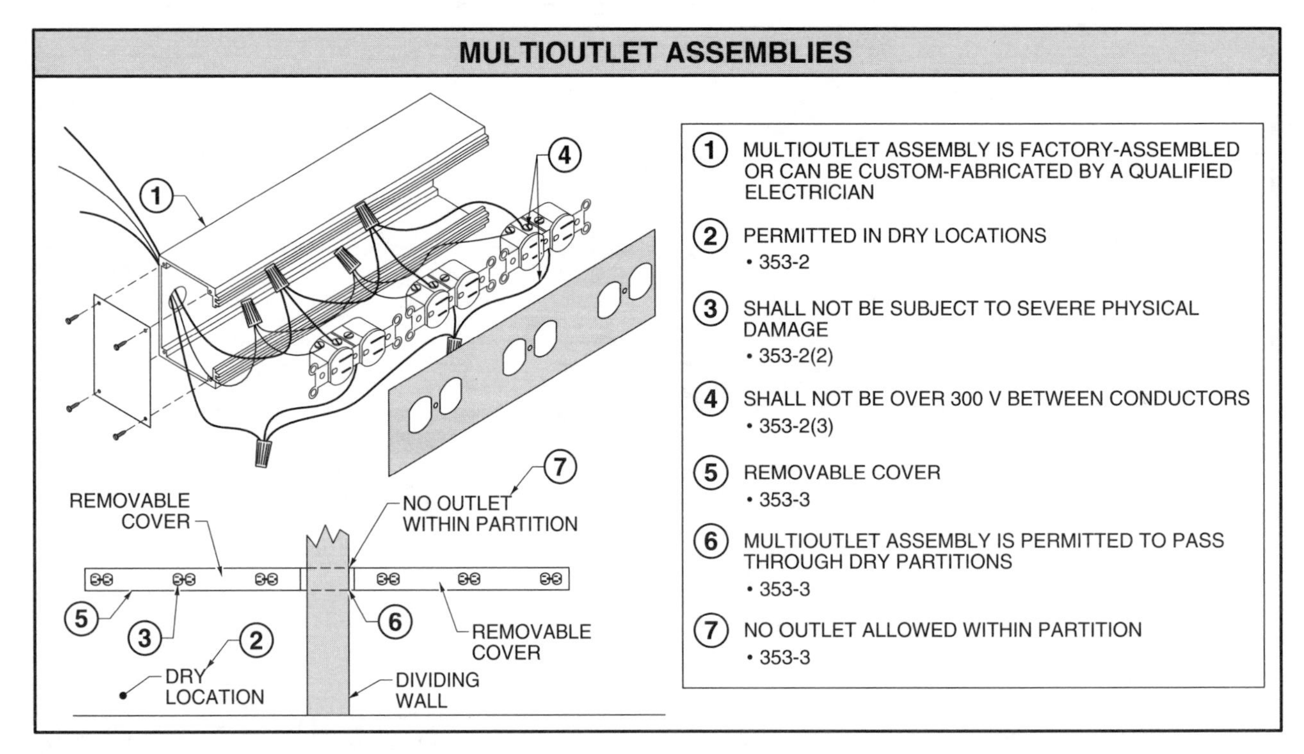

Figure 3-10. A multioutlet assembly is a metal raceway with factory-installed conductors and attachment plug receptacles.

CABLE ASSEMBLIES

A *cable assembly* is a flexible assembly containing multiconductors with a protective outer sheath. The purpose of this outer sheath is to protect the conductors contained therein. Four of the more popular types of cable assemblies are armored cable (AC), metal-clad cable (MC), nonmetallic-sheathed cable (NMC), and service-entrance cable (SE).

Armored Cable (AC) – 333

Armored cable (AC) is a factory assembly that contains the conductors within a jacket made of a spiral wrap of steel. See Figure 3-11. A bare bonding strip is placed inside the armor with the conductors. This bare bonding strip is in intimate contact with the outer metal jacket for its entire length to ensure a low-impedance path on the outer armor in the event of a ground fault. AC cable may only be used in dry locations.

Section 333-7 requires that AC cable be supported every 4½′ and within 12″ of outlets and fittings. Section 333-9 requires that a connector be used when terminating AC cable in a box, cabinet, or piece of equipment. The purpose of this connector is to protect the conductors within the cable from abrasion. An insulating bushing (anti-short) shall be used with the connector. This bushing slips over the conductors and under the armor of the cable for additional protection.

Section 333-10 requires the AC cable be installed at least 1¼″ back from the finished edge of framing members. This protects the cable from nails and drywall screws per 300-4.

Section 333-11 requires that exposed runs of AC cable closely follow the surface of the building finish or of running boards. The three exceptions to this rule are:

(1) AC cable lengths shall not be more than 24″ at terminals where flexibility is necessary.

(2) AC cable may be run on the underside of a joist where supported at each joist and not subject to physical damage.

(3) AC cable may be run in lengths not exceeding 6′ from an outlet for connection within an accessible ceiling to lighting fixtures or other equipment.

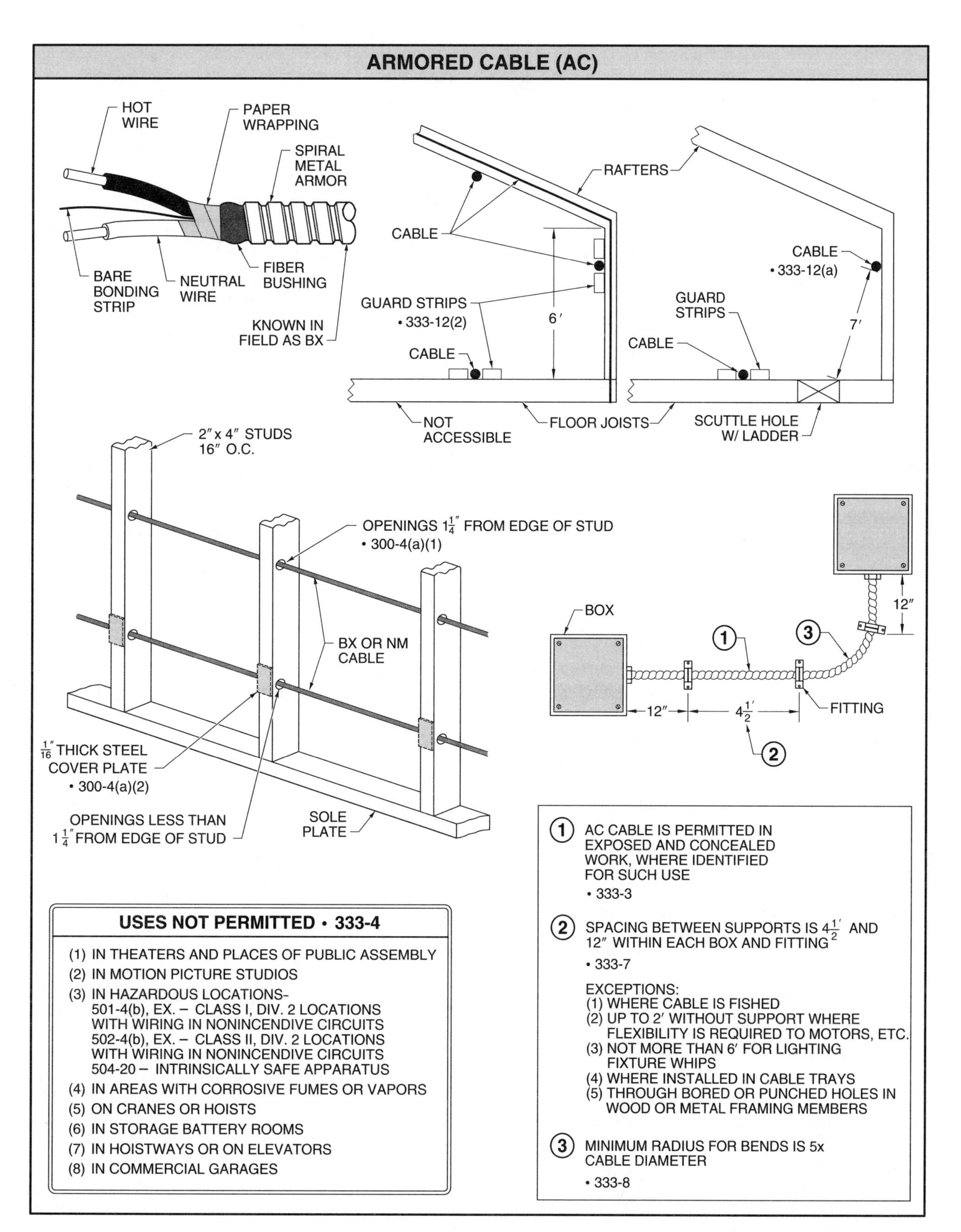

Figure 3-11. Armored cable (AC) is an assembly that contains the conductors within a jacket made of a spiral wrap of steel.

Section 333-12 permits the use of AC cables in accessible attics. AC cables installed across the top of floor joists require protection by using guard strips that are at least as high as the cable. Guard strips shall also be used when AC cable is installed across the face of rafters or studs that are less than 7′ above the attic floor.

AC cable installed in attics that are inaccessible (without stairways or pull-down ladders) only require guard strips within 6′ of the scuttle-hole opening. Within inaccessible attics, AC cables may be run along the sides of joist, rafters, and studs or fished through drilled holes without guard strips. Guard strips are usually 1″ × 2″ wooden strips nailed directly to the structure.

USING AC CABLE AS AN EGC

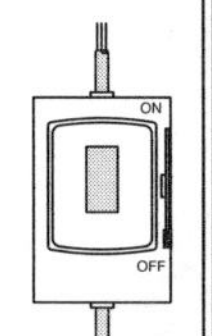

The outer armor of AC cable has the same electrical characteristics as a wound coil. The armor offers inductive reactance (opposition to current flow) to an AC current trying to flow around the armor. Thus, the armor alone cannot be used as an EGC since it would oppose the flow of large currents in the event of a ground fault and prevent the operation of the overcurrent protection device.

An internal bonding strip is in intimate contact with each spiral of the outer armor. The outer armor no longer has the characteristics of a coil, but rather a solid piece of metal. The outer armor, with the bonding strip, provides a low-impedance path for high fault-currents to flow, which, in turn, allows the overcurrent protection device to open the circuit in the event of a ground fault.

Metal-Clad Cable (MC) – 334

Metal-clad cable (MC) is a factory assembly of one or more conductors with or without fiber-optic members. See Figure 3-12. MC cable does not have an internal bonding strip in intimate contact with the outer armor as does AC cable. The conductors are enclosed in a metal sheath with one of the following configurations per 334-1: smooth sheath, corrugated sheath, or interlocked sheath.

Smooth Sheath – The outer metallic sheath is smooth and continuous. This type of sheath makes an excellent EGC provided it is terminated with connectors listed for this type of cable. Since the outer sheath is metallic, smooth, and continuous, it is a little difficult to bend.

Corrugated Sheath – This type has some of the metallic sheath grooved out to make it more flexible. It looks like a piece of flexible gas tubing used to hook up gas appliances. The outer sheath is a continuous piece of metal and makes an excellent EGC.

Interlocking Sheath – This type has an outer jacket which looks like AC cable. The outer jacket is spiraled and has the electrical characteristics of a coil. When used in an AC circuit, it offers inductive reactance which causes a high impedance, thus opposing the flow of ground-fault current. With this type of MC cable, the manufacturer always provides an insulated green EGC for the grounding.

Type MC cable is manufactured with conductor sizes of No. 14 AWG up to No. 1000 kcmil. MC cable is listed up to 5000 V by UL. MC cable is available with insulations rated as high as 15,000 V. Per 334-4, MC cable shall not be used where exposed to destructive corrosive conditions. It shall not be directly buried in earth or concrete, and it shall not be exposed to cinder fills. MC cable shall not be used where exposed to strong chlorides, caustic alkalis, chlorine vapors, or hydrochloric acid vapors.

Section 334-3 permits MC cable to be used for services, feeders, and branch circuits. It may be used for power, lighting, control and signal circuits, indoors or outdoors, exposed or concealed, in any approved raceway, as open runs, as aerial cable on a messenger, in hazardous locations per 501, 502, 503, and 504, and in any wet location (provided the cable is approved for that location) with a metallic covering that is impervious to moisture or with insulated conductors under the metallic covering approved for wet locations.

Section 334-4 prohibits the use of MC cable where exposed to destructive corrosive conditions such as direct burial in the earth, in concrete, or where exposed to cinder fills, strong chloride, caustic alkalis, or vapors of chlorine or hydrochloric acids. Section 334-10 requires MC cable to be supported and secured at intervals not exceeding 6′. Cables containing four of fewer conductors, sized no larger than No. 10 shall be secured within 12″ of every box, cabinet, or fitting.

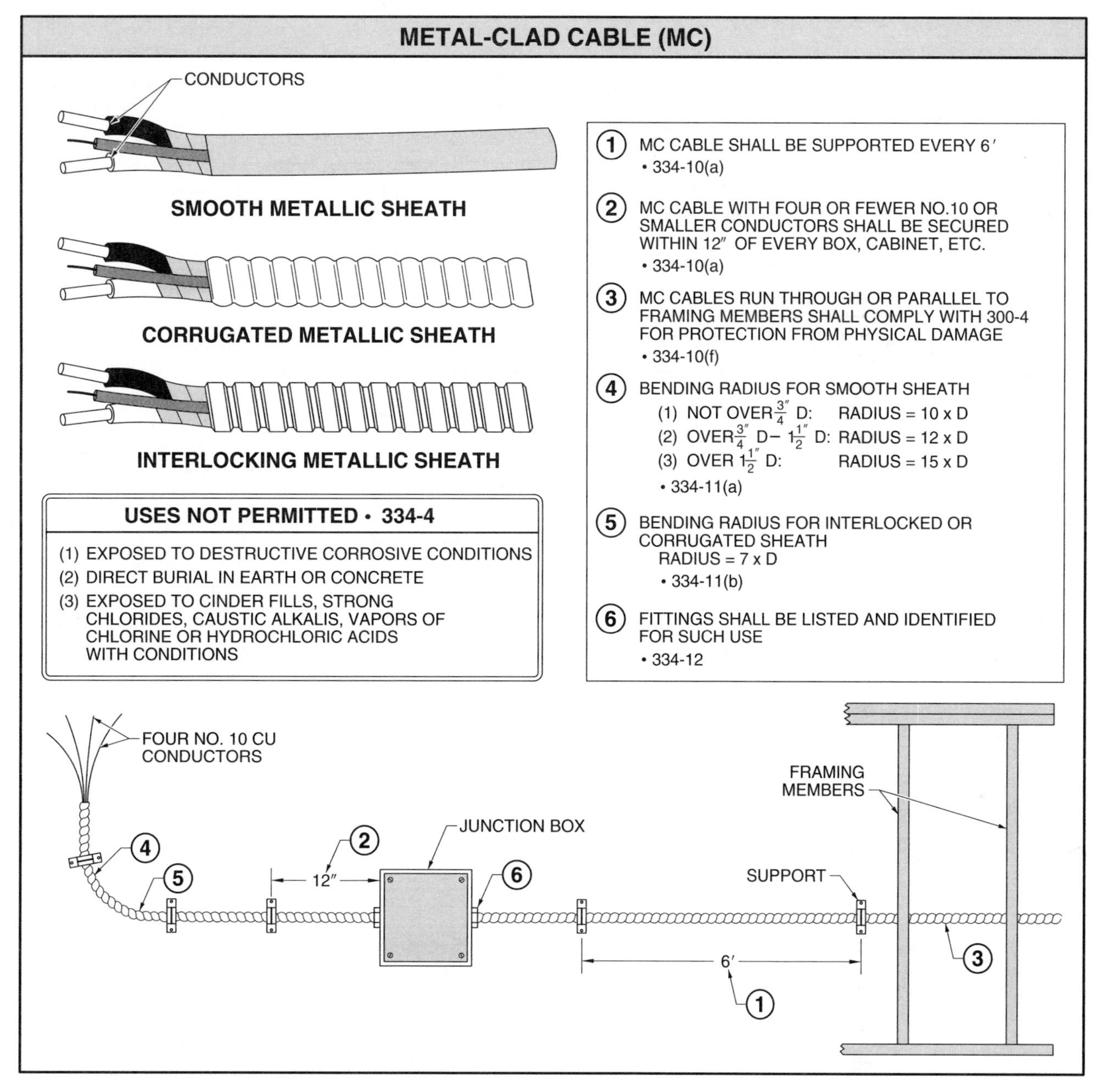

Figure 3-12. Metal-clad cable (MC) is a factory assembly of one or more conductors with or without fiber optic members.

The radius of the curve of the inside bend of MC cable varies with the type of outer covering and the diameter of the cable per 334-11.

(a) Smooth sheath – (1) Not more than 3/4″ in diameter, shall have a bending radius of a minimum of 10 times the diameter of the cable. (2) More than 3/4″ in diameter but not more than 1½″ in diameter, shall have a bending radius of a minimum of 12 times the diameter of the cable. (3) More than 1½″ in diameter, shall have a bending radius of a minimum of 15 times the diameter of the cable.

(b) Interlocked or corrugated sheath – Shall have a bending radius of not less than seven times the external diameter of the metallic sheath.

(c) Shielded conductors – Shall have a bending radius of twelve times the overall diameter of one of the individual conductors or seven times the overall diameter of the multiconductor cable, whichever is greater.

REDUNDANT GROUNDING

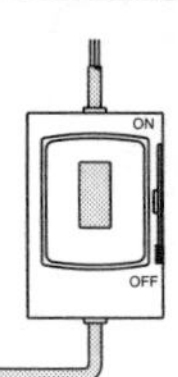

Redundant grounding is required in several places in the NEC®. Redundant grounding is grounding with two separate grounding paths. For example, in patient care areas per 517-13, the metal raceway or recognized outer metallic sheath of a smooth tube or corrugated tube MC cable counts as one grounding path, and an insulated EGC contained within the metal raceway or recognized metallic sheath counts as the second path. Interlocking tape MC cable does not meet NEC® requirements for redundant grounding.

Nonmetallic-Sheathed Cable – 336

Nonmetallic-sheathed cable (NM) is a factory assembly of two or more insulated conductors having an outer sheath of moisture-resistant, flame-retardant, nonmetallic material. See Figure 3-13. NM is an economical wiring method used in residential units and small commercial locations of fewer than four stories. It is known in the field as "Romex."

NM comes with a bare EGC. The circuit conductors are manufactured in sizes No. 14 through No. 2 AWG. It is available in copper, aluminum, or copper-clad aluminum. It is shipped in 250′ rolls.

Until 1984, NM was manufactured with TW insulation which has a temperature rating of 60°C for Type A. Since 1984, the conductors have been made with THHN insulation which has a temperature rating of 90°C for Type B. This change was initiated due to higher R values of insulation being installed in the attics of dwelling units. The Type A NM cable did not dissipate the heat under the thicker type insulation and, as a result, many of these cables failed. Caution should be taken not to take the ampacity from the 90°C column in Table 310-16 when installing Type B NM cable. The ampacity from the 60°C column shall be used. See 336-30(b).

Uses Permitted – 336-4. Nonmetallic-sheathed cable is permitted for use in one-family, two-family, and multifamily dwelling units and other structures except as prohibited by 336-5. The three types of nonmetallic-sheathed cable are NM, NMC, and NMS.

Type NM cable, 336-4(a), has a flame-retardant and moisture-resistant outer jacket that is restricted to inside wiring. Type NMC cable, 336-4(b), has a flame-retardant, moisture-, fungus-, and corrosion-resistant outer jacket. NMC cable is permitted to be installed in outside locations. Type NMS cable is new to the 1996 NEC®. It is used in closed-loop and programmed power distribution systems (Smart House) as permitted in 780.

The use of NM cable in any dwelling or structure more than three floors above grade is prohibited per 336-5(a)(1). The first level is not counted if it is used for vehicle parking, storage, or similar use.

NM, NMC, and NMS shall not be used in corrosive locations, commercial garages, theaters, motion picture studios, hoistways, or hazardous locations. NM, NMC, and NMS shall not be embedded in plaster.

Installation – 336. Section 336-6 defines the provisions for installing NM when exposed. The cable shall closely follow the surface of the building finish or running boards. It shall be protected whenever it passes through a floor in an exposed work area. This protection may be provided by a piece of conduit, EMT, or Schedule 80 PVC. The protective sleeve shall extend a minimum of 6″ above the floor. See Figure 3-14.

Cables in unfinished basements shall be installed through bored holes in joists or on running boards. Three-conductor No. 8 or two-conductor No. 6 or larger cables may be installed on the lower edges of the joist. Per 336-18, NM cable shall be secured every 4½′ and within 12″ of every outlet or fitting.

Two Romex cables shall not be stapled on top of one another. Cables run through holes in wood or metal joists, rafters, or studs shall be considered to be supported and secured.

ROMEX

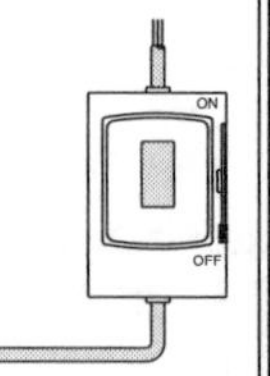

The General Cable Co. developed NM cable in the early 1900s. NM cable was first recognized in the NEC® in 1928. The Rome Cable Co. of Bronx, NY registered this product under the name "Romex." This name has become a generic term for NM cable.

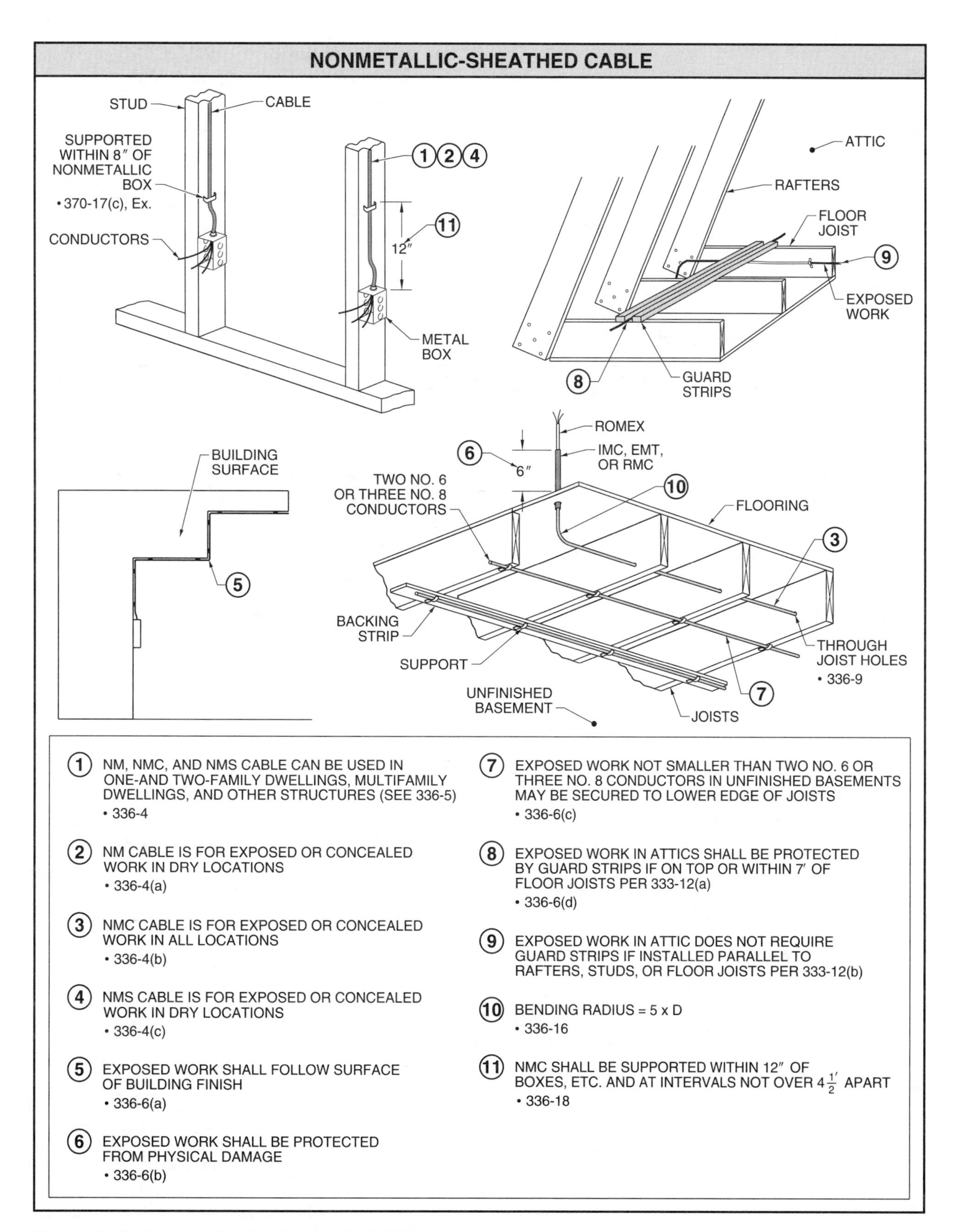

Figure 3-13. Nonmetallic-sheathed cable (NM) is a factory assembly of two or more insulated conductors having an outer sheath of moisture-resistant, flame-retardant, nonmetallic material.

Metal or nonmetallic boxes may be used with NMS cable. The cable shall be supported within 12″ of a metal box. NMS cable shall be supported within 8″ of a nonmetallic box with at least ¼″ of the cable's outer jacket inside the box.

Service-Entrance Cable (SE and USE) – 338

Service-entrance cable (SE) is a single or multiconductor assembly with or without an overall covering. See Figure 3-14. There are two types of service-entrance cable recognized by the NEC®, Type SE and Type USE. Both are rated up to 600 V.

SE cable is available in two styles, U and R. U is a flat cable with an uninsulated concentric grounded conductor. R is a round cable with an insulated grounded conductor. Both styles are listed for aboveground installations.

The outer jacket or finish of SE cable is suitable for use where exposed to the sun. SE cable is available in sizes from No. 12 through No. 4/0. The cable comes with two or three insulated conductors and one bare conductor.

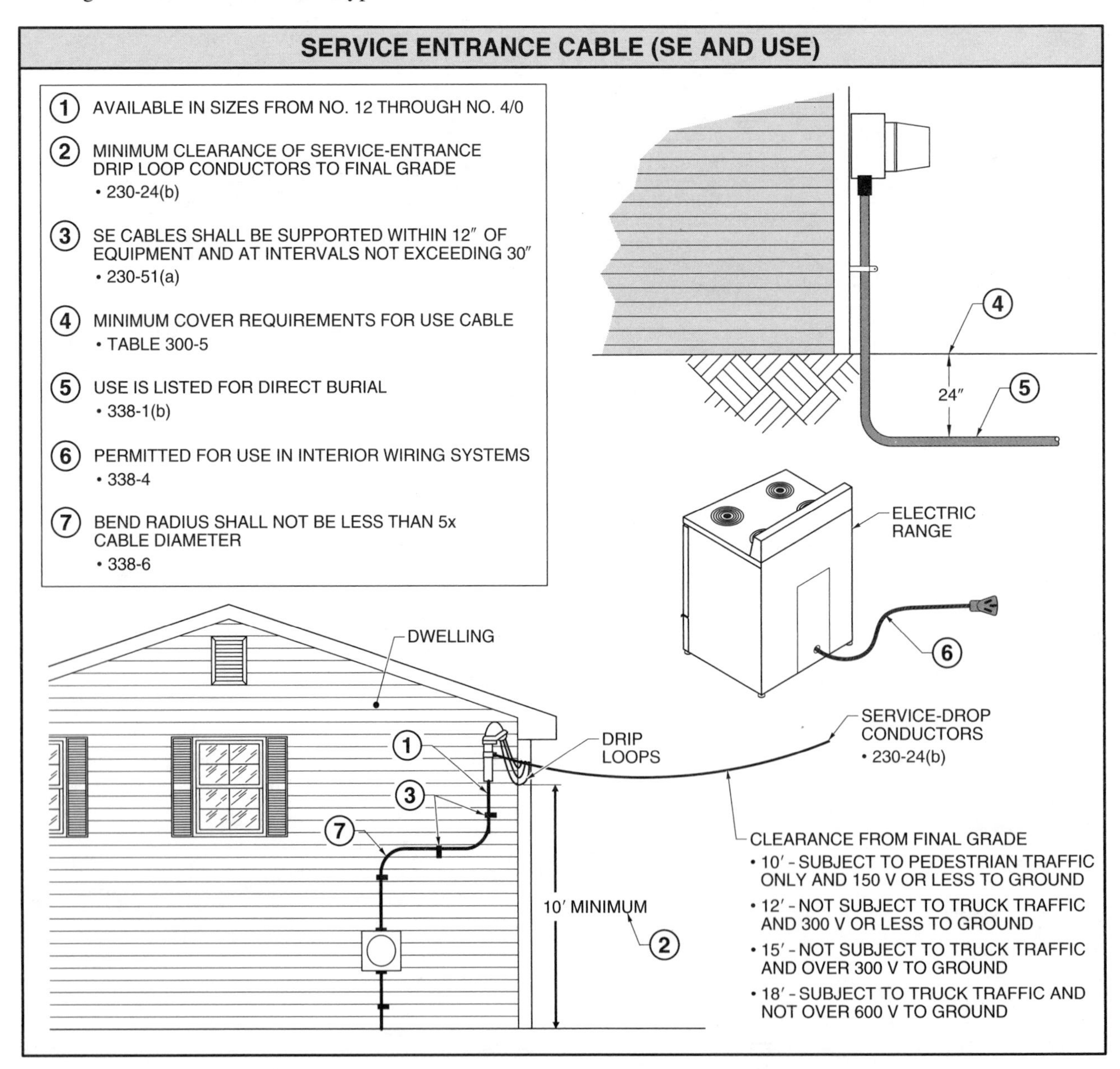

Figure 3-14. Service-entrance cable (SE) is a single or multiconductor assembly with or without an overall covering.

Section 338-2 permits SE cable to be used for service-entrance wiring. When used for service-entrance wiring, the grounded conductor may be insulated or bare. Section 338-4 permits SE cable to be used for general interior wiring in buildings with not more than three floors. SE cable used indoors for feeders or branch circuits shall have an insulated grounded conductor. It is permissible to use the uninsulated conductor for equipment grounding purposes.

Section 338-6 requires that the protective coverings of the cable not be damaged when the cable is bent during installation. The radius of the curve of the inner edge of any bend shall not be less than five times the diameter of the cable.

USE cable has a moisture-resistant outer jacket. It is listed for burial directly in the earth. USE cable is moisture-resistant. It is not fire retardant and is not suitable for use in premises or aboveground except to terminate at the service equipment or metering equipment.

OTHER WIRING SYSTEMS

Cable trays, wireways, and busways are found primarily in commercial and industrial installations. The advantage of these systems, as opposed to a conduit system, is the ease of installation and the ability to alter the loads that are served from these systems.

Cable Tray Systems (CTS) – 318

A *cable tray system (CTS)* is an assembly of sections and associated fittings which form a rigid structural system used to support cables and raceways. See Figure 3-15. These systems are available in different configurations including ladders, troughs, channel, solid bottom, and ventilated troughs.

Cable tray systems are assembled in the field. They are available in aluminum, steel, or nonmetallic. Nonmetallic cable trays shall be of a flame-retardant material. Cable trays are classified as a support system for cables and raceway systems. The cable tray itself is not considered a raceway system.

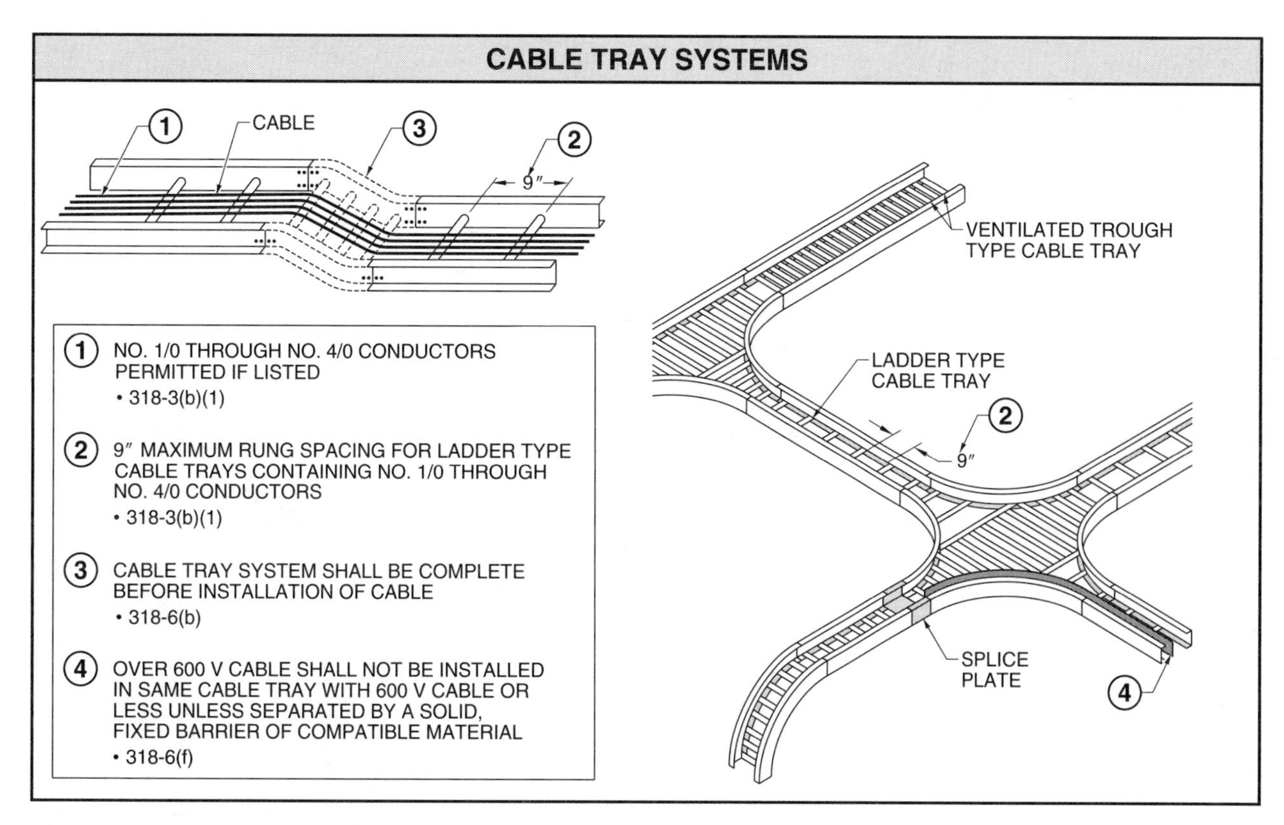

Figure 3-15. A cable tray system (CTS) is an assembly of sections and associated fittings which form a rigid structural system used to support cables and raceways.

Wiring Methods Permitted – 318-3. There are 18 wiring methods permitted in cable tray systems. Section 318-3(b) permits ladder type, ventilated trough type, and ventilated channel type CTS to be installed in industrial establishments when running single and multiconductor cable. Section 318-3(b)(1) permits single conductors in smaller sizes, No. 1/0 through No. 4/0, to be installed in cable trays provided they are marked and listed for such use. Installing conductor sizes No. 1/0 through No. 4/0 in ladder type trays requires that the rungs of the ladder tray be spaced no more than 9″ apart.

INSTRUMENTATION TRAY CABLE

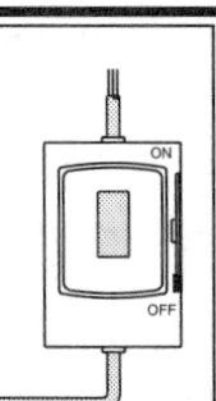

Article 727 is new in the 1996 NEC®. This article applies only to industrial locations where maintenance and supervision assure that only qualified persons service the installation. This cable tray system shall not carry power, lights, Class I, and nonpower-limited circuits. The circuits contained in this system shall not operate at more than 150 V and shall not carry more than 5 A.

Installation as a Complete System – 318-6(a). CTS shall be installed as complete systems. The electrical continuity of the tray system shall be maintained. Section 318-7 permits the use of steel and aluminum cable tray systems as the EGC when the cable tray and fittings are identified for the purpose and the total cross-sectional area of both side rails meet the requirements of Table 318-7(b)(2). The concern here is that there is enough mass of metal to carry the available fault current.

Sizing Cable Trays for Multiconductor Cables – 318-9. Multiconductor cables rated 2000 V or less are permitted to be installed in ladder, ventilated trough, solid bottom, or ventilated channel cable trays. See Figure 3-16.

Section 318-9(a)(1) requires that ladder or ventilated trough cable trays have a width equal to or greater than the sum of all the diameters of the cables No. 4/0 or larger. These cables shall be installed side by side in a single layer. Section 318-9(a)(2) requires that cables smaller than No. 4/0 in size meet the maximum allowable fill requirements of Column 1 in Table 318-9. These smaller cables do not have to be placed side by side in a single layer. Section 318-9(a)(3) requires installations that have a combination of cable sizes smaller and larger than No. 4/0 to meet the maximum allowable fill requirements of Column 2 in Table 318-9. Section 318-9(b) requires that signal and control cables shall not exceed 50% of the cross-sectional area of cable trays that have an inside depth of 6″ or less.

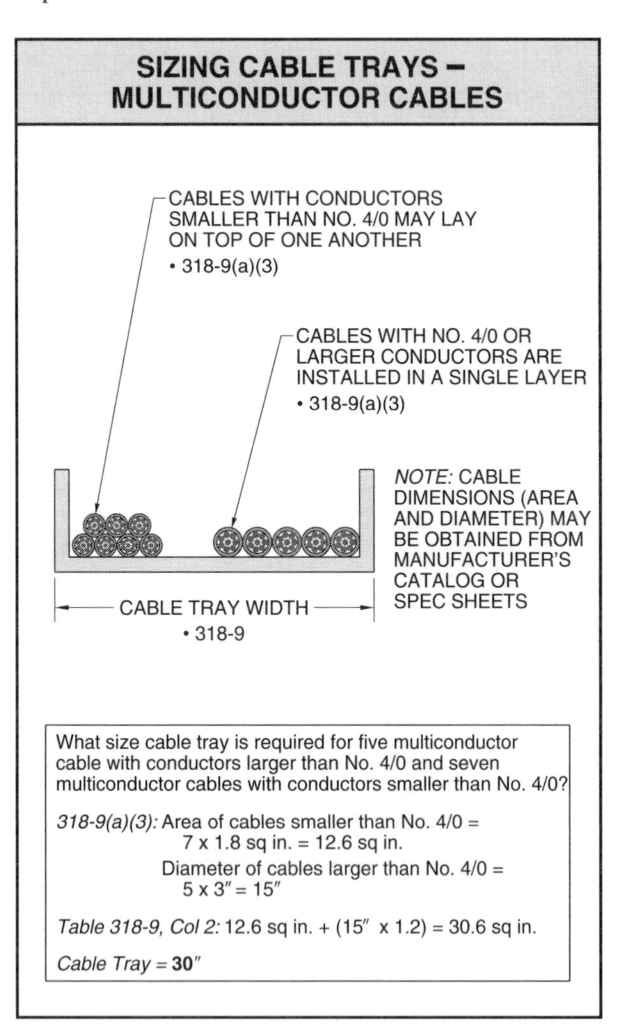

Figure 3-16. Cable trays for multiconductor cables are sized by Table 318-9.

For solid-bottom cable trays, 318-9(c)(1) requires that the diameter of all cables No. 4/0 and larger shall not exceed 90% of the cable tray width and that the cables shall be installed in a single layer. Section 318-9(c)(2) requires that cables smaller than No. 4/0 in size shall meet the maximum allowable fill requirements of Column 3 in Table 318-9. These cables do not have to be installed in a single layer. Section 318-9(c)(3) requires installations containing cables smaller and larger than No. 4/0 to comply with the maximum allowable fill requirements of Column 4 in Table 318-9. Section 318-9(d) requires that signal and control cables shall not exceed 40% of the cross-sectional area of the cable trays that have an inside depth of 6″ or less.

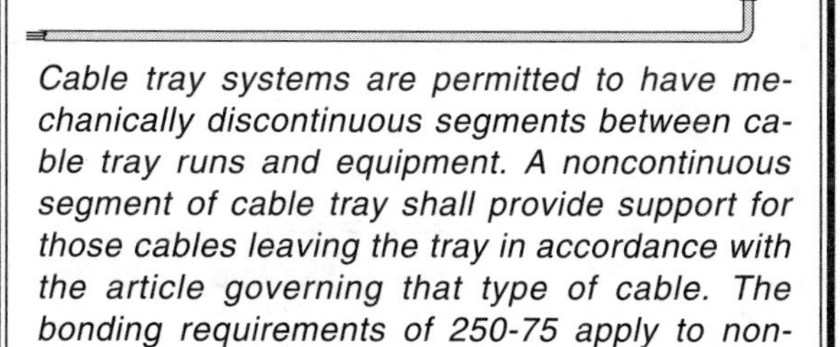

Section 318-9(e) requires that for ventilated channel cable trays, the total cross-sectional areas of all cables shall not exceed 1.3 sq in. for 3″ wide trays, 2.5 sq in. for 4″ wide trays, and 3.8 sq in. for 6″ wide trays.

Sizing Cable Trays for Single Conductor Cables – 318-10. Single conductors rated 2000 V or less are permitted in ladder, ventilated trough, and ventilated channel cable trays. See Figure 3-17. The single conductors shall be evenly distributed across the cable tray. Cable trays are sized from Columns 1 and 2 of Table 318-10 based on the square inch area and diameter of conductors per Chapter 9, Table 5.

Section 318-10(a)(1) requires that ladder or ventilated trough cable trays containing 1000 kcmil or larger cables shall have a minimum width equal to the sum of the diameters of all the cables. Section 318-10(a)(2) requires that ladder or ventilated trough cable trays containing 250 kcmil up to 1000 kcmil cables shall have a minimum width equal to the sum of the cross-sectional areas of all single conductor cables as permitted in Column 1 of Table 318-10. Section 318-10(a)(3) requires that ladder or ventilated trough cable trays containing single conductor 250 kcmil up to and larger than 1000 kcmil cables shall have a minimum width based on the following calculation:

1. Find the sum of the diameters of all single conductors 1000 kcmil and larger.
2. Multiply the sum by 1.1 to convert to square inches.
3. Add the sum of all diameters of 250 kcmil to 1000 kcmil to sum in Step 2.
4. Select cable tray per Table 218-10.

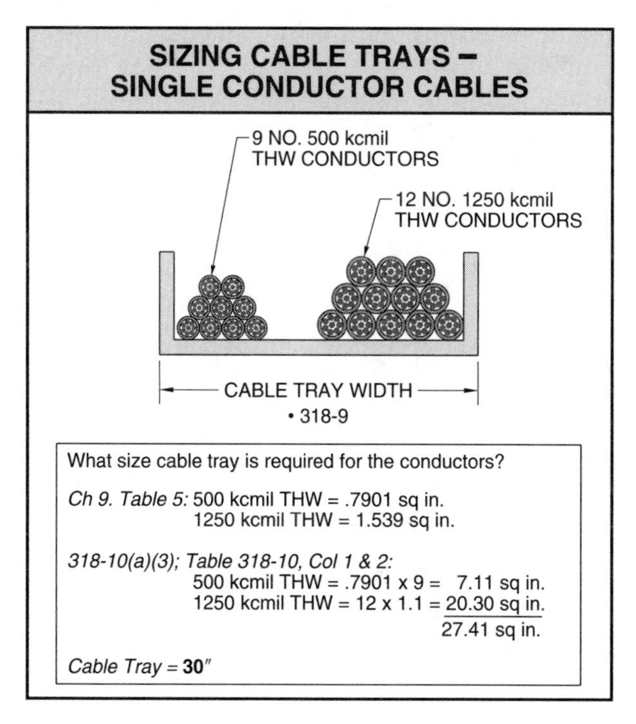

Figure 3-17. Cable trays for single conductor cables are sized by Table 318-10.

Wireways – 362

A *wireway* is a metallic or nonmetallic trough with a hinged or removable cover designed to house and protect conductors and cables. See Figure 3-18. Wireways shall be installed as a complete system before the conductors are installed. They are sized by Chapter 9, Table 5. See Figure 3-19.

Metallic Wireways – 362, Part A. Metallic wireways shall be used for exposed work only per 362-2. Wireways installed in wet locations shall be of raintight construction. There shall be no more than 30 current-carrying conductor at any cross section per 362-5.

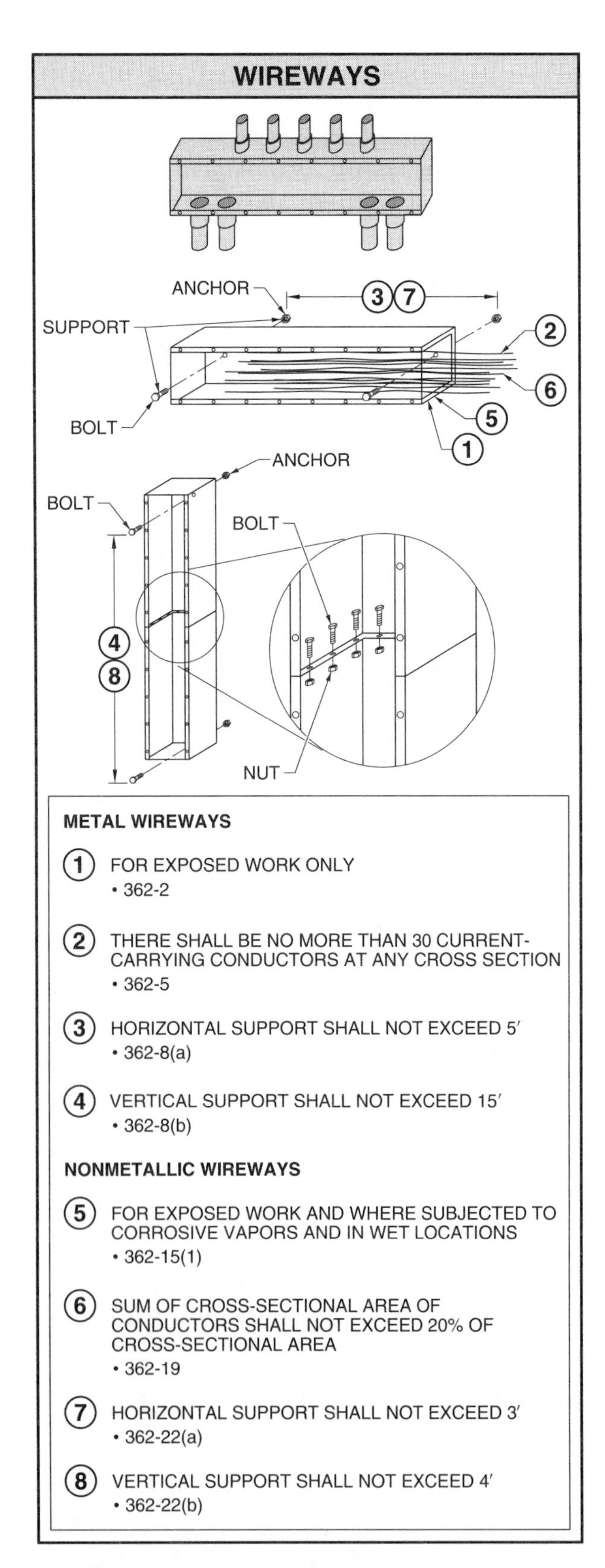

Figure 3-18. A wireway is a metallic or nonmetallic trough with a hinged or removable cover designed to house and protect conductors and cables.

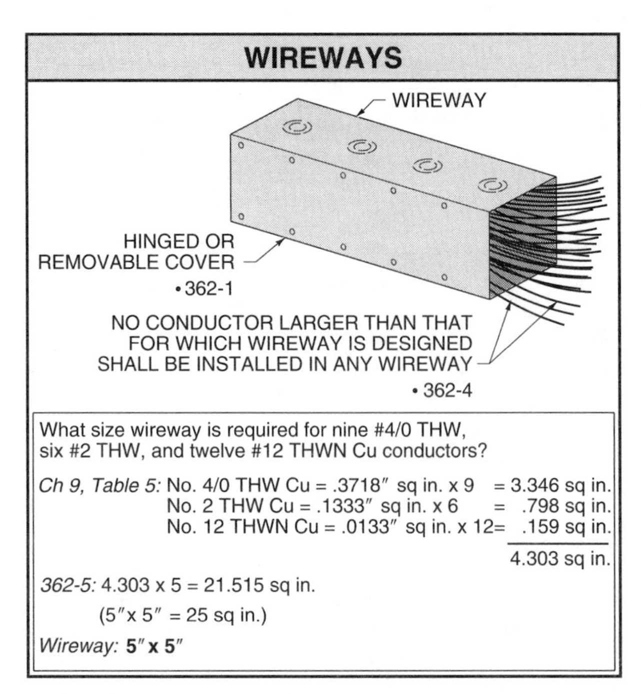

What size wireway is required for nine #4/0 THW, six #2 THW, and twelve #12 THWN Cu conductors?

Ch 9, Table 5: No. 4/0 THW Cu = .3718″ sq in. x 9 = 3.346 sq in.
No. 2 THW Cu = .1333″ sq in. x 6 = .798 sq in.
No. 12 THWN Cu = .0133″ sq in. x 12 = .159 sq in.
4.303 sq in.

362-5: 4.303 x 5 = 21.515 sq in.
(5″ x 5″ = 25 sq in.)

Wireway: **5″ x 5″**

Figure 3-19. Wireways are sized by Chapter 9, Table 5.

Conductors used for signaling circuits or conductors of a motor control circuit used only for starting duty shall not be considered current-carrying conductors. The sum of the cross-sectional area of all of the contained conductors shall not exceed 20% of the interior cross-sectional area of the wireway. The derating factors contained in Note 8 of Notes to Ampacity Tables of 0 to 2000 Volts do not apply to the 30 current-carrying conductors at 20% fill.

Splices and taps in wireways are permitted per 362-7 provided they are accessible. The conductors, including splices and taps, shall not occupy more than 75% of the cross-sectional area at that point.

Section 362-8(a) requires supports of horizontal wireways not to exceed 5′. Wireways longer than 5′ shall have a support at each end or joint. The distance between supports shall not exceed 10′. Section 362-8(b) requires that vertical runs of metallic wireways shall be supported at intervals not exceeding 15′. Vertical runs shall not have more than one joint between supports.

Nonmetallic Wireways – 362, Part B. Nonmetallic wireways shall be used for exposed work only per 362-15. Nonmetallic wireways may be used in areas subject to corrosive vapors and in wet locations where listed for the purpose. Per 362-15, FPN the installer is cautioned that extreme cold may cause nonmetallic wireways to become brittle and more susceptible to damage from physical contact. Section 362-16 prohibits the use of nonmetallic wireways in the following five areas:

(1) Where subject to physical damage.

(2) In any hazardous location, except as permitted in 504-20.

(3) Where exposed to sunlight unless listed and marked as suitable for the purpose.

(4) Where subject to ambient temperatures other than those for which it is listed.

(5) For conductors whose insulation temperature limitations exceed those for which it is listed.

The sum of the cross-sectional area of all the contained conductors in a nonmetallic wireway shall not exceed 20% of the cross-sectional area of the wireway per 362-19. Conductors for signaling circuits or conductors of motor control circuits used only for starting duty shall not be considered current-carrying conductors. The derating factors of Note 8 of Notes to Ampacity Tables of 0 to 2000 Volts shall apply to the current-carrying conductors at the 20% fill.

Section 362-21 permits splices and taps in nonmetallic wireways. The conductors, including splices and taps, shall not fill the nonmetallic wireway to more than 75% of its area at that point.

Horizontal runs of nonmetallic wireways shall be supported at least every 3′, per 362-22(a), unless listed for other support intervals. The distance between supports shall never exceed 10′. Section 362-22(b) requires vertical runs of nonmetallic wireways to be supported at intervals not exceeding 4′, per 362-22(b), unless listed for other support intervals. There shall not be more than one joint between supports.

Section 362-23 requires expansion fittings in straight runs of nonmetallic wireways where the length change is expected to be 0.25″ or greater. Table 10 in Chapter 9 gives the expansion characteristics of PVC rigid nonmetallic conduit. PVC nonmetallic wireway has identical characteristics.

Busways – 364

A *busway* is a sheet metal enclosure that contains factory-assembled aluminum or copper busbars which are supported on insulators. See Figure 3-20. Busways are available for either indoor or outdoor use. If used in an outdoor location, the busway shall be listed for outdoor use. UL lists busways up to 600 V. Their ampacity can be as high as 6500 A for copper bus and 5000 A for aluminum bus. Most systems are in the 200 A to 1000 A range. There are small busway systems available with ranges between 20 A to 250 A. Some systems are designed to accept disconnect switches with overcurrent devices that can be plugged into the busway in order to feed loads as required. The three basic types of busways are busways that provide an opening for plugging in a disconnect with an overcurrent device, busways that will not accept a plug-in disconnect, and trolley busways.

A trolley busway has an open bottom. The trolley runs on wheels and electrical contact is made by brushes or rollers on the busbars. The trolley system feeds portable equipment in industrial installations.

Section 364-4(a) permits busways to be installed only in open locations where they are visible. The only exception is for totally enclosed, nonventilating-type busways. They have to be installed so that the joints between sections and fittings are accessible for maintenance. Section 364-4(a), Ex. permits these busways to be installed behind panels where means of access are provided, and one of two provisions is followed:

(1) The space behind the access panels is not used for air-handling purposes.

(2) The space behind the access panels is used for environmental air, in which case there shall be no provisions for plug-in connections and the conductors shall be insulated.

Section 364-4(b) prohibits the use of busways in the following four locations:

(1) Where subject to severe physical damage or corrosive vapors.

(2) In hoistways.

(3) In any hazardous location, unless approved for such use, such as 501-4(b), Class I, Division 2 (enclosed gasketed busways).

(4) Outdoors or in damp or wet locations unless identified for the use.

Lighting busways and trolley busways shall not be installed less than 8′ above the floor or working platform unless provided with a cover identified for the purpose. Section 364-5 requires that busways shall be supported at intervals not exceeding 5′.

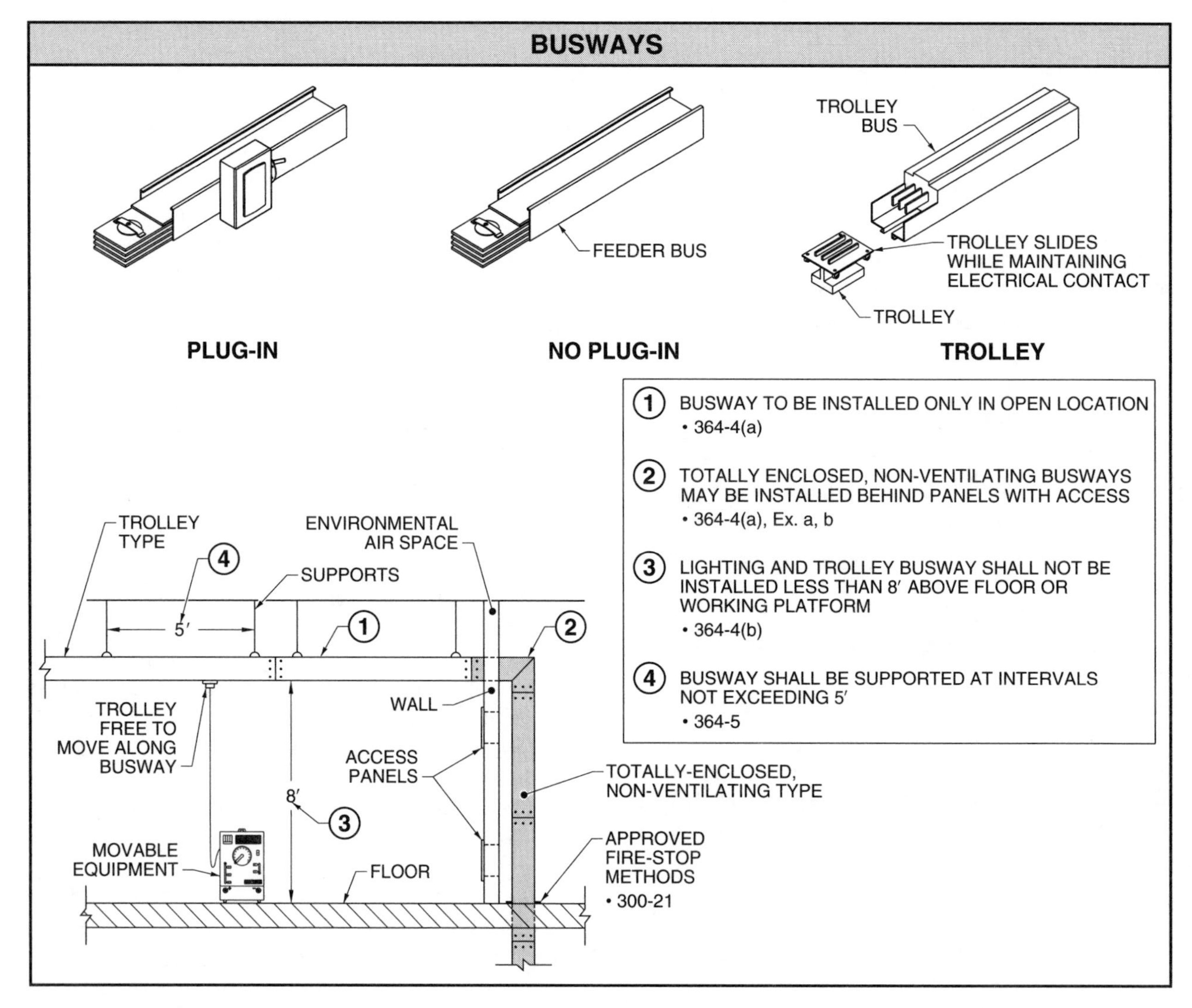

Figure 3-20. A busway is a sheet-metal enclosure that contains factory-assembled aluminum or copper busbars which are supported on insulators.

Busways are permitted to be used for service-entrance conductors where the voltage is 600 V nominal or less per 230-43(9). Unbroken lengths of busways are permitted to extend through dry walls per 364-6. It is permissible to extend busways vertically through dry floors provided the busway is of the totally enclosed type.

Section 364-10 requires that busways have overcurrent protection in accordance with the current rating of the busway. Where the allowable current rating of the busway does not correspond to a standard ampere rating of the overcurrent device, the next higher standard size of the overcurrent device may be used. However, if this rating exceeds 800 A, then the lower size shall be used.

Branches from Busways – 364-8(a). Branches from busways are permitted, provided one of the following methods is used:

- RMC
- IMC
- EMT
- FMC
- MC
- RNMC
- ENT
- LTFMC
- LTFNMC
- Surface metal and nonmetallic raceways and busways

Where a nonmetallic raceway is used, the EGC in the nonmetallic raceway shall be connected to the busway in accordance with 250-113. The concern here is that grounding and bonding conductors are terminated properly by cadwelding or using listed materials. *Cadwelding* is a welding process used to make electrical connections of copper to copper or copper to steel in which no outside source of heat or power is required. Powdered metals (copper oxide and aluminum) are dumped from a container into a graphite crucible and ignited by means of a flint igniter. The reduction of the copper oxide by the aluminum (exothermic reaction) produces molten copper and aluminum oxide slag. The molten copper flows over the conductors in the graphite mold, melting them and welding them together.

Section 364-8(b) permits suitable cord and cable assemblies recognized for hard usage per Table 400-4 and listed bus drop cable to be used for the connection of portable or stationary equipment, provided the following four conditions are met:

(1) The cord or cable shall be attached to the building by an approved means.

(2) There shall be no more than 6′ of cord or cable between the busway plug-in device and the tension take-up device. Section 364-8(b)(2), Ex. permits industrial establishments that have maintenance and supervision performed by qualified persons to exceed this 6′ length. Although there is not a length limitation between the busway plug-in device, and the tension take-up device the cord or cable shall be supported every 8′.

(3) The cord or cable shall be installed as a vertical riser between the tension take-up device and the equipment served.

(4) Strain-relief cable grips shall be provided for the cord or cable at the busway plug-in device and at the equipment terminations.

Reduction in Size of Busway – 364-11. Section 364-11 requires overcurrent protection at the point where busways are reduced in ampacity. See Figure 3-21. Section 364-11, Ex. permits only industrial establishments to reduce the ampacity of busways without providing overcurrent protection provided the following three conditions are met:

(1) The length of the reduced busway does not exceed 50′.

(2) The reduced busway has an ampacity of at least ⅓ of the rating or setting of the overcurrent device ahead of where the reduction takes place.

(3) The reduced busway is free from contact with combustible material.

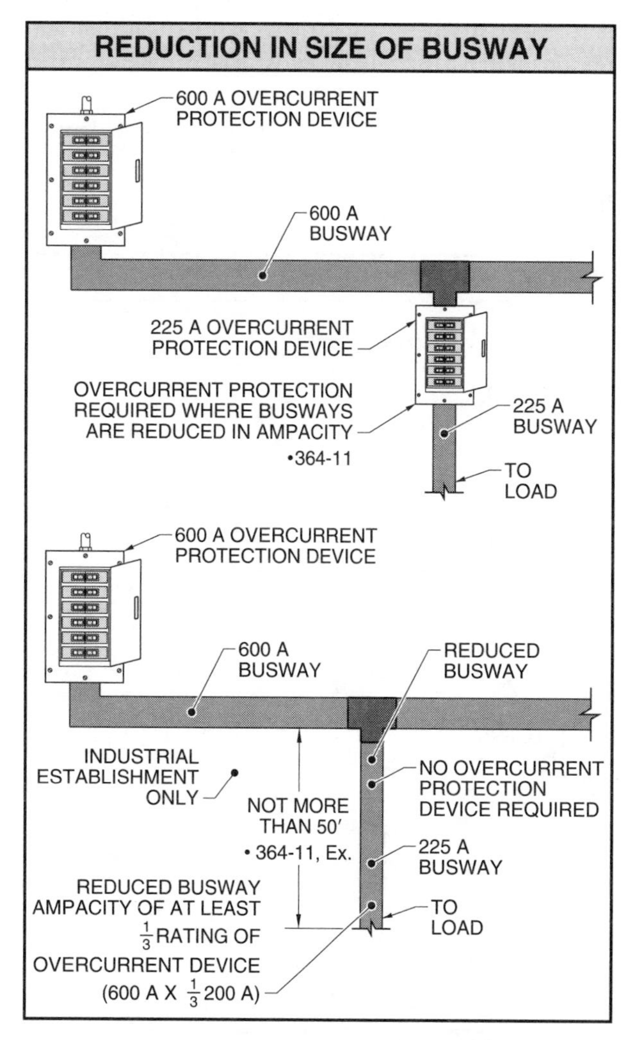

Figure 3-21. Overcurrent protection shall be provided at the point where busways are reduced in ampacity.

Busways are permitted to be used as branch circuits per 364-13. The overcurrent protection device for the busway used as a branch circuit is the same as any other wiring method used to supply a branch circuit. The rating or setting of the overcurrent protection device protecting the busway shall determine the ampere rating of the branch circuit.

Auxiliary Gutters – 374

An auxiliary gutter and associated fittings are identical to wireways and associated fittings. UL registers one listing for both of these products, *Listed Wireway or Auxiliary Gutter.* The only difference between them is their intended use. An *auxiliary gutter* is a sheet-metal enclosure equipped with hinged or removable covers that is used to supplement wiring space. These gutters supplement wiring space at meters, panelboards, switchboards, and similar types of equipment. See Figure 3-22. They are manufactured in 1′ to 10′ lengths in various widths and depths. Different fittings are available such as couplings, elbows, end plates, and tees.

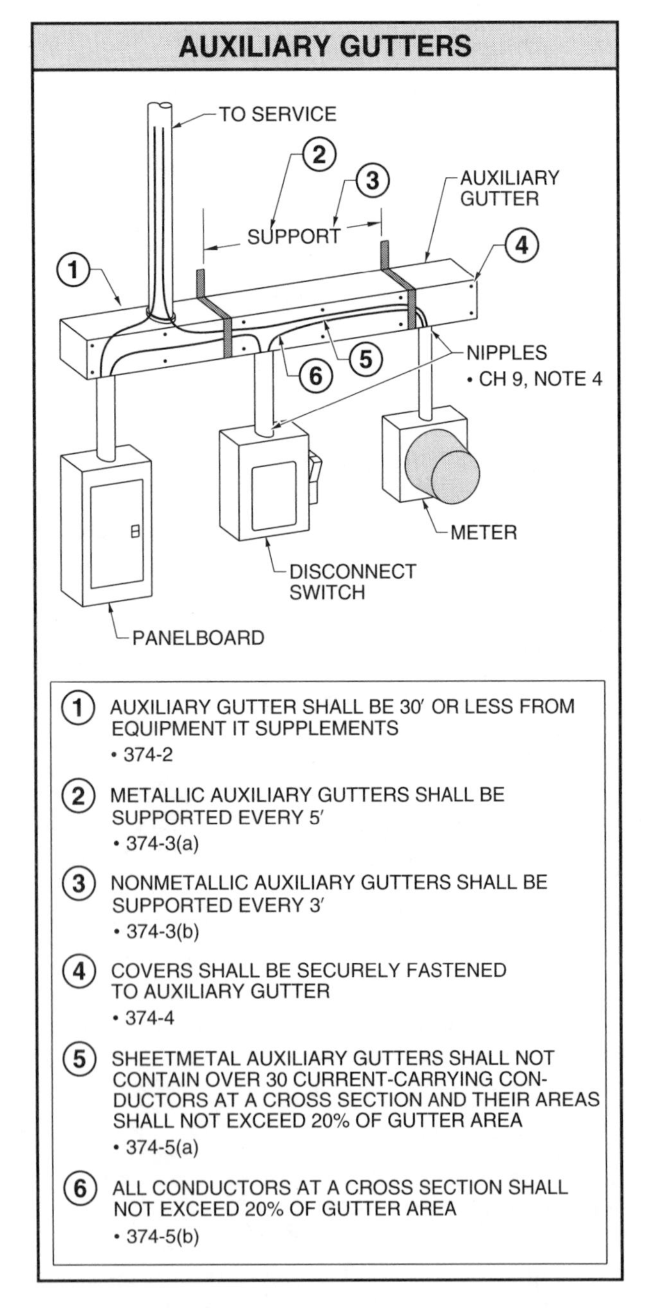

Figure 3-22. An auxiliary gutter is a sheet-metal enclosure equipped with hinged or removable covers that is used to supplement wiring space.

An auxiliary gutter is prohibited by 374-2 from extending more than 30′ beyond the equipment that it supplements. Section 374-2, Ex. eliminates the 30′ extension restriction for elevator installations as permitted in 620-35.

BOXES, CONDUIT BODIES, AND FITTINGS – 370

This article covers the installation and use of all boxes and conduit bodies. Section 370 also includes the installation requirements for fittings used to join raceways and to connect raceways and cables to boxes and conduit bodies. Section 300-15 requires a box or conduit body to be installed at each conductor splice connection point, outlet, switch point, junction point, or pull point.

Types of Boxes

A *box* is a metallic or nonmetallic electrical enclosure used for equipment, devices, and pulling or terminating conductors. See Figure 3-23. An *outlet box* is a box which houses a piece of utilization equipment. For example, an outlet box may contain a lighting fixture. A *device box* is a box which houses an electrical device. For example, a device box may contain a switch or receptacle.

Nonmetallic boxes shall be permitted only when using one of the following wiring methods: open wiring on insulators, concealed knob-and-tube wiring, nonmetallic-sheathed cable, and nonmetallic raceways per 370-3. There are two exceptions to this rule. Section 370-3, Ex. 1 allows metal raceways or metal-jacket cables to be used with nonmetallic boxes so long as an internal bonding means is provided between all entries. Section 370-3, Ex. 2 permits metal raceways or metal-jacket cables to be used with nonmetallic boxes. An integral bonding means with a provision for attaching an equipment grounding jumper inside the box shall be provided between all threaded entries. All metal boxes shall be grounded per 370-4.

An *outlet* is any point in the electrical system where current supplies utilization equipment. An outlet box provides that access point in the electrical system. Wall cases and handy boxes are used to terminate branch circuits of current-consuming devices such as fixtures, appliances, and other pieces of utilization equipment. Boxes are also used for terminations to switches, receptacles, and other non-current consuming devices.

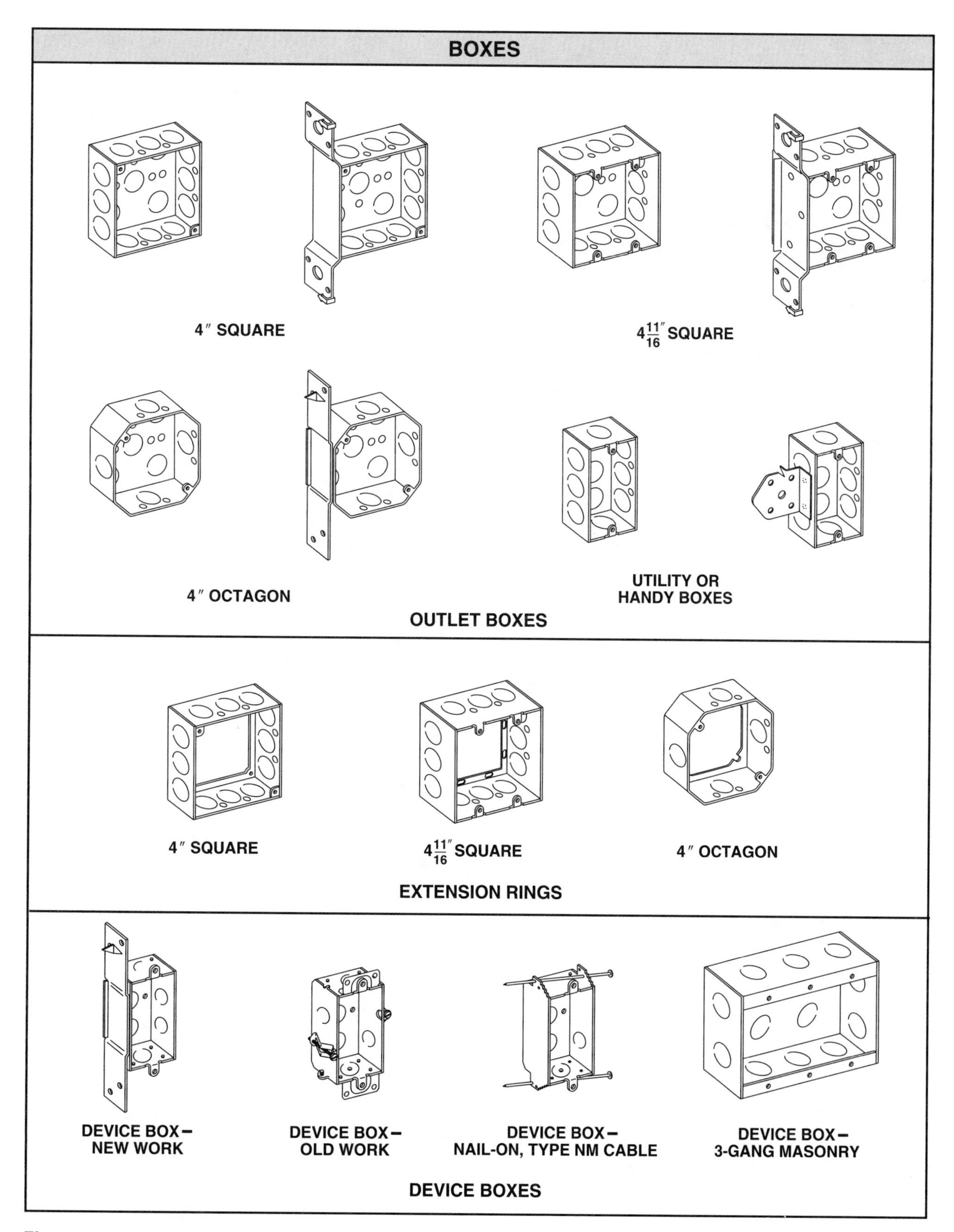

Figure 3-23. A box is a metallic or nonmetallic electrical enclosure used for equipment, devices, and pulling or terminating conductors.

Boxes shall be securely supported.

Outlet and device boxes are available in different sizes for mounting switches, receptacles, and other devices. The mounting holes on these boxes are drilled and tapped with No. 6 × 32 threads which are standard for each box. Handy boxes are used for exposed work. Boxes come in a variety of types and sizes and in various depths ranging from 1¼″ to 3½″. See Table 370-16(a).

Section 370-27(a) requires that boxes used with lighting fixture outlets be designed for the purpose and permits the lighting fixture to be attached to the box. Section 370-27(b) requires that floor boxes be listed for the application of receptacles located in the floor. Section 370-27(b), Ex. allows for the installation of other type boxes in elevated floors of show windows and similar locations. The AHJ shall judge these installations to be free from physical damage, moisture, and dirt.

Section 370-27(c) prohibits outlet boxes from being the sole support for ceiling fans. Section 370-27(c), Ex. permits boxes listed for this application to be used as the sole means of support for ceiling fans. Square and octagonal boxes may also be used as outlet or device boxes. They come in a variety of depths and are available with ½″ and ¾″ knockouts.

Pull boxes and junction boxes are points in the electrical system which provide access to the raceways entering and leaving the boxes. See Figure 3-24. A *pull box* is a box used as a point to pull or feed electrical conductors into the raceway system. A *junction box* is a box in which splices, taps, or terminations are made. Conduits shall not have bends totaling more than 360° between pull points per 346-11. If a conduit run has a total of more than 360° of bends and offsets, a pull box is installed. Vertical runs require support for the conductors per Table 300-19(a). A pull or junction box provides a place to install these supports.

Pull and junction boxes are sized for straight or angle pulls. See Figure 3-25. Calculating the minimum size pull and junction boxes to be used with raceways or cables containing conductors No. 4 or larger shall be determined based on the type of pull. The length of boxes containing straight pulls shall not be less than eight times the trade diameter of the largest raceway per 370-28(a)(1).

Section 370-28(a)(2) considers the requirements for sizing boxes where the conductors are pulled at an angle or in a U configuration. The distance between the side of the box in which the raceways enter and the opposite side of the box shall be at least six times the trade diameter of the largest raceway in that row, plus the diameters of all the other raceways in that row. The distance between raceway entries containing the same conductors shall not be less than six times the trade size diameter of the raceway containing these conductors.

Conduit Bodies – 370-29. A *conduit body* is a conduit fitting that provides access to the raceway system through a removable cover at a junction or termination point. Conduit bodies are named based upon their similarity to letters of the alphabet. See Figure 3-26.

Conduit bodies get their letter designation from their shape. All conduit bodies have an opening to provide access to the conductors. When a conduit body is placed on a flat surface with the opening in the up position, the shape of the letter designation is apparent. For example, the T and the X conduit bodies resemble the letters *T* and *X*.

The conduit bodies that are used as 90° L get their designation in another way. If the hub of the short end of the L is held in the hand like a pistol, the location of the access opening determines the letter designation. If the opening is to the *left*, it is an LL conduit body. If the opening is to the *right*, it is an LR conduit body. If the opening is on top, it is an LB conduit body, meaning the access cover goes on the *back* of the conduit body. The C conduit body has a hub at both ends, therefore a person can *see* right through it. The E designation comes from one entrance hub on the *end*.

Section 370-29 requires that conduit bodies be installed so that the wiring contained in them is accessible without removing any part of the building. Section 370-40(a) requires that conduit bodies be corrosion resistant or galvanized, enameled or otherwise coated inside and out to prevent corrosion. The wall of a conduit body shall not be less than ⅛″ thick.

Number of Conductors Permitted in a Box – 370-16

Boxes shall be large enough to provide enough free space for all of the enclosed conductors. This free space is needed so that any heat given off by the conductors I^2R losses can be dissipated. Box size is calculated by the number and size of conductors contained within the box. See Figure 3-27.

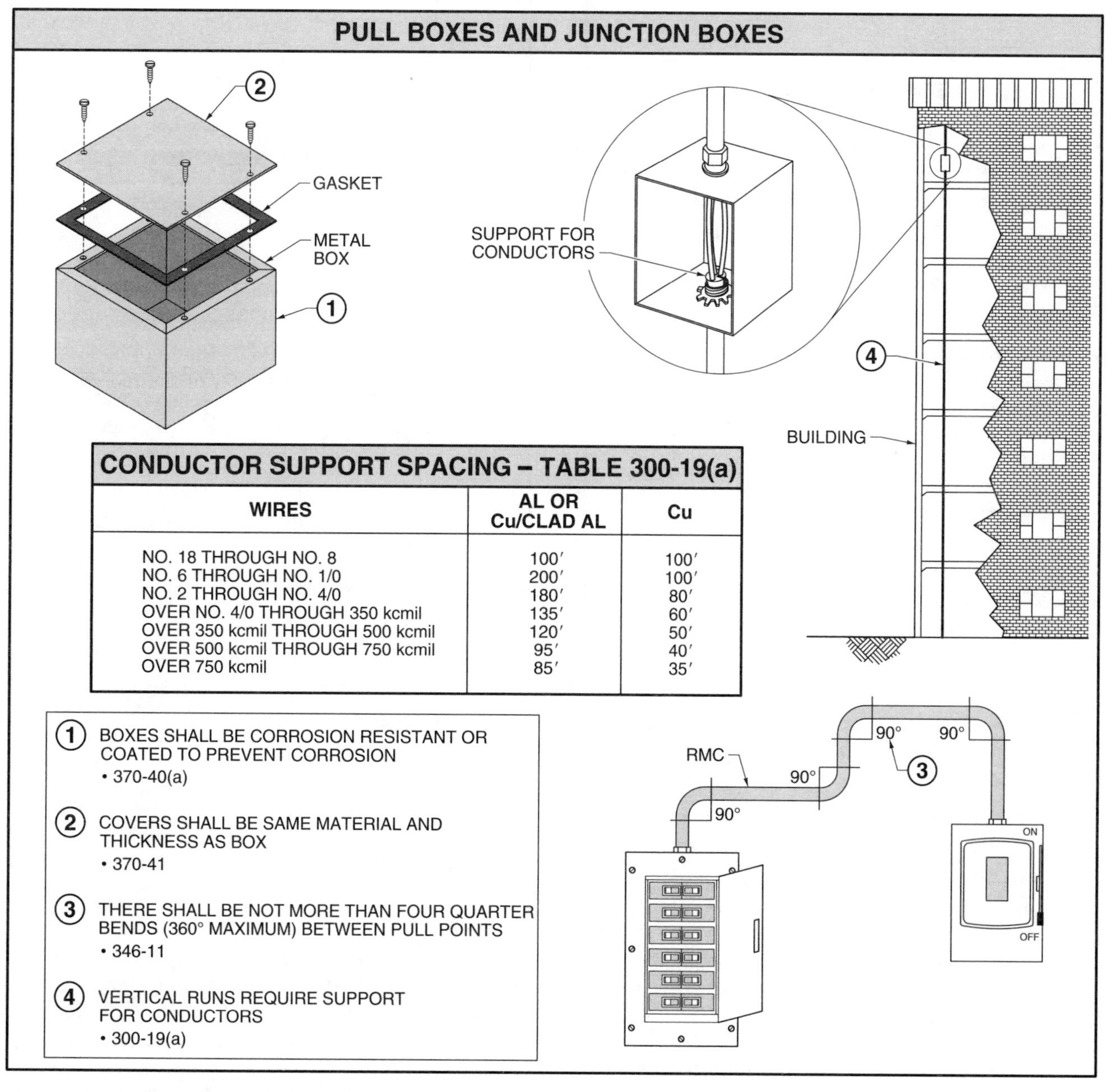

CONDUCTOR SUPPORT SPACING – TABLE 300-19(a)

WIRES	AL OR Cu/CLAD AL	Cu
NO. 18 THROUGH NO. 8	100′	100′
NO. 6 THROUGH NO. 1/0	200′	100′
NO. 2 THROUGH NO. 4/0	180′	80′
OVER NO. 4/0 THROUGH 350 kcmil	135′	60′
OVER 350 kcmil THROUGH 500 kcmil	120′	50′
OVER 500 kcmil THROUGH 750 kcmil	95′	40′
OVER 750 kcmil	85′	35′

Figure 3-24. Pull boxes and junction boxes are points in the electrical system which provide access to the raceways entering and leaving the boxes.

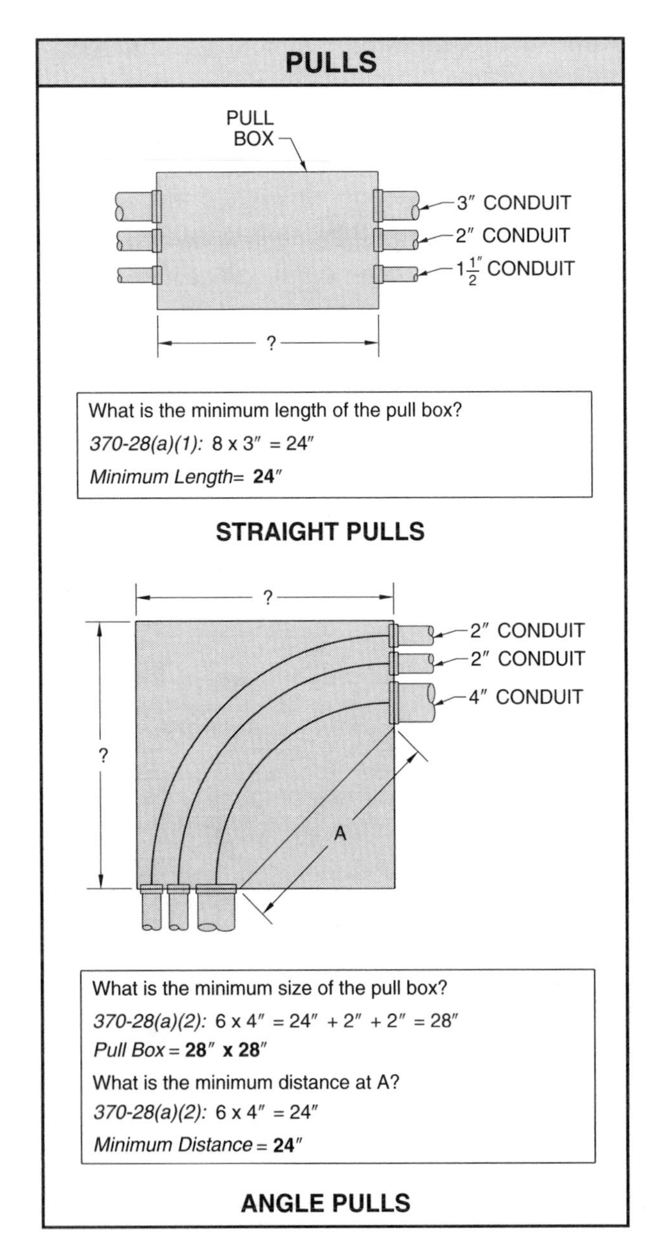

Figure 3-25. Pull and junction boxes are sized for straight or angle pulls.

Table 370-16(a) lists the number of conductors of the same size that may be installed in a certain volume box. For example, a 4″ square box 1½″ deep has a minimum capacity of 21 cu in. This is sufficient to contain ten No. 14 conductors or eight No. 10 conductors, provided there are no cable clamps, support fittings, devices, or EGC contained within the box per 370-16(b)(2), (3), (4), and (5).

To determine the minimum size box needed when there are different-size conductors contained within the box, see Table 370-16(b) and calculate the total cubic inches required for all of the conductors in the box. For example, if a box contained four No. 12 conductors and six No. 10 conductors, the box would require a minimum capacity of at least 24 cu in. (4 × 2.25 cu in. = 9 cu in.; 6 × 2.5 cu in. = 15 cu in.; 9 cu in. + 15 cu in. = 24 cu in.). Refer to table 370-16 (a) to find the proper size box. A 4″square × 2⅛″ deep or any of the 4¹¹⁄₁₆″ square boxes would satisfy this requirement.

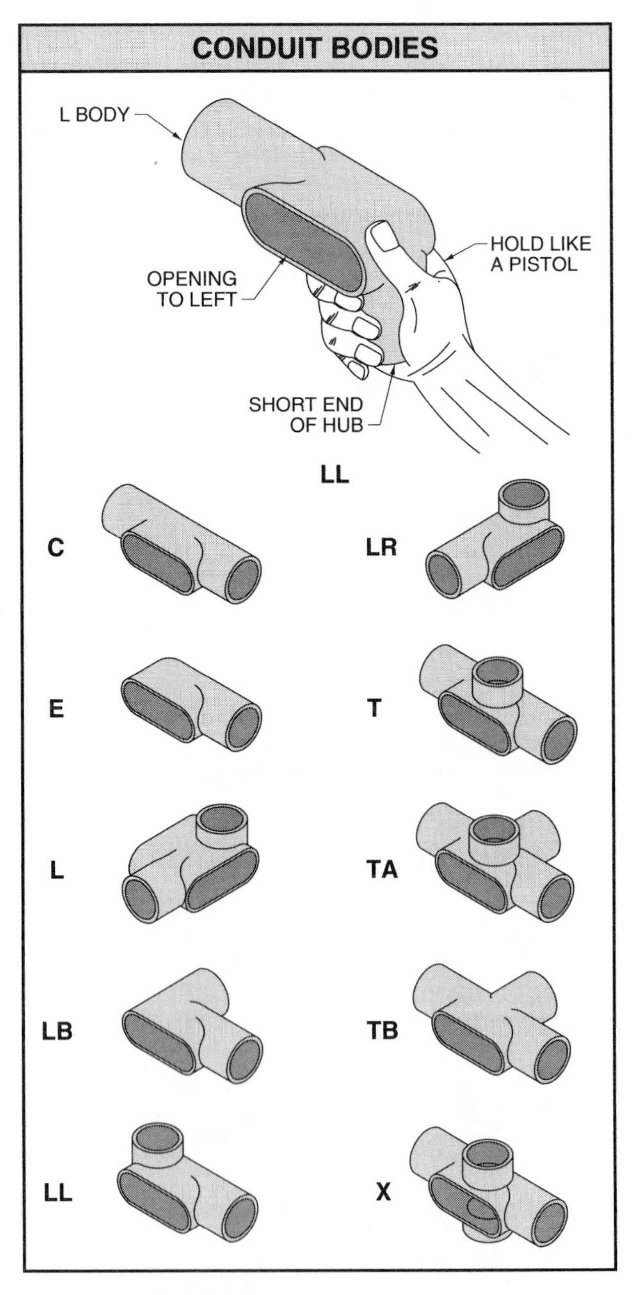

Figure 3-26. A conduit body is a conduit fitting that provides access to the raceway system through a removable cover at a junction or termination point.

CALCULATING BOX FILL

What minimum size 4″ square box is required for four No. 12-3 type AC cables? The box has two internal AC cable clamps.

Table 370-16 (a): 8 No. 12 ungrounded conductors = 8
4 No. 12 grounded conductors = 4
2 cable clamps = 1
13

Table 370-16 (a): Minimum size box for 13 conductors = **4″ x $2\frac{1}{8}$″ square box**

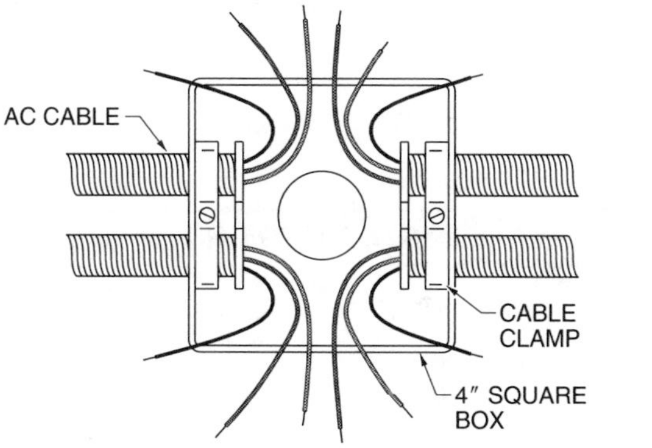

SAME-SIZE CONDUCTORS

What minimum size $4\frac{11}{16}$″ square box is required for six No. 12, four No. 10, and two No. 6 conductors spliced in the box?

Table 370-16(b): 6 No. 12 = 2.25 cu in. x 6 = 13.5 cu in.
4 No. 10 = 2.5 cu in. x 4 = 10.0 cu in.
4 No. 8 = 3.0 cu in. x 4 = 12.0 cu in.
35.5 cu in.

Table 370-16(b): Minimum size box for 35.5 cu in. = **$4\frac{11}{16}$″ x $2\frac{1}{8}$″ square box**

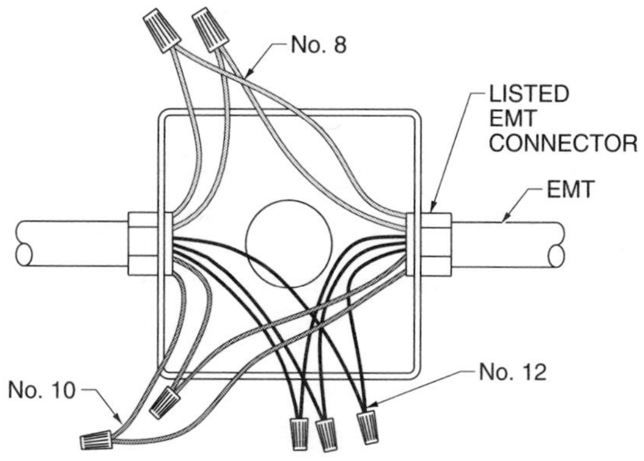

DIFFERENT-SIZE CONDUCTORS

Figure 3-27. Box size is calculated by the number and size of conductors contained within the box.

FINDING BOX CAPACITY

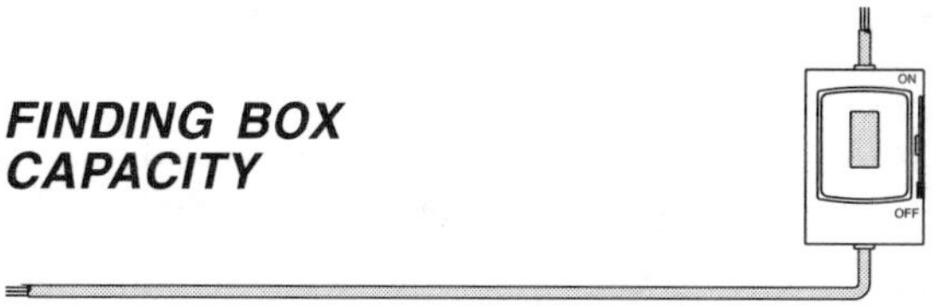

The cubic inch capacity of a box is calculated by finding its volume. To find volume, apply the formula:

$V = l \times w \times t$

where

V = volume

l = length

w = width

t = thickness

For example, what is the cubic inch capacity (volume) of a 6″ × 8″ box which is 4″ deep?

$V = l \times w \times t$

$V = 6 \times 8 \times 4$

V = **192 cu in.**

Conductor Fill – 370-16(b)(1). Each conductor originating outside the box and terminated or spliced inside the box counts as one. Each conductor passing through the box without a splice or termination counts as one. A conductor that does not leave the box is not counted. Section 370-16(b)(1), Ex. permits an EGC and up to four fixture wires smaller than No. 14 which enter a box from a domed fixture or similar canopy and terminate within that box to be omitted from the count.

When different-size conductors are used, the conductor fill in cubic inches shall be computed per Table 370-16(b). After the volume for the conductors is calculated, a box is selected according to its cubic inch capacity listed in Table 370-16(a).

Section 370-16(a) permits the total volume in cubic inches to include the cubic inch volume marked on any assembled sections added to the box, such as plaster rings, domed covers, extension rings, etc. Section 370-16 prohibits using a box with less volume than calculated according to the conductor volume.

Clamp Fill – 370-16(b)(2). A deduction of one conductor from Table 370-16(a) shall be made when one or more internal cable clamps are inside the box. A single volume allowance per Table 310-16(b) shall be made based on the largest conductor in the box.

Support Fitting Fill - 370-16(b)(3). A deduction of one conductor from Table 370-16(a) is made when one or more fixture studs or hickeys are inside the box. A single volume allowance per Table 370-16(b) shall be made based on the largest conductor in the box.

Device or Equipment Fill – 370-16(b)(4). A deduction of two conductors from Table 370-16(a) is made for each yoke or strap containing one or more devices or equipment. A double volume allowance per Table 370-16(b) is made for each yoke or strap based on the largest conductor connected to the device or equipment on that yoke or strap.

EGC Fill – 370-16(b)(5). A deduction of one conductor from Table 370-16(a) is made when one or more EGCs enter a box. Per 250-74, Ex. 4, a second deduction of one conductor is made if an additional set of EGCs are present. This refers to an isolated EGC permitted to reduce electrical noise. A single volume allowance per Table 370-16(b) is made based on the largest EGC in the box. An additional volume allowance is made where an additional set of EGCs are present as permitted by 250-74, Ex. 4. This second allowance is based on the largest EGC of the second set.

Box Supports – 370-23

Boxes and enclosures shall be rigidly and securely fastened in place. Boxes and enclosures may be installed under one of seven different types of installations. The box is considered one of the anchors of the electrical system since raceways and cables terminate at this point. Since many types of raceways and outer sheaths of cable systems act as the EGC, it is imperative that the box is securely anchored in place.

Surface and Structural Mounting – 370-23 (a)(b). Boxes shall be rigidly fastened to the surface upon which they are mounted. Section 370-23(b) requires that enclosures be supported from a structural member of the building or by using a metal, polymeric, or wood brace. Support wires shall provide a rigid support. Metal braces shall be protected against corrosion and not be less than .020″ thick and uncoated. Wood braces shall not be less than 1″ × 2″. Wood braces in wet locations shall be treated for the conditions. Polymeric braces shall be identified suitable for the use.

Nonstructural Mounting – 370-23(c). Framing members of suspended ceiling systems are permitted as support for boxes if the framing member is adequately supported and securely fastened to the building structure. Boxes shall be fastened to the framing member by a mechanical means such as bolts, screws, or clips identified for the use. See Figure 3-28.

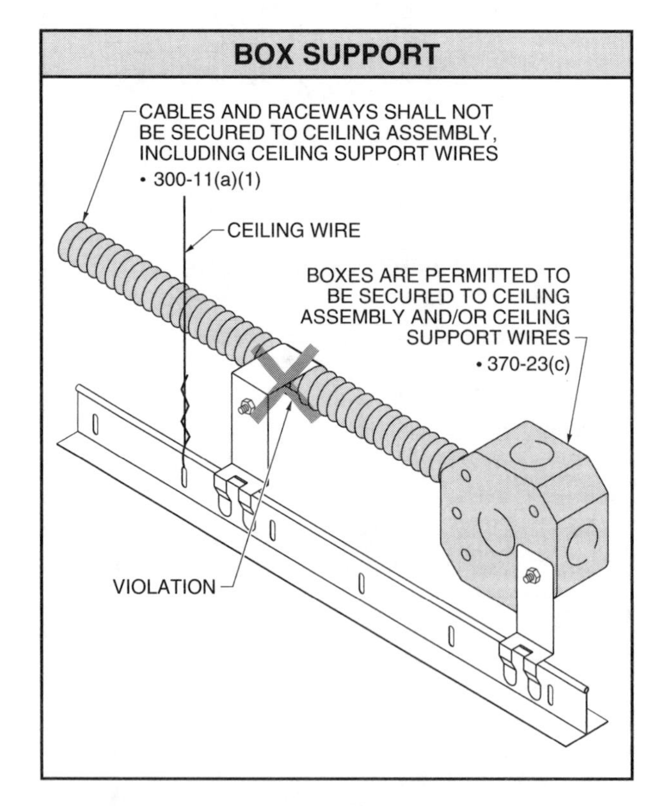

Figure 3-28. Framing members of suspended ceiling systems are permitted as support for boxes if the framing member is adequately supported and securely fastened to the building structure.

Raceway Supported Enclosures Without Devices or Fixtures – 370-23(d). Boxes that are not over 100 cu in. and have threaded entries or hubs may be supported where two or more conduits are threaded wrench tight into the enclosure or hubs. These conduits shall be supported within 3′ of the box on two or more sides so as to provide a rigid and secure installation. See Figure 3-29. Section 370-23(d), Ex. permits RMC, IMC, EMT, and RNMC to support conduit bodies, provided the conduit bodies are not larger than the largest trade size of the conduit or tubing.

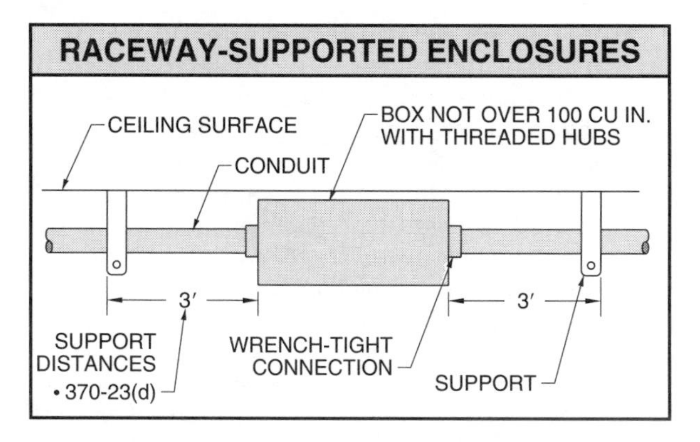

Figure 3-29. Conduits for boxes not over 100 cu in. with threaded hubs shall be supported within 3′ where two or more conduits are threaded wrench tight into the box.

Raceway Supported Enclosures With Devices or Fixtures – 370-23(e). Boxes that are not over 100 cu in. and have threaded entries or hubs are permitted to support fixtures, contain devices, or both, provided they are supported by two or more conduits and are threaded wrench tight into the enclosure or hubs and each conduit is supported within 18″ of the enclosure.

Pendant Boxes – 370-23(g). Boxes shall be permitted to be supported on a pendant by one of two methods. Section 370-23(g)(1) permits boxes to be supported from a multiconductor or cable in an approved manner. The installation shall ensure that the conductors are protected against strain by use of a strain-relief connector threaded into a box with a hub per 370-23(g)(2).

Floor Boxes – 370-27(b). Section 370-27(b) permits receptacles to be installed in floor boxes provided the box is specifically listed for this application. The exception to this rule permits other boxes located in elevated floors of show windows and similar locations to be used provided the AHJ judges them to be free from physical damage, moisture, and dirt.

THREADED WRENCH TIGHT

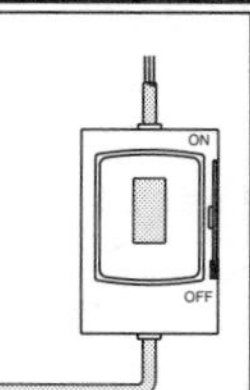

Threaded wrench tight means the threads of the conduit have to be tightened with a wrench into the coupling, hub, or other fitting. Threaded wrench tight connections shall not be made by hand only. RMC and IMC are the only conduits with threads on the ends or capable of being cut and threaded. RNMC does not have threads. A listed glue is used to attach it to couplings, hubs, or other fittings.

Boxes at Fan Outlets – 370-27(c). Section 370-27(c) prohibits outlet boxes from being used as the sole support for ceiling fans. The exception permits boxes that are listed for this application.

A ceiling fan is a dynamic load. A *dynamic load* is a load that produces a small but constant vibration. On a standard ceiling outlet box, the No. 8 × 32 mounting screws may have a tendency to back out or loosen completely due to the dynamic load. Therefore, the NEC® does not permit an outlet box to be the sole support for a ceiling fan.

GENERAL RULES FOR BOXES

General rules pertaining to boxes:

- *110-13(a) prohibits using wooden plugs in masonry, concrete, plaster, or similar materials for securing boxes.*
- *250-91(b) permits metal conduit, bare conductor, or green insulated conductor to be used as a grounding means for metal boxes.*
- *250-114(a) requires that an EGC shall ground a metal box by means of a grounding screw or another approved grounding device.*
- *300-14 requires that at least 6" of free conductor be left at each box for the purpose of connecting fixtures and devices.*
- *370-2 prohibits the use of conduits and locknuts on the sides of round boxes.*
- *370-16(a)(1) permits the volume of a box to be increased by the cubic inches marked on the plaster ring.*
- *370-18 requires that all unused openings in boxes or fittings shall be closed by knockout seals.*
- *370-20 requires that boxes mounted in a wall of combustible material shall be flush with the surface or project from it. Boxes mounted in a wall of noncombustible material may be recessed ¼" from the finished surface of the wall.*
- *370-22, Ex. permits surface extensions to be made from the cover of a concealed box designed for such use. The wiring shall be flexible and arranged so that the required grounding continuity is independent of the connection between the box and the cover.*
- *370-25 requires that each outlet box shall have a cover, faceplate, or fixture canopy to complete the installation.*
- *370-25(a) permits the use of metal or nonmetallic covers and plates with nonmetallic boxes. Metallic covers and plates shall be grounded per 250-42. This requires that all exposed noncurrent-carrying metal parts of fixed equipment likely to become energized shall be grounded.*
- *370-25(b) requires a cover or canopy to cover all exposed space around the edge of the box.*
- *370-29 requires that junction boxes shall be accessible without removing or disturbing any part of the building. However, liftout panels in suspended ceilings are considered accessible. Ceilings with scuttle holes or pull-down ladders are also considered accessible.*

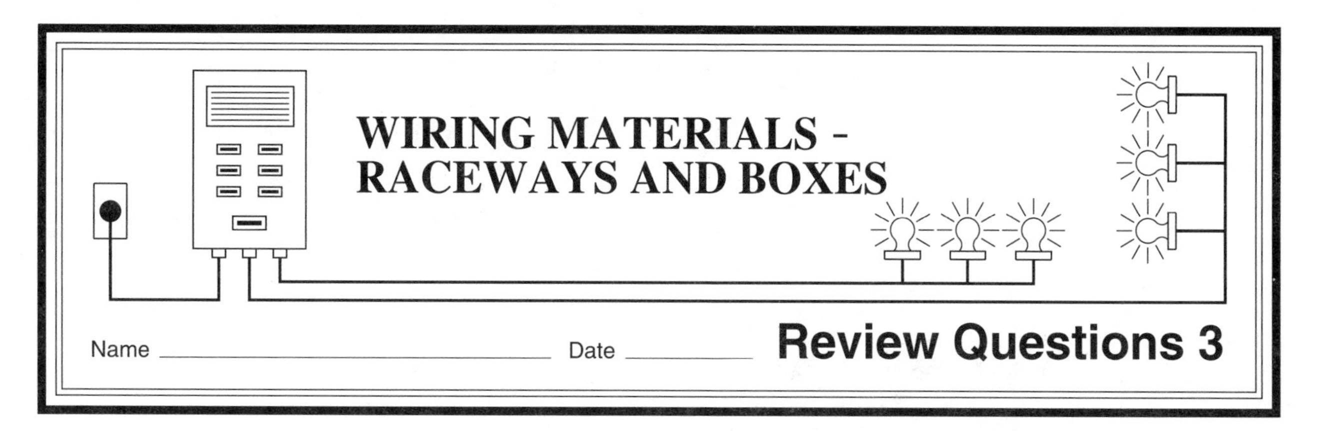

Review Questions

__________ **1.** A(n) ________ is an enclosed channel for conductors.

__________ **2.** A(n) ________ is a short piece of conduit or tubing that does not exceed 24″ in length.

__________ **3.** A total run of PVC shall not exceed ________° of bends.

__________ **4.** A(n) ________ assembly is a metal raceway with factory-installed conductors and attachment plug receptacles.

__________ **5.** A(n) ________ is a metallic or nonmetallic trough with a hinged or removable cover designed to house and protect conductors and cables.

T F **6.** Aluminum fittings and enclosures shall not be used with RMC.

T F **7.** AC cable shall be installed at least 3/4″ back from the finished edge of framing members.

T F **8.** MC cable with a smooth sheath shall have a bending radius of a minimum of 10 times the diameter of the cable.

__________ **9.** The use of NM cable in any dwelling or structure more than ________ floors above grade is prohibited.

T F **10.** A trolley busway has an open bottom.

__________ **11.** A(n) ________ conduit body has a hub at both ends.

T F **12.** USE cable is listed for direct burial.

__________ **13.** An auxiliary gutter shall not extend more than ________′ beyond the equipment it supplements.

__________ **14.** A(n) ________ is a double bend in a raceway.

__________ **15.** MC cable is a factory assembly of ________ or more conductors.

T F **16.** Two Romex cables shall not be stapled on top of one another.

T F **17.** CTS shall be installed as complete systems.

T F **18.** Busways shall be used for indoor work only.

__________ **19.** A(n) ________ is any point in the electrical system where current supplies utilization equipment.

__________ **20.** A(n) ________ box is a box in which splices, taps, or terminations are made.

Support

__________ **1.** MC cable shall be supported within ________″ at A.

__________ **2.** MC cable shall be supported every ________′ at B.

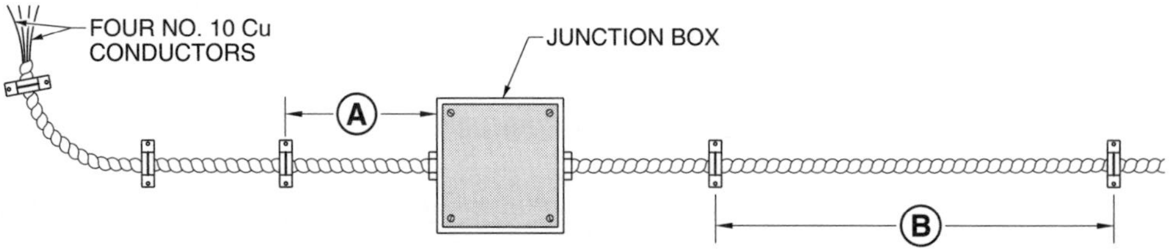

FOUR NO. 10 Cu CONDUCTORS
JUNCTION BOX
A
B

__________ **3.** AC cable shall be supported within ________″ at C.

__________ **4.** AC cable shall be supported every ________′ at D.

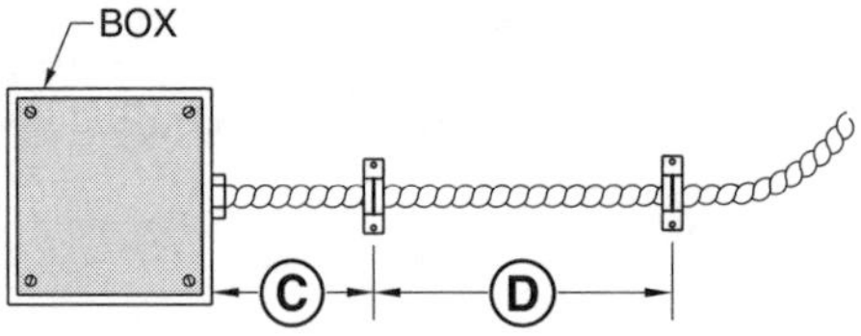

BOX
C
D

__________ **5.** FMC cable shall be supported within ________″ at E.

__________ **6.** FMC cable shall be supported every ________′ at F.

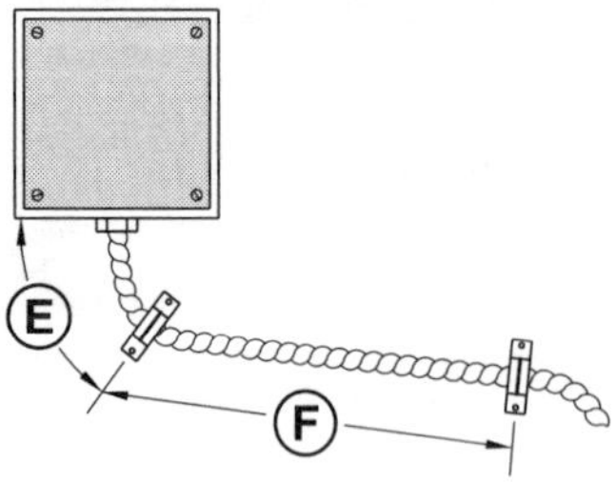

E
F

__________ **7.** EMT shall be supported every ________′ at G.

__________ **8.** EMT shall be supported within ________′ at H.

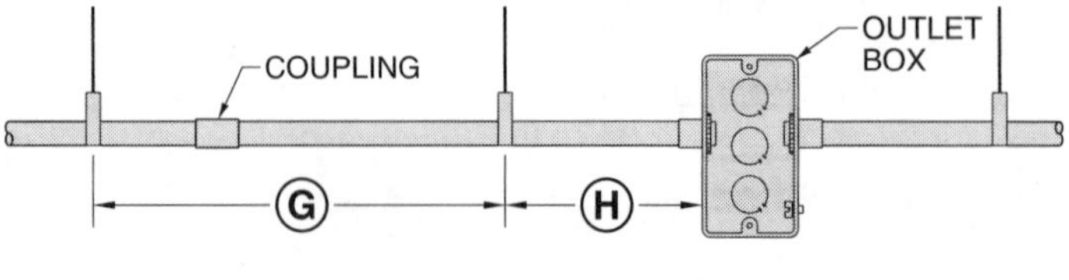

COUPLING
OUTLET BOX
G
H

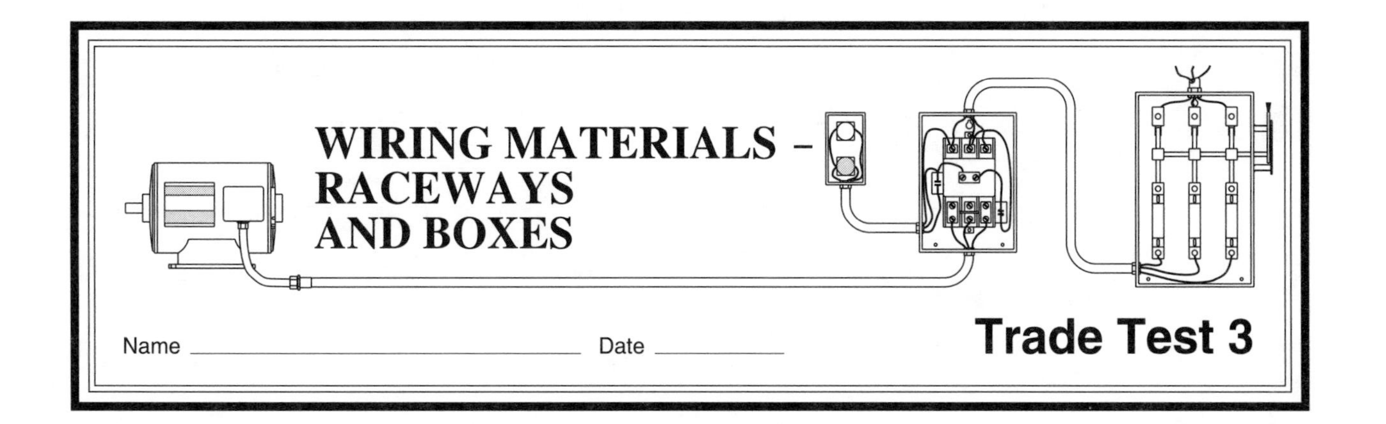

NEC®	Answer	
________	________	1. *See Figure 1.* Determine the minimum size EMT nipple required for the installation.
________	________	2. Determine the minimum cubic inch capacity required for a 4″ square outlet box which contains three No. 12 and six No. 8 THW Cu conductors. No internal clamps are provided in the outlet box.
________	________	3. Determine the minimum size 4″ square box required for four No. 14/3 AC cables. The box has two internal AC clamps.
________	________	4. *See Figure 2.* Determine the minimum size $4^{11}/_{16}$″ box required for the installation.

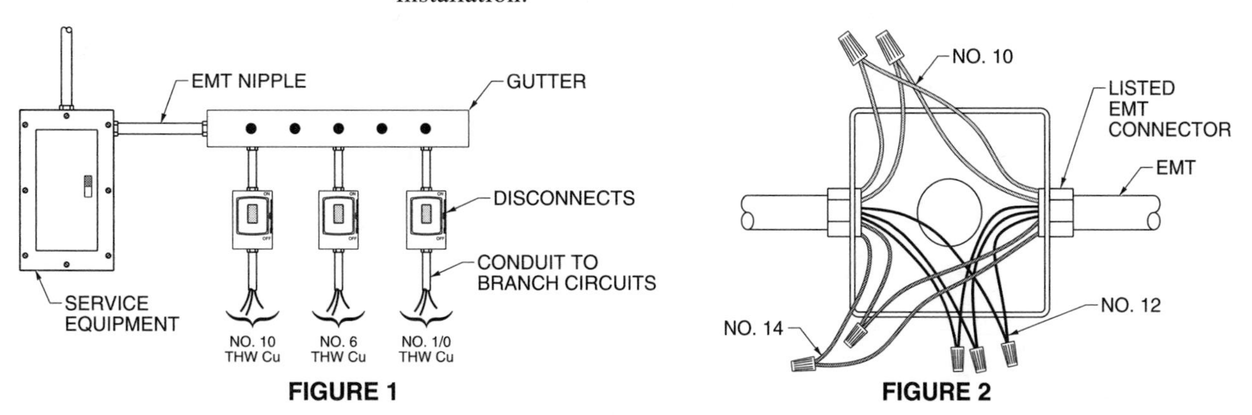

FIGURE 1

FIGURE 2

________	________	5. Determine the minimum length of a pull box which contains a 3″ RMC entering and leaving the box through opposite sides.
________	________	6. *See Figure 3.* What is the minimum length of the pull box?

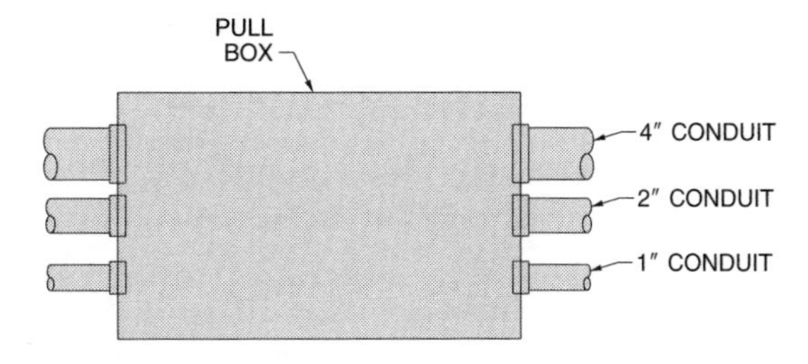

FIGURE 3

______ ______ **7.** *See Figure 4.* What is the minimum size of the pull box?

______ ______ **8.** *See Figure 4.* What is the minimum distance of A?

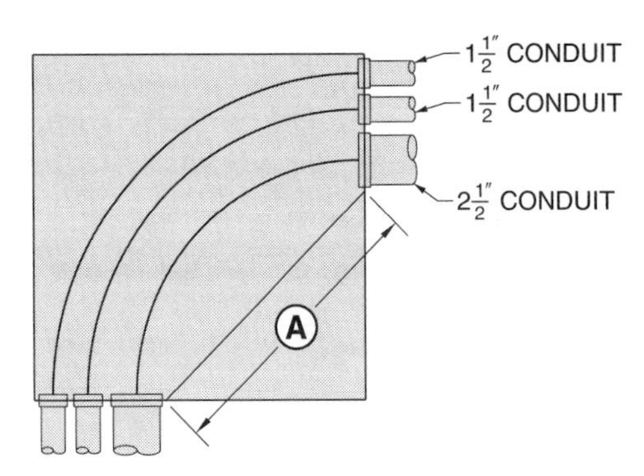

FIGURE 4

______ ______ **9.** Determine the minimum bending radius for a smooth-sheath Type MC cable with an external diameter of 1.25″.

______ ______ **10.** Determine the minimum bending radius for Type AC cable with an external diameter of .80″.

______ ______ **11.** *See Figure 5.* Determine the minimum number of conduit supports required for the installation.

______ ______ **12.** Determine the maximum spacing between conduit supports for 2½″ RNMC installed horizontally on a wall.

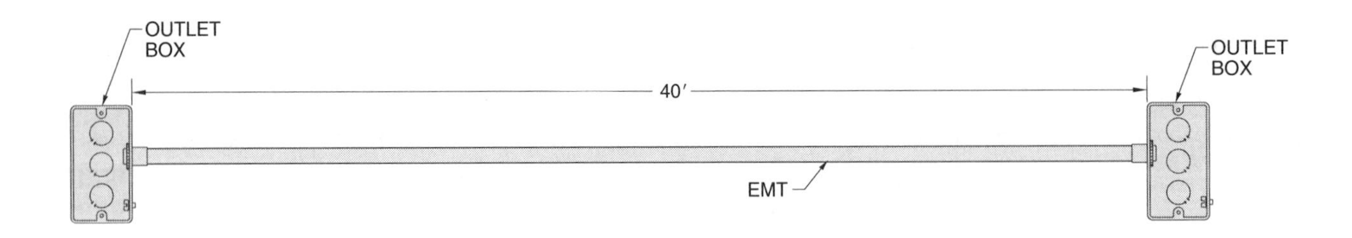

FIGURE 5

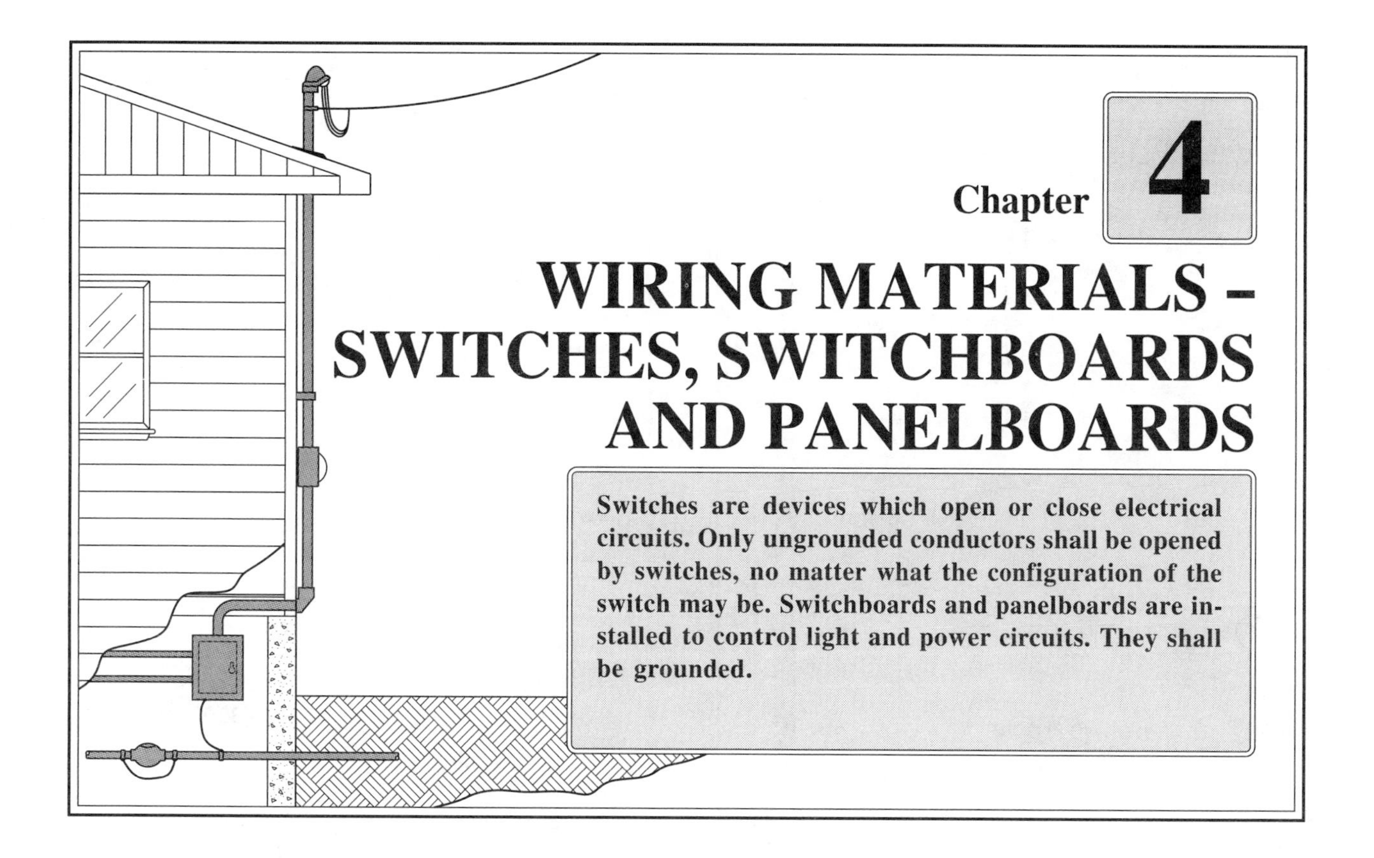

Chapter 4

WIRING MATERIALS - SWITCHES, SWITCHBOARDS AND PANELBOARDS

Switches are devices which open or close electrical circuits. Only ungrounded conductors shall be opened by switches, no matter what the configuration of the switch may be. Switchboards and panelboards are installed to control light and power circuits. They shall be grounded.

SWITCHES – 380

The provisions of 380 apply to all switches and circuit breakers (CBs) used as switches. In 100 there are six definitions for a switch. Switches defined are the bypass isolation switch, general-use snap switch, general-use switch, isolating switch, motor-circuit switch, and transfer switch. A *switch* is a device, with a current and voltage rating, used to open or close an electrical circuit. This definition does not include an isolation switch. A *circuit breaker (CB)* is an overcurrent protection device with a mechanical mechanism that may manually or automatically open the circuit when an overload condition or short circuit occurs.

Switch Connections – 380-2(a)

Only ungrounded conductors shall be opened by switches, no matter what the configuration of the switch might be. A switched conductor run in a metallic raceway or cable assembly shall have the return conductor in the same metallic raceway or cable assembly. The current in each conductor is the same, but the flow of current is in opposite directions, canceling out the inductive heating effect. See Figure 4-1.

A switched conductor run in conduit shall have the return conductor run in the same conduit. Travelers or shunts for 3-way and 4-way switches cannot be run alone in conduit or any metal wiring system. The grounded conductor (neutral) could be considered the return. If so, it is run in the same raceway as the travelers or shunts. With both polarities in the same enclosure or raceway, the inductive heating is minimized.

Grounded Conductors – 380-2(b)

Switching of the grounded neutral conductor is prohibited. There are two exceptions to this rule. Per 380-2(b), Ex. 1, the switching of a grounded neutral conductor is permitted provided all of the circuit conductors are disconnected simultaneously. Per 380-2(b), Ex. 2 a switch or CB is permitted to disconnect a grounded neutral conductor provided the device is arranged so that the grounded conductor cannot be disconnected until all the ungrounded conductors of the circuit have been disconnected. See Figure 4-2.

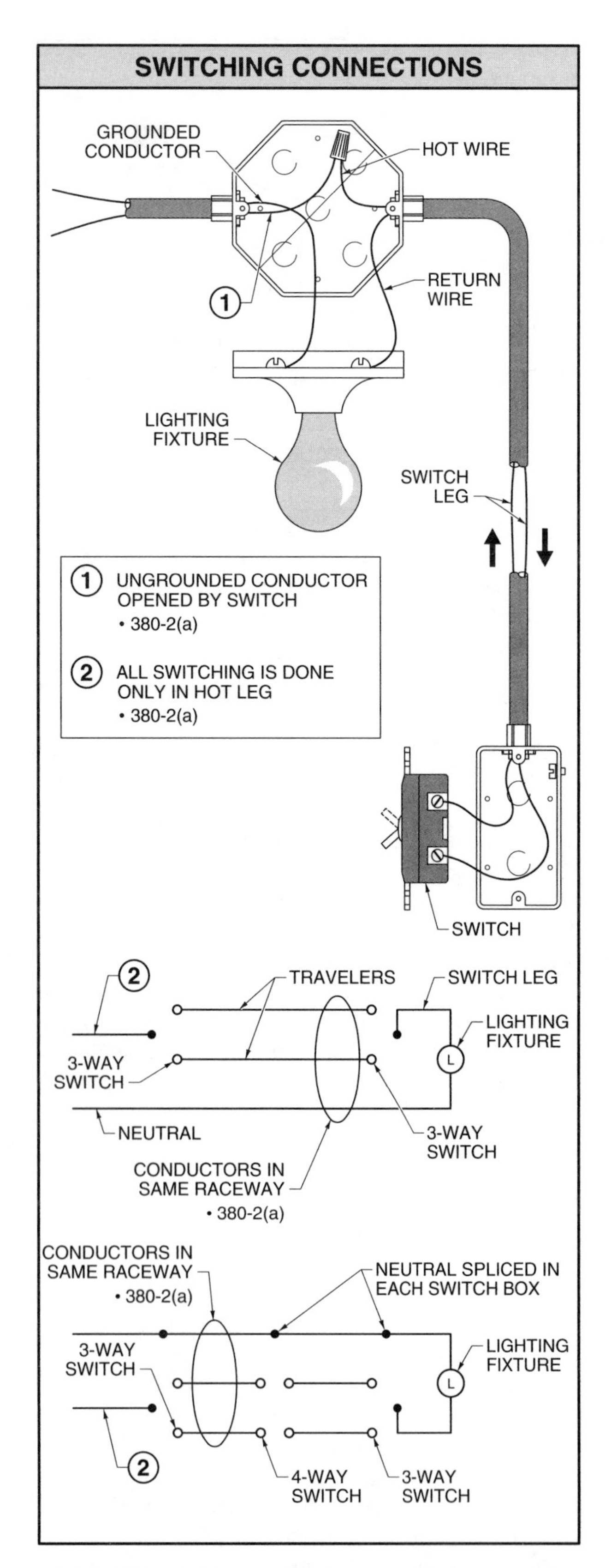

Figure 4-1. Only ungrounded conductors shall be opened by switches.

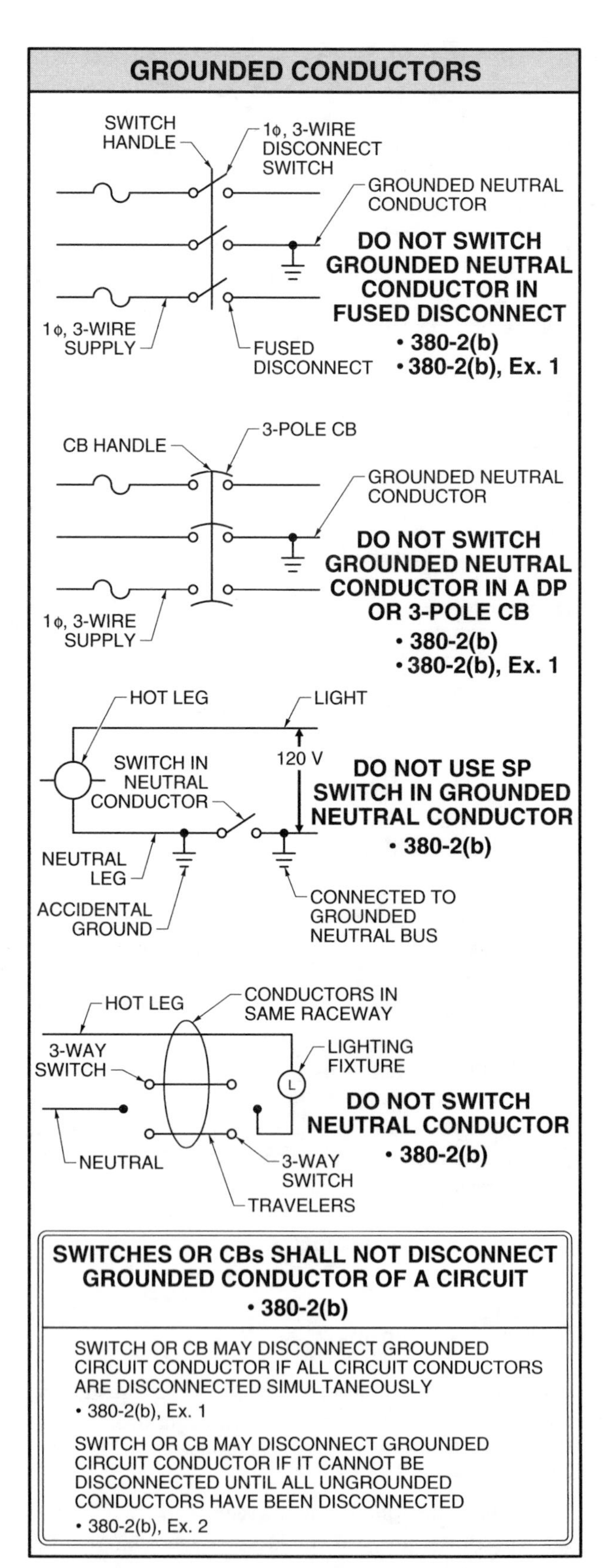

Figure 4-2. A switch or CB shall not be used to open the grounded neutral conductor.

Enclosures – 380-3

Enclosed switches minimize fire hazards and protect switch components from mechanical abuse. Enclosed switches also protect people from electrical shock. The enclosure shall provide adequate wire-bending space at the terminals per Tables 373-6(a) and 373-6(b). The concern here is to provide adequate space within the enclosure to bend the conductors when making terminations. The two types of termination configurations are L bends and S bends. See Figure 4-3.

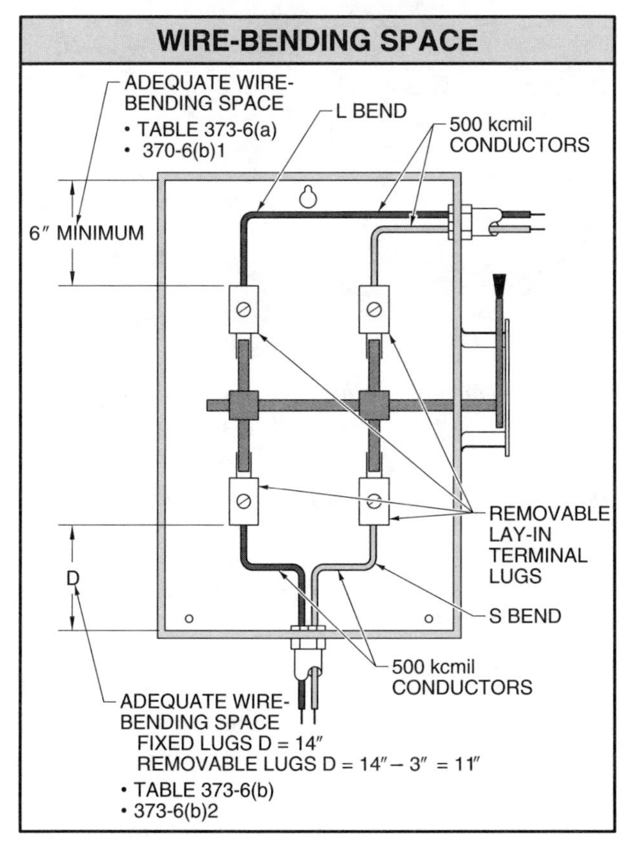

Figure 4-3. Enclosed switches shall provide adequate wire-bending space.

In an L bend, the conductor does not enter or leave the enclosure through the wall opposite its terminals. In this bend, the conductors take a 90° or L bend. For an L bend, the distance between the terminal lugs and the opposite wall of the enclosure shall conform to Table 373-6(a).

In an S bend, the conductor enters or leaves the enclosure through the wall opposite the terminals. In this bend, more space is required to bend the conductors so that they can terminate properly at the terminal lugs. In an offset or S bend, the distance between the terminal lugs and the opposite wall of the enclosure shall conform to Table 373-6(b). For removable and lay-in wire terminals intended for only one wire, bending space shall be permitted to be reduced by the number of inches shown in parentheses in the table.

Wet Locations – 380-4. Switch enclosures installed in wet locations shall be of the weatherproof type. A 1/4″ space shall be provided between the enclosure and the wall to which it is mounted. This 1/4″ airspace prevents the water from accumulating at the back of the enclosure, helping to prevent rust and corrosion. Nonmetallic cabinets and cutout boxes may be installed without the 1/4″ airspace between the enclosures and the wall or other surface used for support per 373-2, Ex. See Figure 4-4.

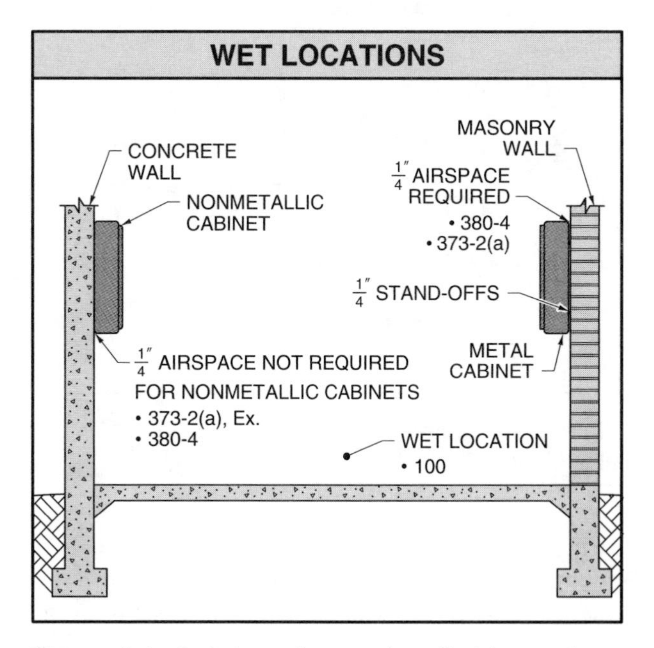

Figure 4-4. Switch enclosures installed in wet locations shall be of the weatherproof type.

Position of Knife Switch – 380-6

An improperly installed knife switch could be inadvertently closed due to gravity or an improperly wired switch could expose live parts in the OFF position, both of which could cause an accident. Per 380-6(a), single-throw knife switches shall be installed with the hinged end down. In this position, gravity will not tend to close the switch blades when left in the open position. Per 380-6(b), double-throw knife switches shall be mounted vertically or horizontally. Double-throw knife switches mounted vertically shall be equipped with a locking device to prevent gravity from closing the blades. See Figure 4-5.

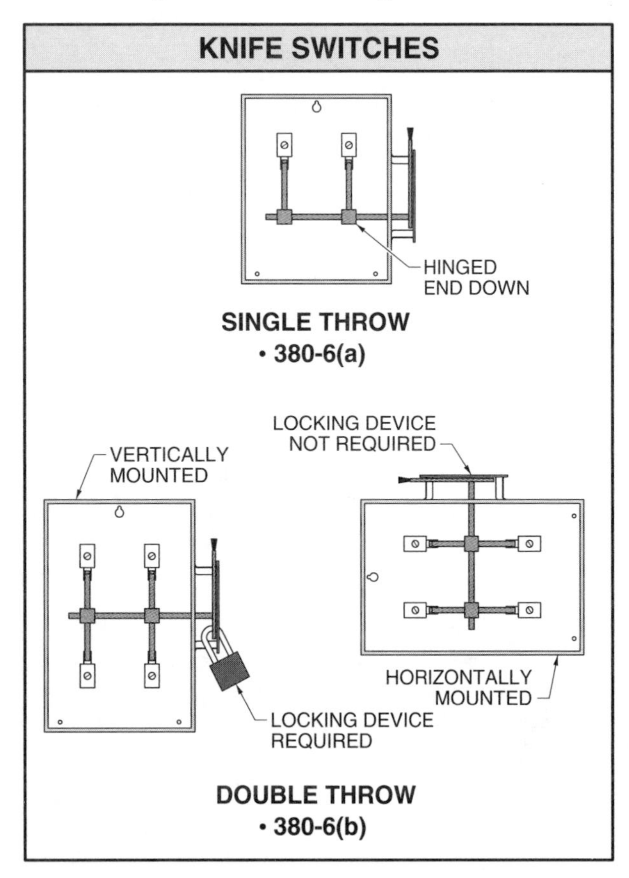

Figure 4-5. Knife switches shall be installed so that gravity does not close the blade.

Per 380-6(c), single-throw knife switches shall be connected in such a way that the blades are de-energized when the switches are in the open position. The load side of switches connected to circuits or equipment which may provide a backfeed source of power shall have a warning sign installed. The sign shall be permanently installed on the switch enclosure or immediately adjacent to open switches. The sign shall read, "WARNING – LOAD SIDE OF SWITCH MAY BE ENERGIZED BY BACKFEED."

ON-OFF Position – 380-7. Switches or CBs mounted in an enclosure shall clearly indicate whether they are in the open (OFF) or closed (ON) position. Where these switches or CB handles are operated vertically, the up position of the handle shall be the ON position.

Vertically-mounted switches shall be ON in the up position and OFF in the down position.

Accessibility and Grouping – 380-8

All switches and CBs used as switches shall be installed so that they are operated from a readily accessible location. The center of the grip of the operating handle, when in the highest position, shall not be more than 6′-7″ above the floor or working platform. See Figure 4-6. There are three exceptions to this general rule.

Per 380-8(a), Ex. 1, fused disconnect switches and CBs shall be mounted at the same level as busways if a suitable means is provided to disconnect the switch or CB from the floor. Per 380-8(a), Ex. 2, motors, appliances, and other equipment mounted higher than 6′-7″ above the floor are permitted to have a switch or CB mounted adjacent to the equipment. The switch or CB shall be accessible by a ladder or other suitable means. Per 380-8(a), Ex. 3, hookstick-operated isolating switches are permitted at heights greater than 6′-7″ above the floor.

Voltage Between Adjacent Switches – 380-8(b). Snap switches are permitted to be grouped or ganged in an enclosure provided the voltage between adjacent snap switches does not exceed 300 V. If the voltage between adjacent snap switches exceeds 300 V, a permanent barrier shall be installed in the enclosure between the adjacent snap switches. See Figure 4-7.

Faceplate for Flush-Mounted Snap Switches – 380-9. Faceplates for flush-mounted snap switches shall be installed so that they are flat against the wall and completely cover the wall opening. Ungrounded metal boxes that can be reached by a person standing on a grounded, conductive floor or near other grounded, conductive items shall be provided with a nonmetallic faceplate. See Figure 4-8.

Grounding Enclosures – 380-12

Metal enclosures for switches and CBs shall be grounded per 250. Exposed noncurrent-carrying metal parts of fixed equipment likely to become energized shall be grounded. See Figure 4-9.

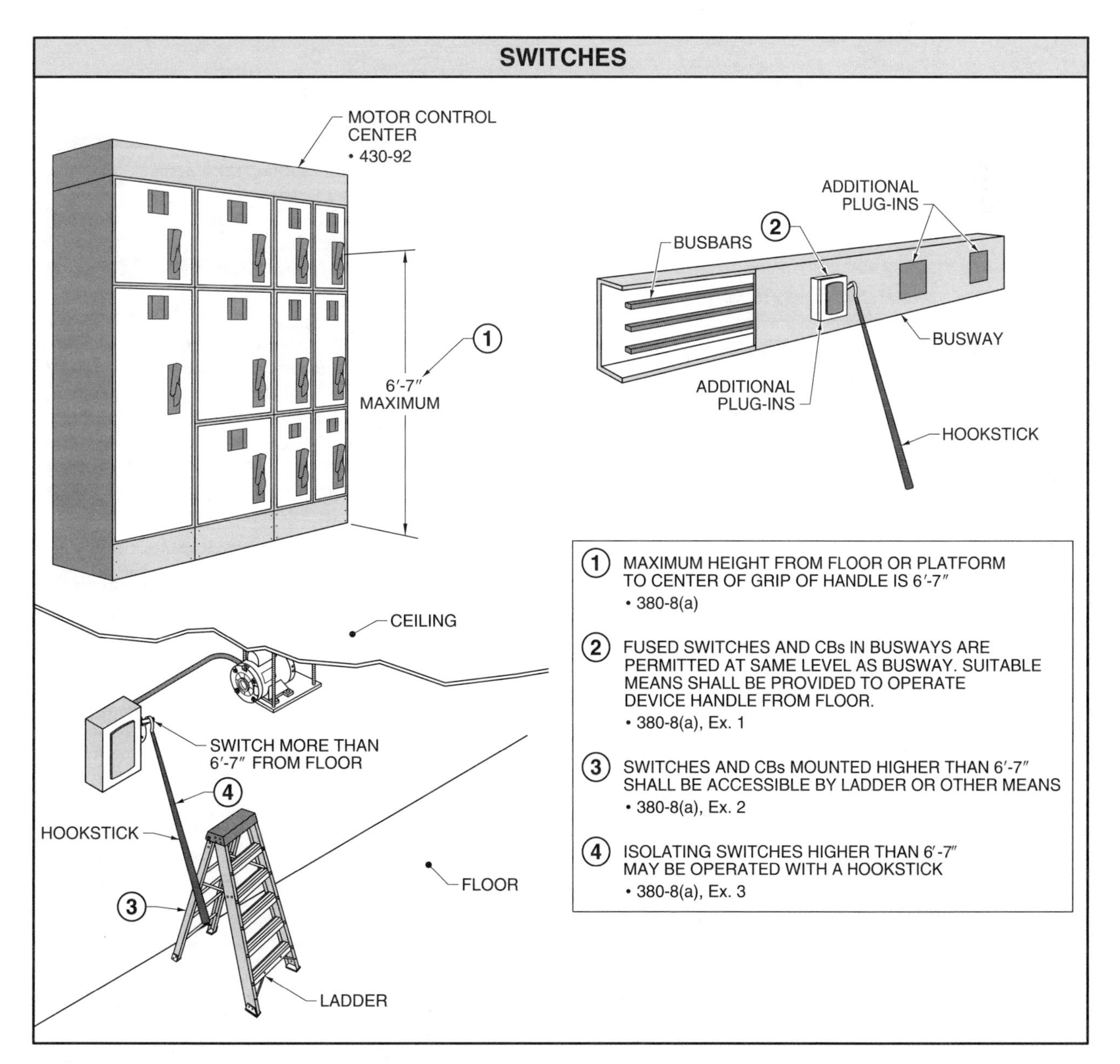

Figure 4-6. All switches and CBs used as switches shall be installed so that they are operated from a readily accessible location.

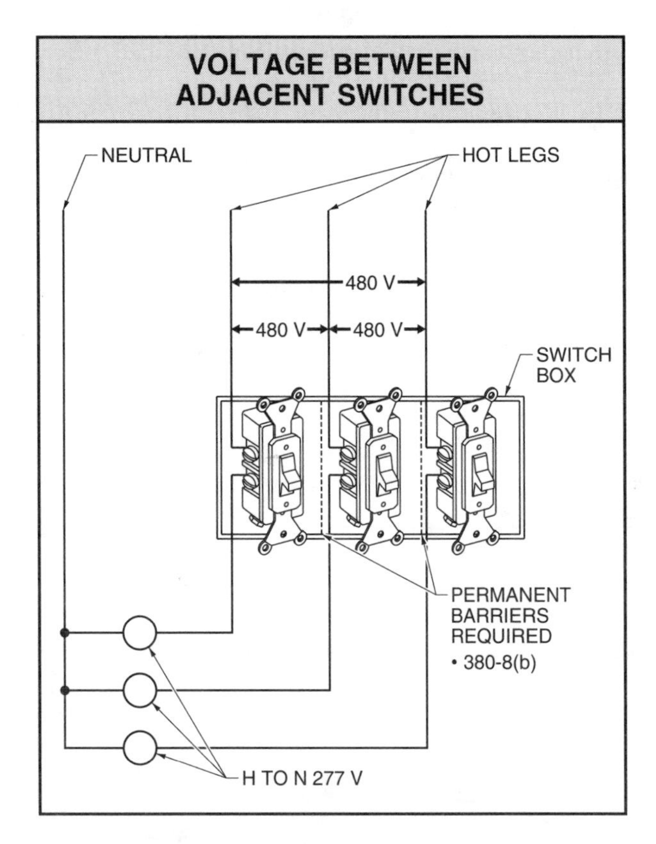

Figure 4-7. Permanent barriers are required if the voltage between adjacent snap switches exceeds 300 V.

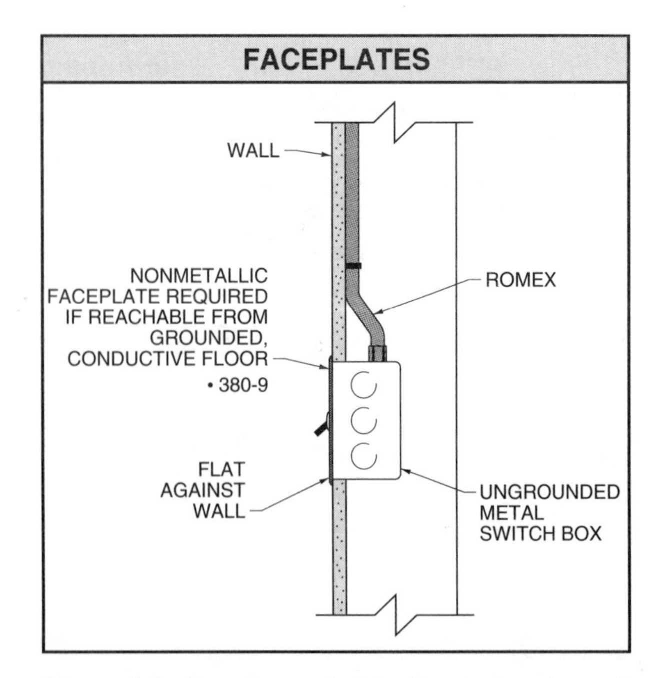

Figure 4-8. Faceplates shall be flat against the wall and completely cover the wall opening.

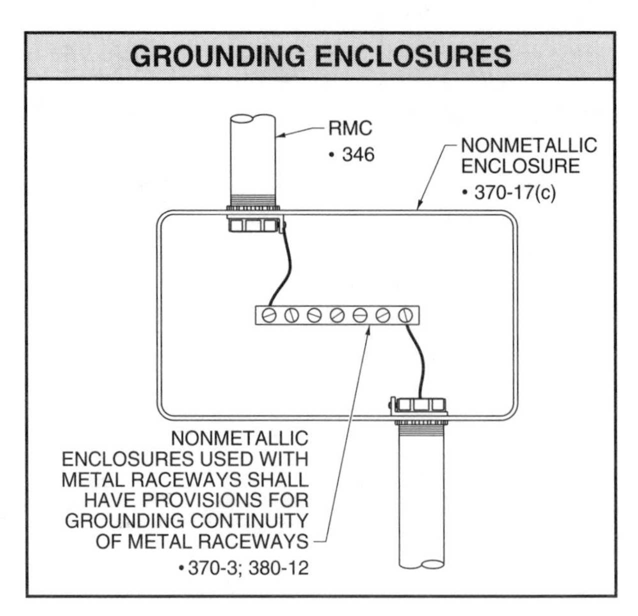

Figure 4-9. Exposed noncurrent-carrying parts of fixed equipment shall be grounded.

Rating and Use of Snap Switches – 380-14

Snap switches shall be used within their ratings. This includes both tumbler and toggle switches. There are four types of snap switches: AC general-use, AC-DC general-use, CO/ALR, and AC specific-use rated for 347 V. See Figure 4-10.

AC General-Use Snap Switch – 380-14(a). AC general-use snap switches are recognized as being suitable for use on AC circuits only and for controlling the following three types of loads:

1. Resistive and inductive loads, including electric-discharge lamps. The load shall not exceed the ampere and voltage rating of the switch.
2. 120 V tungsten-filament lamp loads. The load shall not exceed the ampere and voltage rating of the switch.
3. Motor loads. The load shall not exceed 80% of the ampere rating of the switch and the voltage of the motor shall not exceed the voltage rating of the switch.

AC-DC General-Use Snap Switch – 380-14(b). AC-DC general-use snap switches are suitable for use on AC or DC circuits to control the following three types of loads:

1. Resistive loads. The load shall not exceed the ampere and voltage rating of the switch.
2. Inductive loads. The load shall not exceed 50% of the ampere rating of the switch at the applied volt-

age. Switches rated in horsepower are suitable for controlling motor loads within their horsepower and voltage rating.

3. Tungsten-filament lamp loads. The load shall not exceed the ampere rating of the switch at the applied voltage provided the switch has a "T" rating.

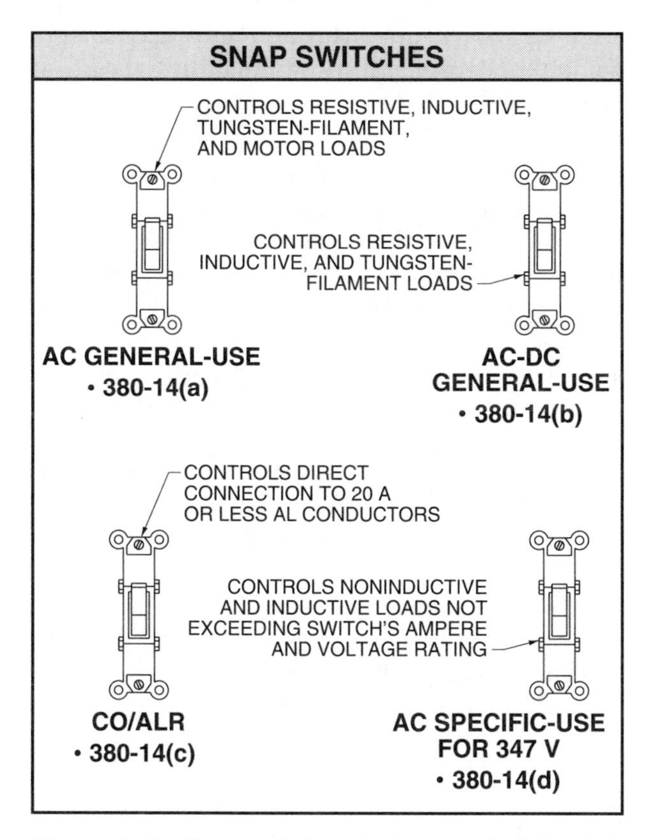

Figure 4-10. Snap switches shall be used within their ratings.

CO/ALR Snap Switches – 380-14(c). Snap switches which are connected directly to aluminum conductors and rated 20 A or less shall be listed and marked CO/ALR. Per UL's *General Information for Electrical Construction Materials*, terminals of 15 A and 20 A switches not marked CO/ALR are for use with copper and copper-clad aluminum conductors only. Terminals marked CO/ALR are for use with aluminum, copper, and copper-clad aluminum conductors.

Screwless pressure terminal connectors of the conductor push-in type are for use only with copper and copper-clad aluminum conductors.

Terminals of switches rated 30 A and above not marked AL/CU are for use with copper conductors only. Terminals of switches rated 30 A and above marked AL/CU are for use with aluminum, copper, and copper-clad aluminum conductors.

AC Specific-Use Snap Switches Rated for 347 V – 380-14(d). This section has been added to the rating and use of snap switches in an effort to harmonize the NEC® with the CEC. Listed snap switches rated 347 V AC are permitted for controlling noninductive and inductive loads not exceeding the ampere and voltage rating of the switch.

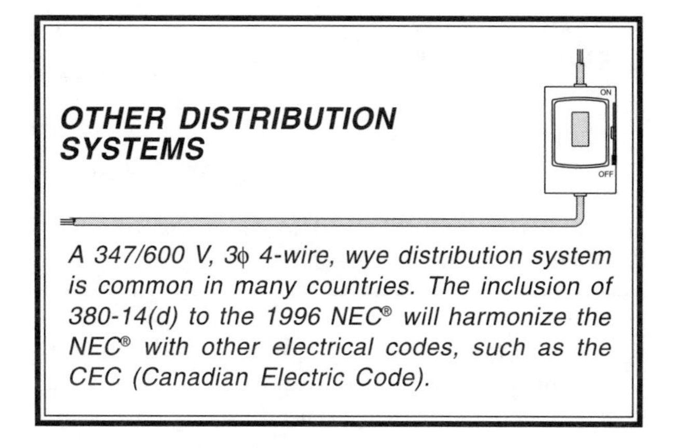

A 347/600 V, 3ϕ 4-wire, wye distribution system is common in many countries. The inclusion of 380-14(d) to the 1996 NEC® will harmonize the NEC® with other electrical codes, such as the CEC (Canadian Electric Code).

SWITCHBOARDS AND PANELBOARDS – 384

Switchboards and panelboards are installed to control light and power circuits. They are designed with busbars and CBs or fuse holders which supply power distribution through feeder circuits and branch circuits. Lighting panelboards include a permanently installed main CB or single set of fuses used as mains. See 384-14.

A *switchboard* is a single panel or group of assembled panels with buses, overcurrent devices, and instruments. Switchboards are not intended to be installed in cabinets. They are commonly accessible from the front and rear.

A *panelboard* is a single panel or group of assembled panels with buses and overcurrent devices, which may have switches to control light, heat, or power circuits. Panelboards are placed in a cabinet or cutout box which is accessible only from the front. Panelboards are mounted in or on partitions or walls.

Support and Arrangement of Busbars and Conductors – 384-3

Busbars and conductors on a switchboard or panelboard shall be installed so as to be free from physical damage. See Figure 4-11. The busbars and conductors shall be held firmly in place. Other than the required interconnections and control wiring, only those conductors that are intended for termination in a vertical section of a switchboard shall be located in that section.

BARRIERS IN SWITCHBOARDS

It is not uncommon to work on switchboards while the main busbars are energized. Barriers provide some protection to those performing maintenance of this equipment.

Conductors and Busbars on a Switchboard or Panelboard – 384-3(a). Barriers are required in all listed switchboards that are used as service-entrance equipment. These barriers isolate the service-entrance conductors or busbars and their termination points from the rest of the switchboard. Load conductors shall exit vertically from the section in which they originate. However, per 384-3(a), Ex., conductors are permitted to travel horizontally through vertical sections of a switchboard where barriers are installed that isolate these conductors from the live busbars. Per 384-3(b), busbars and conductors shall be arranged to avoid overheating from inductive effects.

Overheating and Inductive Effects – 384-3(b). Conductors or busbars carrying AC current produce a fluxing magnetic field around the conductor or busbar. When ferrous metal (iron or steel) surrounds a single current-carrying conductor, there is an interaction between the molecules of the ferrous material and the fluxing magnetic field. The molecules of the ferrous material vacillate back and forth, producing a heating effect on the ferrous material. In some cases, the heat produced may be hot enough to ignite combustible materials.

Used as Service Equipment – 384-3(c). Switchboards or panelboards used for service equipment shall have a grounding bar bonded to the casing of the panelboard or switchboard frame. A main bonding jumper shall be placed in the panelboard or one of the sections of the switchboard. It connects the grounded conductor of the service-entrance conductors to the switchboard or panelboard enclosure frame. See 250-79(d) for sizing the main bonding jumper.

Terminals – 384-3(d). Terminals in switchboards and panelboards shall be arranged so that it is not necessary to reach over or behind an ungrounded line bus in order to make connections. The concern here is safety to the worker who has to modify or make changes to an existing switchboard or panelboard.

High-Leg Marking – 384-3(e). The B phase shall be durably and permanently marked with an orange outer finish or by other effective means. For example, a 4-wire, delta system is used to supply large 3ϕ loads and small 1ϕ loads. These systems are also known as 3ϕ, 4-wire, 120/240 V systems. Per 384-3(f), the B phase is required to be the high-leg (wild leg or stinger leg). In order to accomplish this, the transformer in which two ends go to the A phase and C phase shall be the transformer that is center-tapped and grounded. This causes the A phase to read 120 V to the grounded midpoint and C phase to read 120 V to the grounded midpoint. From B phase to the grounded midpoint is 208 V. This is the vector sum of the 240 V of one full winding and 120 V of the half-winding that is center-tapped and grounded.

To calculate the high-leg voltage, take the voltage between the A phase or C phase and the grounded midpoint (usually 120 V). Multiply this voltage by 1.73 ($\sqrt{3}$) to find the voltage of the high-leg to ground (120 V × 1.73 = 208 V).

Per 384-3(e), the high-leg or B phase, shall be identified by the color orange. This alerts other installers who tap into the 4-wire delta system and take out a 1ϕ, 3-wire, 120/240 V system to avoid the high-leg.

Phase Arrangement – 384-3(f). On a 3ϕ, 4-wire, delta-connected system, the B phase shall be the phase having the higher voltage to ground. The phase arrangement on 3ϕ buses shall be A, B, C from front to back, top to bottom, or left to right as viewed from the front of the switchboard or panelboard.

Wire Bending Space – 384-3(g). The minimum wire bending space at terminals in panelboards and switchboards shall comply with 373-6. The concern is to provide adequate wiring space between the connecting terminal in the equipment and the opposite wall of the enclosure. If there is not adequate room, the installer may have to make sharp bends in the conductors while shaping them to enter the connection terminals.

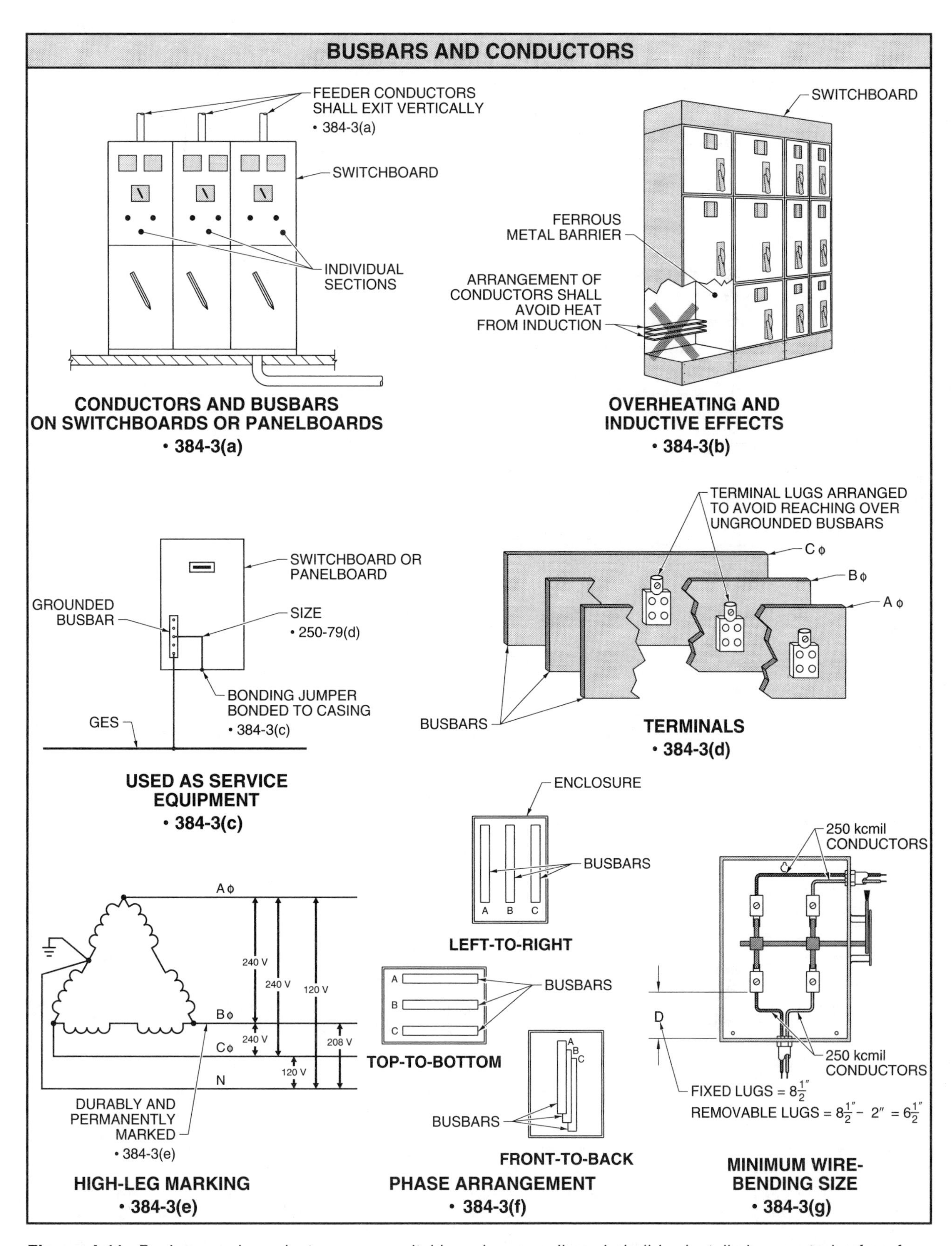

Figure 4-11. Busbars and conductors on a switchboard or panelboard shall be installed so as to be free from physical damage.

Sharp bends in conductors often cause damage to the insulation. Per 373-6(b)(1), the wire bending space in Table 373-6(a) shall be used where the conductor does not enter or leave the enclosure through the wall opposite its terminal. Per 373-6(b)(2), the wire bending space in Table 373-6(b) shall be used where the conductor enters or leaves the enclosure through the wall opposite its terminal.

Installation of Switchboards and Panelboards – 384-4

Switchboards, panelboards, and motor control centers shall be located in dedicated spaces and protected from damage. The installation may be indoors or outdoors. Per 384-4(a), the dedicated space or footprint for switchboards, panelboards, and motor control centers installed indoors shall be equal to the width and depth of the equipment extending from the floor to a height of 25′ or the structural ceiling, whichever is lower. This space shall be used for electrical installation only. No piping, ducts, or equipment foreign to the electrical installation shall be located in this zone. Any space above 25′ is not considered dedicated space. A dropped or suspended ceiling that does not add strength to the building structure is not considered to be a structural ceiling.

Per 384-4(a)(2), the working clearance space in 110-16(a) shall be met. No architectural barriers or other equipment shall be located in this zone. See Figure 4-12.

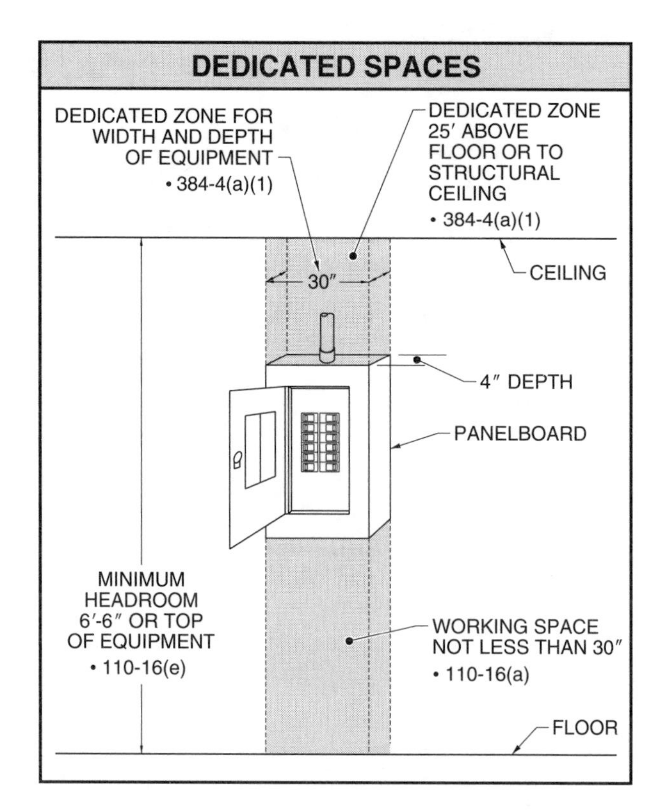

Figure 4-12. No barriers or other electrical equipment shall be located in dedicated spaces of switchboards, panelboards, and motor control centers.

SWITCHBOARDS – 384-5

Switchboards that have any exposed live parts shall be installed in dry locations accessible only to qualified persons. Switchboards shall be so located that the probability of damage from equipment or processes is reduced to a minimum. Per 384-7, switchboards shall not be placed in locations where it is possible for the switchboards to spread fire to adjacent combustible materials.

Clearances – 384-8(a)

A space of not less than 3′ shall be provided between the top of a switchboard and a combustible ceiling. This is not required if a noncombustible shield is provided between the switchboard and the ceiling or if the switchboard is of the totally-enclosed type. See Figure 4-13.

Per 384-8(b), the clearances around switchboards shall comply with 110-16. The minimum space required between the bottom of the switchboard enclosure and the busbars, their supports, or other obstructions is 8″ for insulated busbars and 10″ for uninsulated busbars.

Per 384-10, conduits or other raceways including their fittings shall not rise more than 3″ above the bottom of switchboards or floor-mounted panelboards. Sufficient space shall be provided to permit the installation of conductors when raceways enter the bottom of these enclosures.

Grounding of Switchboards – 384-11

Switchboard frames and structures supporting switching equipment shall be grounded. Per 384-11, Ex., frames of 2-wire, DC switchboards are not required to be grounded if effectively insulated from ground.

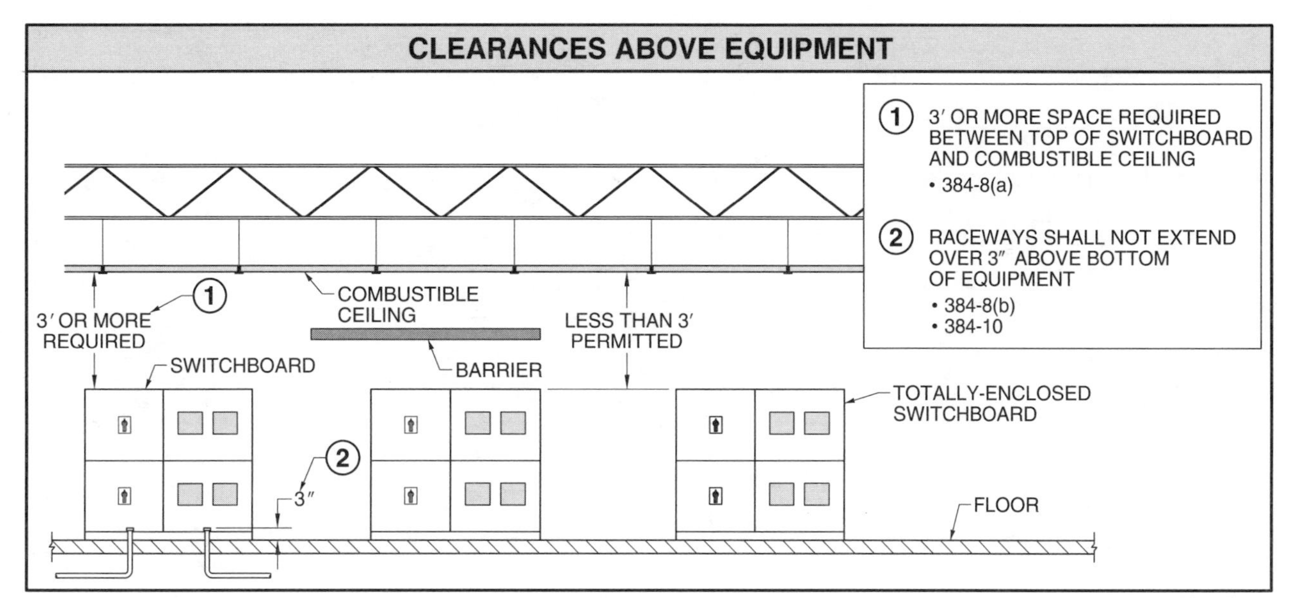

Figure 4-13. Minimum space above the top of a switchboard and a combustible ceiling is based on open or totally-enclosed switchboards and if barriers are present.

PANELBOARDS – 384-13

Panelboards shall be rated equal to or higher than the minimum feeder capacity required for the load. If the computed load on a panelboard is 175 A, the panel shall be sized at a rating of at least 175 A (minimum load). The next available standard size panelboard is 200 A.

Panelboards shall be durably marked by the manufacturer with the voltage and current ratings and the number of phases for which they are designed. The manufacturer's name or trademark shall also be marked on the panel so that it is visible after the installation, without disturbing the interior parts or wiring. All panelboards shall be provided with a circuit directory located on the face or inside cover of the panel door to identify the loads. All panelboard circuit modifications shall be legibly marked on the panelboard directory.

Lighting and Appliance Branch-Circuit Panelboards – 384-14

A *lighting and appliance branch-circuit panelboard* is a panelboard that provides neutral connections with more than 10% of its branch-circuit fuses or CBs rated at 30 A or less (15 A, 20 A, 25 A, or 30 A). The lighting and appliance branch-circuit panelboard shall also provide provisions for neutral connections. See Figure 4-14.

Number of Overcurrent Devices on One Panelboard – 384-15

A lighting and appliance branch-circuit panelboard shall not contain more than 42 overcurrent protection devices. This does not include those provided for in the mains. For the purpose of this rule, a 2-pole CB is considered as two overcurrent protection devices and a 3-pole CB is considered as three overcurrent protection devices. A *power panelboard* is a panelboard with more than 10% of its branch-circuit fuses or CBs rated over 30 A or more. There is no limit to the number of overcurrent protection devices permitted in a power panelboard.

Overcurrent Protection – 384-16(a). All lighting and appliance branch-circuit panelboards shall be provided with overcurrent protection to protect the busbars supplying the connected loads. See Figure 4-15. Every panel shall be protected individually on the supply side by not more than two main CBs or two sets of fuses having a combined rating not greater than the rating of the panelboard. Per 384-16(a), Ex. 1, a lighting and appliance panelboard may be protected by an overcurrent protection device (OCPD) on the feeder side if it does not exceed the panelboard rating. In this case, individual protection is not required for the panelboard. For example, a 600 A main may protect the busbars on two or more different panelboards provided the feeder conductors and the busbars in each panelboard are rated 600 A or more.

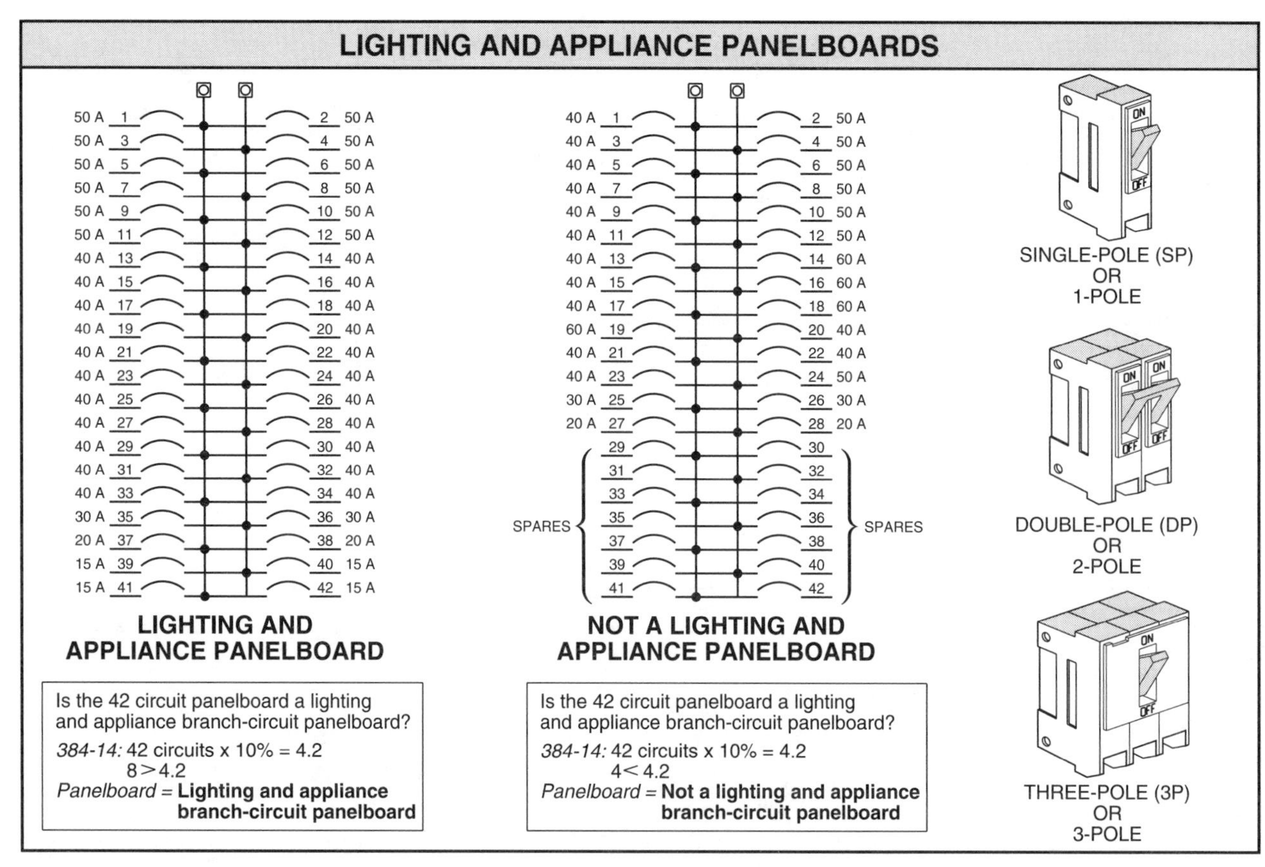

Figure 4-14. A lighting and appliance panelboard has 10% or more of its branch-circuit OCPDs rated at 30 A or less.

Continuous Load – 384-16(c). A continuous load shall not exceed 80% of the rating of the OCPD. A *continuous load* is a load in which the maximum current may continue for three hours or more. For example, a 200 A OCPD could only handle 160 A of continuous load (200 A × 80% = 160 A). Conversely, a 160 A continuous load requires a 200 A OCPD. Per 384-16(c), Ex., an assembly, including the overcurrent protection device, is permitted to be used for continuous operation at 100% of its rating when it is listed for the purpose.

Supplied Through a Transformer – 384-16(d). Panelboards supplied through transformers shall have their overcurrent protection located on the secondary side of the transformer. See Figure 4-16. Per 384-16(d), Ex., a 1ϕ transformer with a 2-wire primary and a 2-wire secondary are permitted to have the OCPD installed on the primary side only.

Back-Fed Devices – 384-16(f). Plug-in OCPDs or plug-in main lug assemblies which are connected for backfeed (the plug-in stabs are the load side of the OCPD) shall be mechanically secured in their installed positions. This requirement covers all plug-in-type protective devices (CBs or fuses). The intent is to eliminate the hazard of exposed energized plug-in stabs of an OCPD that could possibly become dislodged from its plug-in or connected position. See Figure 4-17.

Grounding of Panelboards – 384-20

All panelboard cabinets and frames of panelboards shall be grounded. This requirement may be satisfied if these enclosures are supplied by metal raceways or metal sheath cable assemblies which are recognized as EGCs. If a nonmetallic wiring method is used, such as Romex or PVC, an EGC shall be run with the feeder conductors.

If a panelboard is used as service equipment, the EGC terminal bar and the neutral terminal bar shall be bonded to the panelboard case along with the GEC per 250-50(a). In other than service equipment, the EGC terminal bar and the neutral terminal bar shall not be connected and bonded together in the panel enclosure. The neutral bar shall be isolated from the panelboard case and the EGC terminal bar shall be bonded to the panelboard case. See Figure 4-18.

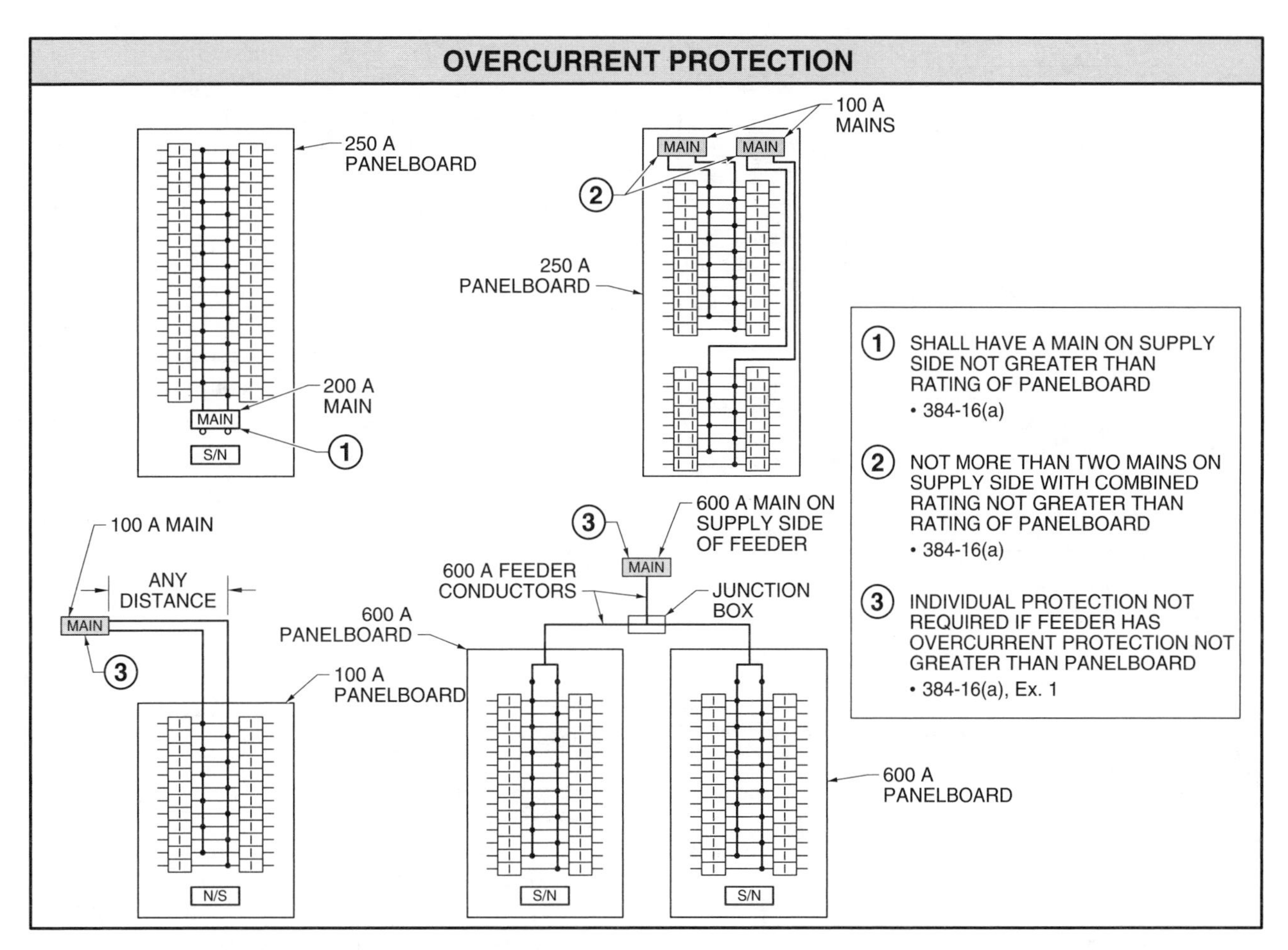

Figure 4-15. All lighting and appliance branch-circuit panelboards shall have overcurrent protection to protect the busbars supplying the connected loads.

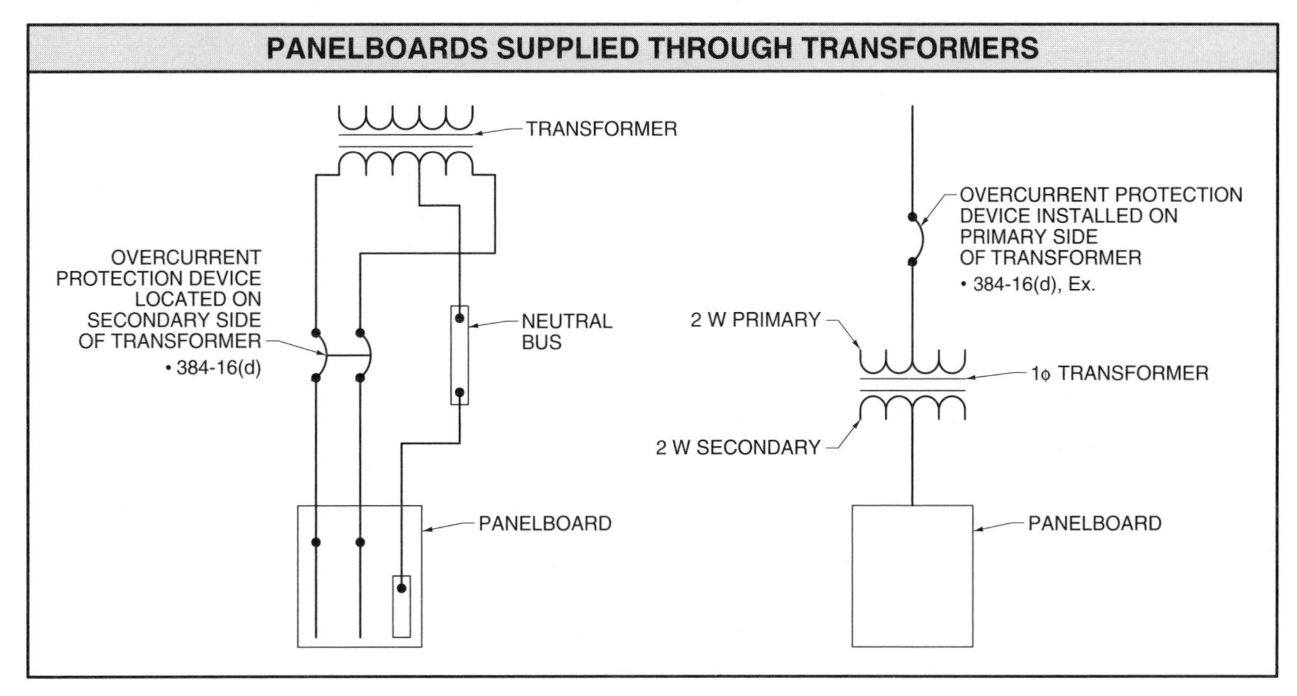

Figure 4-16. Panelboards supplied through transformers shall have OCPDs located on the secondary side.

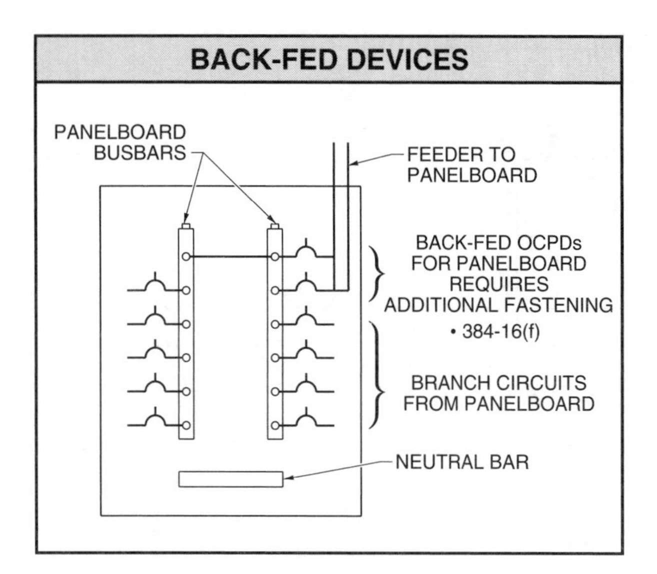

Figure 4-17. Plug-in OCPDs which are connected for backfeed shall be mechanically secured.

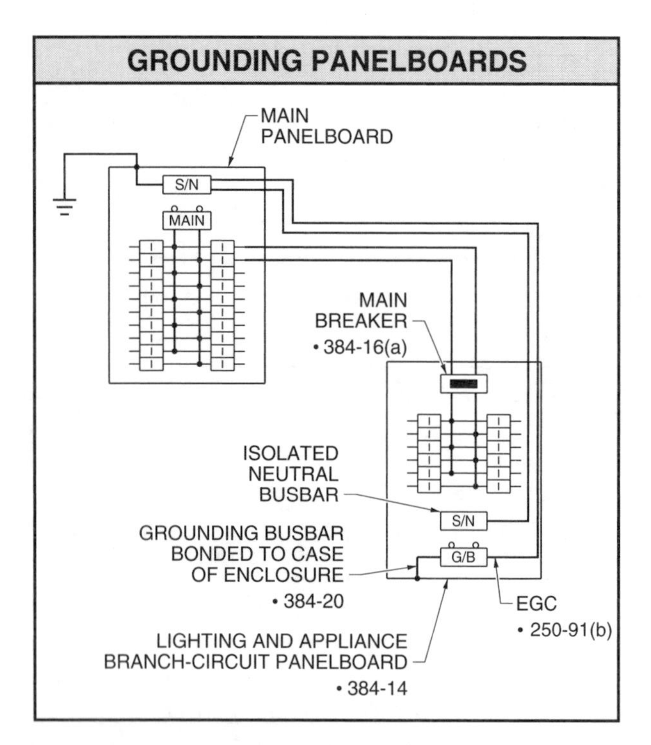

Figure 4-18. All panelboard cabinets and frames shall be grounded.

The neutral (grounded conductor) of an electrical system is always carrying the unbalanced current of that system, while the only time an EGC is expected to carry current is in the event of a ground fault. A ground fault occurs when an ungrounded conductor (hot wire) inadvertently comes in contact with a portion of the metallic electrical system not intended to carry current. Per 250-33, metal enclosures and raceways shall be grounded. Exposed noncurrent-carrying metal parts of fixed equipment likely to become energized shall be grounded per 250-42.

If the EGC system were to touch the grounded conductor (neutral) system, which is carrying current all the time, there would be a parallel path for this current to return to the service equipment bonding point, such as the metallic raceway or the metal armor of an AC cable. This would create the possibility of a potential of voltage to exist between all of the exposed dead metal parts of electrical equipment and any other grounded metallic parts within that facility, such as metallic pipe and appliances within the plumbing system.

Panelboard Construction Specifications – 384-30 through 384-34

Per 384-30, panels of switchboards are required to be made of moisture-resistant, noncombustible material. Insulated or bare busbars shall be rigidly and securely mounted in place per 384-31. Section 384-32 requires circuits which provide switchboards with pilot lights, instrument devices, and switchboard devices with potential coils to be protected by standard overcurrent devices rated 15 A or less.

Per 384-32, Ex. 1, OCPDs rated more than 15 A are permitted where interruption to the circuit may create a hazard. Per 384-32, Ex. 2, special types of enclosed fuses rated 2 A or less are permitted. Section 384-34 requires that exposed blades of knife switches be de-energized when open.

Wire-Bending Space in Panelboards – 384-35. The concern here is the same as in 373-6, which is the deflection of conductors in cabinets and cutout boxes. Adequate space is required where the conductors are shaped and terminated at the equipment. The wire bending space at the top and bottom of a panelboard shall comply with the dimensions in Table 373-6(b), regardless of the position of the conduit entries. The width of the side wiring gutters may be in accordance with the lesser distances of Table 373-6(a), based on the largest conductor to be terminated in that space.

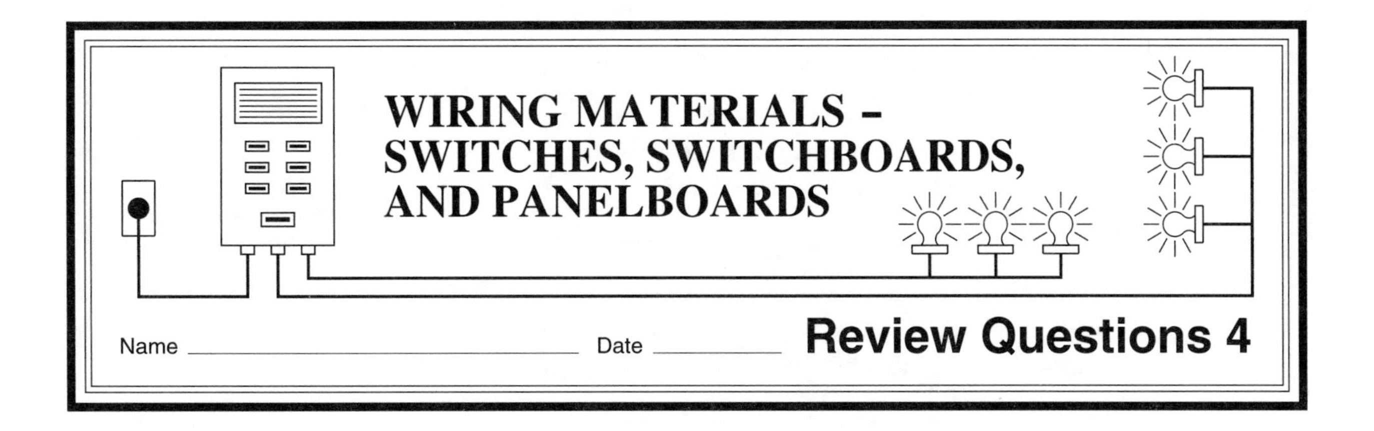

Review Questions

_________ **1.** A(n) ________ is a device, with a current and voltage rating, used to open or close an electrical circuit.

_________ **2.** Vertically-mounted switches shall be ________ in the up position.

_________ **3.** A(n) ________ is a single panel or group of assembled panels with buses, overcurrent devices, and instruments.

_________ **4.** A(n) ________ is a single panel or group of assembled panels with buses and overcurrent devices, which may have switches to control light, heat, or power circuits.

_________ **5.** A space of not less than ________′ shall be provided between the top of a switchboard and a combustible ceiling.

_________ **6.** A lighting and appliance branch-circuit panelboard is a panelboard with more than 10% of its branch-circuit fuses or CBs rated at ________ A or less.

T F **7.** All panelboard cabinets and frames of panelboards shall be grounded.

T F **8.** The neutral of an electrical system is always carrying the unbalanced current of that system.

T F **9.** Only ungrounded conductors shall be opened by switches.

_________ **10.** A(n) ________ load is any load that operates for three or more hours.

_________ **11.** A(n) ________ is an overcurrent protection device with a mechanical mechanism that may manually or automatically open the circuit when an overload condition or short circuit occurs.

_________ **12.** The two types of wire termination configurations are ________.

A. U and L
B. U and S
C. L and S
D. neither A, B, nor C

_________ **13.** Snap switch terminals marked CO/ALR are for use with ________ conductors.

A. aluminum
B. copper
C. copper-clad aluminum
D. A, B, and C

__________ **14.** A(n) ________ panelboard is a panelboard with more than 10% of its branch-circuit fuses or CBs rated over 30 A or more.

__________ **15.** A metal switch enclosure installed in a wet location shall be spaced ________″ off the wall to which it is mounted.

__________ **16.** Double-throw knife switches mounted vertically shall be equipped with a locking device to prevent ________ from closing the blades.

__________ **17.** Conduits or other raceways including their fittings shall not rise more than ________″ above the bottom of switchboards or floor-mounted panelboards.

__________ **18.** Panelboards supplied through transformers shall have their overcurrent protection located on the ________ side of the transformer.

__________ **19.** A permanent ________ is required between adjacent snap switches rated over 300 V.

T F **20.** Switchboards and panelboards are installed to control light and power circuits.

Dedicated Spaces

__________ **1.** The minimum distance at B is ________ or the top of the equipment.

__________ **2.** The minimum working space at A is ________″.

CEILING
4″ DEPTH
PANELBOARD
B
A
FLOOR

Phases

T F **1.** The busbars at A are correctly arranged.

T F **2.** The busbars at B are correctly arranged.

T F **3.** The busbars at C are correctly arranged.

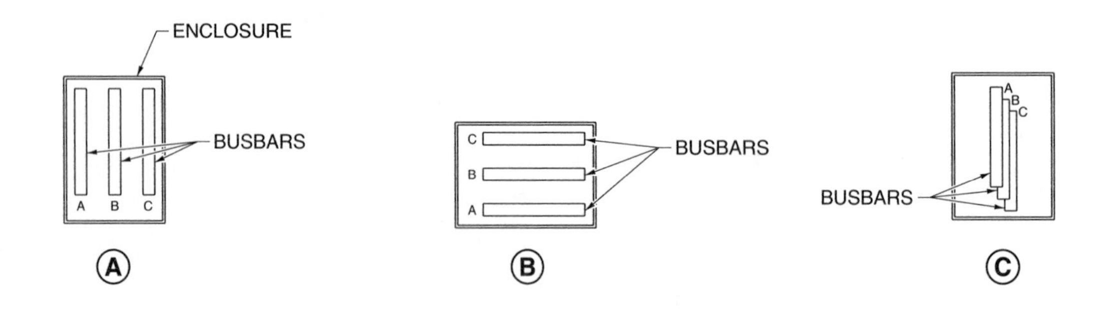

WIRING MATERIALS – SWITCHES, SWITCHBOARDS, AND PANELBOARDS

Name ______________________ Date ____________

Trade Test 4

NEC®	Answer	
______	______	**1.** *See Figure 1.* Determine the minimum wire-bending space required at A.
______	______	**2.** *See Figure 1.* Determine the minimum wire-bending space required at B.
______	______	**3.** *See Figure 1.* Determine the minimum wire-bending space required at B if the installation is done with removable and lay-in wire terminals intended for only one wire.
______	______	**4.** *See Figure 2.* Which phase is required to be identified as the high-leg?

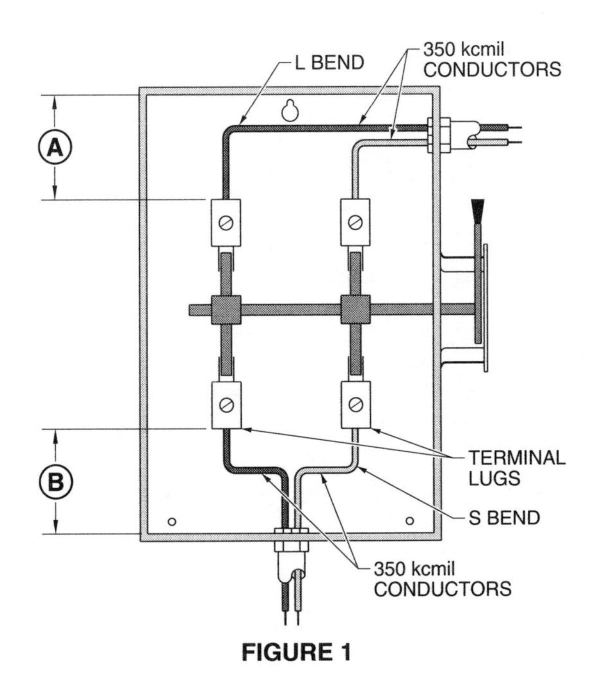

FIGURE 1

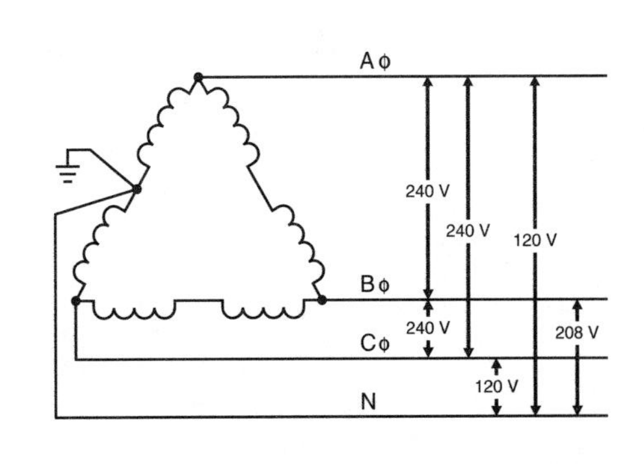

FIGURE 2

______	______	**5.** Determine the maximum permissible height above the finished floor to the center of the handle of a motor disconnect switch which is located in the top row of a motor control center.
______	______	**6.** Determine the maximum voltage between adjacent snap switches which are installed in a ganged enclosure that is not provided with permanent barriers.

_______________ _______________ **7.** *See Figure 3.* Determine the minimum clear space required between the switchboard and combustible ceiling.

_______________ _______________ **8.** *See Figure 4.* Is the panelboard classified as a lighting and appliance branch-circuit panelboard?

_______________ _______________ **9.** A service switchboard is installed in an electrical equipment room to supply a commercial building. The top of the switchboard is 7′ above the finished floor. Determine the minimum headroom about the switchboard per Article 384 of the NEC®.

_______________ _______________ **10.** A 4″ RNMC is run underground into an electrical service room. The conduit emerges from the concrete floor beneath an open-bottom switchboard. Determine the maximum height, including fittings, which the conduit is permitted to rise above the bottom of the enclosure.

_______________ _______________ **11.** What is the minimum space required between the bottom of the switchboard enclosure and the energized busbars for the installation in Problem 10? The busbars are uninsulated.

_______________ _______________ **12.** Determine the minimum rating for the main OCPD of a panelboard which supplies a total-connected load of 160 A. The load is a continuous load.

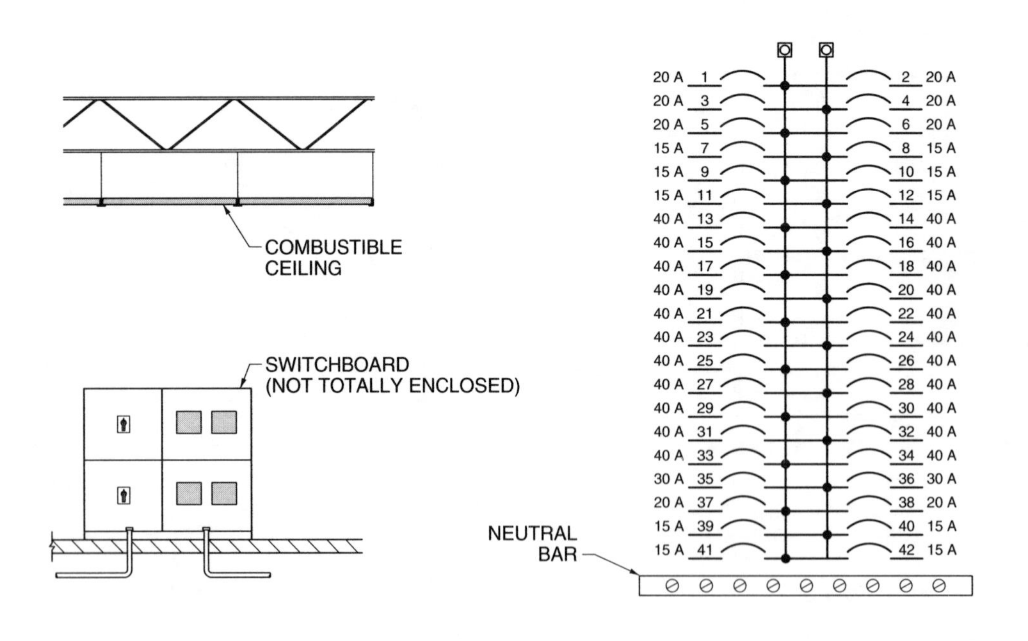

FIGURE 3 **FIGURE 4**

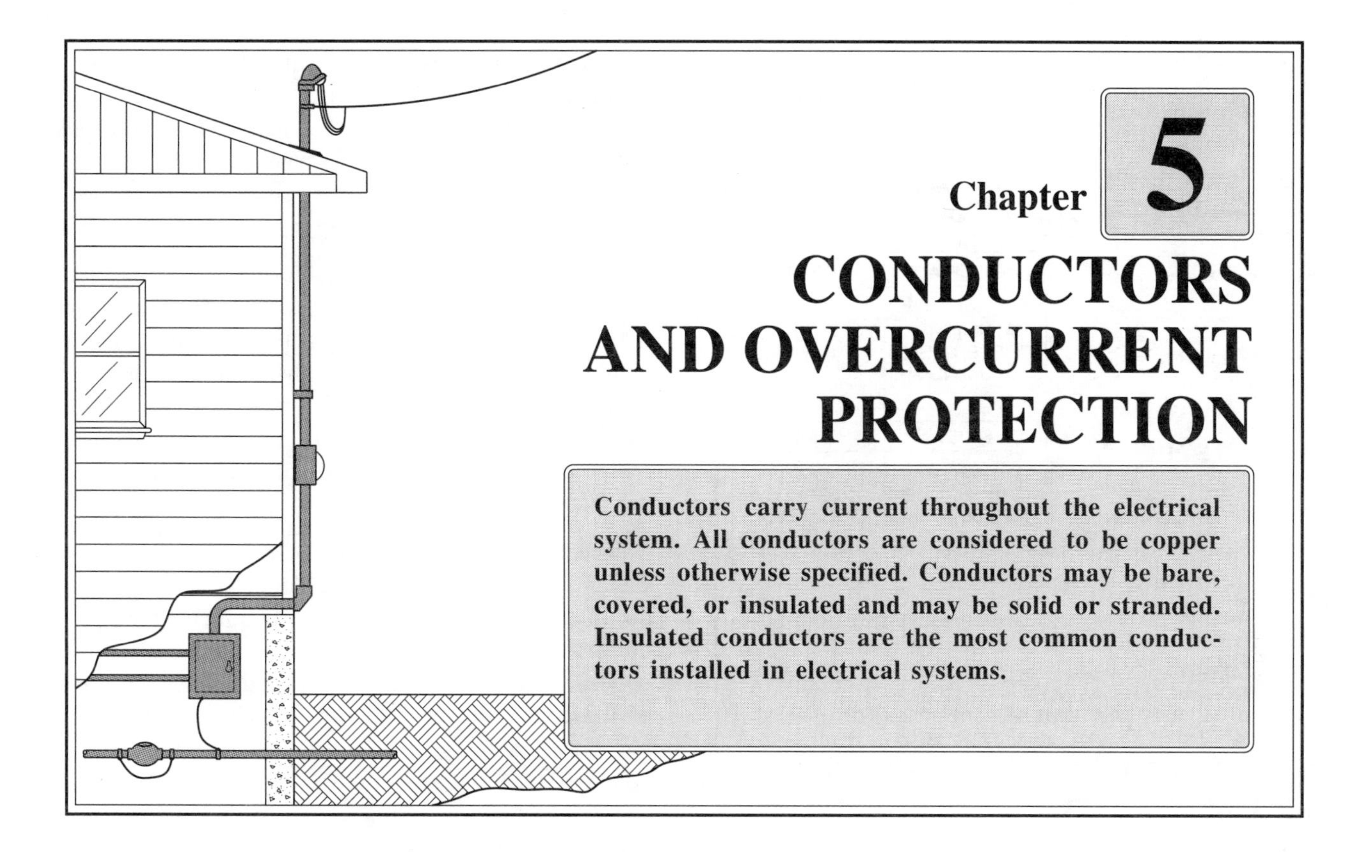

Chapter 5

CONDUCTORS AND OVERCURRENT PROTECTION

Conductors carry current throughout the electrical system. All conductors are considered to be copper unless otherwise specified. Conductors may be bare, covered, or insulated and may be solid or stranded. Insulated conductors are the most common conductors installed in electrical systems.

CONDUCTORS – 100; 310

The heart of all electrical systems is the conductors that carry the current to electrical devices and utilization equipment. There are associated effects whenever current flows through a conductor. For example, a magnetic field that varies in direct proportion to the amount of current flowing in the conductor is formed when current flows.

Heat, another property that is associated with the flow of current in a conductor, is one of the most significant factors to be considered in designing electrical systems. Conductors, in general, are rated on their ability to withstand the effects of heat. The conductor's rating is dependent upon its type of electrical insulation.

Conductors are assigned a specific ampacity that reflects the insulation's ability to handle and dissipate heat under varying conditions. For example, Table 310-16 lists the allowable ampacity of a No. 12 Cu conductor with THHN insulation at 30 A, while a similar No. 12 Cu conductor with TW insulation has a listed allowable ampacity of only 25 A. In general, all conductors shall be protected against overcurrents in accordance with their listed ampacities, at the point where they receive their supply.

Definitions – 100

Conductors used in electrical systems are often defined by the material, if any, that is used to encase the actual electrical conductor. Three classifications exist for electrical conductors: bare, covered, and insulated. See Figure 5-1.

Bare. Bare conductors have no outer covering or insulation encasing the actual conductor material. Bare conductors provide no protection for the conductor material, thus the NEC® has very limited applications for bare conductors.

The most common use for bare conductors is for equipment grounding purposes. A copper or other corrosion-resistant material may serve as an equipment grounding conductor (EGC) per 250-91(b)(1). The conductor is permitted to be bare and in the form of a busbar or wire.

Another application for bare conductors is as a grounding electrode conductor (GEC). The GEC is permitted to be constructed of copper, aluminum, or copper-clad aluminum and it may be a bare conductor per 250-91(a).

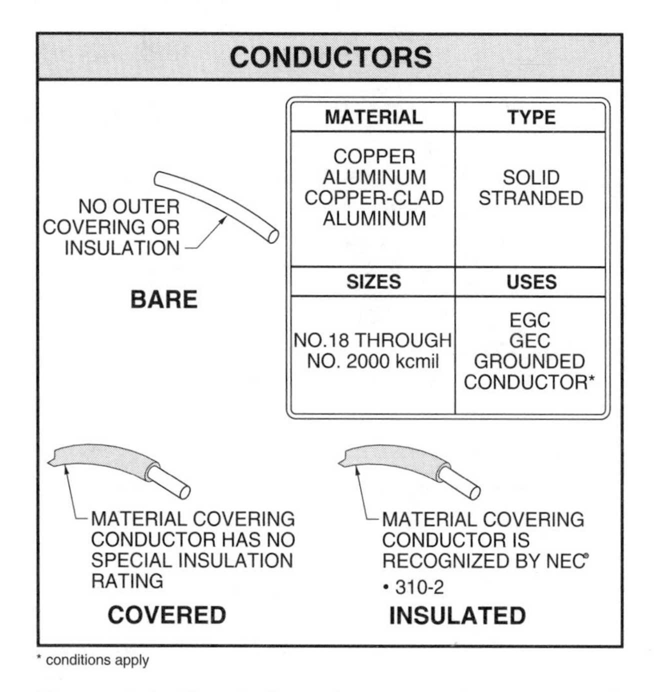

Figure 5-1. Electrical conductors are bare, covered, or insulated.

For the most part, service-entrance (SE) conductors are required to be insulated. However, the grounded conductor (neutral) may be a bare conductor under specific conditions per 230-41, Ex. A common element in all the provisions for the use of bare conductors is that the hazards associated with the potential electrical flow in these conductors have been lessened by isolation or circuit design.

Covered. Of the three basic types of conductors, covered conductors are utilized the most infrequently in electrical systems. By definition, covered conductors are constructed with a material that encases the electrical conductor, but the material has not been evaluated and is not recognized as having a specific insulation rating. Grounding electrode conductors and equipment grounding conductors are both permitted to be covered conductors, but more frequently these are installed in electrical systems as either bare or insulated conductors.

Perhaps the most frequent use of covered conductors is in conductors installed in free air, in particular service-drop conductors. By their nature, service-drop conductors are installed in a manner that does not make them accessible to unqualified persons. Additionally, because these conductors are installed in free air, their capacity to dissipate heat associated with current flow is greatly enhanced. While covered electrical conductors may offer some degree of electrical insulation, the covering has not been evaluated as an electrical insulation recognized by the NEC®.

Insulated. Insulated conductors are the most common conductors installed in electrical systems. These conductors are constructed with a material that has been identified by the NEC® as a recognized electrical insulation. Unless specifically permitted elsewhere in the NEC®, all conductors shall be insulated per 310-2.

There are numerous types of electrical insulations. The type of insulation selected depends upon the conditions of use of the conductor. A conductor for direct burial requires an insulation that is suitable for the conditions it is likely to be exposed to over the life of the conductor. Conductors for direct burial shall be identified for such use per 310-7. For example, UF cable is an underground feeder and branch-circuit cable. The outer covering of UF cable shall be suitable for direct burial in the earth per 339-1(a). *Cable* is a factory assembly with two or more conductors and an overall covering.

Conductors exposed to potentially corrosive conditions shall be provided with an insulation suitable for such exposure. Corrosive conditions include installations where the conductors are exposed to oils, greases, vapors, etc. that break down the integrity of the conductors' insulation. For example, NMC cable is identified in 336-30(a)(2) as being constructed of an overall covering that is corrosion-resistant. See 310-9.

Conductors installed in wet locations require insulation that protects conductors from continuous exposure to moisture. A *wet location* is any location in which a conductor is subjected to excessive moisture or saturation from any type of liquids or water. Unprotected locations outdoors, installations underground or concrete slabs, and installations in direct contact with the earth are classified as wet locations. See 100. Section 310-8(a) permits insulated conductors installed in wet locations to be lead covered with RHW, TW, THW, THHW, THWN, or XHHW or other insulation listed for use in wet locations.

A common element among all electrical insulations is that they provide an opposition or resistance to the flow of electricity. All conductors and cables are marked to identify, among other things, the type of insulation that is used. See 310-11(a)(2). All conductors and cables are required to be marked to identify the maximum rated voltage at which the conductor is listed, the letter or letters necessary to identify the type of wire or cable, the manufacturer's name or trademark or other identifiable markings, and the AWG or circular mil area size of the conductor. See Figure 5-2.

The specific insulation types listed for use in wet locations contain the letter "W" in their prefix. The "W" identifies the conductor as having an insulation that is moisture-resistant and suitable for use in wet locations. Table 310-13 can be used to identify the prefix markings found on conductors.

Another important consideration in selecting conductor insulation type is the temperature limitations of the insulation. Table 310-13 provides the maximum operating temperature for the particular insulation types. The three basic insulation temperature ratings listed for the conductors used for general wiring are 60°C (140°F), 75°C (167°F), and 90°C (194°F). These ratings designate the maximum temperature that the conductor can be exposed to without the danger of encountering insulation breakdown and damage. The maximum temperature that the conductor is exposed to along its entire length shall be used when determining the temperature rating of a conductor.

The common factors that can contribute to conductor insulation degradation because of excessive operating temperature are listed in 310-10. See Figure 5-3. The first factor is the ambient temperature in which the conductor insulation shall operate. The second factor is the internal heat created in the conductor as a result of current flow. The third factor is the dissipation rate of the heat into the surrounding environment. The final factor is the heat generated by adjacent current-carrying conductors. All of these factors shall be considered prior to selecting the type of insulation and determining the conductor's allowable ampacity.

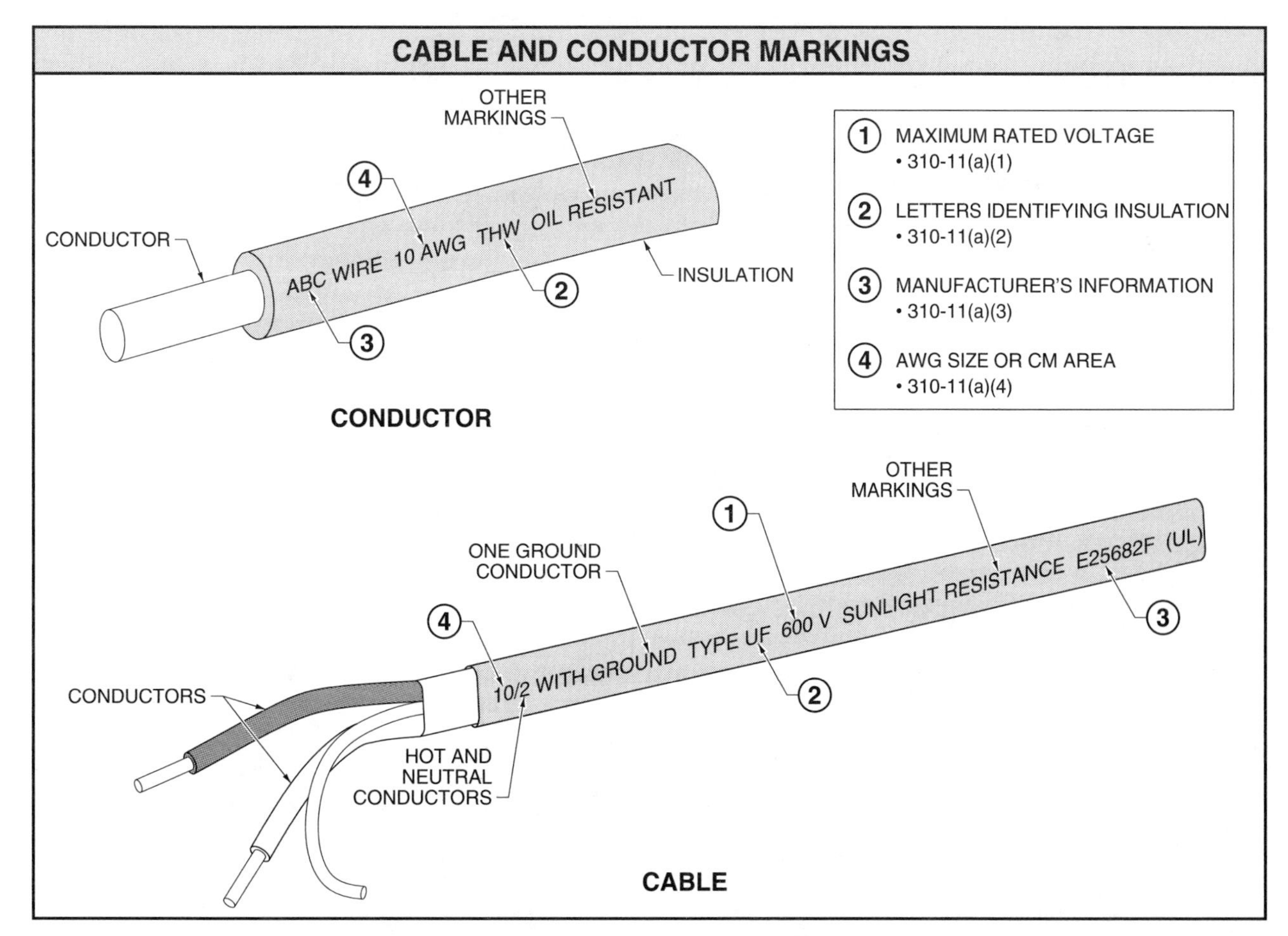

Figure 5-2. All conductors and cables are identified by their markings.

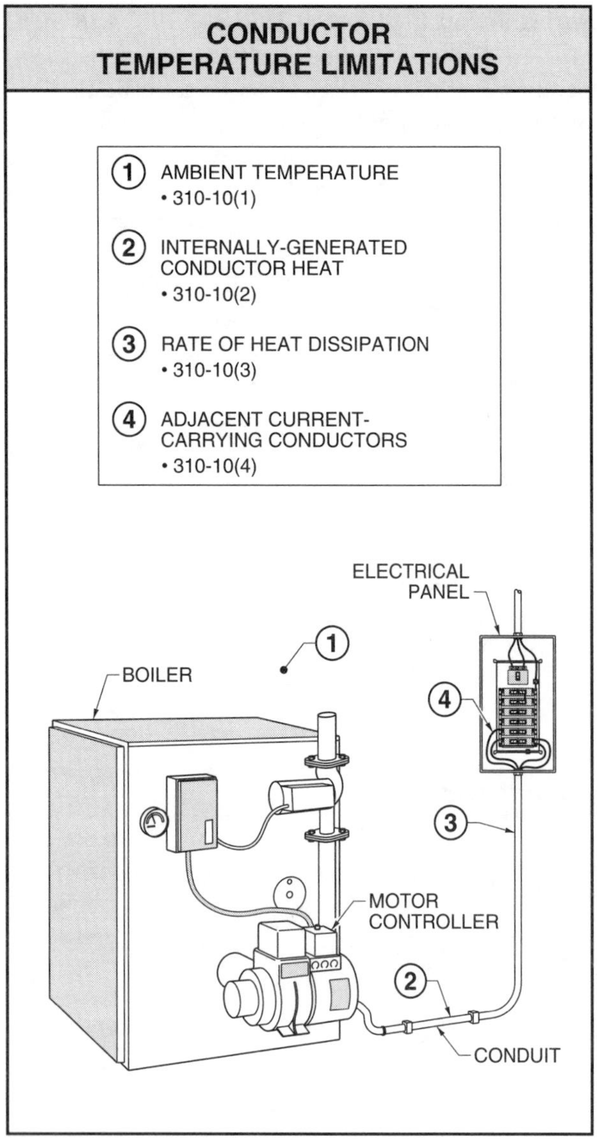

CONDUCTOR TEMPERATURE LIMITATIONS

① AMBIENT TEMPERATURE
• 310-10(1)

② INTERNALLY-GENERATED CONDUCTOR HEAT
• 310-10(2)

③ RATE OF HEAT DISSIPATION
• 310-10(3)

④ ADJACENT CURRENT-CARRYING CONDUCTORS
• 310-10(4)

ELECTRICAL PANEL
BOILER
MOTOR CONTROLLER
CONDUIT

Figure 5-3. Ambient temperature and conductor heat contribute to insulation degradation.

Construction – 310-2; 310-3

The two most common materials used for construction of electrical conductors are copper and aluminum. Copper has excellent conductivity and therefore has a higher allowable ampacity than a comparable aluminum conductor. Aluminum, on the other hand, is not as good a conductor in terms of its allowable ampacity, but is lighter and cheaper to install than copper.

Copper. All conductors used to carry current are copper unless specified otherwise. See 110-5. While copper is most extensively used, its cost is a primary disadvantage. The cost of installing electrical distribution systems with copper conductors, particularly in larger sizes, can be prohibitive. Another factor to consider when installing electrical conductors is the weight of the conductors. Copper conductors installed in vertical raceways require support at shorter intervals than are required for aluminum conductors. Section 300-19 lists the required intervals and methods for vertical supports to ensure that the weight of the conductors is not passed on to the conductor terminations. See Figure 5-4.

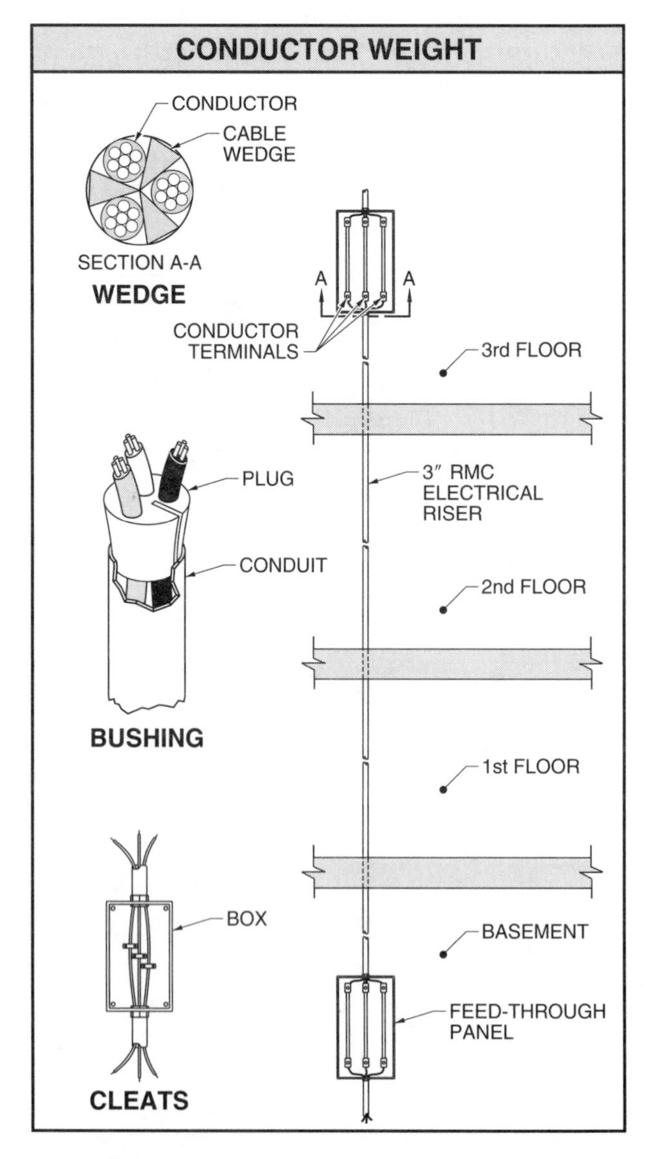

Figure 5-4. The weight of the conductor shall not be passed on to conductor terminations.

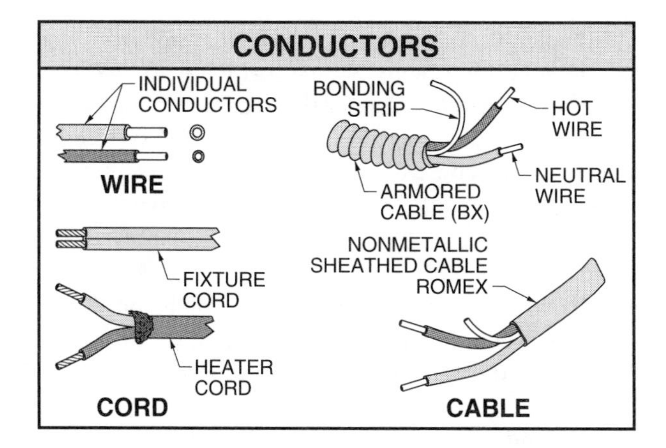

Copper conductors are available as hard-drawn, medium-hard-drawn, and soft-drawn. Hard-drawn copper has the greatest strength, but is difficult to work with. Because of the difficulty in bending and shaping hard-drawn copper, its uses are limited. It is used primarily for utility transmission and some service-drop conductor applications.

Medium-hard-drawn copper is easier to work with than hard-drawn copper but it does not have the tensile strength that hard-drawn copper has. Therefore, it is used primarily for transmission and utility conductors where greater flexibility is required.

Soft-drawn copper is easy to work with and can be installed in many different types of raceways and cable assemblies. Most general building wires used in electrical distribution systems are made of soft-drawn copper.

Aluminum. Aluminum conductors have been used extensively for a number of years in the utility distribution and transmission field. More recently, aluminum conductors have found increasing applications in building electrical distribution systems. Section 310-14 requires solid aluminum conductors in sizes No. 8, 10, and 12 to be constructed of an AA-8000 series electrical grade aluminum alloy. Stranded conductors are permitted in sizes from No. 8 through No. 1000 kcmil if constructed from an AA-8000 series electrical grade aluminum alloy.

Despite the fact that aluminum is not as good a conductor as copper, its reduced conductivity is compensated by the considerable cost savings in using aluminum conductors. For the most part, aluminum conductors can be installed with the same installation procedures as used for copper. The primary difference occurs when the conductors are terminated. Because the surface of aluminum conductors oxidizes readily, terminations are generally made with the aid of joint compounds designed to prevent the oxide from re-forming in the installation process. An *oxide* is a thin, but highly resistive coating that forms on metal when exposed to the air. If the oxide is not prevented from re-forming, a high-resistance connection could occur which could lead to insulation failure at the termination.

Another consideration when installing aluminum conductors is terminations with dissimilar metals. Conductors of different materials, like copper and aluminum, should not be terminated in a manner that causes the dissimilar metals to come in direct contact with each other unless the termination or splicing device is identified for such use. See 110-14. See Figure 5-5.

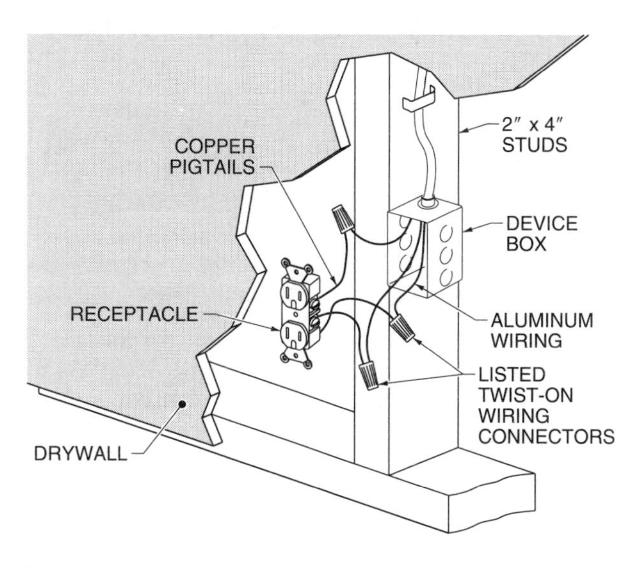

Figure 5-5. Listed twist-on wire connectors are available for directly splicing aluminum to copper.

Torque is a turning or twisting force, typically measured in foot pounds (ft-lb). Manufacturers provide torquing specifications for terminations. The termination should be tightened to those specifications to ensure that the connection is electrically sound. The UL Marking Guide for Molded-Case Circuit Breakers requires that all CBs be marked with their rated tightening torque where field terminations are made. If the tightening torque is dependent on the wire size, the appropriate range of tightening torques shall be provided for each wire size.

Copper-Clad Aluminum. The most recent development in conductor material and construction is the advent of copper-clad aluminum. Copper-clad aluminum offers a compromise between the increased conductivity and termination qualities of copper conductors with the lighter weight and cost efficiency of aluminum conductors. Copper-clad conductors are constructed using a minimum of 10% copper, which is bonded metallurgically to the aluminum. The ampacity of copper-clad aluminum conductors is selected from the same column used for aluminum conductors. See Table 310-16.

Solid or Stranded. Conductors for electrical systems are either solid or stranded. See Figure 5-6. Solid conductors are constructed of a single piece of wire (strand). Stranded conductors are constructed of multiple wires (strands).

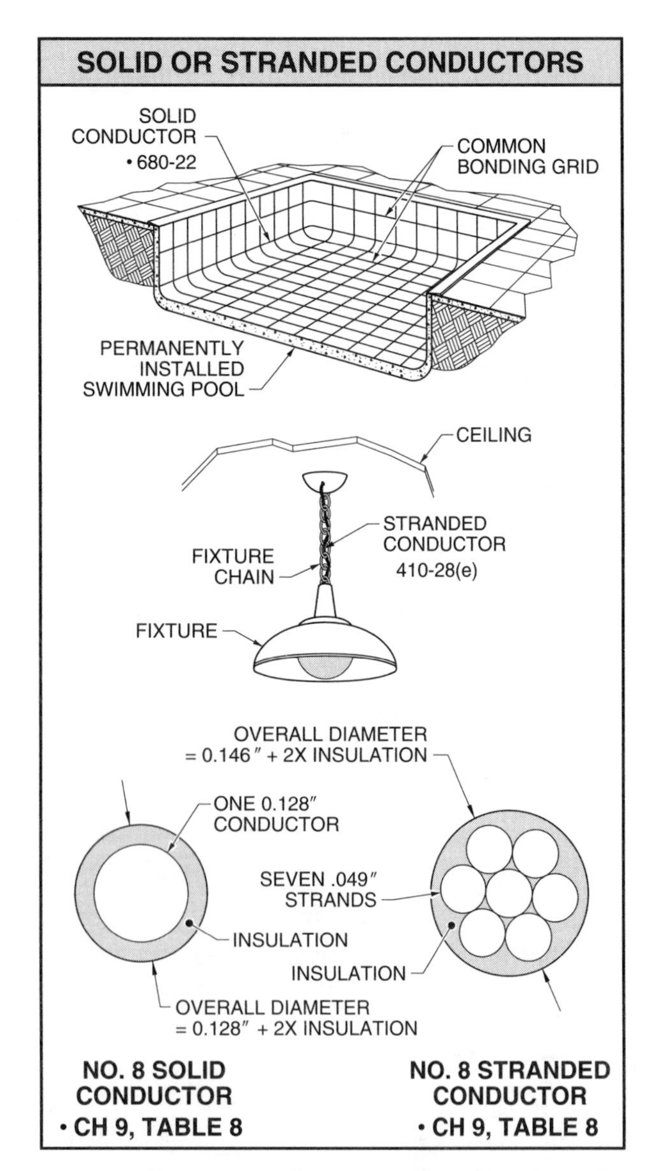

Figure 5-6. Some electrical installations specifically require that solid or stranded conductors be used.

Chapter 9, Table 8 lists conductor properties for common building wire. Notice that there are two listings for No. 8 AWG. Under the heading "conductors" there is a column for stranding. A No. 8 AWG conductor can be constructed from a single strand (solid) or from seven individual strands, each .049″ in diameter (stranded). The choice of solid or stranded conductor lies with the designer and is based largely on the needs of flexibility in the conductor. There are, however, a few provisions in the NEC® that specify either solid or stranded conductors.

Color Code. The NEC® has adopted various color code requirements for many of the conductors used in electrical distribution systems. See Figure 5-7. These requirements ensure proper identification of circuits and conductors which is critical for the safety of those who are required to maintain the electrical system. Section 310-12 contains conductor identification requirements for the grounded, ungrounded, and equipment grounding conductors.

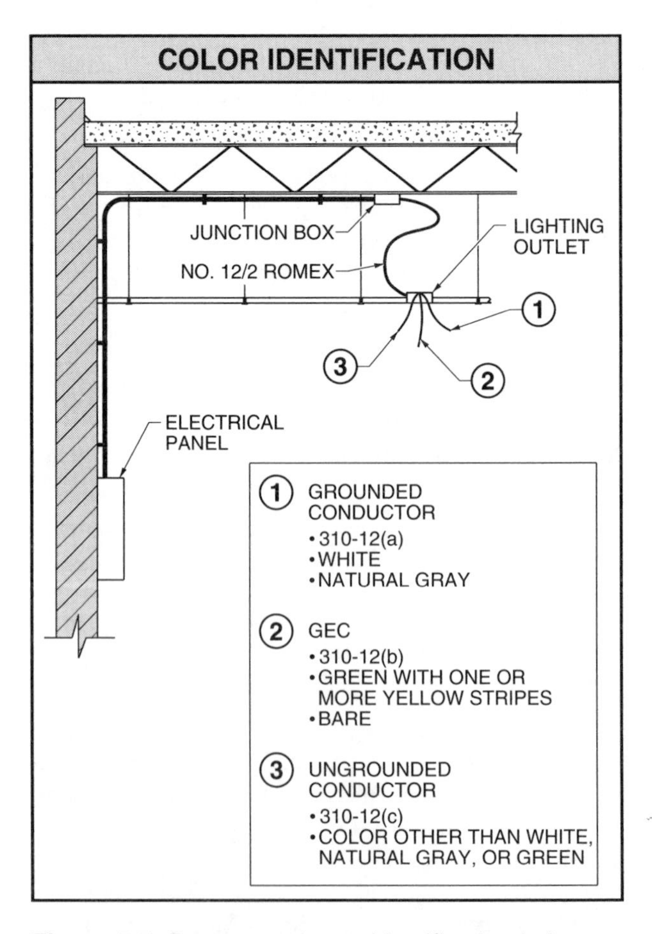

Figure 5-7. Proper conductor identification helps ensure worker safety and maintain the integrity of the electrical system.

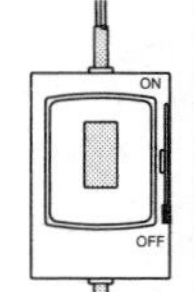

SYSTEM GROUNDED CONDUCTORS

When system grounded conductors are installed in the same raceway, cable, box, enclosure, etc., one of the system grounded conductors shall have a white or natural gray outer covering per 200-6(d). The other grounded conductor shall be white with a stripe or tracer that is not the color green.

A common practice among electricians is to use white for 120/208 V system grounded conductors and gray for 277/480 V system grounded conductors. While this does distinguish between the two systems, it is still a literal violation of the requirement in 200-6(d) for using white with a stripe to identify the second system grounded conductor. When installing conductors of different systems in the same raceway, check with the AHJ for an official interpretation.

In general, the grounded conductor (neutral) is colored white or natural gray. Depending on the size of the conductor, there are provisions for reidentifying a conductor that is not white or gray, at the conductor termination. For example, a No. 2 AWG conductor with black insulation is permitted to be reidentified at the point where it terminates with distinctive white markings to identify it as a grounded conductor. Additionally, where only qualified persons maintain an electrical installation, the grounded conductor in a multiconductor cable is permitted to be reidentified at the termination. See 200-6(a)(b) and 200-6(b), Ex.

When the grounded conductors for two different electrical systems are installed in the same raceway, cable, box, etc., the two system neutrals shall not be mixed. Therefore, the NEC® requires that when grounded conductors are mixed, one system neutral shall be white or natural gray and the other system neutral shall be white with a stripe, other than green, running along the insulation. See 200-6(d).

Ungrounded conductors shall be color coded so they are easily distinguishable from grounded and other grounding conductors. While there are no specific color code requirements for ungrounded conductors, the colors white, natural gray, and green are not permitted. Section 210-4(d) contains requirements for color coding of ungrounded conductors where more than one nominal voltage system is present in a building. Each of the system's ungrounded conductors shall be identified and the means of identification shall be clearly and permanently posted at each branch-circuit panelboard. The means that may be employed for identification of the conductors may be separate color coding, marking tape, tagging, or other equally effective means per 210-4(d), FPN. No particular color code scheme is specified. See Figure 5-8.

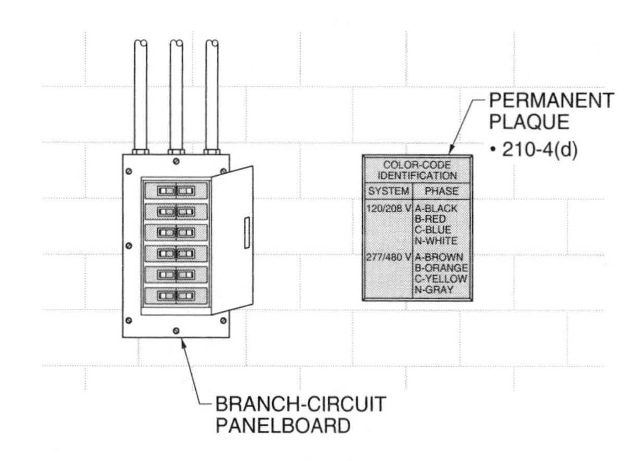

Figure 5-8. Color-code identification directories shall be located at the branch-circuit panelboard.

Another color provision that installers and designers may encounter involves the use of delta 4-wire systems. Sections 215-8 and 384-3(e) require that the high leg of the delta 4-wire system be identified by an outer finish that is orange in color or other equally effective means.

Equipment Grounding Conductors

Equipment grounding conductors, for the most part, are permitted to be bare conductors. When these conductors are required to be covered or insulated, or if the designer or installer chooses to provide insulation, the outer covering shall be green or green with one or more yellow stripes per 310-12(b). As with grounded conductors, there are provisions for reidentification of the equipment grounding conductor under specific conditions. See 310-12(b), Ex. 1 and 2.

Ex. 1 permits reidentification of EGCs, No. 4 and larger, at the time of installation by permanent means such as stripping the insulation or covering from the exposed conductor, coloring the exposed insulation or covering green, or marking the exposed covering or insulation with green tape or green labels. Ex. 2 permits the same methods of reidentification for multiconductor cables provided the conditions of maintenance ensure that only qualified personnel service the installation. See Figure 5-9.

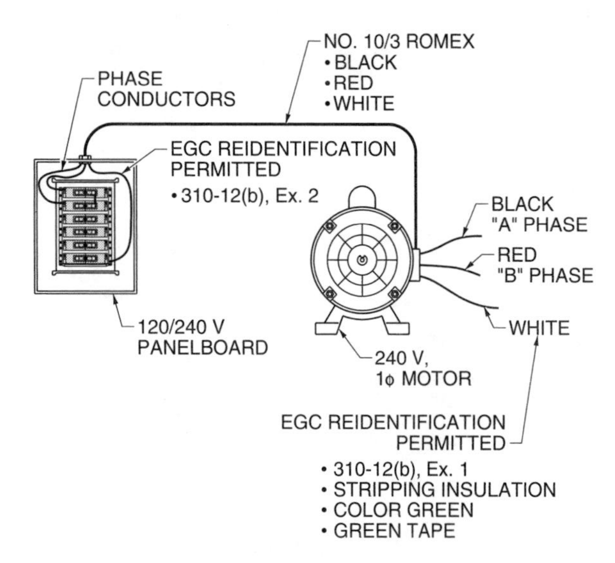

Figure 5-9. Reidentification of multiconductor cables is permitted during installation if only qualified personnel perform service.

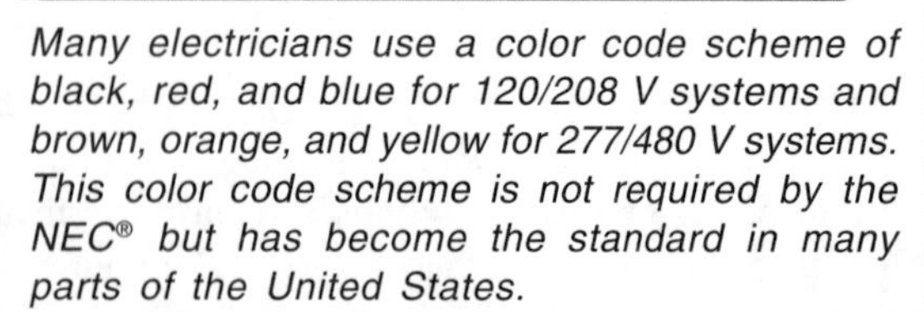

MARKING SYSTEMS

Many electricians use a color code scheme of black, red, and blue for 120/208 V systems and brown, orange, and yellow for 277/480 V systems. This color code scheme is not required by the NEC® but has become the standard in many parts of the United States.

The FPN to 210-4(d) does not require that the marking system be by color code. Other equally effective means, such as tagging and marking tape, are acceptable. Whatever system is utilized, the means used to identify the ungrounded conductors of the electrical systems shall be permanently posted at each panelboard.

Size – 310-5

Table 8, Chapter 9 lists copper, aluminum, and copper-clad aluminum conductors in sizes from No. 18 AWG through No. 2000 kcmil. In general, the NEC® does not permit conductors smaller than No. 14 Cu, No. 12 Al, or copper-clad aluminum to be used for conductors for general wiring in applications up to 2000 V. However, 310-5 contains eight exceptions to this rule. Table 310-5 lists the minimum conductor size. This size changes depending upon the voltage rating of the conductor. These are minimum sizes and there are many instances in which a particular NEC® provision requires a specific size conductor.

Parallel Conductors – 310-4

Parallel conductors are two or more conductors that are electrically connected at both ends to form a single conductor. Despite the wide range of conductor sizes available, there may be installations in which it is desirable to use parallel conductors to meet a specific design consideration or installation requirements. See 310-4 for specific rules concerning the installation of parallel conductors. Copper, aluminum, and copper-clad aluminum conductors are all permitted to be installed in parallel. See Figure 5-10.

Length – 310-4(1). All conductors shall be the same length. Paralleling conductors of different lengths may result in an uneven distribution of current in the conductors resulting in excessive heat and possible insulation degradation. Many cable manufacturers recommend cutting the parallel conductors to the same length prior to installation in a raceway to ensure that the conductors remain the same length.

Material – 310-4(2). All conductors shall be the same material. Aluminum and copper have different properties that make them incompatible for paralleling.

Size – 310-4(3). All conductors shall be the same size. Same size conductors ensure that the current split and the potential voltage drop between conductors remains the same.

Insulation – 310-4(4). All conductors shall be the same insulation type. There are three basic temperature ratings for conductor insulations. Parallel conductors shall have the same temperature rating to ensure that the termination does not reach a temperature beyond the operating temperature of the insulation.

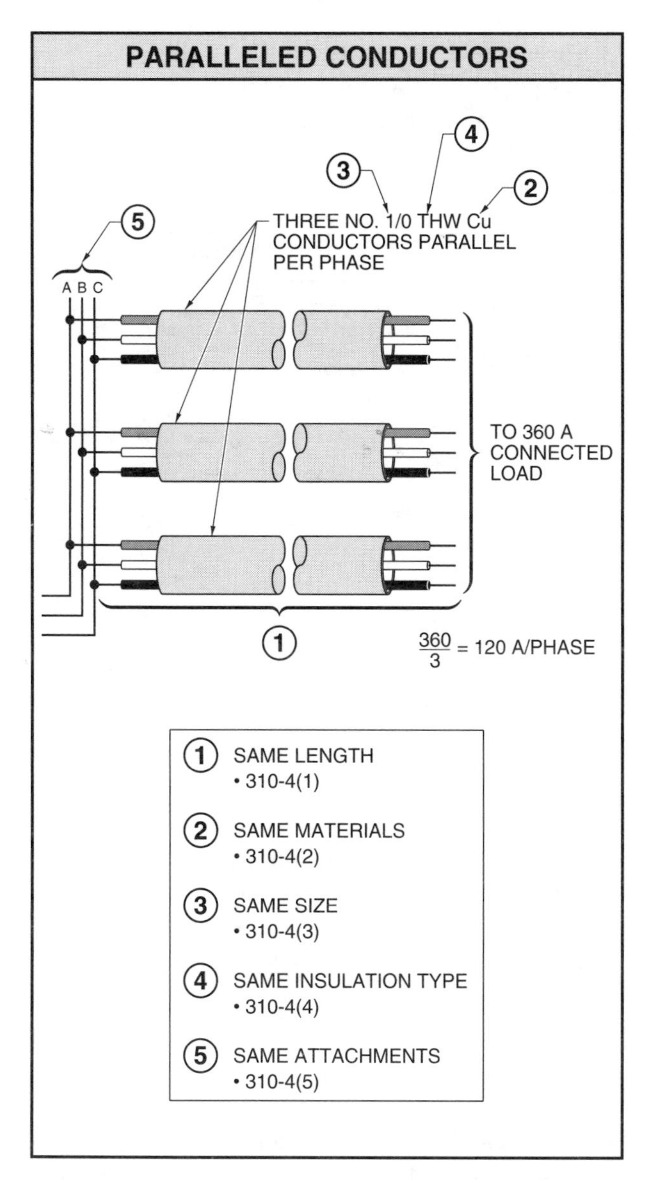

Figure 5-10. Each conductor, of each set of paralleled conductors, shall meet five conditions.

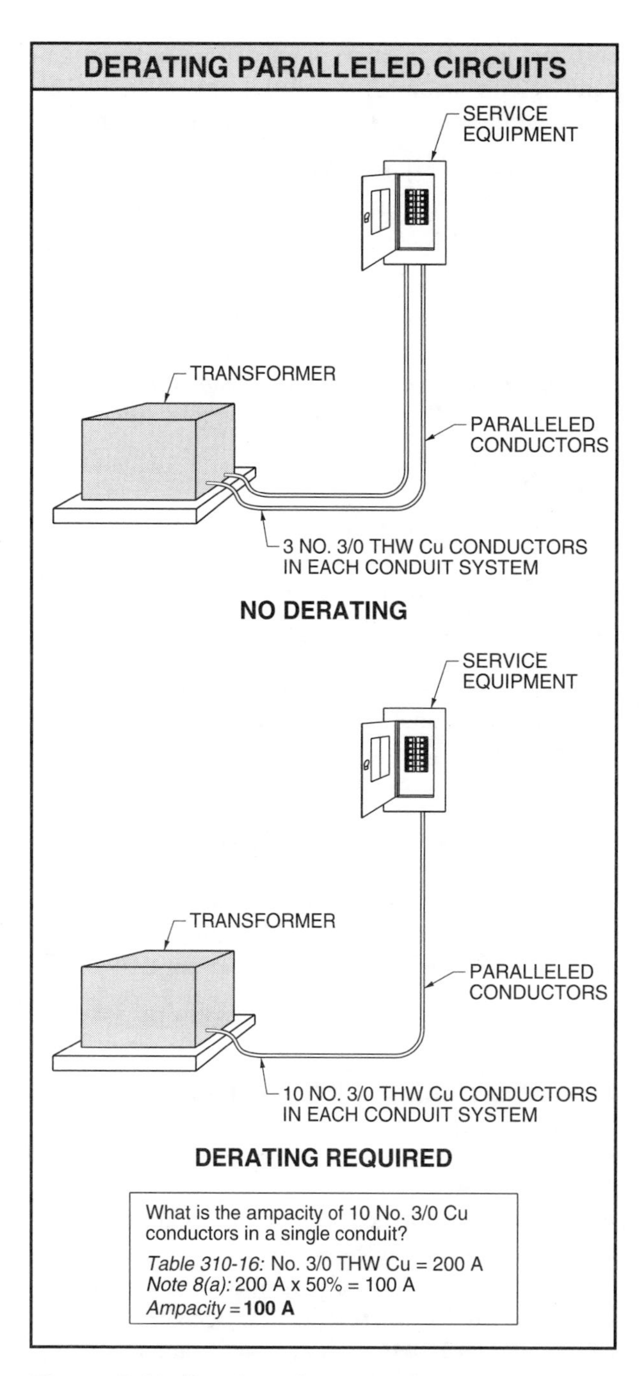

Figure 5-11. Derating of parallel conductors is determined by the design of the electrical distribution system.

Terminations – 310-4(5). All conductors shall be terminated using the same methods or materials. If termination methods are different, such as the when single-barrel lugs and multiple-barrel lugs are used on the same phase, the impedance of the paralleled conductors may be different resulting in uneven current splits between the paralleled conductors.

These requirements apply to the individual conductors that comprise a single-phase or grounded conductor and not to all of the conductors of the circuit. Paralleled conductors are derated per Note 8(a) of Notes to Ampacity Tables of 0 to 2000 Volts. See Figure 5-11.

AMPACITY – 100; 310

Ampacity is a term derived from combining the words ampere and capacity. *Ampacity* is the current that a conductor can carry continuously, under the conditions of use. See 100. Conditions which may affect a conductor's allowable ampacity, and require derating, include the type of insulation, ambient temperature surrounding the conductor, number of current-carrying conductors, and temperature rating.

The conductor insulation type has a direct effect on the allowable ampacity of the conductor. Generally, the higher the temperature rating of the conductor insulation, the higher the allowable ampacity.

Heat is generated when current flows through a conductor. As the ambient temperature is increased, the ability of the conductor to dissipate this heat is greatly reduced. All ambient temperatures above 30°C require a correction of the conductor's allowable ampacity.

Table 310-16 lists allowable ampacities based upon not more than three current-carrying conductors in the same raceway or cable. When additional current-carrying conductors are installed, the additional heat generated by these conductors shall be accounted for. Note 8 of Notes to Ampacity Tables of 0 to 2000 Volts provides adjustment factors for installations with more than three current-carrying conductors.

All electrical equipment has temperature ratings associated with the equipment terminations. The temperature rating of the conductor shall be selected in accordance with the temperature rating of any connected termination in the equipment per 110-14(c).

Insulation Ratings – Table 310-13; Table 310-16

All insulated conductors have a maximum operating temperature at which the insulation of the conductor is not adversely affected. Conductor ampacities are directly related to these operating temperature limitations. Table 310-16 lists conductor ampacities for insulated conductors under 2000 V, installed in a raceway, cable, or earth. These values assume an ambient temperature of 30°C (86°F), and not more than three current-carrying conductors. While there are many different types of insulations listed, all of them are assigned to one of three basic classifications; 60°C (140°F), 75°C (167°F), and 90°C (194°F). See Figure 5-12.

For example, three different ampacities are listed for a No. 8 AWG Cu conductor depending upon the insulation used. Any of the 60°C insulations result in a listed Table ampacity of 40 A. The 75°C insulations permit a 50 A rating, and the 90°C insulations permit a 55 A rating. These ratings are the listed Table values only. The actual allowable ampacity depends on the installation conditions.

COMMON ELECTRICAL INSULATIONS

60° C 140° F	75° C 167° F	90° C 194° F
TW	FEPW	TBS
UF	RH	SA
	RHW	SIS
	THHW	FEP
	THW	FEPB
	THWN	MI
	XHHW	RHH
	USE	RHW-2
	ZW	THHN
		THHW
		THW-2
		THWN-2
		USE-2
		XHH
		XHHW
		XHHW-2
		ZW-2

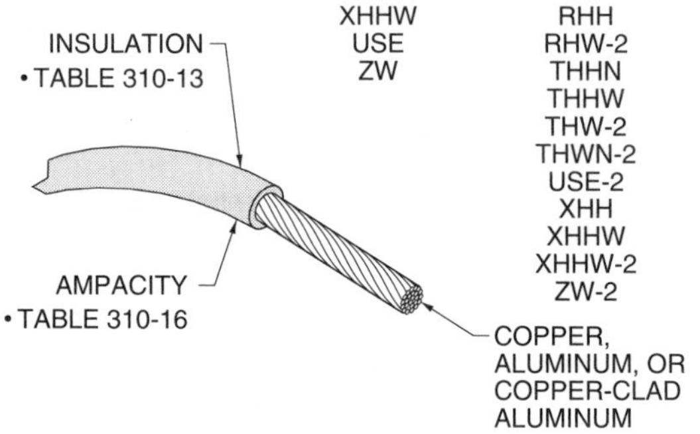

INSULATION
• TABLE 310-13
AMPACITY
• TABLE 310-16
COPPER, ALUMINUM, OR COPPER-CLAD ALUMINUM

Figure 5-12. Electrical insulations are classified according to their temperature rating.

Ambient Temperature – Table 310-16

Table 310-16 ampacities are based on an assumed ambient temperature of 30°C (86°F). *Ambient temperature* is the temperature of air around a piece of equipment. The ambient temperature is the maximum temperature that can be found anywhere along the conductor length. It directly affects the conductor's ability to dissipate heat. If the ambient temperature is too high, the heat dissipation rate decreases, resulting in an increased conductor and insulation temperature.

Generally, the higher the ambient temperature, the more difficult it is for a conductor to dissipate heat. When the ambient temperature is above 30°C, the ampacity of the conductor shall be adjusted to compensate for the increased ambient temperature. The ambient temperature correction factors are included in the ampacity tables. The correction factors depend upon the type of insulation on the conductor, the type of conductor material (copper, aluminum, or copper-clad aluminum), and the ambient temperature. Correction factors are cross-listed for both Celsius and Fahrenheit to make the application easier. There may be instances in which the ambient temperature remains below the table value of 30°C. In these cases, an increased ampacity is permitted by the correction factors.

MULTICONDUCTOR CABLES

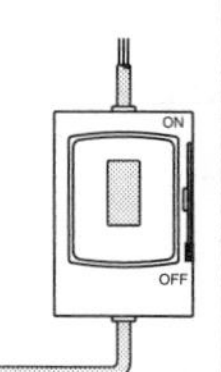

When using multiconductor cables, such as AC (BX), NM (Romex), MC, etc., the derating provisions of Note 8 of Notes to Ampacity Tables of 0 to 2000 Volts apply. Many electrical installations use these types of cables for branch-circuit wiring for both lighting and receptacle circuits. The most common violation of this rule occurs where all of the cable assemblies come together, such as at a panelboard.

Cables entering an electrical room shall have adequate spacing to allow for proper heat dissipation or the conductor's allowable ampacity shall be calculated per provisions of Note 8 of Notes to Ampacity Tables of 0 to 2000 Volts.

Number of Conductors. The values in Table 310-16 are based upon the assumption that no more than three current-carrying conductors are installed in the same raceway, cable, or trench. For installations that require more than three current-carrying conductors, the table values shall be adjusted. See Notes to Ampacity Tables of 0 to 2000 Volts.

Note 10 of Notes to Ampacity Tables of 0 to 2000 Volts covers the conditions under which the neutral conductor shall be counted as a current-carrying conductor. In general, if the neutral carries only unbalanced current from other conductors of the same circuit, it need not be considered a current-carrying conductor. If a major portion of a 3ϕ, 4-wire, wye circuit consists of nonlinear loads such as those associated with electric-discharge lighting and data processing equipment, then the neutral shall be considered a current-carrying conductor. In addition, 3-wire circuits derived from a 3ϕ wye system have neutral currents equivalent to those in the phase conductors and are counted as current-carrying conductors.

Note 8 of Notes to Ampacity Tables of 0 to 2000 Volts lists the required derating percentages for installation which have more than three current-carrying conductors. These percentages reflect the percent of the Table 310-16 values that are permitted to be used in determining the allowable conductor ampacity. Note 8 of Notes to Ampacity Tables of 0 to 2000 Volts also requires that where single conductors or multiconductor cables are bundled together for more than 24″, the conductors shall be derated.

Termination Rating. The temperature rating of the equipment terminations also affects a conductor's allowable ampacity. The ampacity of the conductors shall be coordinated so as not to exceed the lowest temperature rating of any component in the electrical circuit, including the electrical equipment per 110-14(c). Although conductors may be selected that are suitable for 90°C conditions, the termination will probably have a lower temperature rating. In these cases, the ampacity of the conductor shall be selected so that it does not exceed the lowest rating of the equipment in the circuit.

Most electrical equipment, not unlike conductors, is rated at 60°C, 75°C, or 90°C. Some equipment contains dual ratings such as 60°C/75°C. Such equipment is permitted to use the 75°C rating provided all other components of the circuit are suitable for 75°C. When the conductor insulation rating exceeds the equipment termination temperature rating, the ampacity of the conductor shall be based on the equipment rating. At the present time, there is very little equipment available that is suitable for use with 90°C conductors. Conductors rated at 90°C are permitted to be installed with the lower-rated equipment but their ampacities shall be based on the temperature rating of the equipment.

Location. The final factor in determining a conductor's allowable ampacity is the type of location in which the conductor is to be installed. Locations can be classified as either dry, damp, or wet. Conductors installed in dry and damp locations may be subjected to only a moderate degree of moisture and thus no adjustment of the conductor's ampacity is necessary. Conductors installed in wet locations however, may be unprotected from the weather and subject to continuous wet conditions.

Depending on the insulation type, an adjustment to the ampacity may be required. For example, if THHW is installed in a wet location, the ampacity shall be selected from the 75°C column even though it has 90°C insulation. This is because the wet location reduces the conductor's ability to dissipate heat. The NEC® has recently recognized new insulations that are capable of maintaining their higher ampacities in both wet or dry locations. Table 310-13 identifies these insulations with the suffix "-2." For example, RHW-2 and XHHW-2 are capable of maintaining their higher ampacities in both wet or dry locations.

ALLOWABLE AMPACITY

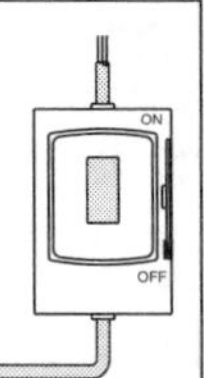

The equipment temperature rating shall be considered when determining a conductor's allowable ampacity. Electrical designers and installers may question the benefit of using 90°C conductors such as THHN if there are no equipment terminations that can take advantage of the higher rating of the conductor.

The advantage of using 90°C conductors is that although the ampacity listed in the 90°C column cannot be used, a 90°C conductor that requires derating for either high ambient temperature, more than three conductors, or both, can be derated starting with the 90°C ampacity. The selected allowable ampacity shall not exceed the temperature ratings of the other components of the electrical circuit.

OVERCURRENT PROTECTION – 240

The purpose of overcurrent protection is to protect the conductors from dangerous overcurrents that could damage the conductors or the conductor insulation per 240-1, FPN. Electrical equipment is also provided with overcurrent protection to protect it from high current levels that may result as a result of short circuits or ground faults.

Protection of Fuses and CBs – 240

In general, all conductors are protected against overcurrents at the point at which they receive their supply. Overcurrent protection is directly related to the ampacity of the conductors. The two most common devices used to provide overcurrent protection are fuses and circuit breakers (CBs). Ratings for standard fuses and CBs are given in 240-6.

Overcurrents. An *overcurrent* is any current in excess of that for which the conductor or equipment is rated. See 100. In terms of application, some equipment is designed to accommodate overcurrents for a period of time depending upon the load. Temporary surge or start-up overcurrents are common for some electrical equipment like motors and generally have no long-term, harmful effects. When overcurrents occur more frequently or persist for longer periods of time, damage to the electrical system may occur. Overcurrents can result from overloads, short circuits, or ground faults. See Figure 5-13.

Overloads. An *overload* is a small-magnitude overcurrent, that over a period of time, leads to an overcurrent which may operate the overcurrent protection device (fuse or CB). It is the lowest magnitude of overcurrent. The least potentially damaging overcurrents occur when there is an overload.

Overcurrents from overloads most commonly occur as a result of connecting too much load to an electrical system component such as a branch-circuit. Assuming the branch-circuit is installed properly, the result of the overload is an opening in the branch-circuit OCPD. As a rule of thumb, overcurrents caused by overloads generally run in the range of one to six times the normal circuit current, and the current flow is confined to the normal current path. For the purposes of Article 240, the term overload does not include short circuits or ground faults. See 100.

FUSES AND ITCBs

Increase	Standard Ampere Ratings
5	15, 20, 25, 30, 35, 40, 45
10	50, 60, 70, 80, 90, 100, 110
25	125, 150, 175, 200, 225
50	250, 300, 350, 400, 450
100	500, 600, 700, 800
200	1000, 1200
400	1600, 2000
500	2500
1000	3000, 4000, 5000, 6000

1 A, 3 A, 6 A, 10 A, and 601 A are additional standard ratings for fuses.

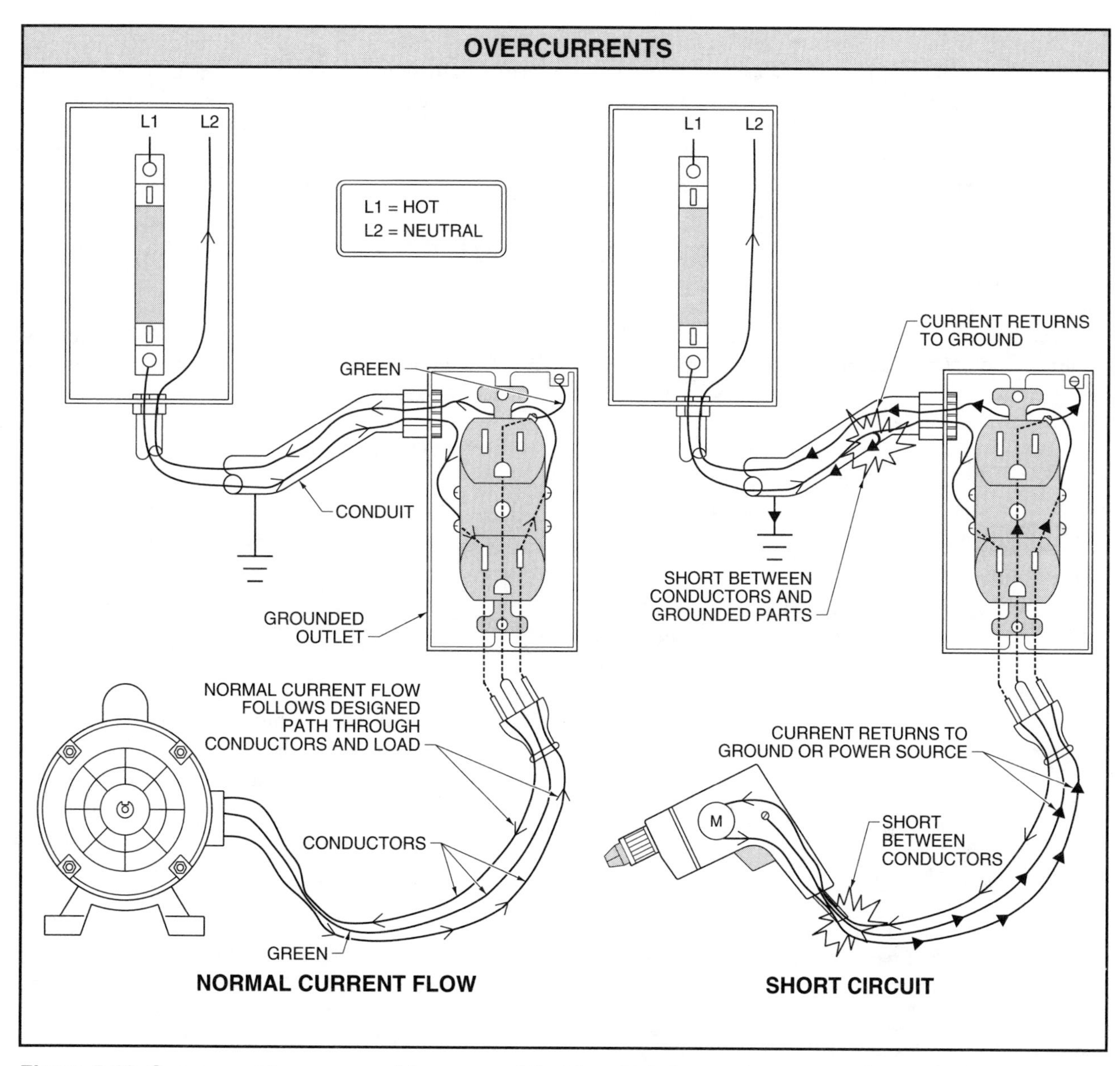

Figure 5-13. Overcurrent is any current in excess of that for which the conductor or equipment is designed.

Short Circuits. A *short circuit* is the condition that occurs when two ungrounded conductors (hot wires), or an ungrounded and a grounded conductor of a 1ϕ circuit, come in contact with each other. The most damaging overcurrents occur as the result of short circuits. For example, if the conductors feeding a 240 V water heater were to come in contact with each other, a short-circuit, as result of this abnormal current path, would result. Short circuits are the largest magnitude overcurrent. If the components of the electrical system are not properly protected, they may be severely damaged by the short circuit.

The amount of short-circuit current that flows depends upon the following factors.

- Amperes available at the source of power. Typically, the source of power is a transformer and the most significant factor is the impedance of the transformer. The higher the impedance of the transformer, the lower the available short-circuit current. Section 450-11 requires all transformers which are 25 kVA and larger to be marked with their impedance ratings.
- Length of the circuit. In general, the farther away from the source, the less available short-circuit current. As the circuit length increases, the distance from the power source also increases. Available short-circuit current declines as the distance from the power supply increases.

- Size of conductors and voltage at which they operate. Usually, the larger the conductor size and the higher the voltage, the higher the available short-circuit current. Larger conductor sizes offer a lower impedance and higher circuit voltages overcome circuit impedances allowing larger short-circuit currents to pass.

Other factors may also be considered in calculating available short-circuit current. These include: impedance of the fault circuit, impedance of the arc, and the type of equipment supplied by the circuit. For example, motors can contribute to available short-circuit current levels and may need to be considered if they are a significant component of the electrical system. Installers of electrical services shall be aware of the available short-circuit current and ensure that the service equipment is suitable for the available short-circuit current per 230-65.

Ground Faults. A *ground fault* is an unintentional connection between an ungrounded conductor and any grounded raceway, box, enclosure, fitting, etc. Like short circuits, ground-fault currents follow an abnormal path. In this case however, the path of current is from an ungrounded conductor to ground. Ground faults can occur due to insulation failures or more commonly at conductor terminations.

Ground-fault currents are generally of a lower magnitude than short circuits. As a rule of thumb, ground-faults generally result in current levels of about 75% of short-circuit currents. Care shall be taken when installing electrical conductors to ensure that insulation is not damaged, as this could contribute to a ground fault.

When No. 4 and larger ungrounded conductors are installed in raceways that enter boxes, enclosures, cabinets, etc., the conductors shall be protected by the use of an insulating fitting per 300-4(f). As with short-circuit currents, the available ground-fault current depends upon the available source current, the impedance of the circuit, the distance from the source that the fault occurs, and the conductor type, size, and voltage rating.

Protection of Equipment – 240-2

High-level fault currents and persistent overloads can cause problems for electrical equipment as well as for electrical conductors. All electrical equipment shall be protected from these potentially damaging currents. The NEC® requires that the type and form of the overprotection provided meet the requirements of the particular article covering the equipment. For example, the provisions for overcurrent protection for appliances are discussed in 422. Overcurrent protection for motors is discussed in 430. See 240-2.

Protection of Conductors – 240-3

Overcurrent protection of conductors ensures that conductors are protected in accordance with their allowable ampacity. This protection is required for all conductors except flexible cords and fixture wires per 240-3. Section 310-15 requires that conductor ampacities be obtained from the Tables 310-16 through 310-19 or under engineering supervision by calculation. The process of selecting overcurrent protection requires that first the conductor's allowable ampacity be calculated and then the overcurrent protection be selected. See 240-3(a–m) for alternate methods to the general rule of protecting conductors in accordance to their allowable ampacity.

Hazardous Shutdown – 240-3(a). There may be instances in which operation of the overcurrent protection device results in a hazard. Section 240-3(a) permits installations of this nature to omit overload protection, but still requires short-circuit protection. For example, while overload protection is important for most motors, fire pumps by their design continue to operate for as long as possible and are not required to have overload protection. Another example is of a material-handling magnet circuit in which a loss of power could create a hazard by dropping the load held by the magnet.

Devices of 800 A or Less – 240-3(b). Overcurrent protection can be selected once the allowable ampacity is determined. Often, the calculated ampacity does not correspond to a standard size fuse or circuit breaker. The next higher standard size device rating may be used in these cases per 240-3(b) provided the conductors are not part of a multioutlet branch circuit for cord- and plug-connected equipment, and the next higher standard device rating does not exceed 800 A. In cases in which the cord- and plug-connected loads are supplied by receptacles, and the ampacity of the conductor does not correspond to a standard rating for a fuse or CB, the next lower standard size device rating shall be selected. See Figure 5-14.

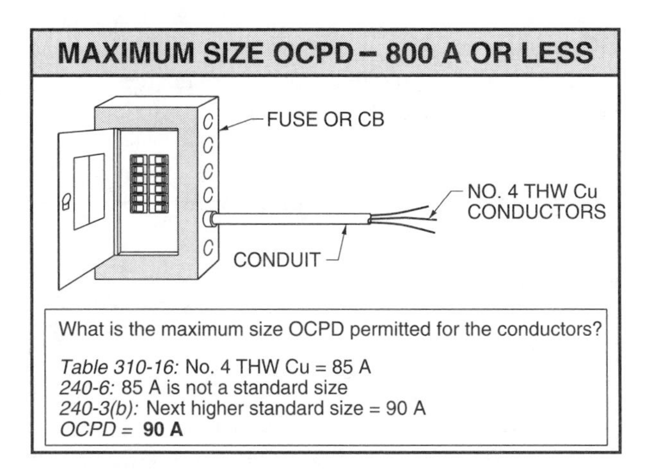

Figure 5-14. OCPDs of 800 A or less which do not correspond to a standard size device are rounded up.

Devices Over 800 A – 240-3(c). In installations where the calculated conductor ampacity does not correspond to a standard size rating for fuses or CBs, and the rating of the overcurrent device is over 800 A, the next lower standard size fuse or CB shall be selected per 240-3(c). See Figure 5-15.

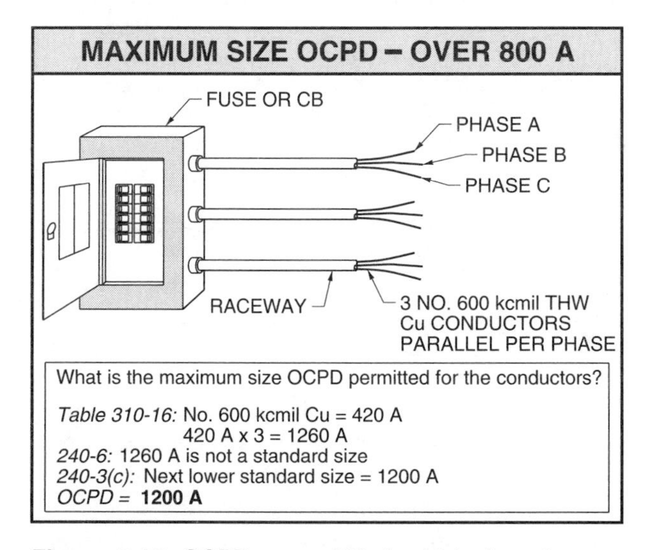

Figure 5-15. OCPDs over 800 A which do not correspond to a standard size device are rounded down.

Conductor Taps – 240-3(d). Smaller conductors are tapped from larger feeder or branch circuit conductors in many installations. In these cases, the conductors are not protected in accordance with their ampacities because the overcurrent protection device is based on the larger conductor. Section 240-3(d) permits taps provided they are in accordance with 210-19(c), 240-21, 364-11, 364-12, and 430-53(d).

Protection of Cords and Equipment – 240-4

Like general building conductors, flexible cords and fixture wires shall be protected against overcurrent in accordance with their ampacities. In the case of these conductors however, ampacities are selected from Tables 400-5(a) and (b) for flexible cords and Table 402-5 for fixture wires. Table 400-4 lists all of the flexible cords and cables which are permitted to be used without being subject to special investigation.

Flexible cords and cables have very specific uses and are not intended to be used as a general wiring method or a replacement for permanent wiring. Only the types of fixture wires listed in Table 402-3 are permitted to be installed. Fixture wires are limited in their use to installation in lighting fixtures where they are suitably protected and for connection to branch circuit conductors serving the fixtures. See 402-10 and 402-11.

Ampere Ratings – 240-6

The standard ratings for overcurrent protection devices are given in 240-6. Part (a) covers the standard ratings for fuses and inverse-time circuit breakers. Part (b) covers the newer adjustable trip circuit breakers.

Fuses. The standard ratings for 15 A to 50 A fuses increase in increments of 5 (15, 20, 25, 30, 35, 40, 45, and 50). For 50 A to 110 A fuses, the ratings increase in increments of 10 (50, 60, 70, 80, 90, 100, and 110). From 100 A to 250 A fuses, the ratings increase in increments of 25 (100, 125, 150, 175, 200, 225, and 250). For 300 A to 500 A fuses, the ratings increase in increments of 50 (300, 350, 400, 450, and 500). Beyond 500 A, the ratings are not as uniform. The maximum standard rating for a fuse is 6000 A. Additional standard ratings for fuses are 1 A, 3 A, 6 A, 10 A, and 601 A per 240-6(a), Ex.

Inverse-Time CBs. The standard ratings for ITCBs are the same as those for fuses with the exception of 1 A, 3 A, 6 A, 10 A, and 601 A which apply only to fuses. See 240-6, Ex. ITCBs are the most commonly used circuit breakers in the electrical industry.

Adjustable Trip CBs. Adjustable trip CBs incorporate an adjustable setting for the long-time pickup or trip setting. In general, the rating of these devices shall be determined from the maximum setting available in the adjustable trip CB per 240-6(b). For example, if the setting is adjustable over a range of 600 A to 800 A, the rating is based on 800 A. The conductors are then selected on that basis. However, 240-6(b), Ex. permits the actual setting, not the maximum setting, to be used provided the CB adjusting means is located behind removable and sealable covers, or is located behind bolted equipment enclosures, or is accessible only to qualified persons. In this example, the actual 600 A setting is permitted to be used to select the conductor size and not the maximum setting if the conditions of the exception are met.

LOCATION OF OVERCURRENT PROTECTION DEVICES – 240-20

The general rule for the location of overcurrent protection devices is to place the OCPD at the point at which the conductor receives its supply. If the OCPD is properly sized in accordance with the conductor's allowable ampacity, and the device is placed at the point where the conductor receives its supply, then the installation should protect both the conductor and the conductor insulation from potentially dangerous overcurrents. In general, overcurrent protection devices are placed in series with the ungrounded conductor and are not used with the grounded conductor.

Ungrounded Conductor – 240-20(a)(b)

Ungrounded conductors shall be installed with an overcurrent protection device installed in series. The overcurrent device may be a fuse, CB, or a combination of an overcurrent relay and current transformer. When CBs are used as the means of overcurrent protection, they shall open all ungrounded conductors of the circuit. Individual single-pole circuit breakers (SPCBs) are permitted for multiwire branch circuits per 210-4(b). Individual CBs with approved handle ties are permitted in grounded systems for line-to-line loads for 1ϕ, 3-wire DC circuits and in 4-wire, 3ϕ systems or 5-wire, 2ϕ systems with a grounded circuit conductor not exceeding the voltage limitations in 210-6.

EXCEPTIONS

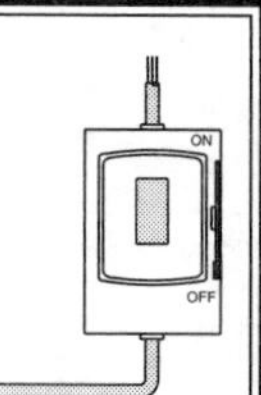

Prior to the 1993 NEC®, 240-21 consisted of eleven exceptions to the general rule requiring overcurrent protection at the point where the conductor receives its supply. Many designers and installers found this important section of the NEC® difficult to apply properly in installing electrical systems. Additionally, many electrical inspectors had a difficult time interrupting the provisions of this section with all of the exceptions.

Remember that an exception is merely an alternate method to meeting some objective. Installations done in accordance with all of the provisions of an exception should result in as good an installation as that of the general rule. Nevertheless, some inspectors were mistakenly not permitting exceptions, fearing the installation would not be adequate. As a result, Code-Making Panel 4, in the 1993 NEC® cycle, revised 240-21 to change the eleven exceptions into general rules with several subsections as they now appear.

Grounded Conductor – 240-22

In the majority of electrical systems, the grounded conductor is the neutral for the system and carries the return circuit current. For this reason, grounded conductors, in general, are not provided with overcurrent protection. There are however, two exceptions to this general rule. The first is for installations in which all of the circuit conductors are opened or disconnected from the supply simultaneously, and the design is such that no pole can operate independently of another. The second is for motor overload protection as provided in 430-36 and 430-37. Some electrical systems, like a 3ϕ, 3-wire, corner-grounded delta, operate with a grounded conductor that is a normal current-carrying conductor. See Figure 5-16.

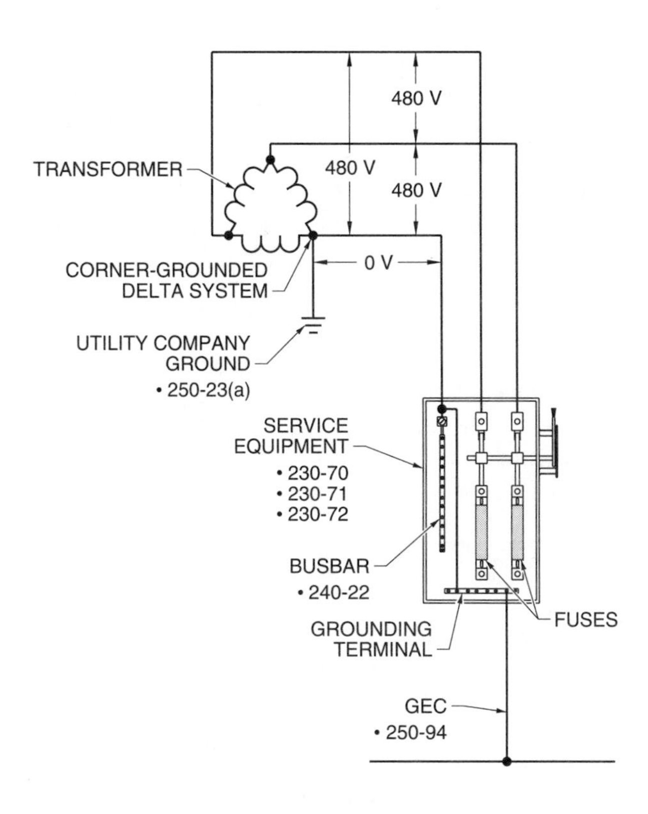

Figure 5-16. All grounded conductors are not neutrals.

Circuit Location – 240-21

Overcurrent protection shall be provided for feeder and branch circuit conductors at the point at which they receive their supply per 240-21(a). Subsections 240-21(b–n) contain provisions which allow conductors to be protected downstream from where they actually receive their supply. These subsections are collectively referred to as "Tap Rules." These rules, which are essentially exceptions to the general rule for overcurrent protection location, list specific sets of conditions that shall be met in order to make taps and protect the tap conductors at a point other than where it receives its supply.

10′ Tap Rule – 240-21(b). The most commonly used tap rule in electrical distribution systems is the 10′ tap rule. Section 240-21(b) sets five conditions which shall be met to permit conductors to be run without overcurrent protection. See Figure 5-17.

The first condition, 240-21(b)(1), requires that the total length of the tap does not exceed 10′. Many of the tap rules contain length requirements that determine how much flexibility the designer or installer has in deviating from the general rule. The second condition, 240-21(b)(2), requires that the ampacity of the tap conductors be at least equal to the computed load(s) on the circuits supplied and not less than the rating of the overcurrent device or the device supplied by the tap conductors. The third condition, 240-21(b)(3), states that the tap conductors cannot continue beyond the device, such as panelboard, switchboard, disconnect, etc., they supply. In other words, it is not permissible to tap a tapped conductor. The fourth condition, 240-21(b)(4), requires that the tap conductors be physically protected, except at the point in which they terminate, by being installed in a raceway. The fifth condition, 240-21(b)(5), requires that the line side OCPD shall not exceed 1000% of the tap conductor's ampacity. This provision does not apply to conductors, no part of which leave the enclosure or vault where the tap is made.

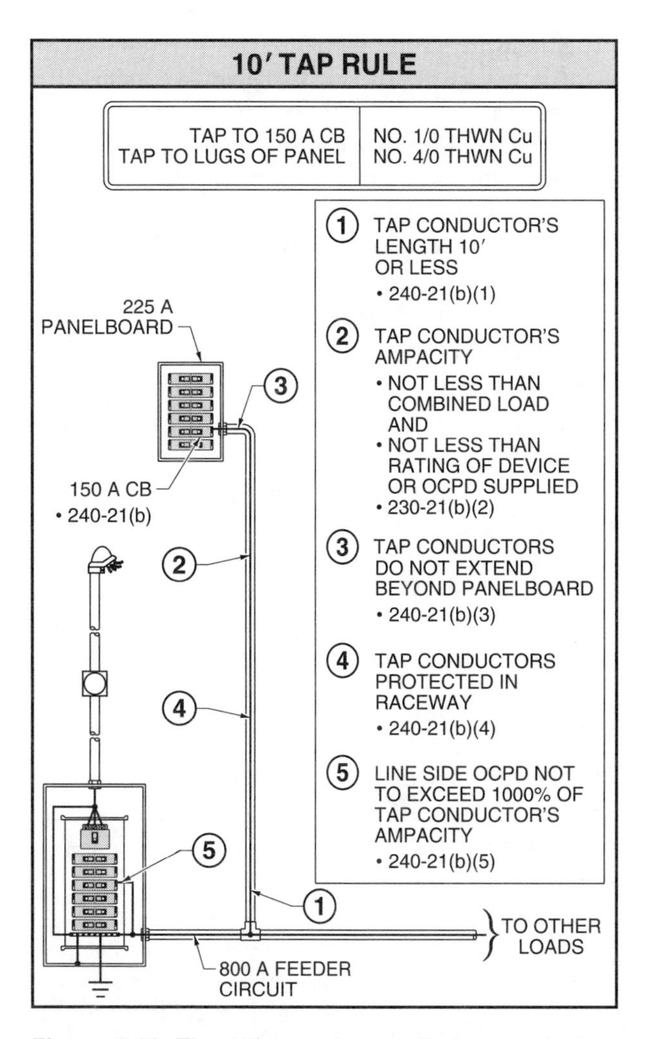

Figure 5-17. The 10′ tap rule permits tap conductors to be supplied from a tap conductor, without overcurrent protection at the point where they receive their supply, provided five conditions are met.

TAP RULES

240-21

- **(b)** Feeder Taps Not Over 10′ Long
- **(c)** Feeder Taps Not Over 25′ Long
- **(d)** Feeder Taps Supplying a Transformer (Primary Plus Secondary Not Over 25′ Long)
- **(e)** Feeder Taps Over 25′ Long
- **(f)** Branch-Circuit Taps
- **(g)** Busway Taps
- **(h)** Motor Circuit Taps
- **(i)** Conductors from Generator Terminals
- **(j)** Transformer Secondary Conductors of Separately Derived Systems for Industrial Installations
- **(m)** Outside Feeder Taps
- **(n)** Service Conductors

25′ Tap Rule – 240-21(c). The 25′ tap rule is also commonly used in electrical distribution systems. The major advantage of the 25′ tap rule over the 10′ tap rule is the additional length that the conductors are permitted to run without overcurrent protection. To compensate for the additional length, the tap conductor's ampacity is more closely matched to the rating of the tapped conductor circuit. There are four conditions that shall be met to use the 25′ tap rule. See Figure 5-18.

The first condition, 240-21(c)(1), limits the total length to 25′. The second condition, 240-21(c)(2), requires that the ampacity of the tap conductor be at least ⅓ the rating of the overcurrent device protecting the tapped conductors. The third condition, 240-21(c)(3), requires the tap conductors to terminate in a single CB or set of fuses that limits the connected load to the ampacity of the tap conductors. The fourth condition, 240-21(c)(4), requires that the tap conductors be installed in a raceway or otherwise protected from physical damage.

25′ Transformer Feeder Tap Rule – 240-21(d). The 25′ transformer feeder tap rule is for feeder taps supplying a transformer in cases in which the total length of the primary plus secondary does not exceed 25′. This tap rule permits the conductors supplying a transformer to be installed without overcurrent protection at the point in which they receive their supply provided five conditions are met. See Figure 5-19.

The first condition, 240-21(d)(1), requires that the ampacity of the primary supply conductors shall be at least ⅓ of the rating of the overcurrent device protecting the feeder conductors. The second condition, 240-21(d)(2), requires that the ampacity of the conductors supplied by the transformer shall be at least ⅓ of the rating of the overcurrent device protecting the feeders when multiplied by the ratio of the secondary to primary voltage. The third condition, 240-21(d)(3), limits the total length of a primary and a secondary conductor to 25′. The fourth condition, 240-21(d)(4), requires that primary and secondary conductors shall be protected from physical damage. The fifth condition, 240-21(d)(5), requires that the conductors shall terminate in a single CB or set of fuses which limits the load to the conductor's allowable ampacity.

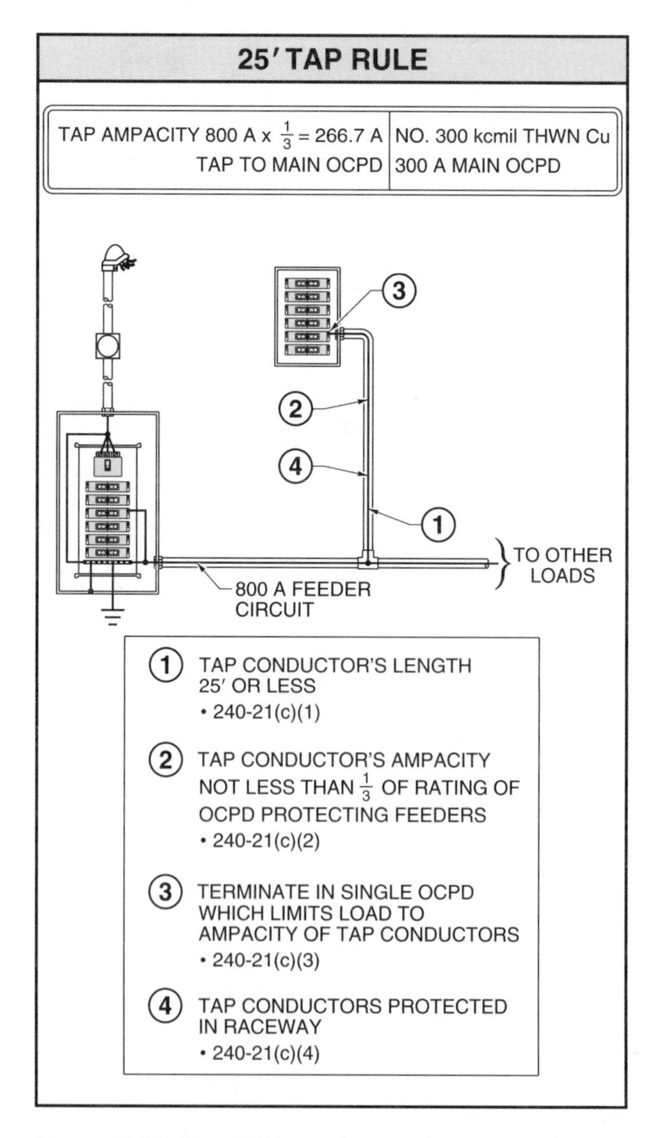

Figure 5-18. The 25′ tap rule permits tap conductors to be supplied from a tap conductor, without overcurrent protection at the point where they receive their supply, provided four conditions are met.

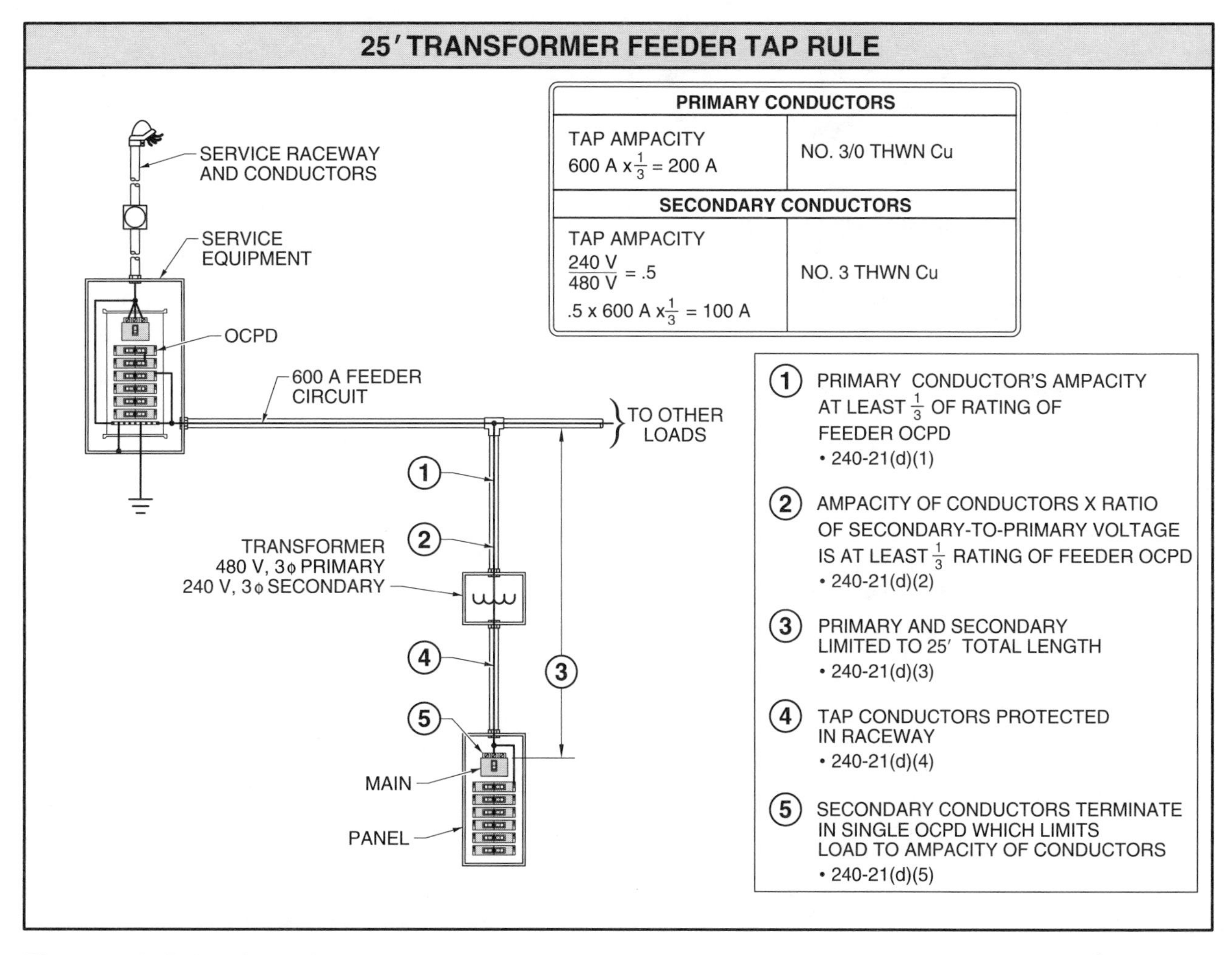

Figure 5-19. The 25′ transformer tap rule permits tap conductors to be supplied from a tap conductor, without overcurrent protection at the point where they receive their supply, provided five conditions are met.

Over 25′ Feeder Tap Rule – 240-21(e). Subpart (e) contains tap rules for a specific set of conditions in a manufacturing building. Overcurrent protection, at the point the conductor receives its supply, can be omitted in installations for high bay manufacturing buildings with walls over 35′ in height. There are seven conditions that shall be met provided the total length of the conductors does not exceed 100′.

Seven conditions shall be met to apply the over 25′ tap rule. See Figure 5-20. The first condition, 240-21(e)(1), requires that the tap conductor's ampacity shall be at least ⅓ the rating of the OCPD protecting the feeder conductors. The second condition, 240-21(e)(2), requires that the tap conductors shall terminate in an OCPD which limits the load to the ampacity of the tap conductors. This ensures that the tap conductors are not overloaded. The third condition, 240-21(e)(3), requires that the tap conductors shall be installed in a raceway or otherwise protected against physical damage. The fourth condition, 240-21(e)(4), requires that the tap conductors cannot contain any splices or taps. It is not permissible to "tap-a-tap." The fifth condition, 240-21(e)(5), requires that the minimum permitted size for the tap conductors shall be No. 6 Cu or No. 4 Al. The sixth condition, 240-21(e)(6), requires that the tap conductors shall not penetrate any floors, ceilings, or walls. The seventh condition, 240-21(e)(7), requires that the location of the tap be made at a point no less than 30′ from the floor. This isolation of the tap enhances the overall safety of the installation.

Outside Feeder Tap Rule – 240-21(m). Outside taps from feeders or a transformer secondary shall be permitted without overcurrent protection at the tap provided five conditions are met. See Figure 5-21. The first condition, 240-21(m)(1), requires that the tapped conductors shall be protected from physical damage by installation in a raceway or other suitable means.

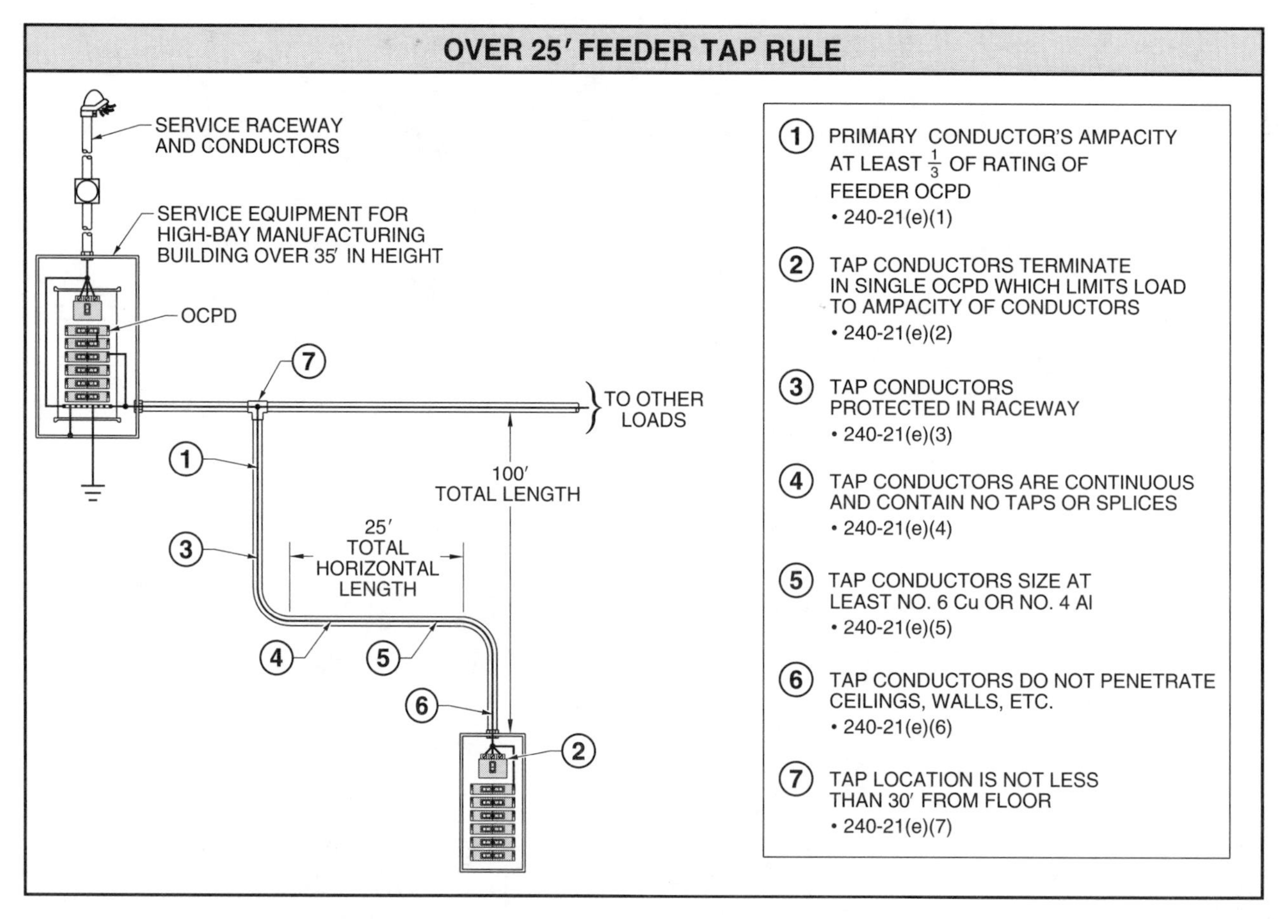

Figure 5-20. The over 25′ tap rule permits tap conductors to be supplied from a tap conductor, without overcurrent protection at the point where they receive their supply, provided seven conditions are met.

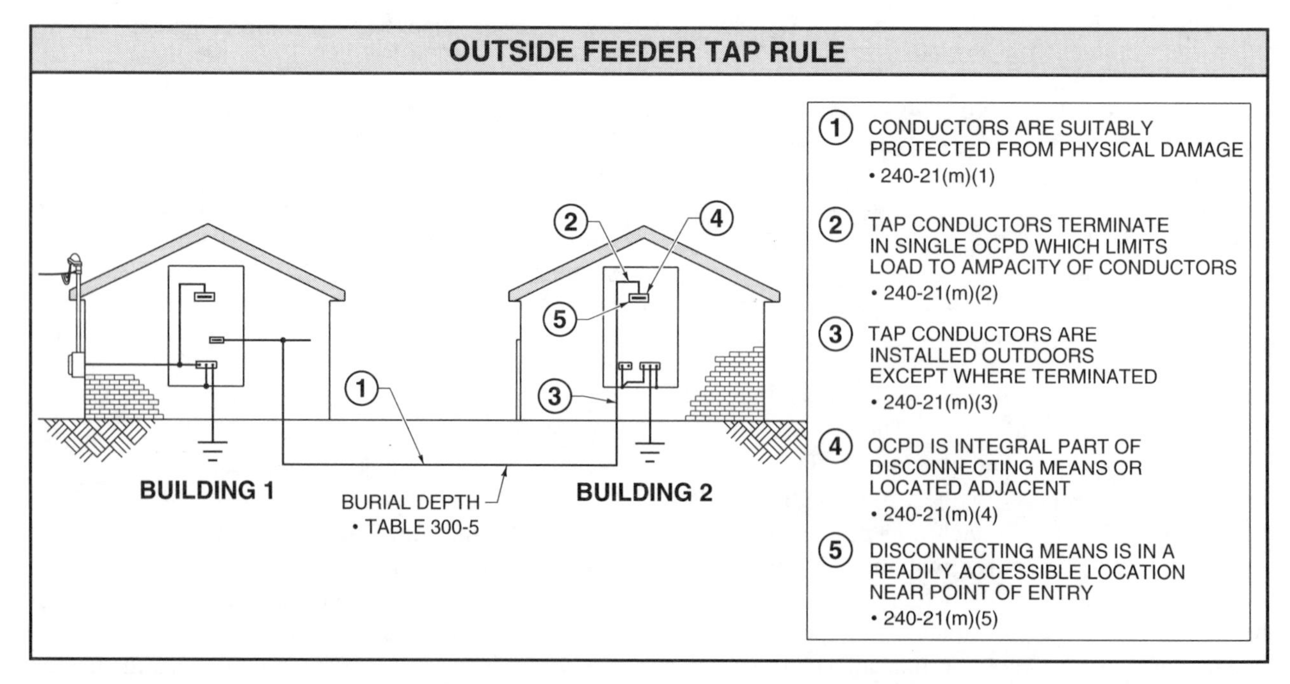

Figure 5-21. The outside feeder tap rule permits tap conductors to be supplied from a tap conductor, without overcurrent protection at the point where they receive their supply, provided five conditions are met.

The second condition, 240-21(m)(2), requires that the tap conductors shall terminate in an OCPD which limits the load to the ampacity of the tap conductors. This ensures that the tap conductors are not overloaded. The third condition, 240-21(m)(3), requires that the tap conductors shall remain outdoors, except at their termination point. The fourth condition, 240-21(m)(4), requires that the disconnecting means for the conductors shall contain the overcurrent protection for the conductors or the OCPD shall be located adjacent to the disconnecting means. The fifth condition, 240-21(m)(5), requires that the location of the disconnecting means shall be in a readily accessible location, either inside or outside the building or structure served by the tap conductors.

Essentially, because the conductors are located outside for their entire length, except where they are terminated, they can be treated like service conductors and no length limitation is imposed. Many of the conditions for outside feeder taps are similar to those imposed upon service conductors in 230. However, like service conductors, these conductors are not suitably protected against short circuits and overloads. The five conditions ensure isolation and protection of the unprotected tap conductors.

Service Conductors – 240-21(n). Subpart (n) permits service conductors to be protected against overcurrents. Section 230-91 requires the service overcurrent protection to be an integral part of the service disconnect or be located adjacent to the service disconnecting means.

Premises Location – 240-24

Overcurrent devices shall be installed in a readily accessible location, free from physical damage, and in a location that ensures that all occupants of a building or structure to have access to the overcurrent devices protecting the supply conductors for their premises. They shall not be near easily ignitible materials or in bathrooms. See Figure 5-22.

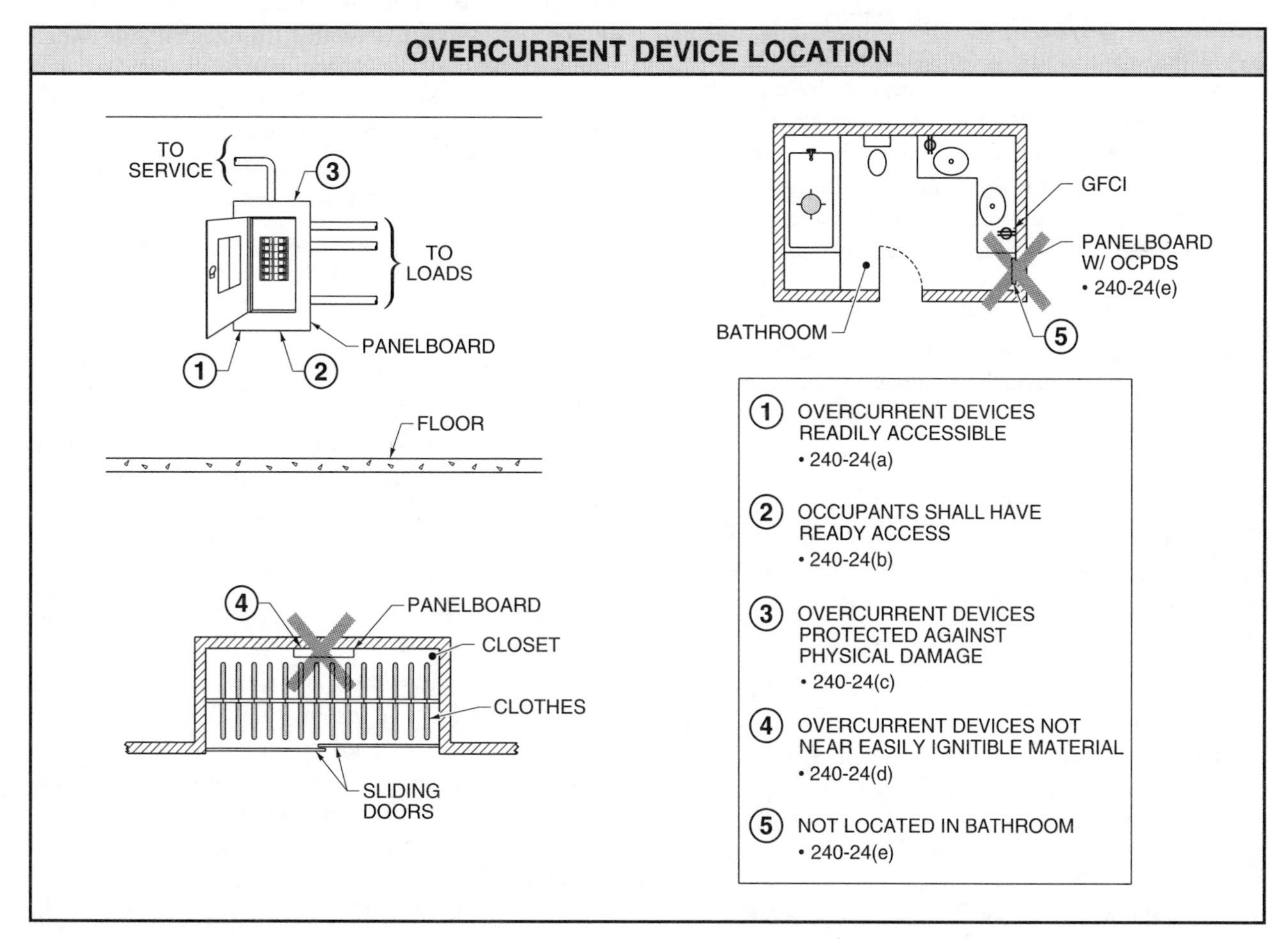

Figure 5-22. Overcurrent devices are not permitted in clothes closets or bathrooms.

Accessibility – 240-24(a). In general, overcurrent protection devices shall be installed in a readily accessible location. Readily accessible indicates that the OCPDs shall be capable of being reached quickly and that there are no obstacles preventing ready access. There are, however, four exceptions to this main rule. Ex. 1 permits the OCPDs to be installed in a location that is not readily accessible if they are for use with busduct per 364-12. Ex. 2 permits supplementary OCPDs to be installed in locations that are not readily accessible. Section 240-10 permits supplementary devices provided they are not used as substitutes for branch-circuit devices. Ex. 3 permits service OCPDs to be installed in locations which are not readily accessible per 225-9(b) and 230-92. Ex. 4 permits OCPDs which are located adjacent to motors and appliances to be installed in locations which are not readily accessible but which are accessible by portable means such as ladders or hook sticks.

Occupant Access – 240-24(b). In multi-occupancy buildings, the OCPDs shall be installed so that each occupant has ready access to the OCPDs which protect conductors supplying the occupants' space. This helps to ensure the safety of occupants by providing them with individual control over the circuits in their space. There are two exceptions to this general rule. Ex. 1 applies to multi-occupancy buildings managed by full-time building supervision. In these types of occupancies, service and feeder OCPDs which supply more than one occupant can be installed where accessible to authorized personnel only. Ex. 2 permits the OCPDs to be installed and accessible to only authorized personnel in occupancies such as motels and hotels and other guest room occupancies.

Physical Damage – 240-24(c). Because OCPDs are a vital link in providing safety to personnel and property, they shall be installed in locations which are not subject to physical damage. This includes locations in which corrosive or other deteriorating agents may be present.

Ignitible Materials – 240-24(d). When OCPDs interrupt the flow of current, an arc can occur depending upon the amount of current flowing in the circuit. Therefore, OCPDs are not permitted to be installed in locations where easily ignitible materials are stored. Examples of such locations are clothes or linen closets.

Bathrooms – 240-24(e). By definition, bathrooms contain a basin and at least a toilet, tub, or shower. Varying degrees of moisture are always present in bathrooms. Therefore, OCPDs are not permitted to be installed in bathrooms in dwelling units or guest rooms of motels or hotels. GFCI receptacles are not OCPDs and are permitted in bathrooms as are supplementary overcurrent devices.

OVERCURRENT DEVICES

Overcurrent devices are designed to open the circuit before damage can occur to either conductors or equipment as a result of abnormally high-current levels. Overcurrent devices shall be capable of protecting the electrical components against overloads, short circuits, and ground faults. Overcurrent devices are placed in series with ungrounded conductors and, by design, they have very little effect on the circuit during normal operation. Overcurrent devices shall have an interrupting rating suitable for the available short-circuit current, and they shall have a voltage and ampere rating suitable for the electrical system in which they are employed. The two most common overcurrent devices used in electrical distribution systems are fuses and circuit breakers.

An overcurrent protection device is used to provide protection from short circuits and overloads. See Figure 5-23. Fuses and circuit breakers (CBs) are overcurrent protection devices designed to automatically stop the flow of current in a circuit that has a short circuit or is overloaded.

A *fuse* is an overcurrent protection device with a fusible link that melts and opens the circuit when an overload condition or short circuit occurs. A *circuit breaker* is an overcurrent protection device with a mechanical mechanism that may manually or automatically open the circuit when an overload condition or short circuit occurs.

Both fuses and circuit breakers are widely used in power distribution systems and individual pieces of equipment. Fuses and circuit breakers have their advantages and disadvantages. The overcurrent protection device selected is normally determined by the application, economics, and individual preference of the person making the choice.

Fuses and circuit breakers include both current and voltage ratings. The listed current rating is the maximum amount of current the overcurrent protection device (OCPD) carries without blowing or tripping the circuit. The current rating of the OCPD is determined by the size and type of conductors, control devices used, and loads connected to the circuit.

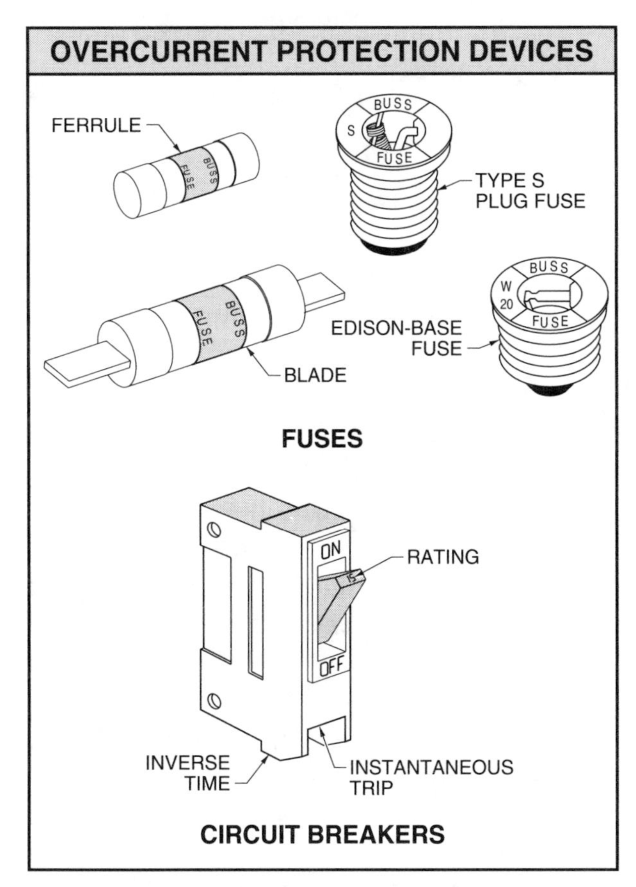

Figure 5-23. Fuses and CBs protect the wiring and components in an electrical circuit from short circuits.

The voltage rating is the maximum amount of voltage applied to the OCPD. It safely suppresses the internal arc produced when the OCPD stops the current flow. The voltage rating of the OCPD is greater than (or equal to) the voltage in the circuit.

Every ungrounded (hot) power line must be protected against short circuits and overloads. A fuse or circuit breaker is installed in every ungrounded power line. One OCPD is required for low voltage, 1ϕ circuits (120 V or less) and all DC circuits. The neutral line (in AC circuits) or the negative line (in DC circuits) does not include an OCPD. Two OCPDs are required for 1ϕ, high-voltage circuits (208 V, 230 V, or 240 V). Both ungrounded power lines include an OCPD. Three OCPDs are required for all 3ϕ circuits (any voltage). All three ungrounded power lines include an OCPD. See Figure 5-24.

Plug Fuses – 240-50

A *plug fuse* is a fuse that uses a metallic strip which melts when a predetermined amount of current flows through it. Plug fuses are inserted in series with ungrounded conductors to protect the conductors against overcurrents which could damage the conductor or the connected equipment. At one time, plug fuses were the most common overcurrent protection device used in residential and light commercial electrical installations.

Plug fuses have a voltage rating of 125 V between conductors in most applications. The maximum voltage rating, however, for circuits supplied from a system utilizing a grounded neutral is 150 V to ground. Thus, for a standard 120/240 V, 1ϕ, 3-wire system, plug fuses can be used for the protection of both 120 V and 240 V circuits.

EDISON-BASE PLUG FUSES

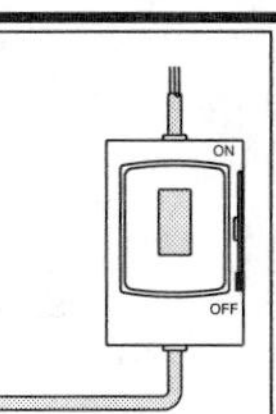

One of the greatest dangers associated with the use of Edison-base plug fuses is that they are interchangeable from one ampere rating to another. The homeowner, often unaware of the potential danger, may replace a fuse that persistently "blows" with another fuse with a greater ampacity. For example, a 15 A circuit installed with No. 14 Cu conductors is overloaded, resulting in frequent blown fuses. The homeowner, in an effort to solve the problem, replaces the properly-sized 15 A fuse with a 30 A fuse. Although the rating is twice that of the 15 A fuse, the 30 A device fits, and the homeowner may mistakenly believe the problem is solved.

Electrical installers, when performing work in occupancies protected by Edison-base plug fuses, should ensure that the homeowner is aware of any overfusing and should encourage the homeowner to consider Type S plug fuses, which do not permit overfusing.

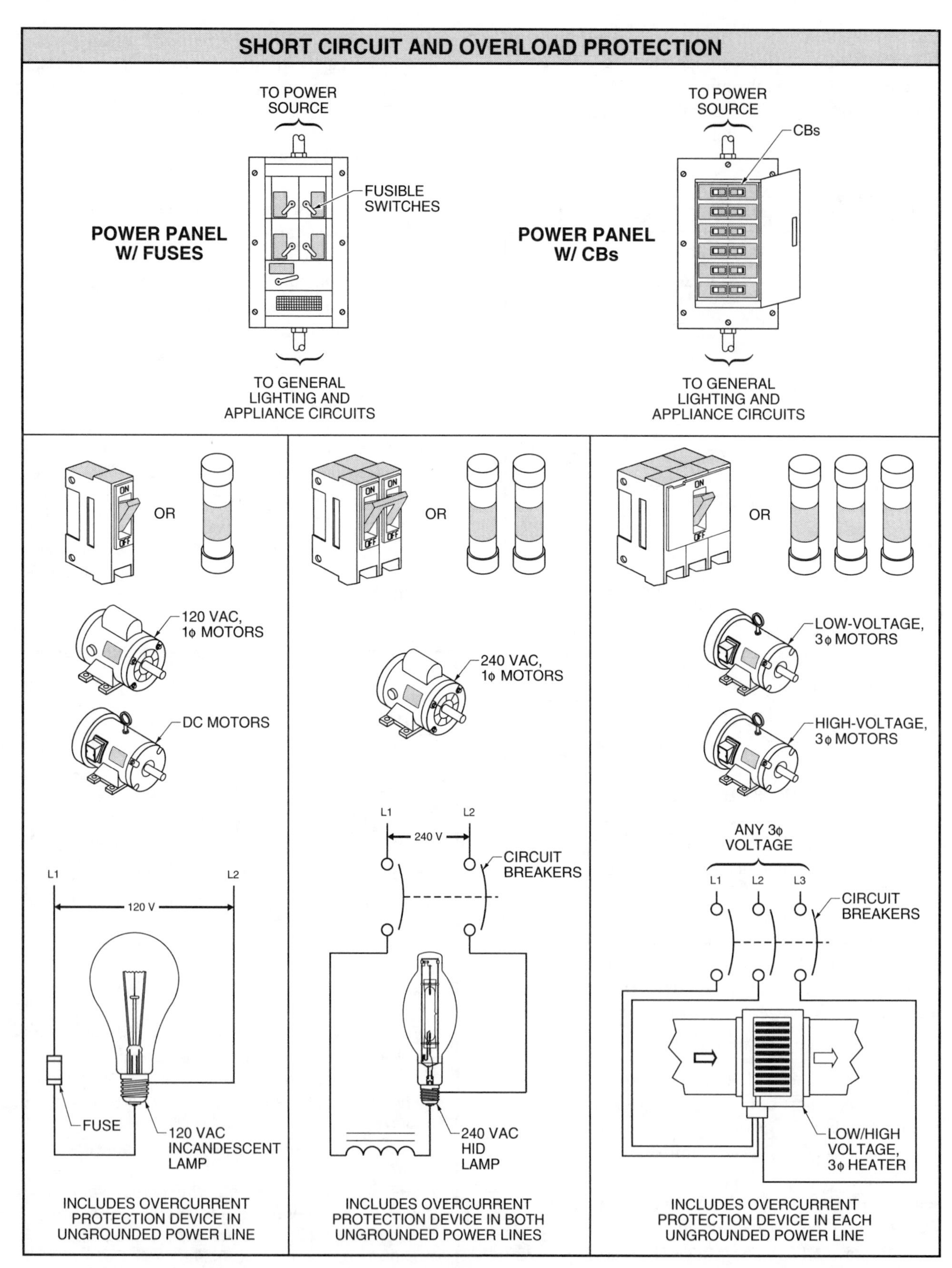

Figure 5-24. All ungrounded (hot) power lines shall be protected against short circuits and overloads.

All plug fuses shall be marked to indicate their ampere rating. By design, the plug fuse ampere rating is identified by the shape of the window on the fuse. Plug fuses with ampere ratings of 15 A or less have a hexagonal window, while those with ratings greater than 15 A have a round window. The two basic types of plug fuses are the Edison-base and the Type S fuse. See Figure 5-25.

Edison-Base Fuses – 240-51. An *Edison-base fuse* is a plug fuse that incorporates a screw configuration which is interchangeable with fuses of other ampere ratings. Edison-base fuses are the older of the plug fuse designs. Edison-base fuses are classified at not over 125 V and are designed to be used in circuits with ampere ratings of 30 A and below. Because the Edison-base fuse is not designed to be noninterchangeable with fuses of other ampere ratings, they are permitted to be installed only as replacements for existing installations where there is no evidence of overfusing.

Type S Fuses – 240-53. A *Type S fuse* is a plug fuse that incorporates a screw and adapter configuration which is not interchangeable with fuses of another ampere rating. Like Edison-base plug fuses, Type S fuses are designed for circuits not exceeding 125 V. Type S fuses have three ampere rating classes: 0 A–15 A, 16 A–20 A, and 21 A–30 A. The fuses are noninterchangeable with ampere ratings of a lower rating, which protects the circuit from the possibility of overfusing. Additionally, Type S fuses are designed with fuseholders and adapters so that the adapters fit Edison-base fuseholders and are nonremovable. Once a Type S adapter is installed, a Type S fuse shall be used.

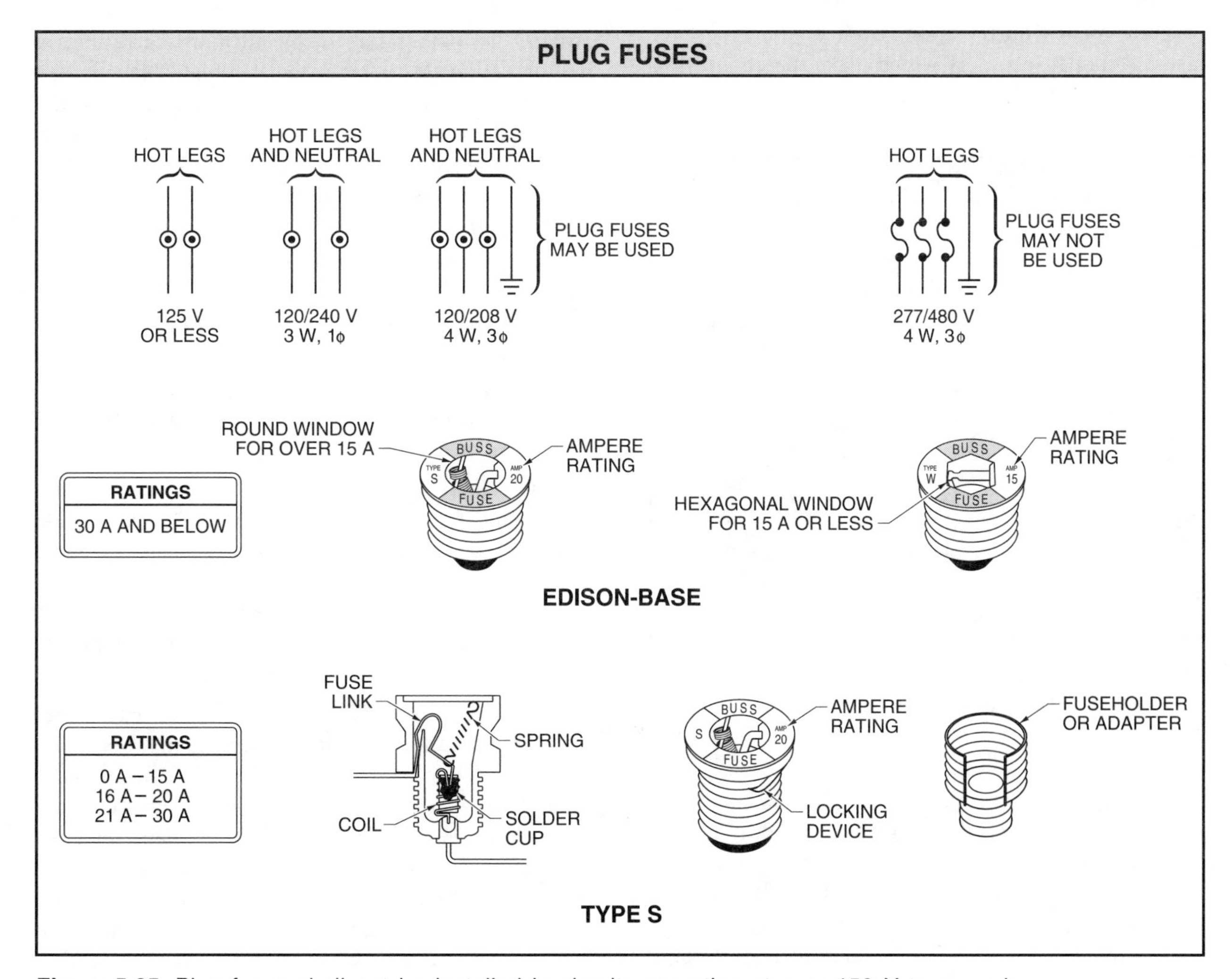

Figure 5-25. Plug fuses shall not be installed in circuits operating at over 150 V to ground.

Cartridge Fuses – 240-60

A *cartridge fuse* is a fuse constructed of a metallic link(s) which is designed to open at predetermined current levels to protect circuit conductors and equipment. Cartridge fuses are classified according to their performance, operation, and construction characteristics. Some of the more common classifications are: RK1, RK5, G, L, T, J, H, and CC.

Cartridge fuses are available in many more types and in more ampere ratings than are plug fuses. Ampere ratings for cartridge fuses are available from 0 A–6000 A. Cartridge fuses are available in either the one-time type or the renewable type. One-time cartridge fuses shall be replaced after they operate, whereas renewable type cartridge fuses contain a fusible link which can be replaced after it operates.

Cartridge fuses can be constructed with a ferrule or knife-blade configuration. Both of the types are inserted into fuseholders which make the connection from the line to the load. See Figure 5-26.

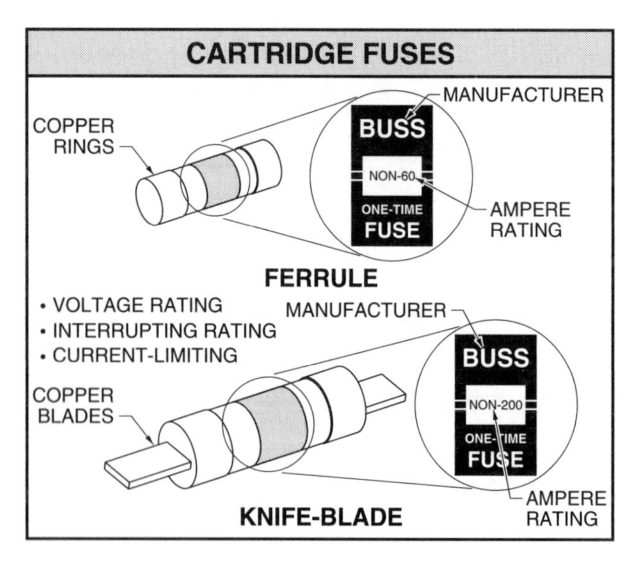

Figure 5-26. Cartridge fuses are designed with either a ferrule or knife-blade configuration.

Cartridge fuses can also be classified as a current-limiting type. Current-limiting fuses are designed to operate very quickly to protect electrical system components. By definition, a current-limiting device shall operate within ½ cycle to prevent high magnitude, short-circuit currents from damaging electrical components. Cartridge fuses can also be classified as to their operational characteristics. Cartridge fuses can be classified as either non-time delay (NTDF) or time delay (TDF).

Non-time delay fuses are fast acting and provide excellent overcurrent protection. Time delay fuses are designed to be used with utilization equipment, like motors, that are subject to temporary start-up or surge currents. Time delay fuses allow overcurrents to exist for short periods of time to avoid nuisance tripping due to equipment start-up characteristics. See Figure 5-27.

Non-time delay fuses (NTDFs) are fuses that may detect an overcurrent and open the circuit almost instantly. They contain a fusible link that melts and opens the circuit at a set temperature. The fusible link is heated by the current. The higher the current, the higher the temperature. Under normal operation and when the circuit is operating at or below its ampere rating, the fusible link simply functions as a conductor. However, if an overcurrent occurs in the circuit, the temperature of the fusible link reaches a melting point at the notched section. The notched section melts and burns back, causing an open in the fuse. If the overcurrent is caused from an overload, only one notched link opens. If the overcurrent is caused from a short circuit, several notched links open. NTDFs are also known as single element fuses.

Low-current rated fuses have only one element, while high-current rated fuses have many elements. Very high-current rated fuses (several hundred amperes) are developed by using many elements connected in parallel inside the same fuse. The heat and pressure created by the arc(s) are safely contained within the fuse.

After removing an overcurrent, the fuse is discarded and replaced. Non-time delay fuses provide a very effective way of removing overloads and short circuits, but may also open on harmless low-level overloads. To prevent this, a time delay may be built into the fuse for use in circuits that have temporary and safe low-level overload currents.

Time delay fuses (TDFs) are fuses that may detect and remove a short circuit almost instantly, but allow small overloads to exist for a short period of time. Time delay fuses include two elements. One element removes overloads, and the other element removes short circuits. The short-circuit element is the same type of fuse link element used in the non-time delay fuse. The overload element is a spring-loaded device that opens the circuit when solder holding the spring in position melts. The solder melts after an overload exists for a short period of time (normally several seconds). The overload element protects against temporary overloads up to approximately 800%. TDFs are also known as dual-element fuses.

TDFs are used in circuits that have temporary, low-level overloads. Motor circuits are the most common type of circuit that includes low-level overloads. Motors draw an overload current when starting. As the motor accelerates, the current is reduced to a normal operating level. Therefore, TDFs are always used in motor applications.

Voltage Ratings – 240-60. All fuses shall have voltage ratings suitable for the electrical circuits in which they are used. Cartridge fuses, in general, shall not be used in circuits exceeding 300 V between conductors. However, 240-60, Ex. permits the maximum voltage to be 300 V line-to-neutral, where the 1ϕ circuit is supplied from a 3ϕ, 4-wire, solidly-grounded neutral system. Additionally, cartridge fuses shall not be used in circuits greater than that for which they are rated. Typical voltage ratings of cartridge fuses are 250 V, 300 V, and 600 V.

Markings – 240-6(c). All cartridge fuses shall be marked, by means of a label or printing on the fuse, to indicate the ampere rating, voltage rating, and the manufacturer's name or trademark. Cartridge fuses shall also be marked to indicate if they are current-limiting.

Current-limiting fuses are fuses that open a circuit in less than ½ of a cycle to protect the circuit components from damaging short-circuit currents. Each cartridge fuse shall also be marked to indicate the interrupting or RMS rating, where other than 10,000 A. The overcurrent device shall have an interrupting rating suitable for the nominal voltage and fault current available at the line side of the terminals per 110-9 and 110-10. Cartridge fuses with no markings are rated for 10,000 A. Other interrupting ratings for cartridge fuses are 100,000 A, 200,000 A, and 300,000 A.

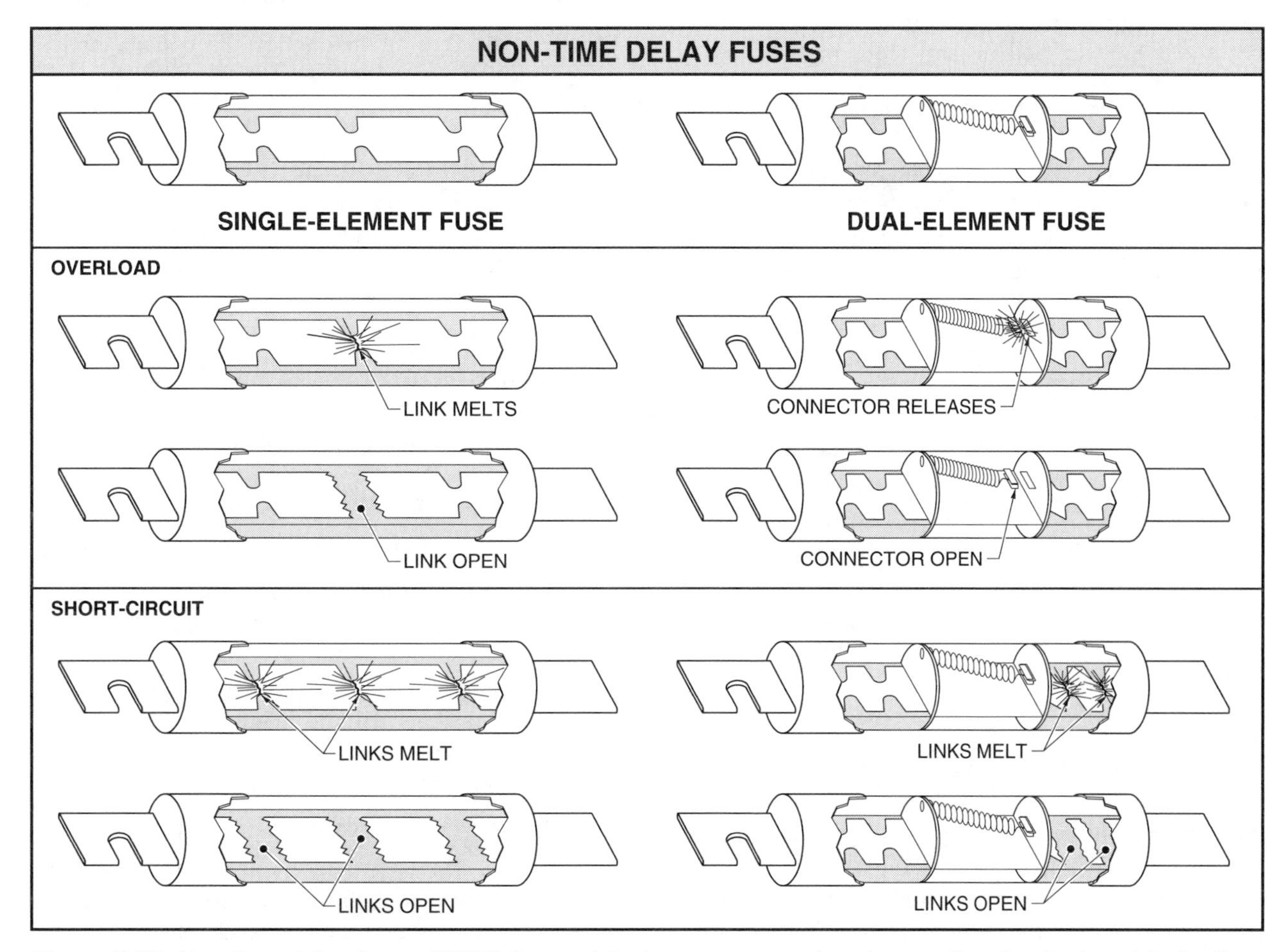

Figure 5-27. Non-time delay fuses (NTDFs) can detect an overcurrent and open the circuit almost instantly.

Circuit Breakers – 240-80

Circuit breakers (CBs) are OCPDs which use a mechanical mechanism to protect a circuit from short circuits and overloads. Like fuses, CBs are connected in series with the circuit's conductors. When the circuit current exceeds the rating of the CB, the CB opens and prevents current from flowing in that part of the circuit.

Circuit breakers are designed to perform the same basic functions as fuses. Circuit breakers offer the added benefit of not damaging themselves and they are resettable after an overload is cleared. CBs that clear short circuits and ground faults can be subjected to extremely high-current levels. Many manufacturers include testing requirements in their instructions for CBs that have cleared short circuits or ground faults. Circuit breakers are designed to be trip free. Essentially, this design permits the internal operation of the CB to occur regardless of whether the handle is locked in the open position or not. CBs are available in four basic types: inverse time, adjustable-trip, nonadjustable-trip, and instantaneous trip.

Operation. Circuit breakers can be designed to be operated either manually, automatically, or both manually and automatically. Although the design characteristics may be different, all CBs contain elements that sense a fault or overload condition and open or interrupt the flow of current. The design elements determine the operational characteristics of the CB.

Thermal CBs contain a spring-loaded electrical contact which opens the circuit. The spring is used to open and close the contacts with a fast snap-action. A handle is added to the contact assembly so the contacts may be manually opened and closed. The contacts are automatically opened on an overcurrent by a bimetal strip and/or an electromagnetic tripping device. The contacts have one stationary contact and one movable contact. The movable contact is attached to a spring that provides a fast snap-action when tripped. See Figure 5-28.

The bimetal strip is made of two dissimilar metals that expand at different rates when heated. The strip bends when heated and opens the contacts. The bimetal strip is connected in series with the circuit and is heated by the current flowing through it. The higher the circuit current, the hotter the bimetal strip becomes. Likewise, the higher the current, the shorter the time required to trip the CB.

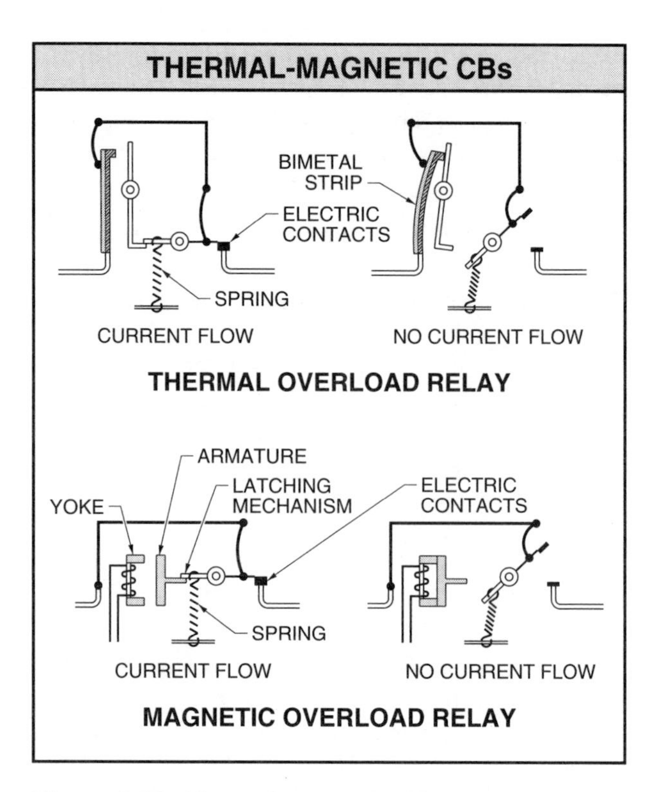

Figure 5-28. Thermal-magnetic CBs contain both a bimetal strip and a magnetic element to sense overloads and faults.

Like the bimetal strip, the electromagnetic device is connected in series with the circuit. As current passes through the coil, a magnetic field is produced. The higher the circuit current, the stronger the magnetic field. When the magnetic field becomes strong enough, the magnetic field opens the contacts.

Whenever the CB contacts are opened, an arc is drawn across the movable and stationary contacts. A short circuit (higher current) produces a much hotter arc than an overload. Therefore, a CB that has tripped numerous times from short circuits may be damaged to the point of requiring replacement.

Magnetic CBs utilize the magnetic effect created when current flows in a conductor. One of the most common CB designs combines a magnetic element with a thermal element. These thermal-magnetic CBs utilize both the thermal and magnetic effects associated with current flow to sense and operate the CB. Generally, the thermal element is used for overload protection while the magnetic element provides faster protection for ground faults and short circuits.

Recent developments in CB design have led to electronically-operated "smart" CBs. Smart CBs incorporate electronic components to accomplish the monitoring, sensing, and tripping mechanisms of the CB. Smart CBs are true rms (root mean square) symmetrical current-sensing devices that utilize a microprocessor to gain closer control over the performance of their operations. Typical settings and adjustments for these CBs include: continuous amperes, long time delay setpoint, short time pickup, short time delay, instantaneous pickup, and ground-fault pickup.

Inverse-Time CBs. *Inverse-time CBs (ITCBs)* are CBs with an intentional delay between the time when the fault or overload is sensed and the time when the CB operates. The relationship between the trip action and the delay is such that the smaller the overload or fault, the longer the delay. Conversely, the larger the overload or fault, the faster the device operates.

Inverse-time CBs are used extensively in electrical distribution systems and for many different types of applications. They are available as single-, two-, and three-pole configurations. ITCBs provide protection from normal overloads and against damaging currents from short circuits and ground faults. This is accomplished in a thermal-magnetic CB by using the thermal element for overload protection and the magnetic element for short-circuit and ground-fault protection. See Figure 5-29.

Adjustable-Trip CBs. *Adjustable-trip CBs (ATCBs)* are CBs whose trip setting can be changed by adjusting the ampere setpoint, trip time characteristics, or both, within a particular range. These CBs are used in industry where utilization equipment may be frequently changed and different overcurrent protection characteristics are required. For example, an ATCB with an 800 A frame size may be installed to protect a feeder for a downstream subpanel. The initial setpoint of the CB is 600 A. If the load on the subpanel increases and the feeder conductors are sized for the additional load, the setpoint of the ATCB can be increased to the new value. This provides great flexibility to the electrical installation at a considerable cost savings.

Nonadjustable-Trip CBs. *Nonadjustable-trip CBs (NATCBs)* are fixed CBs designed without provisions for adjusting either the ampere trip setpoint or the time-trip setpoint. These CBs are used when utilization equipment with known and constant operational characteristics require protection. For example, if a piece of utilization equipment, with a full load ampere rating of 60 A is installed, an NATCB can be installed because the operational characteristics of the equipment are constant, and there is no need for increasing the size or rating of the circuit supplying the equipment.

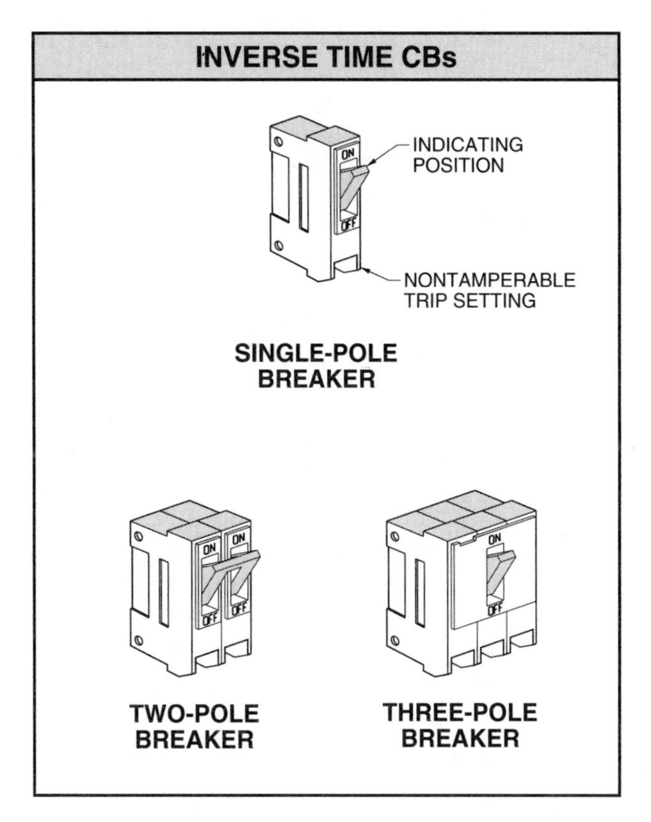

Figure 5-29. Inverse-time CBs are available in single-, two-, and three-pole configurations.

Instantaneous-Trip CBs. *Instantaneous-trip CBs (ITBs)* are CBs with no delay between the fault or overload sensing element and the tripping action of the device. Because a great deal of utilization equipment, such as motors, is subject to surge or start-up currents much higher than normal operating levels, ITBs are sized well above the ampere rating for normal operation. For example, motors with ITBs for the branch-circuit, short-circuit, and ground-fault protection required by 430-52, may have the CB sized at up to 1300% of the FLC when starting current is a problem.

CB Markings – 240-83

All CBs shall clearly indicate whether they are in the open or closed position per 240-81. The open position is OFF and the closed position is ON. For vertically-mounted switches, the ON position of the CB shall be up in order that the CB falls to the open rather than the closed position. CBs shall be marked to indicate the ampere rating, the interrupting rating, and the voltage rating of the device. Additionally, CBs which are permitted to be used as switches, shall be marked to indicate such use. See Figure 5-30.

Ampere Rating – 240-83(a). CBs shall be durably marked to indicate their ampere rating and shall be installed so that the ampere rating is visible. This can be accomplished, for CBs rated at 600 V and 100 A or less, by having the ampere rating marked or printed directly into the handle. For other CBs, it is permissible to make the marking visible by removing a panelboard trim or cover.

Interrupting Rating – 240-83(c). The *interrupting rating* is the maximum amount of current that an OCPD can clear safely. All CBs shall be marked to indicate interrupting rating. All CBs without markings have an interrupting rating of 5000 A. The overcurrent device shall have an interrupting rating suitable for the nominal voltage and fault current available at the line side of the terminals per 110-9 and 110-10.

CBs Used as Switches – 240-83(d). In some electrical installations, it is practical to control the lighting circuits directly from the panelboard or load center. CBs used for switches in 120 V or 227 V fluorescent lighting circuits shall be listed and shall contain the marking "SWD."

Voltage Markings – 240-83(e). A critical requirement for the proper installation of CBs in electrical distribution systems is that the CBs have a voltage rating suitable for the voltages in the system. CBs are generally marked with either a straight or slash voltage rating. CBs with straight voltage ratings are suitable for use only in circuits in which the nominal voltage between two conductors is not greater than the voltage rating of the CB. Slash voltage markings indicate that the CB is suitable for use in electrical systems in which the nominal voltage between two conductors is not greater than the highest value of the voltage slash rating, and the voltage to ground is not greater than the lowest voltage slash rating.

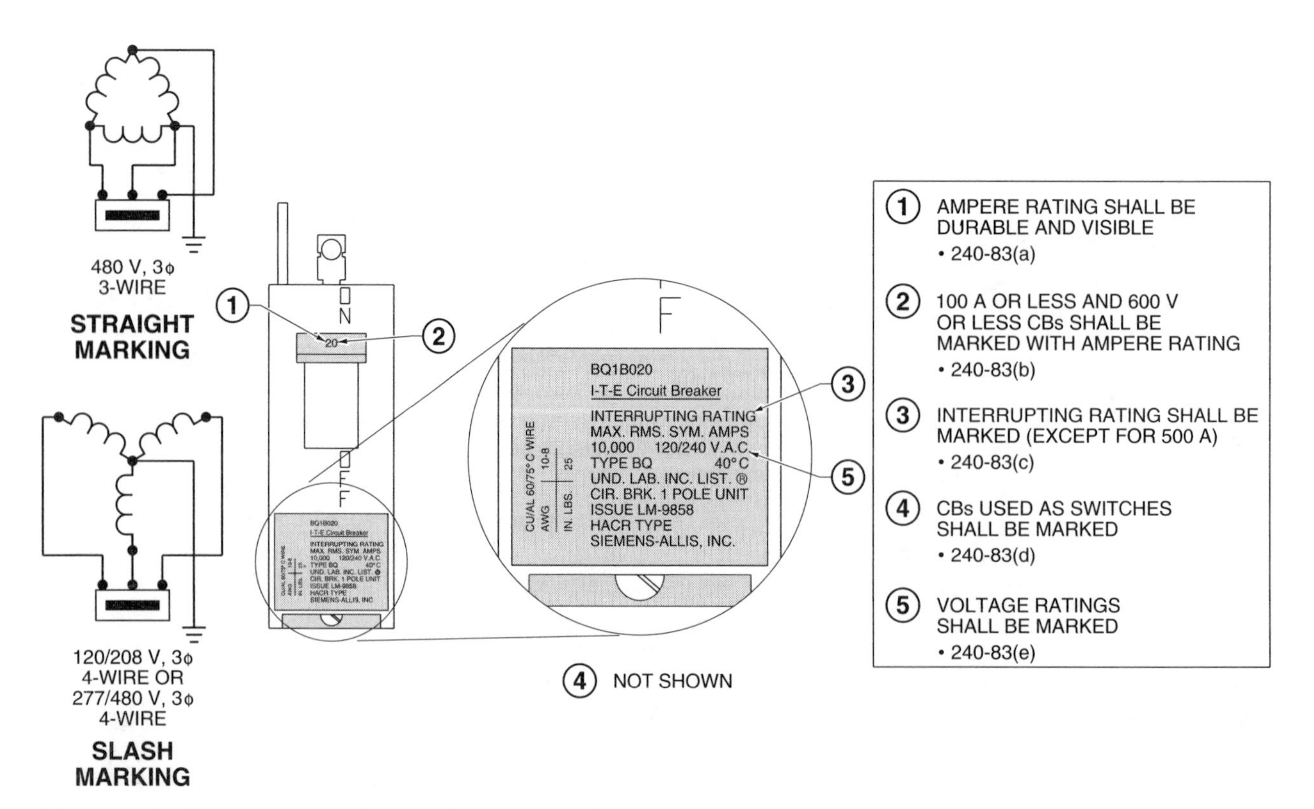

Figure 5-30. CBs are required to be clearly marked with information necessary to ensure their proper application.

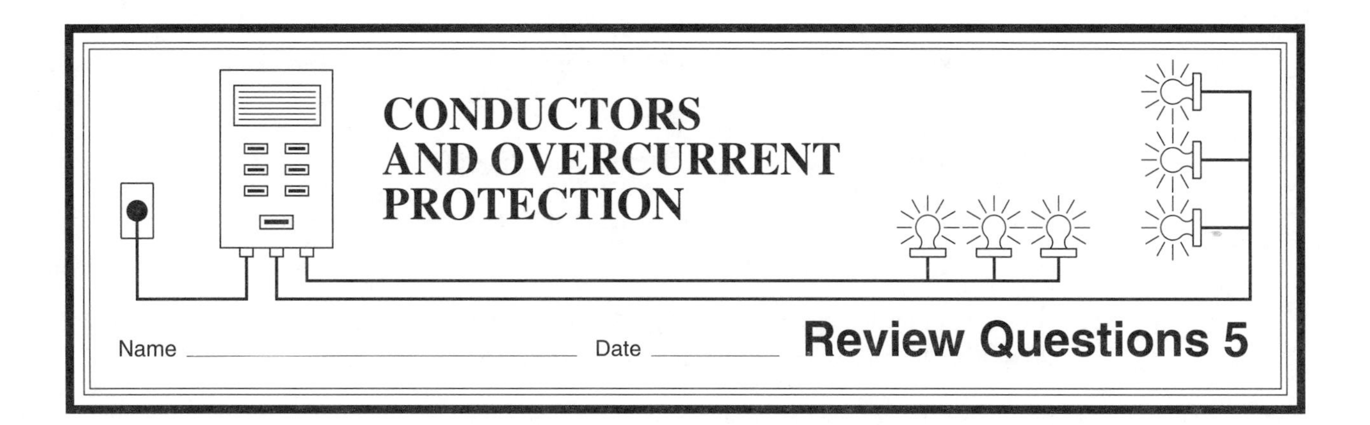

Review Questions

T F **1.** Flexible cords and cables shall be protected against overcurrent in accordance with their ampacities.

__________ **2.** Plug fuses and fuseholders shall not be installed in circuits exceeding ________ V between conductors.

T F **3.** All conductors installed in electrical distribution systems shall be insulated.

T F **4.** THW conductors are permitted to be installed in wet locations.

T F **5.** Conductors used for direct burial shall be identified for such use.

__________ **6.** Conductors used for direct burial shall be ________ for the use.

__________ **7.** Edison-base fuses are for ________ in existing locations only.

T F **8.** When used for equipment grounding purposes, only covered or insulated conductors are permitted.

__________ **9.** No conductor shall be installed in such a manner that the operating temperature of its ________ is exceeded.

__________ **10.** When conductors are run in parallel in separate raceways, the raceways shall have the same ________ characteristics.

__________ **11.** Plug fuses of 15 A and lower are identified by a(n) ________ configuration of the window.

T F **12.** CBs shall clearly indicate whether they are in the ON or OFF position.

T F **13.** Overcurrent protection devices are permitted to be installed in clothes closets if they are located at least 12″ from combustible materials.

T F **14.** CBs used as switches for a 120 V fluorescent lighting circuit shall be listed and marked SWD.

_______ **15.** Fuses rated ________ V or less shall be permitted to be installed in circuits at or below their voltage rating.

_______ **16.** CBs that are not marked have an interrupting rating of ________ A.

_______ **17.** Conductors exposed to corrosive conditions shall have an insulation that is ________ for the application.

_______ **18.** Ampacity of feeder taps not over 25′ shall not be less than ⅓ of the rating of the overcurrent protection device protecting the ________ conductors.

_______ **19.** In general, fuses and CBs shall not be connected in ________.

_______ **20.** Equipment grounding conductors shall be provided with an outer finish that is ________ in color.

_______ **21.** A(n) ________ is a CB with no delay between the fault and the tripping action.

_______ **22.** In general, the grounded conductor (neutral) is colored white or natural ________.

_______ **23.** The three basic classifications of insulation ratings are 60°C (140°F), 75°C (167°F), and ________°C (194°F).

_______ **24.** The next ________ standard size OCPD of 800 A or less is used where the OCPD does not correspond to a standard size.

T F **25.** Overcurrent devices are permitted in clothes closets in bathrooms.

T F **26.** Copper, aluminum, and copper-clad aluminum conductors are all permitted to be installed in parallel.

T F **27.** Paralleled conductors shall all be the same length.

_______ **28.** ________ is generated when current flows through a conductor.

_______ **29.** A(n) ________ is any current in excess of that for which the conductor or equipment is rated.

_______ **30.** No. ________ and larger ungrounded conductors installed in a raceway that enters a box shall be protected by an insulating fitting.

T F **31.** The least potentially damaging overcurrent occur when there is an overload.

T F **32.** The maximum standard rating for a fuse is 5000 A.

T F **33.** Type S fuses are designed for circuits not exceeding 100 V.

T F **34.** Insulated conductors are the most common conductors installed in electrical systems.

_______ **35.** Locations can be classified as either dry, ________, or wet.

Paralleled Conductors

_______ **1.** Same length.

_______ **2.** Same material.

_______ **3.** Same size.

_______ **4.** Same insulation type.

_______ **5.** Same attachments.

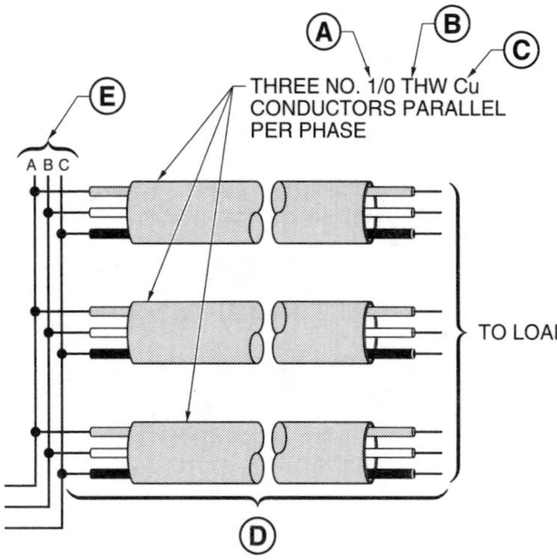

Matching

_______ **1.** Temperature of the air around a piece of equipment.

_______ **2.** Any current in excess of that for which the conductor or equipment is rated.

_______ **3.** A small-magnitude overcurrent which may operate the OCPD.

_______ **4.** Condition that occurs when two ungrounded conductors, or an ungrounded and a grounded conductor of a 1ϕ circuit, come into contact with each other.

_______ **5.** An unintentional connection between an ungrounded conductor and any grounded raceway, box, fitting, etc.

_______ **6.** Any location in which a conductor is subjected to excessive moisture.

_______ **7.** A turning or twisting force, typically measured in foot pounds.

_______ **8.** Two or more conductors that are electrically connected at both ends to form a single conductor.

_______ **9.** The current that a conductor can carry continuously under the conditions of use.

_______ **10.** An OCPD with a link that melts and opens the circuit when an overload or short current occurs.

A. overload

B. ground fault

C. torque

D. wet location

E. ampacity

F. overcurrent

G. short circuit

H. ambient temperature

I. parallel conductors

J. fuse

Cable Markings

__________ **1.** Maximum rated voltage.

__________ **2.** Letters identifying insulation.

__________ **3.** Manufacturer's information.

__________ **4.** AWG size or CM area.

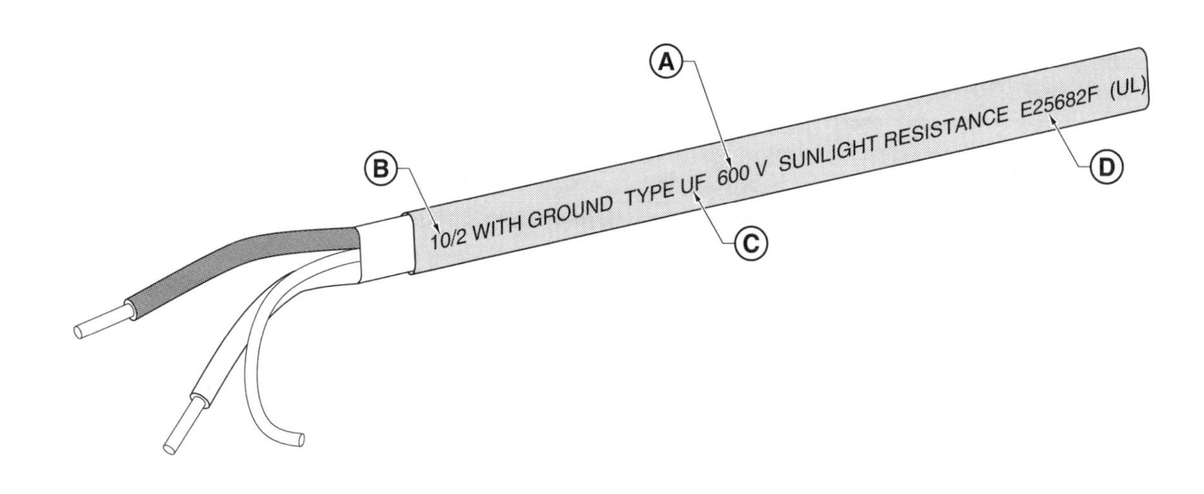

Overcurrent Protection Devices

__________ **1.** Blade fuse.

__________ **2.** Ferrule fuse.

__________ **3.** Type S plug fuse.

__________ **4.** Edison-base fuse.

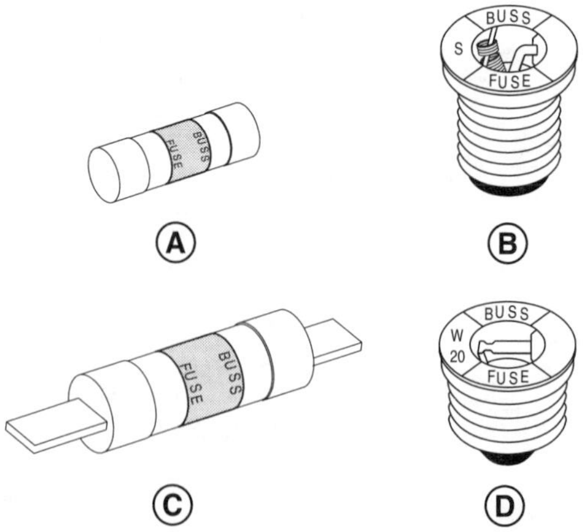

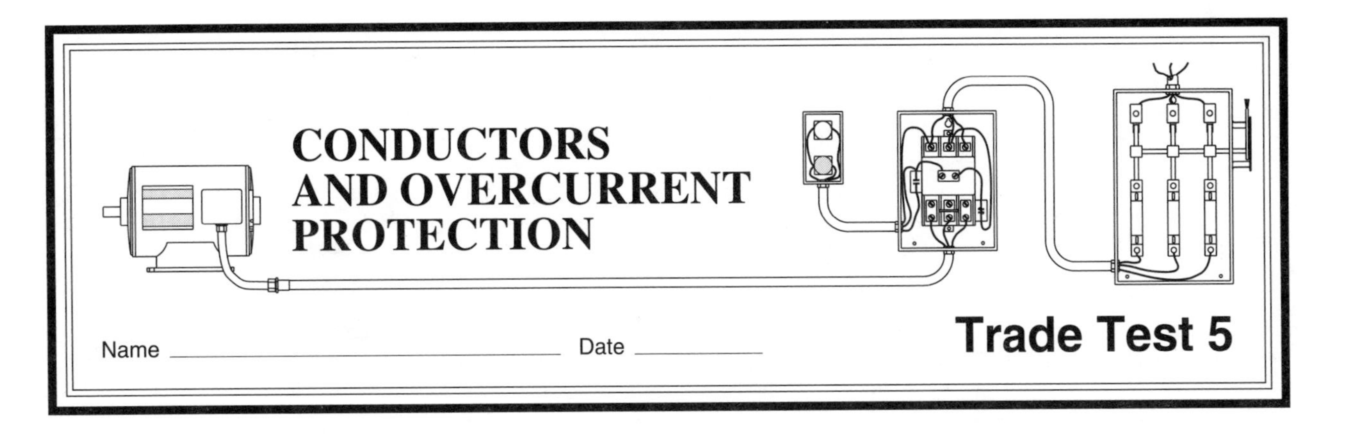

CONDUCTORS AND OVERCURRENT PROTECTION

Name ______________________ Date __________

Trade Test 5

NEC®	Answer	
______	______	**1.** Determine the ampacity of three No. 8 THW Cu conductors installed in the same conduit in an ambient temperature of 30°C. All of the conductors are current-carrying conductors.
______	______	**2.** Determine the ampacity of twelve No. 10 THHN Cu conductors installed in the same conduit in an ambient temperature of 30°C. Nine of the conductors are current-carrying conductors.
______	______	**3.** Determine the ampacity of six No. 6 THWN Al conductors installed in the same conduit in an ambient temperature of 30°C. All of the conductors are current-carrying conductors.
______	______	**4.** Determine the ampacity of three No. 12 THHN Cu conductors installed in the same conduit in an ambient temperature of 104°F. All of the conductors are current-carrying conductors.
______	______	**5.** Determine the ampacity of twelve No. 1/0 THW Cu conductors installed in the same conduit in an ambient temperature of 96°F. All of the conductors are current-carrying conductors.
______	______	**6.** Determine the ampacity of three No. 4 THHN Cu conductors installed in the same conduit in an ambient temperature of 86°F. All of the conductors are current-carrying conductors. The conductors terminate on equipment rated for 75°C terminations.
______	______	**7.** Four No. 6 THW Cu conductors are installed in the same conduit to supply fastened-in-place equipment. Three of the conductors are current-carrying conductors. The ambient temperature around the conduit is 30°C. What is the largest size OCPD permitted to protect these conductors?
______	______	**8.** *See Figure 1.* Determine the maximum size OCPD permitted for the installation.
______	______	**9.** *See Figure 2.* Determine the maximum size OCPD permitted for the installation.

FIGURE 1

FIGURE 2

_______ _______ **10.** *See Figure 3.* Determine the minimum size of the tap conductors for the installation.

_______ _______ **11.** *See Figure 4.* Determine the minimum size of the tap conductors for the installation.

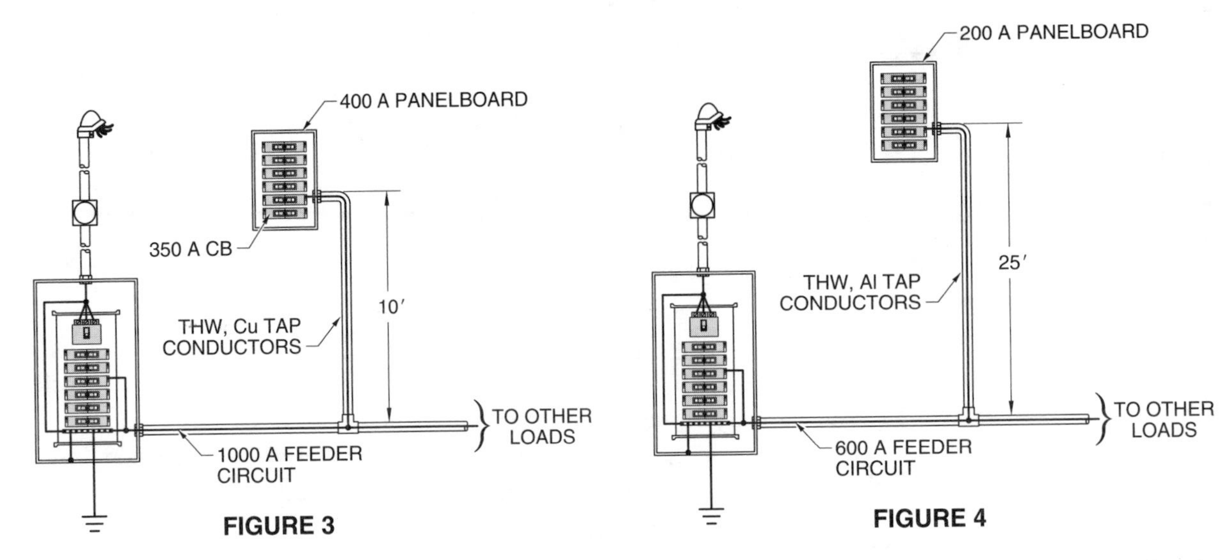

FIGURE 3

FIGURE 4

_______ _______ **12.** *See Figure 5.* Determine the minimum size of the tap conductors for the installation.

_______ _______ **13.** *See Figure 6.* Determine the minimum size of the tap conductors for the installation.

_______ _______ **14.** A 30 A, ITCB is installed as a branch-circuit OCPD in a 200 A branch-circuit panelboard. The interrupting rating of the 30 A CB is not marked on the CB. The available short-circuit current at the line side of the CB is 7500 A. Does this installation violate the overcurrent protection provisions of Article 240?

_______ _______ **15.** An 800 A OCPD protects the installation of three parallel 500 kcmil Cu conductors per phase. At a suitable wireway, a 10′ field tap is made from the 500 kcmil conductors. Determine the minimum size THW Cu conductors required for the tap conductors.

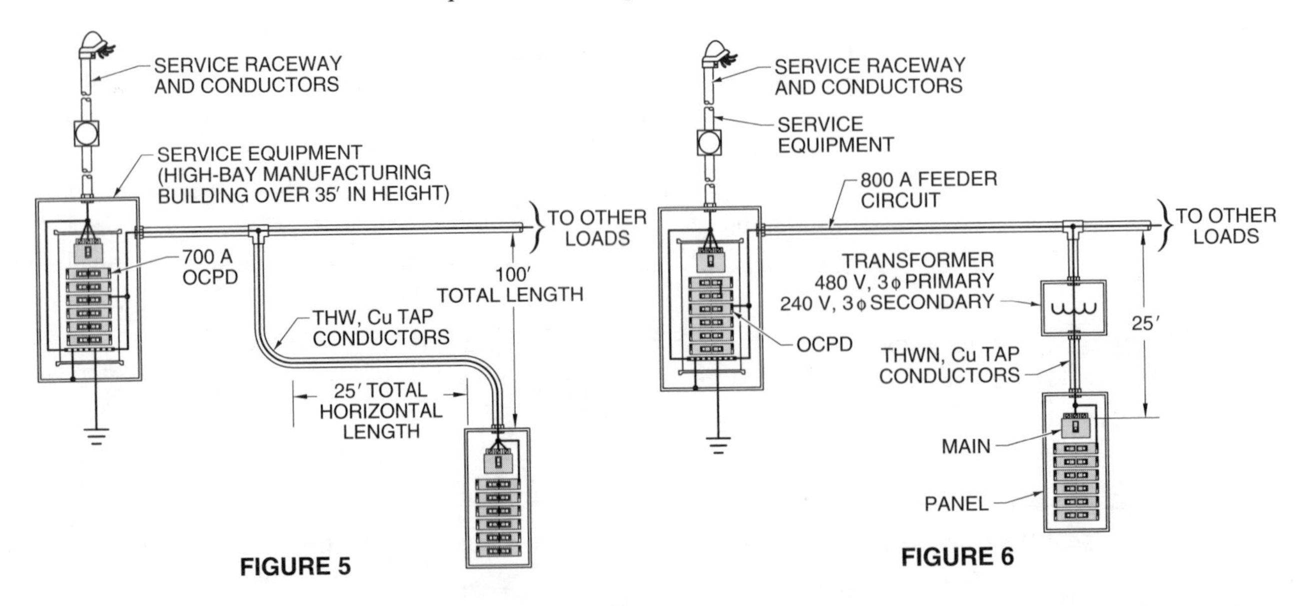

FIGURE 5

FIGURE 6

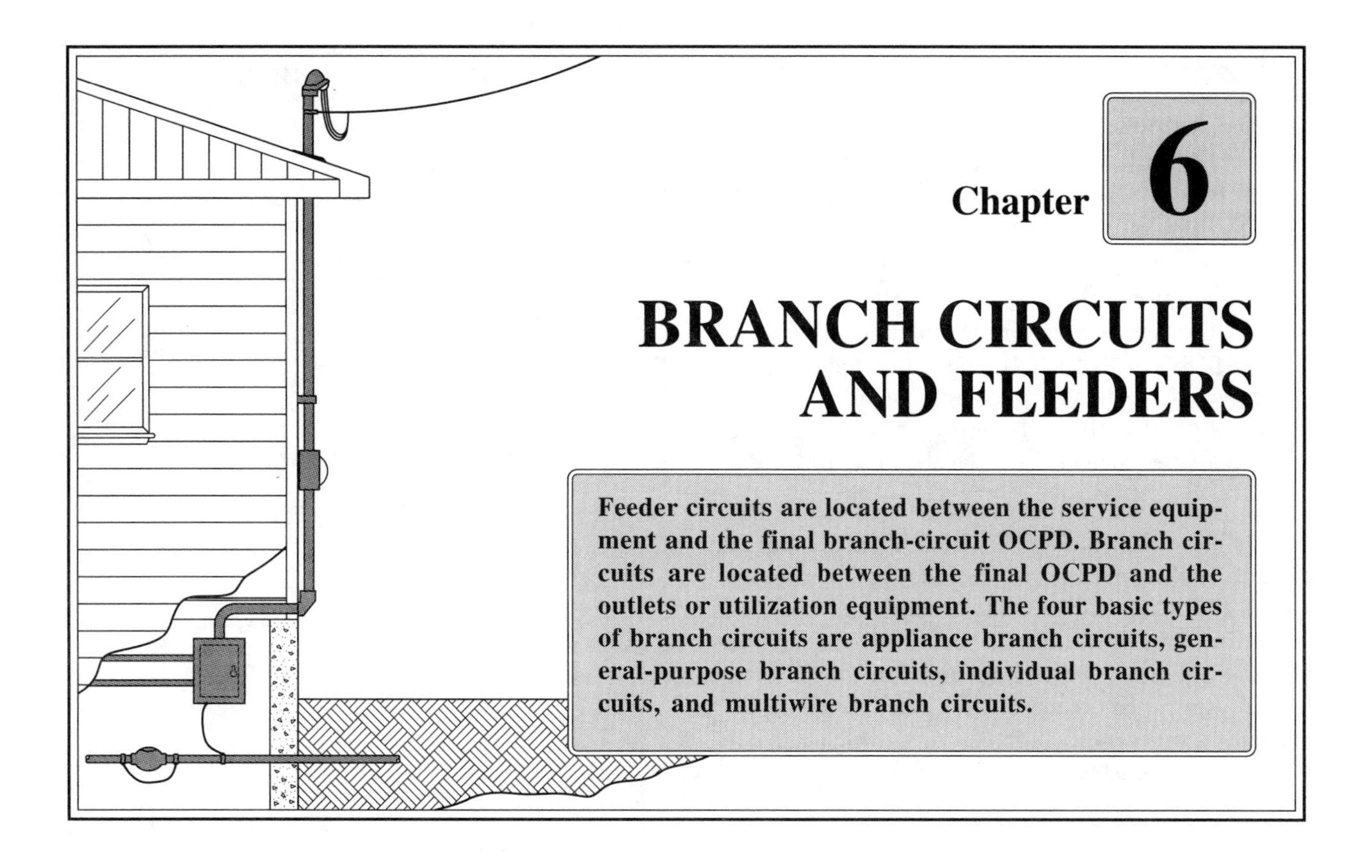

BRANCH CIRCUITS – 210

There are basically three types of conductors in any electrical installation. See Figure 6-1. These are the service-entrance conductors, feeder conductors, and branch-circuit circuits. Article 100 defines *service conductors* as the conductors from the service point or other source of power to the service disconnecting means. Service conductors supply power to the service equipment. Article 100 defines *service equipment* as the necessary equipment, usually consisting of a CB or switch and fuses, located near the point of entrance of supply conductors to a building or other structure, and intended to constitute the main control and means of cutoff of the supply. The wires leaving the service equipment are feeder conductors. Article 100 defines *feeder* as all circuit conductors between the service equipment or the source of a separately derived system and the final branch-circuit overcurrent device.

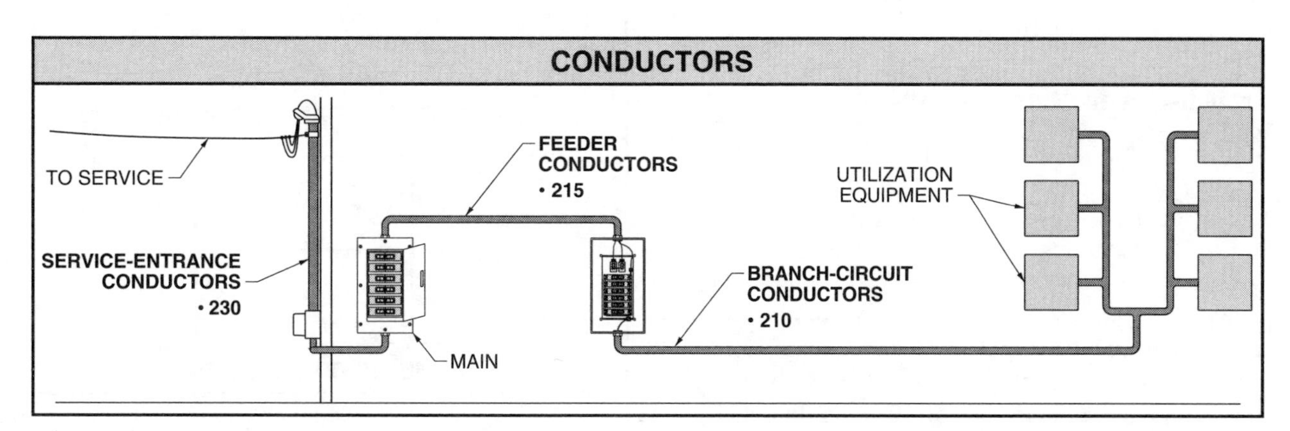

Figure 6-1. Three types of conductors are service-entrance conductors, feeder conductors, and branch-circuit conductors.

The *branch circuit* is that portion of the electrical circuit between the last overcurrent device (fuse or CB) and the outlets or utilization equipment. *Utilization equipment* is equipment that utilizes electric energy for electronic, electromechanical, chemical, heating, lighting, or similar purposes.

Section 210-1 limits the scope of 210 to all branch circuits except branch circuits that supply only motor loads and branch circuits for electrolytic cells as covered in 668-3(c). Branch circuits shall comply with 210 and with applicable provisions of other articles of the NEC® per 210-2. The provisions in the more specific articles amend the provisions of 210.

The four types of branch circuits defined in 100 are appliance branch circuits, general-purpose branch circuits, individual branch circuits, and multiwire branch circuits. See Figure 6-2. An *appliance branch circuit* is a branch circuit that supplies energy to one or more outlets to which appliances are to be connected. Appliance branch circuits shall have no permanently-connected lighting fixtures which are not a part of an appliance.

A *general-purpose branch circuit* is a branch circuit that supplies a number of outlets for lighting and appliances. An *individual branch circuit* is a branch circuit that supplies only one piece of utilization equipment. A *multiwire branch circuit* is a branch circuit with two or more ungrounded conductors having a potential difference between them, and a grounded conductor having equal potential difference between it and each ungrounded conductor, and is connected to the neutral or grounded conductor of the system. The most popular forms of multiwire branch circuits occur in 1ϕ, 3-wire and 3ϕ, 4-wire systems. The main concern in multiwire branch circuits is that the neutral shall never be opened while the circuit is energized.

Branch-Circuit Ratings – 210-3

Branch circuits shall be rated in accordance with the maximum ampere rating or setting of the overcurrent device. These ratings shall be 15 A, 20 A, 30 A, 40 A, and 50 A in other than individual branch circuits. Where conductors of higher ampacity are used, the rating or setting of the overcurrent device shall determine the circuit rating. For example, a branch circuit of No. 8 THW Cu conductors and an ampacity of 50 A per Table 310-16 with a 20 A overcurrent device protecting it is considered a 20 A branch circuit.

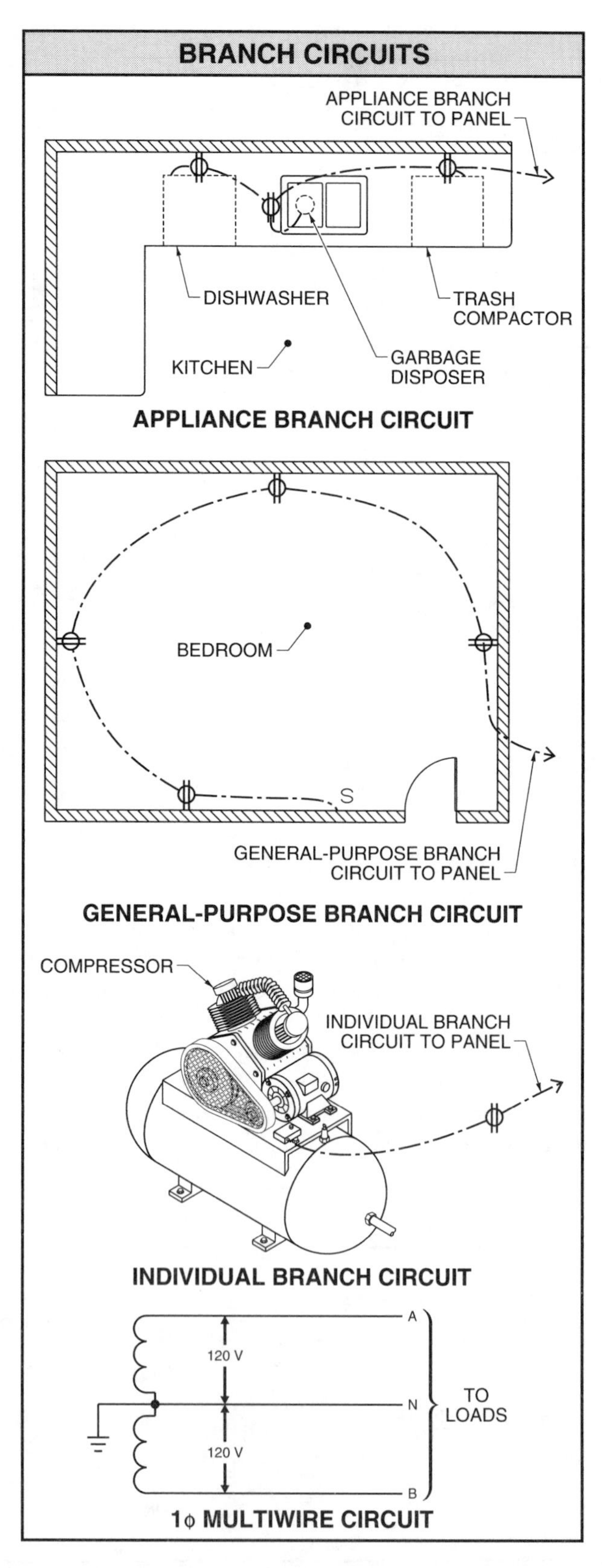

Figure 6-2. Four types of branch circuits are appliance branch circuits, general-purpose branch circuits, individual branch circuits, and multiwire branch circuits.

Multiwire Branch Circuits – 210-4

Section 210-4 permits multiwire branch circuits to be considered as multiple circuits. All of the conductors in a multiwire branch circuit shall originate from the same panelboard. Per 210-4(a), FPN, the neutral conductor in multiwire branch circuits supplying power to nonlinear loads may carry more current than the ungrounded conductors. This is due to the high harmonic currents. A *nonlinear load* is a load where the wave shape of the steady-state current does not follow the wave shape of the applied voltage. See Figure 6-3. Examples of nonlinear loads include electronic equipment, electronic/electric-discharge lighting, adjustable speed drive systems, and similar equipment.

Studies have been made of 3ϕ conductors which are perfectly balanced and it was found that the neutral conductor carried more current. For example, if each phase conductor carried 100 A, it was found that the neutral conductor could carry as much as 130 A to 140 A. Engineering manuals suggest that the neutral conductor be increased in multiwire circuits that supply current to nonlinear loads.

Dwelling Units. A multiwire branch circuit supplying more than one device on the same yoke or strap shall be provided with a means to simultaneously disconnect all ungrounded conductors at the panelboard. In general, multiwire branch circuits shall supply only line-to-neutral loads.

SPLIT-WIRED DUPLEX RECEPTACLES

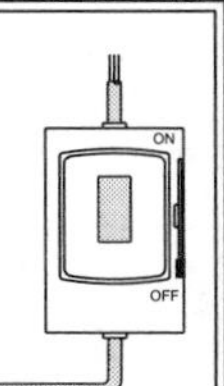

While trying to replace a defective, split-wired duplex receptacle, the homeowner may plug a lamp into one of the receptacle outlets and then open the CBs until the lamp turns OFF in order to determine the circuit on which the receptacle is located. The homeowner may then think that both outlets on the duplex receptacle are de-energized and begin to replace the receptacle. On a split-wired receptacle, each outlet of a duplex receptacle is on a separate circuit. The NEC® requires that the CB be a double-pole type or two SPCBs with a listed tie handle to ensure that they both operate simultaneously.

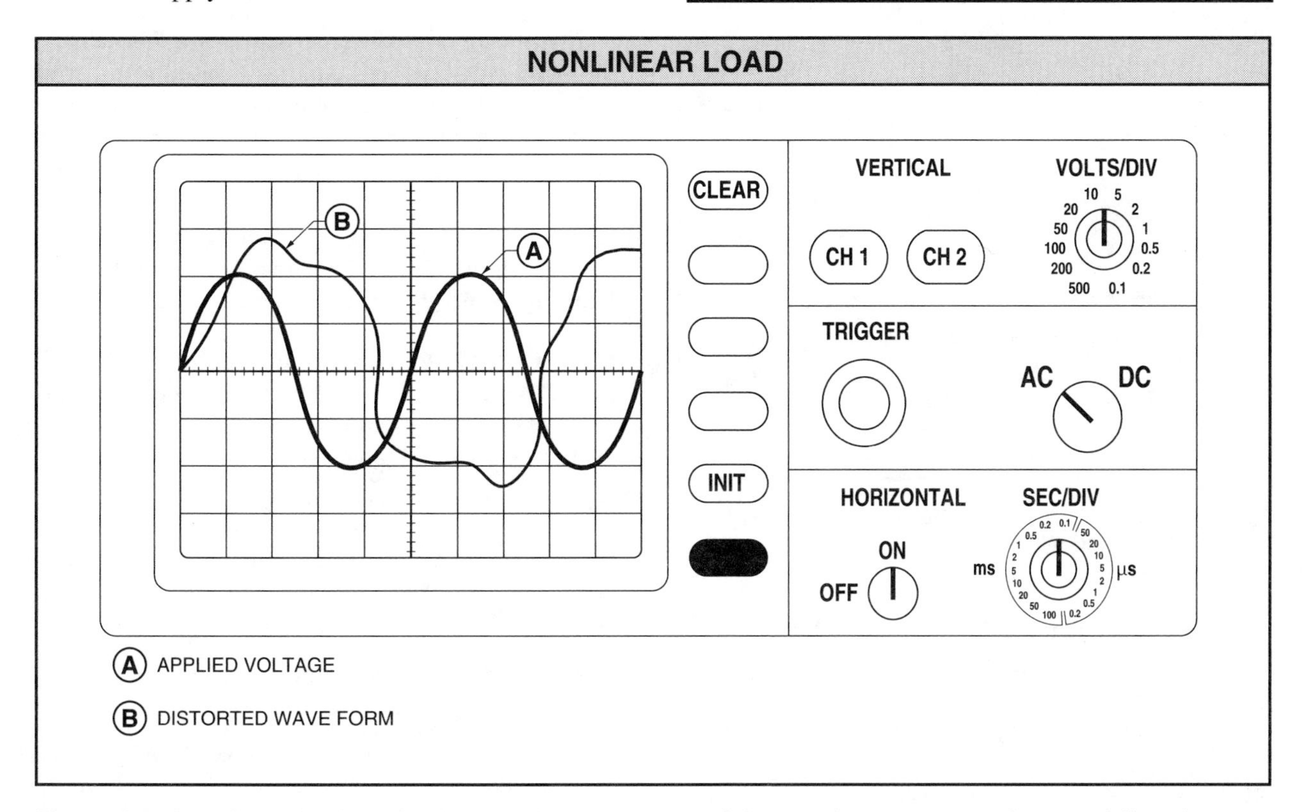

Figure 6-3. A nonlinear load is a load where the wave shape of the steady-state current does not follow the wave shape of the applied voltage.

Identification of Ungrounded Conductors – 210-4(d). In buildings where more than one voltage system exists, each ungrounded system conductor shall be identified by phase and system. Per 210-4(d), FPN, this identification occurs wherever the conductors are accessible, such as panel enclosures, junction boxes, and outlet boxes. Identification may be by separate color coding, marking tape, tagging, or other effective means. See Figure 6-4.

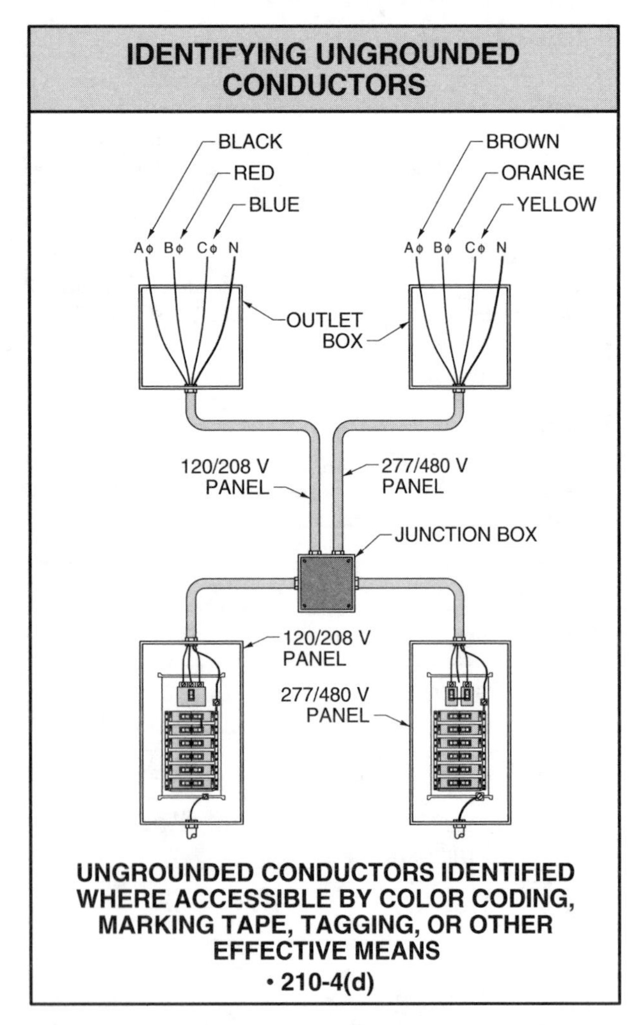

Figure 6-4. Each ungrounded system shall be identified by phase and system.

Color Code for Branch-Circuit Grounded Conductors – 210-5(a). The grounded conductor of a branch circuit shall be identified by the color white or natural gray. See Figure 6-5. *Note:* The wording is natural gray, not gray. Natural gray is an off-color or dirty white. The NEC® does not specify the color gray because there are so many shades of gray. Some shades of gray are so dark that they are almost black.

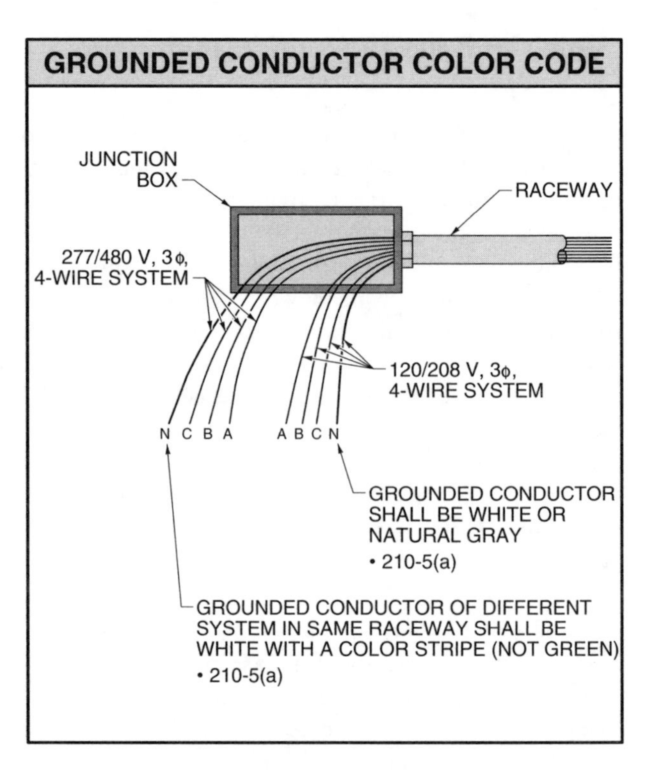

Figure 6-5. The grounded conductor of a branch circuit shall be identified by the color white or natural gray.

If conductors of different systems are installed in the same raceway, box, auxiliary gutter, or other enclosure, the grounded conductors of each system shall be clearly identified. One of the system's grounded conductors (if required) shall be white or natural gray. The other systems' grounded conductors (if required) shall have an outer covering of white with a colored stripe, other than green, on the insulation or other, different means of identification.

Color Code for Branch-Circuit Equipment Grounding Conductors – 210-5(b). Branch-circuit EGCs shall have a continuous outer finish that is either green or green with one or more yellow stripes unless it is bare. Insulated conductors larger than No. 6 are permitted to be permanently identified as an EGC at each end and at every point where the conductor is accessible per 250-57(b). Identification shall be accomplished in one of three ways:

- Stripping the insulation or covering from the full exposed length.
- Coloring the exposed insulation or covering green.
- Marking the exposed insulation or covering with green-colored tape or adhesive labels.

BRANCH-CIRCUIT VOLTAGE LIMITATIONS – 210-6

The voltage limitations for branch circuits is covered by 210-6. Generally, branch circuits supplying lampholders, fixtures, or receptacles of the standard 15 A or less rating are limited for operation on circuits where the voltage rating is not more than 120 V.

Occupancy Limitation – 210-6(a)

This section pertains specifically to dwelling units, motels, hotels, and other occupancies, such as dormitories, nursing homes, and similar residential occupancies. In these types of dwellings, any lighting fixture or receptacle for plug-connected loads rated up to 1440 VA, or less than ¼ HP, shall be supplied at not more than 120 V. High-wattage loads such as clothes dryers, electric ranges, and ACs are permitted to be supplied with 208 V or 240 V.

Plug-connected loads rated up to 1440 VA shall be supplied at not more than 120 V.

120 V Between Conductors – 210-6(b)

Branch circuits with voltages not over 120 V between conductors are permitted to supply three types of loads. See Figure 6-6. These loads are:

(1) Terminals of lampholders applied within their voltage rating.

(2) Auxiliary equipment of electric-discharge lamps.

(3) Cord- and plug-connected or permanently-connected utilization equipment.

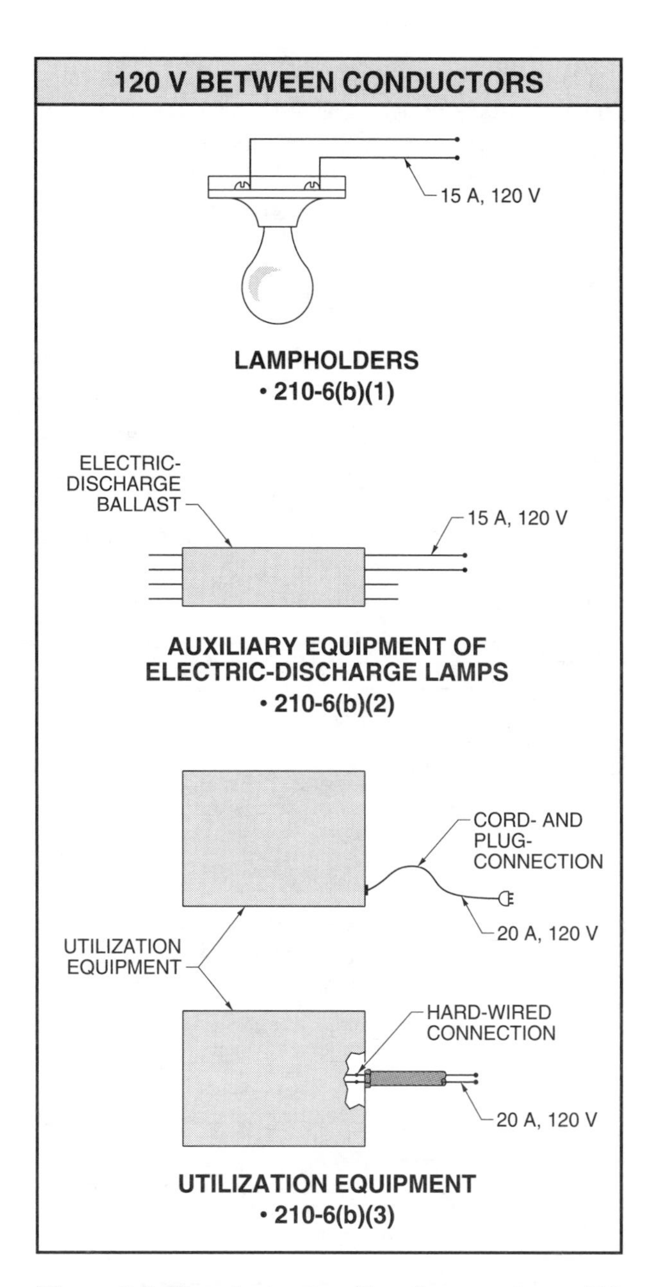

Figure 6-6. Branch circuits with voltages not over 120 V between conductors are permitted to supply three types of loads.

277 V to Ground – 210-6(c)

Circuits exceeding 120 V between conductors, and not exceeding 277 V to ground, are permitted to supply six types of loads. See Figure 6-7. These loads are:

(1) Listed electric-discharge lighting fixtures used within their rating.

(2) Listed incandescent lighting fixtures with step-down autotransformers which are an integral part of the fixture. In this installation, the autotransformer supplies the lampholder with 120 V and the grounded conductor is connected to the screw shell of the lampholder.

(3) Lighting fixtures equipped with mogul-base, screw-shell lampholders.

(4) Other than screw-shell type lampholders applied within their voltage ratings.

(5) Auxiliary equipment of electric-discharge lamps.

(6) Cord- and plug-connected or permanently-connected, equipment. Examples include air-conditioning and heating equipment in commercial locations and electric cooking equipment.

600 V Between Conductors – 210-6(d)

Branch circuits rated over 277 V to ground and up to 600 V between conductors are permitted to supply two types of loads. See Figure 6-8. These loads are:

(1) High-discharge lighting. Examples include fluorescent, mercury-vapor, and sodium fixtures. Installation is limited to outdoor areas such as roads, bridges, athletic fields, and parking lots. The lamps shall be mounted in permanently-installed fixtures on poles not less than 22′ in height or not less than 18′ in height on other structures, such as tunnels.

(2) Cord- and plug-connected or permanently-connected equipment.

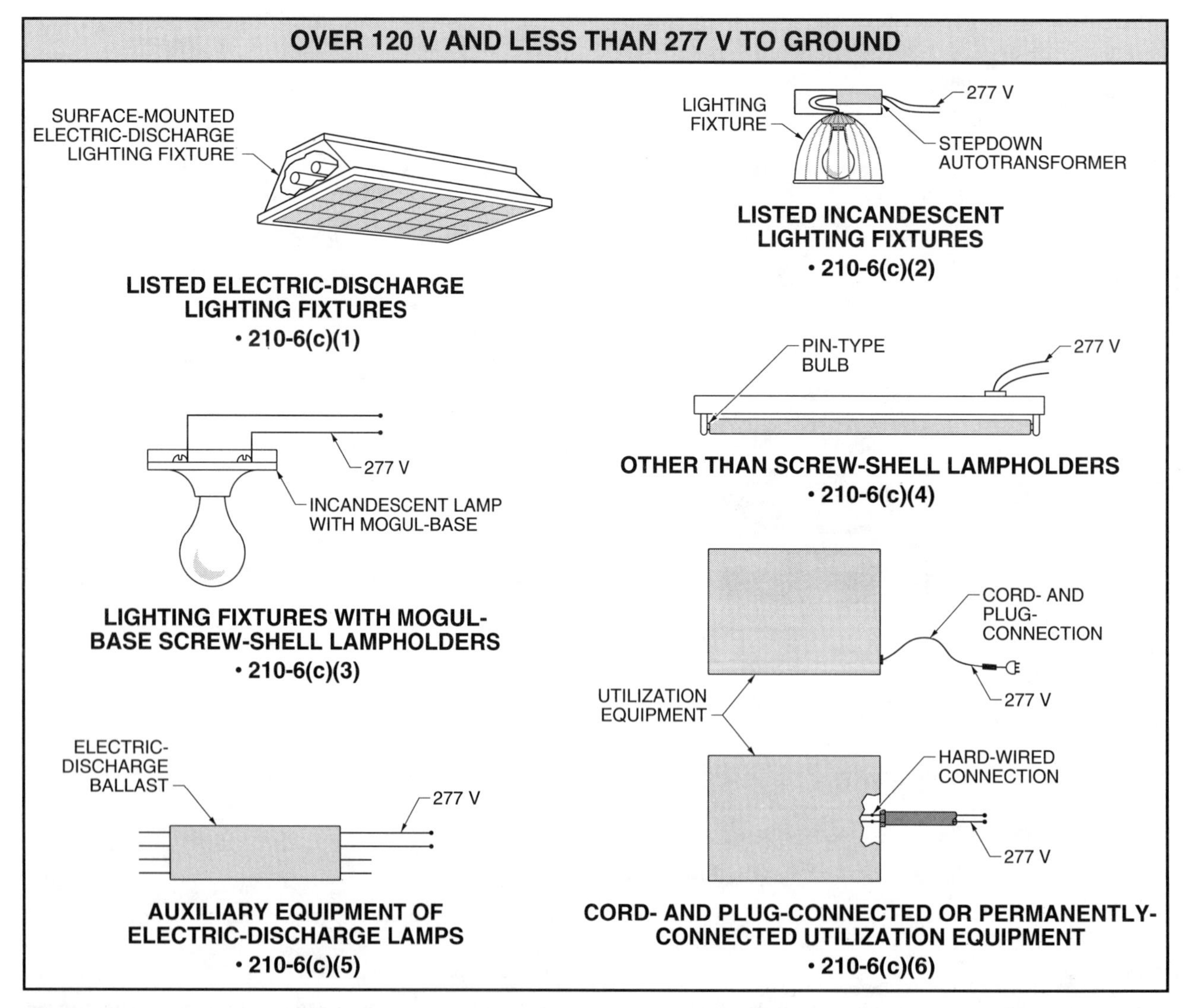

Figure 6-7. Branch circuits with voltages over 120 V, but not more than 277 V to ground, are permitted to supply six types of loads.

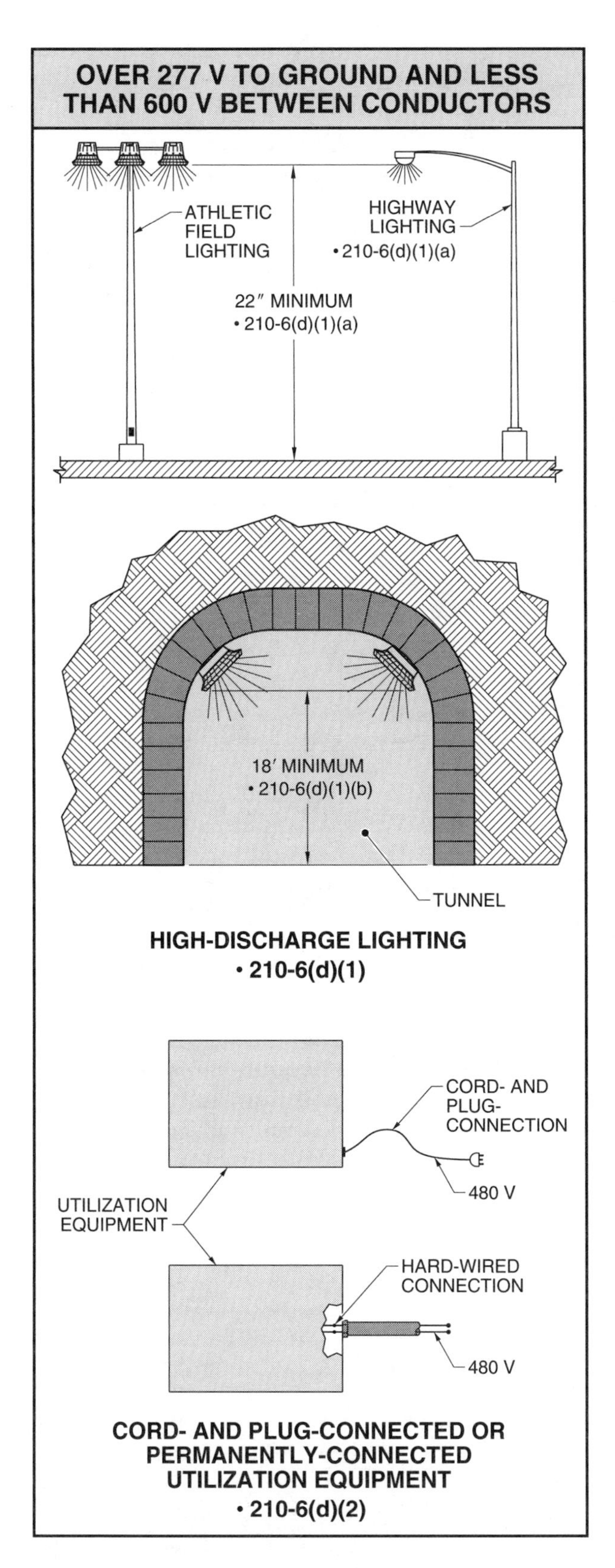

Figure 6-8. Branch circuits with voltages over 277 V, but less than 600 V between conductors, are permitted to supply two types of loads.

RECEPTACLES AND CORD CONNECTORS – 210-7(a)(b)(c)

Receptacles installed on 15 A and 20 A branch circuits shall be of the grounding type per 210-7(a). These receptacles shall be installed only on circuits of the voltage class and current for which they are rated. Section 210-7(b) requires receptacles and cord connectors having grounding contacts to have those contacts effectively grounded. Section 210-7(b), Ex. 1 does not require this on receptacles mounted on portable and vehicle-mounted generators. Per 210-7(c), the grounding contacts of receptacles and cord connectors shall be grounded to the EGC of the circuit which supplies the receptacle or cord connector. The branch circuit wiring method shall provide an EGC for this purpose.

Replacing Receptacles – 210-7(d)

Per 210-7(d)(1), a grounding receptacle shall be used where a grounding means exists in the receptacle enclosure. Per 210-7(d)(2), a GFCI-protected receptacle shall be used where replacements are made at receptacle outlets that are required to be protected elsewhere in the NEC®. For example, if a non-GFCI-protected receptacle in a bathroom becomes defective, it shall be replaced with a GFCI-protected receptacle. This is one of the few retroactive rules found in the NEC®.

Three methods are permitted for the installation of replacement receptacles when grounding does not exist. See Figure 6-9. Per 210-7(d)(3)a, a nongrounding receptacle shall be permitted to be replaced with another nongrounding receptacle. Per 210-7(d)(3)b, a nongrounding receptacle shall be permitted to be replaced with a GFCI receptacle. The GFCI receptacle shall be marked "No Equipment Ground." These stickers are shipped with the GFCI receptacles. An EGC shall not be connected from the GFCI receptacle to any outlet supplied from the GFCI receptacle.

Per 210-7(d)(3)c, a nongrounding receptacle shall be permitted to be replaced with a grounding receptacle provided it is supplied through a GFCI device. This device may be either a GFCI receptacle or a GFCI CB. The replacement grounding receptacle shall be marked "GFCI Protected" and "No Equipment Ground." An EGC shall not be connected between the grounding receptacles.

REPLACEMENT RECEPTACLES			
EXISTING NONGROUNDING RECEPTACLE	REPLACEMENT RECEPTACLE	NEC®	REPLACEMENT PERMITTED
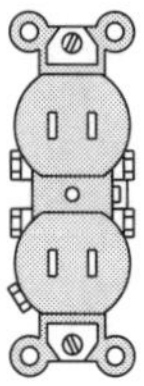	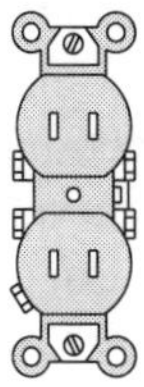	• 210-7(d)(3)a	REPLACE A NONGROUNDING-TYPE RECEPTACLE WITH ANOTHER NONGROUNDING-TYPE RECEPTACLE IF NO GROUNDING MEANS EXIST
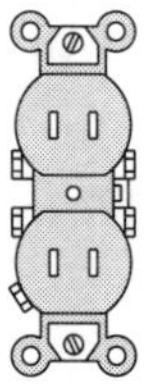	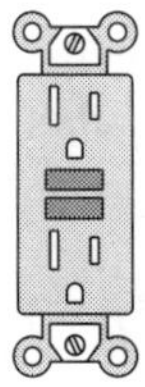	• 210-7(d)(3)b	REPLACE A NONGROUNDING-TYPE RECEPTACLE WITH A GFCI-TYPE RECEPTACLE MARKED "NO EQUIPMENT GROUND"
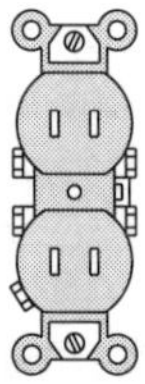	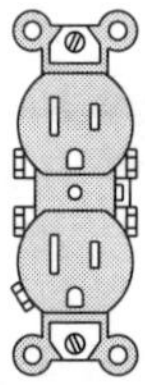	• 210-7(d)(3)c	REPLACE A NONGROUNDING-TYPE RECEPTACLE WITH A GROUNDING-TYPE RECEPTACLE WHEN SUPPLIED BY A GFCI PROTECTIVE DEVICE. REPLACEMENT GROUNDING RECEPTACLE SHALL BE MARKED "GFCI PROTECTED" AND "NO EQUIPMENT GROUND"

Figure 6-9. Three methods are permitted for the installation of replacement receptacles when grounding does not exist.

GROUND-FAULT CIRCUIT-INTERRUPTER PROTECTION FOR PERSONNEL

A GFCI receptacle or CB is set to trip at 5 mA. A *milliampere (mA)* is $\frac{1}{1000}$ of an ampere (1000 mA = 1 A). The effects of current flow on the human body are:

- 1 mA = Threshold of sensation
- 2 mA = Mild shock
- 5 mA = GFCI will trip
- 10 mA = Cannot let go
- 20 mA = Muscles contract. Breathing difficulty begins
- 50 mA = Breathing difficult. Suffocation possible
- 100 mA = Heart stops pumping
- 300 mA = Severe burns. Breathing stops

The first requirements for GFCI protection were included in the 1975 NEC®. Protection was required in dwelling unit bathrooms and in outdoor receptacles of dwelling units. Subsequent editions of the NEC® have continued to expand these requirements. Studies continue to indicate a decreasing trend in the number of electrocutions in the United States since the GFCI was introduced in the 1975 NEC®.

Dwelling Units – 210-8(a)

All 125 V, 15 A and 20 A, 1ϕ receptacles installed in dwelling units at any of seven locations shall have GFCI protection. These seven locations are bathrooms, garages and accessory buildings, outdoors, crawl spaces, unfinished basements, kitchens, and wet bar sinks.

Bathrooms – 210-8(a)(1). All bathroom receptacles shall be GFCI-protected. See Figure 6-10. A *bathroom* is an area with a basin and one or more of a toilet, tub, or shower. See 100. For an area to qualify as a bathroom, a basin shall be present.

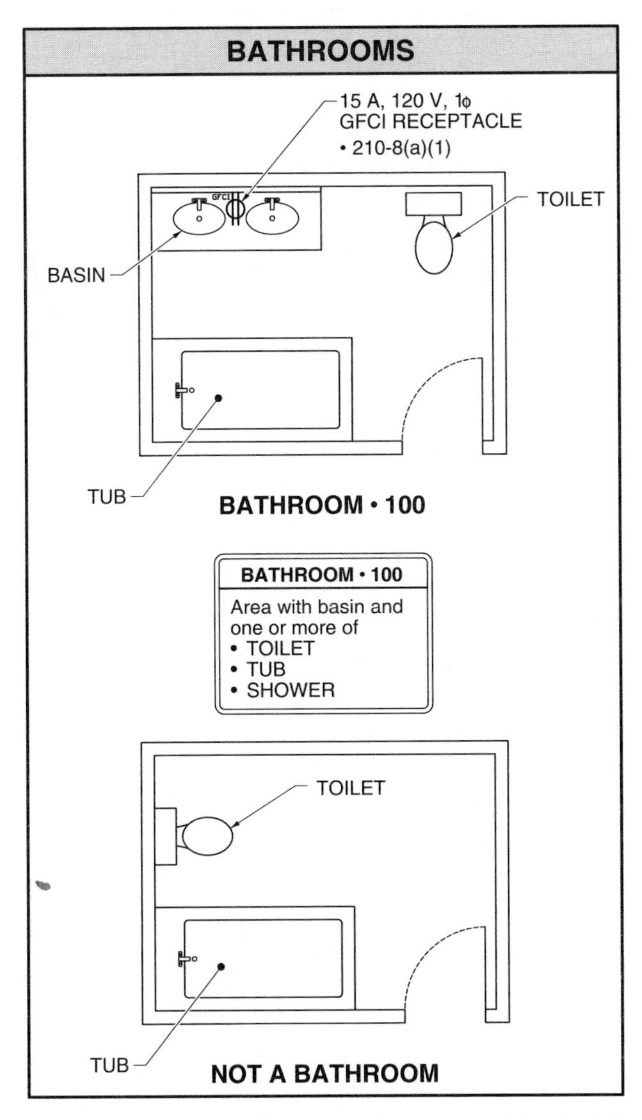

Figure 6-10. All bathroom receptacles shall be GFCI-protected.

Garages and Accessory Buildings – 210-8(a)(2). GFCI-protected receptacles shall be installed in garages and grade-level portions of unfinished accessory buildings used for storage or work areas. See Figure 6-11. Per 210-8(a)(2), Ex. 1, receptacles that are not readily accessible do not require GFCI protection. For example, a receptacle installed in the ceiling for a garage door opener does not require GFCI protection because it is not readily accessible.

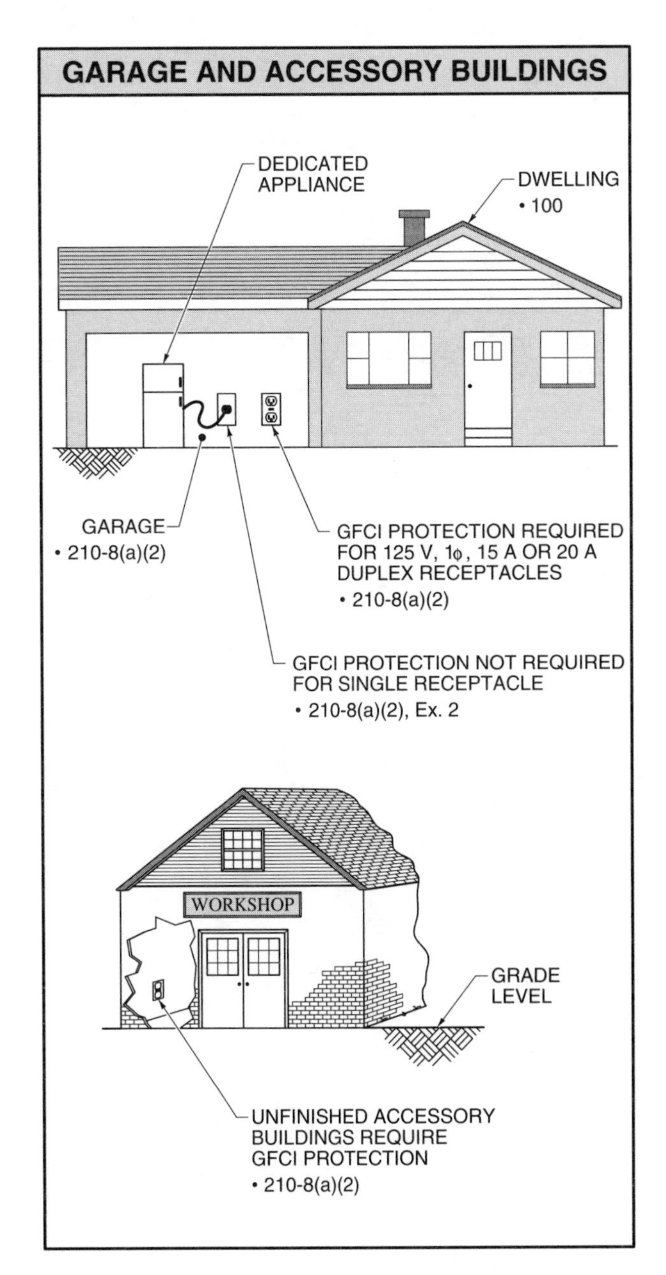

Figure 6-11. GFCI-protected receptacles shall be installed in garages and grade-level portions of unfinished accessory buildings used for storage or work areas.

Per 210-8(a)(2), Ex. 2, a single receptacle installed for an appliance which occupies dedicated space, such as a refrigerator or freezer, does not require GFCI protection. A duplex receptacle is also exempt provided two appliances that occupy dedicated space are cord- and plug-connected into the duplex receptacle. The intent here is to provide GFCI protection to any receptacle outlets available for use for portable hand tools, gardening tools, etc. Receptacles installed under either of these exceptions shall not be considered as meeting the requirements of 210-52(g).

Outdoors – 210-8(a)(3). All outdoor receptacles shall be GFCI-protected. See Figure 6-12. The one exception to this requirement is receptacles which are not readily accessible and are supplied from a dedicated branch circuit to serve electric snow-melting or deicing equipment per 426.

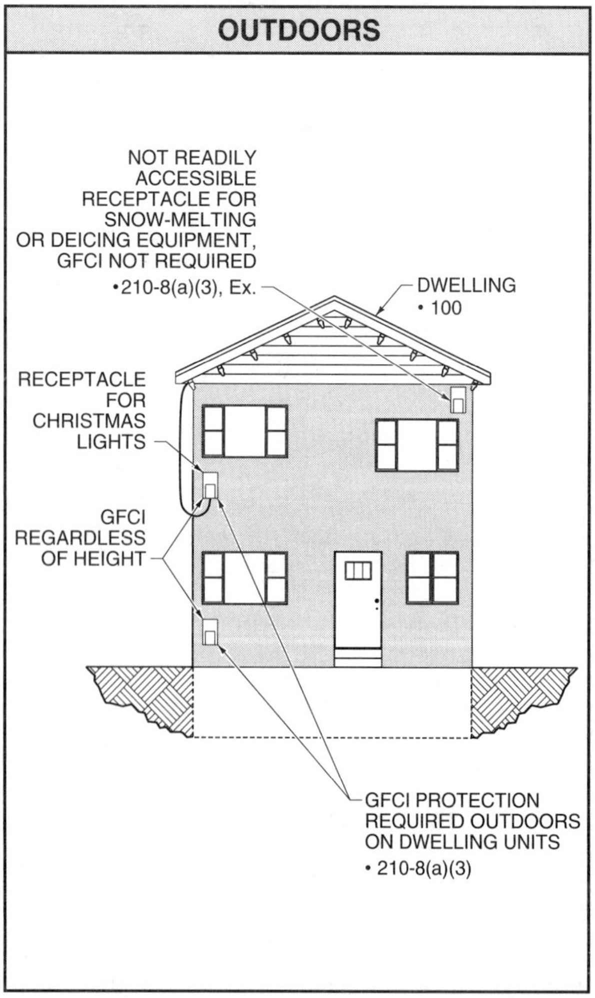

OUTDOORS

NOT READILY ACCESSIBLE RECEPTACLE FOR SNOW-MELTING OR DEICING EQUIPMENT, GFCI NOT REQUIRED
•210-8(a)(3), Ex.
DWELLING
• 100
RECEPTACLE FOR CHRISTMAS LIGHTS
GFCI REGARDLESS OF HEIGHT
GFCI PROTECTION REQUIRED OUTDOORS ON DWELLING UNITS
• 210-8(a)(3)

Figure 6-12. All outdoor receptacles shall be GFCI-protected.

These receptacles are permitted to be installed without GFCI protection for personnel. However, 426-28 requires that ground-fault protection of equipment (GFPE) be provided for branch circuits supplying fixed outdoor electric deicing and snow-melting equipment. GFPEs protect the equipment, not personnel. GFPEs are set to trip at 30 mA, not 5 mA which is necessary to protect personnel.

Crawl Spaces – 210-8(a)(4). All receptacles in crawl spaces where the crawl space is at or below grade level shall be GFCI-protected. See Figure 6-13. *Note:* Crawl spaces at or below grade level may be wet or damp locations.

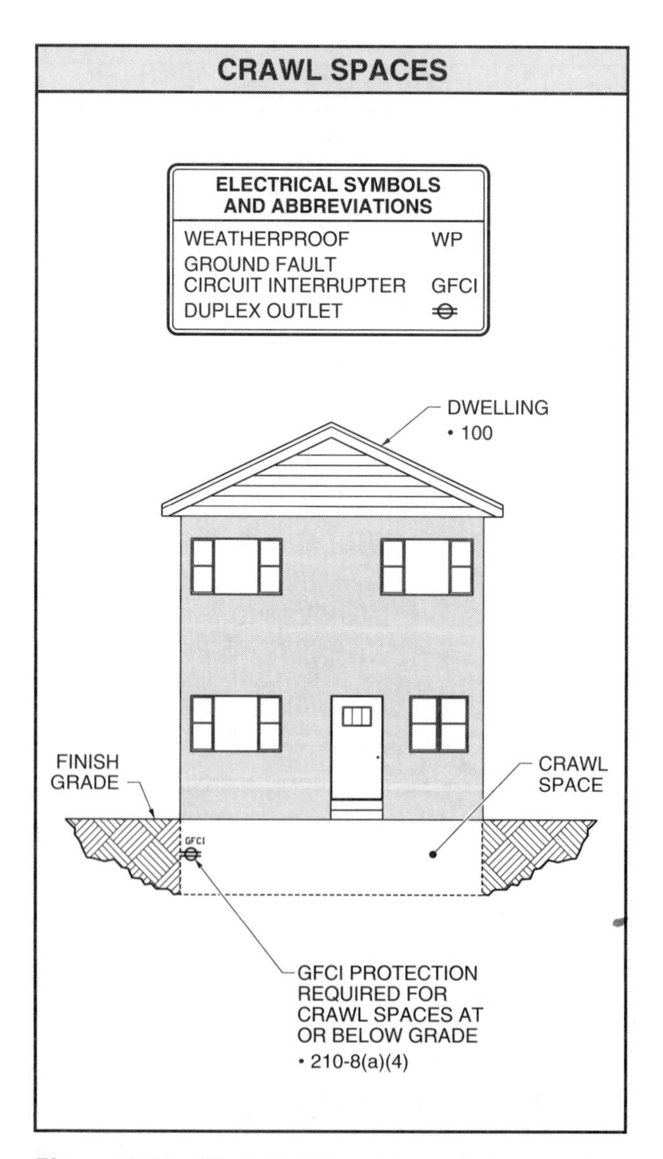

Figure 6-13. All receptacles in crawl spaces at or below grade level shall be GFCI-protected.

Unfinished Basements – 210-8(a)(5). All receptacles in unfinished basements shall be GFCI-protected. See Figure 6-14. An *unfinished basement* is the portion or area of a basement which is not intended as a habitable room, but is limited to storage areas, work areas, etc. *Note:* GFCI protection is not required in finished basements.

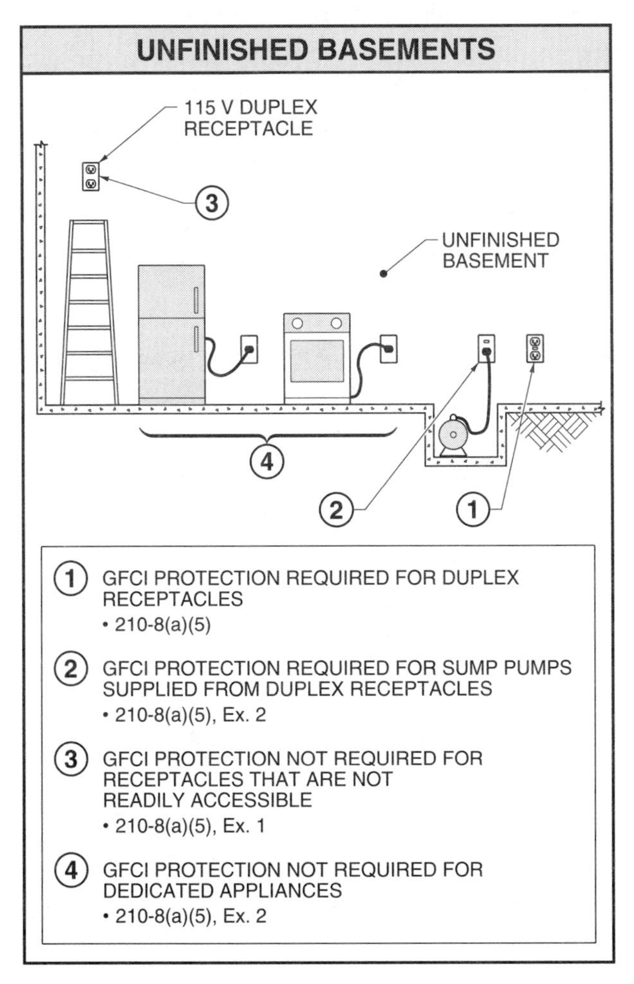

Figure 6-14. All receptacles in unfinished basements shall be GFCI-protected.

Per 210-8(a)(5), Ex. 1, GFCI protection is not required in unfinished basements for receptacles which are not readily accessible. Per 210-8(a)(5), Ex. 2, GFCI protection is not required for a single receptacle or duplex receptacle (for two appliances) located within dedicated space if the appliance, in normal use, is not easily moved from one place to another and is cord- and plug-connected to that receptacle. Receptacles installed under these two exceptions shall not be considered as meeting the requirements of 210-52(g).

Kitchens – 210-8(a)(6). All receptacles which serve countertop surfaces installed in kitchens shall be GFCI-protected. See Figure 6-15. Most appliances intended for countertop use are not equipped with an EGC. Since water and many grounded surfaces are present in kitchens, a shock hazard exists. Receptacles installed for dedicated appliances, such as garbage disposers, dishwashers, and trash compactors are not required to have GFCI protection.

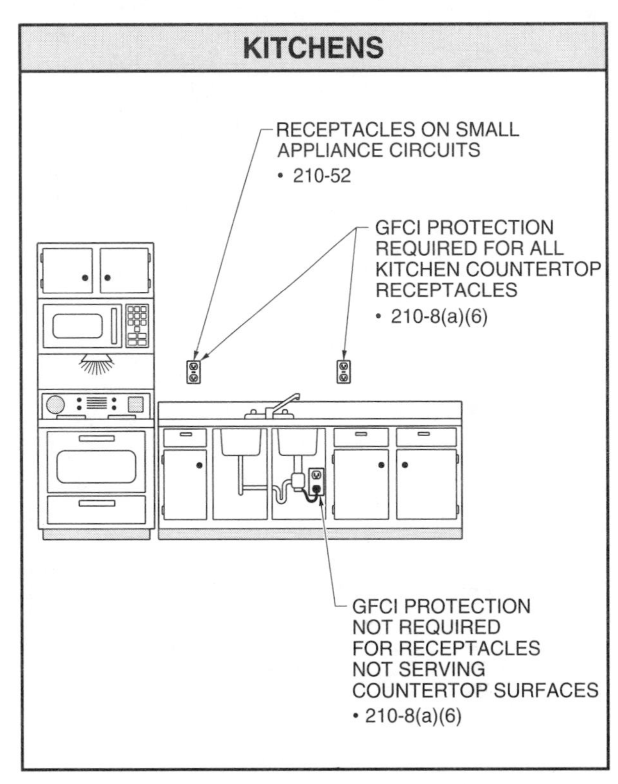

Figure 6-15. All receptacles which serve countertop surfaces installed in kitchens shall be GFCI-protected.

Wet Bar Sinks – 210-8(a)(7). All receptacles installed to serve the countertop surfaces of wet bar sinks and which are installed within 6′ of the outside edge of the sink shall be GFCI-protected. See Figure 6-16. Such receptacles may be used to operate blenders, ice crushers, etc.

Other Than Dwelling Units – 210-8(b)

Section 210-8(b) requires all 15 A and 20 A, 125 V, 1ϕ receptacles installed in other than dwelling units at either of two locations to have GFCI protection. See Figure 6-17. These two locations are bathrooms and rooftops.

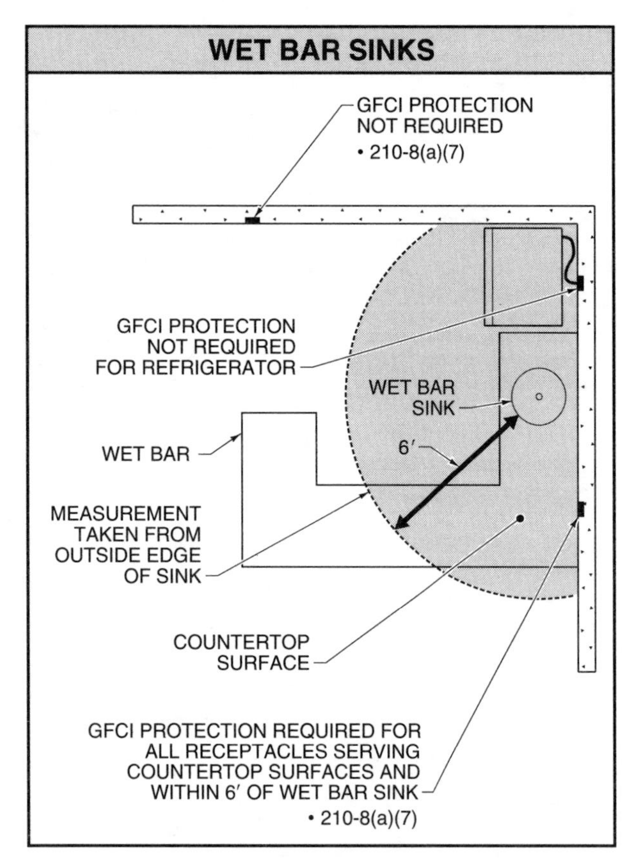

Figure 6-16. All receptacles within 6′ of the outside edge of a wet bar sink shall be GFCI-protected.

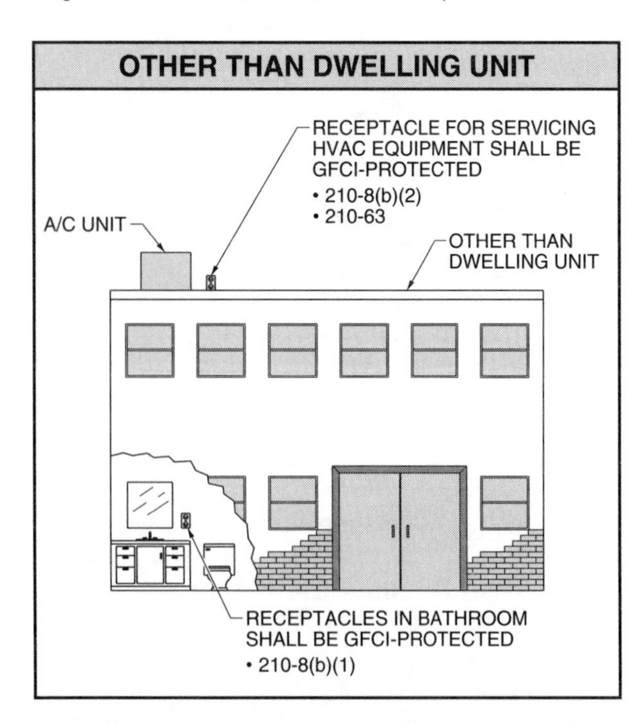

Figure 6-17. All 15 A and 20 A, 125 V, 1ϕ receptacles in bathrooms and on rooftops in other than dwelling units shall be GFCI-protected.

Per 210-8(b)(1), all receptacles installed in bathrooms shall be GFCI-protected. However, 517-21 does not require GFCI protection if the toilet and basin in health care facilities are installed within critical care areas of the patient's room. Per 210-8(b)(2), all receptacles installed on rooftops shall be GFCI-protected.

In addition to the requirements listed in 210-8 for GFCI protection for personnel, the installations requiring GFCI protection include:

Section Location

- 210-7(d) – Replacement
- 215-9 – Feeders
- 305-6 – Temporary Wiring
- 422-8(d)(3) – High-Pressure Spray Washing Appliances
- 511-10 – Commercial Garages
- 517-20 – Health Care Facilities
- 517-21 – Health Care Facilities
- 530-73(a)(1) – Motion Picture and TV Studios
- 550-8(b) – Mobile Homes
- 550-23(d) – Mobile Homes
- 551-41 – Recreational Vehicles
- 551-71 – Recreational Vehicle Parks
- 555-3 – Marinas
- 600-10(c)(2) – Signs (Mobile or Portable)
- 620-85 – Elevators, Escalators, and Moving Walkways
- 625-22 – Electric Vehicle Charging Systems
- 680-6(a)(3) – Pools (Permanently Installed)
- 680-31 – Pools (Storable)
- 680-51(a) – Fountains
- 680-70 – Hydromassage Bathtubs

BRANCH-CIRCUIT RATINGS – 210, PART B

Part B deals with the branch-circuit ratings of conductors, overcurrent devices, and outlet devices. It also deals with maximum and permissible loads. The basic rule is that the branch circuit shall have an ampacity not less than the maximum load to be served. The rating of a branch circuit is determined by the ampere rating or setting of the overcurrent device per 210-3.

Minimum Size Conductors – 210-19(a)

Branch-circuit conductors shall have an ampacity not less than the maximum load to be served. In addition, conductors of multioutlet branch circuits supplying receptacles for cord- and plug-connected portable loads shall have an ampacity of not less than the rating of the branch circuit. This is the rating of the overcurrent device. Multioutlet branch circuits for cord- and plug-connected portable loads have random unpredictable loads, therefore, the conductors shall have an ampacity equal to the rating of the branch circuit overcurrent protective device.

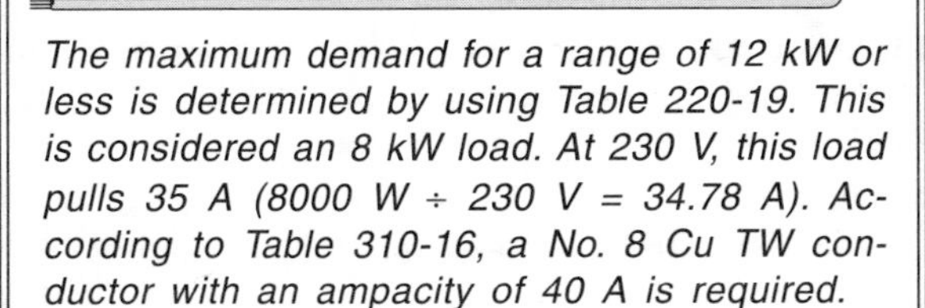

DEMAND LOAD FOR 12 kW OR LESS RANGE

The maximum demand for a range of 12 kW or less is determined by using Table 220-19. This is considered an 8 kW load. At 230 V, this load pulls 35 A (8000 W ÷ 230 V = 34.78 A). According to Table 310-16, a No. 8 Cu TW conductor with an ampacity of 40 A is required.

Household Ranges and Cooking Appliances – 210-19(b). Branch-circuit conductors supplying household ranges, and other cooking appliances, shall have an ampacity not less than the rating of the branch circuit, which is the overcurrent device. For ranges rated 8¾ kW or more, the minimum branch-circuit rating shall be 40 A.

Per 210-19(b), Ex. 1, tap conductors supplying electric ranges and other cooking units from a 50 A branch circuit are permitted to have an ampacity of not less than 20 A. The taps shall not be longer than necessary for servicing the appliance. See Figure 6-18.

Per 210-19(b), Ex. 2, the neutral conductor of a 3-wire branch circuit supplying a household electric range or other cooking units, is permitted to be smaller than the ungrounded conductors. This neutral shall have an ampacity of not less than 70% of the branch-circuit rating and shall not be smaller than No. 10.

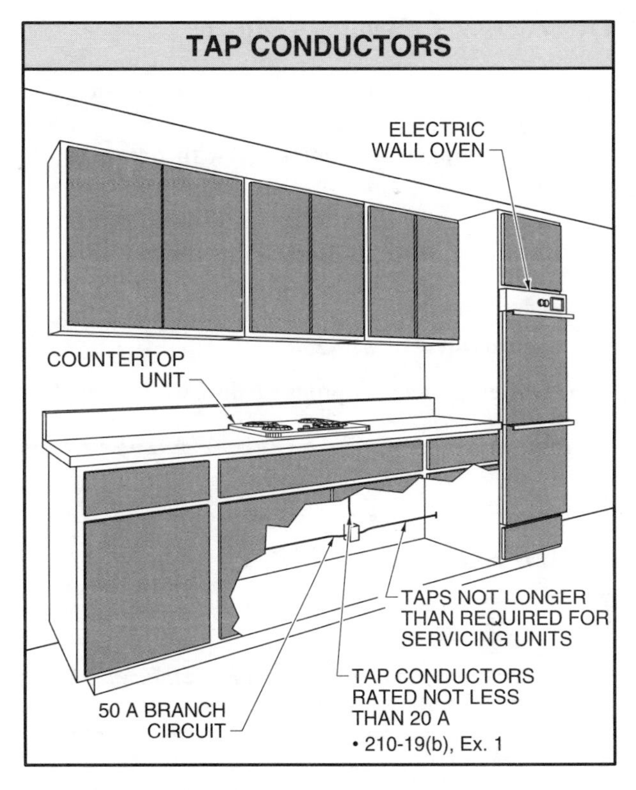

Figure 6-18. Tap conductors supplying electric ranges and other cooking units from a 50 A branch circuit are permitted to have an ampacity of not less than 20 A.

Other Loads – 210-19(c). Branch-circuit conductors shall have an ampacity sufficient for the loads served and shall be No. 14 or larger. There are two exceptions to this requirement. Per 210-19(c), Ex. 1, tap conductors serving one of five type of loads shall have an ampacity not less than 15 A for circuits rated less than 40 A and not less than 20 A for circuits rated at 40 A or 50 A. The five types of loads are:

- Individual lampholders or fixtures with taps extending no more than 18″ beyond the lampholder or fixture.
- A fixture with tap conductors per 410-67.
- Individual outlets, other than receptacle outlets, with taps no more than 18″ long.
- Infrared lamp industrial heating appliances.
- Nonheating leads of deicing and snow-melting cables and mats.

Exception 2 is for fixture wires and cords as permitted in 240-4.

Overcurrent Protection – 210-20

Branch-circuit conductors and equipment shall be protected by overcurrent protective devices with a rating or setting which complies with one of three conditions:

- Shall not exceed conductors specified in 240-3 for conductors.
- Shall not exceed equipment specified in the articles listed in 240-2.
- Outlet devices as provided in 210-21.

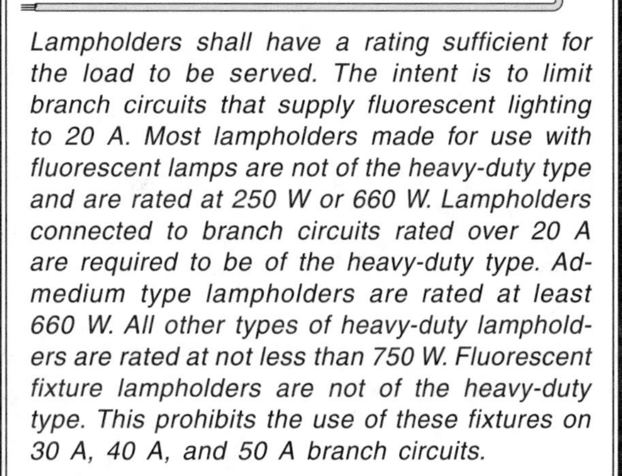

LAMPHOLDER RATINGS

Lampholders shall have a rating sufficient for the load to be served. The intent is to limit branch circuits that supply fluorescent lighting to 20 A. Most lampholders made for use with fluorescent lamps are not of the heavy-duty type and are rated at 250 W or 660 W. Lampholders connected to branch circuits rated over 20 A are required to be of the heavy-duty type. Admedium type lampholders are rated at least 660 W. All other types of heavy-duty lampholders are rated at not less than 750 W. Fluorescent fixture lampholders are not of the heavy-duty type. This prohibits the use of these fixtures on 30 A, 40 A, and 50 A branch circuits.

Outlet Devices – 210-21

Outlet devices shall have an ampere rating not less than the load to be served. Lampholders shall be of the heavy-duty type when connected to a branch circuit in excess of 20 A per 210-219(a). Heavy-duty, admedium lampholders shall have a rating of not less than 660 W. All other lampholders shall have a rating of not less than 750 W.

A single receptacle installed on an individual branch circuit shall have an ampere rating not less than the branch circuit per 210-21(b)(1). See Figure 6-19. The important wording here is "single receptacle," not a duplex receptacle. A *single receptacle* is a single contact device with no other contact device on the same yoke. A *multiple receptacle* is a single device with two or more receptacles. If a single receptacle were on a 20 A branch circuit (protected with a 20 A overcurrent device), the single receptacle would have to be rated 20 A.

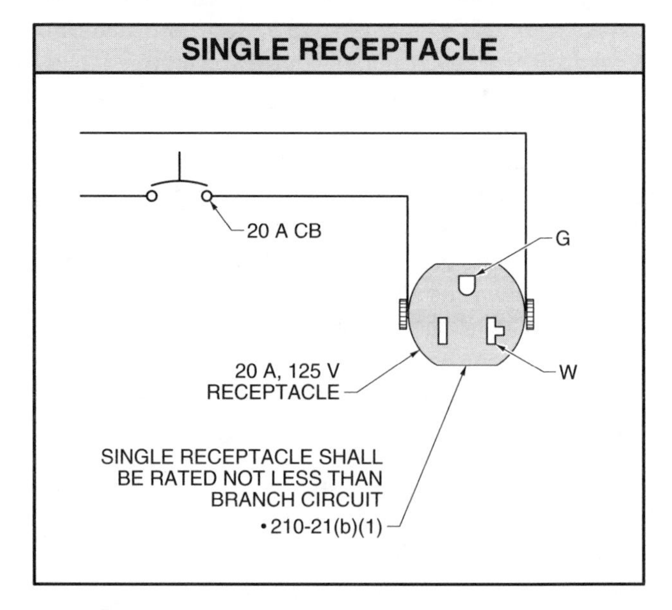

Figure 6-19. A single receptacle on an individual branch circuit shall have an ampere rating not less than the branch circuit.

Per 210-21(b)(2), if a branch circuit supplies two or more receptacles or outlets, the receptacle shall not supply a total cord- and plug-connected load in excess of the amount specified in Table 210-21(b)(2). For example, two or more 15 A receptacles are permitted on a 20 A general-purpose branch circuit.

TABLE 210-21(B)(2)

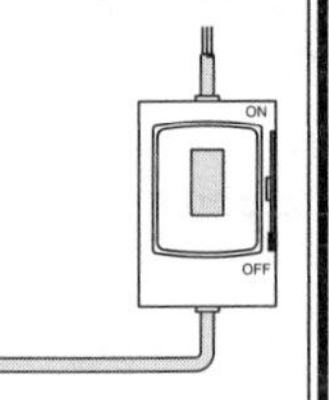

There is some confusion as to the intent of Table 210-21(b)(2). Under the column, Circuit Rating Amperes, 15 A or 20 A is listed. Then, in the next column, Receptacle Rating Amperes, 15 A is listed. The question that arises is whether a 15 A rated receptacle is permitted to be installed on a circuit that is protected for 20 A?

The answer is No, because Table 210-21(b)(2) is applied only when there are two or more receptacles, such as a duplex receptacle. The configuration of a 15 A rated receptacle does not accept a 20 A rated plug. Each receptacle of a duplex receptacle is rated 15 A.

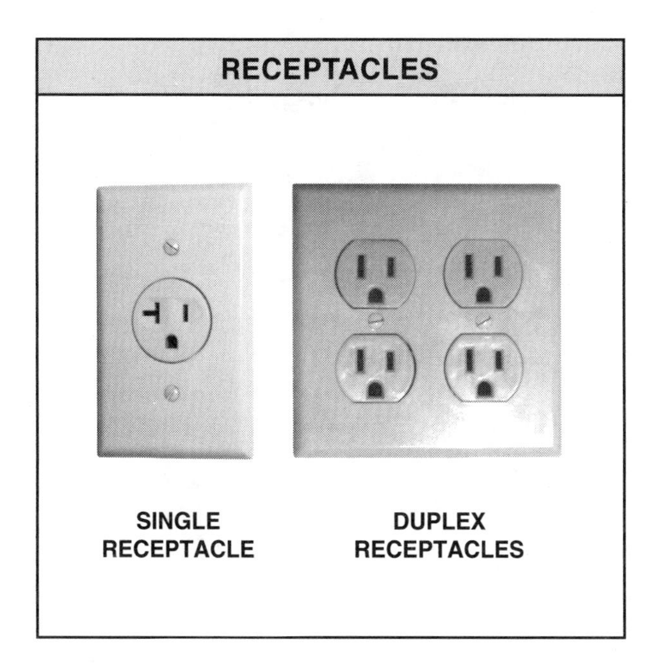

A receptacle is a contact device installed at outlets for the connection of cord-connected electrical equipment.

Maximum Loads – 210-22

The total load shall not exceed the maximum loads specified under three conditions:

Per 210-22(a), branch circuits supplying motor loads only shall be installed per the requirements of 430. Branch circuits supplying only AC and/or refrigeration equipment shall comply with 440. Branch circuits supplying fastened-in-place motor loads larger than 1/8 HP in combination with other loads are calculated at 125% of the largest motor load plus the sum of the other loads.

Per 210-22(b), lighting units having ballasts, transformers, or autotransformers shall be calculated with the current rating on the ballasts, transformers, or autotransformers. Do not use the total wattage of the lamps.

Per 210-22(c), a branch circuit that supplies continuous and noncontinuous loads shall have its overcurrent device sized at not less than the noncontinuous load plus 125% of the continuous load. The minimum branch-circuit conductor size, without applying any adjustment or correction factors, shall be equal to or greater than the noncontinuous load plus 125% of the continuous load. See Figure 6-20.

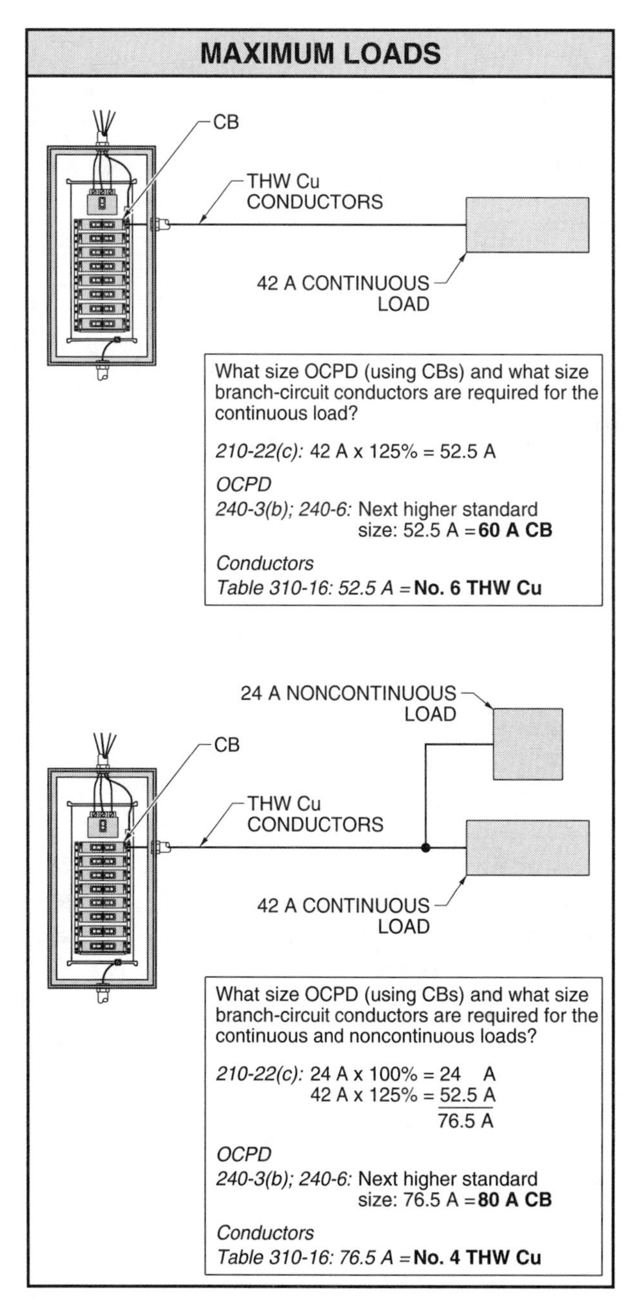

Figure 6-20. The minimum branch-circuit conductor size shall be equal to or greater than the noncontinuous load plus 125% of the continuous load.

Permissible Loads – 210-23

The load of a branch circuit shall not exceed the branch-circuit ampere rating. An individual branch circuit may serve any load for which it is rated. Branch circuits supplying two or more outlets or receptacles shall be subject to four limitations:

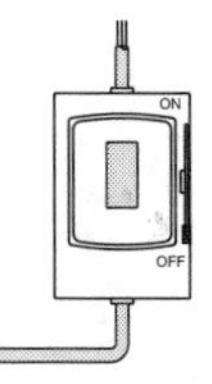

RECEPTACLE OUTLETS ON A SINGLE BRANCH CIRCUIT

It is not uncommon in industrial applications to supply several single receptacle outlets rated 50 A on a single branch circuit. This allows for the relocation of heavy-duty equipment which is used in the production process or for maintenance equipment.

Per 210-23(a), 15 A and 20 A branch circuits are permitted to serve lighting and cord- and plug-connected utilization equipment. The rating of any one cord- and plug-connected piece of utilization equipment shall not exceed 80% of rating of the branch circuit. Fastened-in-place equipment may be connected to a circuit serving lighting fixtures and cord- and plug-connected equipment, provided the total rating of the fixed equipment does not exceed 50% of the branch-circuit rating.

Per 210-23(b), 30 A branch circuits are permitted to supply fixed lighting fixtures with heavy-duty lampholders in other than dwelling units. They can also supply utilization equipment in any occupancy. The rating of any one cord- and plug-connected piece of equipment shall not exceed 80% of the rating of the branch circuit.

Per 210-23(c), branch circuits rated 40 A and 50 A are permitted to supply fixed lighting units equipped with heavy-duty lampholders or infrared heating units in other than dwelling units. They can also serve cooking appliances that are fastened in place in any occupancy.

Per 210-23(d), lighting loads are prohibited on branch circuits larger than 50 A.

REQUIRED OUTLETS – 210-50(b)

Receptacle outlets shall be installed wherever cord- and plug-connected equipment is used. Receptacle outlets installed in a dwelling unit for specific appliances shall be located within 6′ of the intended location of the appliance per 210-50(c). See Figure 6-21.

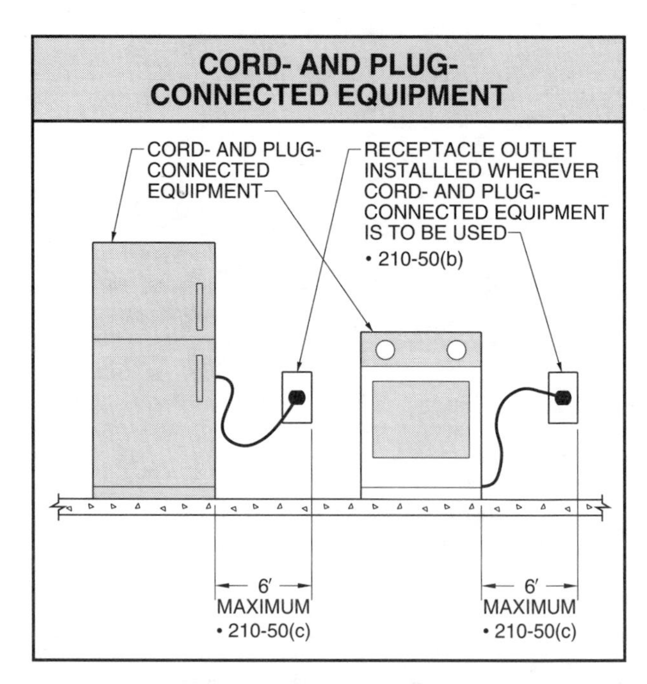

Figure 6-21. Receptacle outlets shall be installed wherever cord- and plug-connected equipment is used.

Dwelling Unit Receptacle Outlets – 210-52(a)

Receptacle outlets shall be installed so that no point along the floor line in any wall space is more than 6′ from an outlet in kitchens, family rooms, dining rooms, living rooms, parlors, libraries, dens, sun rooms, bedrooms, recreation rooms, or similar rooms or areas of dwelling units. See Figure 6-22. Wall space provided by freestanding bar-type counters or railings, shall be included in this 6′ measurement. A *dwelling unit* is a dwelling with one or more rooms used by one or more persons as a housekeeping unit with eating, living, and sleeping space, and permanent cooking and sanitation provisions.

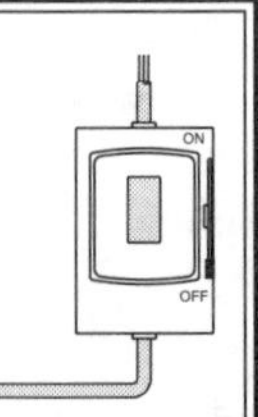

NO POINT MORE THAN 6′ FROM OUTLET

The 6′ dimension is based on the UL requirement that all floor and table lamps have a 6′ cord. The intent is that wherever the homeowner decides to place a floor or table lamp along the floor line there is a receptacle available to supply the lamp without using an extension cord.

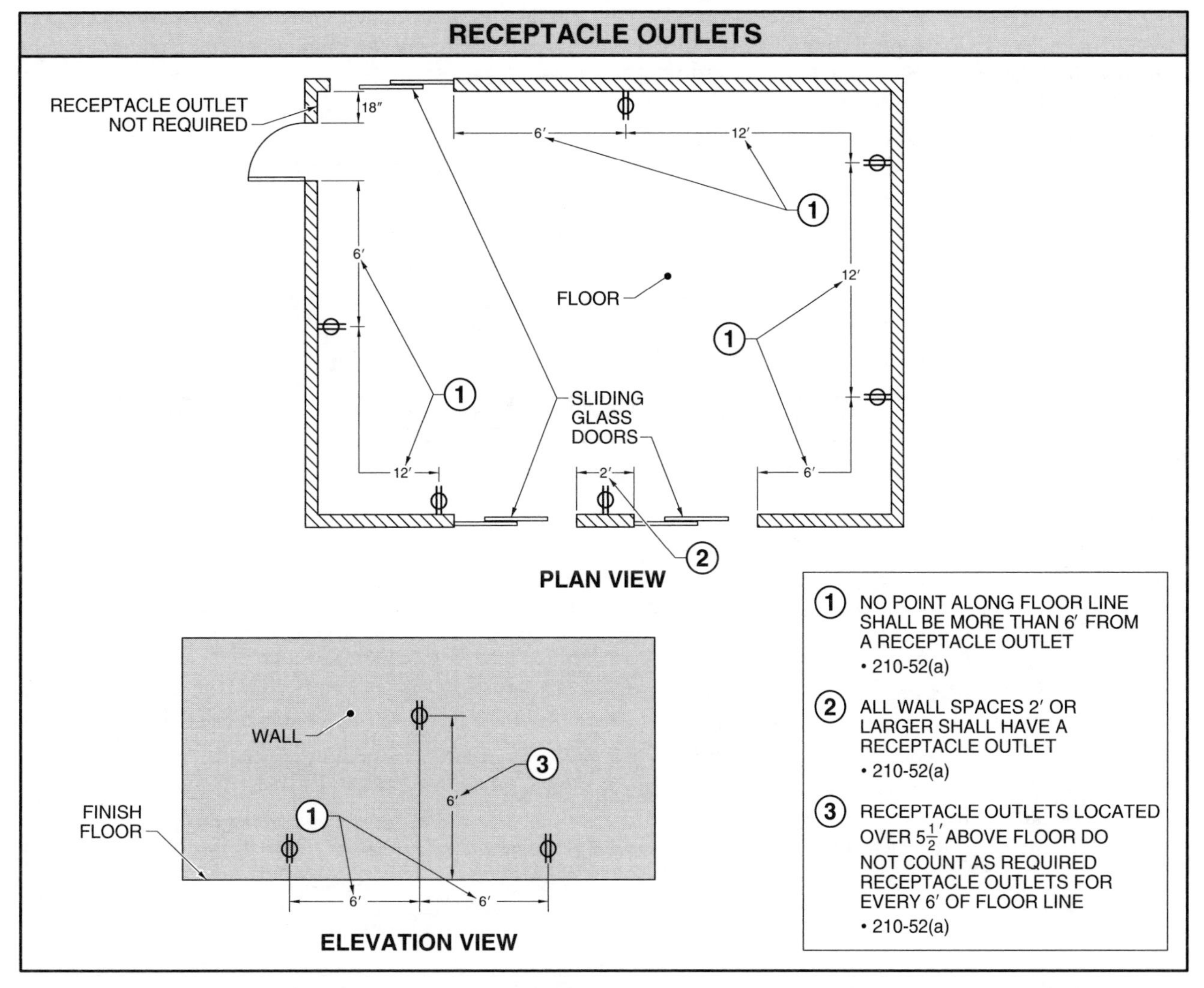

Figure 6-22. No point along the floor line in any wall space shall be more than 6′ from a receptacle outlet.

Receptacles are required for all floor line dimensions 2′ or greater and unbroken by doorways, windows, fireplaces, etc. A wall space may include two or more walls of a room (around corners) if it is unbroken at the floor line. Receptacle outlets in the floor shall not be counted as part of the required number of outlets unless they are located within 18″ of the wall.

Receptacle outlets which are part of a lighting fixture or appliance, located within cabinets or cupboards, or over 5½′ above the floor shall not count as a required outlet. However, permanently-installed electric baseboard heaters with factory-installed receptacle outlets which are not connected to the heater circuits are permitted to count as the required outlet or outlets for the wall space per 210-52(a), Ex.

Dwelling Unit Small Appliances – 210-52(b)

Per 210-52(b)(1), two or more 20 A small appliance branch circuits required by 220-4(b) shall supply all of the receptacle outlets in the kitchen, pantry, dining room, breakfast room, or similar areas of a dwelling unit. Refrigeration equipment is also permitted to be served by these circuits.

However, 210-52(b)(1) permits switched receptacles supplied from a general-purpose branch circuit as defined in 210-70(a). *Note:* This exception does not include the kitchen or bathroom. The receptacle outlet for refrigeration equipment may be supplied from an individual branch circuit rated 15 A or greater per 210-52(b), Ex. 2.

The two small appliance branch circuits required by 210-52(b)(1) shall have no other outlets per 210-

52(b)(2). This requirement has two exceptions. Per 210-52(b)(2), Ex. 1, a receptacle outlet is permitted to be installed for an electric clock. Per 210-52(b)(2), Ex. 2, receptacles are permitted to be installed to provide power for lighting and supplemental equipment of gas-fired ranges, ovens, or counter-mounted cooking units.

Section 210-52(b)(3) requires that at least two of the 20 A small appliance branch circuits shall supply the receptacle outlets that are required to serve the countertop surfaces in the kitchen. Additional small appliance branch circuits, if necessary, may be located in the kitchen and other rooms specified in 210-52(b)(1).

Dwelling Unit Countertops – 210-52(c)

The location of receptacle outlets for counter spaces in kitchen and dining rooms of dwelling units is determined by the wall counter space, island counter space, peninsular counter space, and separate spaces in these rooms. See Figure 6-23.

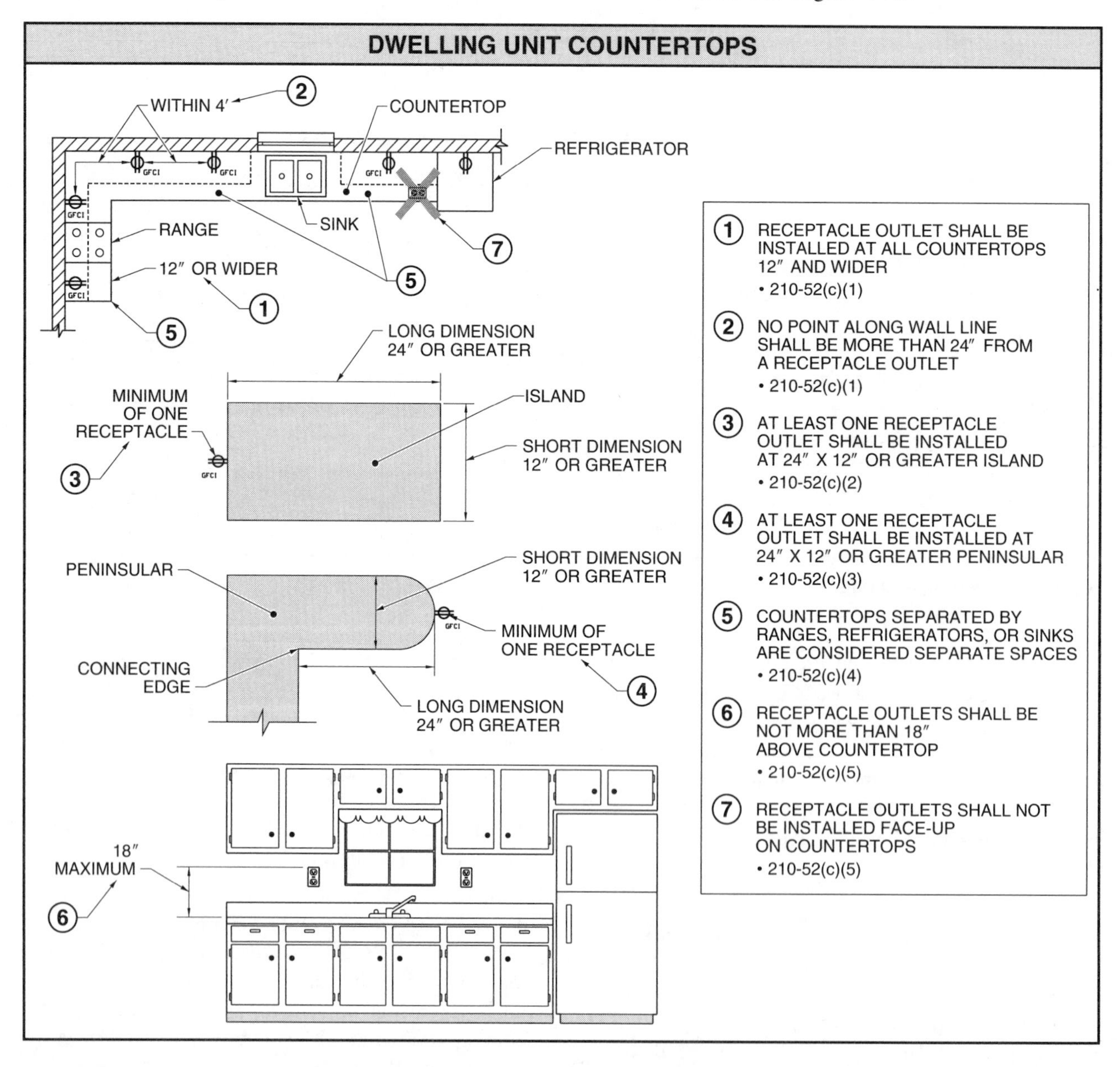

Figure 6-23. The location of receptacle outlets for counter spaces in kitchen and dining rooms of dwelling units is determined by the wall counter space, island counter space, peninsular counter space, and separate spaces in these rooms.

Wall Counter Space – 210-52(c)(1). Receptacle outlets shall be installed at each wall counter space 12″ or wider. No point along the wall line shall be more than 24″ measured horizontally from a receptacle outlet in that space.

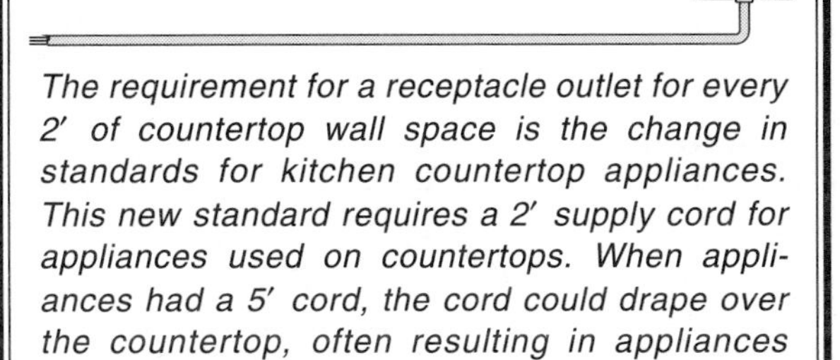

2′ SUPPLY CORDS

The requirement for a receptacle outlet for every 2′ of countertop wall space is the change in standards for kitchen countertop appliances. This new standard requires a 2′ supply cord for appliances used on countertops. When appliances had a 5′ cord, the cord could drape over the countertop, often resulting in appliances pulled off of the countertop.

Island Counter Space – 210-52(c)(2). At least one receptacle outlet shall be installed to serve each island countertop space that is 24″ × 12″ or larger. The long dimension is 24″ or larger and the small dimension is 12″ or larger.

Peninsular Counter Space – 210-52(c)(3). At least one receptacle outlet shall be installed to serve each peninsular countertop space, provided the peninsular has a long dimension of 24″ or greater and a short dimension of 12″ or greater. A peninsular countertop is measured from where it connects to the countertop.

Separate Spaces – 210-52(c)(4). Countertop spaces which are separated by range tops, refrigerators, or sinks shall be considered as separate countertop spaces in applying the requirements for placing receptacles on wall counter space, island counter space, or peninsular counter space.

Receptacle Outlet Location – 210-52(c)(5). The location of receptacle outlets shall not be more than 18″ above the countertop surface. Wall cabinets are generally 18″ above the countertops of the base cabinets. Receptacle outlets shall not be installed in a face-up position in the work surfaces or countertops. Receptacle outlets which are not readily accessible, due to fastened-in-place appliances which prevent ready access, shall not be counted as required receptacle outlets.

Per 210-52(c)(5), Ex., outlets shall be permitted to be mounted not more than 12″ below the countertop when acceptable to the AHJ. The construction shall be for the physically impaired or where the island or peninsular counter space construction prevents practical mounting above the countertop.

All counter-mounted receptacles in dwelling unit kitchens shall be GFCIs.

Dwelling Unit Bathrooms – 210-52(d)

At least one wall receptacle outlet shall be installed adjacent to each basin in all dwelling unit bathrooms. See Figure 6-24. The intent of the CMP in using the word adjacent was to allow one wall receptacle outlet between two basins to satisfy this requirement. These receptacle outlets shall be supplied by at least one 20 A branch circuit and the circuit shall have no other outlets. This 20 A circuit may supply receptacle outlets for other bathrooms within the dwelling unit. Receptacle outlets shall not be installed in a face-up position in the countertop in a bathroom basin location.

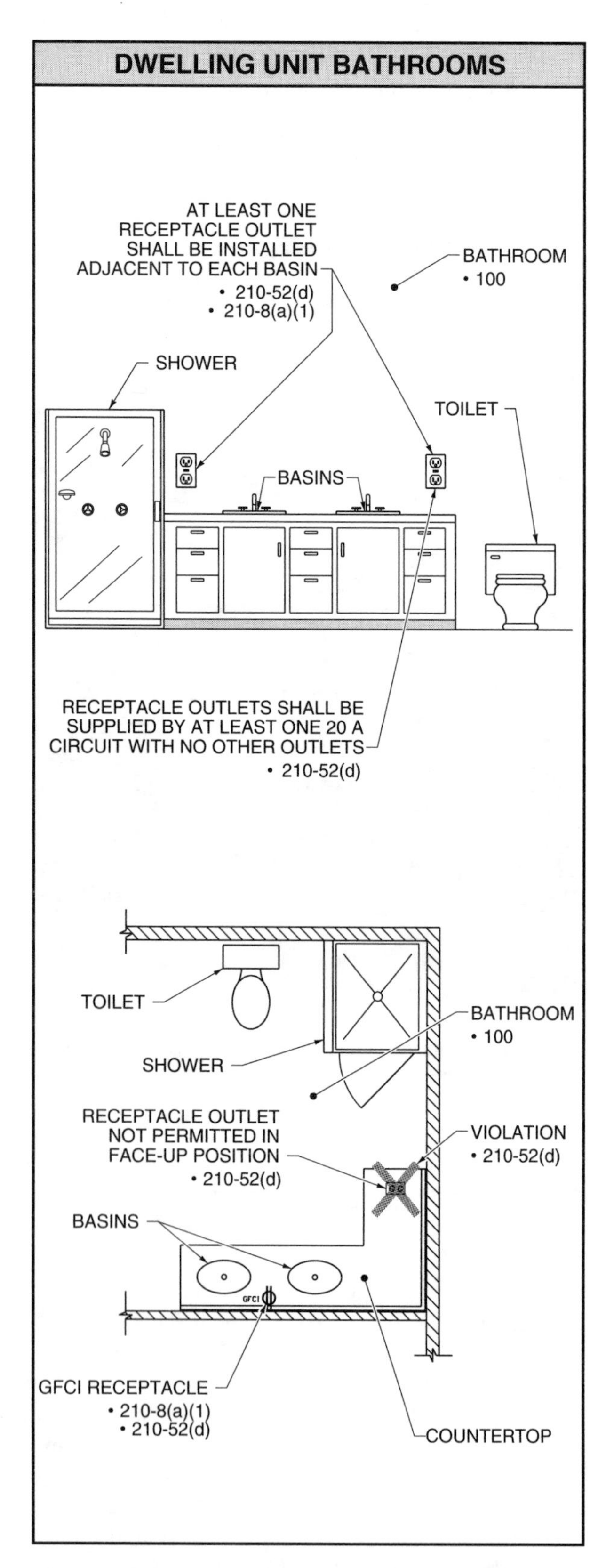

Figure 6-24. At least one wall receptacle shall be installed adjacent to each basin in all dwelling unit bathrooms.

Dwelling Unit Outdoor Outlets – 210-52(e)

At least one receptacle outlet shall be installed outdoors at grade level at the front and the back of each one-family dwelling and each unit of a two-family dwelling. These receptacle outlets shall be accessible at grade level and not more than 6′-6″ above grade. See Figure 6-25.

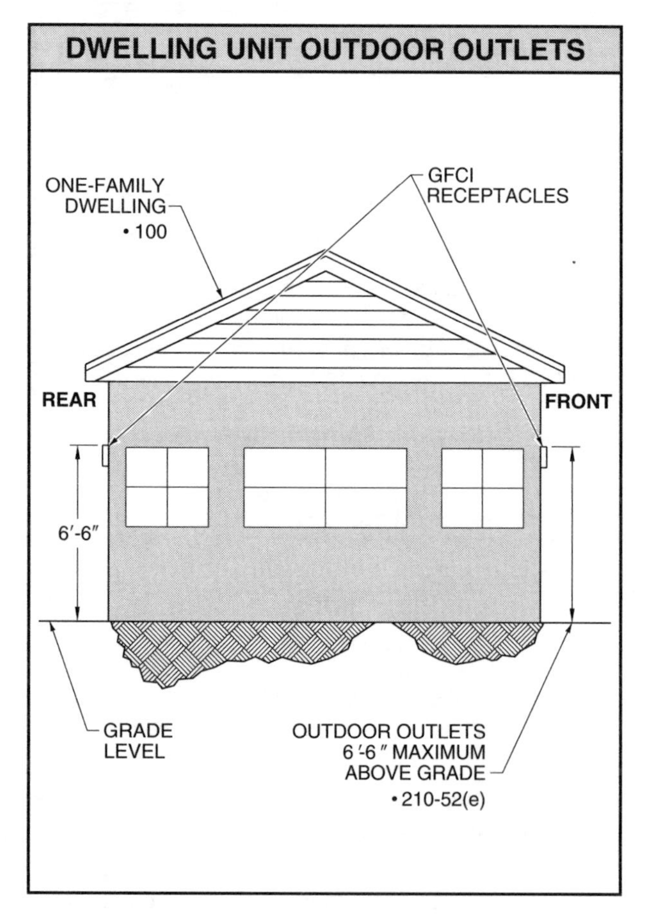

Figure 6-25. At least one receptacle shall be installed outdoors at grade level at the front and back of each one-family dwelling and each unit of a two-family dwelling.

Dwelling Unit Laundry Areas – 210-52(f)

At least one receptacle outlet shall be installed in dwelling units for the laundry. Per 210-52(f), Ex. 1, the laundry receptacle is not required in each unit of an apartment or living area in a multifamily building where laundry facilities are provided on the premises that are available to the building occupants. Per 210-52(f), Ex. 2, the laundry receptacle is not required in other than one-family dwellings where laundry facilities are not to be installed or permitted.

Dwelling Unit Basements and Garages – 210-52(g)

At least one receptacle outlet shall be installed in the basement of a one-family dwelling. This outlet is in addition to any outlets provided for laundry equipment. Also, at least one receptacle outlet shall be installed in each attached garage and in each detached garage which has electric power.

Dwelling Unit Hallways – 210-52(h)

At least one receptacle outlet shall be installed in 10′ or longer hallways in dwelling units. The hallway length is considered the length along the centerline of the hall without passing through a doorway. This outlet prevents the need to use extension cords for hallway table lamps, vacuum cleaners, etc.

Guest Rooms – 210-60

Guest rooms in hotels, motels, and similar occupancies shall have receptacle outlets installed per 210-52(a). The number of outlets required is based upon the general rule that no point along the floor line in any wall space shall be more than 6′, measured horizontally, from an outlet in that space. However, per 210-60, Ex., the required number of receptacle outlets is permitted to be located conveniently for the permanent furniture layout and to be readily accessible.

Heating, Air-Conditioning, and Refrigeration Equipment Outlets – 210-63

A 15 A or 20 A, 125 V, 1ϕ receptacle outlet shall be installed at an accessible location for the purpose of servicing HACR equipment on rooftops, in attics, and crawl spaces. On rooftops, this receptacle shall be located on the same level of the roof as the HACR equipment and within 25′ of the equipment. The receptacle outlet shall not be connected to the load side of the equipment disconnect. The exception to this requirement is for HACR equipment on rooftops of one- and two-family dwellings.

LIGHTING OUTLETS REQUIRED – 210-70

A *lighting outlet* is an outlet intended for the direct connection of a lampholder, lighting fixture, or pendant cord terminating in a lampholder. Section 210-70 is the section that requires lighting outlets for safety. Lighting outlets are required for safety in dwelling units, guest rooms, and other locations.

Dwelling Units – 210-70(a)

At least one wall switch-controlled lighting outlet shall be installed in the following locations of each dwelling unit:

- Every habitable room
- Bathrooms
- Hallways
- Stairways
- Attached garages and detached garages with electric power
- Near the exterior side of outdoor entrances or exits. A vehicle door for a garage is not considered as an outdoor entrance or exit.

At least one lighting outlet controlled by a light switch is controlled at the point of entry to the attic, under-floor space, utility room, and basement if these spaces are used for storage or contain equipment requiring servicing. The lighting outlet shall be located at or near the equipment requiring servicing.

A lighting outlet shall be installed for interior stairways with six or more steps between floor levels. There shall be a wall switch at each floor level to control the lighting outlet.

Per 210-70(a), Ex. 1, in habitable rooms other than kitchens and bathrooms, one or more receptacles controlled by a wall switch shall be permitted instead of lighting outlets. The intent is to comply with the demands of homeowners. Most people prefer switch-controlled table and floor lamps as opposed to switch-controlled ceiling lighting outlets.

Per 210-70(a), Ex. 2, remote, central, or automatic lighting control is permitted in hallways, stairways, and at outdoor entrances. Per 210-70(a), Ex. 3, lighting outlets may be controlled by occupancy sensors provided the occupancy sensor is in addition to the wall switch and it is located at a customary wall switch location. It is also equipped with a manual override which allows the sensor to function as a wall switch.

Guest Rooms – 210-70(b)

At least one wall switch-controlled lighting outlet or wall switch-controlled receptacle shall be installed in guest rooms in hotels, motels, or similar occupancies. This is generally located on the inside wall near the entrance door for the convenience of guests.

Other Locations – 210-70(c)

At least one wall switch-controlled lighting outlet shall be installed at or near equipment requiring servicing in attics or under-floor spaces. The wall switch shall be located at the point of entry to the attic or under-floor space.

FEEDER LOADS

The actual load on a feeder depends upon the total load supplied by the feeder, multiplied by the demand factor. If an entire connected load is operating at the same time, the demand factor is 100%. The maximum load, or maximum demand, is equal to the total connected load. If the maximum operating load is only three-quarters of the total connected load, then the demand factor is considered to be 75%.

FEEDERS – 215

A *feeder* is all of the circuit conductors between the service equipment or the source of a separately derived system and the final branch-circuit overcurrent device. The actual load on a feeder depends on the total load connected to the feeder and the demand factor. *Demand factor* is the ratio of the maximum demand of a system, or part of a system, to the total connected load of a system or the part of the system under consideration. See 100.

Minimum Rating and Size – 215-2

Feeder conductors shall have an ampacity not less than that required to supply to load. Feeder conductors for dwelling units and mobile homes are not required to be larger than the service-entrance conductors. Note 3 of Notes to Ampacity Tables of 0 to 2000 Volts is permitted to be used in sizing some feeder installations. See Figure 6-26.

AMPACITIES FOR 1ϕ, 3-WIRE DWELLING SERVICES AND FEEDERS		
CONDUCTOR	TABLE 310-16	NOTE 3 OF NOTES TO AMPACITY TABLES OF 0 TO 2000 VOLTS
NO. 4 THW Cu	85 A	100 A
NO. 2 THW Cu	115 A	125 A

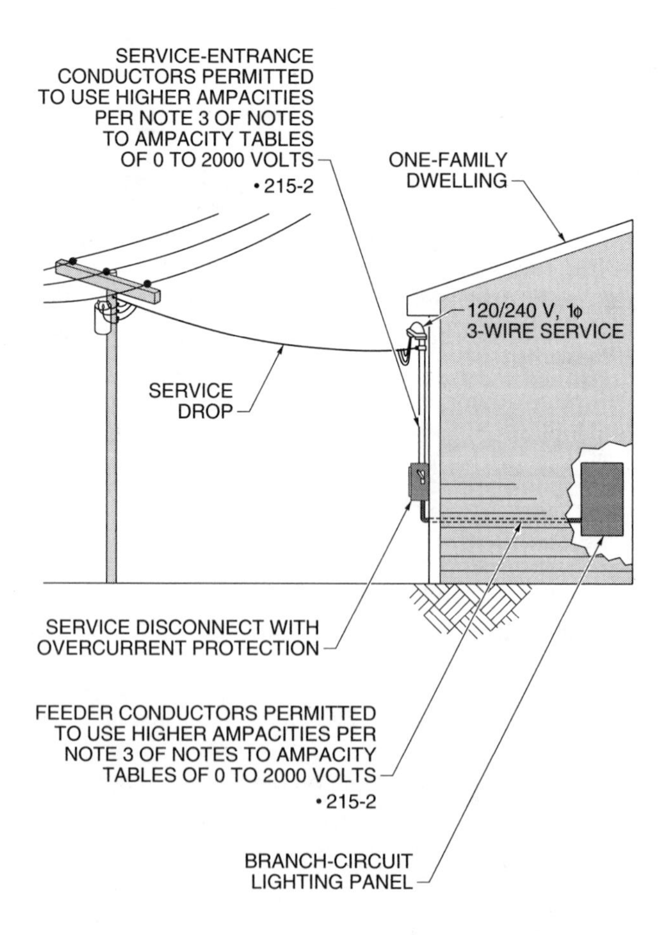

Figure 6-26. Note 3 of Notes to Ampacity Tables of 0 to 2000 Volts is permitted to be used in sizing conductors.

Feeders with Common Neutral – 215-4(a)

Feeders containing a common neutral are permitted to supply two or three sets of 3-wire feeders, or two sets of 4-wire or 5-wire feeders. Per 215-4(b), feeder conductors, including the neutral supplying AC current, shall be grouped together when run in a metallic raceway or enclosure. This avoids induction heating of the surrounding metal. Parallel conductors run through multiple raceways shall contain the conductors from each phase plus the neutral.

Identifying High-Leg in Delta 4-Wire Systems – 215-8

The circuit conductor with the higher voltage-to-ground in a delta 4-wire system shall be identified. The means of identification shall be by orange finish, tagging, or other effective means. This identification shall occur wherever the high-leg conductor and the neutral are accessible, such as junction boxes, pull boxes, troughs, or other accessible locations. See Figure 6-27.

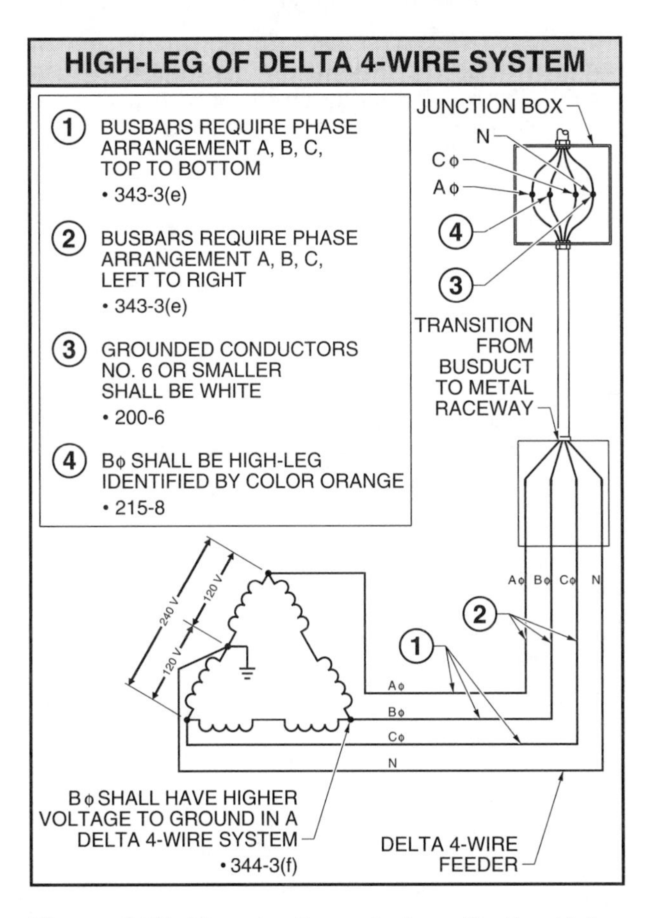

Figure 6-27. The circuit conductor with the higher voltage-to-ground in a delta 4-wire system shall be identified.

The two most common 3φ, 4-wire, wye systems used to feed AC lighting and AC motor loads are 120/208 V and 277/480 V systems. The neutral conductor does not carry any current for the power or motor portions of these loads. The current flowing through the neutral conductor is due to the lighting loads or circuits where the neutral is used. On these 3φ, 4-wire, wye systems, 220-22 permits a 70% demand factor to be used for that portion of the neutral load in excess of 200 A. For example, if the maximum unbalanced load that the neutral carries is calculated to be 400 A, the neutral would have to be sized to carry 140 A (400 A – 200 A = 200 A × 70% = 140 A). The first 200 A is figured at 100% and the remaining 200 A at 70% to equal 340 A (200 A × 100% = 200 A + 200 A × 70% = 140 A. 200 A + 140 A = 340 A). This derating factor does not apply to that portion of the load that supplies non-linear loads, such as electric discharge lighting, personal computers, and variable speed controllers.

Section 220-22 also permits a 70% demand factor to be applied to the maximum unbalanced load for the ungrounded conductors of feeders supplying clothes dryers, household ranges, wall-mounted ovens, and counter-mounted cooking units.

A delta 4-wire system is popular where there is a large 3φ load and a small 1φ load. See Figure 6-28. By mid-tapping off one of the secondary windings, a 1φ, 3-wire system can be provided for the 1φ load.

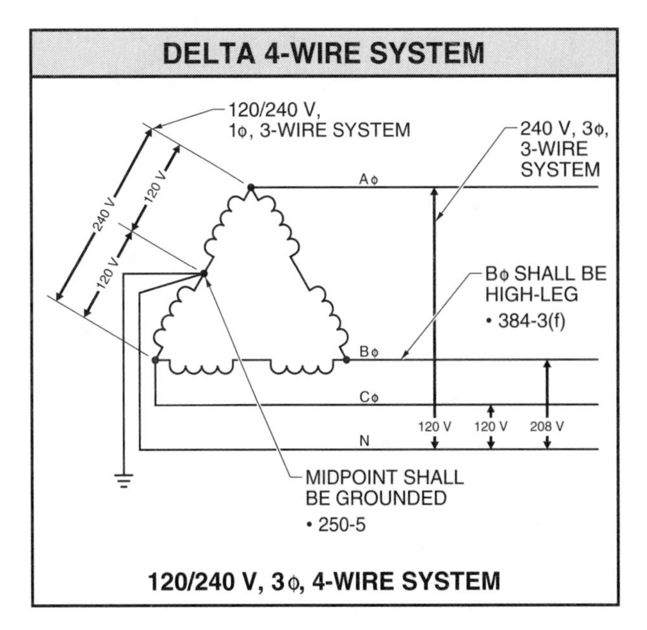

Figure 6-28. A delta 4-wire system is popular where there is a large 3φ load and small 1φ load.

The conductor coming off this midpoint tap then acts as a neutral and 250-5(b)(3) and 250-25(5) require this point to be grounded. The voltage between this midpoint and the two ends of the secondary winding is 120 V. The voltage across the secondary winding that was midpoint-tapped is 240 V. The remaining point of the 3ϕ system (B phase) has a voltage-to-ground of 208 V (120 V × 1.73 = 208 V).

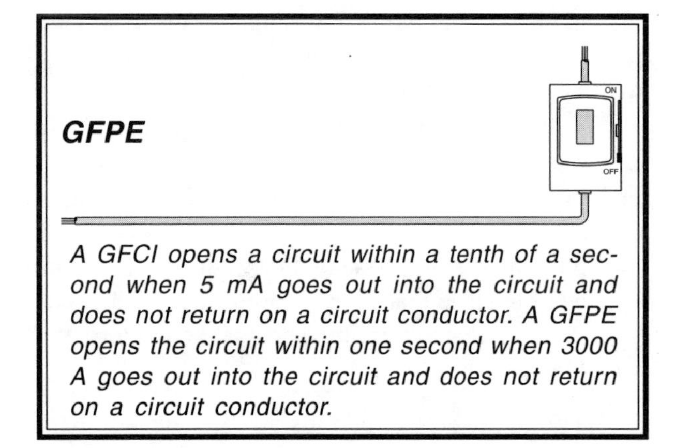

A GFCI opens a circuit within a tenth of a second when 5 mA goes out into the circuit and does not return on a circuit conductor. A GFPE opens the circuit within one second when 3000 A goes out into the circuit and does not return on a circuit conductor.

When tapping into an existing delta 4-wire system in order to take off a 120/240 V, 1ϕ, 3-wire system, it is imperative to know which phase conductor is the high-leg. This is the B phase, which is the conductor with a voltage-to-ground of 208 V. This is the reason that 215-8 requires the phase conductor with the higher voltage-to-ground to be identified by the color orange. If the B phase were tapped into while trying to create a 1ϕ, 3-wire system, half of the 1ϕ load would be destroyed due to the 208 V.

Ground-Fault Protection of Equipment – 215-10

Ground-fault protection of equipment (GFPE) is required for feeder disconnects rated at 1000 A or more. This protection is only required in solidly-grounded wye systems where the voltage-to-ground is more than 150 V and the phase-to-phase voltage does not exceed 600 V. For example, protection is required for 277/480 V, 3ϕ, 4-wire systems. A similar requirement is found in 230-95 for services.

The intent of this requirement is to prevent help burndowns on feeders and services in this voltage range. GFPE is not required on a feeder if this protection is provided on an upstream feeder, or at the service per 215-10, Ex.

Review Questions

________ **1.** ________ conductors are the conductors from the service point to the service disconnecting means.

________ **2.** ________ conductors are all circuit conductors between the service equipment and the final branch-circuit overcurrent device.

________ **3.** The ________ is that portion of the electrical circuit between the last overcurrent device and the outlets or utilization equipment.

________ **4.** A(n) ________ load is a load where the wave shape of the steady-state current does not follow the wave shape of the applied voltage.

________ **5.** The grounded conductor of a branch circuit shall be identified by the color white or natural ________.

T F **6.** Branch circuits with not over 120 V between conductors are permitted to supply auxiliary equipment of electric-discharge lamps.

T F **7.** A GFCI receptacle or CB is set to trip at 10 mA.

________ **8.** A bathroom is an area with a ________ and one or more of a ________.

A. toilet; basin, tub, or shower
B. basin; toilet, tub, or shower
C. tub: toilet, basin, or shower
D. shower; tub, toilet, or basin

________ **9.** A GFCI-protected receptacle is required within ________′ of the outside edge of a wet bar sink.

T F **10.** All receptacles installed in dwelling bathrooms shall be GFCI-protected.

T F **11.** All receptacles which serve countertops in dwelling kitchens shall be GFCI-protected.

________ **12.** Branch-circuit conductors shall be No. ________ or larger.

________ **13.** Receptacle outlets installed in a dwelling for specific appliances shall be within ________′ of the appliance.

________ **14.** The required dwelling unit outdoor receptacles shall be not more than ________ above finished grade level.

______ **15.** At least one receptacle outlet shall be installed in ________′ or longer hallways of dwelling units.

______ **16.** The required 15 A or 20 A, 125 V, 1ϕ, receptacle on rooftops for HACR servicing shall be within ________′ of the equipment.

______ **17.** A lighting outlet shall be installed in dwellings for interior stairways with ________ or more steps between floor levels.

______ **18.** ________ is the ratio of the maximum demand of a system to the total connected load.

______ **19.** Branch circuits are rated in accordance with the maximum ampere rating or setting of the ________.

______ **20.** Branch-circuit EGCs shall have a continuous outer finish that is either green or green with one or more ________ stripes unless it is bare.

______ **21.** High-discharge lighting, operating at 480 V in a tunnel shall be mounted at least ________′ in height.

______ **22.** A current flow of ________ mA will cause the heart to stop pumping.

T F **23.** All receptacles in dwelling crawl spaces at or below grade shall be GFCI-protected.

T F **24.** No point along the wall line for a dwelling unit countertop shall be more than 48″ from a receptacle outlet.

T F **25.** At least one wall receptacle outlet shall be installed adjacent to each basin in all dwelling unit bathrooms.

Wall Receptacles

______ **1.** The maximum dimension at A is ________′.

______ **2.** The maximum dimension at B is ________′.

______ **3.** The maximum dimension at C is ________′.

______ **4.** The maximum dimension at D is ________′.

T F **5.** A receptacle is required at E.

______ **6.** The maximum dimension at F is ________′.

______ **7.** The maximum dimension at G is ________′.

T F **8.** A receptacle is required at H.

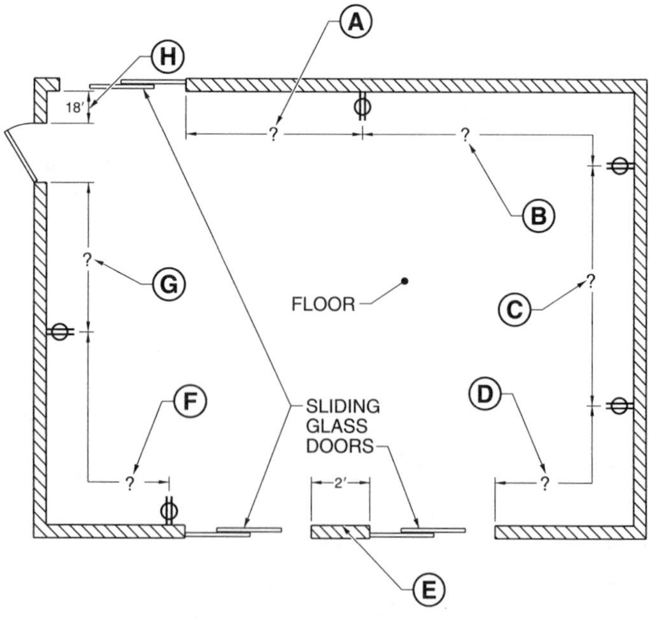

DWELLING PLAN VIEW

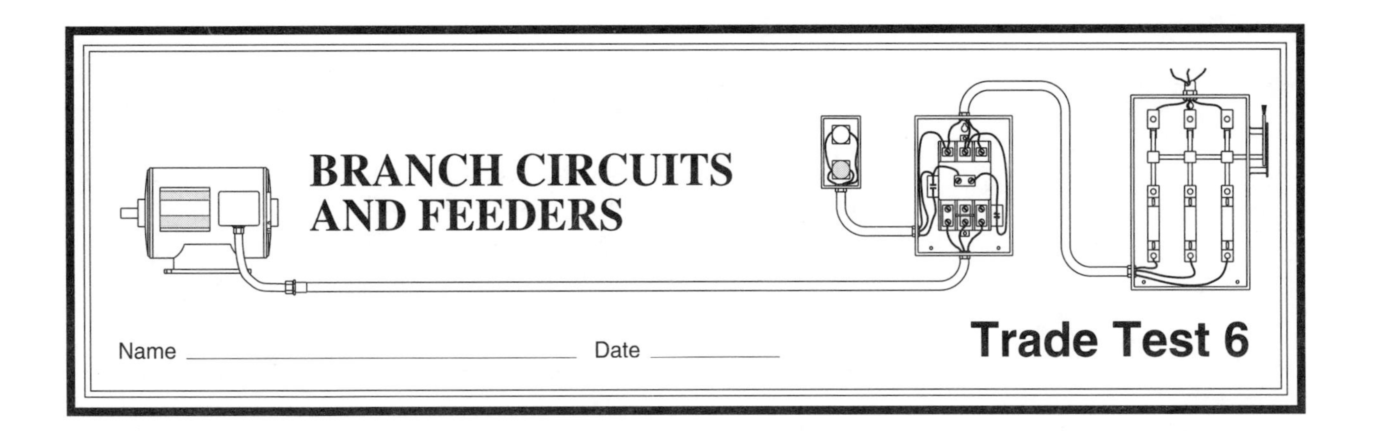

NEC®	Answer	
______	______	**1.** A parking lot lighting installation requires the use of No. 10 THW conductors with a listed ampacity of 35 A to be installed to limit the voltage drop on the circuit. In the panelboard, the No. 10 conductors are terminated on a 20 A OCPD. Determine the branch-circuit rating.
______	______	**2.** *See Figure 1.* Does the multiwire branch circuit run to supply lighting in a commercial installation violate Article 210 of the NEC®?

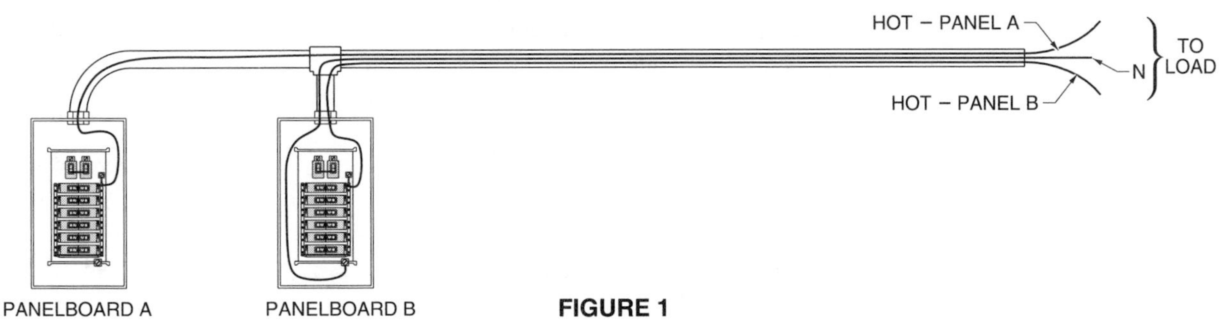

FIGURE 1

______	______	**3.** Determine the minimum distance from ground permitted for auxiliary equipment of electric-discharge lighting fixtures used to illuminate a parking lot. The lighting fixtures operate at 480 V to ground.
______	______	**4.** *See Figure 2.* Does the installation violate the provisions of Article 210 of the NEC®?

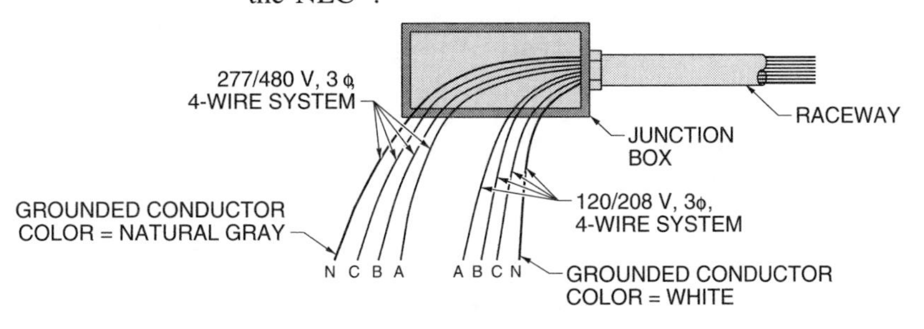

FIGURE 2

______	______	**5.** A contractor installs a 15 A, 125 V, 1φ receptacle in the bathroom of a one-family dwelling. The receptacle is not GFCI-protected, but it is installed in a location which is not readily accessible. Does this installation violate Article 210 of the NEC®?

_______________ _______________ **6.** A 240 V, 6 kW household electric range is installed in the basement of a one-family dwelling. Determine the minimum branch-circuit rating for the circuit which supplies the range.

_______________ _______________ **7.** *See Figure 3.* What size OCPD (using CBs) and what size branch-circuit conductors are required for the continuous load?

_______________ _______________ **8.** *See Figure 4.* What size OCPD (using CBs) and what size branch-circuit conductors are required for the continuous and noncontinuous loads?

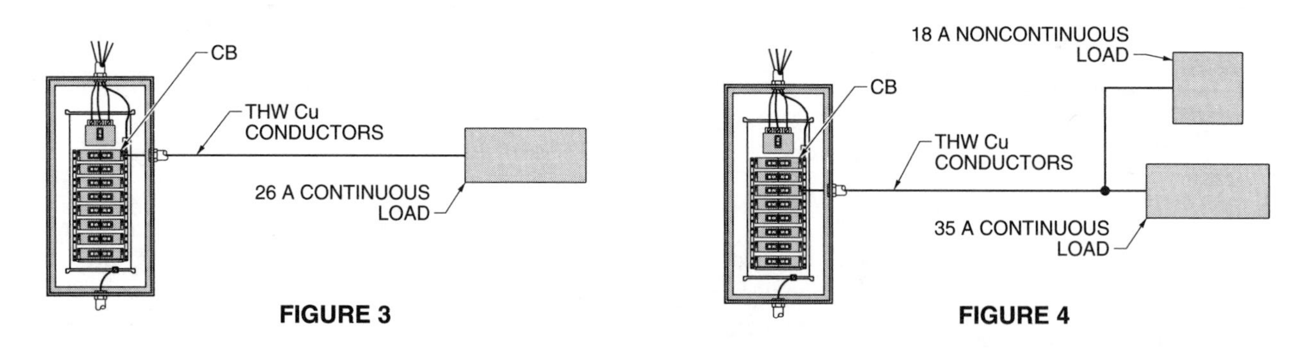

FIGURE 3

FIGURE 4

_______________ _______________ **9.** Determine the maximum rating for a single cord- and plug-connected piece of utilization equipment which is installed on a 30 A branch circuit.

_______________ _______________ **10.** Determine the maximum rating for a single cord- and plug-connected piece of utilization equipment which is fastened-in-place and installed on a 20 A branch circuit. The branch circuit also supplies lighting units in addition to the utilization equipment.

_______________ _______________ **11.** A 40 A branch circuit is installed in a dwelling unit to supply lighting fixtures equipped with heavy-duty lampholders. Does this installation violate Article 210 of the NEC®?

_______________ _______________ **12.** *See Figure 5.* Determine the minimum number of receptacles required to be installed in the dwelling unit den.

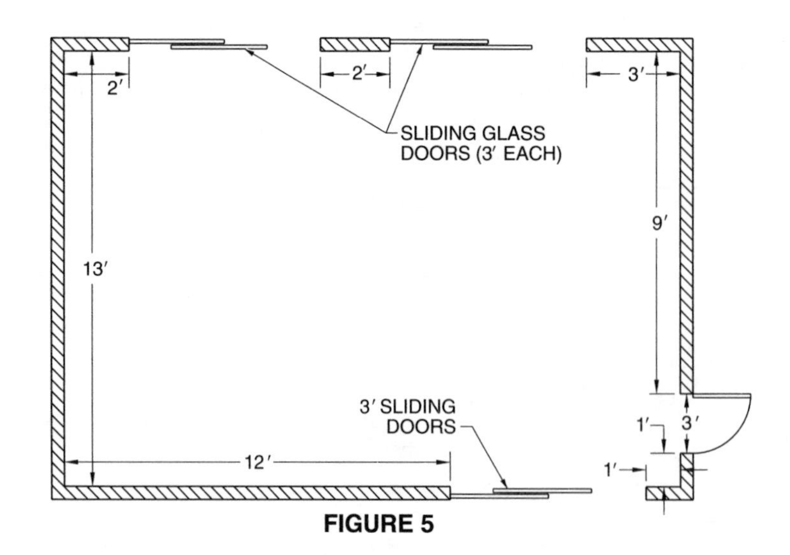

FIGURE 5

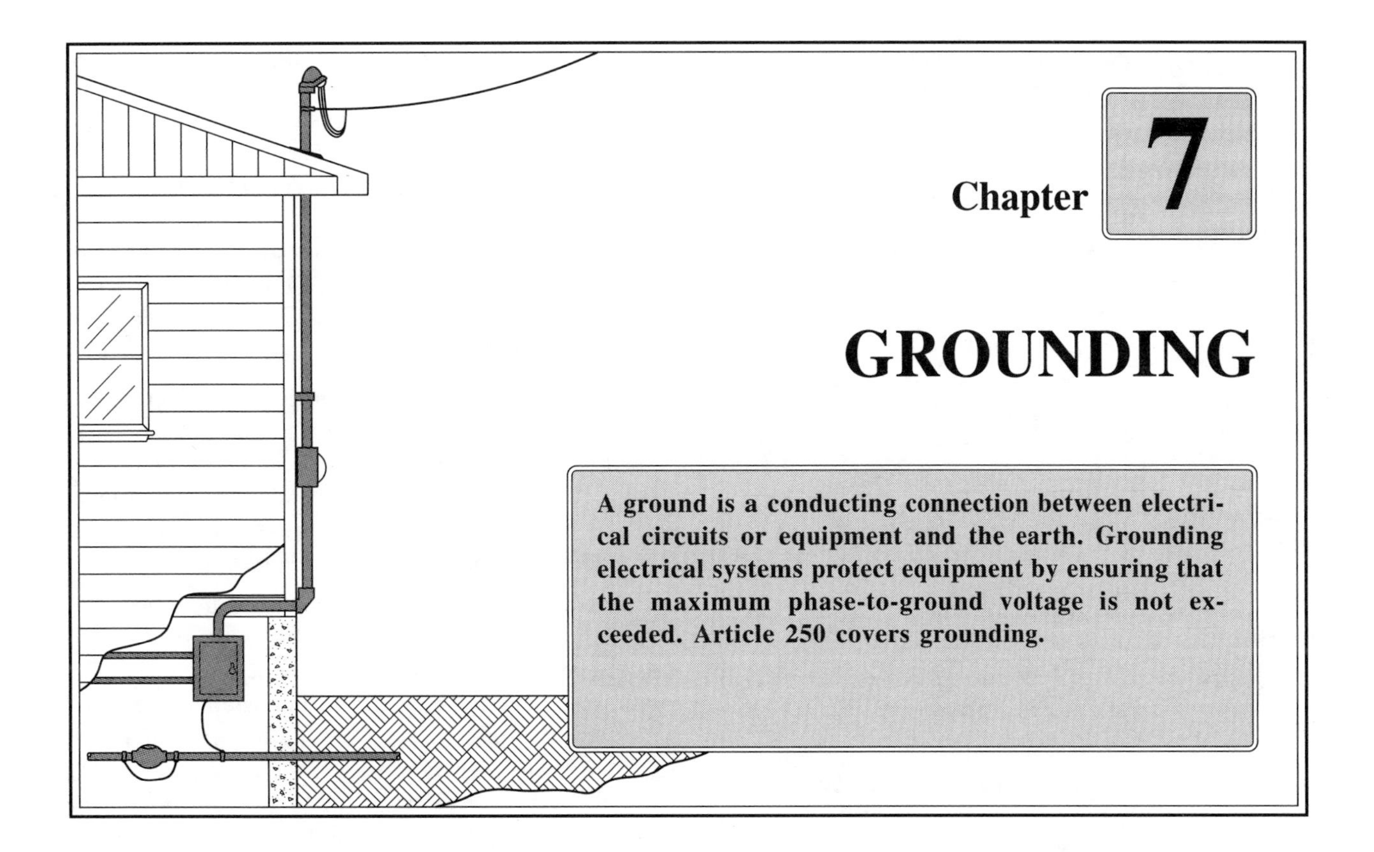

Chapter 7

GROUNDING

A ground is a conducting connection between electrical circuits or equipment and the earth. Grounding electrical systems protect equipment by ensuring that the maximum phase-to-ground voltage is not exceeded. Article 250 covers grounding.

GROUNDING – 250

Proper grounding of electrical systems is one of the most critical concerns facing installers and designers of electrical systems. Despite this, no other conductor is so often installed improperly. Section 250-1, and the accompanying FPNs list the major reasons why electrical systems and equipment are grounded.

Electrical systems are grounded to limit and stabilize voltages to ground. Unintentional line surges, lightning strikes, or contact with higher voltage lines may result in voltages being placed on the electrical system that could damage or destroy electrical components and equipment. A *line surge* is a temporary increase in the circuit or system voltage or current that may occur as a result of fluctuations in the electrical distribution system. Grounding electrical systems protects equipment by ensuring that the maximum phase-to-ground voltage is not exceeded. *Phase-to-ground voltage* is the difference of potential between a phase conductor and ground. For example, the phase-to-ground voltage in a 120/240 V, 1ϕ, 3-wire system is 120 V. *Phase-to-phase voltage* is the maximum voltage between any two phases of an electrical distribution system. In the 120/240 V, 1ϕ, 3-wire system, the phase-to-phase voltage is 240 V.

Electrical systems are grounded to ensure that a low-impedance ground path for fault current is present. A low-impedance ground path is a ground path that contains very little resistance or opposition to the flow of fault current. The impedance of the ground path is kept low by ensuring that all electrical connections and any other connections or terminations in the ground path are electrically continuous and done in a manner that ensures very little opposition to the flow of current. The low-impedance ground path helps to ensure that overcurrent protection devices open under ground-fault conditions.

Ohm's Law states that the amount of current (I) that flows in any circuit is determined by the voltage (E) and resistance (R) of the circuit ($I = \frac{E}{R}$). Ground-fault circuits are no different. If there is a high resistance or impedance in the ground-fault path, the amount of current that flows is smaller and it may not permit the overcurrent protection device to operate. See Figure 7-1.

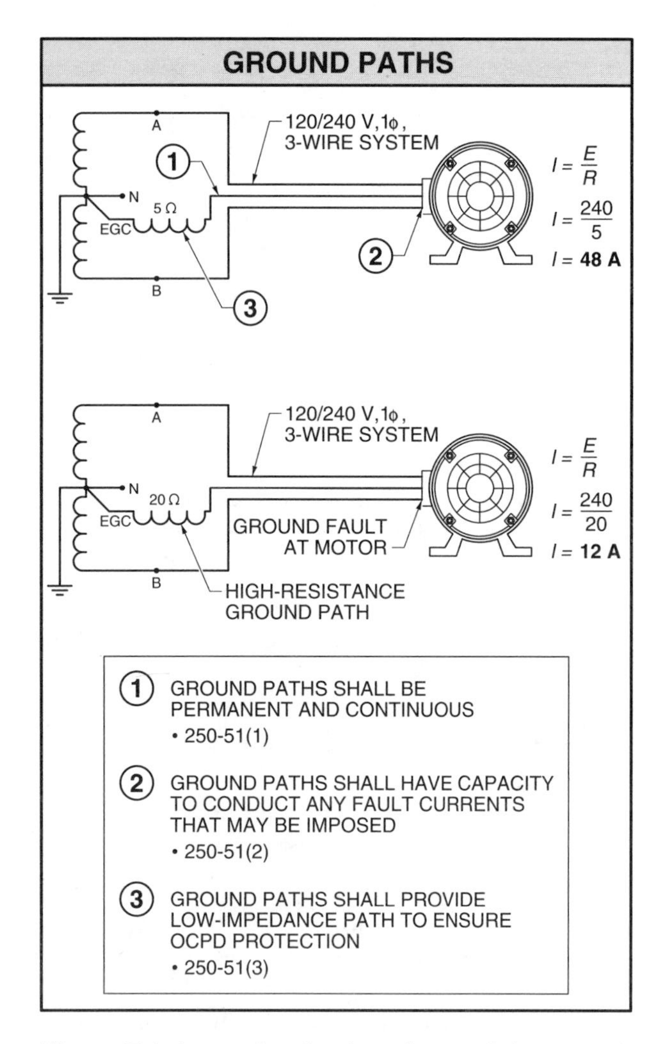

Figure 7-1. Increasing the impedance of the ground-fault path reduces the amount of current that flows in the circuit.

Ground faults can, and often times do occur. Proper equipment grounding is required to limit the duration of the fault and protect electrical equipment and personnel. Three provisions are necessary to ensure that an effective grounding path is present per 250-51. The grounding path shall be permanent and continuous per 250-51(1). Splices are permitted in grounding conductors but they must be done in a manner that maintains a low-circuit impedance. The grounding path shall be capable of handling any fault currents that are likely to be placed upon it per 250-51(2). The size of the grounding conductor shall be adequate for the available fault current. The ground path shall be of a sufficiently low impedance to permit the ready operation of the OCPD per 250-51(3). This is the primary purpose of EGCs and the most important factor in ensuring personnel safety.

Grounding Terminology – 100

Terms used in grounding are similar and may be inadvertently misused or misinterpreted if care is not taken. For example, ground, grounded, effectively grounded, grounded conductor, grounding conductor, equipment grounding conductor (EGC), and grounding electrode conductor (GEC) are grounding terms that, while similar, are distinct. Each of these terms is defined in 100.

Ground. A *ground* is a conducting connection between electrical circuits or equipment and the earth. See Figure 7-2. This connection can be an intentional connection or it can result accidentally. Intentional connections are made to achieve the benefits of grounding as noted in 250-1, FPNs.

Accidental grounds result in an unintentional connection to the earth or something connected to the earth. For example, an ungrounded conductor's insulation may get nicked while being installed in a metal raceway. An unintentional ground occurs, and a fault current flows, depending upon the impedance of the ground path. *Impedance* is the total opposition to the flow of current in a circuit. In AC circuits, impedance consists of the total resistance, inductance, and capacitance of the circuit.

Often, the connection for a ground is not made directly to the earth but by means of an electrode that is, in turn, connected to the earth. In some parts of the world, the ground connection is known as earthing.

Grounded. *Grounded* is connected to the earth or a conducting body connected to the earth. Circuits, systems, or equipment that have been grounded have been connected to the earth directly or to some type of electrode which is, in turn, connected to the earth. For example, 250-32 requires that service raceways and enclosures shall be grounded. These raceways and enclosures shall be connected to the earth or a conducting body connected to the earth.

Effectively Grounded. *Effectively grounded* is grounded with sufficient low impedance and current-carrying capacity to prevent hazardous voltage buildups. This term was added to the definitions in 100 in the 1993 NEC®. Several provisions in the NEC® require equipment to be effectively grounded. For example, per 250-81(b), the metal frame of a building is required to be part of the grounding electrode system (GES) where it is effectively grounded. See Figure 7-3.

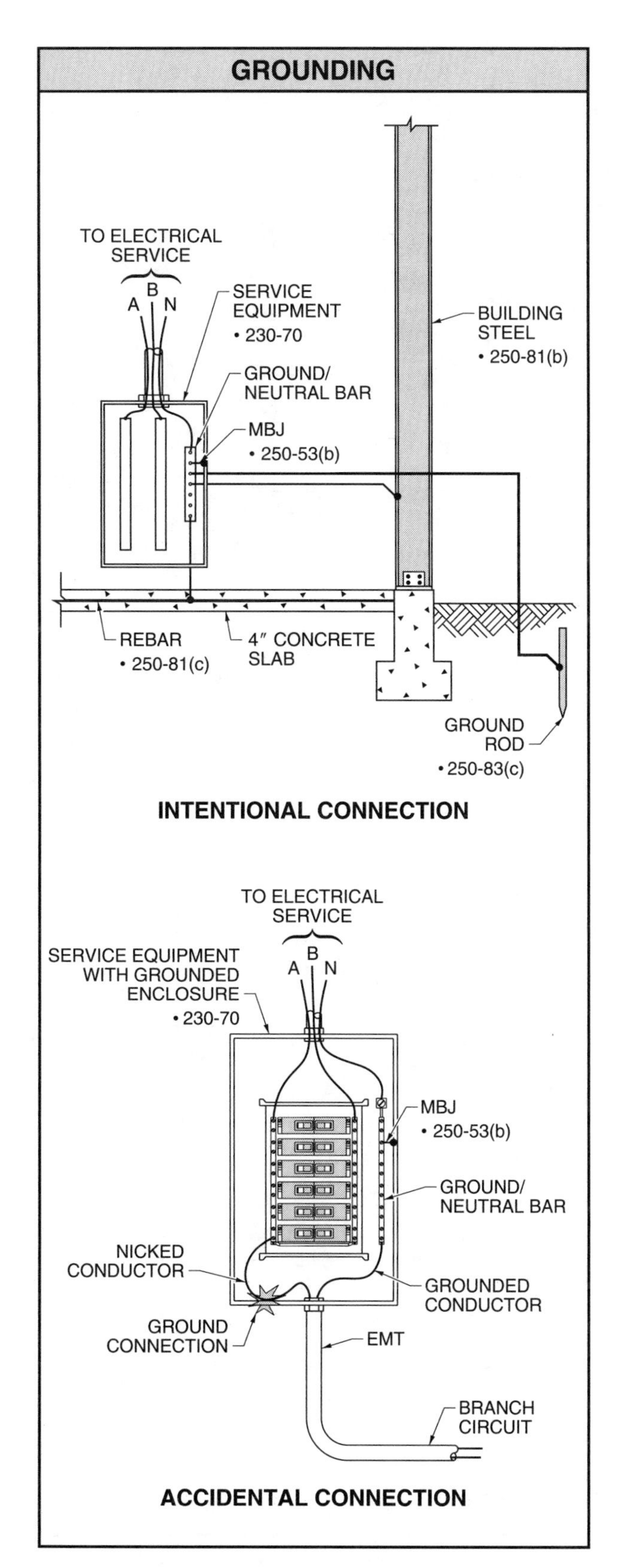

Figure 7-2. A ground is a conducting connection between electrical circuits or equipment and the earth.

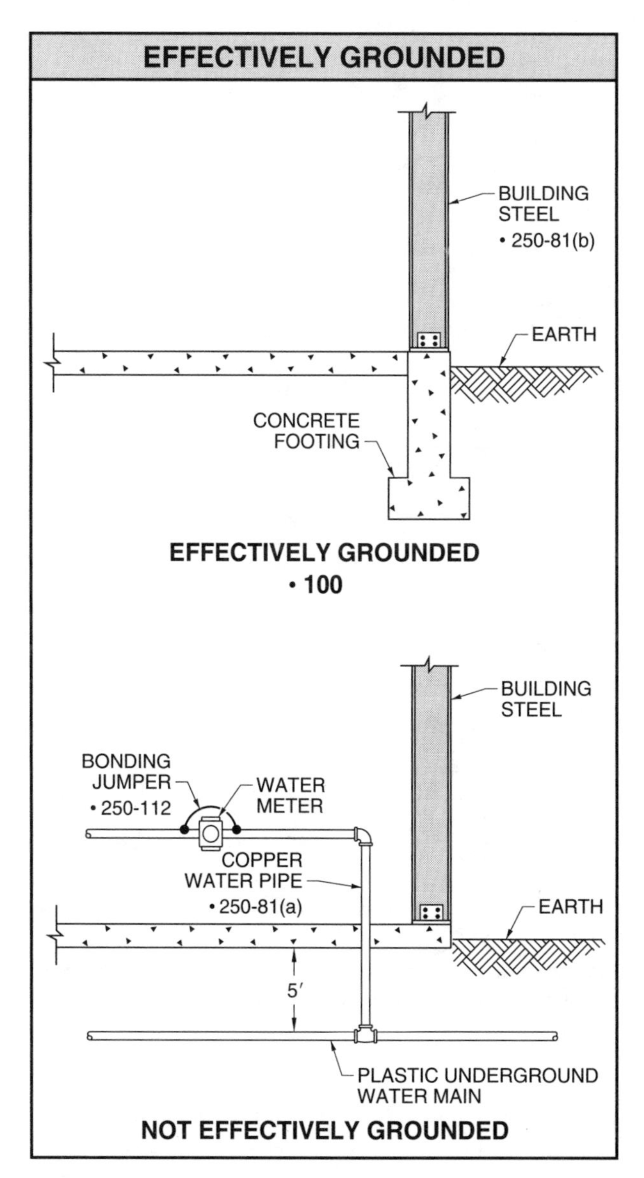

Figure 7-3. Electrodes that are not effectively grounded are not required to be connected to the GES.

Grounded Conductor. A *grounded conductor* is a conductor that has been intentionally grounded. The grounded conductor can be a system conductor or a circuit conductor. See Figure 7-4. The grounded conductor is commonly a neutral conductor. Not all electrical distribution systems, however, use the grounded conductor as a neutral. For example, corner-grounded delta systems contain a grounded conductor that is not a neutral conductor. Therefore, it is incorrect to refer to all grounded conductors as a neutral conductor, although that is the case in the majority of electrical distribution systems.

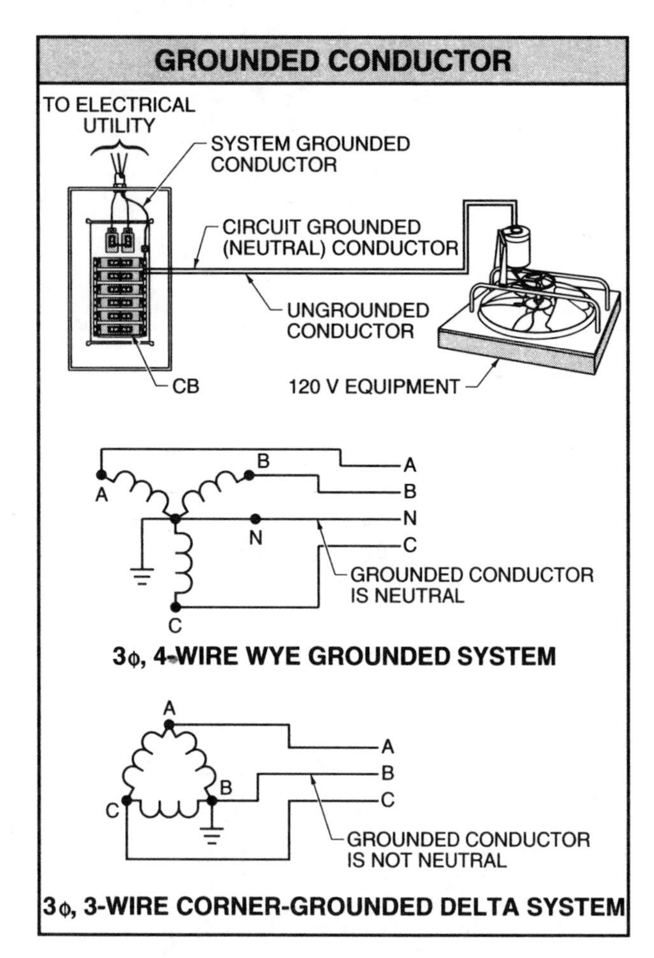

Figure 7-4. A grounded conductor is a conductor that has been intentionally connected or grounded.

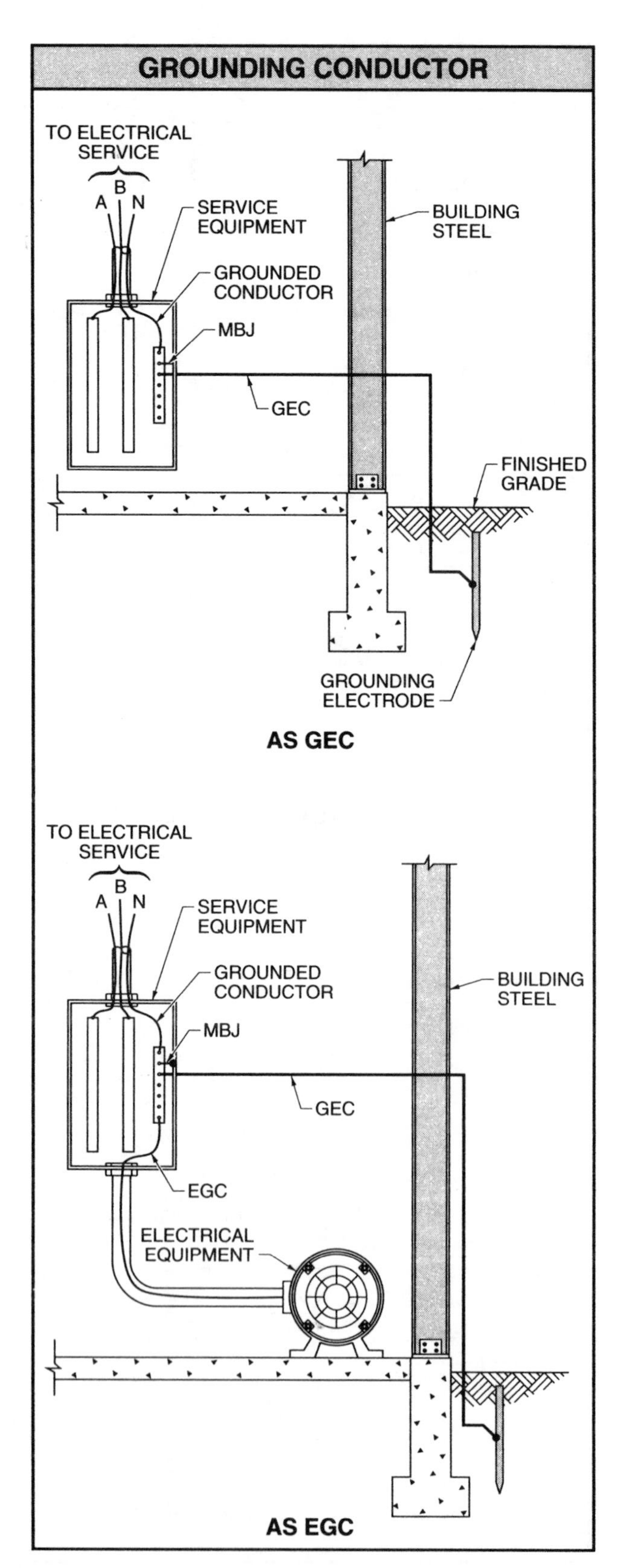

Figure 7-5. A grounding conductor is the conductor that connects electrical equipment or the grounded conductor to the grounding electrode.

Grounding Conductor. A *grounding conductor* is the conductor that connects electrical equipment or the grounded conductor to the grounding electrode. The grounding conductor can be a GEC or an EGC. In either case, the connection is made between electrical equipment or the grounded conductor and the grounding electrode(s). See Figure 7-5.

Equipment Grounding Conductor (EGC). An *equipment grounding conductor (EGC)* is an electrical conductor that provides a low-impedance path between electrical equipment and enclosures and the system grounded conductor and GEC. This connection usually occurs at the service equipment or at the source of power and helps to ensure the operation of the OCPD when fault conditions occur. EGCs may be in the form of a separate wire, conduit, tubing, cable sheath, etc.

The most important factor in reducing the risk of shock and electrocution in electrical distribution systems is the proper installation of the EGC. The EGC connects metal frames and enclosures of equipment and other noncurrent-carrying metal parts of equipment to the grounded conductor and/or the EGC. This connection can occur at the service equipment or at the source of a separately derived system, such as a transformer.

As noted in 250-1, FPN 1, the EGC is connected to the grounded system conductor to facilitate the operation of the OCPD by providing a low-impedance path for fault current to flow. This conductor is essential for life safety, yet most equipment can function quite normally with a poor EGC or even without an EGC. See Figure 7-6.

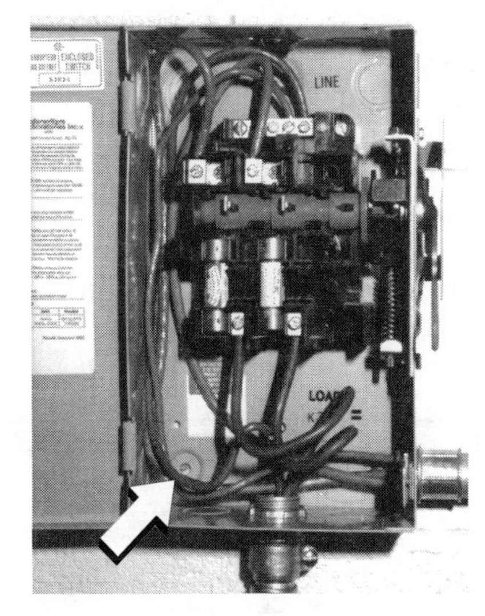

The EGC is essential for life safety.

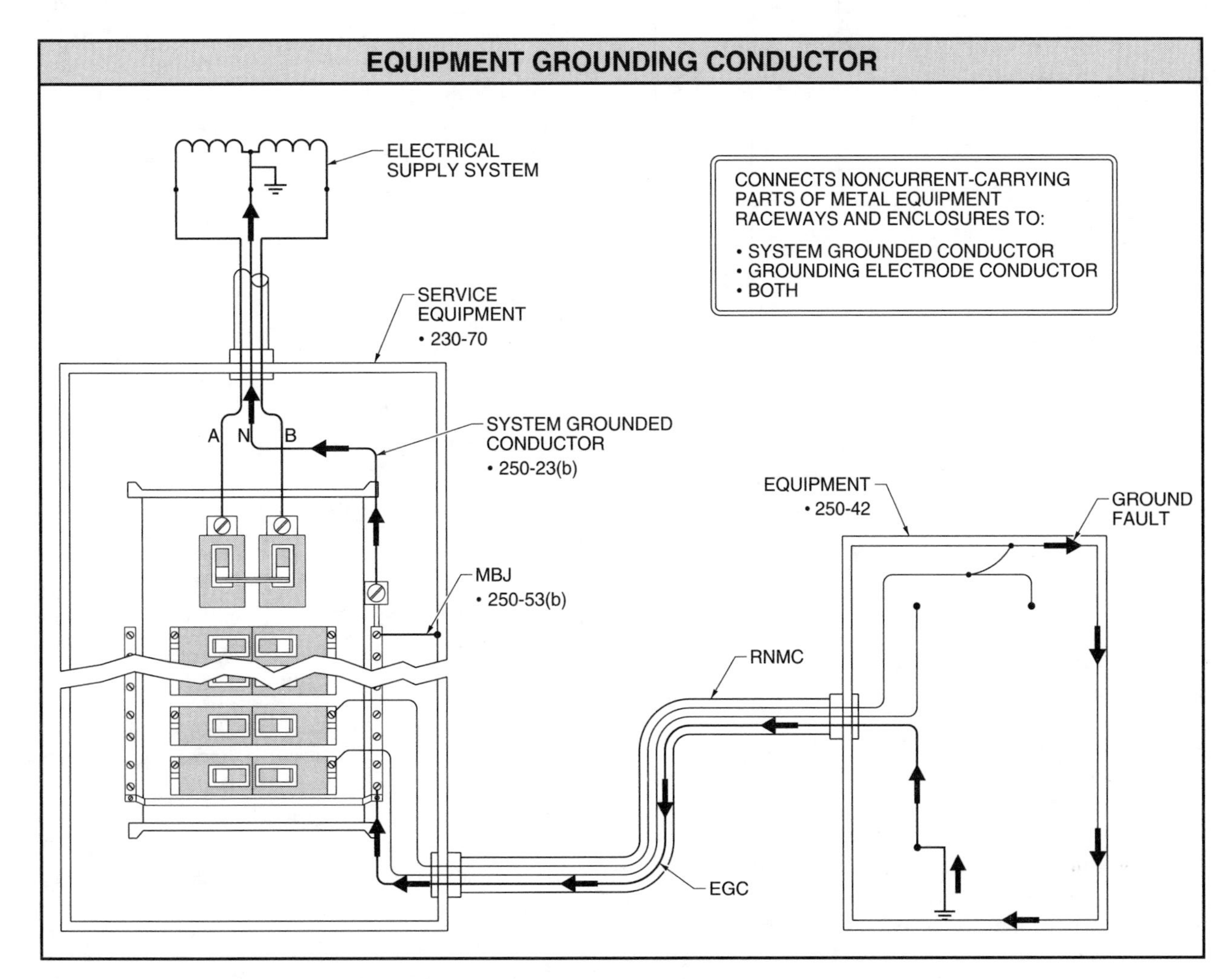

Figure 7-6. An equipment grounding conductor is an electrical conductor that provides a low-impedance path between electrical equipment and enclosures and the system grounded conductor and GEC.

Grounding Electrode Conductor (GEC). A *grounding electrode conductor* is the conductor that connects the grounding electrode(s) to the grounded conductor and/or the EGC. Like the EGC, this connection can occur at the service equipment or at the source of a separately derived system. The GEC may be constructed of aluminum, copper, or copper-clad aluminum; may be solid or stranded; and either insulated, covered, or bare per 250-91(a). See Figure 7-7.

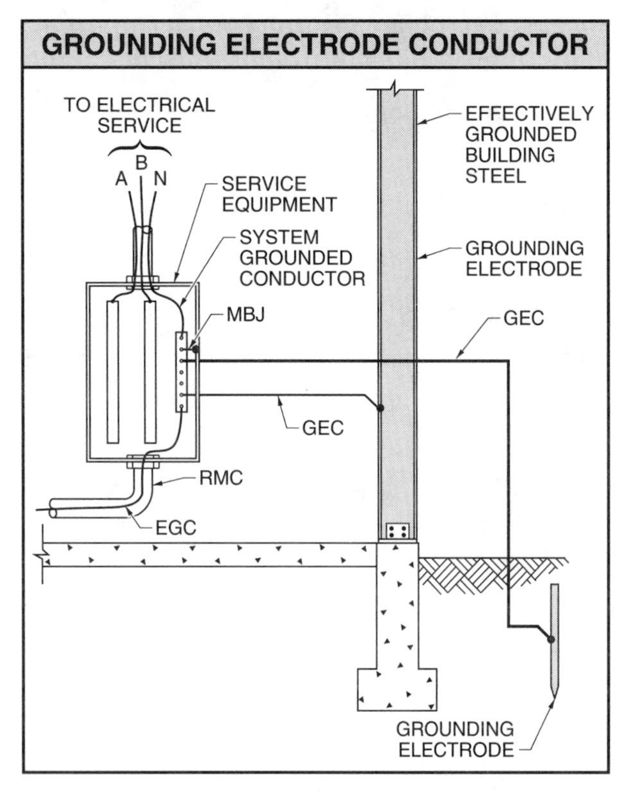

Figure 7-7. A GEC is the conductor that connects the grounding electrode(s) to the grounded conductor and/or the EGC.

GROUNDING SYSTEMS

Despite the benefits of grounding, there are electrical systems and circuits that are not grounded. Electrical systems can be classified as those that are required to be grounded, those that are permitted to be grounded, and those that are prohibited from being grounded. The first step in designing electrical systems is to determine if the system is to be grounded. Sections 250-3 for direct current (DC) systems and 250-5 for alternating current (AC) systems contain the requirements for system grounding. Section 250-7 lists the electrical circuits that are not permitted to be grounded. Once a determination has been made that the system is to be grounded, the NEC® is used to determine the manner in which the grounding occurs. Systems are grounded as follows:

- AC Systems – 250-5; 250-23
- DC Systems – 250-22
- Separately Derived AC Systems – 250-26
- Generators – 250-6

AC Systems – 250-5(a)(b); 250-23(a)

AC systems can be installed as either grounded or ungrounded systems. The voltage at which the system operates plays an important part in determining if the system is required to be grounded. Section 250-5(a) contains three conditions under which AC systems operating at less than 50 V are required to be grounded. The first condition involves systems that are supplied by transformers of circuits that operate at over 150 V to ground. These systems shall be grounded per 250-5(a)(1) because if a fault in the transformer windings occurs, the higher voltages of the primary could be imposed on the secondary side of the transformer.

The second condition under which an AC system of less than 50 V is required to be grounded occurs when the system is supplied from a transformer that is ungrounded. Per 250-5(a)(2), the low voltage system is required to be grounded because the supply system is ungrounded. Section 250-5(a)(3) requires AC systems of less than 50 V to be grounded when they are installed outside of a building as overhead conductors.

The NEC® does not require other AC systems operating at less than 50 V to be grounded. The decision as to whether to ground the system is left to the designer of the electrical system.

Section 250-5(b) provides the requirements for grounding AC electrical systems operating between 50 V and 1000 V. AC systems shall be grounded when the system can be grounded in a manner that the maximum voltage to ground does not exceed 150 V per 250-5(b)(1). Examples of these types of systems are 120 V, 1ϕ, 2-wire systems; 120/240 V, 1ϕ, 3-wire systems; and 120/208 V, 3ϕ, 4-wire systems.

Section 250-5(b)(2) requires 3ϕ, 4-wire, wye-connected AC systems in which the neutral is used as a circuit conductor to be grounded. The most common electrical system to fall under this condition is the 277/480 V, 3ϕ, 4-wire system. This system is frequently used to distribute power in commercial and industrial locations such as high-rise office buildings and manufacturing plants.

Section 250-5(b)(3) requires 3ϕ, 4-wire, delta-connected systems with the midpoint of one phase used as a circuit conductor to be grounded. This condition requires 120/240 V, 3ϕ, 4-wire delta systems to be grounded. These systems are commonly used in commercial and light industrial installations where the 3ϕ power requirements are large and the 120 V lighting and receptacle requirements are small.

Under some conditions, such as when part of a multiconductor cable, or where suitable for direct burial installations, the neutral conductor is permitted to be installed as an uninsulated conductor. These installations shall be grounded per 250-5(b)(4). See Figure 7-8.

Some highly specialized electrical installations permit AC systems which operate between 50 V and 1000 V to be installed without grounding. Section 250-5(b), Ex. 1 permits electric systems used exclusively to supply electric furnaces used in industry for the purposes of melting, refining, or tempering to be ungrounded. Another exemption from the grounding requirements is for industrial, adjustable-speed drive systems.

Section 250-5(b), Ex. 2 permits systems supplying rectifiers which supply these adjustable drive systems to be ungrounded if they are a separately derived system. Separately derived systems supplied from transformers with a primary voltage of less than 1000 V may also be installed ungrounded. Section 250-5(b), Ex. 3 permits this type of installation if the system is used solely for control circuits, only qualified persons service the installation, loss of the control power cannot be tolerated, and ground detector systems are installed.

High-impedance grounded neutral systems and isolated power systems are two other conditions under which AC systems of 50 V to 1000 V can be installed ungrounded. See 250-5(b), Ex. 4 and 5. The isolated power systems shall be in accordance with 517 and 668. The high-impedance grounded neutral system shall be installed for only line-to-line loads where continuity of power is required. Ground detection monitors are required for these systems and they shall be serviced by qualified personnel only.

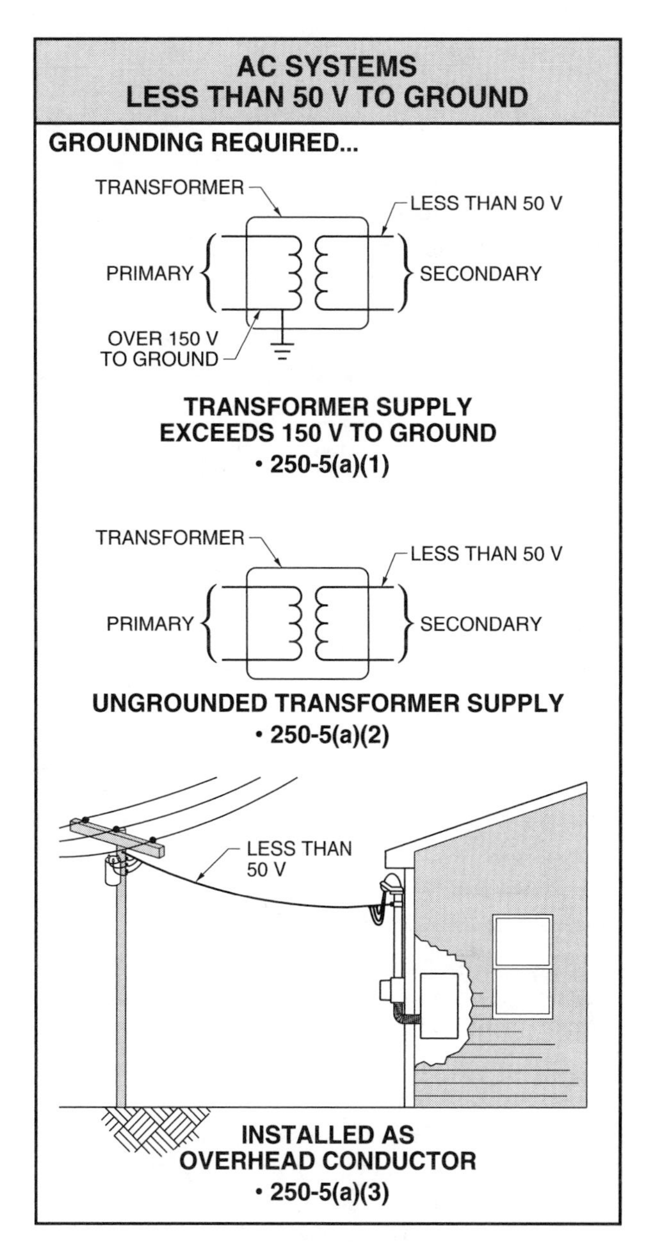

Figure 7-8. Three AC systems of less than 50 V are required to be grounded.

The majority of the electrical systems in the United States are required to be grounded per 250-5(b). Section 250-23 specifies the required grounding connections for these AC-supplied systems. Section 250-23(a) requires that all AC-supplied systems which are required to be grounded shall be connected to a GES meeting the requirements of 250, Part H. Therefore, each AC system which is required to be grounded shall have a GES.

Section 250-23(a) also requires that the system grounded conductor shall be connected to the GEC at any accessible point on the load end of the service drop or lateral up to and including the grounded conductor terminal in the service disconnecting means. See Figure 7-9.

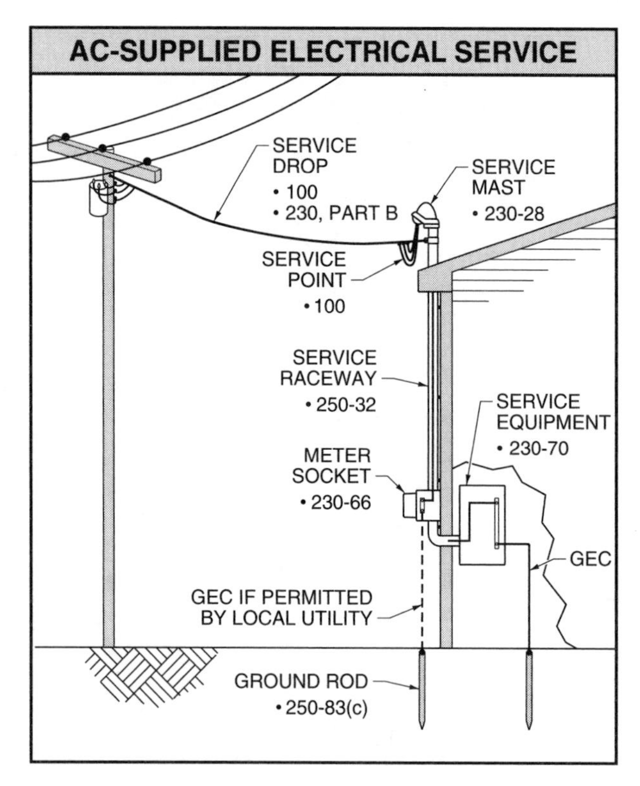

Figure 7-9. GECs may be connected to the grounded conductor within the meter enclosure if the local utility permits connections within the metering equipment.

Section 250-23(a) also requires that an additional connection be made between the system grounded conductor and the GEC where the service is supplied by a transformer located outside the building. Connection of the grounded conductor and the GEC on the load side of the service disconnecting means is prohibited. If a connection were to be made on the load side of the service disconnecting means, a parallel connection between the grounded conductor and the grounding conductor would occur. Since some current flow always occurs in a parallel circuit, this connection might result in an undesirable current flow on the grounding conductor. Under some conditions, however, this connection is permitted to be made. For example, 250-23(a), Ex. 1 permits this connection for separately derived systems. This exception is necessary because when a separately derived system is installed, a new ground reference must be established. The load side connection is permitted when a common service is used to supply two or more separate buildings per 250-23(a), Ex. 2. Under the requirements of 250-24, this connection may need to be re-established at the remote building. The most common installation, however, in which a connection is made between the grounded conductor and the grounding conductor on the load side of the service disconnecting means is for electric ranges, clothes dryers, counter-mounted ovens, and metering enclosures per 250-23(a), Ex. 3. This permission, except for the meter enclosures, has been restricted by the 1996 NEC® to new installations only.

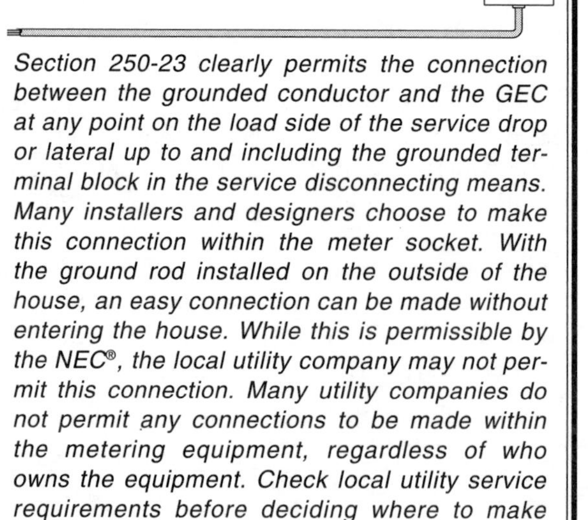

GROUNDING CONNECTIONS

Section 250-23 clearly permits the connection between the grounded conductor and the GEC at any point on the load side of the service drop or lateral up to and including the grounded terminal block in the service disconnecting means. Many installers and designers choose to make this connection within the meter socket. With the ground rod installed on the outside of the house, an easy connection can be made without entering the house. While this is permissible by the NEC®, the local utility company may not permit this connection. Many utility companies do not permit any connections to be made within the metering equipment, regardless of who owns the equipment. Check local utility service requirements before deciding where to make this connection.

Section 250-23(b) contains installation provisions for the grounded conductor of AC systems operating at less than 1000 V. The grounded conductor shall be run to each service disconnecting means and it shall be bonded to the service disconnecting means enclosure. The initial determination of the size of the grounded conductor is done per 220-22 based on the neutral load. After that calculation is performed, the size is checked against the calculation required by 250-23(b).

On the line side of the service disconnecting means, the grounded conductor performs the additional task of the EGC. Therefore, the EGC shall be of a sufficient size that it can adequately handle any line side fault current that may occur. The size of the grounded conductor is determined per Table 250-94. Unlike the GEC however, there is no point at which a maximum conductor size is reached. For example, if the size of the largest service-entrance conductor exceeds No. 1100 kcmil Cu or No. 1750 kcmil Al, the GEC stops at a maximum size of No. 3/0 Cu and No. 250 kcmil Al. The grounded conductor, however, shall never be smaller than 12½% of the largest service-entrance conductor, no matter how large the service-entrance conductors may be. See Figure 7-10.

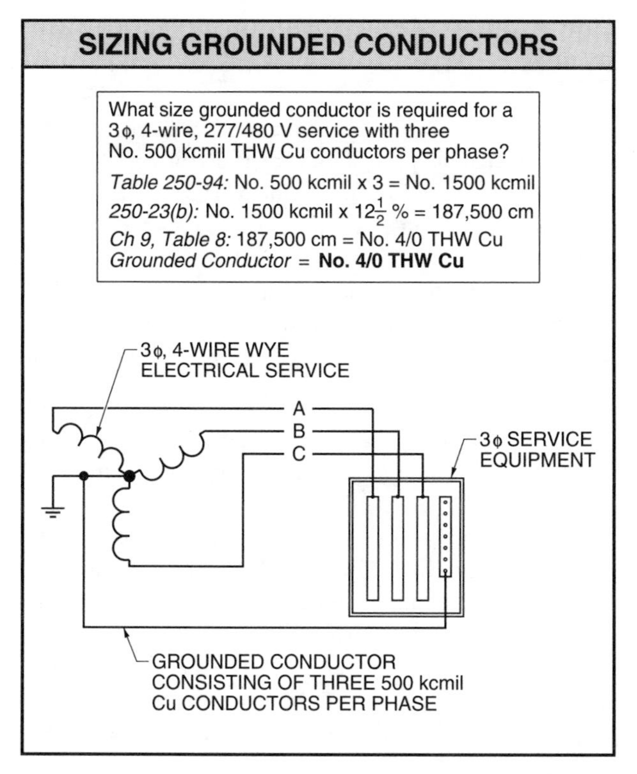

Figure 7-10. The grounded conductor shall never be smaller than 12½% of the largest service-entrance conductor.

DC Systems – 250-22

DC systems are treated separately by the NEC®. Sections 250-3(a) and (b) contain the provisions for DC systems that are required to be grounded. As a general rule, all 2-wire DC systems that supply premises wiring shall be grounded. *Premises wiring* is basically all interior and exterior wiring installed on the load side of the service point or the source of a separately derived system. See 100. There are, however, five exceptions to 250-3(a) for grounding two-wire DC systems.

DC systems installed with ground detection equipment that serve only industrial equipment in limited areas need not be grounded per 250-3(a), Ex. 1. There are several types of ground detectors available, but they all work on the principle that a ground detection system lamp arrangement identifies when an unintentional ground occurs. Ground detection system lamp arrangements vary but they all employ a light bank, consisting of individual light bulbs representing each system phase, that changes in lamp intensity or turns the lamps ON and OFF, depending upon the status of the system.

Per 250-3(a), Ex. 2, two-wire DC systems are permitted to be operated ungrounded if the system operates at 50 V or less between conductors. These low-voltage (24 V) DC systems do not pose any significant safety concerns. They are used extensively for control purposes, such as with high-voltage switchboards and metering equipment.

Per 250-3(a), Ex. 3, the system is permitted to operate without being grounded if the system voltage is in excess of 300 V between conductors. Higher-voltage DC systems are often used to supply individual equipment such as motors. In these installations, where the phase-to-phase voltage exceeds 300 V, the system can operate ungrounded.

Per 250-3(a), Ex. 4, specialized installations are permitted in which a rectifier-derived system is supplied by an AC system installed per 250-5. For example, a 2-wire DC rectifier-derived circuit could operate ungrounded provided that the AC supply system is installed per 250-5. Per 250-3(a), Ex. 5, a 2-wire DC system is permitted to operate ungrounded if it supplies fire alarm circuits installed per 760, Part C. These circuits are limited to a maximum current of 30 mA.

The provision for grounding 3-wire DC systems is contained in 250-3(b). A neutral of a 3-wire DC system that supplies premises wiring shall be grounded. Once a determination has been made to ground a DC system, 250-22 is reviewed to determine how and where the grounding connection shall be made.

The general rule for the grounding connection for DC systems, unlike AC systems, is that a grounding connection shall not be made at the building or structure. See 250-22. The grounding connection is required to be made at the source of supply only, and connections are not permitted at the service or at any point on the premises wiring. However, per 250-22, Ex., the connection is permitted to be made at the source or at the first disconnecting means or OCPD, if the supply system is located on the premises. The second part of 250-22, Ex. permits the connection to be made on the premises if listed and identified equipment is used, and there is a determination that an equivalent system protection is provided.

DC supply systems are quite rare today. The AC system is almost universally accepted as the best means for distributing electrical power. There are, however, more and more applications for DC circuits, but the vast majority of these circuits are derived or converted from AC sources.

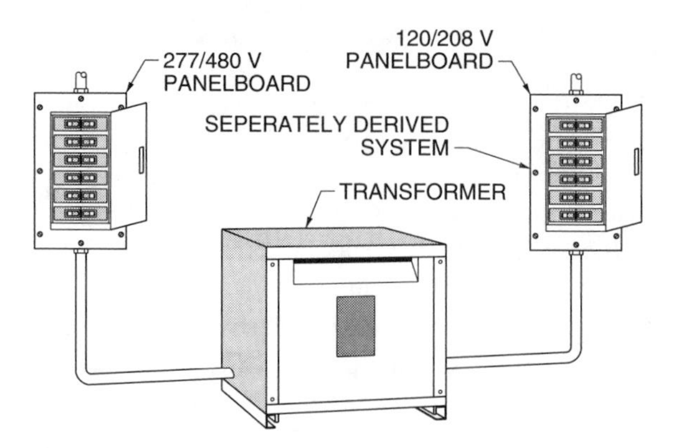

The 120/208 V panel is a separately derived system.

Separately Derived Systems – 250-26

Separately derived systems are a very common component of today's electrical distribution systems. A *separately derived system* is a system that supplies premises with electrical power derived or taken from storage batteries, solar photovoltaic systems, generators, transformers, or converter windings. See 100. By definition, a separately derived system can have no direct electrical connection to the conductors of the supplying system. This also applies to the grounded conductor. If the grounded conductor is solidly connected, then there is not a separately derived system. See Figure 7-11.

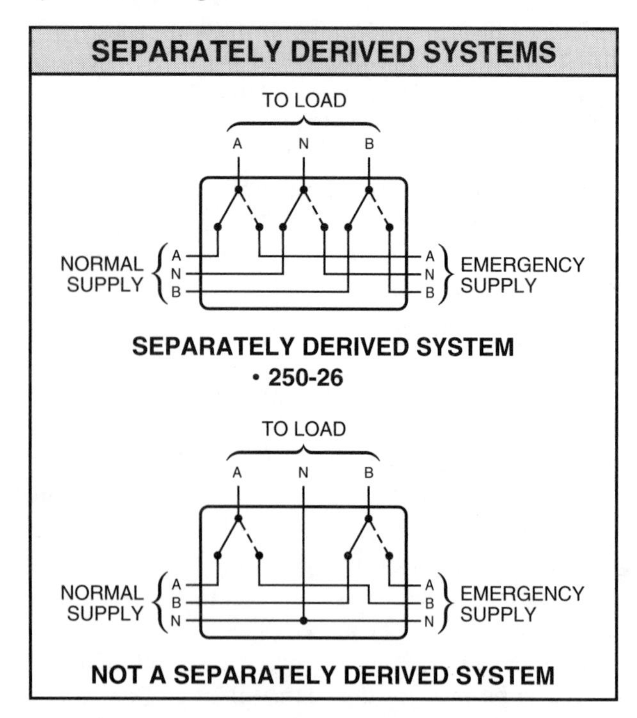

Figure 7-11. Separately derived systems do not have any direct electrical connection between the supply system and the separately derived system.

The vast majority of separately derived systems which electricians work on are transformers. Transformers are available in a wide variety of types for various applications. A *transformer* is a device that converts or "transforms" electrical power at one voltage or current to another voltage or current. Transformers accomplish this task without any rotating parts and essentially without the use of significant amounts of energy.

No matter what the source of the separately derived system is, the grounding requirements are very similar. As with AC and DC systems, 250-5 contains the requirements for grounding separately derived systems. Subsection 250-5(d) states that a separately derived system that is required to be grounded by 250-5(a) or (b) shall also be grounded. A separately derived system that operates outside the requirements of 250-5(a) or (b) is not required to be grounded, but is permitted to be grounded. See 250-26.

Ground references cannot be transferred through windings of any type. The grounding connections that were in place for the service must be re-established. Per 250-26(a), a bonding jumper is required be installed between the EGC of the derived system and the grounded conductor. This connection shall be made at a point from the source of the separately derived system up to and including the first system disconnecting means or OCPD. See Figure 7-12.

The bonding jumper shall be sized per 250-79(d). It shall not be smaller than the sizes shown in Table 250-94 for GECs. Unlike the GEC, the size of the bonding jumper does not stop increasing at a specified point. For conductors larger than No. 1100 kcmil Cu or No. 1750 kcmil Al, the bonding jumper is required to be not less than 12½% of the area of the largest derived phase conductor. See Figure 7-13.

Per 250-26(b), a GEC is required to connect the newly established grounded conductor to the grounding electrode. As with the bonding jumper, this connection occurs at any point of the separately derived system from the source to the first disconnecting means or OCPD. In some installations, such as a panelboard with main lugs only, the connection is made at the source of the separately derived system. The GEC is sized per 250-94. Instead of basing the calculation on the size of the largest service-entrance conductors, the calculation is based on the size of the largest derived phase conductors. See Figure 7-14.

The selection of the grounding electrode for separately derived systems is somewhat different than that of AC-supplied services. Instead of four components, which must be connected together, if available, the GEC for separately derived systems can consist of any one of three components. Section 250-26(c) requires that for whichever method is selected, the grounded electrode shall be as near as practicable to and preferably in the same areas as the GEC connection to the system. Per 250-26(c)(1), the nearest effectively grounded building steel is permitted to be used as the grounding electrode. A determination is made to see if the building steel qualifies as a suitable grounding electrode per 100.

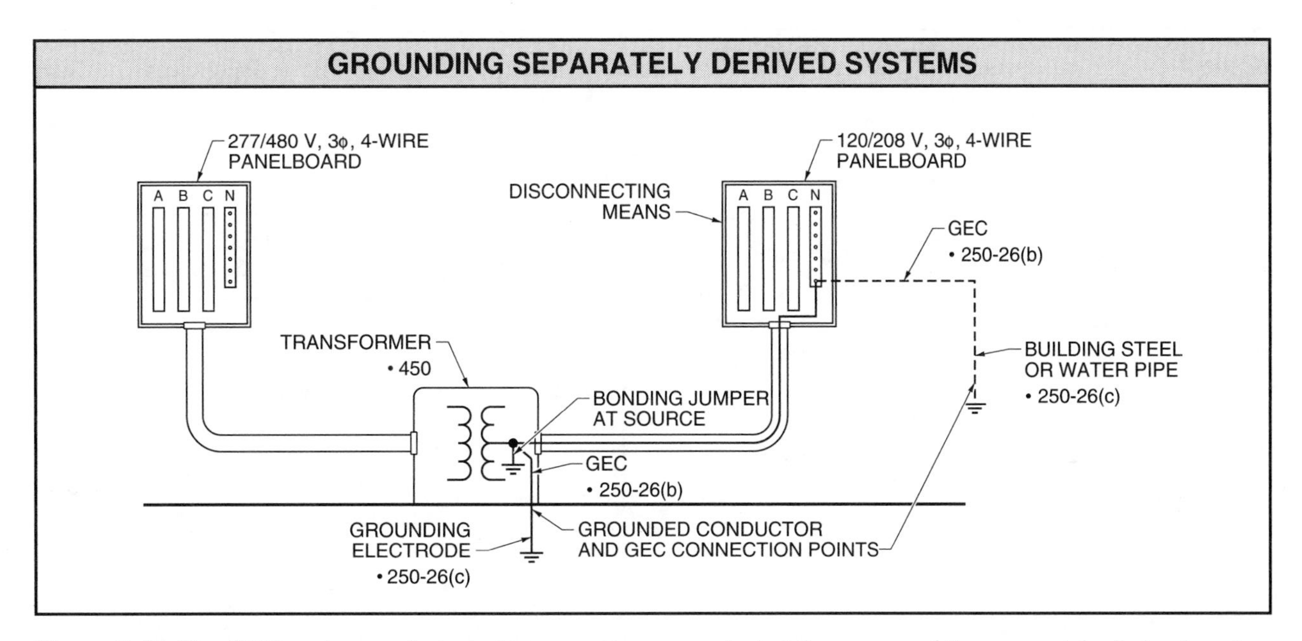

Figure 7-12. The GEC and grounded conductor can be connected at the source of the separately derived system or at the first disconnecting means, but not at both locations.

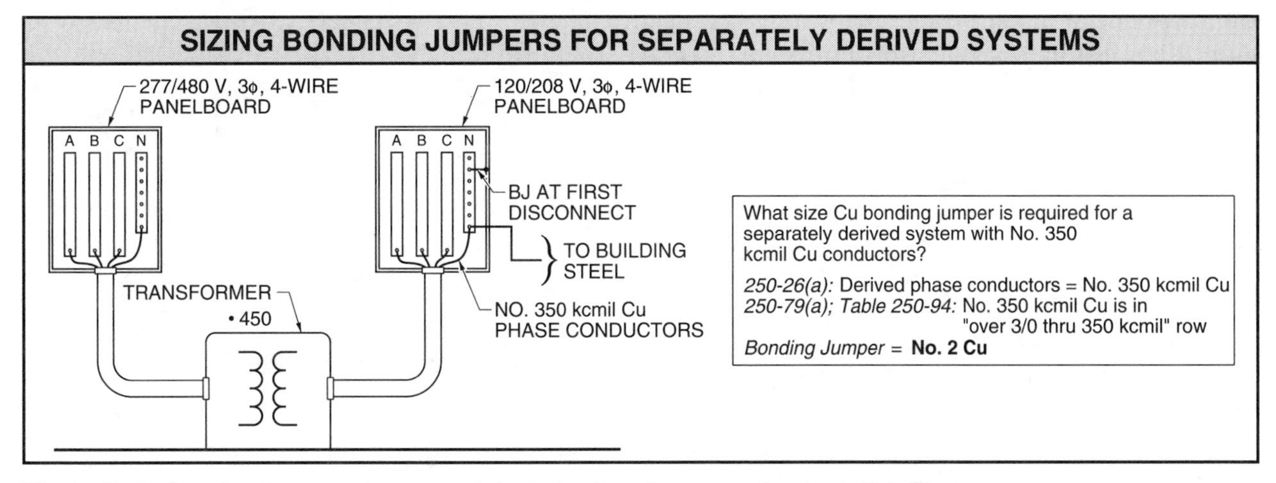

Figure 7-13. Bonding jumpers for separately derived systems are sized per 250-94.

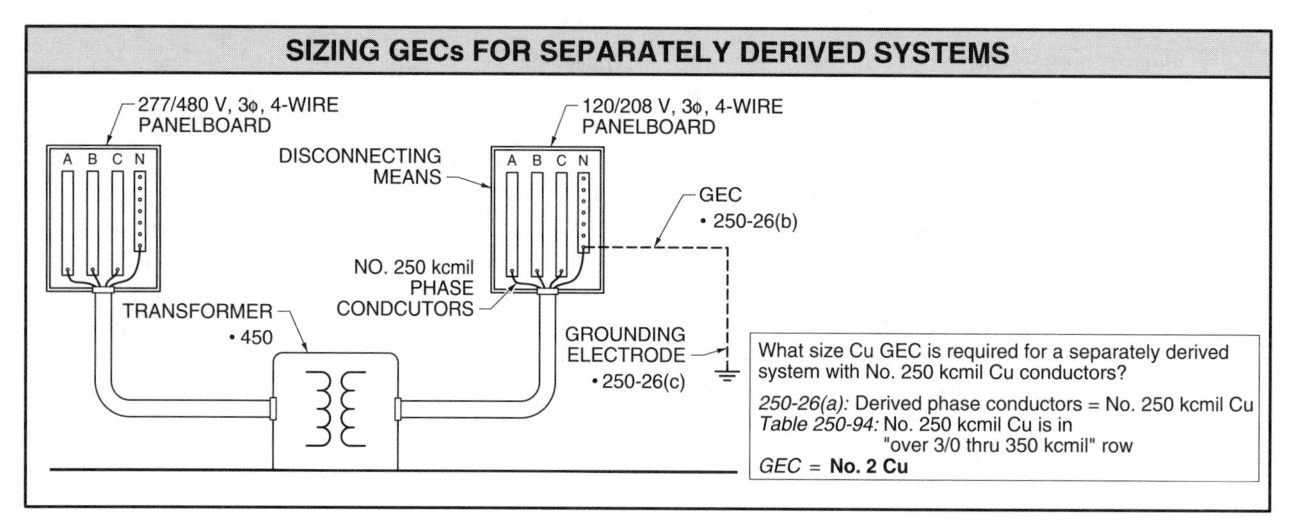

Figure 7-14. GECs for separately derived systems are sized on the largest derived phase conductors and Table 250-94.

If the building steel does not meet the definition, the second option is for the nearest metal water pipe which also is effectively grounded. If neither the steel nor the water piping proves to be suitable, other electrodes such as concrete-encased electrodes, ground rings, or rod and pipe electrodes shall be used. No matter what type of grounding electrode is selected, it shall be as near as possible to, and preferably in the same area as, the point where the connection is made to the system. See Figure 7-15.

Generators – 250-6

A *generator* is a device that is used to convert mechanical power to electrical power. Generators are another source of power for separately derived systems. While not as common as transformers, they are commonly used for supplying emergency power to buildings where required. For example, buildings in which the interruption of power could pose a life safety concern frequently are required to used emergency sources of power. Examples of these types of buildings are schools, office buildings, supermarkets, restaurants, etc. See 445.

Portable Generators – 250-6(a). The frame of a portable generator is not required to be grounded. This is because the frame of the generator, which is in direct contact with the earth, acts like a grounding electrode. The frame is permitted to serve as the grounding electrode if two conditions are met.

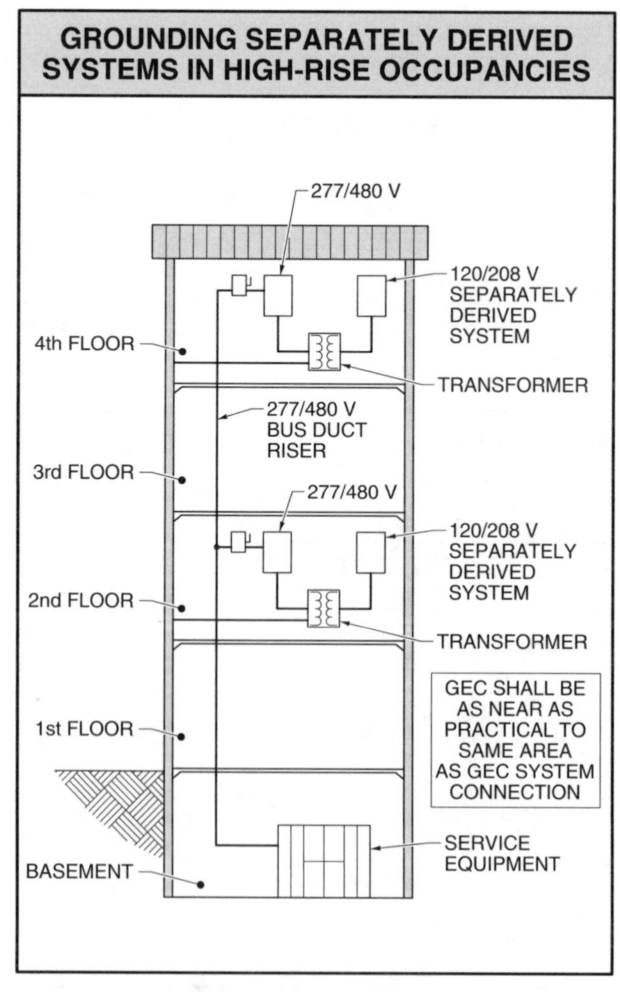

Figure 7-15. A grounding electrode system shall be created for each separately derived system.

Per 250-6(a)(1), the portable generator shall supply loads or equipment that are either directly connected to the frame of the generator or are cord- and plug-connected through receptacles which are directly connected to the frame of the generator. In either case, the frame of the generator is in direct contact with the earth and is permitted to serve as the GEC. If necessary, both conditions are permitted to occur at the same time. In this case, equipment mounted directly on the generator frame and equipment which is cord- and plug-connected to receptacles mounted on the generator can be supplied from the portable generator.

Per 250-6(a)(2), the frame of the generator shall be bonded to the equipment grounding terminals of the receptacles and to any noncurrent-carrying parts of equipment that is served from the generator. Bonding the receptacle terminals to the frame of the generator ensures that any equipment which is cord- and plug-connected to the generator is also connected to the generator frame which serves as the GEC. See Figure 7-16.

Figure 7-16. Portable generators are considered a separately derived system.

Vehicle-Mounted Generators – 250-6(b). Vehicle-mounted generators are treated in a similar fashion to portable generators. The main difference is that the frame of the generator is not in direct contact with the earth. The frame of the vehicle shall be permitted to serve as the grounding electrode under four conditions.

Per 250-6(b)(1), the vehicle frame shall be bonded to the frame of the generator. Per 250-6(b)(2), as with portable generators, the generator shall supply loads or equipment that are either directly connected to the frame of the vehicle or are cord- and plug-connected through receptacles which are directly connected to the frame of the vehicle. Equipment can be supplied by cord- and plug-connected means and direct connection to the vehicle frame at the same time.

Per 250-6(b)(3), the frame of the generator shall be bonded to the equipment grounding terminals of the receptacles and to any noncurrent-carrying parts of equipment that is served from the generator. All provisions of 250 shall be adhered to per 250-6(b)(4).

GROUNDING EQUIPMENT AND ENCLOSURES – 250-32; 250-42

The primary reason enclosures and noncurrent-carrying metal parts of equipment are grounded is to limit the voltage to ground and facilitate the operation of the OCPD under ground-fault conditions. See 250-1, FPN 2. Most equipment operates regardless of the state of the grounding connection. Grounding of enclosures and equipment is done for safety considerations and is the primary factor in reducing the risk of electrical shock.

Enclosures

An *enclosure* is the case or housing of equipment or other apparatus which provides protection from live or energized parts. See 100. Outdoor equipment is sometimes provided with fences or walls to accomplish the same goal. In these cases, the fences and walls are also considered to be enclosures. Enclosures also protect equipment and apparatus from physical damage.

Service Enclosures – 250-32. Service enclosures provide protection from severe physical damage and protection from accidental contact with energized parts that comprise part of the electrical service. All service enclosures and raceways for the protection of service conductors and service equipment shall be grounded.

A *service conductor* is a conductor that extends from the service point to the service disconnecting means. *Service equipment* is all of the necessary equipment to control the supply of electrical power to a building or structure. See 100. These definitions, taken together, require that essentially all of the components of the electrical service are required to be grounded.

Grounding on the line (supply) side of the service equipment is accomplished differently from grounding of equipment on the load side. Grounding on the load side is primarily through the use of an EGC. There is no separate EGC on the line side. Per 250-61(a), the grounded conductor is permitted to be used to ground any noncurrent-carrying part of equipment and enclosures on the supply side of the service. The grounded conductor performs the dual function of serving as the neutral conductor and the EGC.

There is however, one exception to the requirement that all service enclosures and raceways be grounded. Per 250-32, Ex., the use of a metal elbow to be installed for underground installations without being grounded is permitted. This is commonly done where nonmetallic raceways are installed and a metal elbow is used to ensure that the raceway is not damaged during installation of the conductors. The elbow shall be installed so that no part of it is within 18″ of finished grade. See Figure 7-17.

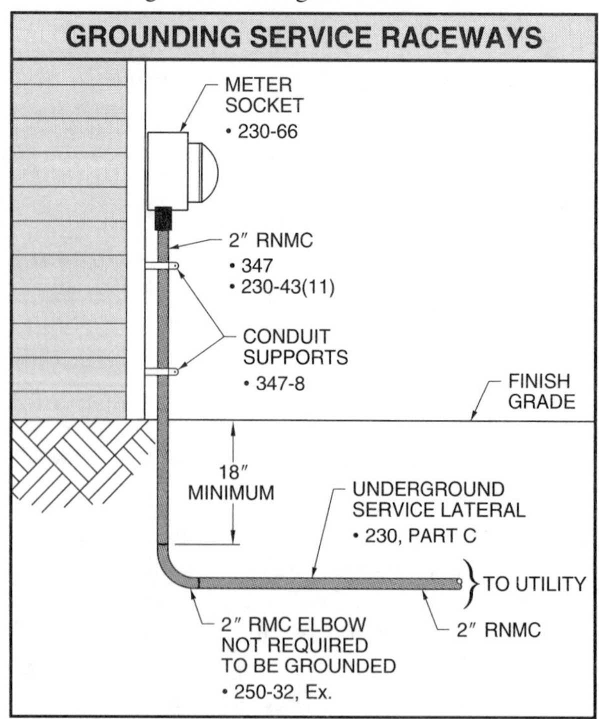

Figure 7-17. Metal elbows installed in nonmetallic raceways are not required to be grounded when they are at least 18″ below grade.

Other than Service Enclosures – 250-33. This section requires all metal enclosures and raceways, other than those covered under 250-32, to be grounded. There is no limitation to service conductors only. Any enclosure or raceway that houses or protects a branch-circuit or a feeder shall be grounded. This protects personnel from an electrical shock hazard. For example, if a raceway protecting a branch circuit is not grounded and, because of an insulation failure, it becomes energized, there is a good probability that the impedance of the ground-fault path would be too great to facilitate the operation of the OCPD. The earth shall never be used as the sole EGC. See Figure 7-18.

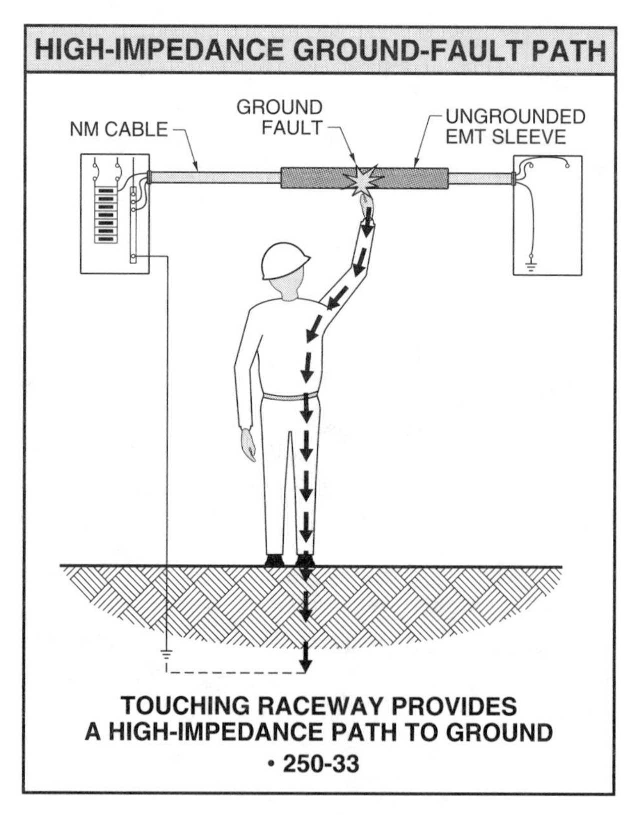

Figure 7-18. The earth shall never be used as the sole EGC.

Per 250-33, Ex. 1, metal enclosures and raceways that are installed as part of an addition to an existing installation of open wiring, knob-and-tube, or NM cables that do not provide an equipment ground are permitted to operate ungrounded. The total length of the circuit shall not exceed 25′, the wiring method shall be free from contact with the ground or other grounded materials, and the installation shall be guarded from contact.

Per 250-33, Ex. 2, short sections of metal enclosures or raceways are permitted to be installed without being grounded. This gives relief from grounding to those installations in which the enclosure or raceway is merely used to provide support or mechanical protection. However, the term "short section" is not defined, and this could lead to confusion in the field and, more importantly, installations that pose a threat to life safety. Always check with the AHJ if there are any questions concerning grounding. See Figure 7-19.

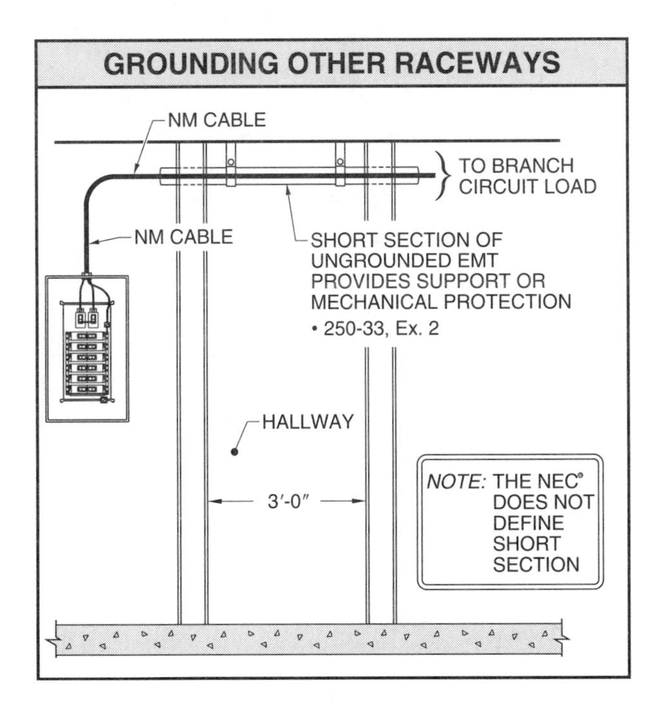

Figure 7-19. Short sections of metal enclosures or raceways are permitted to be installed without being grounded.

Fixed Equipment – 250-42

Equipment is any device, fixture, apparatus, appliance, etc., used in conjunction with electrical installations. See 100. Grounding of equipment is divided into two basic sections depending on whether the equipment is fixed in place or cord- and plug-connected.

Section 250-42 lists six conditions where equipment, which contains exposed noncurrent-carrying metal parts and which is fastened in place or connected by a permanent wiring method, shall be grounded. No specific equipment is mentioned. The requirements apply generally to all electrical installations. See Figure 7-20.

The first condition, 250-42(a), establishes a zone measuring 5′ horizontally and 8′ vertically from ground or grounded objects in which exposed equipment shall be grounded. These dimensions establish a "reach zone" under which someone could contact an exposed energized metal part of a piece of equipment and ground or any other grounded metal objects. Objects beyond this zone do not pose a significant risk to personnel safety because they cannot be reached while maintaining contact with an energized metal part.

The second condition, 250-42(b), requires that all equipment located in a wet or damp location shall be grounded unless the equipment is isolated. Equipment is considered to be isolated when it is not readily accessible to persons without the use of special means. A *wet location* is any location in which a conductor is subjected to excessive moisture or saturation from any type of liquids or water. A *damp location* is a partially protected area subject to some moisture. See 100. Equipment installed in a damp or wet location poses an increased hazard for persons who may use the equipment or otherwise come in contact with the equipment.

The third condition, 250-42(c) requires that all fixed electrical equipment in contact with metal be grounded. If the metal were to become energized, it would sustain a voltage to ground and pose a threat of electrical shock for persons in contact with the equipment.

The fourth condition, 250-42(d), requires that all fixed equipment located in hazardous locations shall be grounded per 500 through 517. Because of the presence of, or the potential for, hazardous atmospheres, special grounding requirements may exist when installing electrical equipment in hazardous locations. Always check the requirements of the specific occupancy to determine if grounding is required.

The fifth condition, 250-42(e), requires that all fixed equipment supplied from metallic wiring methods be grounded. For example, fixed equipment supplied by metal-clad cable or metal raceways shall be grounded. This requirement does not apply to short sections of metal enclosures where their function is support or mechanical protection. See 250-33, Ex. 2.

The sixth condition, 250-42(f), requires grounding of fixed equipment when it operates with any terminal at over 150 V to ground. This is a broadly worded provision. There are four exceptions for specialized electrical installations to this general rule.

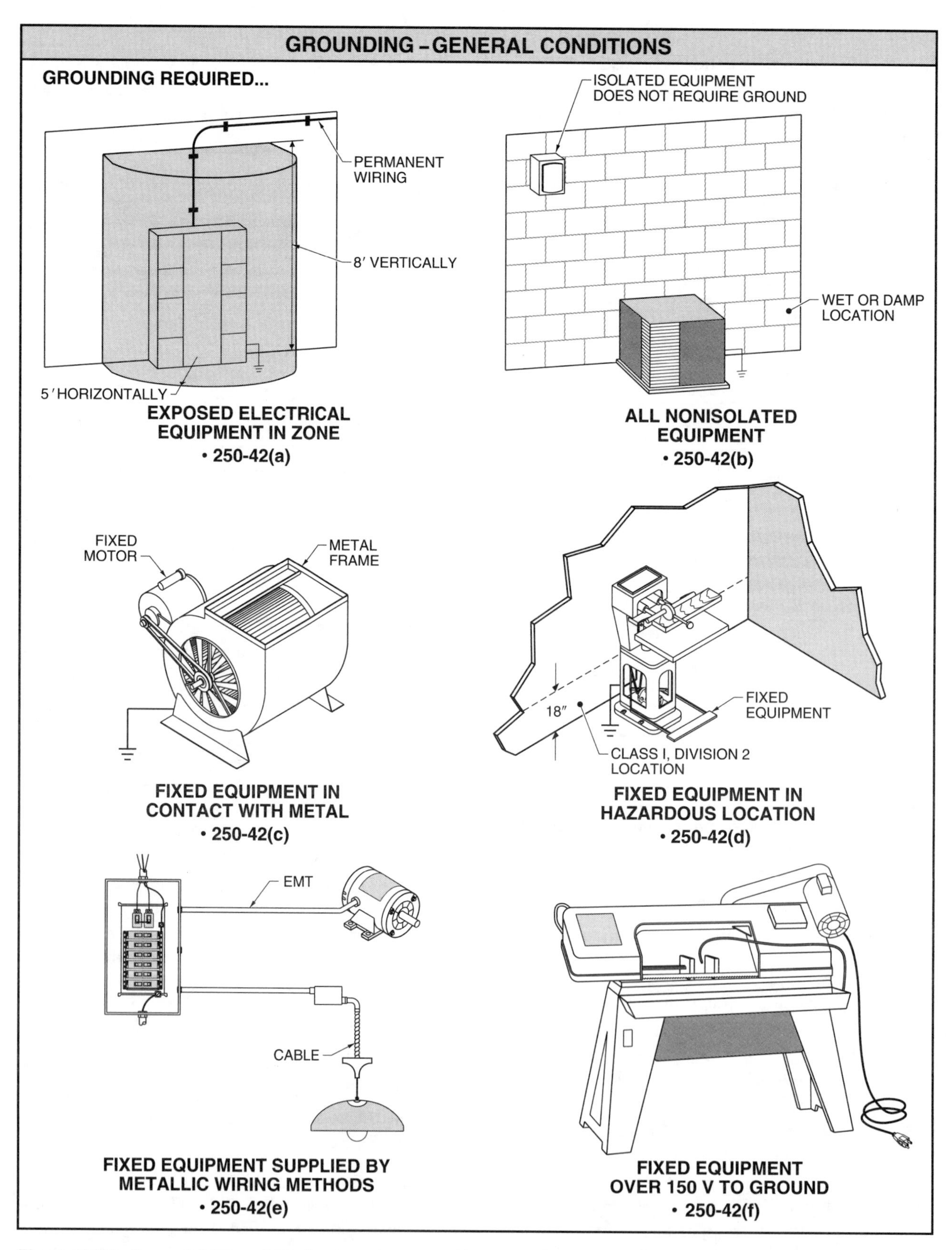

Figure 7-20. General fastened-in-place equipment, with exposed noncurrent-carrying metal parts, which is permanently wired shall be grounded.

Specific Conditions – 250-43. This section lists specific equipment that shall be grounded if it contains exposed, noncurrent-carrying metal parts that are likely to become energized. There are no voltage limitations to these requirements. See Figure 7-21. The following shall be grounded, regardless of the voltage at which they operate:

- Switchboard frames and structures
- Pipe organs
- Motor frames
- Motor controller enclosures
- Elevators and cranes
- Garages, theaters, and motion picture studios
- Electric signs
- Motion picture equipment
- Power-limited remote control, signaling, and fire alarm circuits
- Lighting fixtures
- Motor-operated water pumps
- Metal well casings

Cord- and Plug-Connected Equipment – 250-45. Grounding cord- and plug-connected equipment poses some design considerations beyond what is required for fixed or fastened-in-place equipment. Cord- and plug-connected equipment is more susceptible to physical damage, which could affect the integrity of the equipment grounding conductor. One potential solution to this problem has been the advent of double-insulated equipment, which in some cases, can be used in lieu of a separate EGC. Subsection 250-45(a–d) contains four conditions under which cord- and plug-connected equipment shall be grounded. See Figure 7-22.

The first two conditions are the same as those for fixed equipment. Cord- and plug-connected equipment located in hazardous locations shall be grounded per 250-45(a). Cord- and plug-connected equipment operating at over 150 V to ground shall be grounded per 250-45(b). The second and third provisions apply to cord- and plug-connected equipment installed in residential occupancies (c) and other than residential occupancies (d).

Subsection 250-45(c) lists five groups of specific equipment that shall be grounded when installed in residential occupancies. Per 250-45(c), Ex., appliances which employ a system of double insulation do not require grounding. The double insulation or equivalent system shall be listed and shall be distinctively marked. While the grounding requirement for any tool or appliance meeting this exception are removed, double-insulated tools are not permitted to be used in lieu of GFCI protection where so required. For example, a 120 V, listed, double-insulated, hand-held, motor-operated tool used for construction purposes on a residential jobsite is still required to be protected with GFCI protection per 305-6.

Subsection 250-45(d) covers requirements for grounding cord- and plug-connected equipment in other than residential occupancies. These provisions apply to commercial and industrial classifications as well as any other classification the AHJ may adopt. While there is a similar exception for listed, double-insulated tools and appliances, the scope of 250-45(d), Ex. 2 is greater than that for the residential occupancies. It applies to listed stationary and fixed motor-operated tools and to listed light industrial motor-operated tools as well.

It is permissible to use ungrounded portable handlamps and tools in other than residential occupancies per 250-45(d), Ex. 1. These ungrounded handlamps and tools may be used in wet or corrosive locations and if they are supplied from an isolating transformer. The ungrounded secondary of the transformer is not permitted to exceed 50 V.

Cord- and plug-connected equipment is susceptible to physical damage.

GROUNDING MEANS – 250, PART F

Once a determination has been made as to whether a system is to be grounded, 250, Part F should be reviewed to determine the method by which the service should be grounded. An effective grounding path, per 250-51, shall be permanent and continuous, it shall have sufficient capacity for the available fault current, and it shall have a low enough impedance to permit the OCPD to operate. These guidelines apply for systems and equipment.

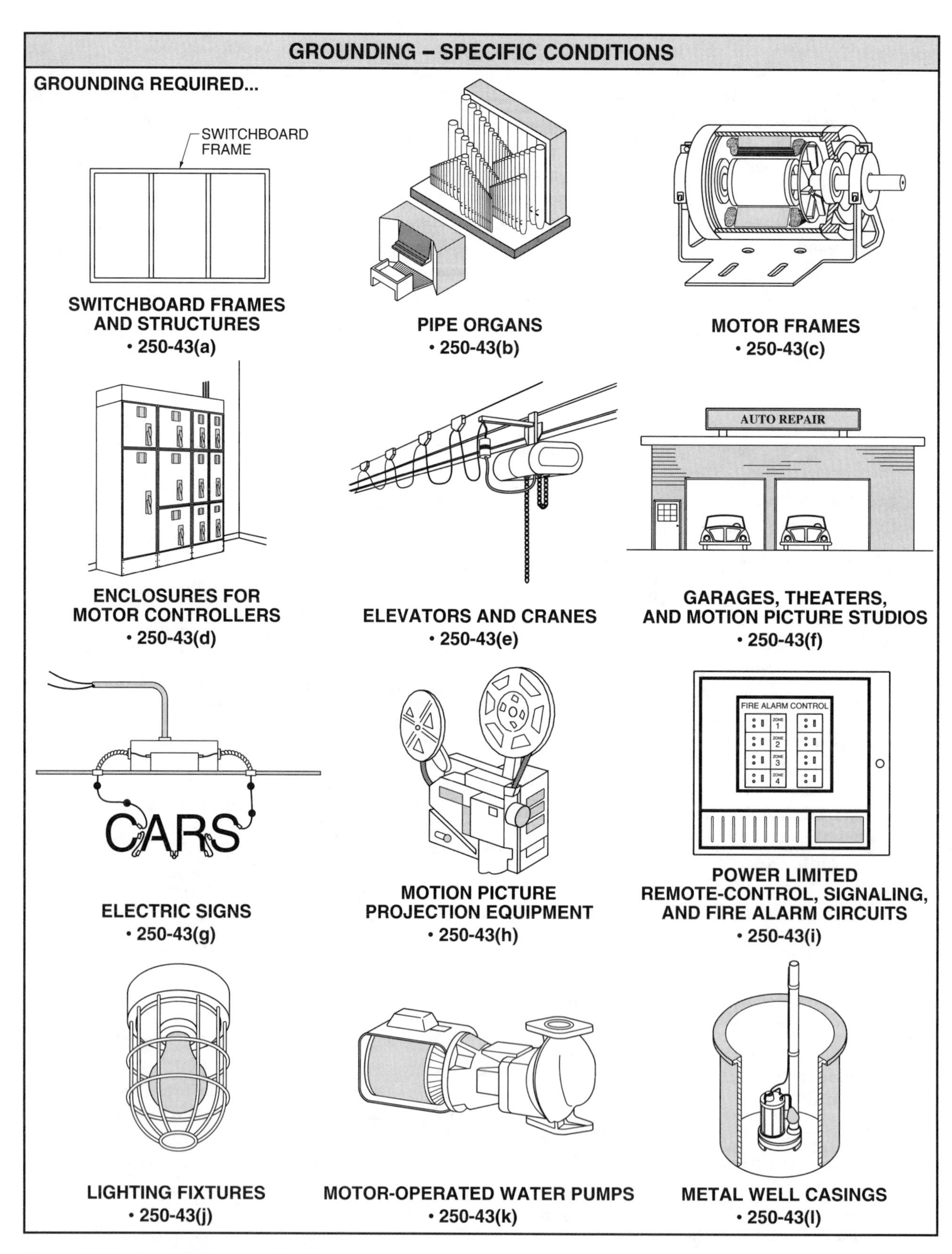

Figure 7-21. Specific fastened-in-place equipment, with exposed noncurrent-carrying metal parts, which is permanently wired shall be grounded regardless of voltage.

Services

Electrical services are grounded by several connections that usually are made within or at the service equipment. The EGC shall be bonded to the grounded service conductor per 250-50(a). The GEC shall also be bonded to the system grounded conductor. See Figure 7-23.

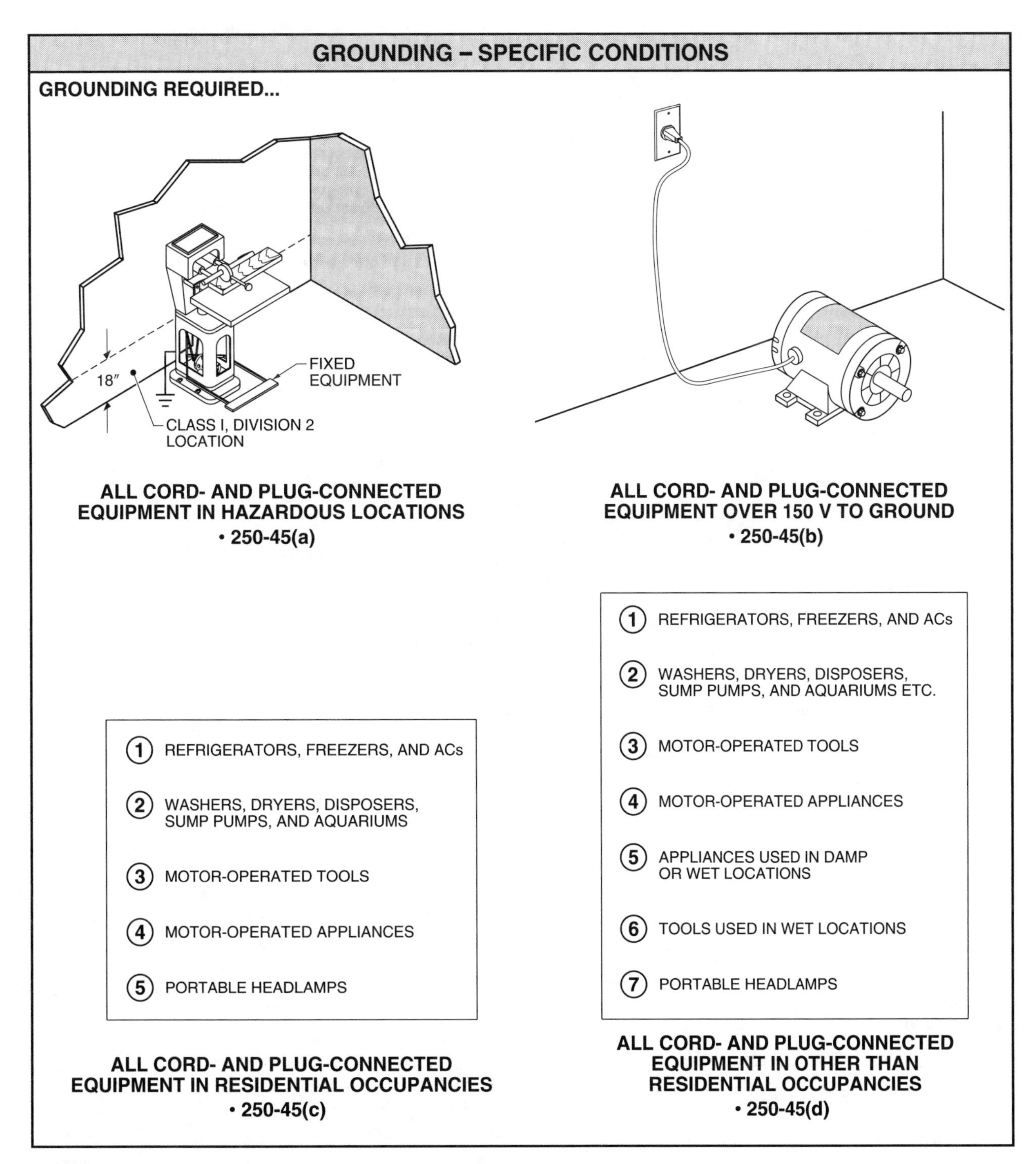

Figure 7-22. Cord- and plug-connected equipment may be used in hazardous locations, over 150 V to ground, in residential occupancies, and in other than residential occupancies.

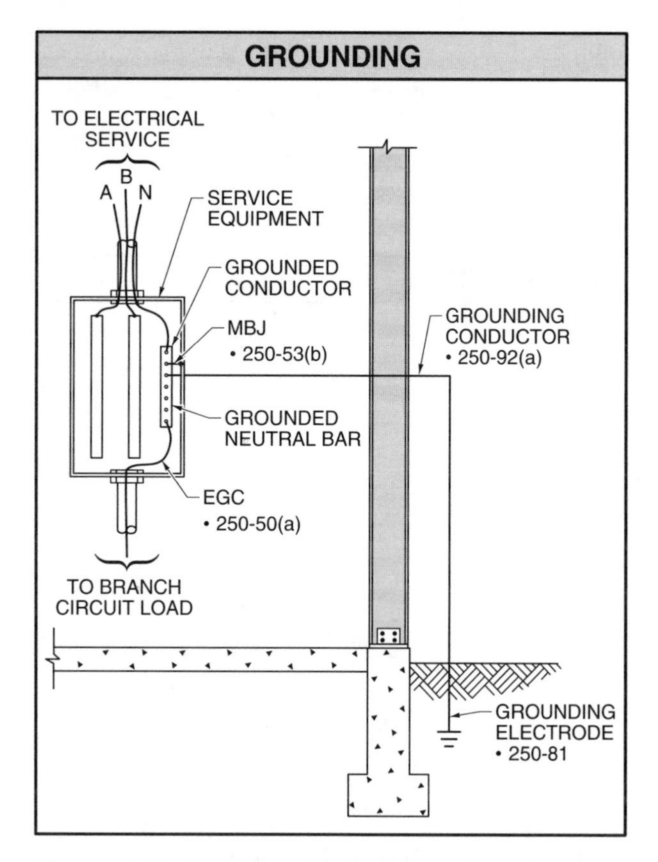

Figure 7-23. Most electrical system grounding connections are made at or within the service equipment.

Section 250-53(a) reiterates the connection of the GEC, EGC, and the grounded service conductor at the service equipment. This connection is made with an unspliced MBJ per 250-53(b). The MBJ connects these conductors and the service disconnecting means together at the service equipment. The MBJ is the heart of the grounding system at the service equipment.

Equipment

Sections 250-57, 250-59, and 250-60 cover requirements for grounding different types of equipment. Equipment, much like systems, can be installed in conditions that require it to be grounded, permit it to be grounded, or even prohibit it from being grounded.

Fixed Equipment – 250-57. This section requires that all noncurrent-carrying metal parts of equipment, enclosures, etc., which are required to be grounded, shall be grounded by one of two methods. The first method, 250-57(a), permits the use of a recognized EGC. See 250-91(b). See Figure 7-24. Eleven different listings under this section could be considered the EGC for fixed equipment. For example, a fixed piece of equipment which operates at 240 V is fed by two ungrounded conductors installed in a ¾″ run of electrical metallic tubing. EMT is listed as item (4) and is permitted to serve as the EGC for this circuit.

The second method, 250-57(b), permits the use of an EGC which is run with or contained within the same circuit. All circuit conductors, including the EGC where so required, shall be run in the same raceway, cable, trench, cord, etc. per 300-3(b). This helps to ensure that the impedance of the ground path remains low to permit operation of the OCPD.

The EGC may be a bare, covered, or insulated conductor. In general, if the conductor is insulated, the color shall be green or green with one or more yellow stripes. Two exceptions permit the reidentification of the EGC at the time of termination if all of the conditions of the exceptions are met. See 250-57(b), Ex. 1 and 4.

IDENTIFYING EGCs

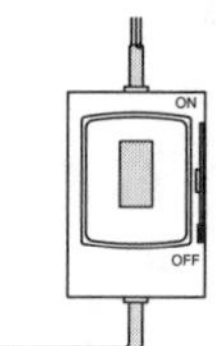

Electrical designers and installers should be careful when applying and identifying the EGC. Per 250-57(b), the EGC may be bare, covered, or insulated. If the EGC is covered or insulated, the outer finish shall be green or green with one or more yellow stripes. The trouble often comes in applying the provisions of 250-57(b), Ex. 1 and 4. Ex. 1 permits reidentification of the EGC only for conductors larger than No. 6 Cu or No. 6 Al. Ex. 1 cannot be applied to conductors or cable assemblies with conductors smaller than No. 4. Ex. 4 is often cited as giving permission for reidentification of conductors of any size. Ex. 4 can only be applied if qualified persons service the installation. See 100.

While 250-57(b), Ex. 4 applies to multiconductor cables, it does not apply to any conductors installed in raceways. Whenever 250-57(b), Ex. 1 and 4 are used, follow the marking or reidentification procedures as detailed.

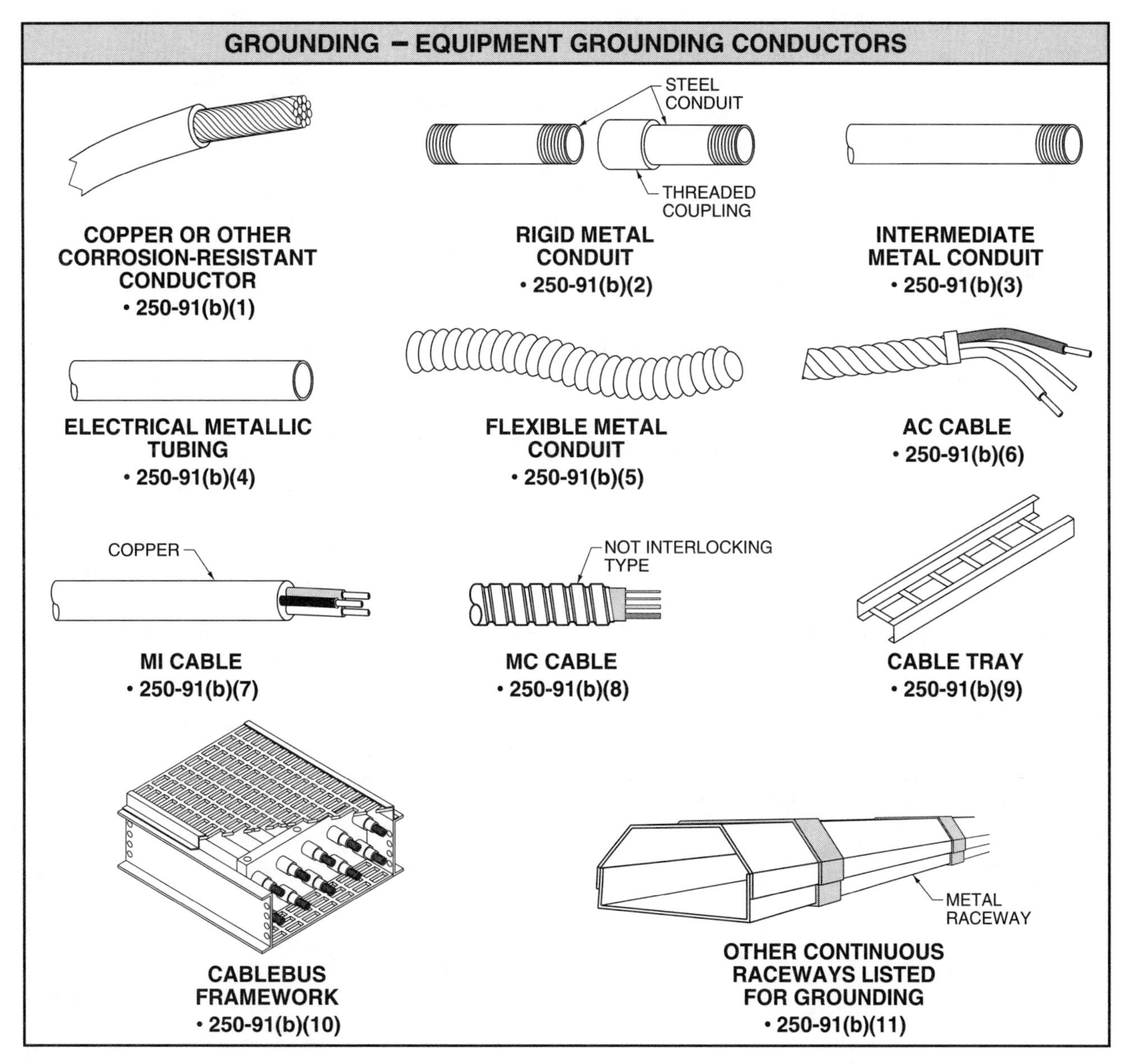

Figure 7-24. Eleven different listings may be considered the EGC for fixed equipment per 250-57.

Cord- and Plug-Connected Equipment – 250-59. Three methods may be used to ground metallic non-current-carrying parts of cord- and plug-connected equipment. Per 250-59(a), metal enclosure of the supplying conductors may serve as the grounding means if a grounding-type attachment plug provided with a fixed grounding contact to ground the metal enclosure is provided. The metal enclosure of the conductors shall, in turn, be connected to the attachment plug and the equipment with approved connectors.

The second and most common method employed to ground cord- and plug-connected equipment is by use of a separate EGC run with the supply conductors per 250-59(b). The EGC is part of the cable assembly or cord and terminated in a grounding-type attachment plug. For fixed equipment, the EGC is permitted to be bare, covered, or insulated, and the same color code requirements shall be followed.

The last method listed for grounding cord- and plug-connected equipment is by use of a separate strap or flexible wire per 250-59(c). The wire or strap may be insulated or bare. If the wire or strap is part of the equipment, it shall be protected against physical damage.

Ranges and Clothes Dryers – 250-60. The 1996 NEC® contains a major change in 250-60, "Frames of Ranges and Clothes Dryers." Since 1947, the NEC® has permitted the frames of ranges, including counter-mounted cooking units, wall-mounted ovens, and clothes dryers to be grounded by the use of the grounded conductor provided all of the following conditions were met:

- The supply circuit is 120/240 V, 1ϕ, 3-wire or 120/208 V, 3ϕ, 4-wire wye
- The grounded conductor is at least No. 10 Cu or No. 8 Al
- The grounded conductor is insulated, or if uninsulated, is part of a SE cable branch-circuit originating at the service equipment
- The equipment is bonded to the grounding contacts of any receptacle which is part of the equipment

The change no longer permits the use of the grounded conductor for this application. The change does, however, only apply to new branch-circuit installations. New installations require the frames to be grounded per 250-57 and 250-59. See Figure 7-25.

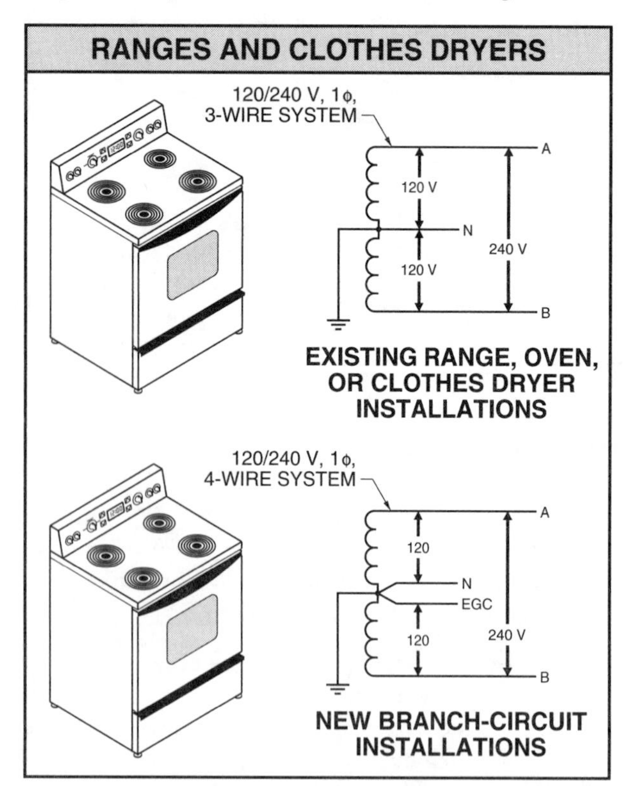

Figure 7-25. New branch-circuit installations for ranges, clothes dryers, and ovens require an EGC.

BONDING – 250, PART G

Part G contains the bonding requirements for service equipment, over 250 V installations, main and equipment bonding jumpers, interior metal piping, and structural steel. Bonding is sometimes used interchangeably with grounding. The two terms should not be used interchangeably because they both seek to accomplish different goals.

Five terms in 100 use "bonding" as a root word. *Bonding* is joining metal parts to form a continuous path to conduct safely any current that is commonly imposed. The definition of bonding contains several components. Bonding is a connection between various metal parts. This connection shall be electrically continuous and shall have sufficient capacity to handle any fault currents that may occur. Although these requirements are similar to those of an effective grounding path, the difference lies in the purpose of the connection. Bonding is done to form an equipotential plane among all metal parts. An *equipotential plane* is an area in which all conductive elements are bonded or otherwise connected together in a manner which prevents a difference of potential from developing within the plane. Bonding ensures that there cannot be a difference of potential between metal parts.

INSTALLING OLD APPLIANCES

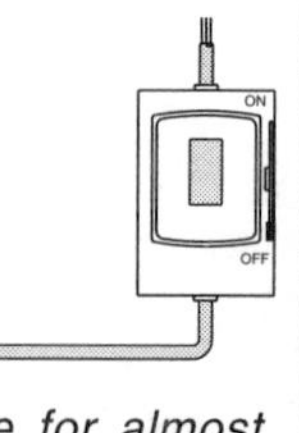

Because 250-60 has been in place for almost 50 years, some time may be required before designers and installers begin to recognize the branch-circuit requirements for new installations. Installers should be particularly concerned when asked to install branch-circuits or hook up any of these appliances. For example, if someone moves to a new house and brings their old appliances with them, the new installation is to be wired for a 4-wire hookup while the appliances are wired for a 3-wire hookup. Depending upon how the branch-circuit is run and whether the appliance is hard-wired or cord- and plug-connected, the installer may need to reconfigure the appliances, devices, and attachment plugs for the new hookup.

Services

Electrical services have special bonding requirements which are contained in 250-71 and 250-72. The primary reason for these special requirements for services is the absence of meaningful overcurrent protection on the line-side of service equipment. Electrical utility power systems frequently do not provide secondary overcurrent protection. The only protection provided by the electrical utility is for the primary of the transformer. Care should be exercised when bonding electrical service equipment because the available fault current is typically greater there than anywhere else in the electrical distribution system.

Bonding Service Equipment – 250-71(a). Part (a) lists the components of the service which shall be bonded together. These include:

- All service raceways or cable assemblies employing metallic sheath
- All enclosures for service conductors, including meter sockets, boxes, etc.
- All metallic raceways used to provided protection for the grounding electrode conductor

Bonding to Other Systems – 250-71(b). Part (b) requires that a means be provided at the service for bonding other systems to the service grounding electrode system. The connection point shall be accessible. Three means listed for this connection point are:

- Metal service raceways shall be exposed
- The GEC shall be exposed
- Other approved means

For example, see 250-71(b)(3), FPN 1. An approved means is a 6″ pigtail of a No. 6 Cu conductor which can be bonded to service equipment or raceway and left accessible on the outside wall for connection of other systems.

Bonding Service Equipment – 250-72. This section lists five methods that can be used to accomplish the required bonding for service equipment. See Figure 7-26. The five methods are:

- Grounded service conductor. See 250-61 and 250-113.
- Threaded connections. Fittings, such as threaded hubs, are permitted for bonding purposes when the fittings are made wrenchtight. Threaded connections are commonly used with RMC and IMC.
- Threadless connections. The use of standard locknuts or bushings are strictly prohibited as a means for bonding at service equipment.
- Bonding jumpers. Properly sized and installed bonding jumpers may be used to bond around concentric or eccentric knockouts. This is the most common bonding method for service equipment.
- Other devices. Bonding is permitted to be accomplished by use of bonding or grounding type locknuts.

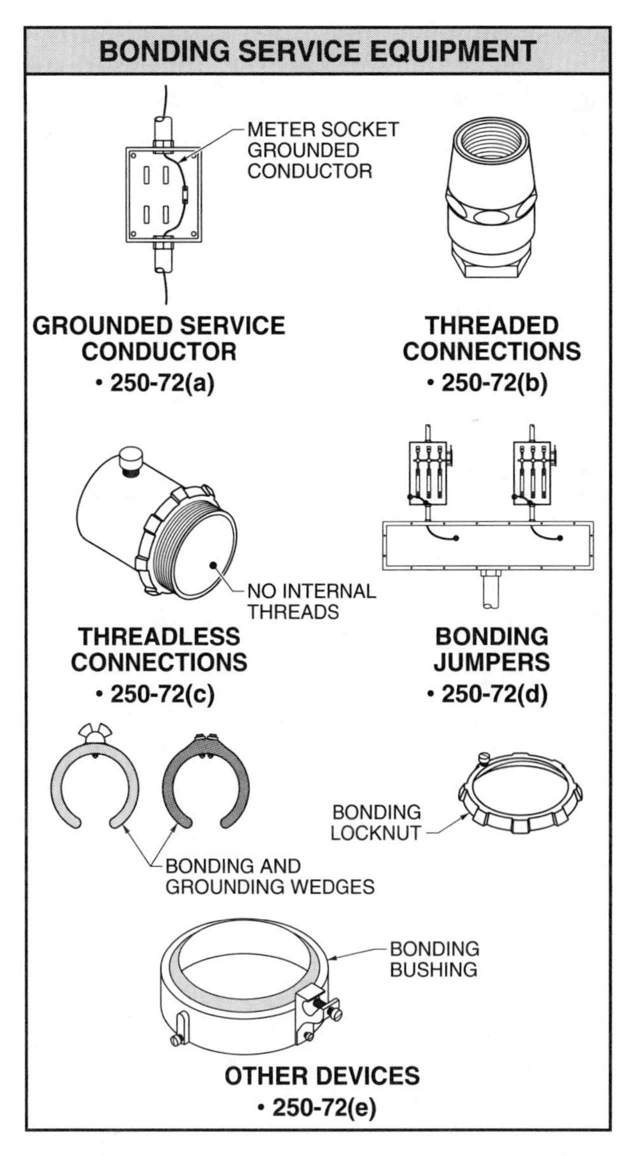

Figure 7-26. Five methods may be used to bond service equipment.

Bonding Over 250 Volts – 250-76

An often overlooked requirement is the bonding of metal raceways and cable for circuits of over 250 V to ground. This section requires that electrical continuity be ensured by either threaded connections, threadless connections, bonding jumpers, or other devices. These are the same methods used for bonding the service equipment with the exception of the grounded conductor. See 250-72. See Figure 7-27.

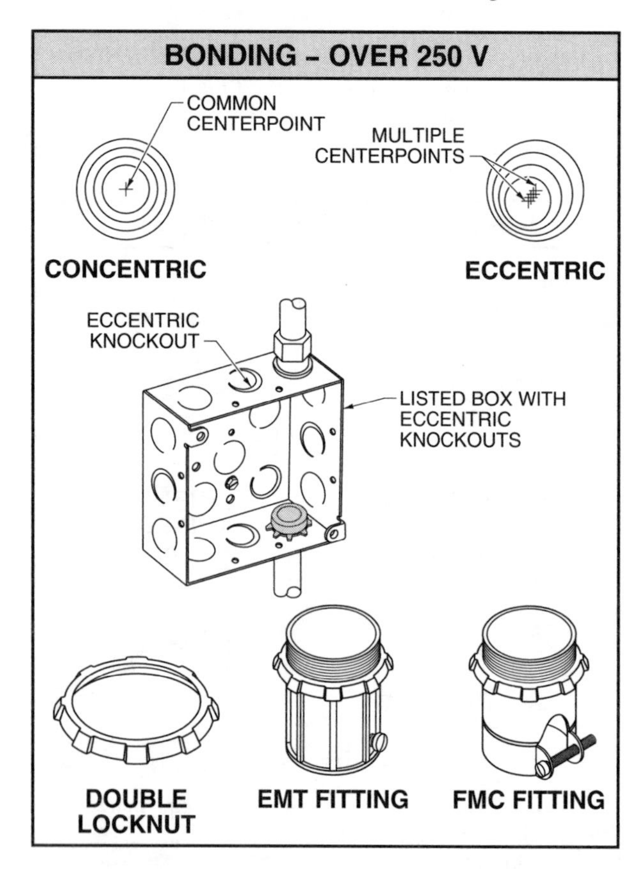

Figure 7-27. Listed boxes, with concentric or eccentric knockouts, may be used for bonding over 250 V circuits.

There is, however, an exception for bonding over 250 V circuits. There are four alternative means to accomplish bonding for circuits which operate at over 250 V to ground per 250-76, Ex. These means are permitted provided that oversized, concentric, or eccentric knockouts are not encountered or when the knockouts are encountered but have been tested and the box or the enclosure is listed for the use. The first means is to use threadless couplings and cable connectors for cables with metallic sheaths. Cable assemblies such as MC, AC, and MI are covered by this section. The second method is to use double locknuts, one on the inside and one on the outside of the box or enclosure. This provision applies to RMC and IMC only. The third option is to use fittings that are designed with a shoulder that seats firmly against the box or enclosure. Fittings for EMT, FMC, and MC are covered by this section. The last alternative is to use listed fittings. This category covers some of the newer fittings that incorporate a "snap-tite" or "speed-lock" design which does not use a separate locknut.

Main and Equipment Bonding Jumpers – 250-79

The *main bonding jumper (MBJ)* is the connection at the service equipment that ties together the EGC, the grounded conductor, and the GEC. The MBJ can be a simple machine screw, as in the case of a 100 A panelboard, or it can be a No. 4/0 conductor for a 1000 A service.

The MBJ may be constructed of copper or other corrosion-resistant materials per 250-79(a). It should not be installed downstream or in subpanels located in the same building as this could result in multiple or parallel grounding paths posing a threat of electrical shock.

When the MBJ is a screw only, it shall be colored green. This distinguishes it from the other screws on the grounded conductor terminal block. The MBJ is attached by a green screw or exothermic welds, listed pressure connectors, listed clamps, or other listed means. See 250-113. See Figure 7-28.

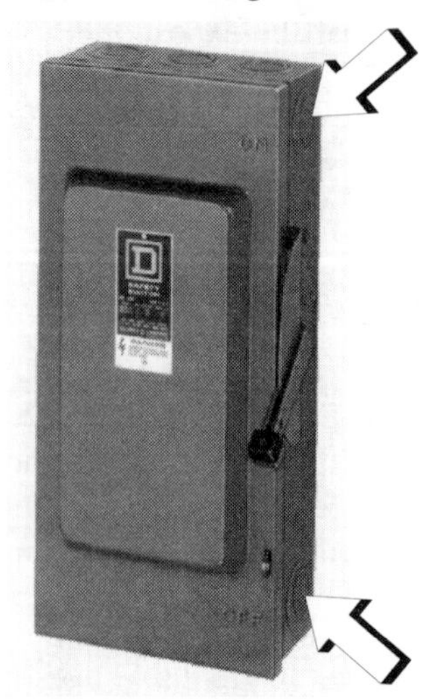

Square D Company

Knockouts provide points to easily connect the raceway to the enclosure.

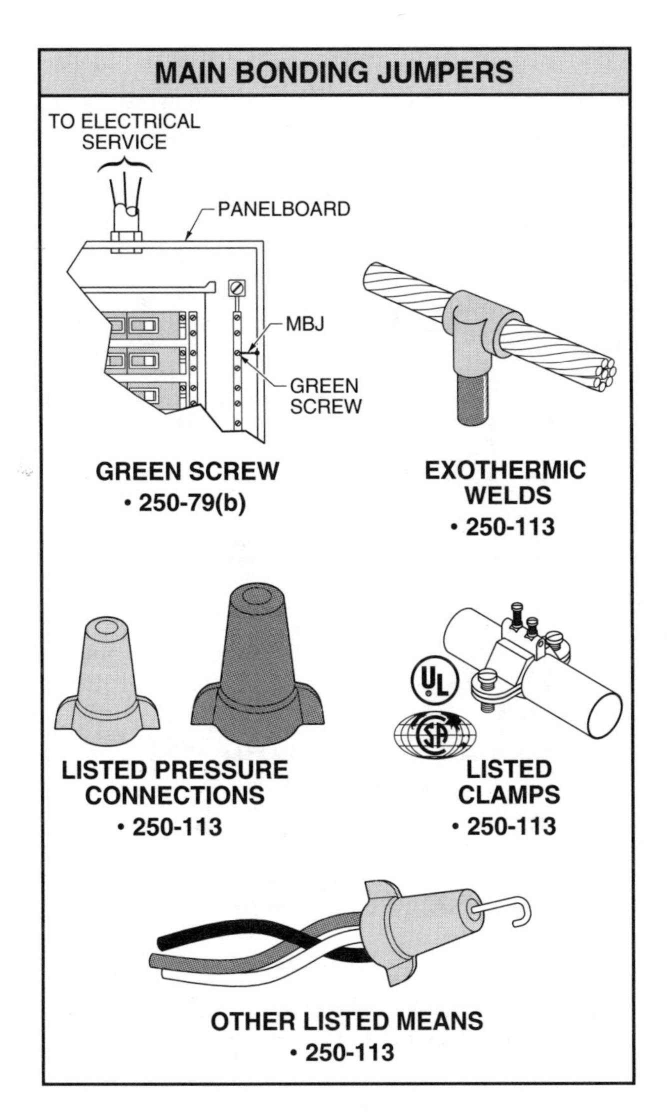

Figure 7-28. The main bonding jumper (MBJ) is the connection at the service equipment that ties together the EGC, the grounded conductor, and the GEC.

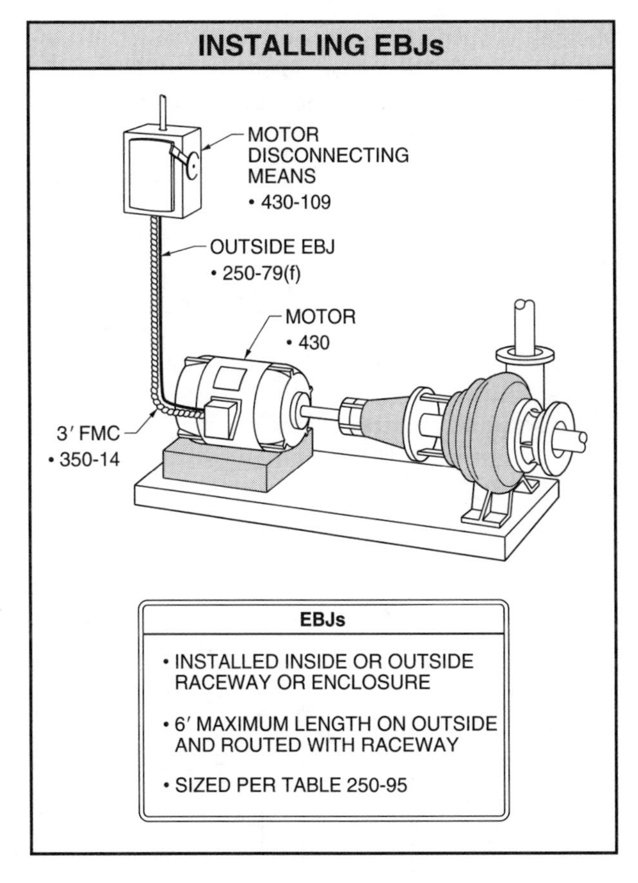

Figure 7-29. The equipment bonding jumper (EBJ) is a conductor that connects two or more parts of the EGC.

The *equipment bonding jumper (EBJ)* is a conductor that connects two or more parts of the EGC. See 100. See Figure 7-29. Like MBJs, EBJs are permitted to be constructed of copper or other corrosion-resistant materials.

EBJs are permitted to be installed either on the inside or the outside of the raceway or enclosure per 250-79(f). The total length shall not exceed 6′ and they shall be installed so that they closely follow the raceway. EBJs should not be installed by wrapping the jumper around a raceway. Such installations would add to the overall inductance of the circuit and act like a choke on any potential current flow.

Size – Supply Side and Main Bonding Jumpers – 250-79(d). EBJs on the line side or the supply side of the service and the MBJ are sized per Table 250-94. The size of the EBJ is based upon the size of the largest phase conductor for the service.

Where the size of the service-entrance conductors exceeds No. 1100 kcmil Cu and No. 1750 kcmil Al, the size of the line side EBJ or MBJ shall not be smaller than 12½% of the phase conductors. If the service-entrance conductors are run in parallel, then the EBJ shall also be run in parallel. Unlike grounded conductors, EBJs, when run in parallel, shall be full-sized conductors based on the largest conductor in each raceway or cable. See Figure 7-30.

Size – Load Side Bonding Jumpers – 250-79(e). Load side EBJs are sized per 250-79(e) and selected based upon the values listed in Table 250-95. The bonding jumper size is determined by the size or rating of the OCPD ahead of the equipment. If the EBJ is to be used for more than one raceway or cable, the size is selected from Table 250-95 based upon the largest OCPD protecting a circuit within the raceways or cables. See Figure 7-31.

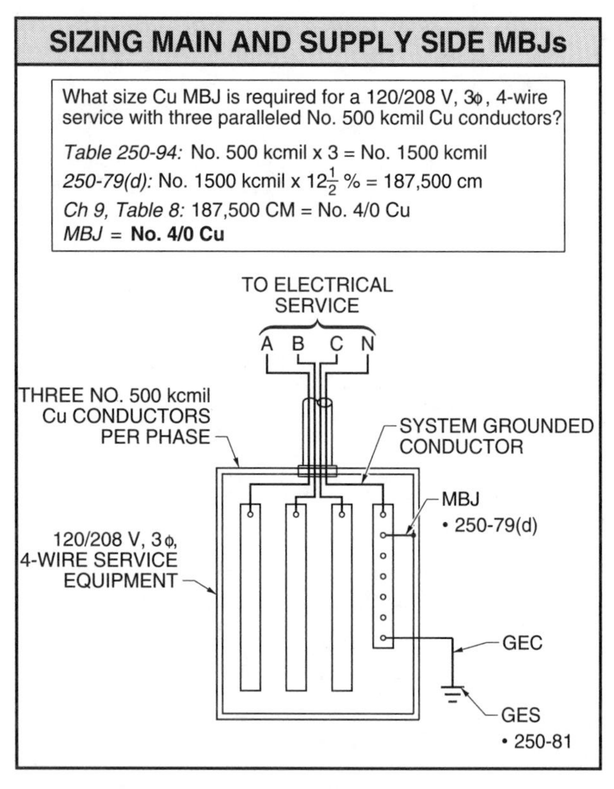

Figure 7-30. Line side and MBJs are sized to handle the available fault current that the system can deliver.

rately derived system. Part (b) of this section contains the additional requirement that other interior metal piping systems shall also be bonded to the service equipment enclosure, service grounded conductor, or the GEC. The size of the bonding jumper for these applications is based on Table 250-95.

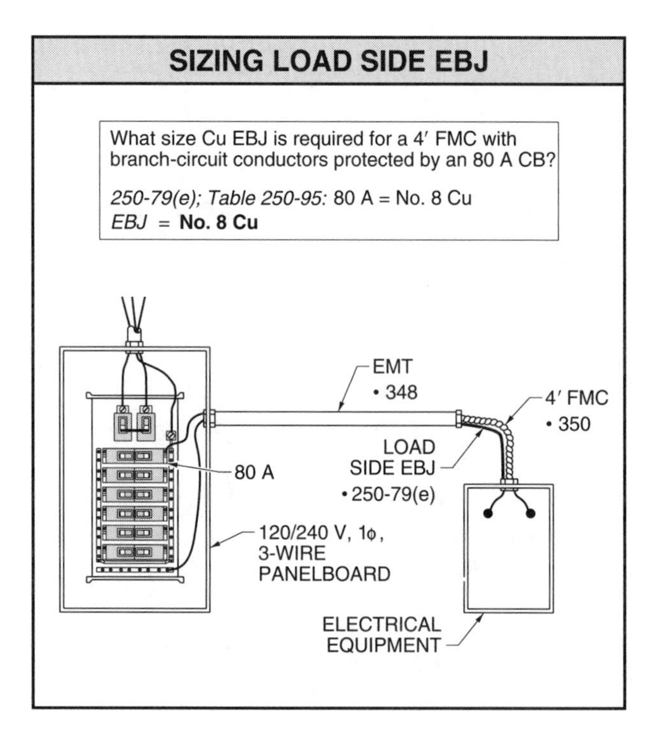

Figure 7-31. Load side EBJs are sized by the rating or setting of the OCPD in the circuit ahead of equipment, conduit, etc.

Interior Metal Piping and Structural Steel – 250-80. Metal underground water piping with at least 10′ of piping in direct contact with the earth shall be part of the GES. If the water piping is not effectively grounded, such as when plastic piping is used outside the building, 250-80 still requires that the interior metal water piping be bonded to the grounded conductor and GEC at the service equipment. The point at which the bonding jumper is attached shall be accessible. As with line side bonding jumpers, the size of the bonding jumper shall be based on Table 250-94.

A 1996 NEC® change requires that when separately derived systems are installed, and the metal water piping is not part of the GES, then the nearest point on the interior metal water piping system shall be bonded to the grounded conductor of the separately derived system.

The OCPD rating is selected based upon the rating of the circuit that may energize the piping. For example, if a piping system contains motor-operated valves, then that piping system could become energized and it should be bonded. Bonding all piping and metal duct work provide an additional level of safety per 250-80(b), FPN.

Bonding requirements have been added in 250-80(c) for exposed interior structural steel. This steel, if it forms a metal frame of a building, shall be bonded to the service equipment enclosure, the service grounded conductor, or the GEC. This section applies to structural steel that is interconnected to form a metal frame. The provisions of this section do not apply to a steel beam supported by concrete columns in a one-family dwelling. See Figure 7-32.

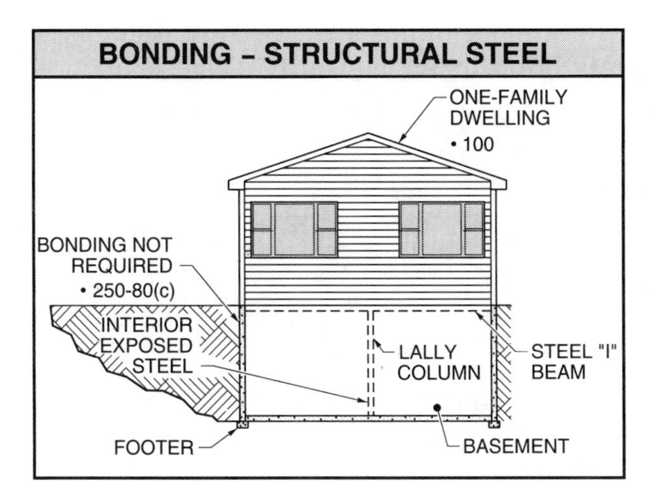

Figure 7-32. Single beams in dwelling units are not required to be bonded.

GROUNDING ELECTRODE SYSTEM – 250, PART H

Section 250-23(a), which details the grounding connections required for an AC service that is to be grounded, requires that the GEC be connected to a grounding electrode that complies with 250. Part H contains the requirements for the grounding electrode system. See Figure 7-33.

Construction – 250-81

Four items are connected together to form the GES; metal underground water pipe, metal building steel, concrete-encased electrode, and ground ring. Each of these items, if available, shall become part of the GES. For example, concrete-encased electrodes would be connected to the other electrodes if the concrete had not yet been poured. It is not the intent of this rule to require that concrete be jack-hammered to gain access to existing concrete-encased electrodes. Additionally, unlike the GES for separately derived systems in 250-26(c), all four shall be connected together if they are available.

Interior metal wiring piping shall not be used as a part of the GES unless it is the first 5′ of piping that enters the building. This ensures that downstream electrical systems are not isolated by the insertion of nonmetallic piping.

Metal Water Piping – 250-81(a). The first component that is connected to GES is metal underground water piping. The piping shall be in direct contact with the earth for a minimum distance of 10′. With the increasing popularity of nonmetallic water piping, always verify that there is at least 10′ in contact with the earth.

All insulating fittings or joints in the piping systems require bonding to ensure electrical continuity. Water meters, filtering devices, and similar equipment can interrupt the continuity of the piping system and require proper bonding. Unlike other electrodes, the metal underground water pipe is required to be supplemented by an additional electrode. The electrode can be one of the other three listed in 250-81 or it can be one of the made electrodes in 250-83. If a supplemental electrode is used, 250-81(a) requires that it be bonded to the GEC, the system grounded conductor, the grounded service raceway, or any grounded service enclosure.

Structural Steel – 250-81(b). This section contains the requirements for the second component of the GES, the metal building frame. The metal structural steel that forms the building frame is required to be part of the GES if it is effectively grounded. A determination is made as to when the metal frame of the building is included in the GES. A key factor in making this determination is whether the metal frame of the building is effectively grounded. See 100 to aid in making this determination. The building steel should be cleaned with a wire brush to remove any nonconductive coatings and to ensure a low-impedance connection per 250-118.

Concrete-Encased Electrode – 250-81(c). A concrete-encased electrode or reinforcing bar (rebar) is one of the best grounding electrodes. The rebar shall be encased in at least 2″ of concrete, be located at the bottom of the pour so as to maintain good contact with the earth, be at least 20′ long, and be at least ½″ in diameter. This section also permits 20′ of bare copper at least No. 4 in size as a substitute for the rebar.

Some rebars may be coated with nonconductive coverings which make them unsuitable as a grounding electrode. The connection to the concrete-encased electrode is not required to be accessible, but the connection to the electrode shall be made by means of exothermic weld, listed lugs, listed pressure connectors, listed clamps, or other listed means. Connectors for connection to a buried electrode shall be listed for direct burial uses. See 250-115.

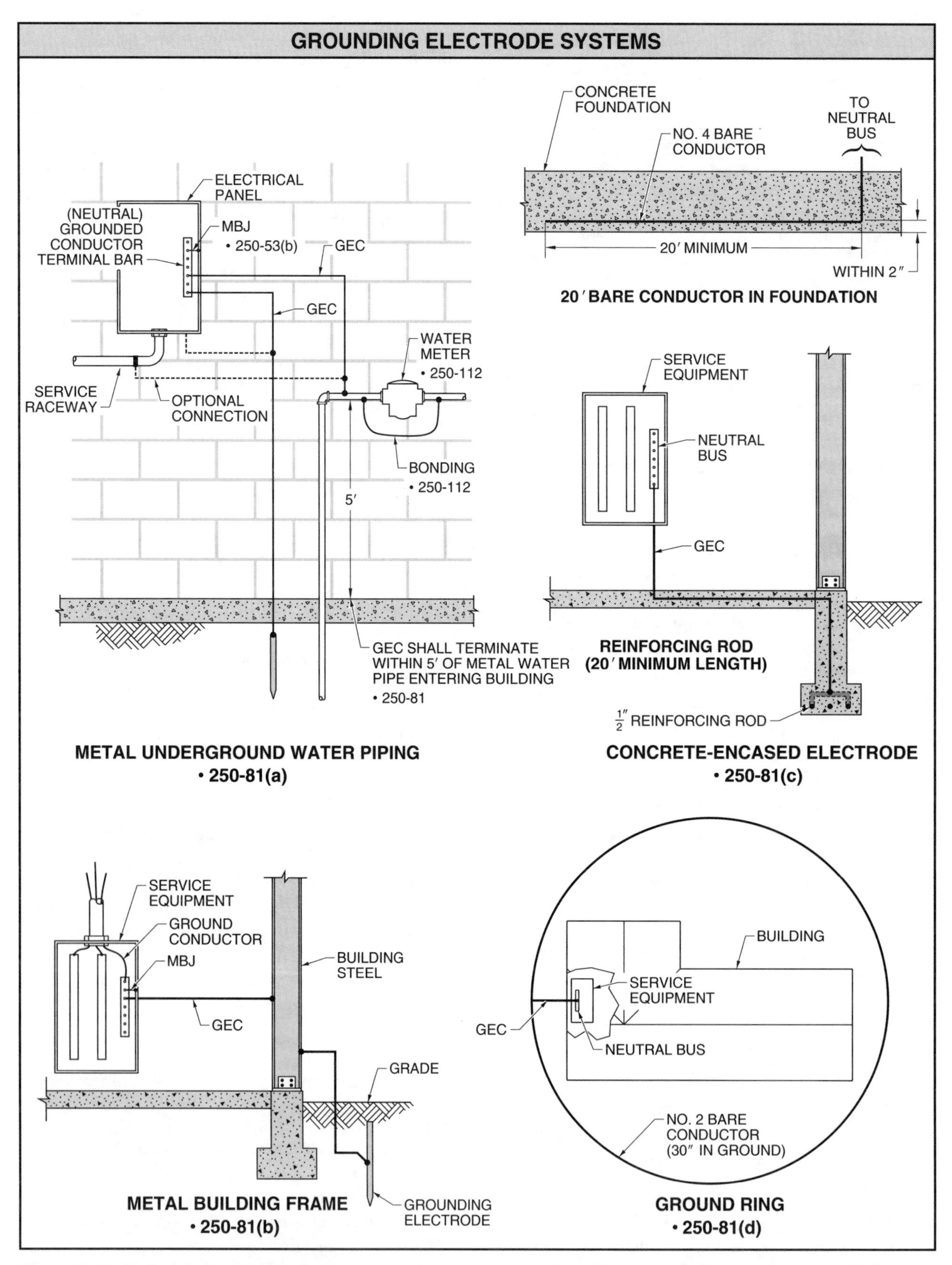

Figure 7-33. Each AC electrical system that is required to be grounded shall be connected to a GES per 250, Part H.

Ground Ring – 250-81(d). This section lists the requirements for ground rings. Ground rings shall consist of at least 20′ of No. 2 or larger bare copper. The ring shall encircle the building and be buried in a trench at least 2½′ in depth. The most common application for this type of electrode is in supplemental grounding systems such as those frequently used for grounding computer systems.

Made Electrodes – 250-83. There may be buildings in which none of the electrodes in 250-81 are available, or in which the only electrode available is the metal underground water pipe. In these cases, a supplemental electrode shall be installed. Section 250-83 contains the requirements for made or other electrodes. An electrode is simply a means to permit current to enter the earth. Made electrodes can be designed to function much like the electrodes listed in 250-81. As with these electrodes, made electrode connections shall be free of nonconductive coverings that may affect the impedance of the grounding path. See Figure 7-34.

Rod and Pipe Electrodes – 250-83(c). The most commonly encountered made electrodes are those constructed of rod and pipe. The minimum length of the ground rod shall not be less than 8′. The minimum diameter is ½″ for stainless steel or other nonferrous rods and ⅝″ for iron or steel rods.

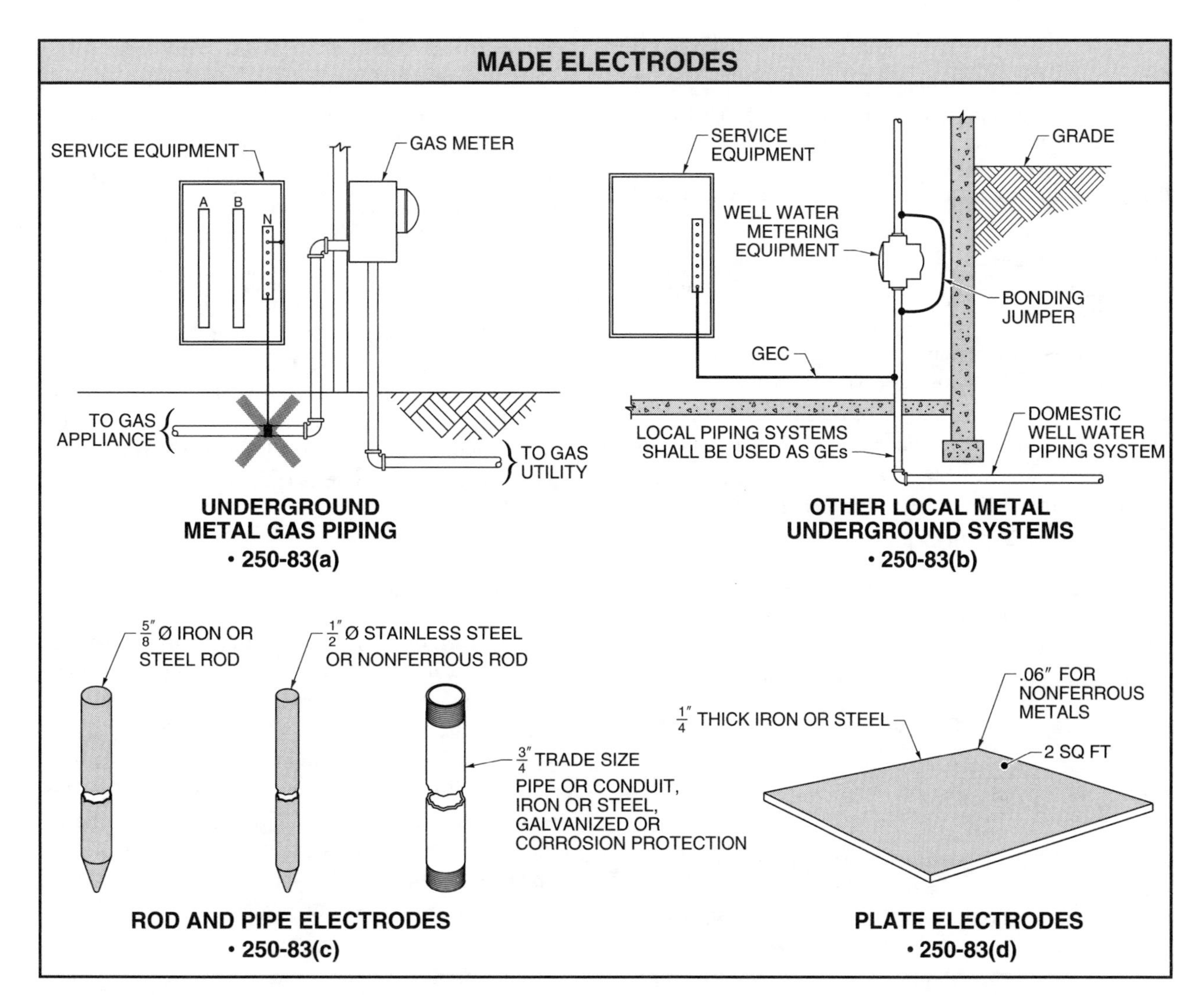

Figure 7-34. Made electrodes are required when none of the electrodes listed in 250-81 are available.

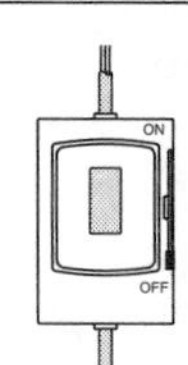

ROD AND PIPE ELECTRODES

The length required for rod and pipe electrodes is 8′ per 250-83(c). At least 8′ of the electrode shall be in contact with the earth per 250-83(c)(3). Therefore, the electrode is driven to a point at least flush with the finished grade. The depth to which the rod or pipe electrode is driven affects the impedance of the ground path. The deeper the electrode is driven, the lower the impedance. Do not cut off a ground rod when rock or other obstructions is encountered. This can have a serious effect on the integrity of the ground path.

The GEC connection is not required to be accessible per 250-112. However, the connection shall be made with a listed fitting per 250-115. Installers often mistakenly use indoor water pipe ground clamps on the ground rod. This is a violation of the NEC®. The listed acorn connector is required for this installation.

Pipe electrodes can be constructed of iron or steel provided they are galvanized or otherwise provided with corrosion protection. Pipe electrodes shall be a minimum ¾″ trade size. At least 8′ of the electrode shall be in contact with the earth. If rock is encountered during the installation, precluding driving the rod straight down, the rod or pipe can be installed on an angle, not to exceed 45° from vertical. If this is not possible, the rod or pipe can be buried in a trench at least 2½′ in depth.

Plate electrodes shall be at least 2 sq ft in area and at least ¼″ thick for iron or steel electrodes and .06″ thick for nonferrous metals. No minimum burial depth is required, but the 2½′ required for the rod or pipe electrode would be adequate.

Prohibited Electrodes

Section 250-83(a) specifically prohibits metal underground gas piping systems from being used as a grounding electrode. Under fault conditions, the GEC and the electrode itself will carry current. The gas piping is required to be bonded per 250-80(b). It is not, however, permitted to be used as a grounding electrode. Per 250-83(e), aluminum electrodes are not permitted in the grounding electrode system.

GROUNDING ELECTRODE CONDUCTORS – 250, PART J

The GEC is used to connect the EGC and the grounded conductor at the service or separately derived system to the grounding electrode. Four major items should be considered when installing or designing the GEC. The first is the construction of the GEC. The second is the installation requirements for the GEC. The third is the sizing calculations for the GEC, and the fourth is the methods of connection which are permitted.

Construction – 250-91

The GEC is constructed of copper, aluminum, or copper-clad aluminum per 250-91(a). The solid or stranded conductors are permitted to be insulated, covered, or bare. The conductor shall have good corrosion-resistant properties and should be suitable for the conditions under which it may be installed.

The GEC shall be one continuous length per 250-91(a), but three exceptions allow splices. Per 250-91(a), Ex. 1, splices in busbars are permitted. Per 250-91(a), Ex. 2, the GEC may be spliced or tapped when multiple service enclosures are used in accordance with 230-40, Ex. 2. GECs may be spliced if the splicing method consists of either exothermic welding or irreversible compression-type connectors per 250-91(a), Ex. 3.

Installation – 250-92

Per 250-92(a), the GEC is not required to be installed in a raceway if it is of adequate size and not subject to severe physical damage. GECs which are subject to severe physical damage shall be a No. 4 Cu or No. 4 Al, or larger, conductor or else it shall be protected. If not subject to severe physical damage, No. 6 and larger conductors shall be permitted to be installed along the surface of the building if adequately supported. RMC, IMC, RNMC, EMT, or cable armor is permitted to be used for protection against severe physical damage.

An aluminum or copper-clad aluminum GEC shall not be used within 18″ of the earth. Additionally, aluminum or copper-clad aluminum cannot be used in direct contact with masonry or other corrosive conditions.

Per 250-92(b), the metal enclosures or raceways in which a GEC is installed shall be electrically continuous. This requires that both ends of the GEC enclosure be bonded to either the equipment or electrode or to the GEC itself. See Figure 7-35.

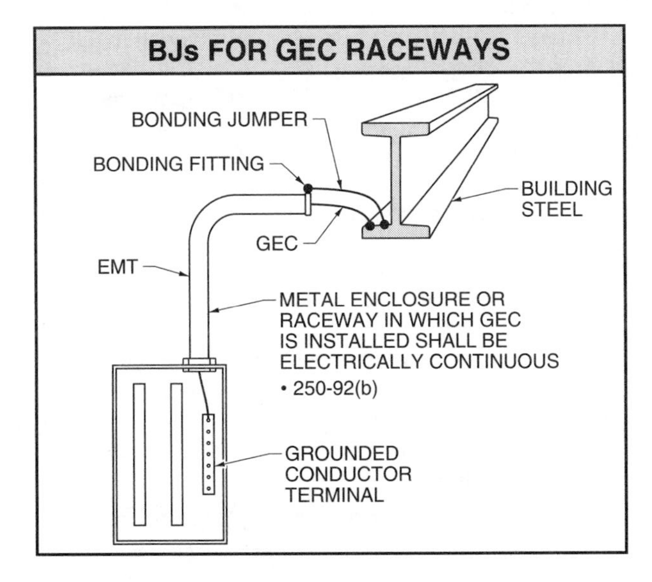

Figure 7-35. The bonding jumper for a grounding electrode conductor raceway shall be the same size or larger than the required GEC.

Size – 250-94

The size of the GEC is selected from Table 250-94 and is based upon the largest service-entrance conductor for 1ϕ. If the service-entrance conductors are paralleled, the equivalent area for 1ϕ conductors shall be used as a basis for sizing the GEC. See Figure 7-36.

While the purpose of the grounding electrode is to provide a path into the earth for current flow, many factors affect its ability to dissipate current flow into the earth. Some electrodes dissipate current flow better than others. For example, rebar is typically a better electrode than a ground rod. The rebar provides a greater surface area from which the electrons can be dissipated. Other conditions may also effect the rate of dissipation. With ground rods, the deeper the ground rod enters the earth, the greater the rate of dissipation. In addition, the higher the moisture content of the earth, the higher the rate of dissipation from the ground rod.

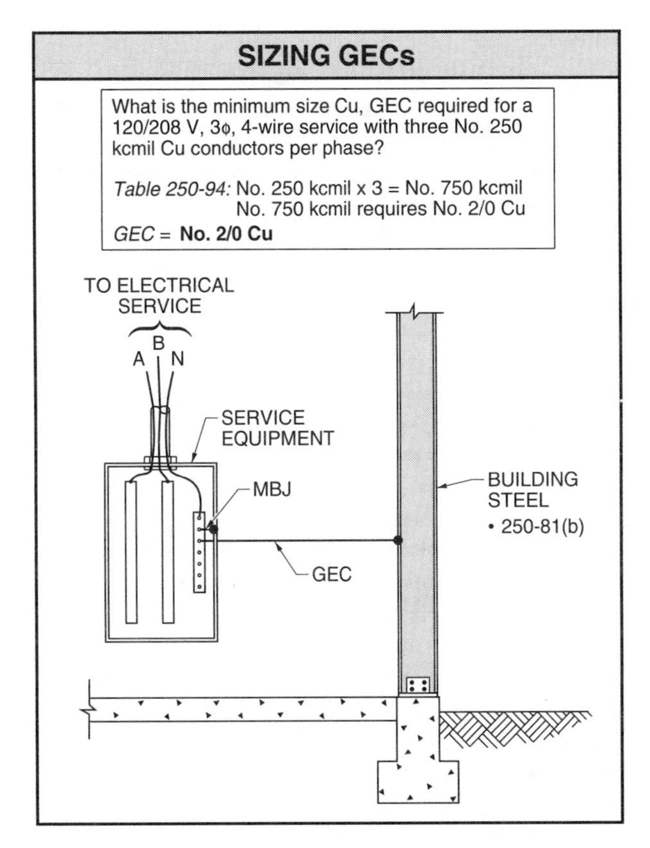

Figure 7-36. Grounding electrode conductors are never required to be larger than No. 3/0 Cu or No. 250 kcmil Al.

The size of the GEC reflects, to some degree, the ability of the grounding electrode to dissipate current flow into the earth. For example, 250-95, Ex. permits smaller size GECs for specific types of grounding electrodes. For sole connections to made electrodes, 250-94, Ex. a, permits the GEC to be No. 6 Cu or No. 4 Al. The GEC is not required to be larger than No. 4 Cu per 250-94, Ex. b for sole connections to concrete-encased electrodes. For ground rings, 250-94, Ex. c permits the GEC, which is a sole connection, to be the same size as the conductor used for the ground ring. See Figure 7-37.

Grounding Electrode Connections – 250, Part K

The provisions for making GEC connections are given in 250, Part K. Per 250-112, the GEC connection shall be accessible.

Per 250-112, Ex., the connection for concrete-encased, driven, or buried electrodes is not required to be accessible. Bonding jumpers around insulated joints to ensure electrical continuity where equipment is installed shall be of sufficient length to permit the removal of the equipment without removal of the bonding jumper. This safety requirement protects service personnel from possible current flow on the water piping system. Connections to electrodes shall be made with listed fittings per 250-115. The connection shall be protected against severe physical damage per 250-117.

EQUIPMENT GROUNDING CONDUCTORS – 250-91

The same items are considered before designing or installing EGCs in electrical distribution systems as are considered when designing or installing GECs. The EGC is used to connect the noncurrent-carrying metal parts of equipment, raceways, enclosures, etc., to the grounded conductor and the GEC at the service equipment or source of a separately derived system. This connection provides a low-impedance path back to the source to permit the operation of the OCPD.

BONDING GEC RACEWAYS

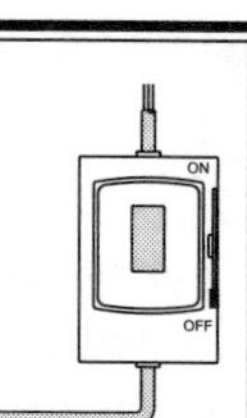

Failing to bond both ends of the GEC raceway is a common NEC® violation which may dramatically affect the integrity of the grounding path. Studies have shown that when a raceway is not properly bonded, and a fault current is applied to the GEC, the current does not split evenly between the GEC and the raceway. The net effect is the fault current is choked by the inductive effect of a single conductor in a metal raceway. GECs, however, may be sized as large as No. 3/0 Cu and No. 250 kcmil Al. With these larger sizes, bonding the raceway can be especially difficult. One solution to the problem is to use RNMC as the raceway for enclosing the GEC. The nonmetallic raceway alleviates any induced current problem resulting from single conductors in a raceway. Make sure that any nonmetallic raceway is not subject to severe physical damage.

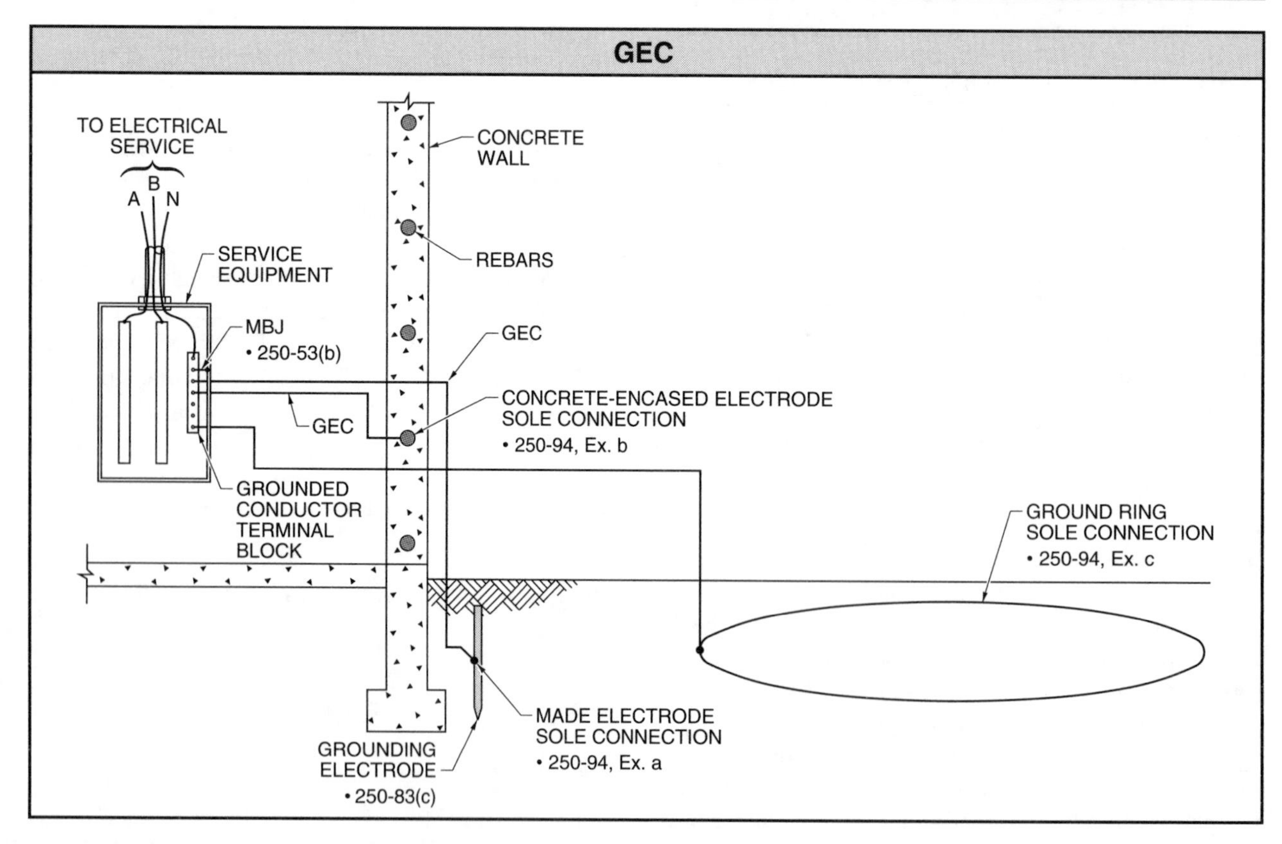

Figure 7-37. The GEC to a ground rod or other made electrode need not be larger than No. 4 Cu when it is the sole connection to the grounding electrode.

Construction – 250-91(b)

Section 250-91 lists all of the permissible types of EGCs. EGCs are constructed of copper or other corrosion-resistant materials. The conductors, which can be in the form of a wire or busbar, may be solid or stranded and either insulated, covered, or bare. However, there are several instances where the EGC is not permitted to be a bare conductor. For example, the EGC to a remote building that houses livestock shall be insulated or covered per 250-24(a), Ex. 2.

Section 250-91(b) contains two exceptions which permit FMC, FMT, and LTFMC under specific conditions to be used as EGCs. Ex. 1 permits FMC, FMT, or LTFMC in the 3⁄8″ through 1 1⁄4″ trade sizes to be used as an EGC when three conditions are met. The total length of the conduits or tubing, in the same ground path, shall be 6′ or less. The conduit or tubing shall be terminated in fittings listed for grounding, and the circuit conductors installed in the conduit or tubing shall be protected by 20 A or less OCPDs.

Ex. 2 permits listed LTFMC in 3⁄4″ through 1 1⁄4″ trade sizes to be used as the EGC, provided four conditions are met. First, the circuit conductors installed within the LTFMC shall be protected by OCPDs rated at more than 20 A but not more than 60 A. The second condition requires that the total length of the LTFMC shall not exceed 6′ in any ground path. The LTFMC shall be terminated in fittings listed for grounding, and there may not be any other FMC, FMT, or LTFMC in the 3⁄8″ and 1⁄2″ trade sizes serving as an EGC in the ground path. See Figure 7-38.

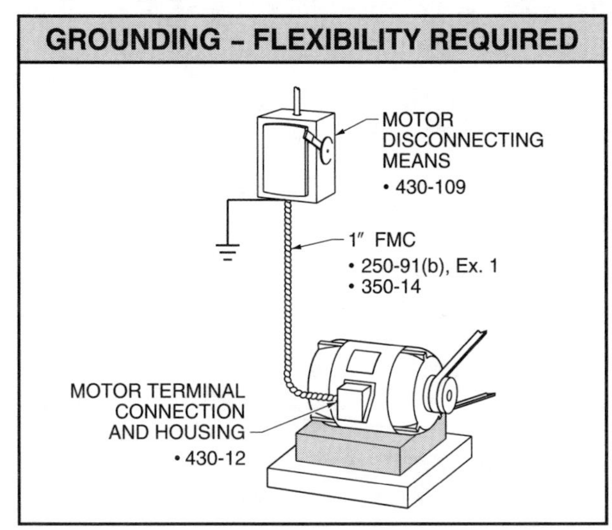

Figure 7-38. A separate EGC shall be provided when FMC is used to connect equipment where flexibility is required.

Installation – 250-92(c)

When the EGC consists of a raceway, cable tray, cable armor, cable sheath, or a wire which is part of a cable or installed in a raceway, the EGC shall be installed in a manner which meets all of the applicable provisions for the type of raceway or cable that is used. Care shall be taken to ensure that the EGC terminations are made with fittings suitable for the raceway or cable system used.

Section 250-92(c)(2) requires that when the EGC is run as a separate conductor, as permitted by 250-50 (a)(b) and 250-57, Ex. 2, it shall be protected against physical damage. Separate EGCs, No. 4 or larger, shall be protected when exposed to physical damage. The separate EGC is permitted to be run along the surface of the building without protection if it is securely fastened in place. When the separate EGC is aluminum or copper-clad aluminum, it shall not be installed in any of the following conditions:

- Direct contact with masonry or the earth
- Where subject to corrosive conditions
- Outdoors within 18″ of the earth

Size – 250-95

The size of the EGC shall be based upon the size or rating of the OCPD ahead of the equipment which is supplied. Table 250-95 lists the OCPD ratings and the appropriate size EGC for copper, aluminum, or copper-clad aluminum. For example, for a 30 A branch-circuit run in NMC to serve equipment, the EGC is based upon the 30 A OCPD rating and requires either a No. 10 Cu or No. 8 Al EGC.

The EGCs shall be run in parallel when conductors are run in parallel in multiple raceways or cables per 310-4. Each of the EGCs shall be a full-sized conductor based on the ampere rating of the OCPD protecting the circuit conductors. If multiple circuits are run in the same raceway, the size of the EGC shall be based on the largest OCPD protecting the circuit conductors per 250-95. See Figure 7-39. If the circuit conductors are adjusted in size for the purposes of voltage drop, a similar adjustment shall be made in the size of the EGC.

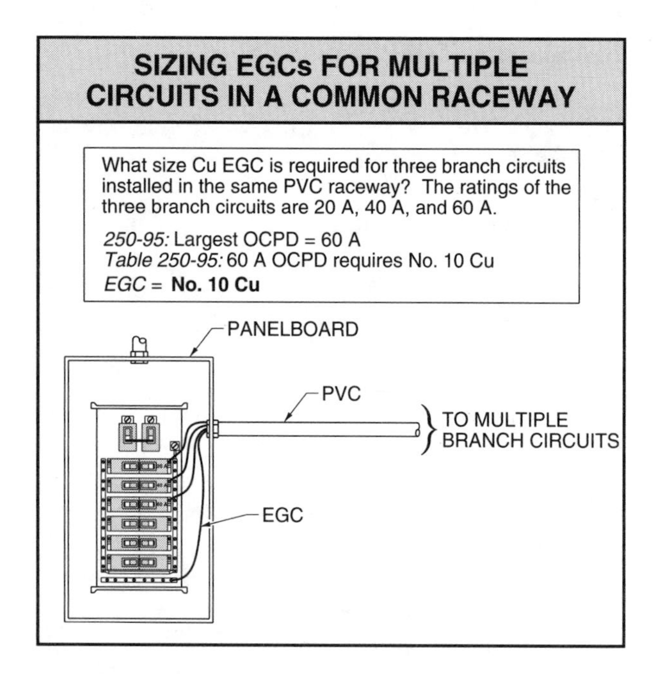

Figure 7-39. If multiple circuits are run in a common raceway, only one EGC, based on the largest size OCPD protecting the conductors, is required.

USING MC CABLE AS EGC

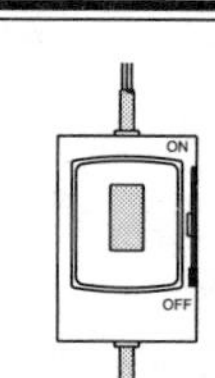

Installers should be very careful when using the outer sheath of MC cable as an EGC. Per 250-91(b)(8), the metallic sheath or the combined metallic sheath and grounding conductors for MC cable may be used as an EGC.

MC cable is constructed in three basic types; interlocking metal tape, smooth, and corrugated metallic sheath. Only the smooth and corrugated types are suitable for equipment grounding purposes. The interlocking tape type requires the use of a separate EGC. Do not use the interlocking tape type in applications which require two EGCs such as for isolated ground receptacles or patient care receptacles in health care facilities.

Connections – 250-114

Whenever several EGCs are present in a box or an enclosure, the conductors shall be spliced together or connected within the box with devices suitable for the use. If devices or fixtures are fed from the box, care shall be taken to ensure that the continuity of the EGCs is not interrupted when the device or fixture is removed. Additionally, if a metal box is used, provisions shall be made to connect the one or more EGCs to the box by means of a dedicated grounding screw or a listed grounding device.

If nonmetallic boxes are used, the EGCs shall be arranged so that any device or fitting in the box can be grounded if so required. Terminals for the EGC in wiring devices shall be identified by either a green-colored screw, green-colored terminal screw, or a green-colored pressure wire nut. If the terminal point is not visible, it shall be marked with either the word "green," the letters "G" or "GR," or the grounding symbol.

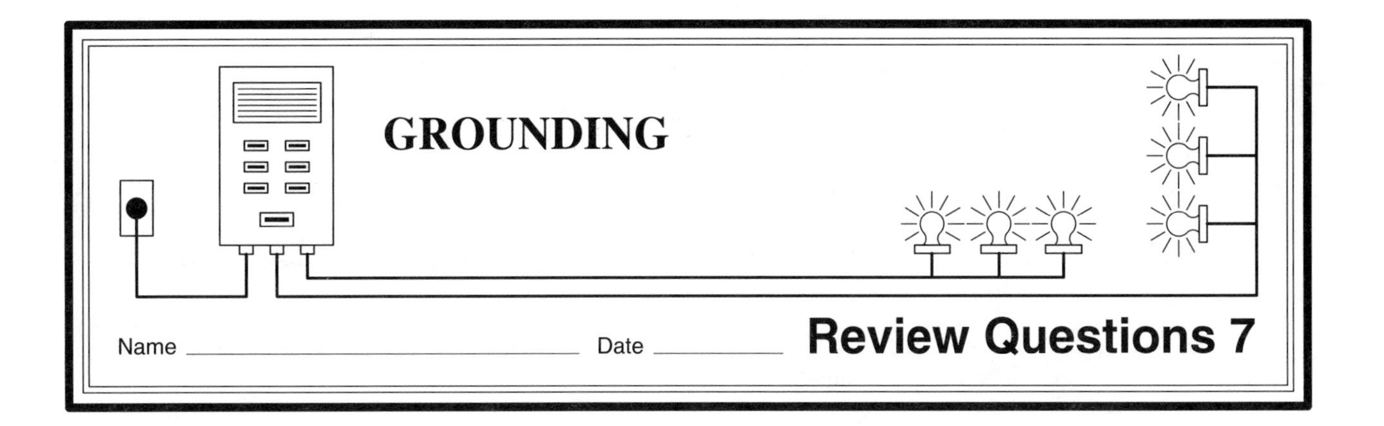

Review Questions

T F **1.** Electrical systems are grounded to limit and stabilize voltages to ground.

T F **2.** The amount of current that flows in any circuit is determined by the voltage and resistance of the circuit.

__________ **3.** A(n) ________ conductor is the conductor that connects electrical equipment or the grounded conductor to the grounding electrode.

__________ **4.** AC systems of less than ________ V to ground shall be grounded when installed outside of a building as overhead conductors.

__________ **5.** ________ wiring is basically all interior and exterior wiring installed on the load side of the service point or the source of a separately-derived system.

__________ **6.** The ________ conductors extends from the service point to the service disconnecting means.

__________ **7.** A wet location is a(n) ________.

A. installation in concrete slabs contacting earth
B. underground installation
C. location saturated with water or other liquids
D. A, B, and C

__________ **8.** A damp location is a(n) ________.

A. partially protected area
B. area subject to some moisture
C. both A and B
D. neither A nor B

__________ **9.** Electrical service equipment which shall be bonded together includes ________.

A. service raceways with metallic sheath
B. service conductor enclosures
C. metallic raceways protecting the GEC
D. A, B, and C

__________ **10.** The ________ is the connection at the service equipment that ties together the EGC, the grounded conductor, and the GEC.

__________ **11.** The ________ connects the EGC and the grounded conductor at the service or separately-derived system to the grounding electrode.

_________ 12. The minimum length of a ground rod is ________′.

_________ 13. The minimum diameter of a nonferrous ground rod is ________″.

_________ 14. The minimum diameter of a ferrous ground rod is ________″.

T F 15. The grounded conductor shall never be smaller than $10\frac{1}{2}\%$ of the largest service-entrance conductor.

T F 16. The first choice for the grounding electrode of a separately-derived system is for effectively grounded building steel.

T F 17. The EGC may be a bare, covered, or insulated conductor.

_________ 18. Metal underground water piping with at least ________′ of piping in direct contact with the earth shall be part of the GES.

_________ 19. The size of the EGC is based upon the rating of the ________ ahead of the supplied equipment.

T F 20. Aluminum electrodes may be used as part of the GES.

Matching

_________ 1. A temporary increase in the circuit or system voltage or current.

_________ 2. The maximum voltage between any two phases of an electrical system.

_________ 3. Grounded with sufficient low impedance and current-carrying capacity to prevent hazardous voltage buildups.

_________ 4. A conducting connection between electrical circuits or equipment and the earth.

_________ 5. A device that converts electrical power at one voltage or current to another voltage or current.

_________ 6. A device that converts mechanical power to electrical power.

_________ 7. The case or housing of equipment or other apparatus which provides protection from live or energized parts.

_________ 8. Any device, fixture, apparatus, appliance, etc. used in conjunction with electrical installations.

_________ 9. Joining metal parts to form a continuous path to conduit safely any current path is commonly imposed.

_________ 10. A conductor that has been intentionally connected or grounded.

A. phase-to-phase voltage

B. ground

C. generator

D. bonding

E. line surge

F. transformer

G. effectively grounded

H. enclosure

I. equipment

J. grounded conductor

Grounding Electrode Systems

__________ **1.** The GEC at A shall terminate within ________′ of the metal water pipe entering the building.

T F **2.** An optional connection point for the GEC at A could be the service raceway.

__________ **3.** The minimum length of the bare conductor at B is ________′.

__________ **4.** The minimum length of the reinforcing rod at C is ________′.

__________ **5.** The minimum diameter of the reinforcing rod at C is ________″.

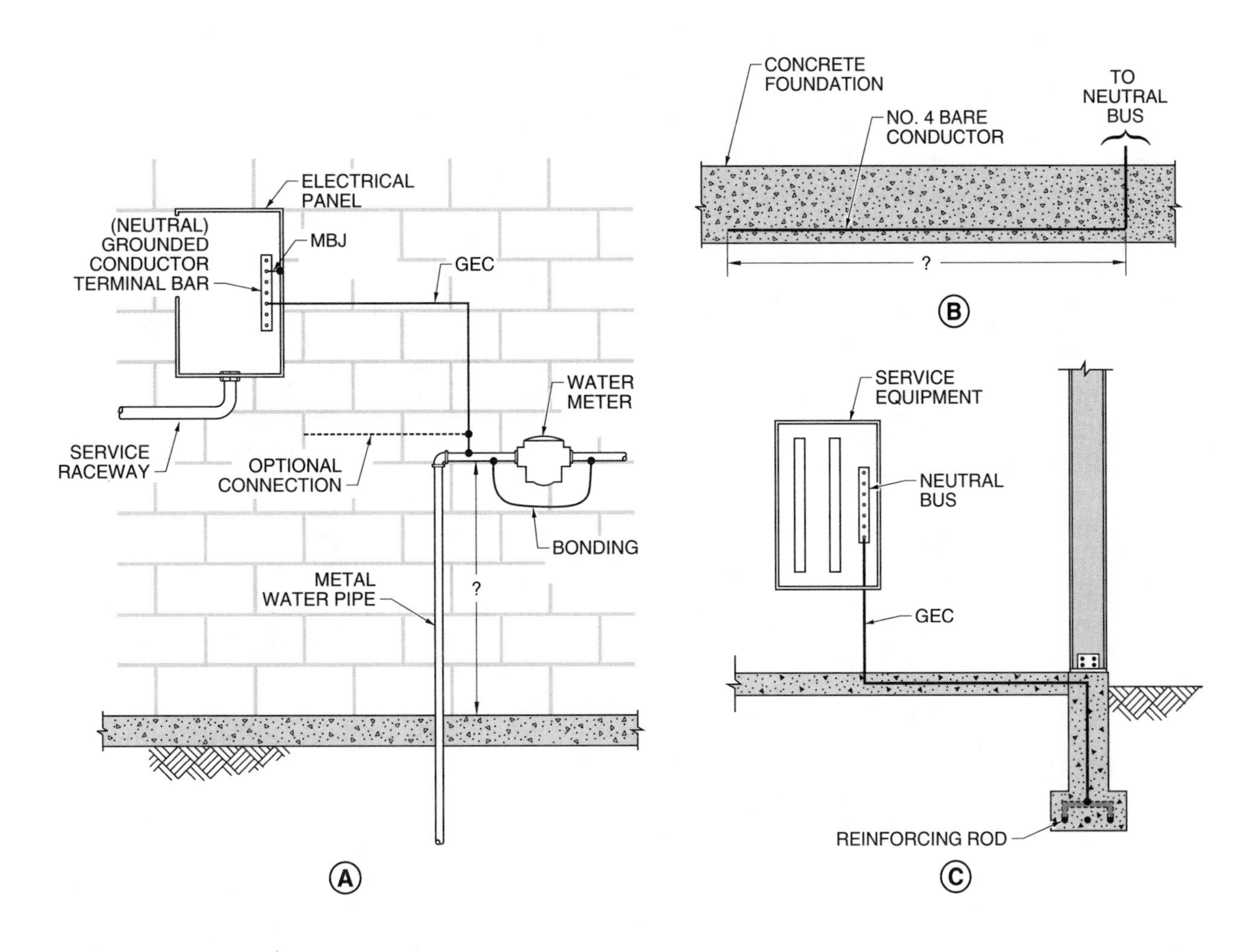

Made Electrodes

T F **1.** The made electrode at A is permitted.

T F **2.** The made electrode at B is permitted.

__________ **3.** The minimum diameter at C is ________″.

__________ **4.** The minimum diameter at D is ________″.

__________ **5.** The minimum diameter at E is ________″.

__________ **6.** The made electrode at F shall be at least ________ sq ft in area.

__________ **7.** The made electrode at F shall be a minimum of ________″ thick when made of iron or steel.

__________ **8.** The made electrode at F shall be a minimum of ________″ when made of nonferrous metals.

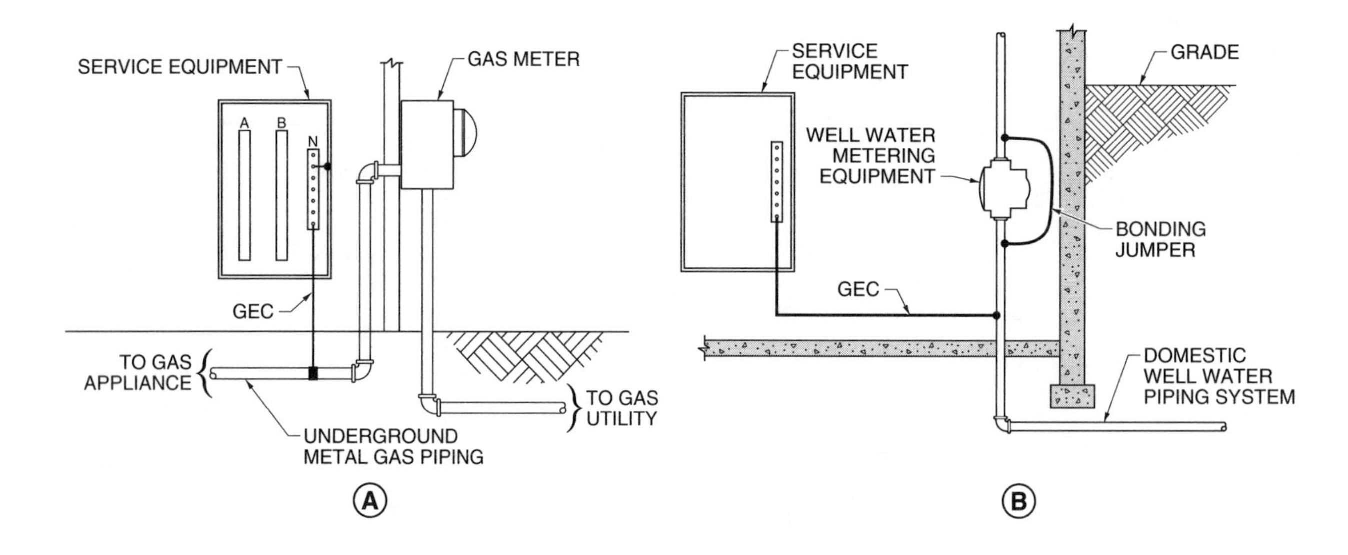

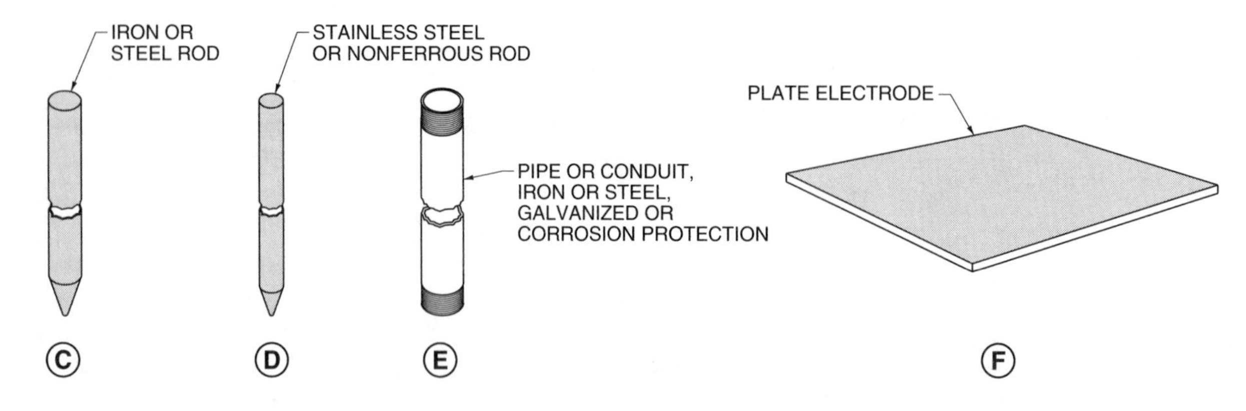

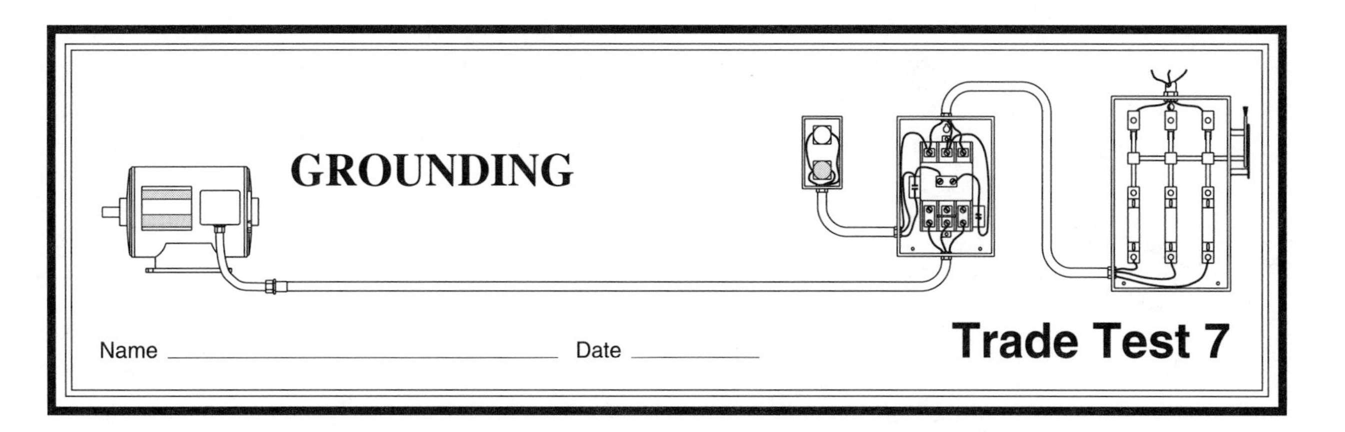

NEC®	Answer	
______	______	**1.** Determine the minimum size Cu grounded conductor required for a 120/208 V, 3ϕ, 4-wire service which consists of two parallel No. 1/0 THW Cu conductors per phase?
______	______	**2.** What size grounded conductor is required for a 277/480 V, 3ϕ, 4-wire service with three No. 400 kcmil THW Cu conductors per phase?
______	______	**3.** *See Figure 1.* Determine the minimum size bonding jumper required for the installation.
______	______	**4.** Determine the minimum size Al main bonding jumper required for a 277/480 V, 3ϕ, 4-wire service which consists of two parallel No. 4/0 THW Cu conductors per phase?
______	______	**5.** *See Figure 2.* Determine the minimum size Cu main bonding jumper required for the installation.
______	______	**6.** Determine the minimum size Cu GEC required for a 120/208 V, 3ϕ, 4-wire service which consists of four No. 1/0 THW Cu conductors per phase. The grounding electrode is effectively-grounded building steel.
______	______	**7.** Determine the minimum size Cu GEC required for a 120/240 V, 1ϕ service which consists of three No. 4/0 Al service-entrance conductors. The grounding electrode is a 5/8″ Cu ground rod. The GEC is the sole connection to the ground rod.
______	______	**8.** What is the minimum size Cu GEC required for a 120/208 V, 3ϕ, 4-wire service with three No. 300 kcmil Cu conductors per phase?

277/480 V, 3ϕ, 4-WIRE PANELBOARD
A B C N
TRANSFORMER • 450
120/208 V, 3ϕ, 4-WIRE PANELBOARD
A B C N
BJ AT FIRST DISCONNECT
TO BUILDING STEEL
NO. 4/0 Cu PHASE CONDUCTORS

FIGURE 1

TO ELECTRICAL SERVICE
A B C N
THREE NO. 600 kcmil Cu CONDUCTORS PER PHASE
SYSTEM GROUNDED CONDUCTOR
MBJ • 250-79(d)
120/208 V, 3ϕ, 4-WIRE SERVICE EQUIPMENT
GEC
GES • 250-81

FIGURE 2

_______ _______ **9.** *See Figure 3.* Determine the minimum size Cu GEC for the separately derived system.

_______ _______ **10.** *See Figure 4.* Determine the size of the Al EBJ required for the installation.

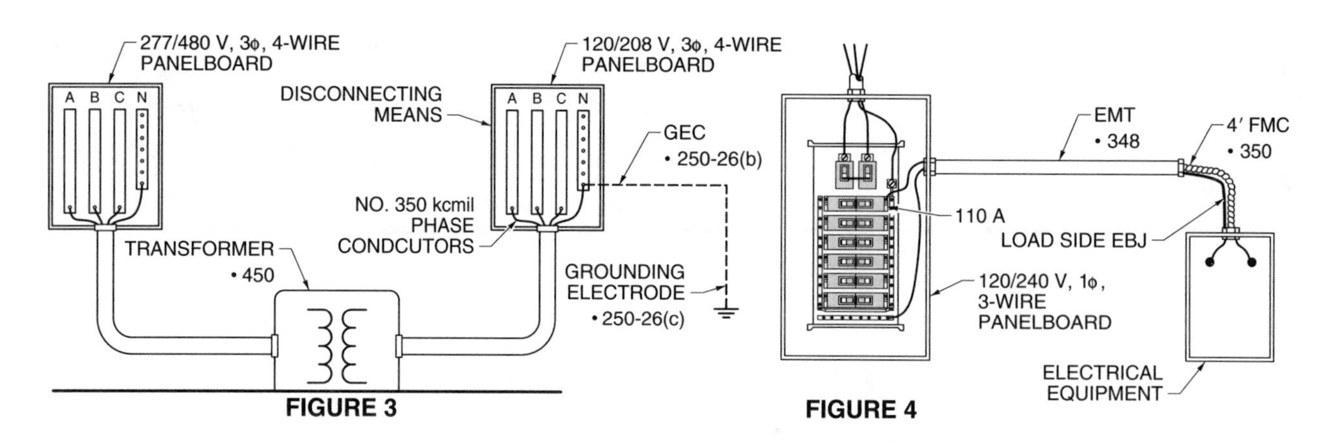

FIGURE 3 FIGURE 4

_______ _______ **11.** Determine the minimum size Cu EGC required for a 50 A branch circuit installed in RNMC.

_______ _______ **12.** Determine the minimum number and minimum size Al EGCs required for an installation which consists of two parallel 2″ RNMCs, each which has the conductor backed up by a 225 A OCPD.

_______ _______ **13.** What size Cu EGC is required for three branch circuits installed in the same PVC raceway? The ratings of the three branch circuits are 20 A, 60 A, and 100 A.

_______ _______ **14.** *See Figure 5.* Determine the minimum size Cu bonding jumper required for the GEC raceway.

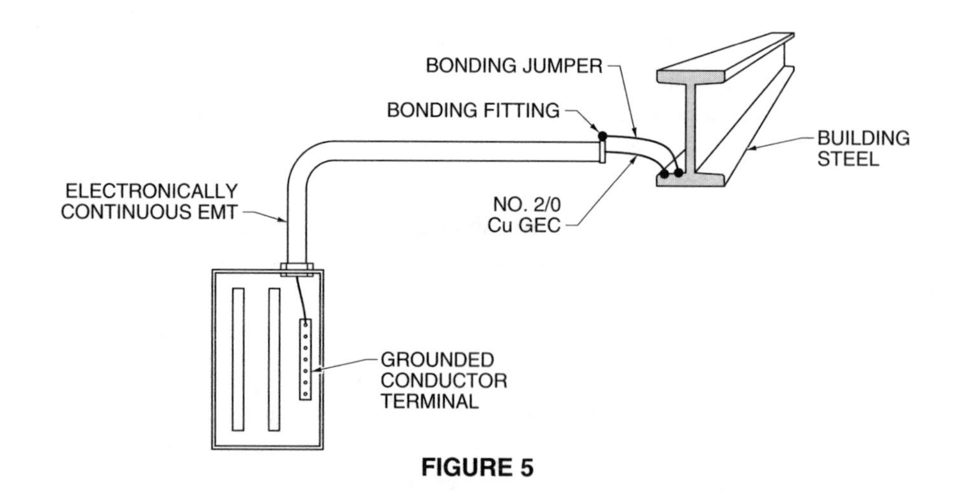

FIGURE 5

_______ _______ **15.** A 4′ piece of 1″ FMC is installed to supply a motor protected by a 20 A OCPD. All fittings used are listed for grounding and the total length of FMC in the circuit is less than 6′. A separate EGC is not provided for the FMC. Does this installation violate the requirements of the NEC®?

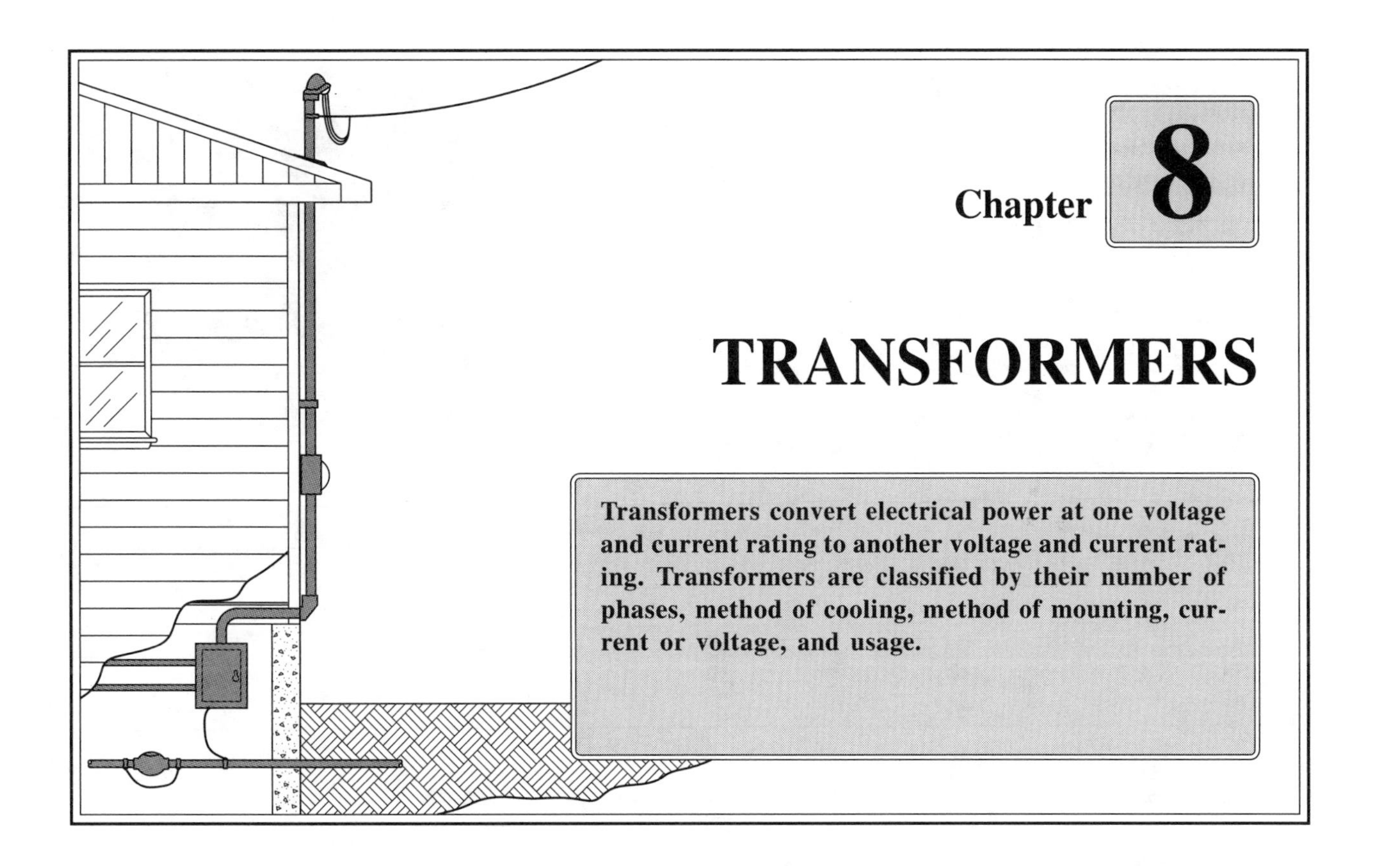

TRANSFORMER CONSTRUCTION AND TYPES

Transformers are one of the most common components of electrical distribution systems. *Transformers* are electrical devices that contain no moving parts and are used primarily to convert electrical power at one voltage and current rating to another voltage and current rating.

Transformers operate on the principle of magnetic induction. Two separate coils or windings are wound around a steel or iron core. When a voltage is impressed upon one of the windings, a strong magnetic field is created by the magnetizing of the core. This magnetic field, in turn, induces a voltage in the second winding. The wire in the first (primary) winding, which is connected to the power supply, usually consists of smaller, heavily-insulated wire to accommodate the applied voltage. The wire in the second (secondary) winding, which delivers the power to the load, is constructed from heavier gauge wire which can accommodate the increased currents.

The arrangement of the windings determines whether the transformer is classified as a step-up transformer or a step-down transformer. A *step-up transformer* is a transformer with more windings in the secondary winding, which results in a load voltage which is greater than the applied voltage. A *step-down transformer* is a transformer with more windings in the primary winding, which results in a load voltage which is less than the applied voltage. See Figure 8-1. Either the primary winding or the secondary winding can be the high-voltage winding depending upon the transformer application.

Transformers can be classified in many ways. Common classifications include number of phases, method of cooling, method of mounting, current or voltage, and usage.

Dry-Type Transformers

Because of the close concentration of the conductors in the windings, heat is a factor that cannot be ignored in transformer installation and design. If excessive heat is permitted to remain in or around the transformer, the insulation on the conductors that comprise the windings would be subject to damage and perhaps failure.

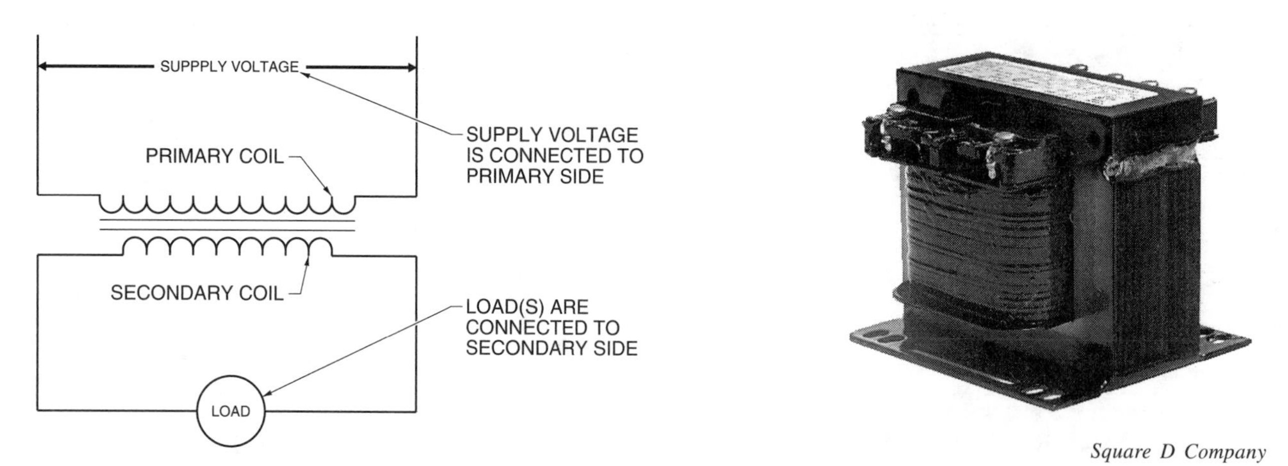

Square D Company

TRANSFORMER	SECONDARY			
	VOLTAGE	AVAILABLE CURRENT	POWER	TERMINAL MARKINGS
STEP-UP	↑	↓	—	PRIMARY = X1 , X2 , ETC. SECONDARY = H1 , H2 , ETC.
STEP-DOWN	↓	↑	—	PRIMARY = H1 , H2 , ETC. SECONDARY = X1 , X2 , ETC.

Figure 8-1. Transformers are classified as step-up transformers or step-down transformers.

Transformers are often classified by the type of coolant or method of cooling that is employed. The NEC® classifies transformers that do not rely upon any type of liquid for cooling purposes as dry-type transformers. A *dry-type transformer* is a transformer which provides air circulation based on the principle of heat transfer. These air-cooled transformers are generally smaller in size. Larger transformers of this type are available and they may incorporate a fan or a system of fins or radiators to aid in the dissipation of heat. Dry-type transformers are constructed as one of three major types: epoxy encapsulated, ventilated, and totally-enclosed. See Figure 8-2.

There are four recognized NEMA insulation classifications for dry-type transformers. These insulation classifications are Class A, Class B, Class F, and Class H. Each classification has a maximum permitted temperature rise for the windings. The type of insulation used and the design of the transformer contribute to the transformer classification. Dry-type transformers are commonly used for the purposes of medium voltage distribution.

Class A transformer insulation operates at not more than a 55°C temperature rise on the windings if the transformer is not overloaded and the ambient temperature does not exceed 40°C. Class A transformers are often selected for use as control-type transformers.

Class B transformer insulation operates at not more than an 80°C temperature rise on the windings provided the transformer is not overloaded and the ambient temperature does not exceed 40°C. Class B transformers are commonly used for distribution purposes.

DRY-TYPE TRANSFORMERS			
Class	**Temperature Rise – °C**	**Ambient Temperature – °C**	**Common Use**
A	55	40	Control
B	80	40	Distribution
F	115	40	Distribution
H	150	40	Large distribution

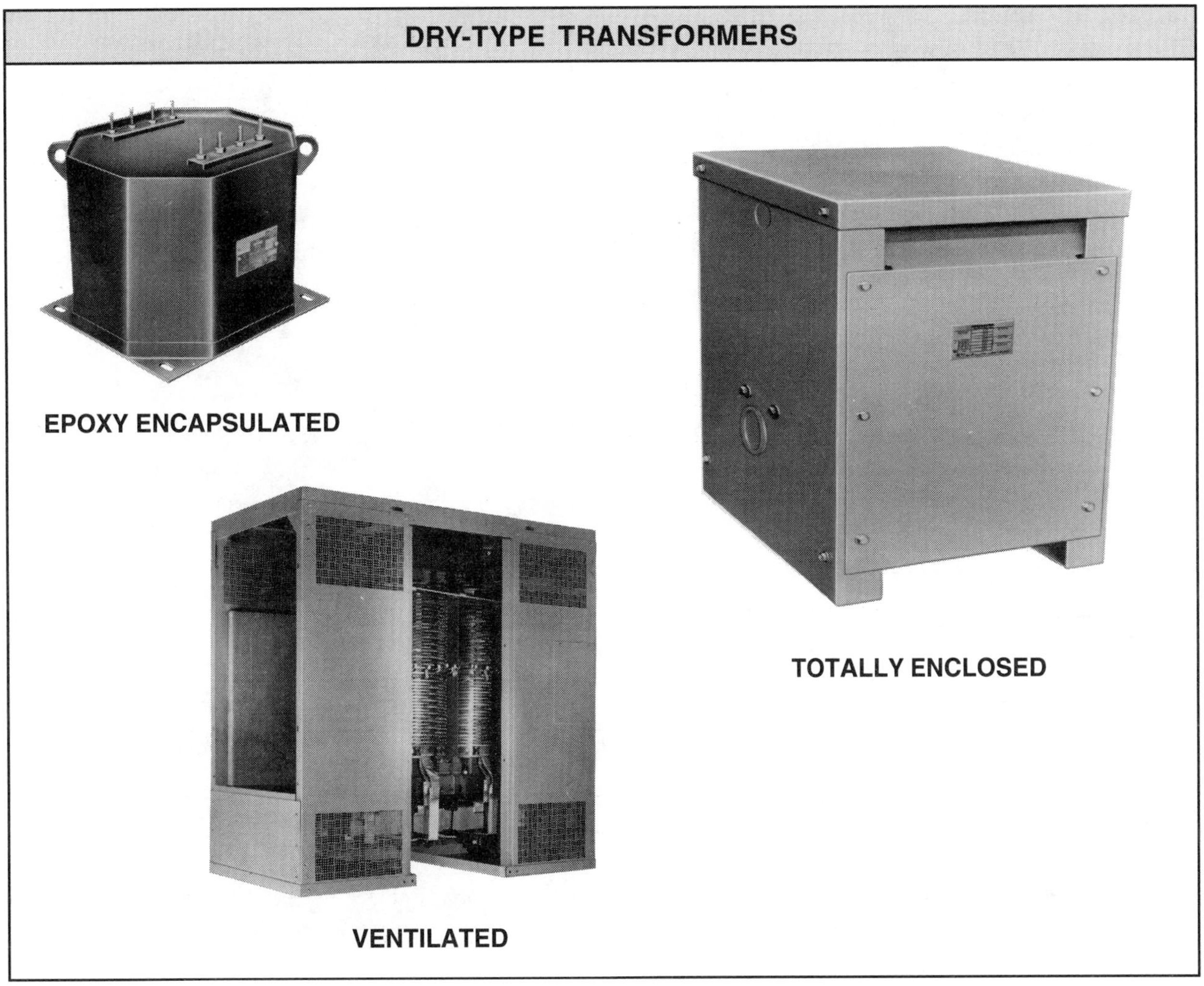

DRY-TYPE TRANSFORMERS

EPOXY ENCAPSULATED

TOTALLY ENCLOSED

VENTILATED

ABB Power T&D Company Inc.

Figure 8-2. Dry-type transformers are constructed as one of three major types: epoxy encapsulated, ventilated, and totally enclosed.

Class F transformers operate at not more than a 115°C temperature rise on the windings if the transformer is not overloaded and the ambient temperature does not exceed 40°C. Class F transformers are rapidly replacing Class B transformers for distribution applications because they are smaller in size and have a higher insulation rating.

Class H transformers operate at not more than a 150°C temperature rise on the windings if the transformer is not overloaded and the ambient temperature does not exceed 40°C. Class H transformer insulation is used primarily with larger distribution transformers.

Dry-type transformers are available in either single- or poly-phase configurations. Single-phase transformers consist of a single core with a primary and secondary winding around the core. Poly-phase transformers have a separate core, each with its own primary and secondary winding for each of the connected phases. See Figure 8-3.

Single-phase transformers are often used to construct poly-phase transformers and in control circuit applications which are commonly found in motor control circuits. Control power transformers are used extensively as a source of supply for motor control circuits. These control transformers are frequently enclosed within combination starters and use the motor's full voltage to derive a new low voltage control circuit. Multitap control transformers offer the same advantages, but are designed with multiple transformer tap locations to permit their use with a variety of supply voltages. Secondary-protected control transformers incorporate overcurrent protection, most often by use of control fuses, into the secondary or output of the control transformer. This offers added protection against short-circuits or other overcurrents that may occur on the secondary of the transformer. High-efficiency control transformers are designed to be used when voltage fluctuation must be held to a minimum and proper voltage regulation is required. These transformers operate at very low temperatures with very little energy loss. See Figure 8-4. Poly-phase transformers are commonly used in electrical distribution systems because the poly-phase (most often 3ϕ) transformer is generally lighter and more efficient than a constructed transformer bank. See Appendix.

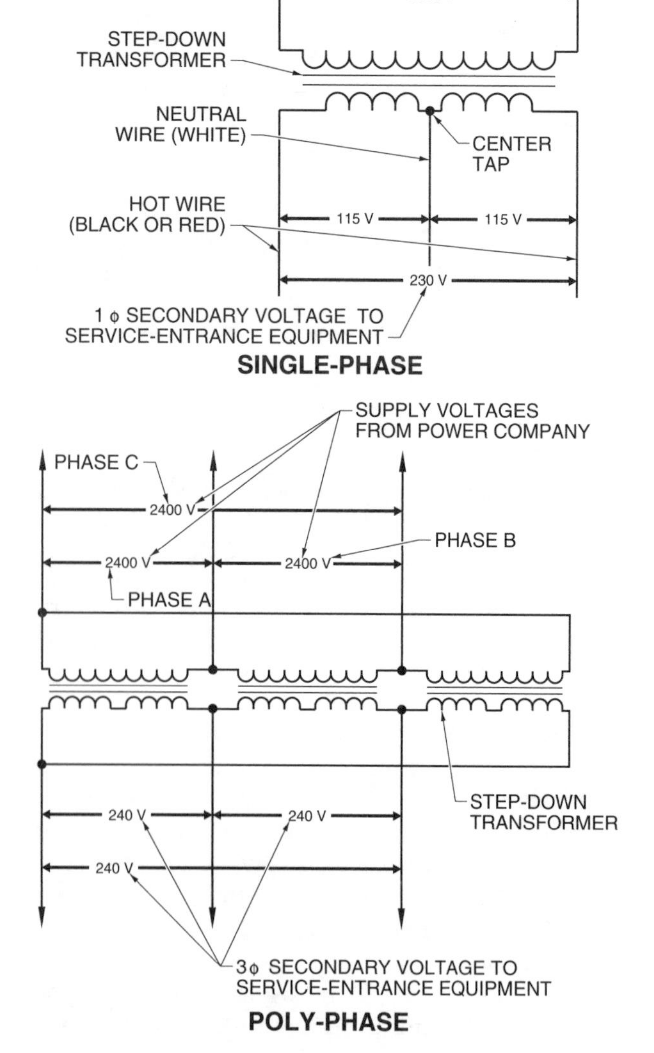

Figure 8-3. Dry-type transformers are available in either single- or poly-phase configurations.

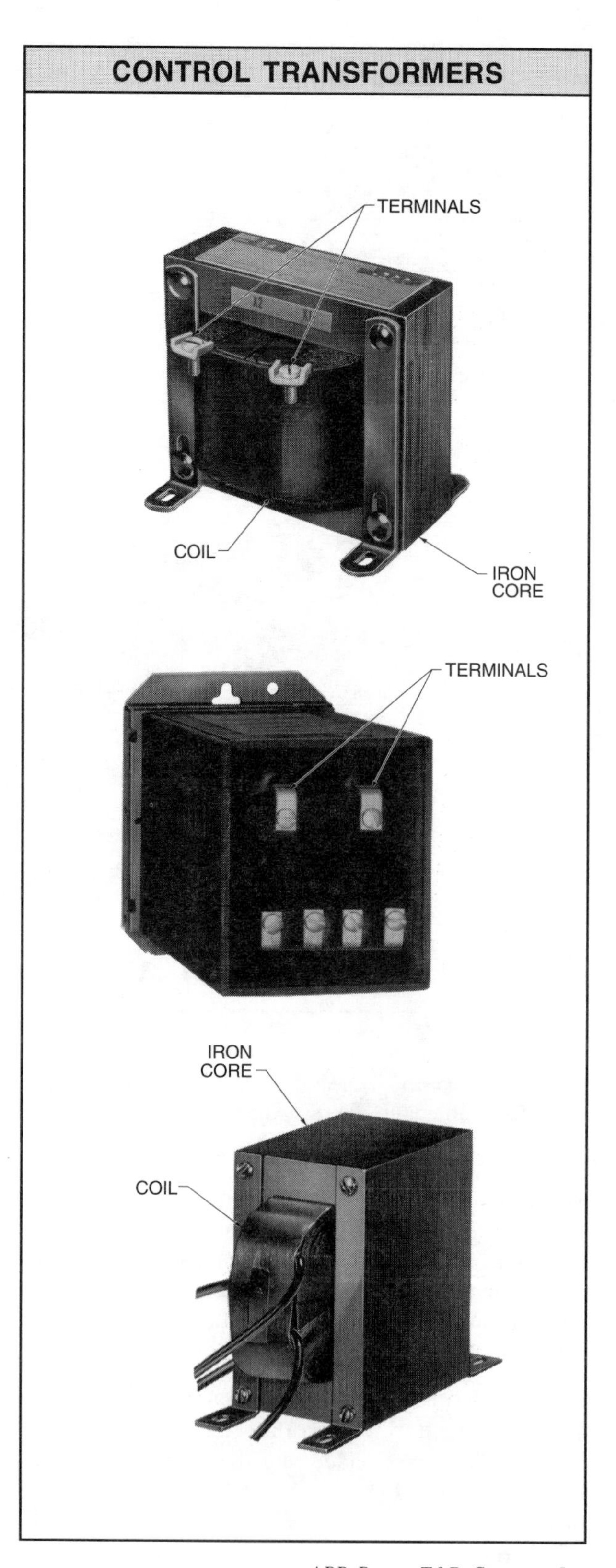

ABB Power T&D Company Inc.

Figure 8-4. Control transformers can be used to provide voltage to operate contactors and relays in various motor control applications.

Liquid-Filled Transformers

As the size of the electrical distribution system increases, the size of the transformers for the system also increases. As the transformers increase in size, natural convection and air-cooled transformers are not capable of avoiding excessive temperature rise in the transformer windings. Therefore, many larger distribution transformers are cooled by a liquid.

A *liquid-filled transformer* is a transformer that utilizes some form of insulating liquid to immerse the core and windings of the transformer. Insulating liquid around the core and windings aids in the removal of heat generated by the transformer windings. Liquid-filled transformers are commonly used in kVA ratings from 150 kVA to 3000 kVA and in primary voltages from 2400 V to 13,800 V. The four types of liquid-filled transformers are oil-insulated, askarel, less-flammable liquid-insulated, and nonflammable fluid-insulated. See Figure 8-5.

ABB Power T&D Company Inc.

Figure 8-5. A liquid-filled transformer is a transformer that utilizes some form of insulating liquid to immerse the core and windings of the transformer to aid in the removal of heat generated by the transformer windings.

The windings and core of liquid-cooled transformers are immersed in liquid. Some common liquids used are mineral oil, insulating oil, askarel, and less-flammable, and nonflammable liquids. *Mineral oil* is a chemically untreated insulating oil that is distilled from petroleum. *Askarel* is a group of nonflammable synthetic chlorinated hydrocarbons that were once used where nonflammable insulating oils were required. A *less-flammable liquid* is an insulating oil which is flammable but has reduced flammable characteristics and a higher fire point. A *nonflammable liquid* is a liquid that is noncombustible and does not burn when exposed to air. Nonflammable liquids do not have a flash or fire point.

Liquid is used in the transformer to insulate the transformer windings and core from each other and for cooling purposes. The liquid provides a medium which enhances heat transfer and dissipation. Liquid-cooled transformers can operate at higher voltage and current levels without damage to, and deterioration of, conductor insulations.

Oil-Insulated. When liquid-filled transformers were introduced, the most common insulation used was mineral oil. This oil is thin enough to circulate freely through the transformer. It does a good job of providing the necessary insulation between the transformer windings and the core. Mineral oil, however, is subject to oxidation and if any moisture enters the oil, the insulating value is dramatically reduced. *Oxidation* is the process by which oxygen mixes with other elements and forms a type of rust-like material.

Another problem associated with the use of mineral oil is that the oil is flammable. Therefore, oil-insulated transformers should not be located near combustible materials indoors or outdoors.

Askarel. Askarel has been banned by the Environmental Protection Agency and its use as a transformer coolant is being phased out. Askarel transformers, however, are still found throughout the electrical industry. Askarel is a nonflammable liquid that was used for many years for liquid-filled transformers which were designed to be installed indoors. Because the liquid is nonflammable, askarel transformers were permitted to be installed outside of transformer vaults. Unfortunately, one of the chief components of askarel is polychlorinated biphenyl (PCB). So, although the liquid has excellent characteristics for use in transformers, the environmental and health concerns about PCBs have led to the withdrawal of askarel for use in liquid-filled transformers. In fact, many askarel transformers are presently being retrofitted with newer less-flammable liquids. The NEC® still contains provisions for installing askarel transformers.

Less-Flammable Liquids. Because of the problems associated with the use of askarel, liquids for use in transformers have been designed which are flammable but do offer a higher flash point. A *flash point (fire point)* is the temperature at which liquids give off vapor sufficient to form an ignitable mixture with the air near the surface of the liquid. The NEC® recognizes transformers with these liquids as less-flammable liquid-insulated transformers. Less-flammable liquids are defined in 450-23 as those liquids which are listed and have a fire point of not less than 300°C.

LIQUID-FILLED TRANSFORMERS

Type	Liquids Used
Oil-Insulated	Chemically untreated insulating oil
Askarel	Nonflammable insulting oil
Less-Flammable Liquid-Insulated	Reduced flammable insulating oil
Nonflammable Fluid-Insulated	Noncombustible liquid

Listed liquids are those liquids which have been suitably evaluated and meet appropriate designated standards. These liquids do not contain any PCBs but they can burn. Therefore, the NEC® has special requirements for less-flammable liquid-insulated transformers. The installation requirements for these transformers depend on whether the transformer is located inside or outside.

Nonflammable Liquids. Because of the added benefit of having nonflammable liquid-filled transformers, the search continues for liquids that have the nonflammability of askarel without the use of PCBs. The NEC®, in 450-24, recognizes nonflammable fluid-insulated transformers. These transformers are filled with a liquid that has no flash or fire point and does not burn in air.

Autotransformers

Autotransformers are single-winding transformers that share a common winding between the primary and secondary circuits. They also work on the principle of magnetic induction. These transformers are often used to "buck" or "boost" the supply voltage. See Figure 8-6. See Appendix.

The ratio of transformer windings is used to determine if the autotransformer is a step-up (boost) or step-down (buck) transformer. The voltage is increased by a boost transformer and decreased by a buck transformer. Autotransformers are used in industry for a wide variety of applications ranging from doorbell power supplies to deriving a 3ϕ, 4-wire, grounded distribution system from a 3ϕ, 3-wire, ungrounded distribution system.

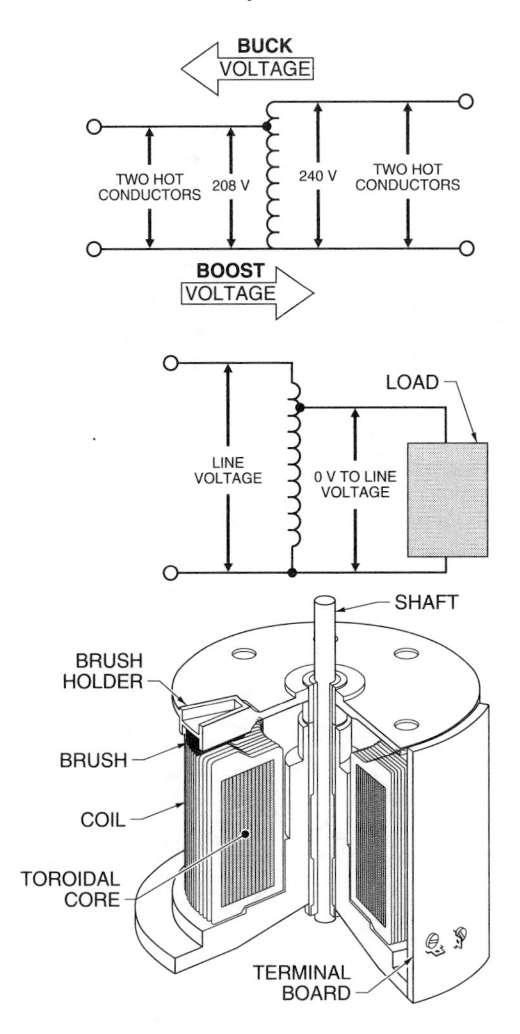

Figure 8-6. Autotransformers are single-winding transformers that share a common winding between the primary and secondary circuits.

Mounting and Use

Transformers are often classified according to the method by which they are mounted. Some common transformer mountings are pole, pad, platform, and vault. Outside electrical distribution systems commonly use pole-mounted transformers, which are quite suitable for overhead connections. Indoor electrical distribution systems use floor- or platform-mounted transformers. Wall-mounted or ceiling-hung transformers are permitted in some installations. Some large capacity, high-voltage transformers are designed to be installed in transformer vaults. The NEC® sets installation requirements for transformers in vaults and the construction of the vault.

Finally, transformers can be classified by their usage. There are many special application transformers in wide use throughout the electrical industry. These transformers include instrument, potential, current, and isolation transformers. See Figure 8-7.

TRANSFORMER INSTALLATION

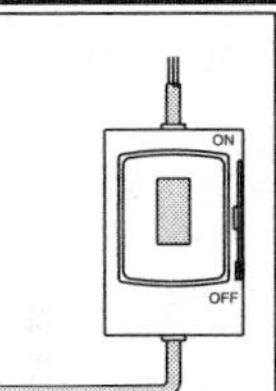

Due to severe space constraints, many electrical designers are requiring that transformers be hung from ceilings. The support system should be adequate for the weight of the transformer. Be particularly careful of modifications to transformers that are not designed to be hung. In addition, access to the space is required. In general, transformers should be installed in a readily accessible location. Although not readily accessible, 450-13, Ex. 2 permits installation of transformers above ceilings under some conditions.

Floor-mounted transformers have similar installation requirements that need to be considered when designing and installing electrical distribution systems. Perhaps the biggest pitfall occurs when installers place transformers below panelboards which are either supplied by or from the transformers. Check the work space clearance requirements of 110-16(a) to ensure that floor-mounted transformers do not impinge upon the working space needed for other electrical equipment.

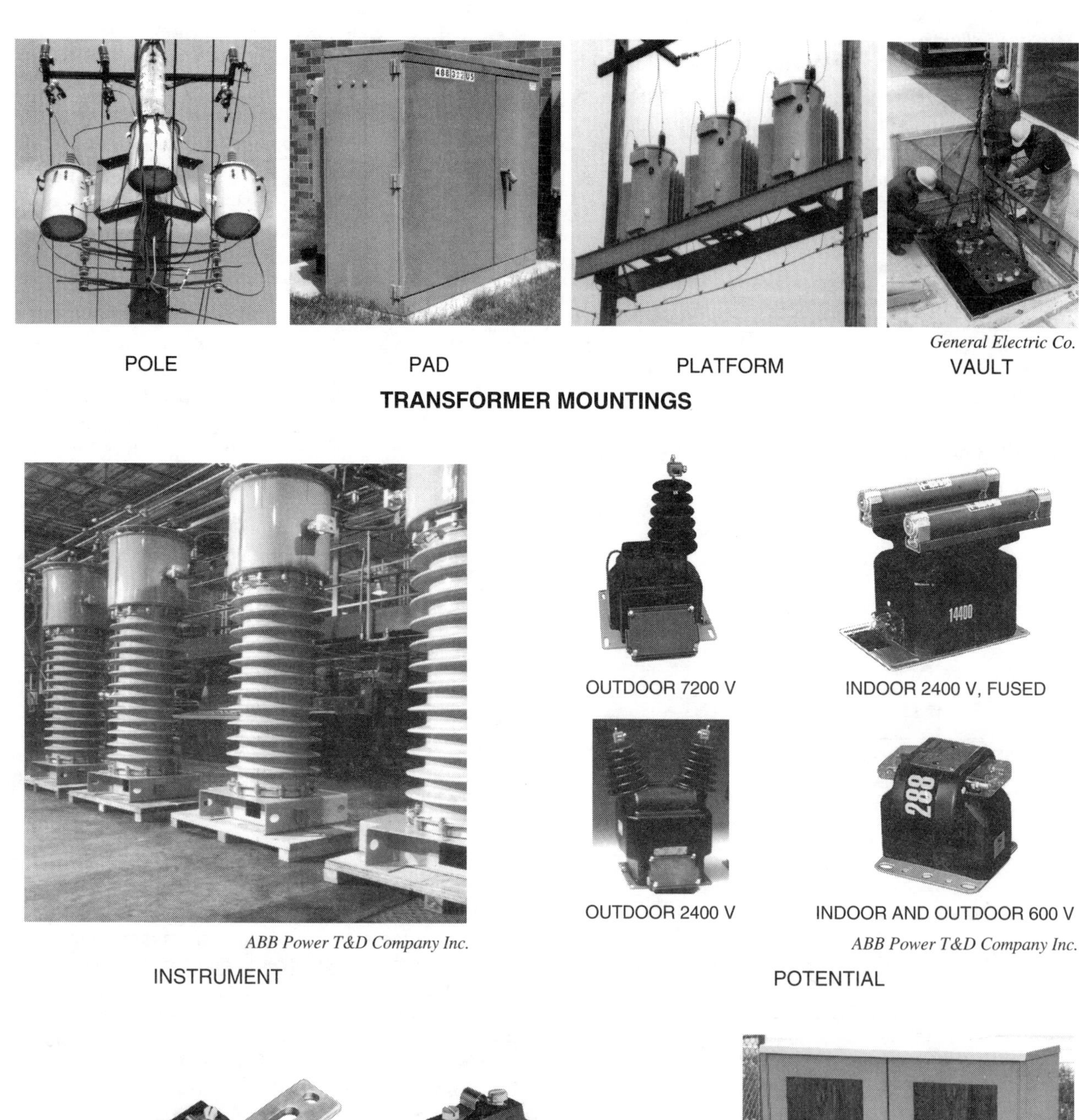

ABB Power T&D Company Inc.

CURRENT

ISOLATION

USAGE

Figure 8-7. Electrical distribution and control applications require transformers of various types and configurations.

An *instrument transformer* is a transformer used to reduce higher voltage and current ratings to safer and more suitable levels for the purposes of control and measurement. A *potential transformer* is a transformer which steps down higher voltages while allowing the voltage of the secondary to remain fairly constant from no-load to full-load conditions.

A *current transformer* is a transformer that creates a constant ratio of primary to secondary current instead of attempting to maintain a constant ratio of primary to secondary voltage. Current transformers are used extensively throughout the industry to measure large values of current that would normally be very difficult to measure.

An *isolation transformer* is a transformer that utilizes a shield between the primary and secondary windings and a transformer ratio of 1:1 to ensure that the load is separated from the power source. Isolation transformers are constructed with a shield between the primary and secondary windings. Isolation transformers can be used to provide "clean" power for information technology system equipment.

TRANSFORMER INSTALLATION

As with most electrical equipment, proper consideration must be given to several important factors before deciding on the best installation procedures for transformers. Since the transformer is in integral part of the electrical distribution system, planning for factors like physical protection, ventilation, grounding, and location helps to ensure the reliability of the electrical system. Transformers are a widely used electrical device and for the most part they are extremely reliable. Transformers can create a problem for the electrical distribution system when proper installation procedures are not followed or required maintenance practices are neglected.

Guarding – 450-8

Physical protection of transformers can be provided by covering, shielding, fencing, enclosing, or otherwise prohibiting the approach or contact by persons or objects. Transformers are required to be protected from physical damage by 450-8(a). Dry-type transformers shall be enclosed in a case or enclosure to protect against the insertion of foreign objects per 450-8(b).

In general, equipment such as switches, shall not be installed within a transformer enclosure unless the equipment operates at 600 V, nominal, or less and only qualified persons have access to such equipment. In addition, 450-8(c) requires that all energized parts within the enclosure be suitably guarded. Guarding can be accomplished by limiting access to the equipment, by use of partitions or barriers, or by elevation to restrict access to the energized parts. See 110-17 and 110-34. In the event that exposed live parts are present, 450-8(d) requires that the operating voltage of the live parts be clearly indicated by visible markings either on the equipment or on adjacent structures.

All energized parts within the transformer enclosure shall be suitably guarded.

Energized parts of transformers shall be guarded per 110-17 and 110-34. Guarding transformers accomplishes two objectives. First, it protects the transformer from objects that could cause physical damage and threaten the reliability of the electrical system. Next, it protects workers and personnel in the vicinity of the transformer from inadvertent contact that could result in physical harm. All transformers are required to be guarded and marked by signs or other markings on the equipment indicating the operating voltage of the transformer. See Figure 8-8.

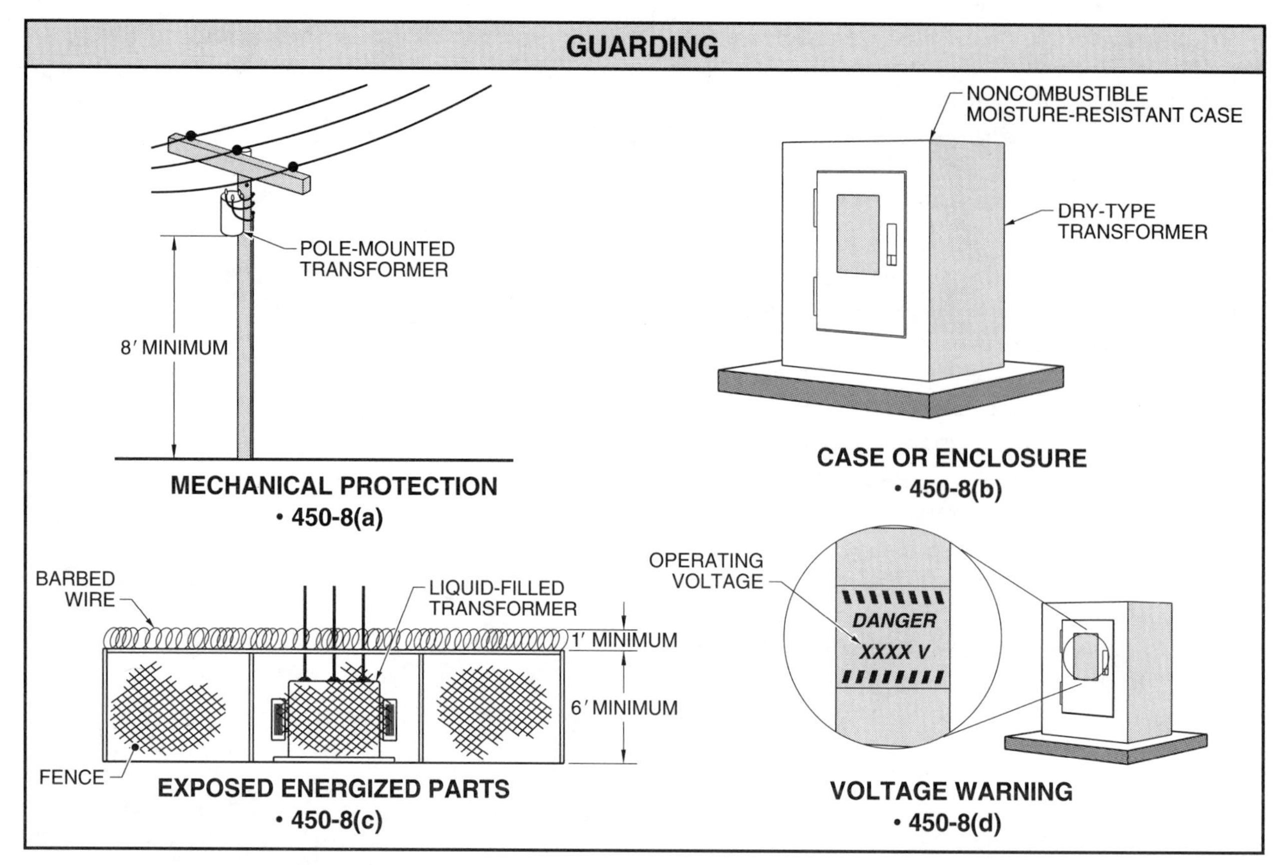

Figure 8-8. All transformers are required to be guarded.

Ventilation – 450-9

Whenever electrical conductors or equipment are installed, care must be taken to protect them from the effects of heat. Heat is a natural by-product of current flow. Excessive heat, or any heat above the temperature rating of the conductor or equipment, can lead to deterioration of the conductors' insulation. Excessive heat can damage the insulation of transformer windings leading to transformer failure. Adequate ventilation shall be provided for all transformer installations so that heat cannot lead to an excessive temperature rise per 450-9. Transformers, like motors, are rated with a maximum temperature rise. *Temperature rise* is the amount of heat that an electrical component produces above the ambient temperature. See Figure 8-9.

In the 1990 NEC®, a further requirement was added to the ventilation requirements for transformers. Transformers shall be installed in a manner which does not block or obstruct openings which are designed for cooling purposes. In addition, transformers are now required to be marked with a minimum distance or clearance from walls or other obstructions to facilitate the dissipation of heat.

SEPARATELY DERIVED SYSTEMS

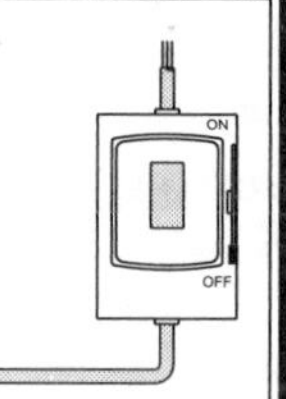

Designers and installers of transformers should consider the requirements of 250-26 when installing electrical distribution systems. Specific components shall be used when grounding separately derived systems. If possible, try to install the transformers in locations that make meeting the requirements of 250-26 easier. For example, 250-26(c) lists building steel as the preferred grounding electrode for separately derived systems. The steel should be as near as practical to, and preferably in the same area as, the transformer. This can reduce the length of the run for installing the GEC.

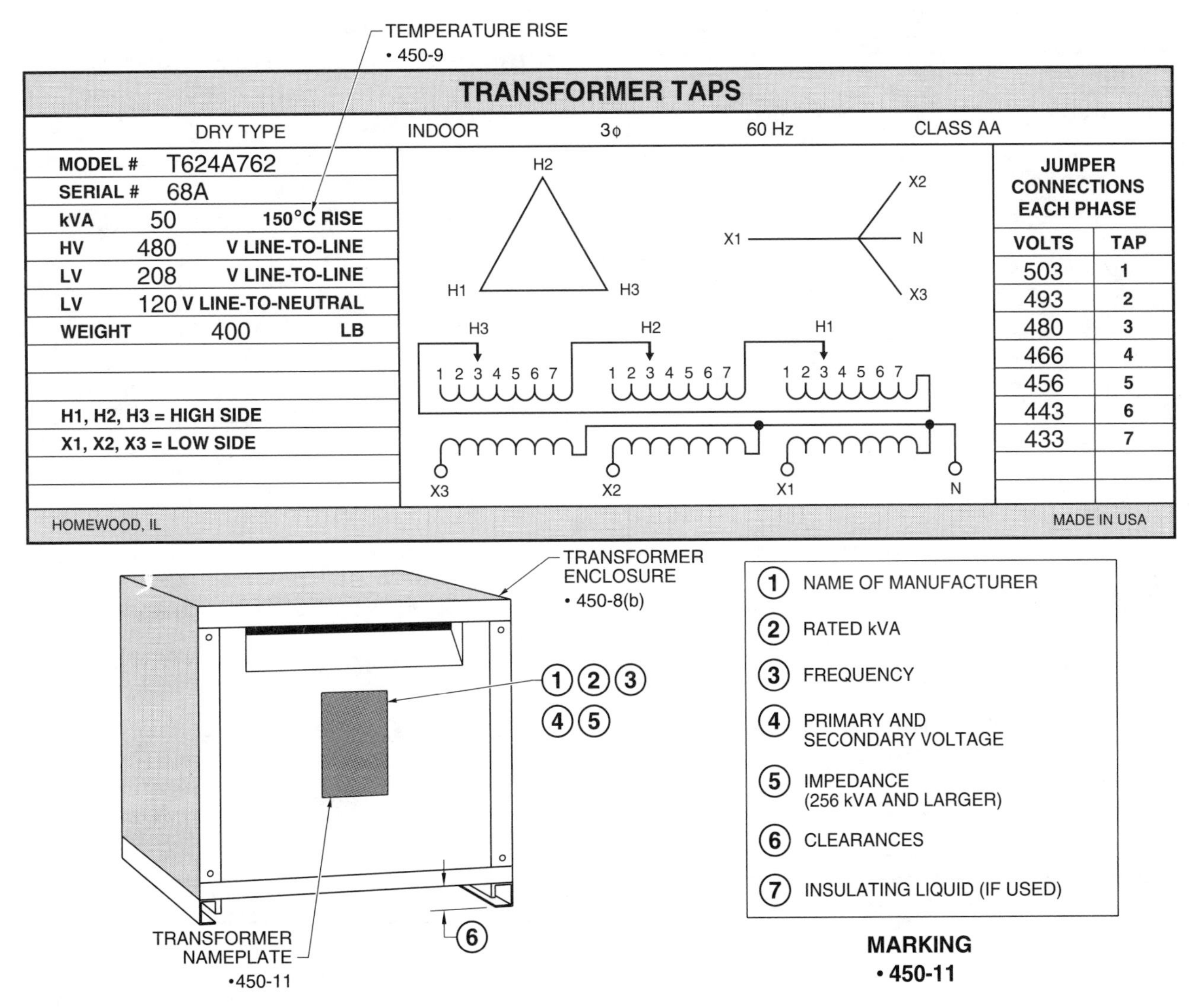

Figure 8-9. All transformers are required to be ventilated.

Grounding – 450-10

Transformers are required to be grounded per 250. All metal, noncurrent-carrying parts of the transformer and surrounding metal objects, like fences and other guarding means, shall also be grounded. When the transformer is used as a source of a separately derived system, it shall be grounded per 250-26.

Location – 450-13

Transformers can be installed ether indoors or outdoors. Because of the potential hazards associated with some types of transformers, special installation requirements apply when transformers are installed indoors. Some transformers, such as the LFLF type, also have special requirements when installed outdoors. In general, transformers and transformer vaults, shall be installed in locations which are readily accessible to personnel. Transformers require some maintenance and may even need to be replaced. For this reason, they cannot be installed where they are blocked or otherwise obstructed in a manner that quick access is denied.

All transformers shall be installed in a readily accessible location per 450-13. Dry-type transformers 600 V, nominal, or less, are not required to be installed in a readily accessible space per 450-13, Ex. 2. This section permits installation in spaces above hung ceilings, etc. provided the transformers are rated 50 kVA or less and adequate ventilation is provided. See Figure 8-10.

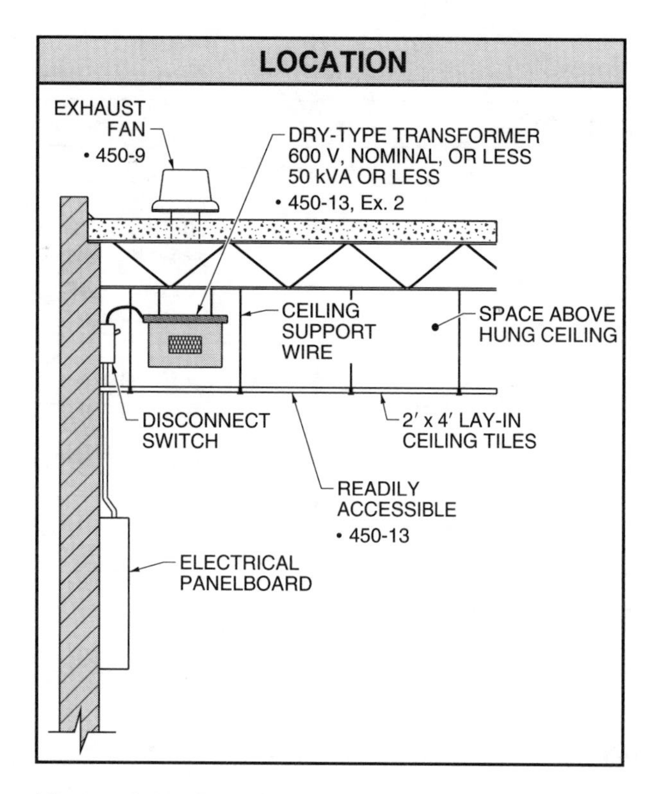

Figure 8-10. Transformers are permitted to be installed in spaces above hung ceilings provided adequate ventilation is provided.

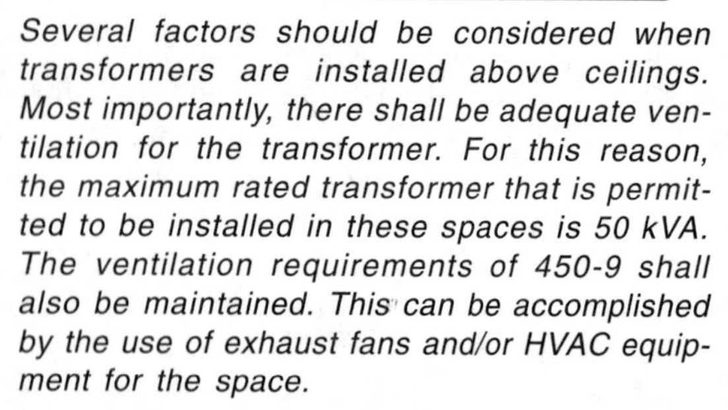

TRANSFORMERS ABOVE CEILINGS

Several factors should be considered when transformers are installed above ceilings. Most importantly, there shall be adequate ventilation for the transformer. For this reason, the maximum rated transformer that is permitted to be installed in these spaces is 50 kVA. The ventilation requirements of 450-9 shall also be maintained. This can be accomplished by the use of exhaust fans and/or HVAC equipment for the space.

Secondly, be sure to provide adequate access to service the transformers. Notice that 450-13, Ex. 2 permits these transformers in spaces that are not permanently closed in by the structure. This permits installation above ceilings with tiles designed to be removed, but prohibits the installation above drywall ceilings that are permanently closed in.

Dry-Type – 450-21; 450-22. Indoor requirements for dry-type transformers depend on the size of the transformer. Dry-type transformers installed indoors, and not over 112½ kVA, shall have a clearance from combustible materials of at least 12″ per 450-21(a). See Figure 8-11.

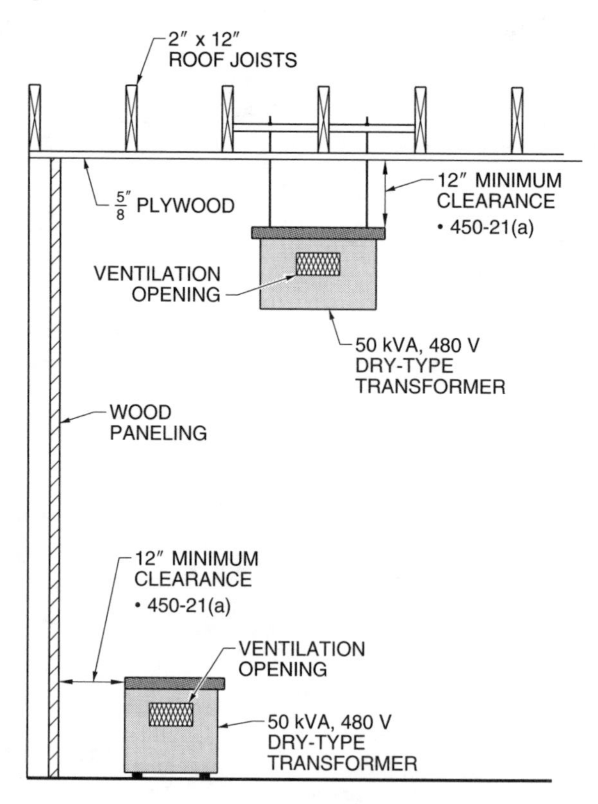

Figure 8-11. Dry-type transformers, less than 112½ kVA, installed indoors shall have a minimum clearance of 12″ clearance from combustible materials.

Two exceptions permit installation within 12″ of combustible materials. Per 450-21(a), Ex. 1, a separation of less than 12″ from combustible materials is permitted when a fire-resistant, heat-insulating barrier is installed between the transformer and the combustible material. For the purposes of this section, fire-resistant means the equivalent of a 1-hour minimum fire rating. Per 450-21(a), Ex. 2, a separation of less than 12″ is permitted when the transformer operates at 600 V, nominal, or less and is completely enclosed, with or without ventilation openings. See Figure 8-12.

In general, transformers rated over 112½ kVA are required to be installed in a fire-resistant transformer room per 450-21(b). Any construction methods or materials that provide at least a 1-hour minimum fire rating are suitable.

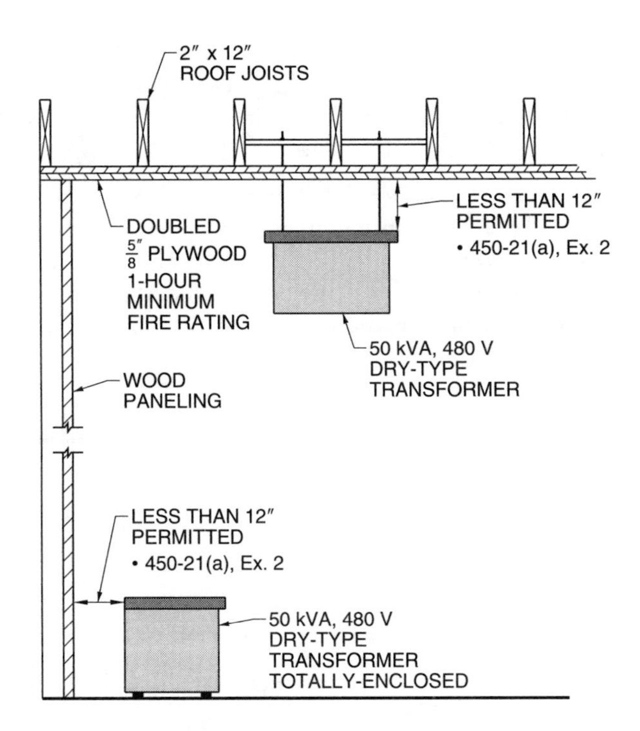

Figure 8-12. Clearance from combustible materials is not required when a fire-resistant barrier is provided or the transformer is the totally-enclosed type.

Two exceptions permit transformers rated over 112½ kVA to be installed outside of fire-resistant transformer rooms. Per 450-21(b), Ex. 1, transformers with 80°C or higher temperature rise ratings are permitted to be installed outside of fire-resistant rooms if the transformers are separated from combustible materials by a fire-resistant, heat-insulating barrier. Per 450-21(b), Ex. 2, transformers with the 80°C ratings or higher, are also permitted to be installed outside of a fire-resistant room if the transformers are separated from combustible materials by a fire-resistant, heat-insulating barrier, or by a minimum clearance of 6′ horizontally and 12′ vertically. See Figure 8-13.

Per 450-21(c), dry-type transformers rated over 35,000 V shall be installed in transformer vaults to reduce the hazard to building personnel and occupants. The transformer vault shall comply with 450, Part C.

Dry-type transformers installed outdoors shall meet the installation requirements of 450-22. In general, the transformer enclosure for transformers installed outdoors shall be weatherproof. The enclosure shall be constructed so that exposure to weather does not interfere with the transformer operation. Additionally, if the transformer rating exceeds 112½ kVA, the transformer is not permitted to be installed within 12″ of any combustible materials used in the construction of the building. However, per 450-22, Ex., if the transformer is provided with a temperature rise of 80°C or higher and is completely enclosed, other than the ventilation openings, the 12″ separation is not required. See Figure 8-14.

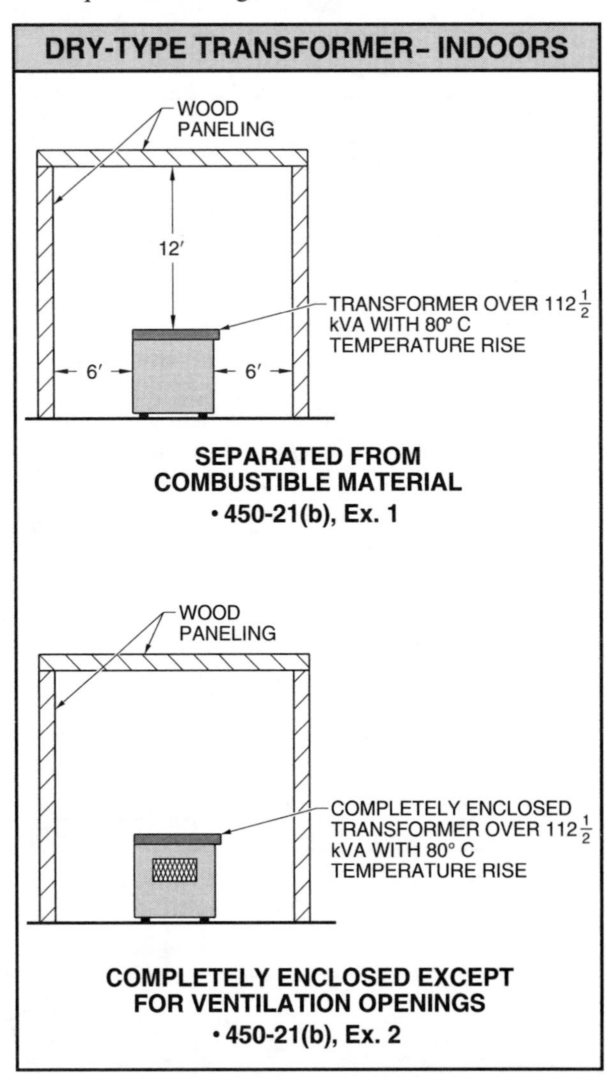

Figure 8-13. Transformers larger than 112½ kVA, with 80°C rise or higher, are permitted to be installed indoors if separated from combustible materials or completely enclosed.

Less-Flammable Liquid-Insulated (LFLI) Transformers – 450-23. LFLI transformers are filled with a liquid that has reduced flammability characteristics but is not completely nonflammable. LFLI transformers installed indoors shall meet one of three general requirements. See Figure 8-15.

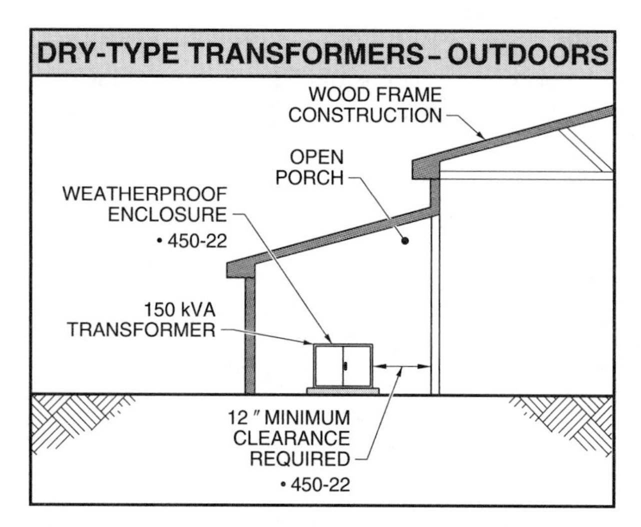

Figure 8-14. Transformers larger than 112½ kVA, installed outdoors shall be located at least 12″ from combustible materials unless the transformer has a 80°C rise or higher and is totally enclosed.

Per 450-23(a)(1), LFLI transformers are permitted to be installed indoors in Type I or Type II buildings when the transformer is rated at 35,000 V or less, the liquid is listed and installed per the listing, a liquid confinement area is provided for leaks or spills, and no combustible materials are stored in the area. Type I and II buildings generally are constructed of noncombustible materials or methods. See *ANSI/NFPA 220-1992, Types of Building Construction*.

A *Type I building* is a building in which all structural members (walls, columns, beams, girders, trusses, arches, floors, and roofs) are constructed of approved noncombustible or limited-combustible materials. Type I buildings have fire resistance ratings that range from 0 to 4 hours.

A *Type II building* is a building that does not qualify as Type I construction in which the structural members (walls, columns, beams, girders, trusses, arches, floors, roofs, etc.) are constructed of approved noncombustible or limited-combustible materials. Type II buildings have fire resistance ratings that range from 0 to 2 hours.

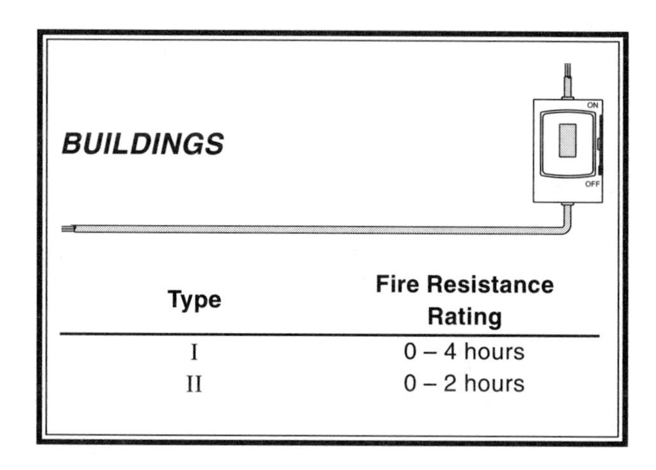

Type	Fire Resistance Rating
I	0 – 4 hours
II	0 – 2 hours

Per 450-23(a)(2), LFLI transformers are permitted to be installed indoors when the transformer is rated at 35 kVA or less, a liquid confinement area is provided, and the area in which the transformer is located is protected with an automatic fire extinguishing system.

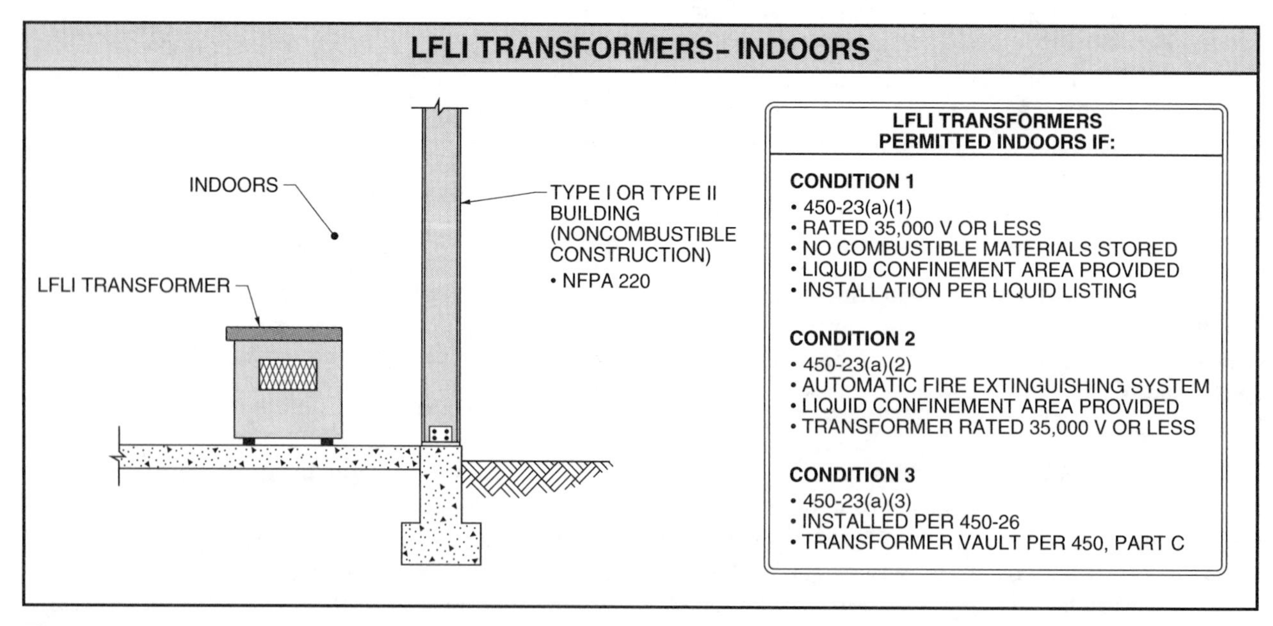

Figure 8-15. LFLI transformers installed indoors shall meet one of three general requirements.

Per 450-23(a)(3), LFLI transformers are permitted to be installed indoors in accordance with the provisions for oil-insulated transformers. See 450-26. In general, the LFLI transformer has to be installed in a transformer vault meeting the requirements of 450, Part C.

LFLI transformers are installed outdoors per 450-23(b). In general, these transformers can be installed outdoors where attached to, adjacent to, or even located on the building roof if one of two conditions is met. See Figure 8-16. Per 450-23(b)(1), LFLI transformers are permitted to be installed outdoors if the building is classified as a Type I or Type II building and the installation complies with all of the instructions included in the listing of the liquid.

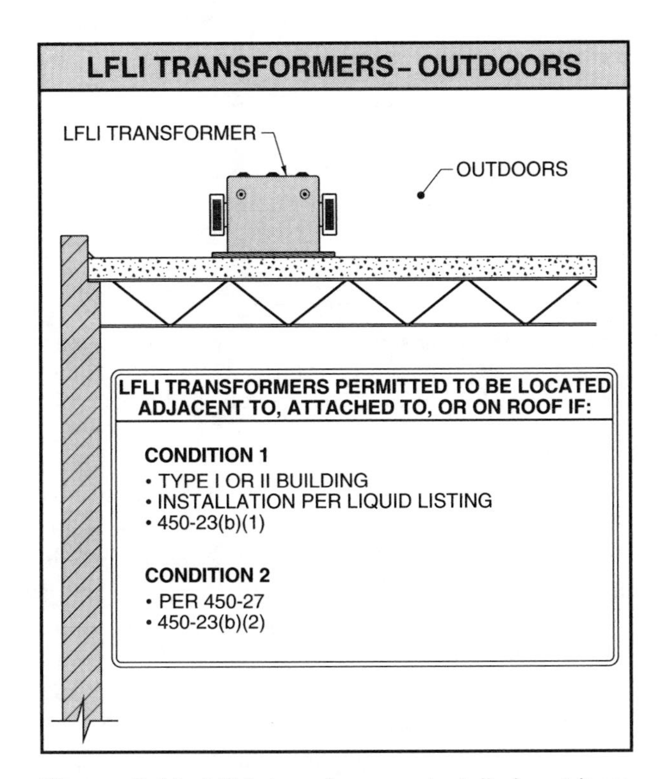

Figure 8-16. LFLI transformers installed outdoors shall meet one of two general requirements.

Additional safeguards may be required if the transformer is installed adjacent to window and door openings, fire escapes, and other combustible materials per 450-23(b)(1), FPN. While FPNs are not mandatory, in this case, additional consideration should be given because of the potential hazards involved.

Per 450-23(b)(2), the second condition under which LFLI transformers can be installed outdoors is when they are installed in accordance with the requirements for oil-insulated transformers installed outdoors. See 450-27. Basically, LFLI transformers can be installed outdoors with one of the following as appropriate for the hazard involved: space separations, fire-resistant barriers, automatic water spray systems, or enclosures.

Nonflammable Fluid-Insulated (NFFI) Transformers – 450-24. NFFI transformers contain a liquid that is nonflammable. Such a liquid does not have a flash or fire point, therefore, NFFI transformers can be installed in either indoor or outdoor locations with very few limitations.

Limitations only apply to NFFI transformers rated over 35,000 V. These transformers, when installed indoors, shall be installed in a vault, a liquid confinement area shall be provided, and the transformer shall be provided with a pressure-relief vent. Gases can be a by-product of these types of transformers, so the transformer shall provide means for absorbing any of these gases, or the pressure-relief vent shall be connected to a chimney or flue for removal purposes. See Figure 8-17.

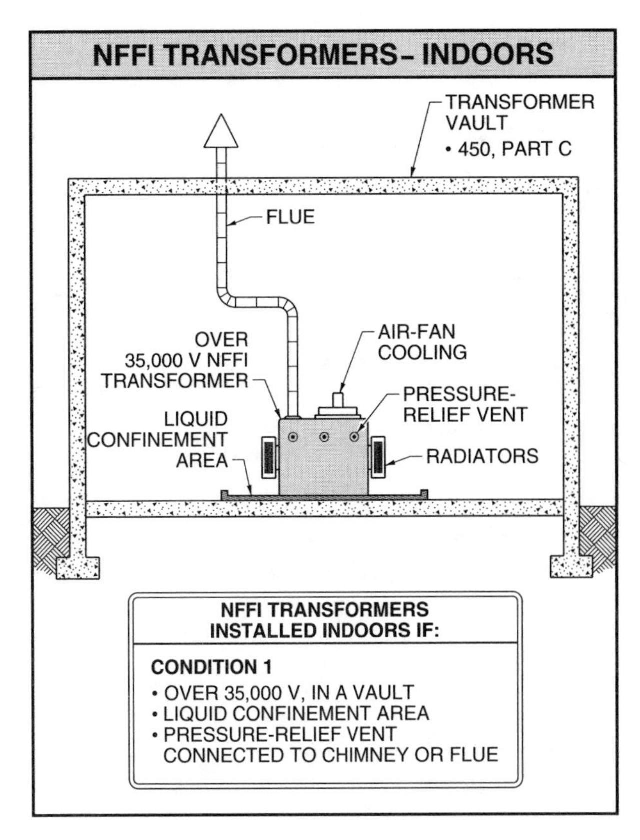

Figure 8-17. Nonflammable fluid-insulated transformers over 35,000 V and installed indoors shall be installed in a transformer vault.

Askarel-Insulated Transformers – 450-25. Askarel-insulated transformers can be installed indoors per 450-25. Askarel is a nonflammable liquid that contains PCBs which are harmful to both persons and the environment. Great care and proper disposal procedures should be followed when working with askarel-insulated transformers.

Askarel-insulated transformers that are rated over 25 kVA and installed indoors shall be equipped with a pressure-relief vent to relieve gases which may be generated from arcing inside the transformer. If the area in which the transformer is installed is poorly ventilated, then the pressure-relief vent shall be connected to a chimney or flue to remove the gases from the building. Askarel-insulated transformers, installed indoors and rated over 35,000 V, shall be installed in a transformer vault. See Figure 8-18.

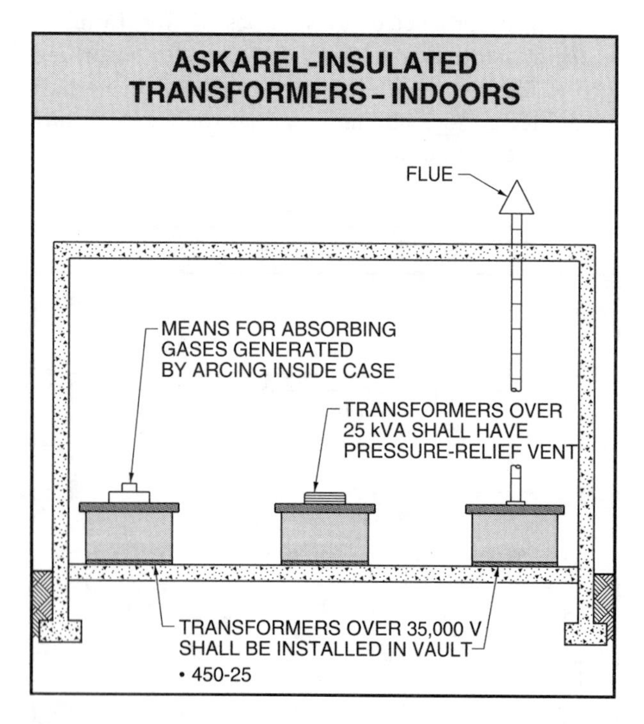

Figure 8-18. Askarel-insulated transformers installed indoors and rated over 35,000 V, shall be installed in a transformer vault.

Oil-Insulated Transformers – 450-26; 450-27. Oil-insulated transformers contain mineral oil. They shall be installed in a transformer vault when installed indoors. There are, however, several installation methods that do not require these transformers to be installed in a transformer vault constructed in accordance with 450, Part C.

Per 450-26, Ex. 1, oil-insulated transformers, rated up to 112½ kVA are permitted to be installed indoors in a vault constructed of reinforced concrete, not less than 4″ thick. Vaults constructed under 450, Part C generally require a larger wall thickness than 4″. See Figure 8-19.

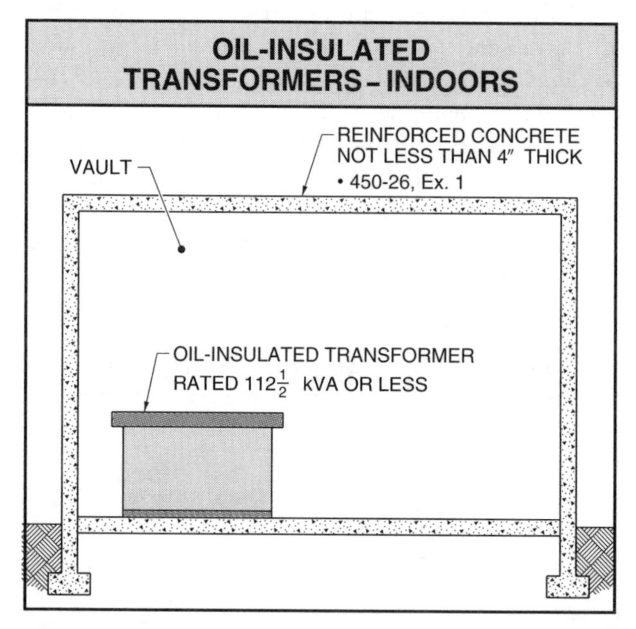

Figure 8-19. Oil-insulated transformers less than 112½ kVA can be installed indoors in a transformer vault constructed of reinforced concrete not less than 4″.

Per 450-26, Ex. 2, oil-insulated transformers rated 600 V, nominal, or less, are not required to be installed in a transformer vault if the installation is designed so that a fire in the transformer oil will not ignite other materials. The total rating of the transformer cannot exceed 10 kVA in areas of the building classified as combustible, or 75 kVA in areas of the building classified as fire-resistant.

Per 450-26, Ex. 3, electric-furnace transformers, with ratings of 75 kVA or less, are allowed to be installed outside of a transformer vault if measures are taken to prevent a transformer oil fire from spreading to other materials.

Per 450-26, Ex. 4, oil-insulated transformers are permitted to be installed in detached buildings not constructed under the requirements for transformer vaults. The detached building shall not present a threat of fire to any other building or structure. Such a detached building shall be used for the purposes of electrical service or supply and it shall be accessible to qualified persons only.

Per 450-26, Ex. 5, oil-insulated transformers are permitted to be installed indoors without a transformer vault if they are used for portable or mobile mining equipment. Leaking fluid shall be drained to the ground, personnel shall have safe egress, and a minimum 1/4″ steel barrier shall be provided for protection of personnel.

Oil-insulated transformers installed outdoors shall meet the installation requirements of 450-27. Care shall be taken to safeguard combustible buildings, parts of buildings, and other life safety apparatus, such as fire escapes, door openings, and window openings from the hazards associated with the transformer installation. At least one safeguard shall be provided to reduce the hazard when the transformer installation presents a risk of fire. See Figure 8-20.

TRANSFORMER VAULTS

The requirements for constructing a transformer vault are covered by 450, Part C. Transformer vaults serve two purposes. First, they provide a means to isolate potentially hazardous electrical components from unqualified personnel. Second, transformer vaults are, by design, intended to contain, or at least slow down, any fire or combustion that may occur as the result of a transformer malfunction. In order to accomplish these objectives, three factors should be considered in designing transformer vaults. The first concern is with the actual location of the transformer vault. The second concern involves the actual construction of the vault. The third concern is somewhat related, and ensures that adequate ventilation is provided for the transformer vault. See Figure 8-21.

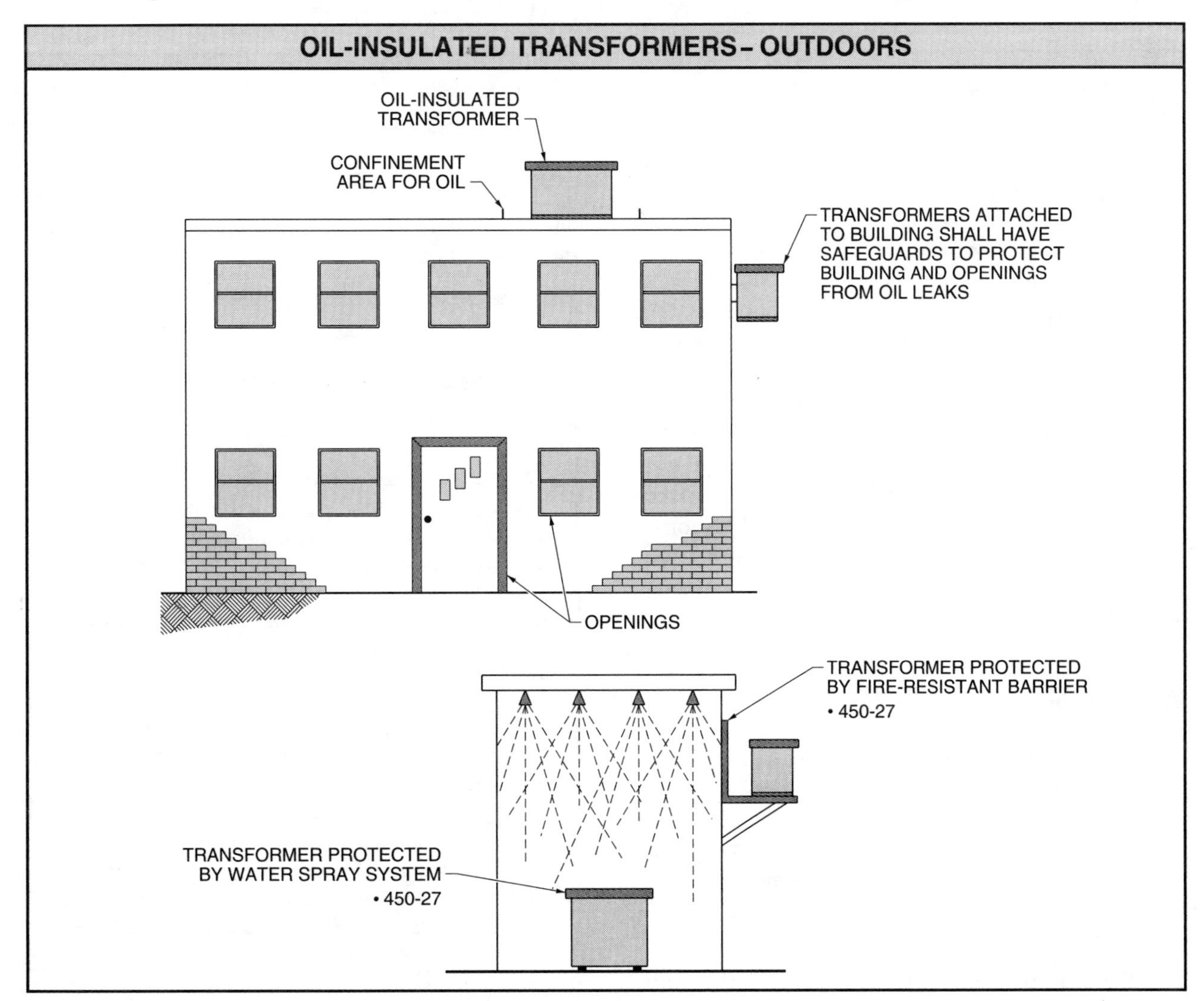

Figure 8-20. To reduce the risk of fire, additional safeguards are required when installing oil-insulated transformers outdoors.

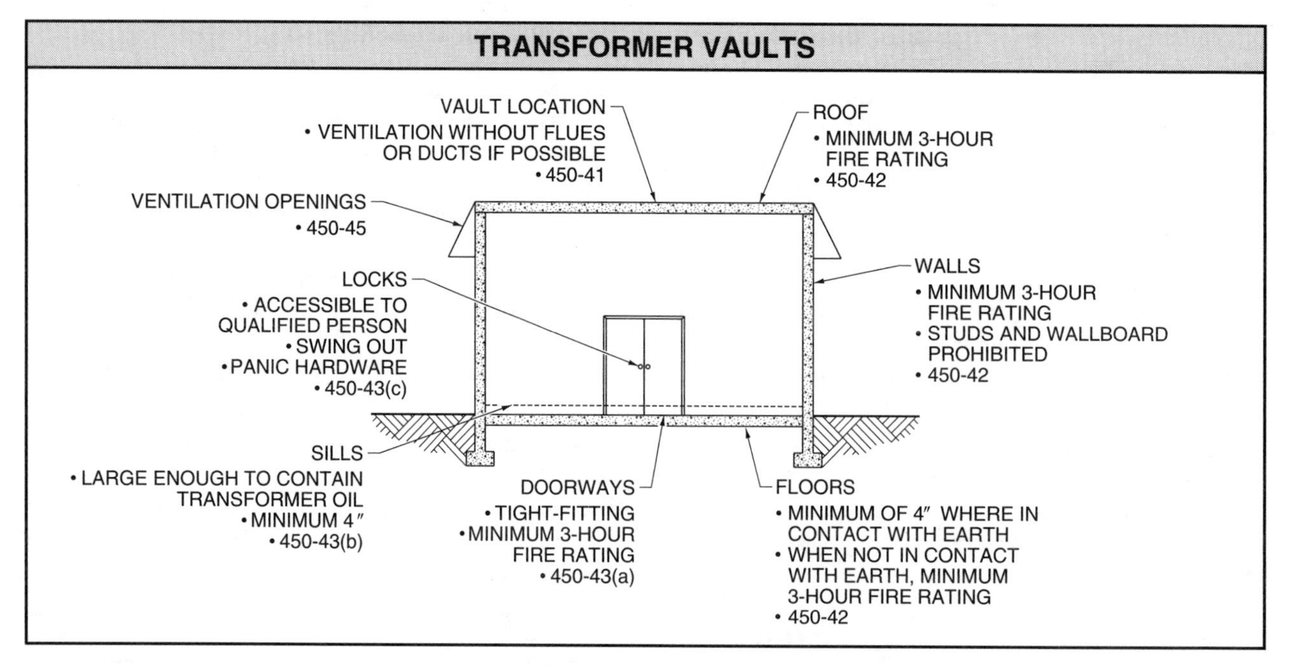

Figure 8-21. Transformer vaults help to ensure that hazards from explosion or fire are not transmitted to adjacent buildings or structures.

Location – 450-41

Where possible, transformer vaults should be located where they can be easily vented in accordance with their ventilation requirements. The use of flues and ducts should be avoided wherever practical. Another important consideration, which is not required by the NEC®, is the placement of the vault within the building. Due to the increase of acts of terrorism and the severity of recent storms, tornadoes, hurricanes, etc., careful consideration should be given to the placement of transformer vaults. Although no location is immune from these occurrences, every possible precaution should be taken to minimize the potential damage.

Walls, Roof, and Floor – 450-42

Because a primary objective is to contain any products of combustion within the transformer vault, construction materials which ensure a minimum fire resistance of three hours for walls, roof, and floor shall by used per 450-42. Standard framing members and drywall are not permitted to be used to achieve the 3-hour rating.

If the vault floor is in direct contact with the earth, the floor shall be constructed of at least 4″ of concrete. If the vault floor is not in direct contact with the earth, it shall be constructed for the load imposed on it and it must have a minimum fire resistance rating of 3 hours.

TRANSFORMER VAULT MATERIALS

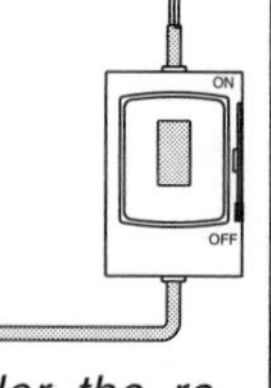

Transformer vaults constructed under the requirements of 450, Part C shall be constructed with materials that ensure a minimum fire resistance of 3 hours. Designers and installers should be careful when selecting the appropriate building materials for the construction of transformer vaults. Some have mistakenly assumed that standard building construction materials, such as wood framing members and drywall could be used, provided the 3-hour fire rating was achieved.

The 1990 NEC® was revised to clarify that studs and drywall are not acceptable methods for meeting the 3-hour fire rating. NFPA and ANSI standards that can be used in selecting building construction materials to achieve the 3-hour rating are listed in 450-42, FPN 1. A typical 3-hour construction consists of 6″ of reinforced concrete per 450-42, FPN 2.

Doorways – 450-43

Doors for transformer vaults shall have a minimum three-hour fire rating unless the vault is protected with an automatic sprinkler system, water spray, carbon dioxide, or halon system. If such a system is provided, the doors are only required to have a 1-hour fire rating. A door sill or curb, of at least 4″, but large enough to contain all of the oil from the largest transformer, shall be included in the vault design.

Another design consideration for transformer vaults is the door construction. For safety considerations, doors shall prohibit the entrance to unqualified personnel. Doors in transformer vaults are required to swing out and shall include panic hardware. *Panic hardware* is door hardware designed to open easily in an emergency situation. For example, a panic bar may simply be pushed to open a door quickly.

TRANSFORMER VAULT DOORS

The 1990 NEC® added 450-43(c), which requires that the doors in transformer vaults swing out and be fitted with panic hardware. However, CMP-1 has continually rejected a similar proposal for panic hardware and doors that swing out for all electrical rooms. CMP-1 has maintained that it is difficult to define "electrical room." While they raise a valid point, look for further proposals to seek similar protection for workers in all electrical rooms.

Ventilation – 450-45

Ventilation facilitates the dissipation of heat from the transformer vault. Ventilation openings shall be located as far as possible from door openings, window openings, fire escapes, and other combustible material per 450-45(a). This helps ensure that if a fire occurs in the transformer vault, it is not spread to other areas by way of the ventilation openings.

Per 450-45(b), if the means of ventilation is natural circulation, ventilation openings should be designed so that one-half of the required opening area is provided at or near the floor level, and the other half is provided in or near the roof. As an alternative, all of the required ventilation openings can be provided in one or more openings in the roof of the transformer vault.

The minimum size ventilation opening is 1 sq ft for all transformers less than 50 kVA per 450-45(c). Ventilation openings for all other transformers are calculated at 3 sq in. per kVA of transformer capacity in service less the area occupied by grates, screens, etc. For example, the ventilation openings of a transformer vault housing a 75 kVA transformer shall be no less than 225 sq in. ($75 \times 3 = 225$) after deducting the area for grates, screens, etc.

Per 450-45(d), ventilation openings shall be covered or otherwise protected with a durable covering, such as grates, screens, or louvers to prevent unsafe conditions. All ventilation openings to indoors shall be provided with automatic smoke or fire dampers that will respond to a fire in the vault per 450-45(e). The dampers shall have a minimum fire rating of 1½ hours. All ventilating ducts are required to be constructed of fire-resistant material per 450-45(f).

TRANSFORMER VENTILATION

Ventilation shall dispose of a transformer's full-load losses without creating a temperature rise in excess of the transformer's rating per 450-9. Ventilation openings shall be provided per 450-45. Location, arrangement, size, covering, dampers, and ducts shall be considered when designing ventilation openings for transformers.

TRANSFORMER OVERCURRENT PROTECTION

Transformer overcurrent protection is required to protect the primary windings from short-circuits and overloads and the secondary windings from overloads. Section 450-3 contains the requirements for overcurrent protection of transformers. Overcurrent protection requirements depend upon several factors. First is the voltage at which the transformer operates. Section 450-3(a) contains the rules for transformers rated over 600 V, nominal. Section 450-3(b) contains the rules for transformers operating at 600 V, nominal or less. The second factor is the location of the overcurrent device. Requirements vary depending on whether the primary winding only is protected, or both the primary and the secondary windings are protected. The last factor applies to transformers rated over 600 V only. If conditions of maintenance ensure that only qualified personnel will work on the transformers, 450-3(a)(2) permits primary protection only for these supervised installations.

The rules of these sections are intended to provide protection for the transformers only. Conductors on both the primary and secondary side will still need to be protected in accordance with the provisions of 240.

Over 600 V, Nominal – 450-3(a)

The requirements for transformer overcurrent protection for transformers rated over 600 V depend upon whether the installation is supervised. A *supervised installation* is an electrical installation in which the conditions of maintenance are such that only qualified persons monitor or service the electrical equipment.

Primary and Secondary – 450-3(a)(1). For nonsupervised installations, each transformer shall have primary and secondary OCPDs in accordance with Table 450-3(a)(1). The Table is organized around the permissible percentages for fuses and circuit breaker ratings on both the primary and secondary sides and the impedance of the transformer. For the purposes of this section, electronically actuated fuses are rated in accordance with the CB percentages. Per 450-3(a)(1), Ex. 1, the next higher standard rating of a fuse of CB is permitted when the calculated value does not correspond to a standard rating of a fuse or CB. See Figure 8-22.

STANDARD AMPERE RATINGS

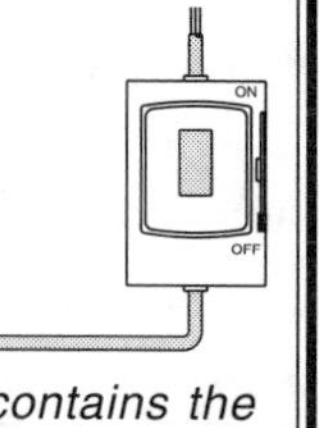

Article 240 states that 240, Part H contains the requirements for overcurrent protection over 600 V. Interestingly, there are no standard ampere ratings provided in Part H. The standard ampere ratings in 240-6 are therefore only applicable to under 600 V installations. Yet, 450-3(a)(2), Ex. 1, makes reference to standard ampere ratings. It appears that CMP-10 intends for the values listed in 240-6 to be used.

Supervised Locations – 450-3(a)(2). For supervised installations, there are two options for providing the required overcurrent protection. The first is based on primary protection only. In these installations, each transformer shall be protected with a primary overcurrent device that does not exceed 250% for fuses or 300% for CBs of the rated primary current of the transformer. If the calculated value does not correspond to a standard rating of a fuse or CB, the next higher standard size shall be permitted. See Figure 8-23. As with transformer installations of 600 V or less, individual overcurrent protection is not required at the transformer if the primary circuit OCPD provides the necessary protection.

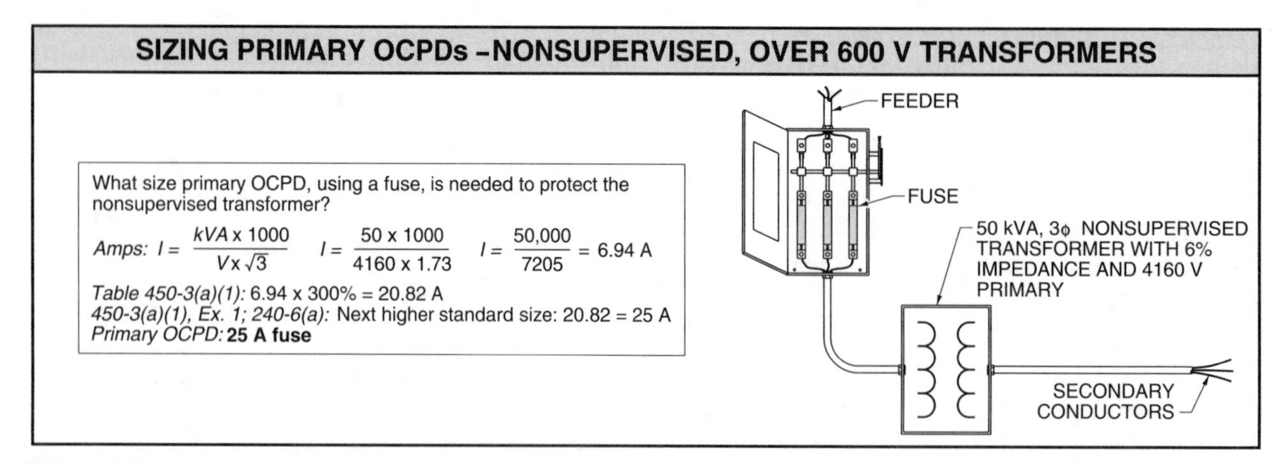

Figure 8-22. Nonsupervised transformers over 600 V shall be protected with primary and secondary OCPDs.

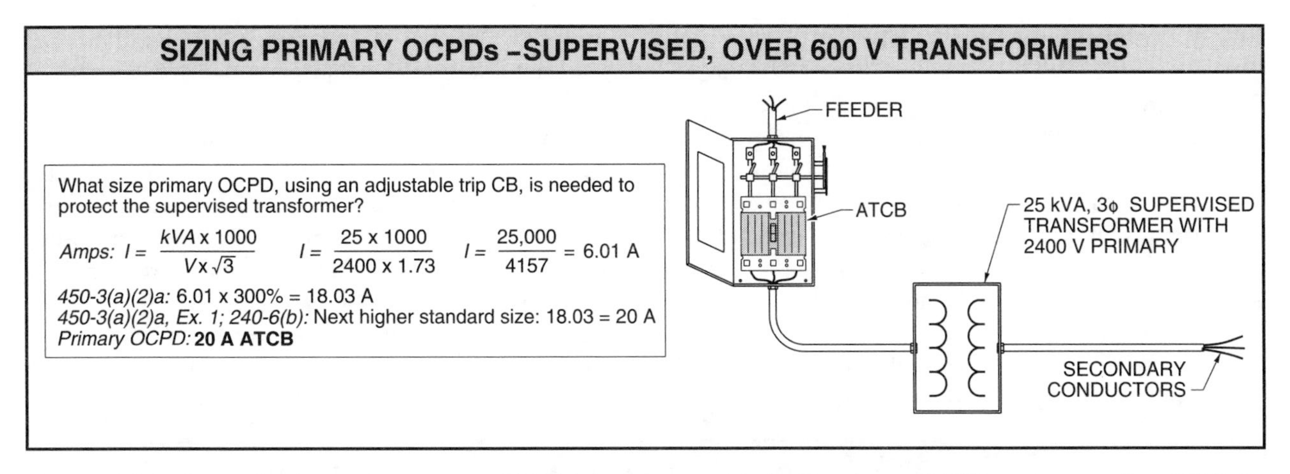

Figure 8-23. Supervised transformers over 600 V shall be provided with primary OCPDs not exceeding 250% for fuses and 300% for CBs.

The second option for transformers rated over 600 V is to provide both primary and secondary protection. Table 450-3(a)(2)b provides the permissible percentages for fuses and CB ratings on both the primary and secondary side based upon the impedance of the transformer. For transformers with an impedance of not more than 6%, the maximum rating or setting of the primary feeder OCPD can range from 250% to 600%, depending on the voltage of the primary and the secondary and the type of OCPD used, either fuse or CB. For transformers with an impedance of more than 6% and less than 10%, the maximum rating or setting of the primary feeder OCPD can range from 225% to 400%, depending on the voltage of the primary and the secondary and the type of OCPD used, either fuse or CB. For example, if a transformer with a rated impedance of 3% operates with a primary and secondary of over 600 V, and fuses are selected as the OCPD, primary overcurrent protection for the transformer is not required if the secondary OCPD is set at not more than 250% of the secondary current and the primary feeder OCPD is set at not more than 300% of the rated primary current of the transformer. See Figure 8-24.

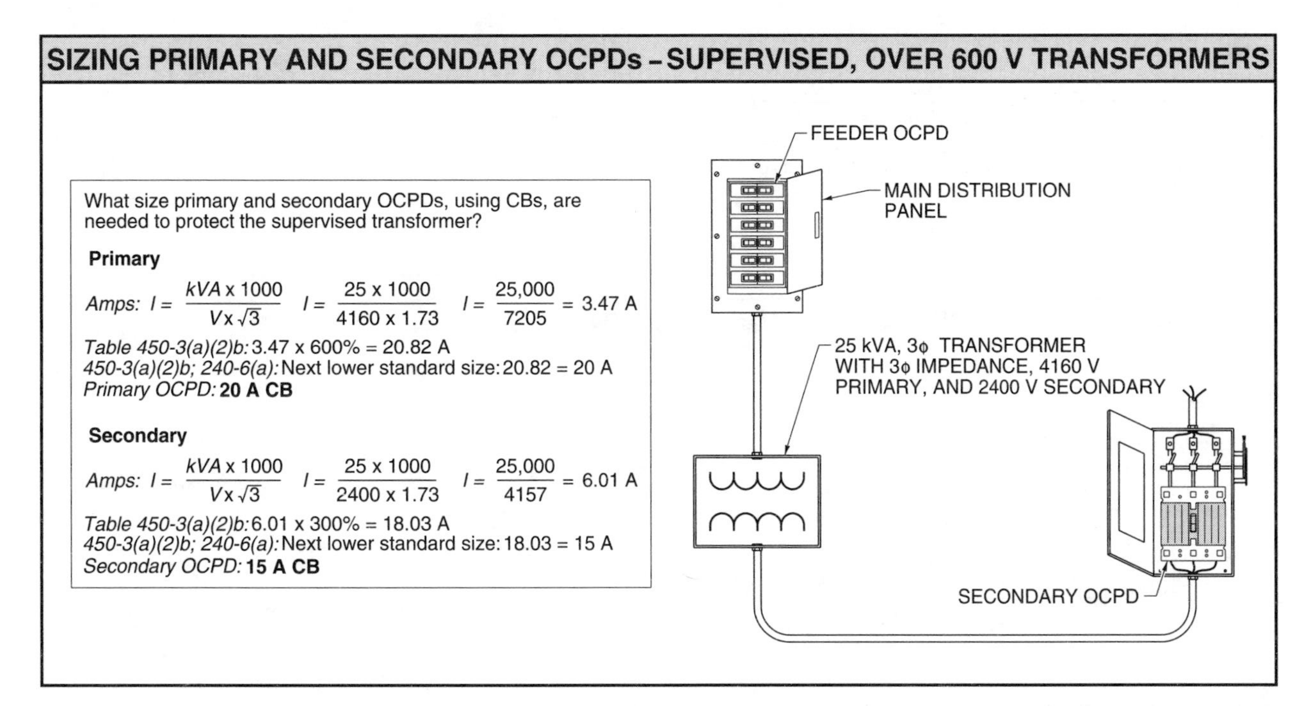

Figure 8-24. Supervised transformers over 600 V shall be provided with primary and secondary OCPDs per Table 450-3(a)(2)b.

600 V, Nominal, or Less – 450-3(b)

The requirements for overcurrent protection for transformers rated at 600 V, nominal, or less are given in 450-3(b). The protection is necessary to protect the windings of the transformer and is independent of the overcurrent protection required for conductors. The secondary overcurrent device is permitted, as it is with services, to consist of not more than six fuses or CBs grouped in one location. The total rating of the six OCPDs cannot exceed that of the required value for a single OCPD. In installations where the six OCPDs consist of both fuses and CBs, the rating shall not exceed the value permitted for fuses.

Primary – 450-3(b)(1). The general rule for transformers rated 600 V, nominal, or less is to protect the primary windings of the transformer at not more than 125% of the rated primary current of the transformer. Where this calculated value does not correspond to a standard rating for a fuse or a CB, per 240-6, and the rated primary current of the transformer is 9 A or more, 450-3(b)(1), Ex. 1 permits the next higher standard rating to be used. See Figure 8-25.

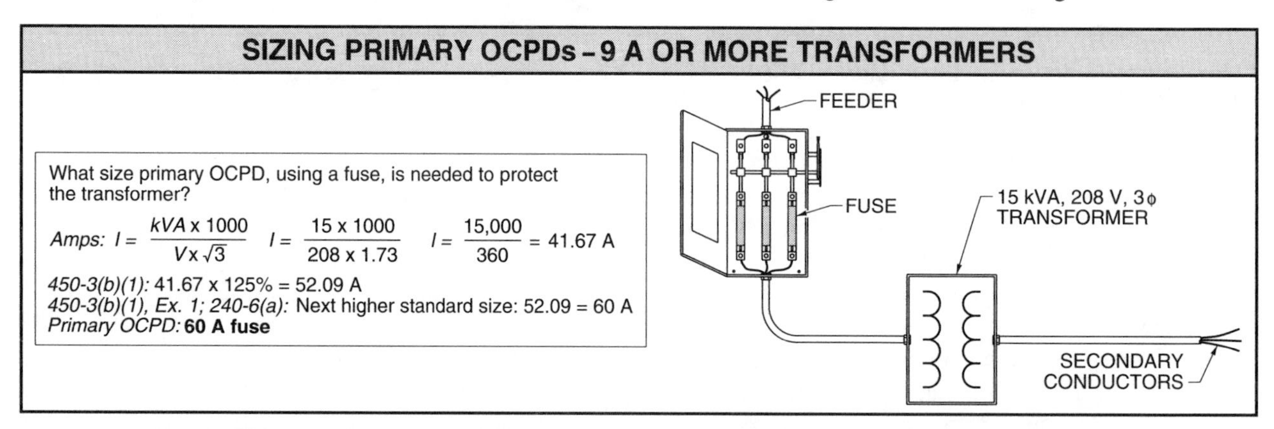

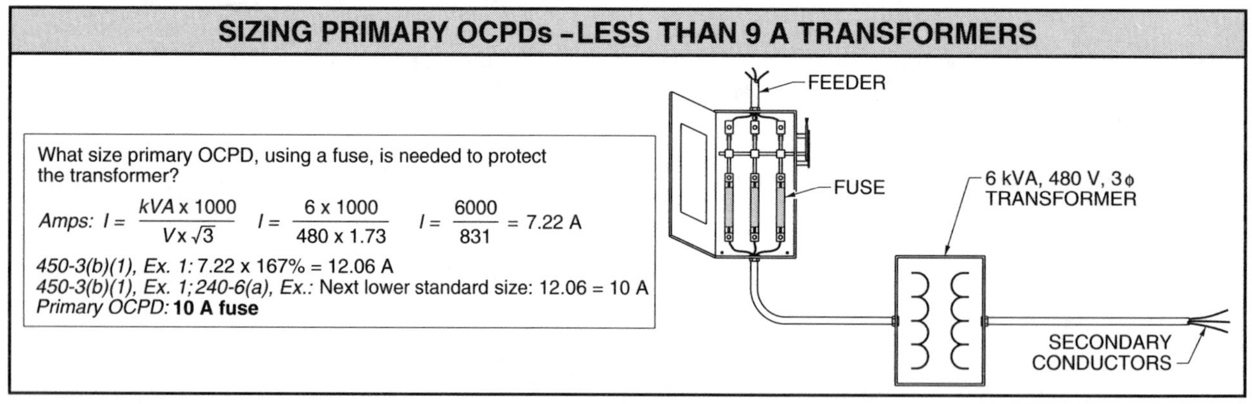

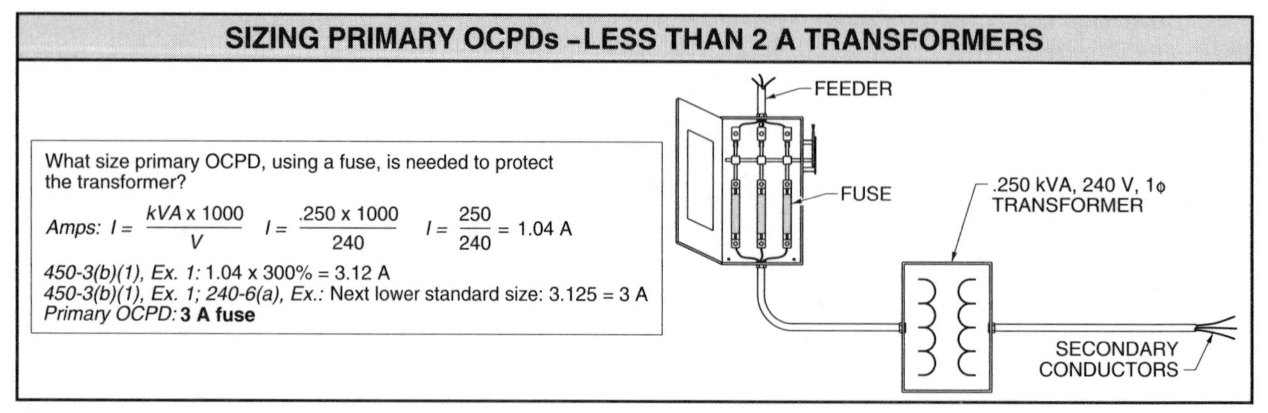

Figure 8-25. Transformers less than 600 V shall be provided with primary OCPDs at not more than 125% of the rated primary current, with exceptions.

If the rated primary current of the transformer is less than 9 A, the rating of the OCPD is permitted to be set at not more than 167% of the rated primary current. For transformer installations where the rated primary current is less than 2 A, the rating of the OCPD shall be permitted to be set at not more than 300% of the primary current. With the lower primary current values, the impedance of the transformer will act to limit potential fault current and the rating of the primary OCPD can be increased.

Per 450-3(b)(1), Ex. 2, transformers are permitted to be installed without individual primary overcurrent protection when the circuit overcurrent device provides the protection required by this section. In these cases, the protection is duplicated at the transformer and such protection is unnecessary. See Figure 8-26.

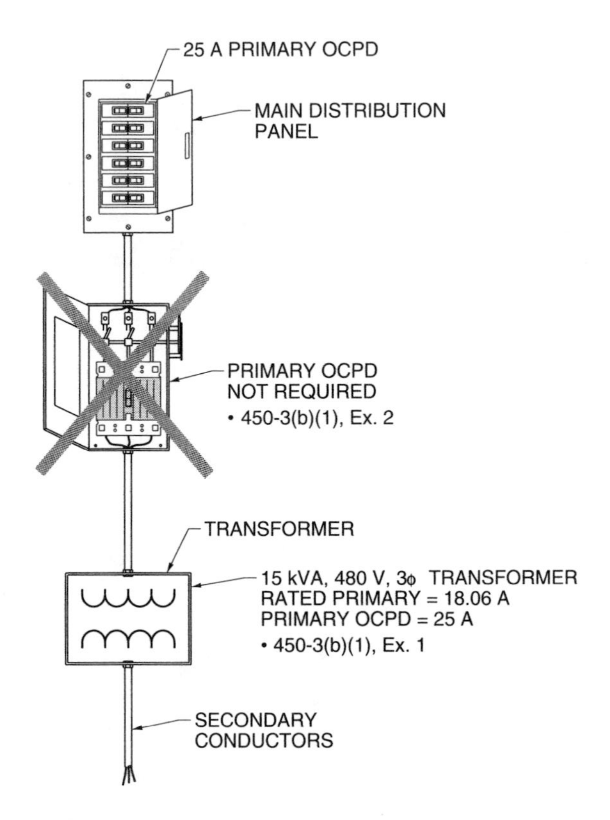

Figure 8-26. Transformers less than 600 V are permitted to be installed without individual primary OCPDs when the circuit OCPDs provide this protection.

Motor-control circuit transformers installed per the exceptions to 430-72(c) are permitted to be installed without primary overcurrent protection. These transformers are generally under 50 VA and have low or limited primary current rating. See Figure 8-27.

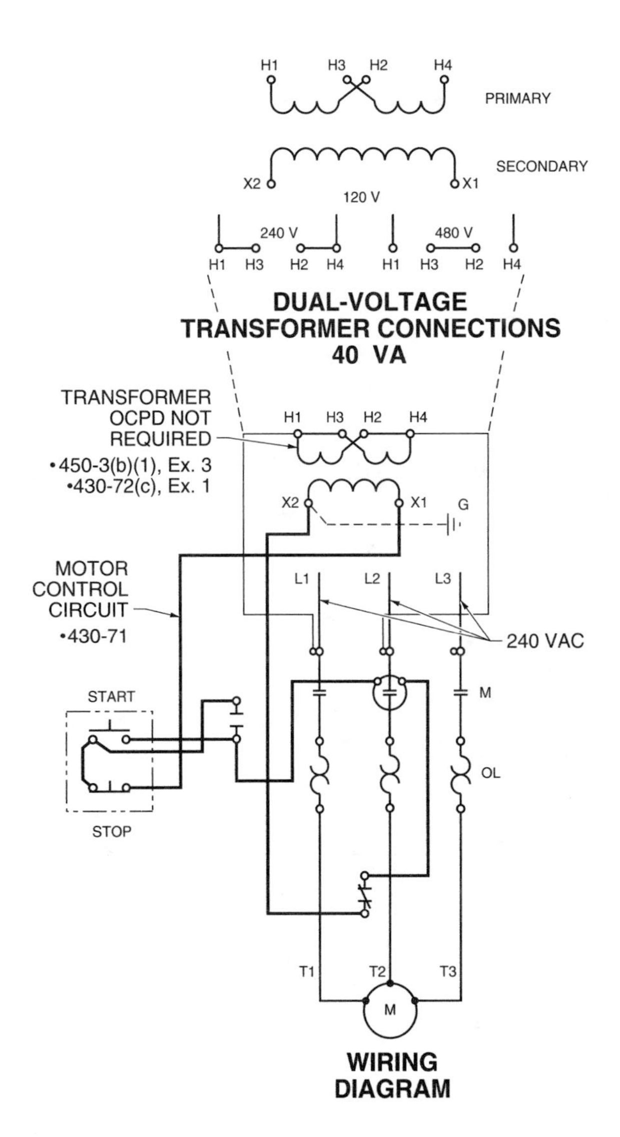

Figure 8-27. Certain motor-control circuit transformers are permitted to be installed without primary overcurrent protection.

Primary and Secondary – 450-3(b)(2). Another option in providing overcurrent protection for transformers rated 600 V, nominal, or less is to protect both the primary and the secondary of the transformer. In these cases, 450-3(b)(2) permits the primary feeder OCPD, set at not more than 250% of the rated primary current of the transformer, to protect the transformer primary, provided the secondary OCPD is set at a value which does not exceed 125% of the rated secondary current of the transformer. As with the primary protection only requirements, if the rated secondary current of the transformer is 9 A or more and the calculated value does not correspond to a standard rating of a fuse or CB from 240-6, the next higher standard rated OCPD is permitted. See Figure 8-28.

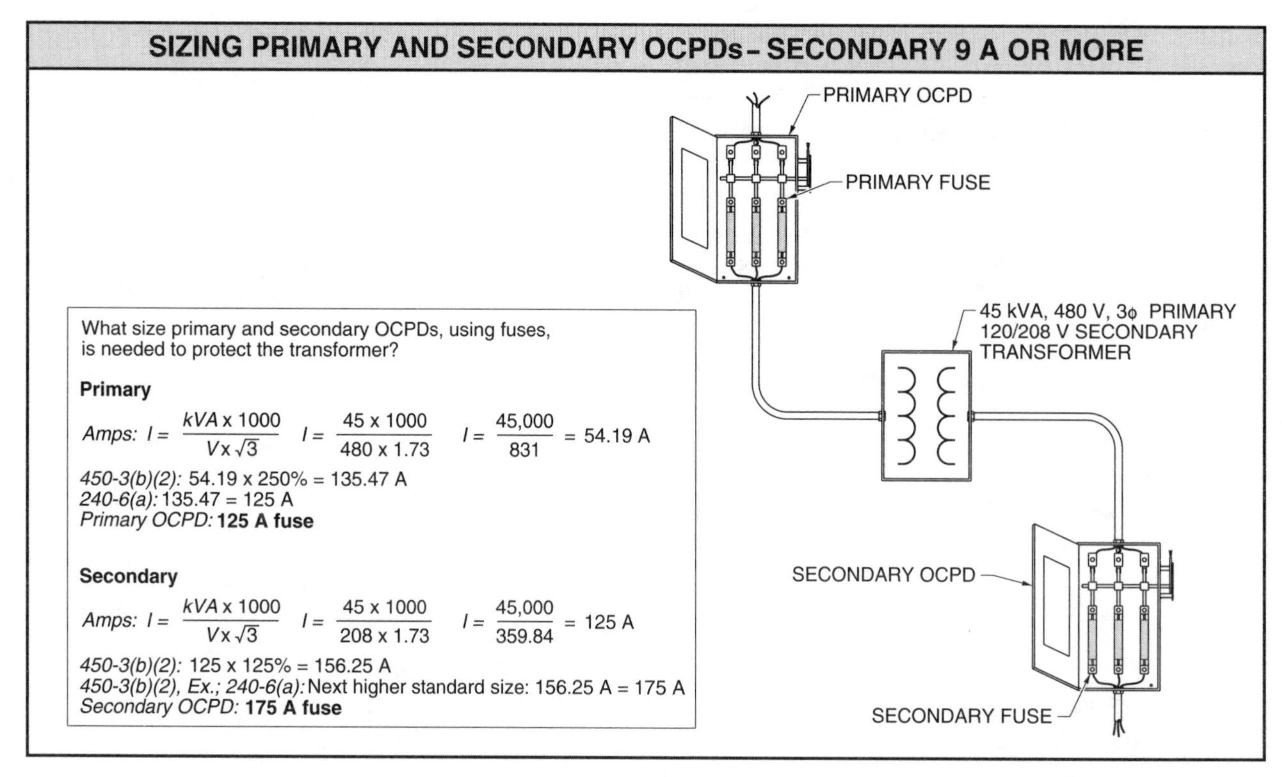

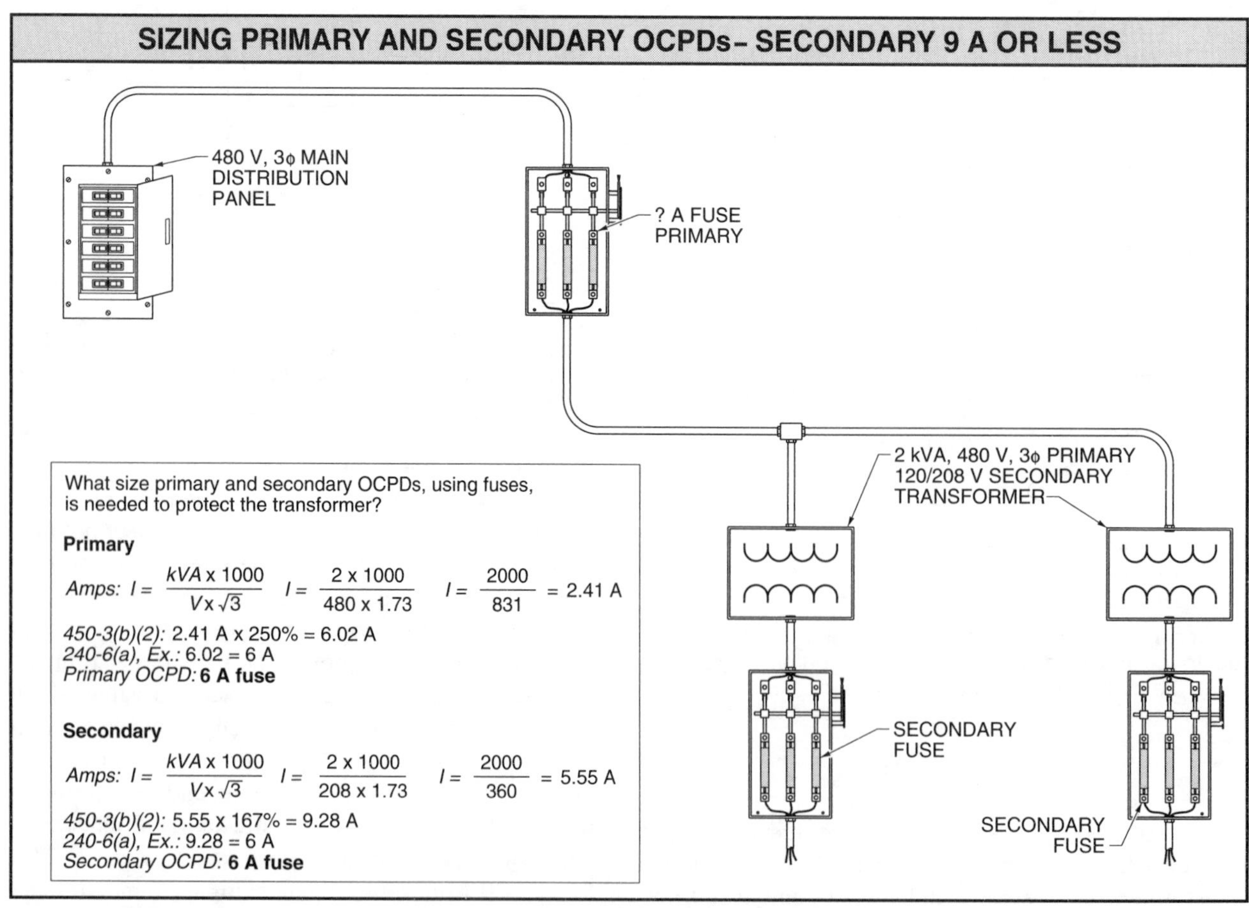

Figure 8-28. Transformers under 600 V may be protected by primary and secondary OCPDs.

Unlike the requirements for primary only protection, there is no permission to increase the rating of the primary OCPD when it does not correspond to a standard rating of a fuse or CB. If the calculated value at 250% does not correspond to a standard rating, the next lower standard rating shall be used. If the rated secondary is less than 9 A, the rating of the secondary OCPD is permitted to be increased to a maximum value of 167% of the rated secondary current of the transformer. This rule is useful for applications in which several transformers are supplied from a single primary OCPD. Sizing at 250% prevents nuisance tripping due to the transformer current in-rush characteristics.

Transformer and Feeder Taps – 240

Closely related to the application of transformer overcurrent protection are the provisions of 240 covering conductor overcurrent protection. Electrical installations are frequently designed in which taps from the secondary of the transformer are made to supply panelboards or other equipment. See 240-21.

The selection of the appropriate tap rule depends largely upon the length of the tap. For example, 240-21(b) contains the requirements for 10′ taps from the secondary of transformers. The length of the tap conductors shall not exceed 10′. In addition, all of the other conditions of the tap rule shall be met. This is a common application where the requirements of 450-3 overlap with the requirements of 240.

A 10′ transformer feeder tap is frequently used to supply panelboards on the secondary side of transformers. Although the tap conductors are not required to terminate in a single CB or single set of fuses, 384-16(a) requires each lighting and appliance branch-circuit panelboard to be protected with a main CB or set of fuses. Per 384-14, a lighting and appliance branch-circuit panelboard has more than 10% of its OCPDs rated 30 A or less, for which neutral connections are provided. If the secondary of the 10′ transformer tap supplies a lighting and appliance branch-circuit panelboard, main protection is required. See Figure 8-29. Notice that nothing in the tap rule removes or amends the transformer overcurrent protection required by 450-3.

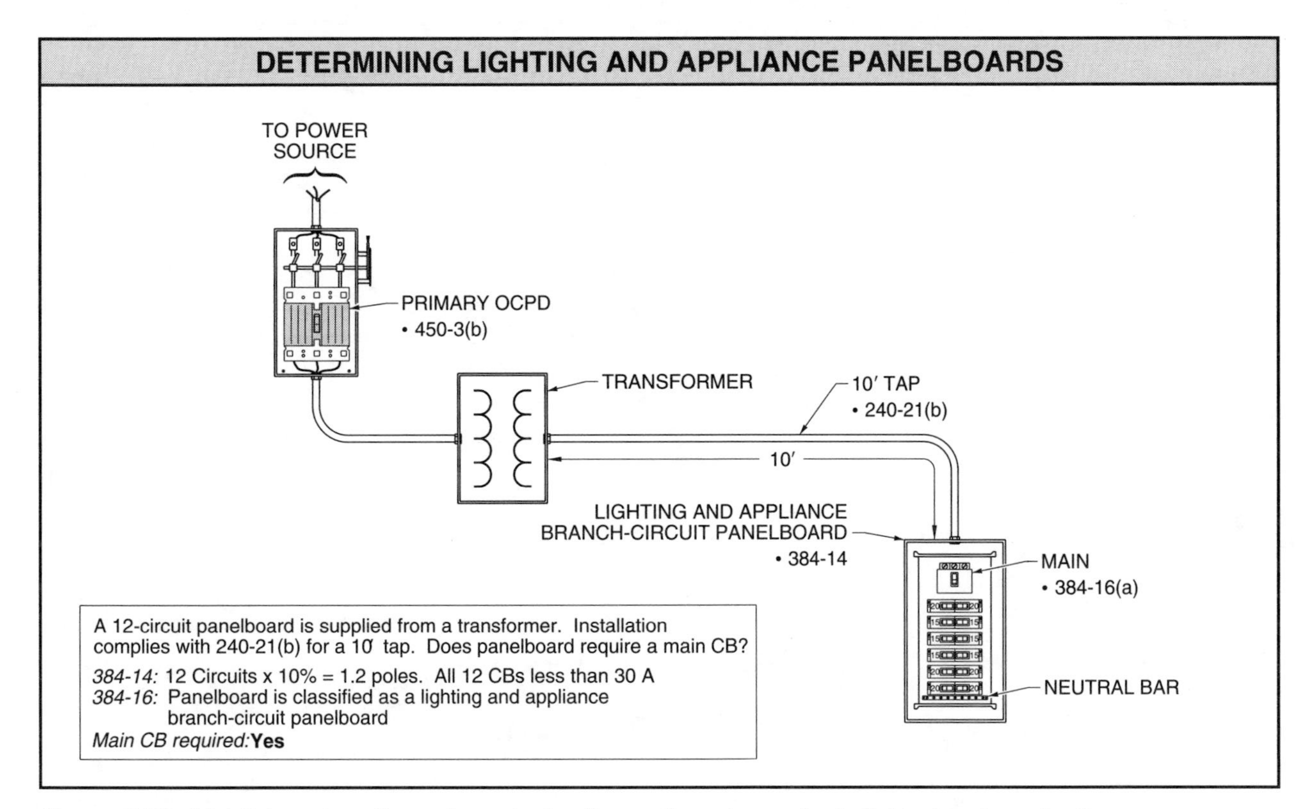

Figure 8-29. Lighting and appliance branch-circuit panelboards require individual main protection.

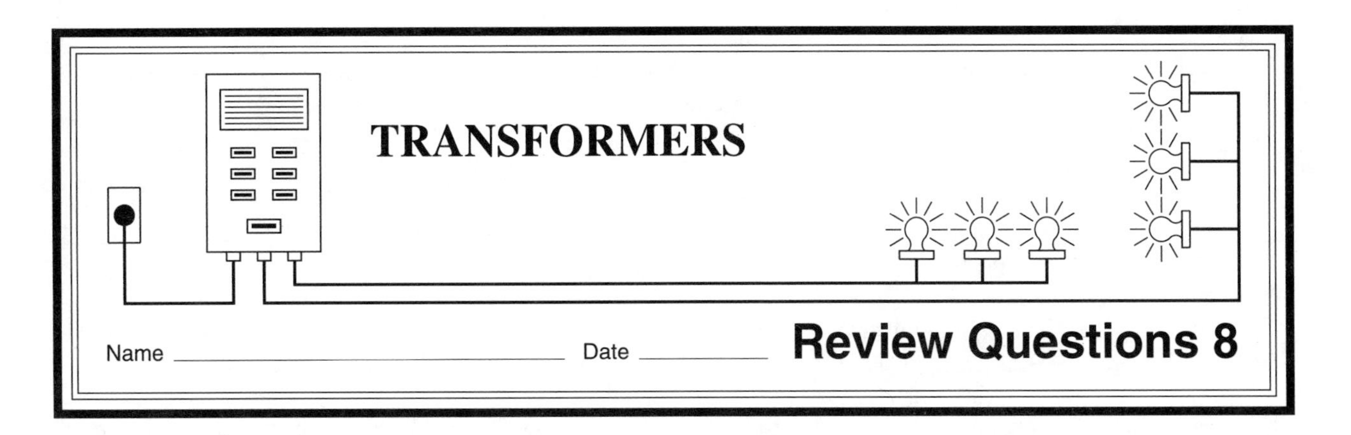

Review Questions

______ **1.** Transformers are electrical devices that contain no ________ parts.

______ **2.** A(n) ________ transformer has more windings in the primary than in the secondary.

______ **3.** The NEMA insulation classifications for dry-type transformers are Classes ________.

A. A, B, C, and D
B. A, B, C, and F
C. A, B, F, and H
D. neither A, B, nor C

______ **4.** ________ is the process by which oxygen mixes with other elements and forms a type of rust-like material.

T F **5.** A less-flammable liquid has a fire point of not less than 300°C.

T F **6.** An autotransformer may be used to buck or boost the supply voltage.

______ **7.** Temperature ________ is the amount of heat that an electrical component produces above the ambient temperature.

______ **8.** Transformer vaults shall be constructed of materials with a minimum fire resistance of ________ hours.

______ **9.** ________ hardware is door hardware that is designed to open easily in an emergency situation.

______ **10.** Liquid-filled transformers are commonly used in ________.

A. 150 kVA to 3000 kVA ratings
B. 2400 V to 13,800 V voltages
C. both A and B
D. neither A nor B

______ **11.** A ________ point is the temperature at which liquids give off vapor sufficient to form an ignitable mixture with the air near the surface of the liquid.

A. flash
B. fire
C. either A or B
D. neither A nor B

T F **12.** Transformers are now required to be marked with a minimum distance or clearance from walls or other obstructions to facilitate the dissipation of heat.

T F **13.** Transformers rated over 112½ kVA shall be enclosed in a fire-resistant room with materials having a 3-hour minimum fire rating.

________ **14.** Type ________ buildings have fire resistance ratings from 0 to 4 hours.

________ **15.** A vault floor in direct contact with the earth shall be constructed of at least ________″ of concrete.

________ **16.** The minimum size of ventilation openings for transformers rated less than 50 kVA is ________ sq ft.

________ **17.** Ventilation openings for transformers rated over 50 kVA is calculated at ________ sq in. per kVA less the area for grates, etc.

________ **18.** A(n) ________ installation has conditions of maintenance which allow only qualified persons to monitor or service the electrical equipment.

________ **19.** The general rule for transformers rated 600 V, nominal, or less is to protect the primary windings at not more than ________% of the rated primary current.

T F **20.** Askarel has been banned by the Environmental Protection Agency and its use as a transformer coolant is being phased out.

T F **21.** All transformers are required to be insulated.

T F **22.** Transformers are required to be installed outdoors only.

T F **23.** All transformers shall be installed in a readily accessible location.

________ **24.** Type ________ buildings have fire resistance ratings from 0 to 2 hours.

T F **25.** Doors for transformer vaults shall swing out.

________ **26.** Dry-type transformer insulation is based on an ambient temperature of ________°C.

________ **27.** Larger distribution transformers are commonly cooled by a(n) ________.

________ **28.** Common transformer mountings include ________.

A. pole or vault
B. floor or platform
C. both A and B
D. neither A nor B

T F **29.** In general, transformers are required to be grounded.

T F **30.** Askarel is a flammable liquid that contains PCBs.

Guarding

__________ **1.** The minimum distance at A is ________′.

__________ **2.** The minimum distance at B is ________′.

__________ **3.** The minimum distance at C is ________′.

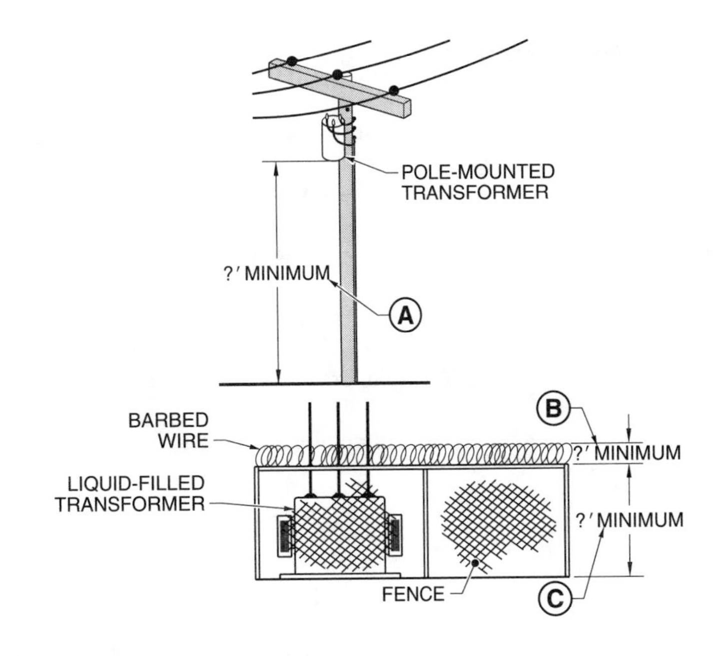

Transformer Vaults

__________ **1.** The minimum fire rating at A is ________ hours.

__________ **2.** The minimum fire rating at B is ________ hours.

__________ **3.** The minimum concrete thickness (when in contact with earth) at C is ________″.

__________ **4.** The minimum fire rating (when not in contact with earth) at D is ________ hours.

__________ **5.** The minimum fire rating at E is ________ hours.

__________ **6.** The minimum height at F is ________″.

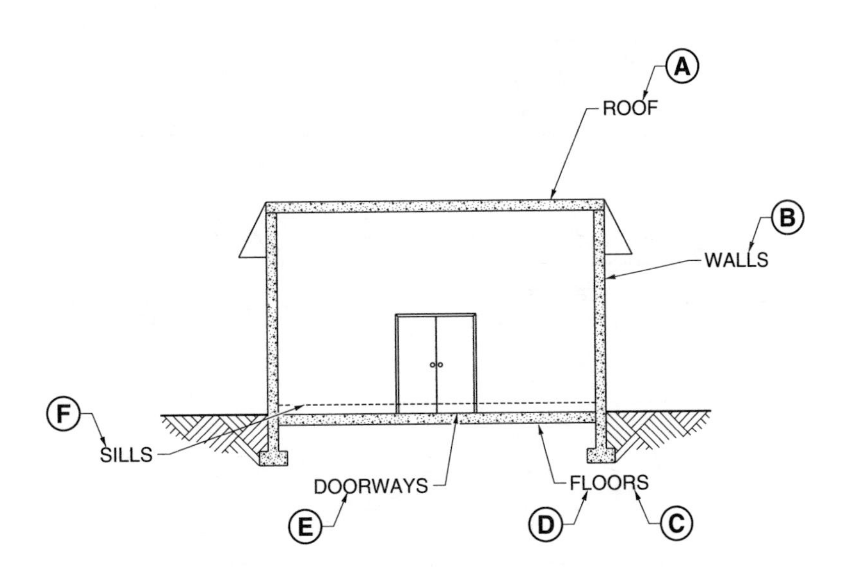

Dry-Type Transformers

__________ 1. The minimum clearance at A is ________″.

__________ 2. The minimum clearance at B is ________″.

__________ 3. The temperature rise at C is ________°C.

__________ 4. The temperature rise at D is ________°C.

__________ 5. The temperature rise at E is ________°C.

__________ 6. The temperature rise at F is ________°C.

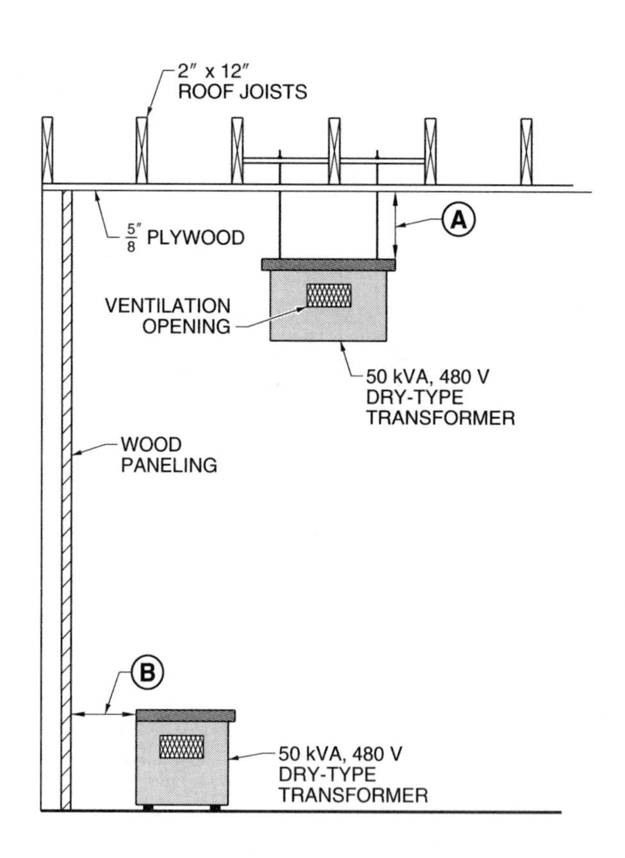

DRY-TYPE TRANSFORMERS			
CLASS	TEMPERATURE RISE – °C	AMBIENT TEMPERATURE –°C	COMMON USE
A	Ⓒ	40	Control
B	Ⓓ	40	Distribution
F	Ⓔ	40	Distribution
H	Ⓕ	40	Large distribution

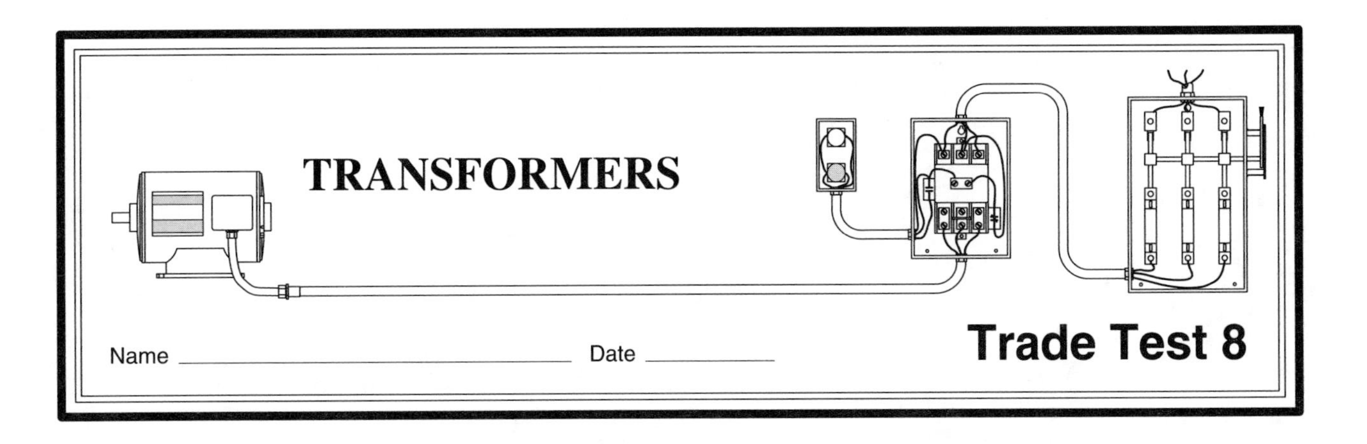

NEC®	Answer	
______	______	**1.** Determine the maximum size fuse permitted to protect the primary of a 480 V, 1ϕ, 15 kVA dry-type transformer. Secondary overcurrent protection is not provided.
______	______	**2.** Determine the maximum size OCPD permitted to protect the primary of a 75 kVA, 3ϕ nonsupervised transformer with 6% impedance and a 4160 V primary.
______	______	**3.** Determine the maximum size ATCB permitted to protect a 50 kVA, 3ϕ supervised transformer with a 2400 V primary.
______	______	**4.** *See Figure 1.* Determine the maximum size OCPD permitted to protect the primary of the supervised transformer.

FEEDER OCPD
MAIN DISTRIBUTION PANEL
45 kVA, 3ϕ TRANSFORMER WITH 3ϕ IMPEDANCE, 4160 V PRIMARY, AND 2400 V SECONDARY
SECONDARY OCPD

FIGURE 1

NEC®	Answer	
______	______	**5.** The rated primary current of a 480 V transformer is 8.75 A. Determine the maximum size OCPD permitted to protect the primary of the transformer. Secondary overcurrent protection is not provided.
______	______	**6.** *See Figure 2.* Determine the maximum size OCPD permitted to protect the primary of the transformer.
______	______	**7.** *See Figure 3.* Determine the maximum size OCPD permitted to protect the secondary of the transformer.

277/480 V, 3ϕ PANEL
120/208 V, 3ϕ SUB-PANEL
PRIMARY OCPD
480/120 V, 6 kVA TRANSFORMER
FUSED DISCONNECT

FIGURE 2

277/480 V, 3ϕ PANEL
FUSED DISCONNECT
480/120 - 208 V, 25 kVA TRANSFORMER
PRIMARY OCPD
SECONDARY CONDUCTORS
FUSED DISCONNECT

FIGURE 3

__________ __________ **8.** *See Figure 4.* Determine the maximum size CB for the primary of the transformer.

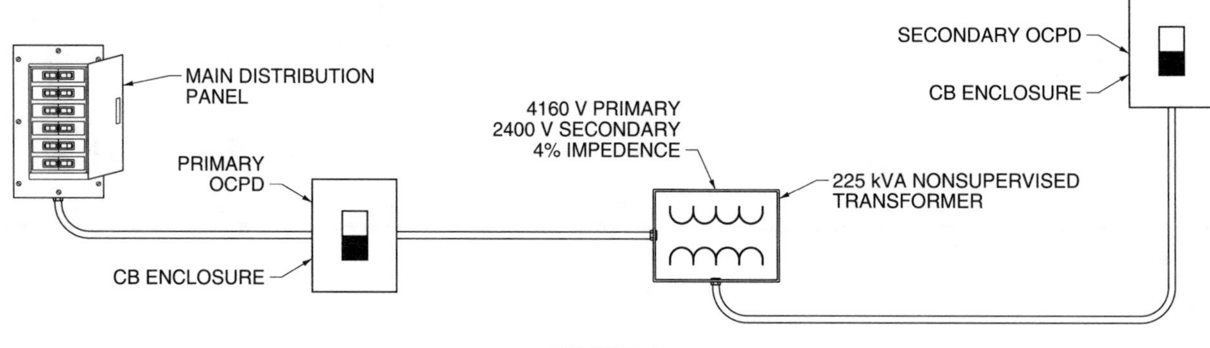

FIGURE 4

__________ __________ **9.** *See Figure 5.* Determine the maximum size fuses permitted for the secondary of the transformer.

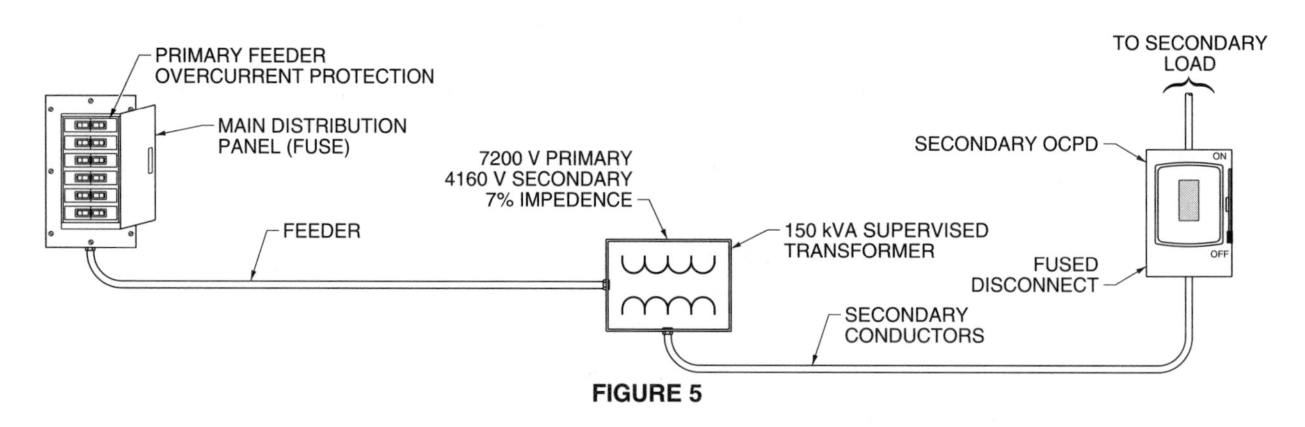

FIGURE 5

__________ __________ **10.** A 480 V transformer is installed without secondary overcurrent protection. The rated primary current of the transformer is 1.875 A. Determine the maximum rating of the OCPD permitted to protect the primary.

__________ __________ **11.** A 30 kVA, 3ϕ transformer operates at 480 V. Secondary overcurrent protection for the transformer is provided at not more than 125% of the rated secondary current of the transformer. Determine the maximum size OCPD permitted to protect the primary.

__________ __________ **12.** A 4160 V, 50 kVA, 3ϕ transformer is installed in a supervised location. Determine the maximum size CB permitted to protect the primary. Secondary overcurrent protection is not provided.

__________ __________ **13.** What size primary OCPD, using a fuse, is required to protect a 30 kVA, 208 V, 3ϕ transformer?

__________ __________ **14.** What size primary OCPD, using a fuse, is required to protect a 15 kVA, 480 V, 3ϕ transformer?

__________ __________ **15.** What size primary OCPD, using a fuse, is required to protect a .500 kVA, 240 V, 1ϕ transformer?

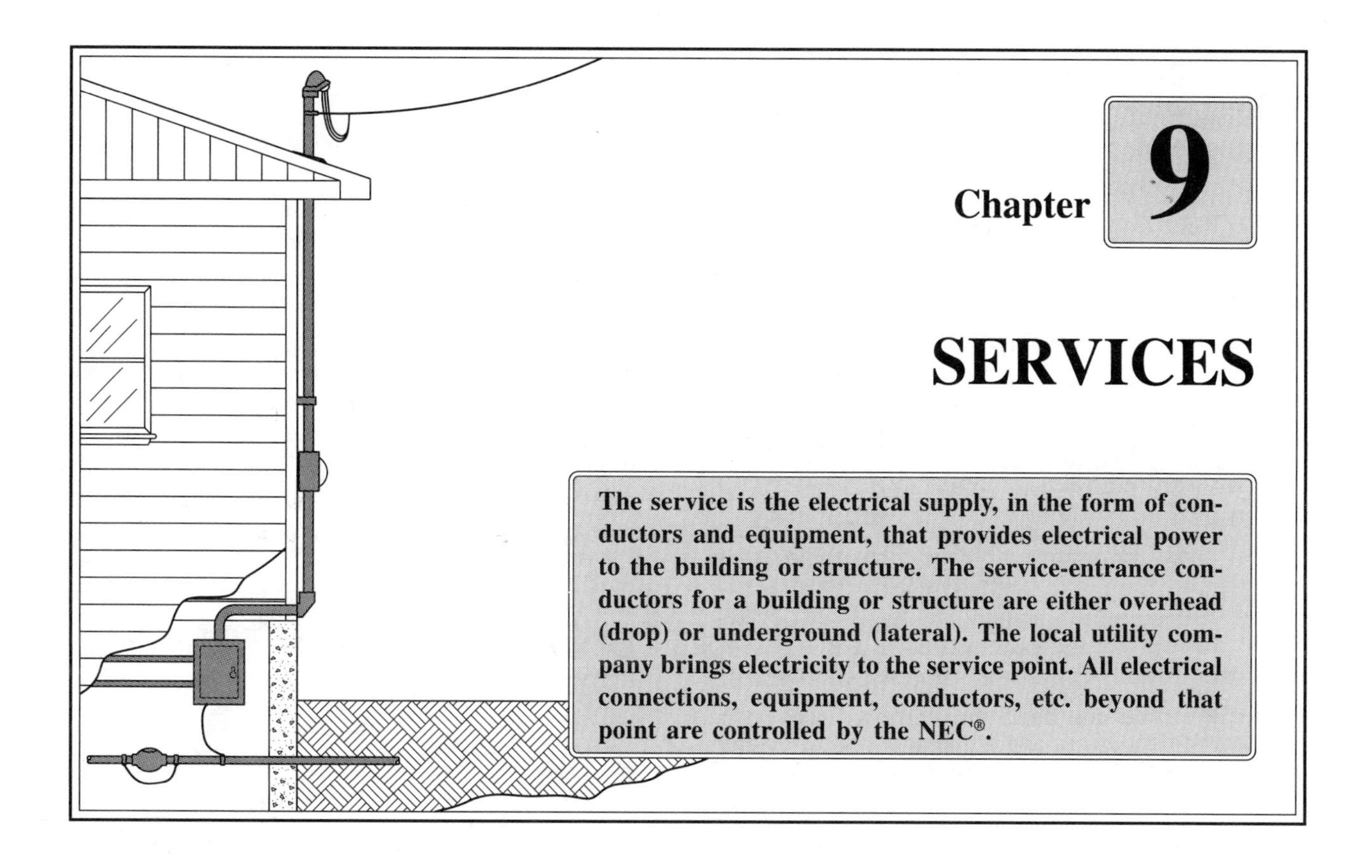

Chapter 9

SERVICES

The service is the electrical supply, in the form of conductors and equipment, that provides electrical power to the building or structure. The service-entrance conductors for a building or structure are either overhead (drop) or underground (lateral). The local utility company brings electricity to the service point. All electrical connections, equipment, conductors, etc. beyond that point are controlled by the NEC®.

SERVICE

The heart of the electrical distribution system is the service. Article 230 of the NEC® contains requirements for installing service conductors and service equipment in all types of occupancies. Parts A through G contain the requirements for services, 600 V and less and Part H contains the requirements for services exceeding 600 V, nominal.

Designers and installers of electrical services may have other requirements to consider in addition to those contained in 230. Most services are supplied by the local power utility company. The local power utility may have additional, and sometimes different, requirements for the installation of electrical services. For example, the NEC® does not address the location of meter sockets. Per 230-66, the meter enclosures need not even be considered service equipment. Yet, most electrical utilities specify where the meter is located and the height at which it is mounted. Designers and installers, therefore, shall comply with the NEC® and the local power utility company. Contact the local utility to obtain a copy of their requirements before planning or installing electrical services.

Definitions – 100

Article 100 contains several important definitions that relate to service. For example, to determine the minimum size of the service conductors, several NEC® sections should be considered. Service-drop conductors are given in 230-23(b), service-lateral conductors in 230-31(b), and service-entrance conductors in 230-42. The minimum size service conductor therefore, depends on the type of service that is used. The requirements for service-lateral conductors contained in Part C cannot be applied to the installation of service-drop conductors which are contained in Part B.

Service. The *service* is the electrical supply, in the form of conductors and equipment, that provides electrical power to the building or structure. See Figure 9-1. Service is the root word for nine terms that are defined in 100. It is a general term and usually includes the service-lateral or service-drop conductors, service equipment, and any associated metering equipment.

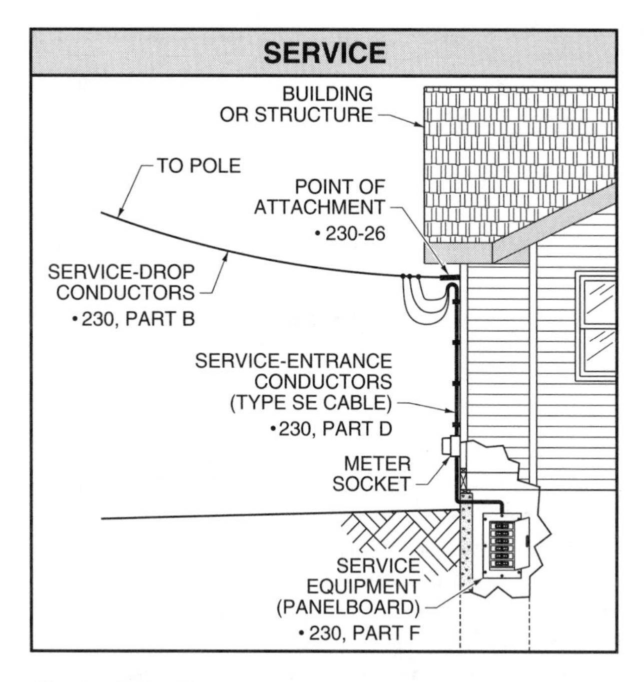

Figure 9-1. The service is the electrical supply, in the form of conductors and equipment, that provides electrical power to a building or structure.

> ***AUTHORITY HAVING JURISDICTION***
>
> *The AHJ has the final say regarding electrical installations no matter what the NEC® states. Due to the possibility of local jurisdictions requiring installation practices that are different from the NEC®, designers and installers should ensure, at the outset of the project, that their intended installation practices meet local as well as NEC® requirements.*
>
> *For example, many local AHJs require that all electrical services be installed in metal conduit despite the list of possible wiring methods that the NEC® recognizes in 230-43. Other AHJs may prohibit any connections within the metering equipment. Always check with the AHJ to ensure that there are no costly surprises at the end of the job.*

Service Conductors. *Service conductors* are the conductors from the service point or other source of power to the service disconnecting means. This is a general term. Service conductors can either be part of an overhead supply system or part of an underground supply system.

From the design and installation point of view, service conductors are the most important part of the service. This is because while other components of the service may be installed or designed by the utility, designers and installers are almost always responsible for the installation of these conductors.

Service-Entrance Conductors – Overhead Systems. *Service-entrance conductors – overhead systems* are conductors that connect the service equipment for the building or structure with the electrical utility supply conductors. This connection usually occurs outside of the building or structure. See Figure 9-2.

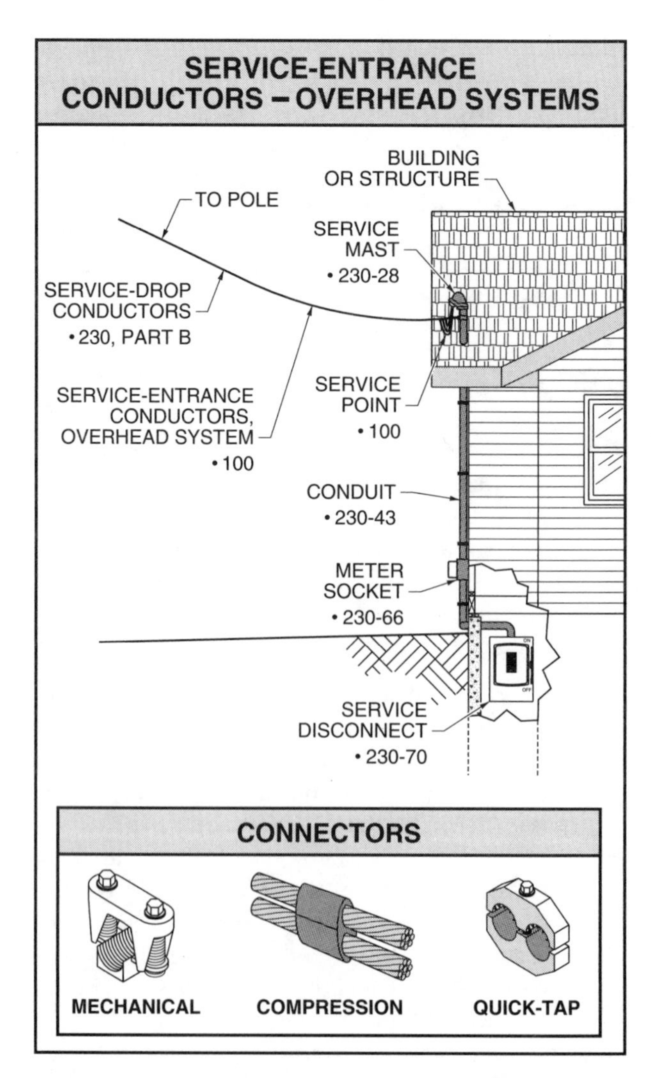

Figure 9-2. Service-entrance conductors – overhead systems are conductors that connect the service equipment for the building or structure with the electrical utility supply conductors.

The method of connection is usually by splice or tap. Devices such as irreversible compression connectors, service lugs, or quick-taps are often used. Overhead systems are the most common system today, however utility companies are moving away from overhead systems and prefer to install underground electrical distribution systems whenever possible.

Service-Entrance Conductors – Underground Systems. *Service-entrance conductors – underground systems* are conductors that connect the service equipment with the service lateral. See Figure 9-3. The FPN, following the definition of underground service-entrance conductors in 100, points out that in some electrical service installations there may not be any service-entrance conductors. In other installations, the service-entrance conductors may not even enter the building. This depends on the configuration of the service and the location of the service disconnecting means.

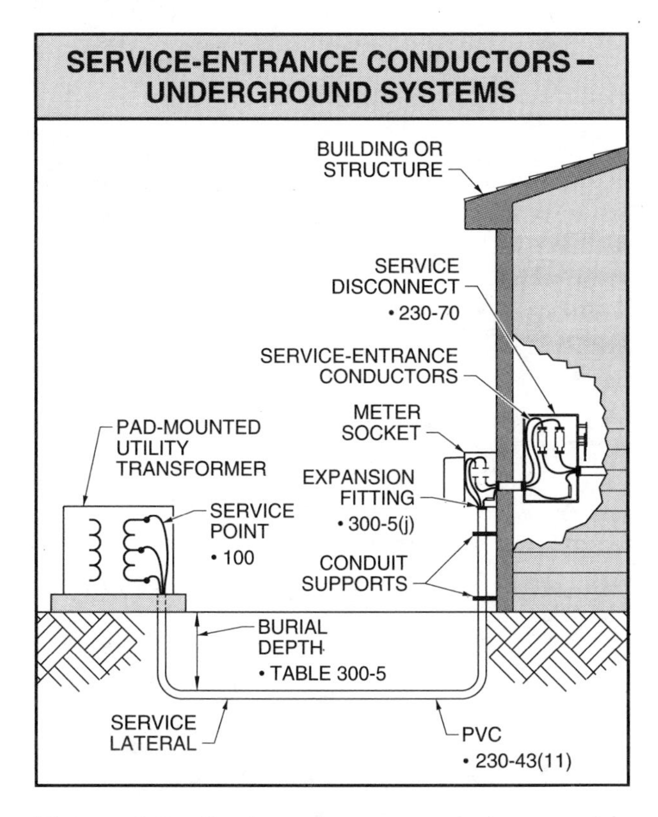

Figure 9-3. Service-entrance conductors – underground systems are conductors that connect the service equipment with the service lateral.

Service Drop. The *service drop* is the conductors that extend from the overhead utility supply system to the service-entrance conductors at the building or structure. See Figure 9-4. Usually, the service-drop conductors originate at the last utility pole or aerial support and continue to their point of attachment on the building or structure. Utility companies frequently use triplex or messenger cable assemblies as the service-drop conductors to the building. In many of these systems, the steel messenger cable is used as the grounded or neutral conductor for the premises wiring system.

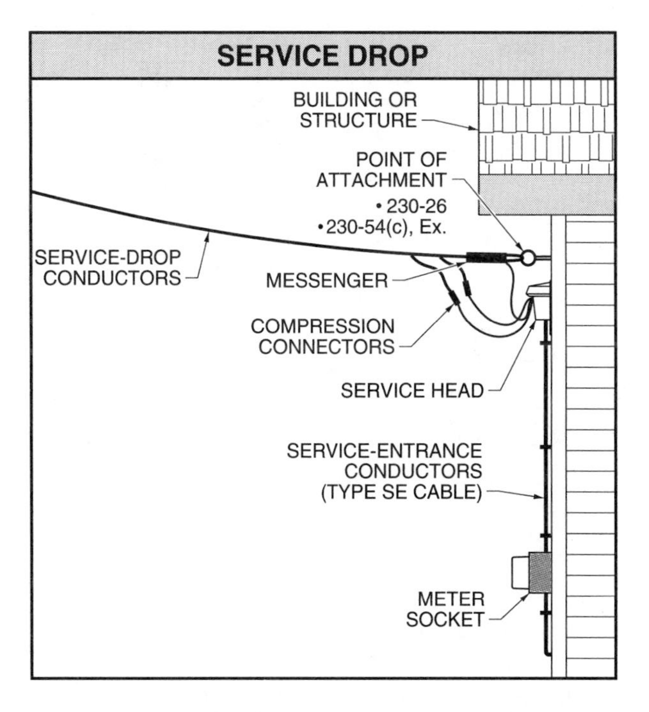

Figure 9-4. The service drop is the conductors that extend from the overhead utility supply system to the service-entrance conductors at the building or structure.

Service Lateral. The *service lateral* is the underground service conductors that connect the utility's electrical distribution system with the service-entrance conductors. See Figure 9-5. The utility supply system can originate from a pad-mounted transformer or an overhead supply system with the conductors emerging from the ground and continuing up the pole.

The point of connection is between the underground service conductors and the service-entrance conductors. It can occur at a terminal box, meter enclosure, or other type of enclosure provided there is adequate space for the connection. The connection point may occur inside or outside the building or structure. In the event there is no terminal box or meter enclosure, the point of connection is considered to be the point at which the service conductors enter the building or structure.

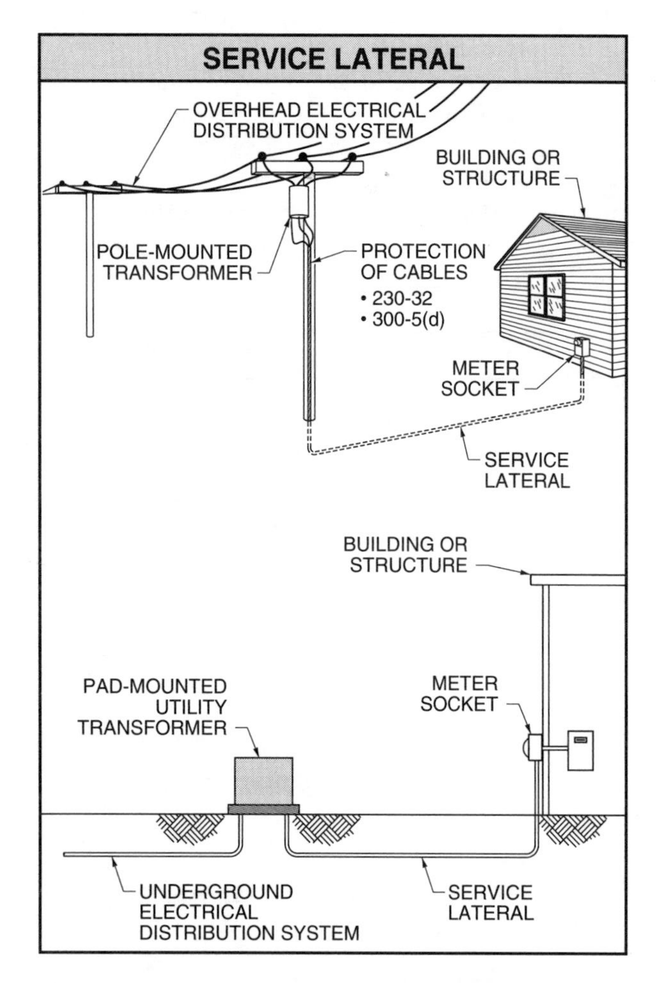

Figure 9-5. The service lateral is the underground service conductors that connect the utility's electrical distribution system with the service-entrance conductors.

Service Point. The *service point* is the point of connection between the local electrical utility company and the premises wiring of the building or structure. See Figure 9-6. An important design consideration in any electrical distribution system is the point at which the utility supply ends and the customer-owned or premises wiring begins. The NEC® typically is only applied to the premises wiring.

In addition to the obvious design concerns, such as who must purchase and maintain the equipment, the service point is important for several legal determinations. The utility company may use its own standards for installation or it may operate under the guidelines established in the *National Electrical Safety Code*. Wiring and equipment connected to the load side of the service point is covered by the NEC®, while conductors located on the line or supply side of the service point are not covered by the NEC®. Unfortunately, as with many of these service-related terms, the location of the service point can change depending upon the design and configuration of the distribution system. Prior to the installation, installers and designers should determine the exact location of the service point and the applicable installation requirements.

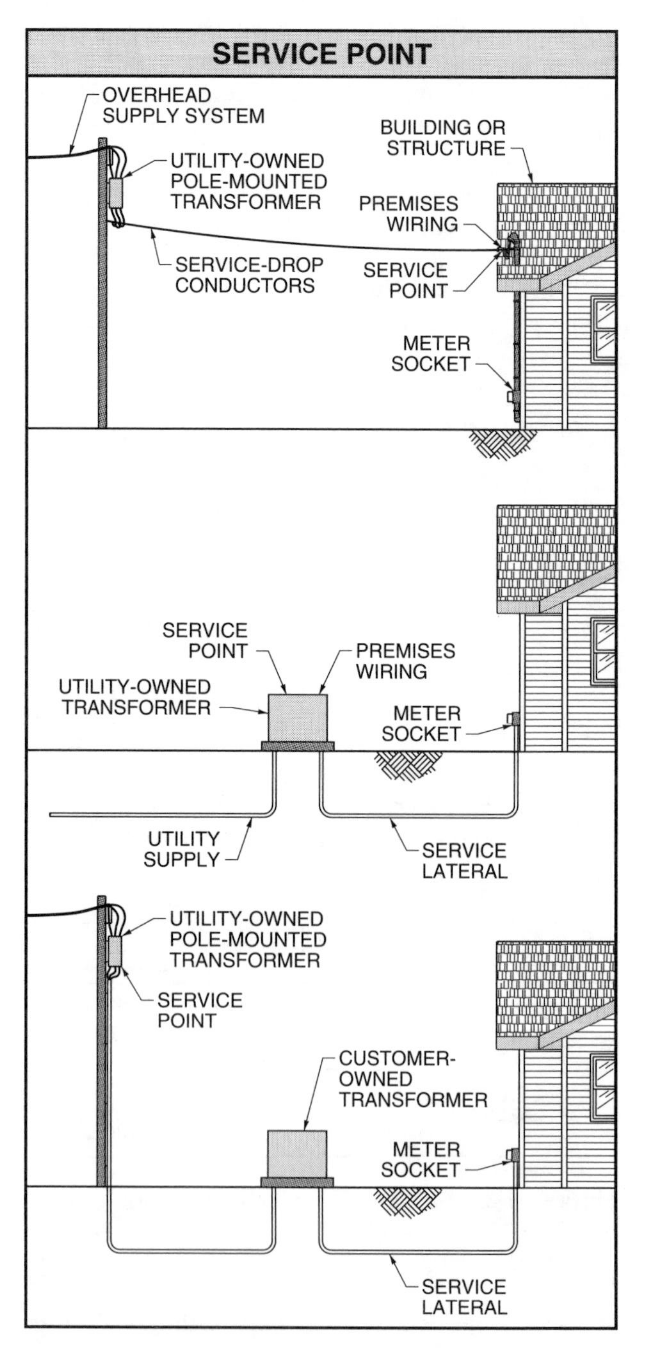

Figure 9-6. The service point is the point of connection between the local electrical utility company and the premises wiring of the building or structure.

SUITABLE FOR USE

Installers of electrical service equipment should verify that the equipment is suitable for use as service equipment. There are many types of identification used to mark the suitability of service equipment. Pay particular attention to equipment that is marked "suitable only for use as service equipment." This marking is associated with panelboards which are supplied with the neutral factory-bonded to the enclosure. These panelboards are not intended for and not suitable for use in subpanels. Use of a panelboard with this marking violates 250-61(b) and could introduce the possibility of a fire and/or shock hazard.

Service Equipment. *Service equipment* is all of the necessary equipment to control the supply of electrical power to a building or a structure. It includes CBs, fuses, disconnecting means, panelboards, etc. See Figure 9-7. All equipment used as service equipment shall be identified as being suitable for use as service equipment per 230-66. Individual meter socket enclosures are not considered service equipment and need not be identified.

The utility-owned transformer provides electrical service through service-lateral conductors through the meter to the customer-owned service equipment.

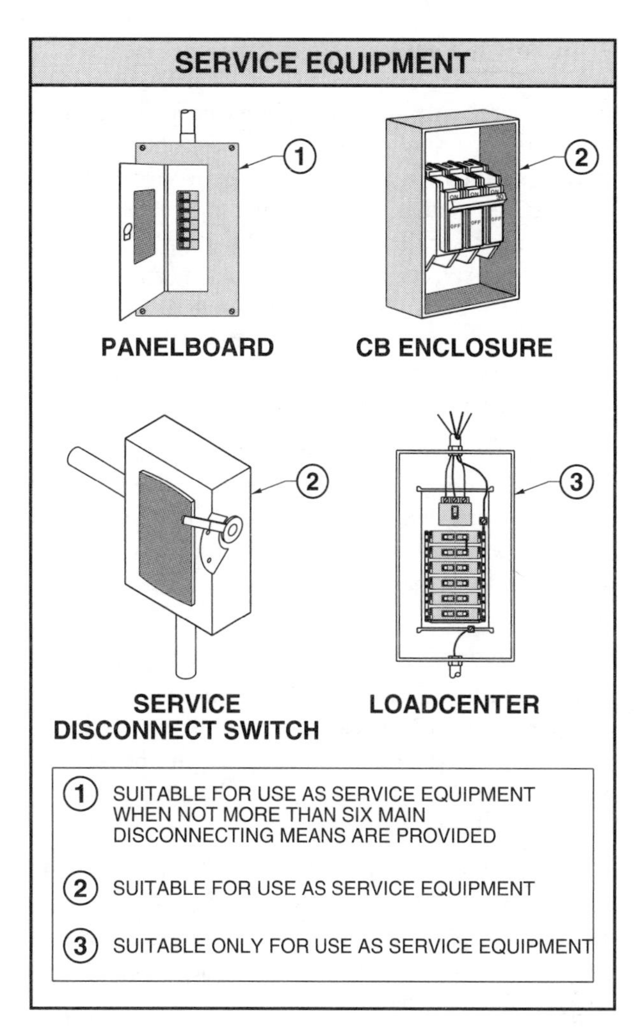

Figure 9-7. Service equipment is all of the necessary equipment to control the supply of electrical power to a building or structure.

SERVICE LIMITATIONS – 230, PART A

Part A contains all of the general requirements applicable to services rated under 600 V, nominal. The general provisions of Part A ensure that adequate protection is provided for the unprotected service conductors. This can be accomplished by maintaining minimum clearances from window and door openings, stairs, balconies, etc. by limiting service raceways to service conductors only, by protecting service conductors which are installed within a building or structure, and by limiting each building or structure to a single service.

Number of Services – 230-2

The general rule of 230-2 limits each building or structure to a single service. Because the definition of service includes the service-drop or service-lateral conductors, the basic rule permits only a single service drop or service lateral to be supplied to each building or structure. This does not mean, however, that each building is limited to a single set of service-entrance conductors. Other rules permit multiple sets of service-entrance conductors to be connected to each service drop or lateral.

There are seven exceptions to the general rule limiting each building to a single service. These exceptions are intended to provide some latitude for the designer and installer for a particular type of occupancy or for a particular type of electrical system. See Figure 9-8.

Per 230-2, Ex. 1, occupancies which utilize a fire pump for life safety protection are permitted to have a separate service provided solely for the fire pump. This helps to ensure that a failure of the primary building service will not remove the life safety protection in the event of a fire.

Emergency or standby systems are permitted by 230-2, Ex. 2. These systems provide power to essential building systems, such as exit lights, emergency lighting, HVAC equipment, etc. in the event of a primary electrical service failure.

Designers of electrical systems may also encounter buildings in which there is not adequate space for service equipment which is accessible to all occupants. Multiple services are permitted to these multiple-occupancy buildings per 230-2, Ex. 3 if special permission is granted. *Special permission* is the written approval of the AHJ. See 100.

Another frequent design problem occurs when the electrical load requirements of the occupancy are so large that it makes the installation of a single service impracticable. Per 230-2, Ex. 4, additional services are permitted to be installed when the capacity requirements of the building or structure exceed 2000 A at 600 V or less. This exception also applies when the 1ϕ load requirements exceed that for which the local utility normally supplies a single service. For example, if the occupancy requires a 1ϕ service of 1500 A and the utility company's largest service is 1200 A, a second service can be supplied. This exception also permits additional services for capacity when special permission is obtained.

Similarly, there may be large area buildings that make the utilization of a single service impracticable from the point of view of the distribution system. Per 230-2, Ex. 5, additional services for these large area single buildings or structures is permitted if special permission is first obtained.

There also may be design considerations regarding the voltage and frequency requirements of the services or different utility rate schedules for the services. Per 230-2, Ex. 6, additional services are permitted for these reasons. For example, a local utility company may offer different usage rates for their customers for off-peak services.

Per 230-3, Ex. 7, underground sets of conductors are permitted to be run to the same building or structure in accordance with the provisions of 230-40, Ex. 2. The conductors must be sized at least No. 1/0 and may be connected at their supply end but not at their load end. Essentially, this provision permits multiple service laterals to be considered a single service lateral.

Conductors – Outside of Buildings – 230-6

Although service conductors are not provided with overload or short-circuit protection, there may be installations in which there is a need to have the service conductors enter the building. Many high-rise distribution systems, with large power needs, utilize 230-2, Ex. 4 to bring a second service into the building. Often, the second service is run to an upper floor electrical room for distribution to the upper levels of the building. In these types of installations, the service conductors enter the building without overcurrent protection. This is accomplished by meeting one of three provisions of 230-6. See Figure 9-9.

Per 230-6(1), service conductors are permitted to be routed within the building if they are installed under not less than 2″ of concrete beneath the building or structure. Per 230-6(2), the service conductors shall be enclosed in a raceway that is encased within 2″ or more of concrete or brick. Per 230-6(3), the service conductors are permitted to enter a transformer vault, provided the vault is constructed per 450, Part C.

In each of these cases, the conductors are considered to be outside of the building even though they are physically located within the building. This is permitted by 230-6 because the conductors are suitably protected from physical damage or are inaccessible to unqualified personnel.

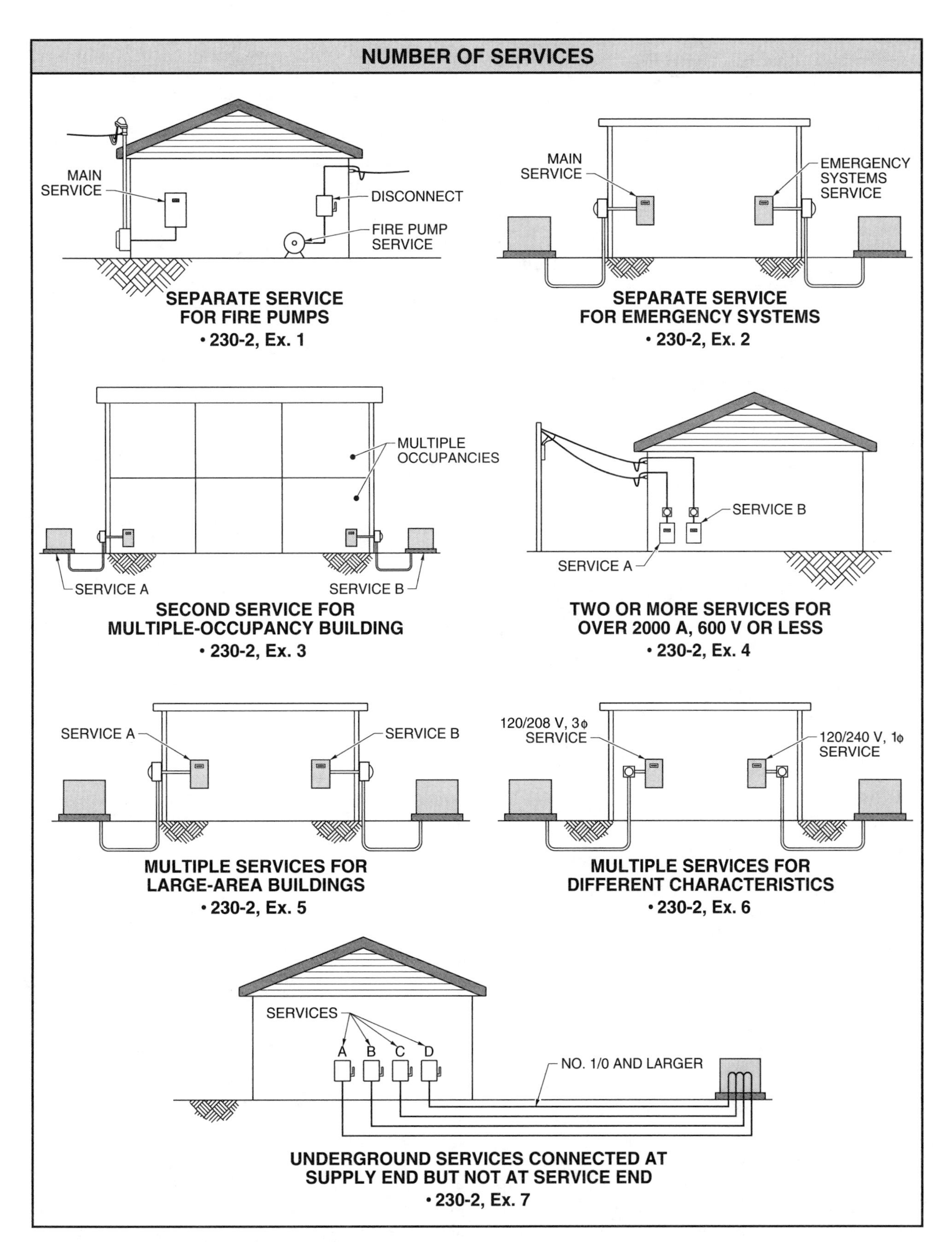

Figure 9-8. Multiple sets of service-entrance conductors may be connected to each service drop or lateral.

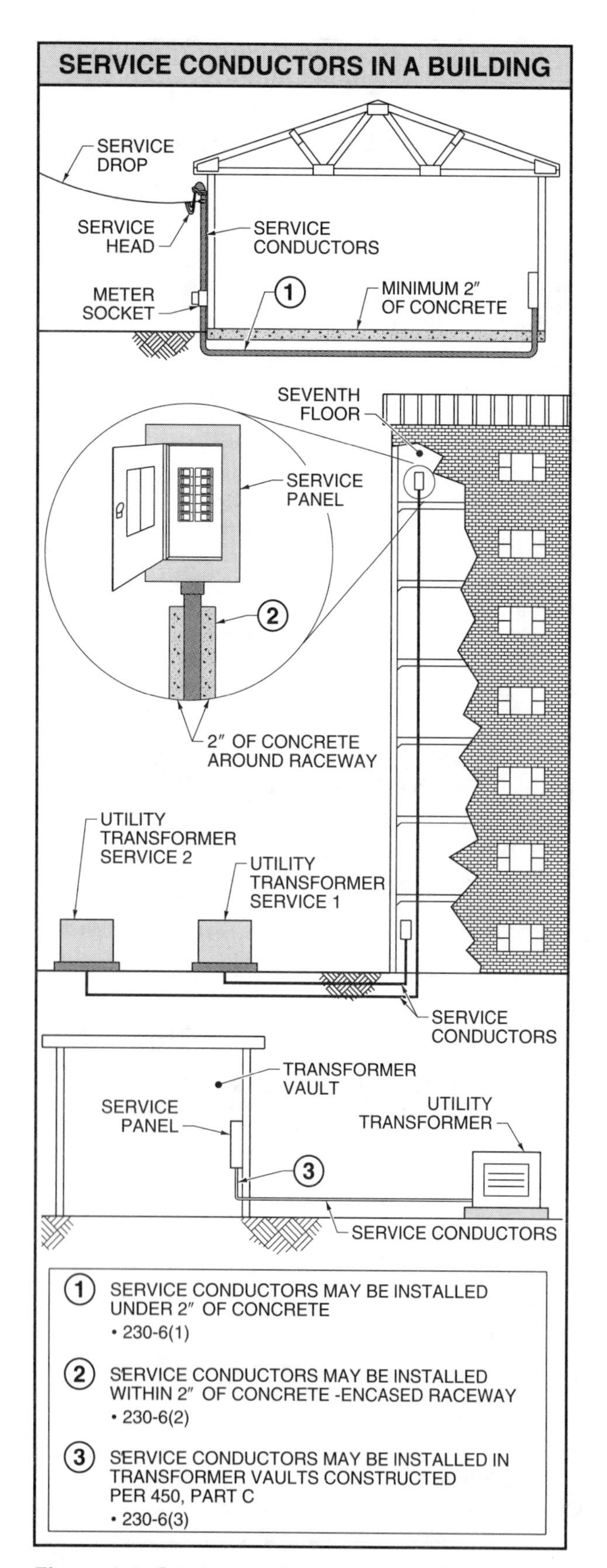

Figure 9-9. Service conductors may enter a building.

Service Raceways and Seals – 230-7; 230-8

Only service conductors shall be installed in service raceways. Service raceway is not defined in 100, but is considered to be any raceway which contains service conductors. Allowing only service conductors to be installed in service raceways ensures that a fault of any type in the service raceway will not affect other conductors. Per 230-7, Ex. 1, this does not apply to grounding or bonding conductors which are permitted within the service raceway. Per 230-7, Ex. 2, load management control conductors are also exempt from the general rule. These conductors are permitted within the service raceway if they are provided with overcurrent protection. See Figure 9-10.

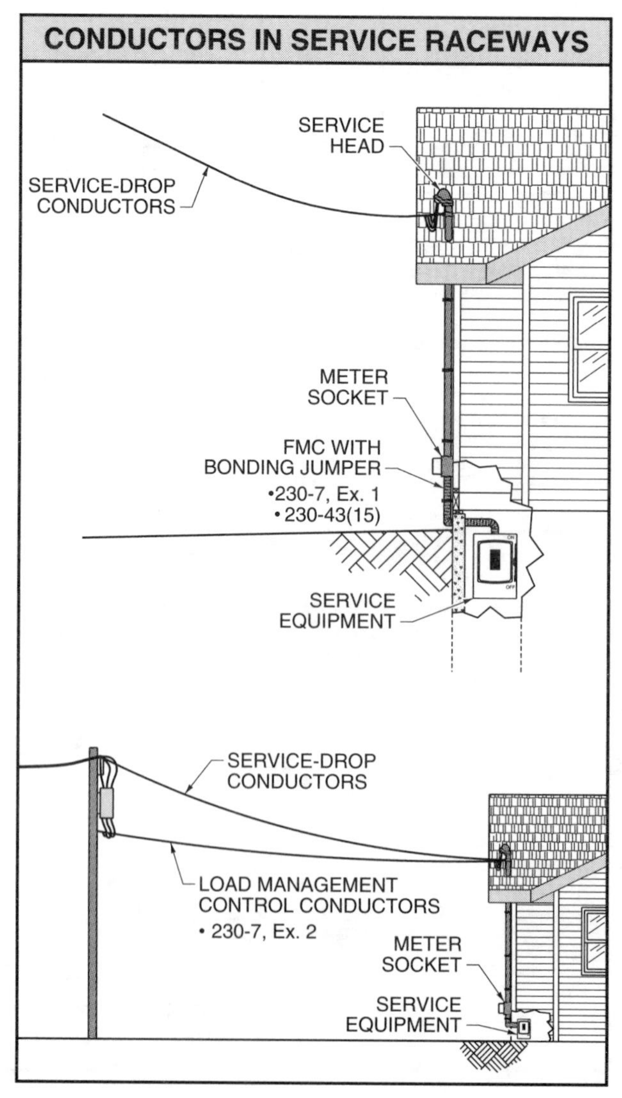

Figure 9-10. Only service conductors shall be installed in service raceways.

Clearance from Openings – 230-9

Because service conductors are not protected by overcurrent protection devices, great care is taken to ensure that the conductors are not accessible to the general public. Per 230-9, whenever service conductors are installed as open conductors or as multiconductor cable without an overall outer jacket, a clearance of 3′ shall be provided from windows that can be opened, doors, porches, balconies, ladders, stairs, fire escapes, or similar locations. See Figure 9-11. The intent of this is to protect people from possible contact with these conductors by maintaining a 3′ zone around the conductors.

Per 230-9, Ex., conductors are permitted to be run above windows within the 3′ zone. The likelihood that the conductors can be reached is reduced by installing them above the window. This only applies to multiconductor cables without an overall outer jacket. It does not apply to Type SE cable with an overall outer jacket. Type SE cable can be installed within the 3′ zone.

The last consideration when installing service conductors near building openings is to ensure that they are not installed beneath openings through which materials may pass. If such openings exist, the service conductors cannot be installed in a manner which obstructs such openings.

OVERHEAD SERVICE-DROP CONDUCTORS – 230, PART B

The service-drop conductors are the overhead service conductors from the utility system to a building or structure. An example of a structure that could be supplied by an overhead system is a pole. In this case, the service-drop conductors terminate at a meter enclosure or disconnecting means which is mounted on the pole. See Figure 9-12.

Service-drop conductors are required to be properly sized for the intended load. In addition, minimum size conductors are specified. Several other safety concerns are associated with service-drop conductors. Service-drop conductors, in most cases, are required to be covered or insulated to protect personnel who might inadvertently come in contact with them.

Service-drop conductors are required to be insulated or covered with an extruded thermoplastic or thermosetting insulating material by 230-22. Per the definition in 100, "covered" means encasement in a material which is not recognized by the NEC® as an insulating material. In this case, however, the NEC® specifies that the covering shall consist of at least an extruded thermoplastic or thermosetting material. This is because there have been several fatalities resulting from contact with service-drop conductors.

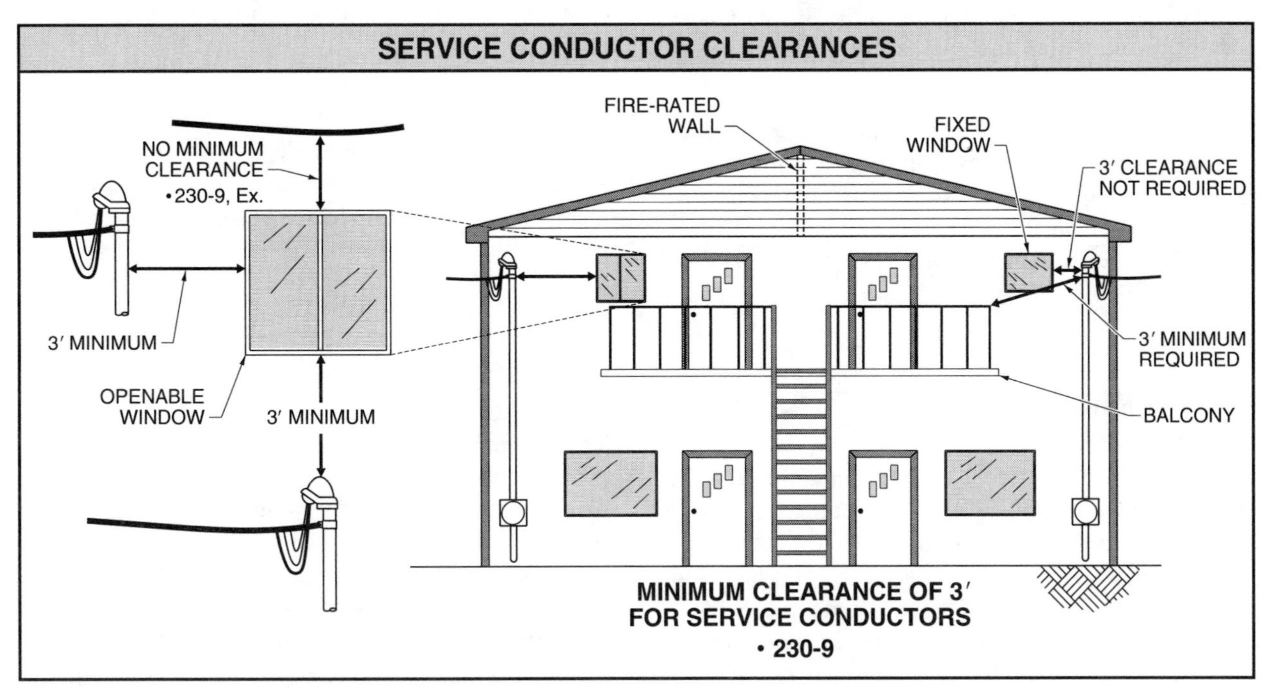

Figure 9-11. Service conductors installed as open conductors or multiconductor cable without an overall outer jacket shall maintain 3′ clearance.

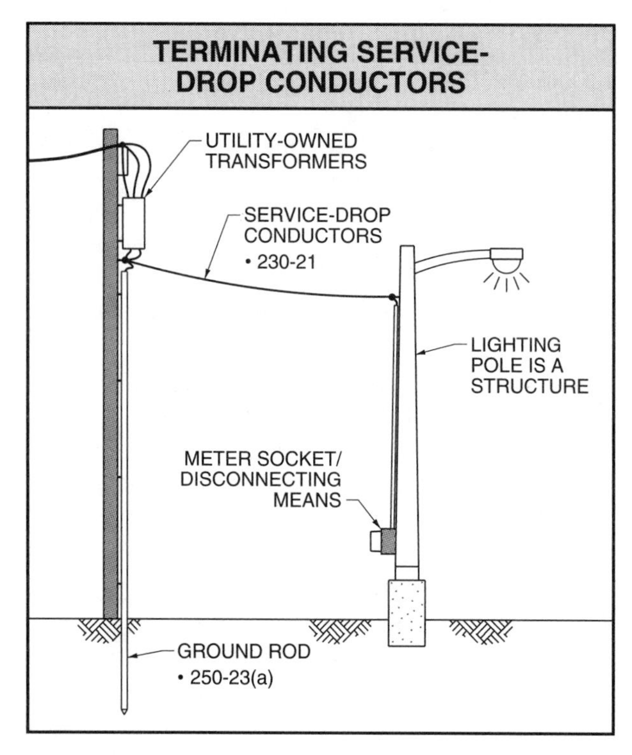

Figure 9-12. Service-drop conductors terminate at a building or structure.

The neutral of service-drop conductors, when installed as multiconductor cable, is permitted to be a bare conductor per 230-22, Ex. For example, a weatherproof, triplex cable which is commonly used for service-drop conductors for 1ϕ, 120/240 V, 3-wire services is permitted to have a bare neutral.

To avoid contact with personnel, the NEC® requires minimum clearances for service-drop conductors from grade, rooftops, building openings, and swimming pools. Service-drop conductors are required to be adequately supported to minimize lines failing in poor weather conditions, such as severe wind or snow/ice storms.

Size and Rating – 230-23

Service-drop conductors shall have an ampacity that is adequate for the intended load. The load is determined based on the calculations required by 220. Another general consideration for service-drop conductors is that they have adequate mechanical strength. Service-drop conductors are often subject to severe weather conditions which can put a great strain on the conductors due to the additional weight. Likewise, service-drop conductors are sometimes subject to large spans which also increase the stress on the conductors. For this reason, 230-23(b) sets a minimum size of No. 8 Cu or No. 6 Al for most service-drop conductors. While some installations for limited loads of a single branch-circuit allow the size of the service-drop conductors to be reduced, they are not permitted to be smaller than No. 12 hard-drawn copper or the equivalent per 230-23(b), Ex. See Figure 9-13.

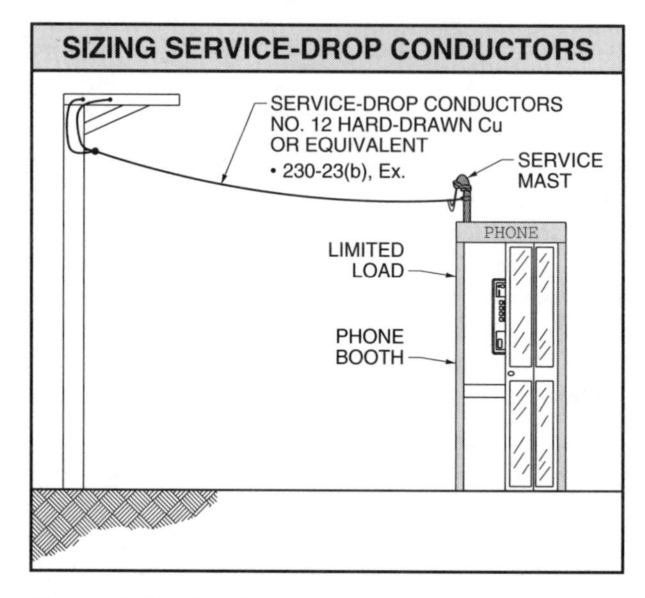

Figure 9-13. Service-drop conductors shall not be smaller than No. 12 hard-drawn Cu or its equivalent.

The grounded conductor is permitted by 230-23(c) to be sized per 250-23(b). Essentially, the grounded conductor is sized per Table 250-94, based on the size of the largest service-drop conductor. In no case shall the grounded conductor be smaller than that required by 250-23(b).

Clearances – 230-24; 230-26

The clearance requirements of 230 are provided to protect the service-drop conductors from physical damage and to protect personnel from contact with the conductors. The clearances listed in 230-24 are based on a set or prescribed conditions. These include no wind condition and a conductor temperature of 60°F (15°C). Designers and installers of service-drop conductors in conditions other than these shall calculate the additional stress on the conductors as a result of the varying conditions.

The general rule for service-drop conductors requires that they be installed so that they are not readily accessible. Service-drop conductors shall not be easily reached. They shall not be installed so that they can be reached without the use of ladders or other portable means. This requirement does not pose a problem for most installations. One area of concern however, is when the service-drop conductors pass over roofs. Per 230-24(a), service-drop conductors shall have a minimum vertical clearance of not less than 8′ above rooftops. See Figure 9-14.

The 8′ clearance shall be maintained in all directions for a minimum distance of 3′. This safety provision ensures that the conductors can not be reached by anyone who happens to be on the roof.

If a roof is subject to pedestrian or vehicular traffic, such as in the case of rooftop parking garages, 230-24(a), Ex. 1 requires that the minimum clearance from the roof surface conform to the requirements of 230-24(b). Therefore, the minimum clearance above the rooftop is 10′. In some cases, the clearances are greater. For services of less than 300 V between conductors, the required clearance is permitted to be reduced to 3′ per 230-24(a), Ex. 2, provided the slope of the roof is not less than 4″ in 12″. See Figure 9-15.

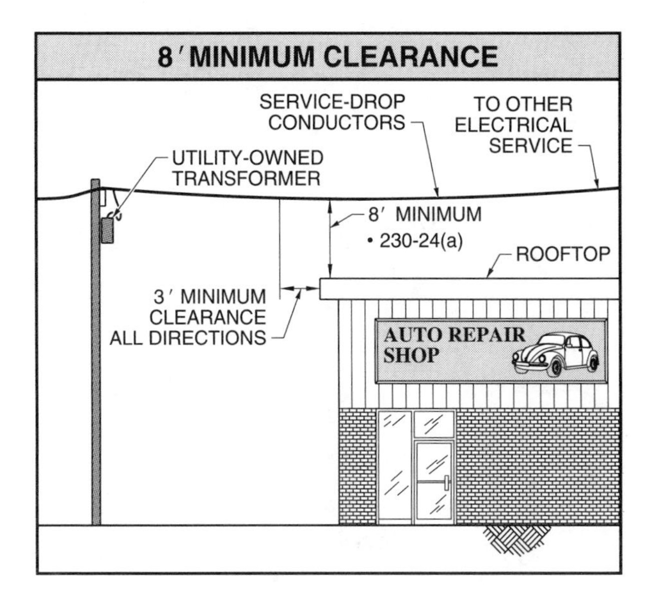

Figure 9-14. Service-drop conductors shall have a minimum vertical clearance of 8′ above rooftops.

Per 230-24(a), Ex. 3, a further clearance reduction is permitted for services not exceeding 300 V between conductors when not more than 6′ of service-drop conductor passes 4′ or less horizontally above a roof overhang and the conductors are terminated in a through-the-roof raceway or approved support. If both of theses conditions are met, the minimum clearance can be reduced to 18″ above the overhanging portion of the roof only. See Figure 9-16.

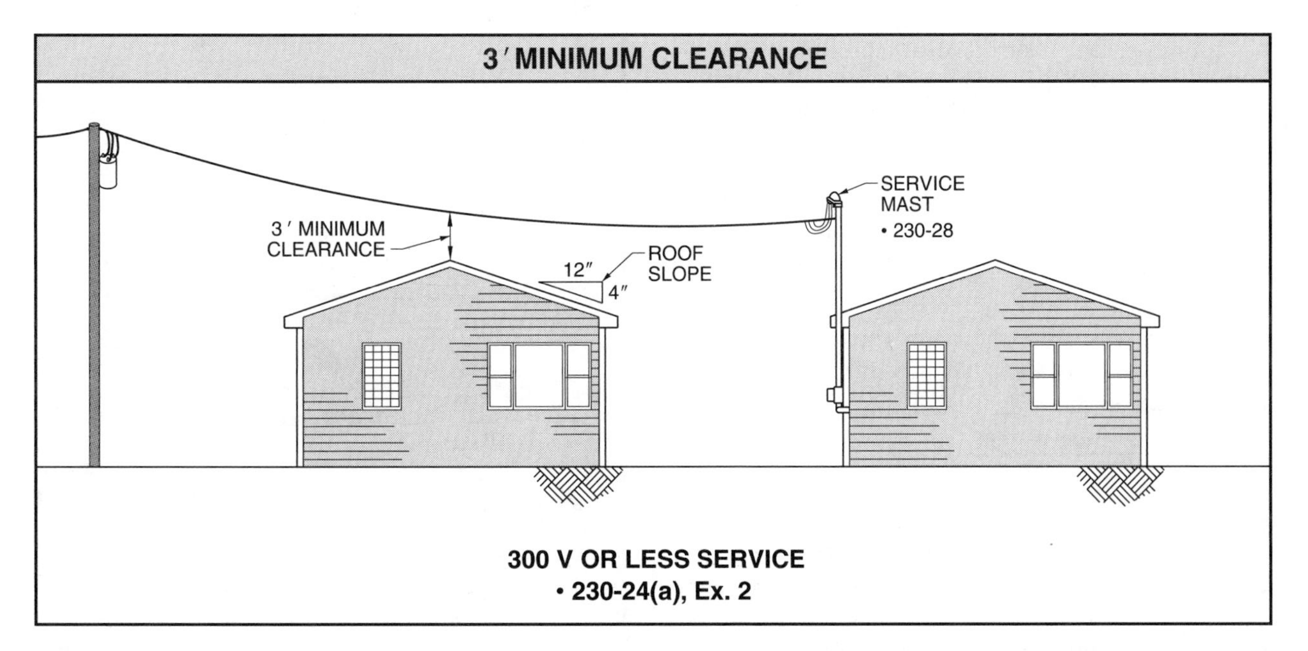

Figure 9-15. The minimum clearance above rooftops is 3′ provided the roof slope is not less than 4″ in 12″ and the voltage does not exceed 300 V.

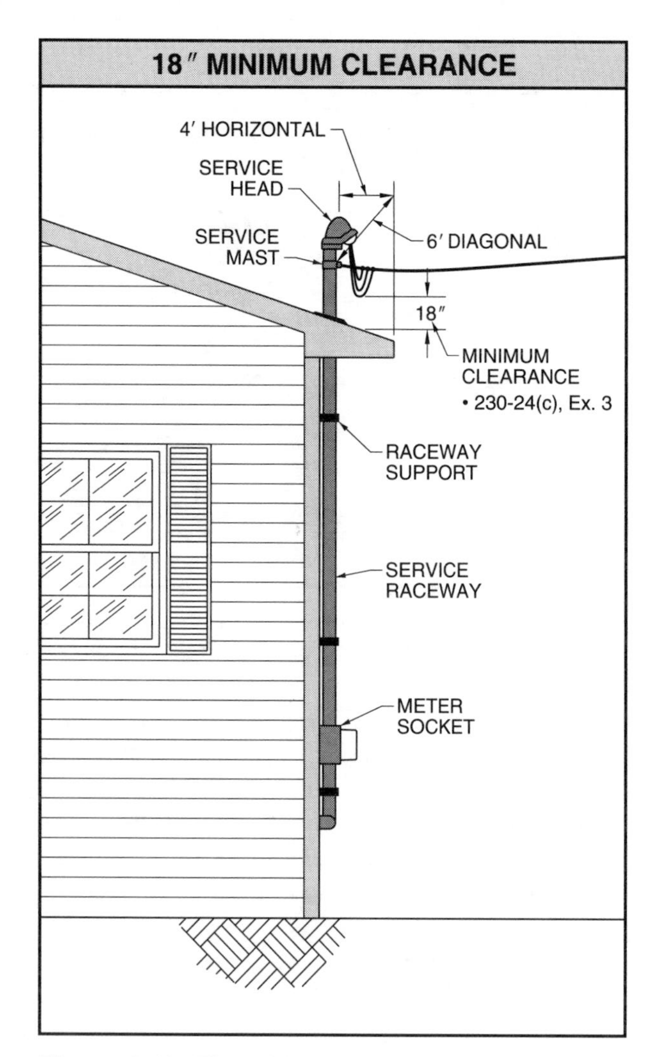

Figure 9-16. The minimum clearance for 300 V or less service-drop conductors terminating in a through-the-roof raceway is 18″ if not more than 6′ of service-drop conductors passes 4′ or less horizontally above the overhang.

Per 230-24(a), Ex. 4, the 3′ vertical clearance from the edge of the roof does not apply to the final span of service-drop conductors when the conductors are attached to the side of the building. This exception is necessary for installations in which the service-drop conductors attach to the side of the building at an angle which does not permit the 3′ clearance in all directions to be maintained. See Figure 9-17.

The second type of clearance that shall be maintained for service-drop conductors is clearance from ground. The requirements of 230-24(b) establish four different clearances based on the voltage of the service-drop conductors and the type of traffic area over which the conductors are installed. See Figure 9-18.

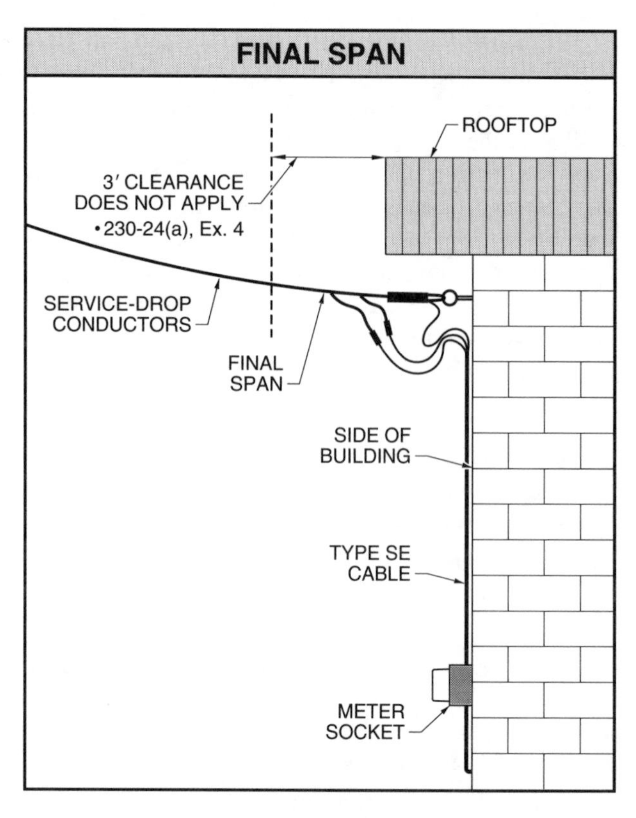

Figure 9-17. The 3′ clearance does not apply to the final span of service-drop conductors attached to the side of a building.

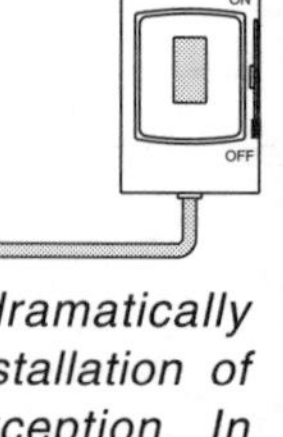

INSTALLING SERVICE-DROP CONDUCTORS

Service requirements can differ dramatically from one utility to another. The installation of service-drop conductors is no exception. In most areas, the local utility company has the responsibility to determine the location of and install service-drop conductors, although this is not always the case. Always contact the local utility company early in the design stage of a job to determine who is responsible for installing service-drop conductors.

Electrical design and installation personnel are usually required to submit an application for an electrical service. The service application asks questions about the type of service required and the type of load to be connected. After receiving the application, the local utility company determines the type of service and the location of the service-drop to the building or structure.

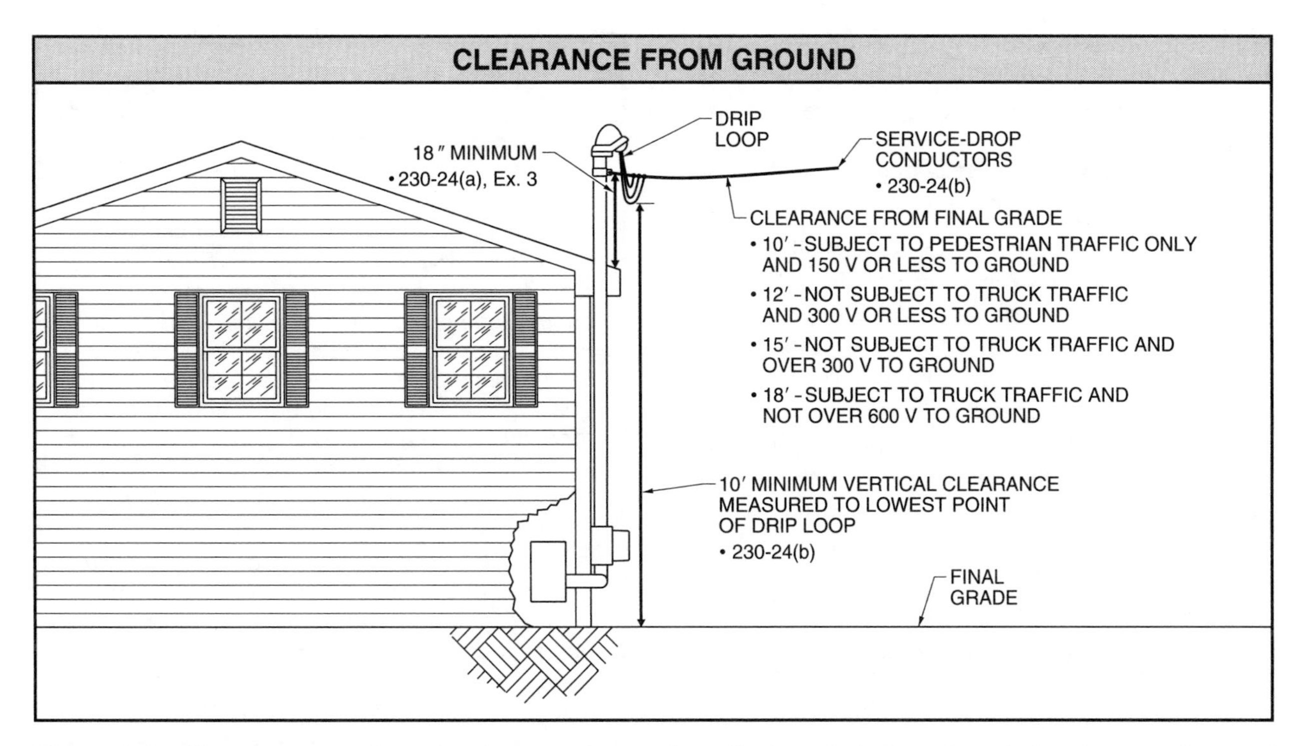

Figure 9-18. The clearance of service-drop conductors from final grade is based on their voltage and type of traffic.

As a general rule, there shall be a 10′ minimum clearance from final grade for all service-drop conductors. The measurement shall be taken from the lowest point of the drip loop of the service-entrance conductors. The 10′ clearance applies to systems with a grounded bare messenger wire where the voltage to ground does not exceed 150 V and the conductors extend over sidewalks and areas accessible to pedestrians only.

Service-drop conductors are required to be insulated or covered.

When the conductors extend over residential properties and driveways or commercial areas not subject to truck traffic and the voltage to ground does not exceed 300 V, the minimum clearance from ground shall be 12′. If a similar installation to that of the 12′ condition exists, but the system voltage exceeds 300 V to ground, the clearance from ground shall be increased to 15′. For those installations in which the service-drop conductors extend over public streets, alleys, roads, parking lots subject to truck traffic, etc., the minimum clearance shall be increased to 18′.

Service Mast – 230-28

Overhead service conductors can terminate at the building or structure in several different ways. Often, eyebolts or rafter attachments are used to attach the service-drop conductors. In some installations however, a through-the-roof assembly is used. In these installations, the service-drop conductors are attached directly to a service mast.

A *service mast* is an assembly consisting of a service raceway, guy wires or braces, service head, and any fittings necessary for the support of service-drop conductors. See Figure 9-19. Service masts are designed to support the weight of the service-drop conductors under varying conditions including those when snow and ice may add additional weight to the conductor span.

Fittings for use with service masts shall be identified for the use. Designers and installers of service masts should check with the local utility company to see if they have additional requirements for the installation of service masts. Finally, service masts shall be used only for the termination and support of service-drop conductors. Other system conductors, such as CATV or communication systems, are not permitted to be attached to service masts.

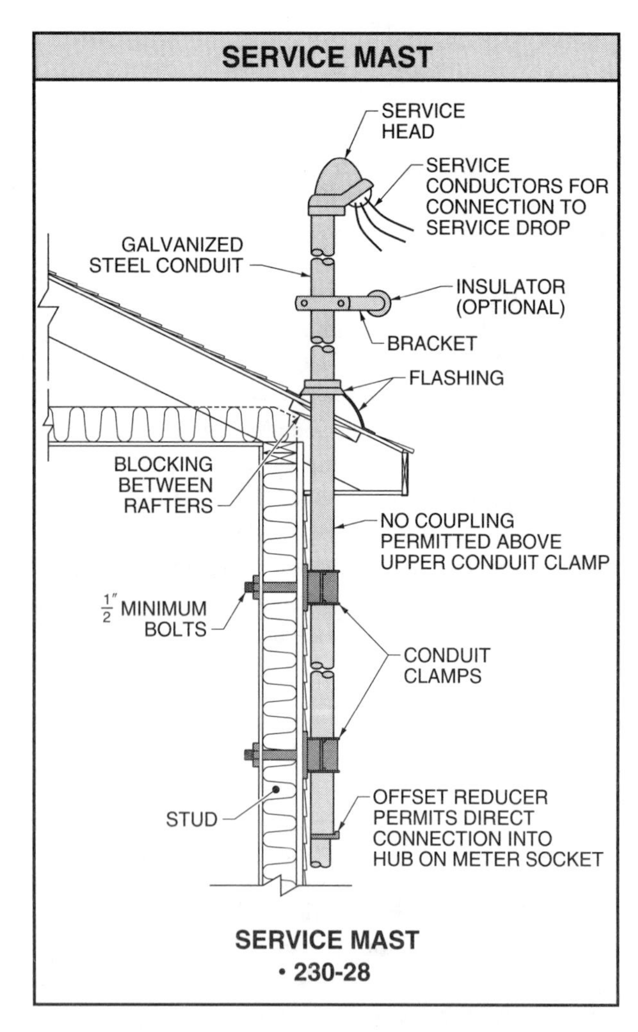

Figure 9-19. A service mast is an assembly consisting of a service raceway, guy wires or braces, service head, and any fittings necessary for the support of service-drop conductors.

SERVICE MASTS

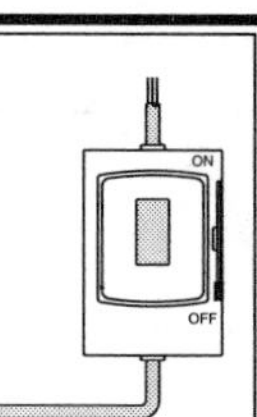

A 1993 NEC® proposal requested that permission to use service masts for other system conductors, such as CATV and communications, be permitted provided the service mast was designed to handle the additional load. The CMP rejected this proposal and generated their own proposal resulting in a FPN in the 1993 NEC® which prohibited these other system conductors from being connected to power service masts.

The FPN was incorporated into the actual text of 230-28 for the 1996 NEC®, leaving no doubt that only power conductors can be attached to the service masts. This does not, however, prohibit separate masts from being installed to handle these other types of systems.

UNDERGROUND SERVICE-LATERAL CONDUCTORS – 230, PART C

In general, service-lateral conductors shall be insulated per 230-30. Service-lateral conductors shall be suitable for and capable of withstanding any atmospheric conditions, without insulation degradation which could lead to current leakage. All service conductors, for this reason, shall be insulated for the applied system voltage. Covered conductors are not permitted. Bare conductors are permitted per 230-30, Ex., provided (a) they are installed in a raceway; (b) judged suitable for direct burial; (c) part of a cable assembly which is identified for use underground; and (d) aluminum or copper-clad aluminum and part of a cable assembly which is identified for use underground either in a raceway or for direct burial.

Like service-drop conductors, service-lateral conductors shall be protected against physical damage. Instead of accomplishing this by maintaining clearances, 230-32 requires that the service-lateral conductors be protected against damage just as any other underground conductors are per 300-5. Per 300-5(d), direct-buried conductors emerging from the ground shall be protected against physical damage from the minimum cover distance required by 300-5(a), up to a point which is at least 8′ above finished grade.

Service-entrance conductors are required to be protected against physical damage by use of service raceway, listed in 230-43, or by encasement in concrete per 230-6.

Size and Rating – 230-31

Service-lateral conductors shall be sized to handle the computed load per 230-31. The load shall be computed per 220. In addition, the service-lateral conductors shall have adequate mechanical strength for the application. To ensure that the conductors have adequate mechanical strength, 230-31(b) requires that the minimum size for service-lateral conductors be No. 8 Cu or No. 6 Al or copper-clad aluminum. Per 230-31(b), Ex., the minimum size conductor is permitted to be reduced to No. 12 Cu or No. 10 Al or copper-clad aluminum when the load served consists of a single branch circuit only. See Figure 9-20.

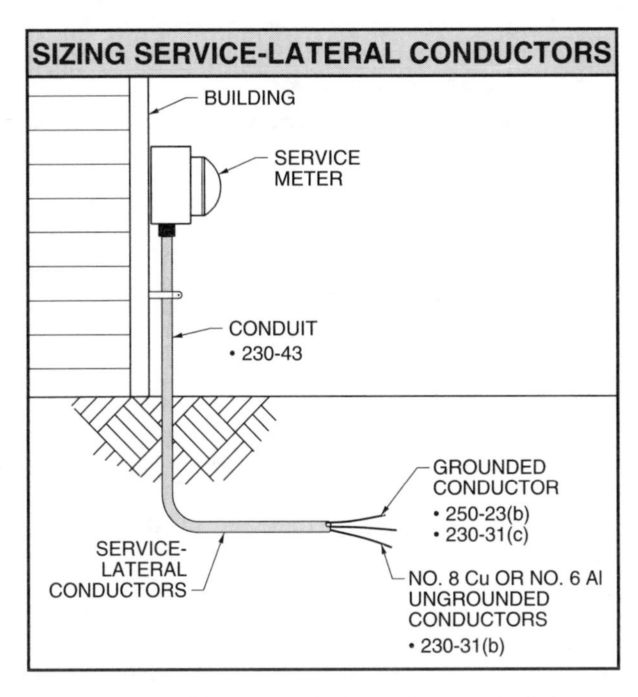

Figure 9-20. The grounded conductor shall be no smaller than that required by 250-23(b).

The grounded conductor is permitted by 230-31(c) to be sized per 250-23(b). Essentially, the grounded conductor is sized per Table 250-94, based on the size of the largest service-drop conductor. In no case shall the grounded conductor be smaller than that required by 250-23(b).

SERVICE-ENTRANCE CONDUCTORS – 230, PART D

Unlike service-drop and service-lateral conductors, service-entrance conductors are almost always the responsibility of the electrical designer and installer. The utility service and responsibility usually ends at the termination or connection to the service-drop or service-lateral conductors. Part D of 230 contains the requirements for installing service-entrance conductors. An important consideration for any electrical system designer is determining the size and rating of the service-entrance conductors, whether they are supplied from a service drop or a service lateral. For the installer of the electrical system, an important consideration is the selection of the wiring method for the service-entrance conductors. Requirements for both of these are contained in 230, Part D.

Size and Rating – 230-42

The general rule for sizing service-entrance conductors is that the conductors shall have an ampacity per 310-15 which is suitable for the computed load. The load for the service-entrance conductors shall be calculated per 220. The minimum sizes for service-entrance conductors is based upon one of three conditions. See Figure 9-21.

Per 230-42(b)(1), a 100 A service is required for any one-family dwelling that has six or more 2-wire branch circuits. Per 230-42(b)(2), a 100 A service is required for any one-family dwelling with an initial net computed load of 10 kVA or more. Per 230-42(b)(3), a 60 A service is permitted for all other types of loads. These ungrounded service-entrance conductors are minimum ampacities. Service calculations, as per 220, may result in ungrounded conductors much larger than these.

There are also some limited installations where the size of the ungrounded conductors may be smaller. Per 230-42(b), Ex. 1, a minimum size of No. 8 Cu or No. 6 Al or copper-clad aluminum is permitted for ungrounded conductors that supply loads consisting of not more than two 2-wire branch circuits. Per 230-42(b), Ex. 2, No. 8 Cu or No. Al or copper-clad aluminum ungrounded conductors are permitted to be used for service-entrance conductors by special permission for loads limited by demand or by the source of supply.

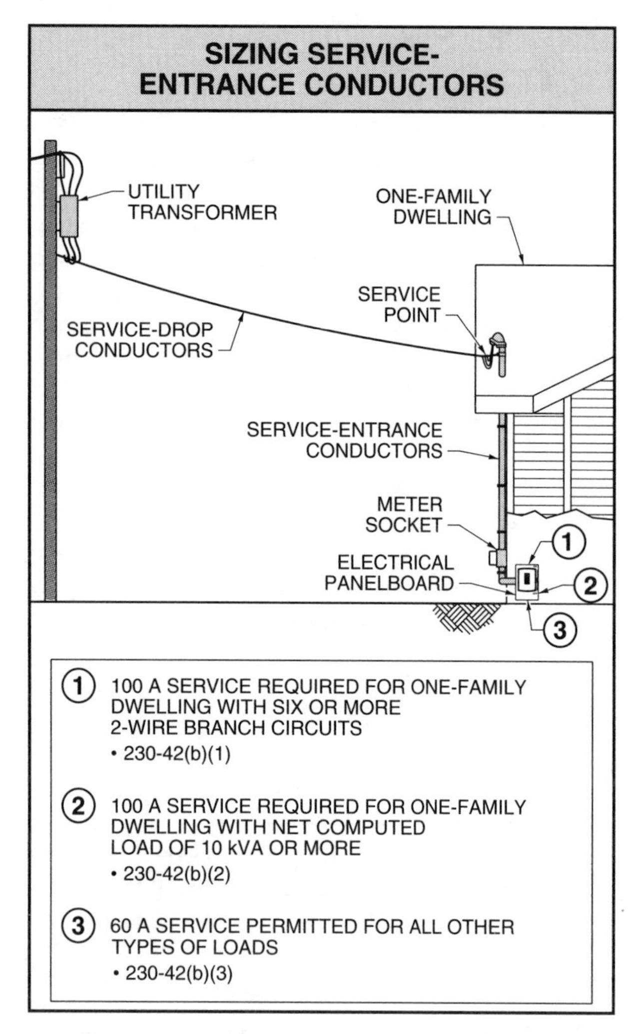

Figure 9-21. The minimum size for service-entrance conductors is based upon one of three conditions.

Per 230-42(b), Ex. 3, smaller ungrounded conductor sizes are permitted for limited loads. In this case, the load shall not exceed a single branch circuit. If the load is only a single branch circuit, the minimum size of the ungrounded conductor is permitted to be reduced to No. 12 Cu or No. 10 Al. In no case, however, shall the minimum size of the ungrounded conductors be smaller than the branch circuit conductors.

The grounded conductor is permitted by 230-42(c) to be sized per 250-23(b). Essentially, the grounded conductor is sized per Table 250-94, based on the size of the largest service-drop conductor. In no case shall the grounded conductor be smaller than that required by 250-23(b).

UNGROUNDED SERVICE CONDUCTORS

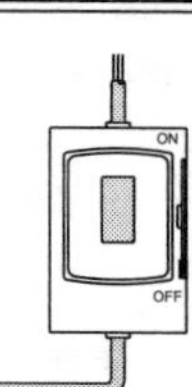

The question of the minimum size necessary for ungrounded service conductors is often used for journeyman wireman and electrical contractors examinations. Three specific conditions in 230-42 require a specific service conductor size and three exceptions permit smaller conductor sizes.

For most one-family dwellings, 100 A is the minimum size required. That is not to say, however, that a small dwelling with less than six branch-circuits could not be designed. A 60 A service, while not common, is still permitted if the conditions of 230-42(a)(1)(2) do not exist. Likewise, the exceptions cover specialized applications with limited loads. Be sure to check this section very carefully when answering any questions concerning the minimum size of ungrounded service conductors.

Sets of Service-Entrance Conductors – 230-40

Just as 230-2 limits the maximum number of services for a building or structure to one, the general rule of 230-40 limits the maximum number of sets of service-entrance conductors that can be connected to a single service drop or service lateral to one. There may be, however, installations where it is desirable to supply additional sets of service-entrance conductors from a single service drop or lateral. See Figure 9-22.

Multi-occupancy buildings for example, are permitted by 230-40, Ex. 1 to have each occupancy or each group of occupancies supplied by a separate set of service-entrance conductors. Per 230-40, Ex. 2, a separate set of service-entrance conductors is permitted to be run to each of the two to six service disconnects that may be used, provided the service disconnects are in separate enclosures and are grouped at one location. Per 230-40, Ex. 3, an additional set of service-entrance conductors is permitted to be connected to a single set of service-drop or service-lateral conductors when used to supply a one-family dwelling and a separate structure.

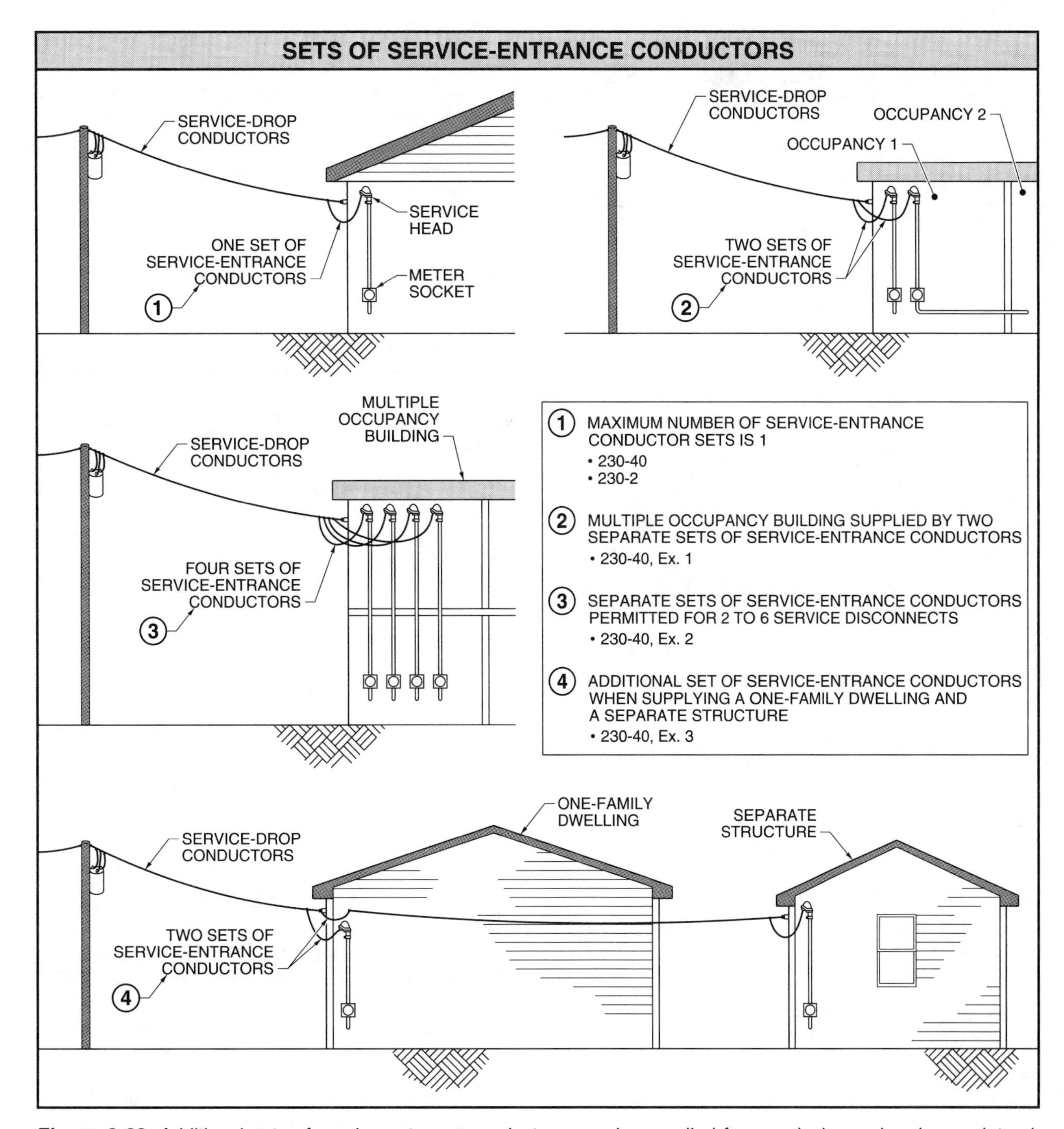

Figure 9-22. Additional sets of service-entrance conductors may be supplied from a single service drop or lateral.

Article 230-41 contains specific requirements for the insulation of service-entrance conductors. See Figure 9-23. Per 230-41, all ungrounded conductors which are used as service-entrance conductors shall be insulated. The service-entrance conductors shall be suitable for the atmosphere in which they are installed. Service-entrance conductors are not permitted to be covered conductors.

Bare, grounded conductors, however, are permitted if they are (a) copper and installed in a raceway or part of a service cable assembly; (b) copper and judged suitable for direct burial; (c) copper and part of a cable assembly which is identified for use underground; or (d) aluminum or copper-clad aluminum and part of a cable assembly which is identified for use underground either in a raceway or for direct burial.

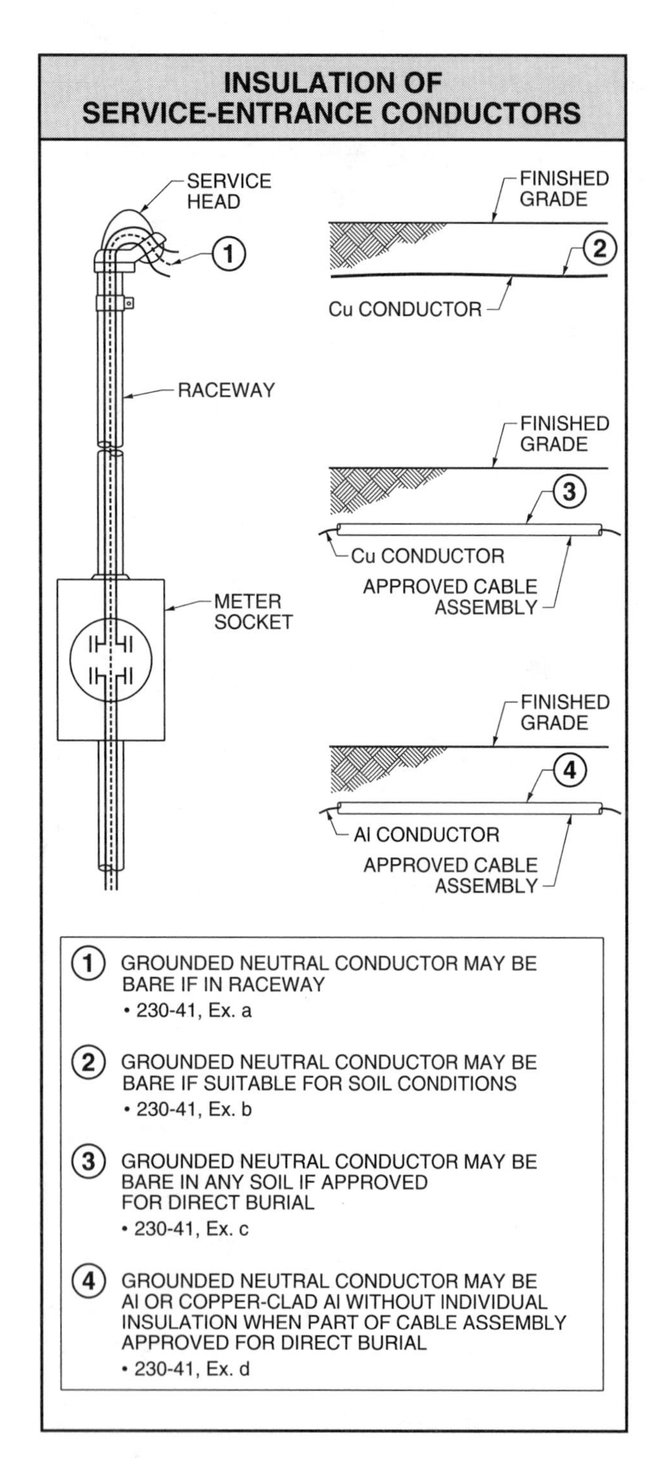

Figure 9-23. Bare, grounded conductors are permitted for service-entrance conductors with conditions.

Wiring Methods – 230-43

Sixteen listed wiring methods are permitted for use as service-entrance conductors in 230-43. See Figure 9-24. Wiring methods that do not appear in 230-43, such as Type NM cable, are not permitted to be used for service-entrance conductors. These 16 wiring methods consist of various types of raceways, cable assemblies, open wiring, and approved cable tray systems. No matter what wiring method is selected, the service-entrance conductors shall be installed per the specific NEC® requirements for that particular wiring method. For example, if Type MC cable is selected as the wiring method, the service-entrance conductors shall be installed according to all of the provisions of 334, Metal-Clad Cable.

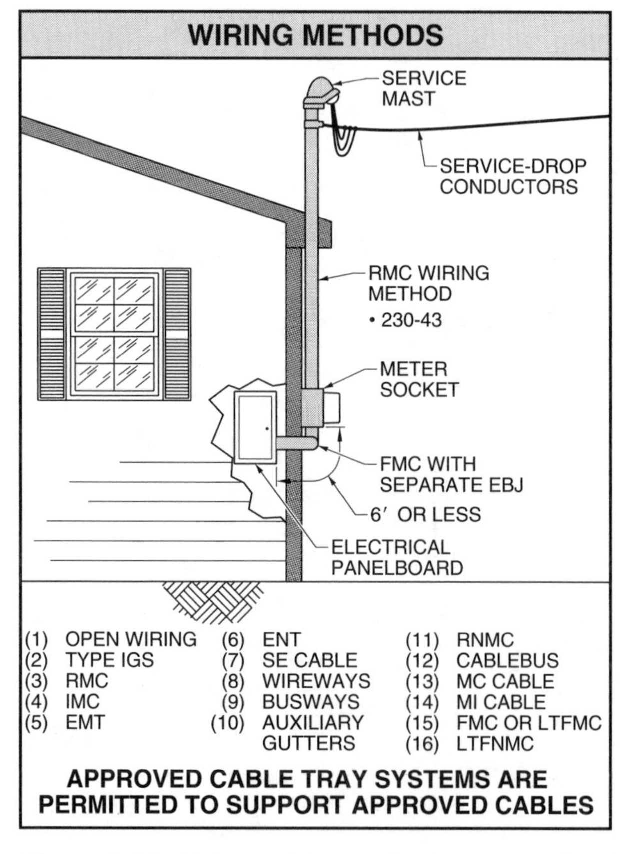

Figure 9-24. Sixteen wiring methods are permitted for use as service-entrance conductors.

Some of the wiring methods have additional limitations when they are used for service-entrance conductors. For example, FMC and LTFMC are permitted to be used for service-entrance conductors, but only in lengths not exceeding 6′. In addition, FMC and LTFMC shall be installed with an EBJ per 250-79.

Protection – 230-50

Type SE, service-entrance cable, is the most commonly used wiring method for 120/240 V, 1ϕ, 3-wire services in one-family and multifamily dwellings. The primary reasons for this are the ease of installation and the relatively low cost of the wiring method. However, there may be local codes and standards or local utility requirements which require all service-entrance conductors to be installed in metal conduits.

Type SE cable is a factory-assembled cable assembly, consisting of one or more conductors, with a flame-retardant, moisture-resistant covering used primarily for services. There are other types of SE cables available, such as those without an overall covering and those for use underground, but Type SE cable is by far the most prevalent. See 338 for other requirements for installing service-entrance cables.

Type SE cable is constructed with a flame-retardant and moisture-resistant covering to protect the conductors within the cable assembly. This covering, however, does not provide a great deal of protection against physical damage. All service-entrance conductors are required to be protected against physical damage per 230-50(a). Specific protection is provided by RMC, IMC, RNMC, EMT, or other approved means. See Figure 9-25.

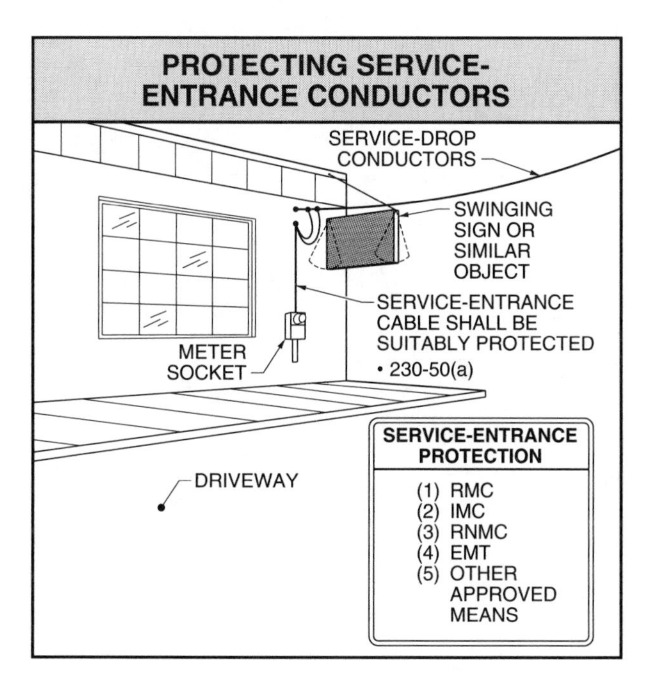

Figure 9-25. Service-entrance cables shall be suitably protected.

When service-entrance cables, such as Type SE cable, are installed in locations subject to physical damage they shall be suitably protected. Suitable protection can be accomplished by use of RMC, IMC, RNMC, EMT, or other approved means. Per 230-50(a), installations near driveways, walkways, sidewalks, or coal chutes require protection as well as installations where the service-entrance cables are subject to contact with awnings, swinging signs, shutters, and similar objects. Per 230-50(b), open wiring and cables other than service-entrance cables shall not be installed within 10′ of grade level or in locations where they would be subject to physical damage.

TYPE SE CABLE

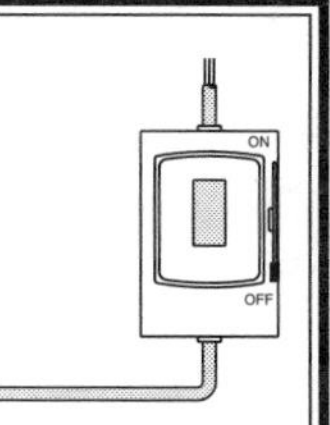

The requirements of 230-50(a) are difficult to apply based on the language of the text. The provision that Type SE cable be protected when it is subject to physical damage, such as near sidewalks, driveways, etc., is vague. Near is not a clearly defined term.

Several areas may not permit Type SE cable for service-entrance conductors, but in those areas where Type SE cable is permitted, electrical installers and designers of electrical systems should check with the AHJ before selecting a wiring method for service-entrance conductors subject to these types of conditions.

Supports – 230-51

Specific support requirements shall be followed when cable assemblies or open wiring are used for service-entrance conductors. Often, these wiring methods are subject to severe weather conditions which may cause damage to the electrical components and the threat of a power outage if the system is not securely fastened to and supported by the building or structure to which it is mounted. Per 230-51(a), service-entrance cables shall be supported at intervals not exceeding 30″. In addition, the cables shall be supported within 12″ of every service head, gooseneck, or point of connection to a raceway or enclosure. The means of support shall be by a cable strap or other approved means. See Figure 9-26.

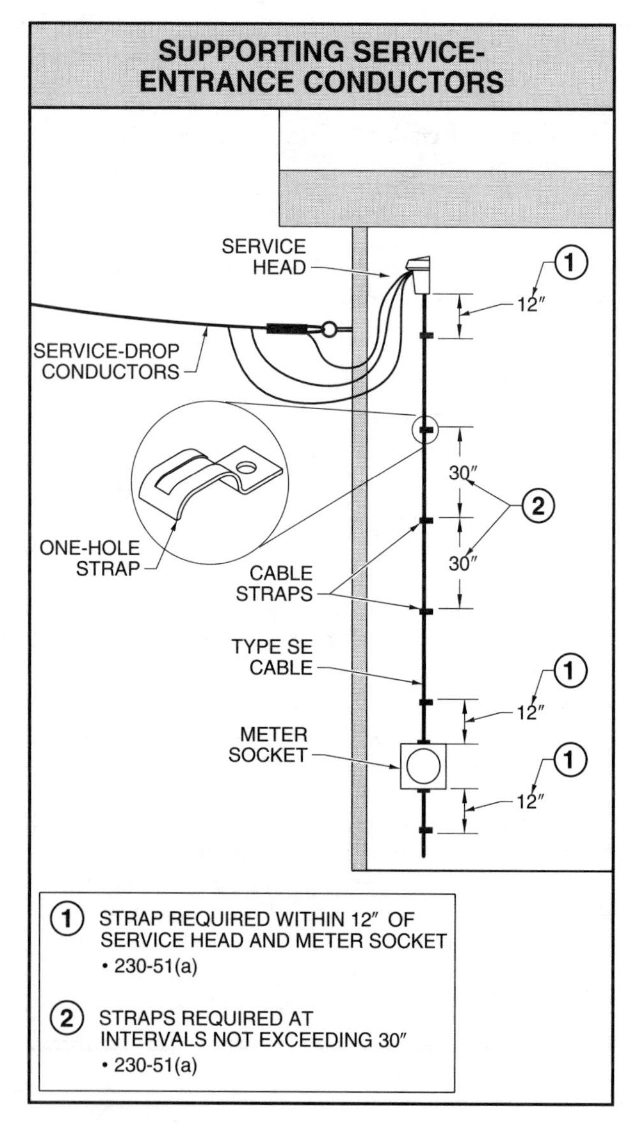

Figure 9-26. Service-entrance cables shall be supported by cable straps or other approved means.

When other types of cables that are not designed for mounting in direct contact with the building are used, 230-51(b) requires that the cables shall be mounted on insulating brackets installed at intervals which do not exceed 15′. The cables shall be installed in a manner which maintains a 2″ minimum clearance from the surface over which they pass.

If the service-entrance conductors consist of individual open conductors, 230-51(c) requires that the conductors be installed per Table 230-51(c). This table establishes maximum support distances and minimum clearances between the conductors, and from the surfaces, depending on the maximum voltage of the conductors.

Overhead Service Locations – 230-54

The overhead service location shall be raintight. See Figure 9-27. To avoid the entrance of water and moisture, service raceways are required by 230-54(a) to be provided with a raintight service head. The raintight service head shall be installed at the point of connection to the service-drop conductors.

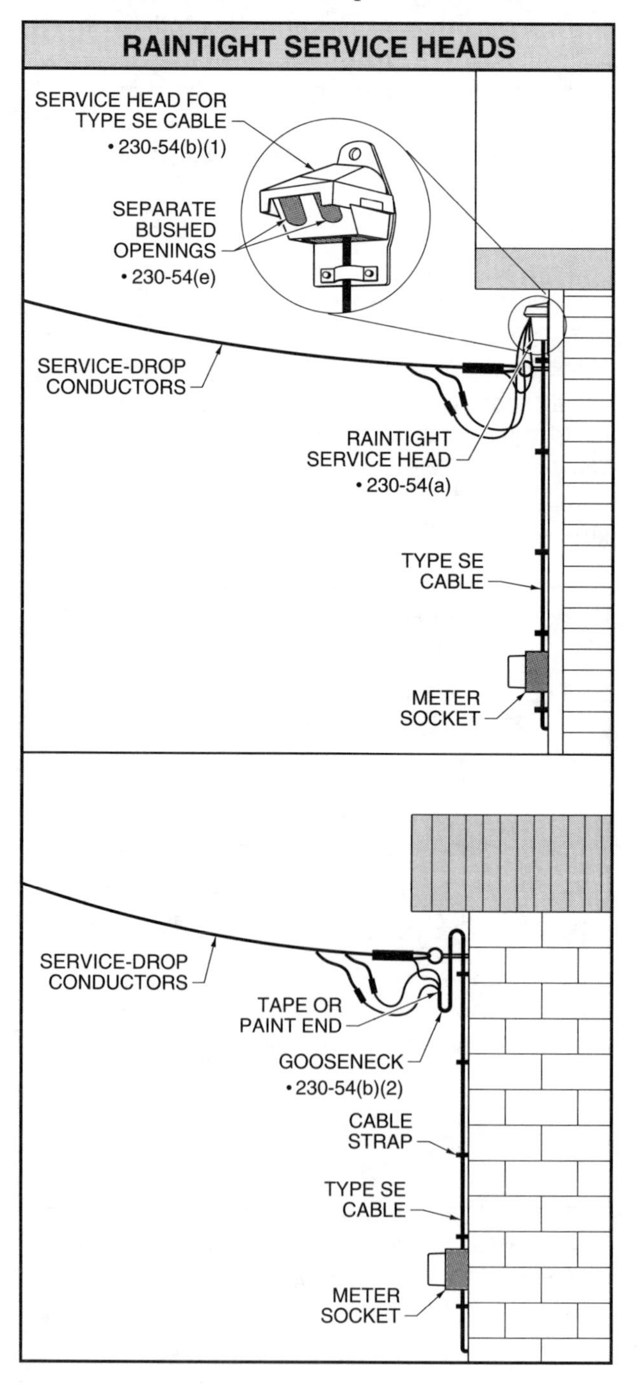

Figure 9-27. The overhead service location shall be raintight.

LOCATION OF SERVICE EQUIPMENT

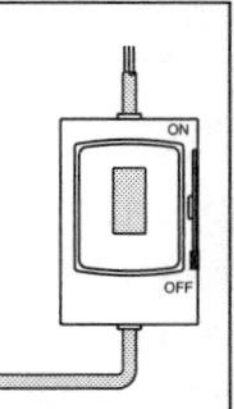

The local utility company determines the point of attachment of the service-drop conductors to the building or structure. Always check with the local utility prior to locating service equipment.

When service-entrance cables are used in lieu of service raceways, similar protection against the entrance of water and moisture shall be provided. Two options are permitted to provide protection from water and moisture for service-entrance cables. The first is to use a raintight service head per 230-54(b)(1). The service head consists of a cap and openings for the conductors. The second option is to form the service-entrance cable into a gooseneck and tape and paint the end, or tape the end with a self-sealing, weather-resistant thermoplastic.

Point of Attachment – 230-54(c). The service head or gooseneck shall be installed at a point which is above the point of attachment for the service-drop conductors. This ensures that water does not enter the service raceway or service-entrance cable. However, there may be installations where it is impracticable to locate the service head or gooseneck above the point of attachment. In these cases, 230-54(c), Ex. permits the point of attachment to be located above the service head or gooseneck. In no case, however, is the service head or gooseneck permitted to be located more than 24″ below the point of attachment. See Figure 9-28.

Drip Loops – 230-54(f). The final requirement to prevent the entrance of water into the service raceway or service-entrance cable is the formation of drip loops on all of the individual service-entrance conductors. In addition, 230-54(f) requires that the point of connection between the service-drop conductors and the service-entrance conductors be either below the level of the service head or below the level of the termination point of the cable sheath. While the NEC® does not contain any specific length requirement for the drip loops, installers of services should check the installation requirements for their particular electrical utility company. Many utility companies require a minimum of 3′ for the drip loop and for connection to the service-drop conductors.

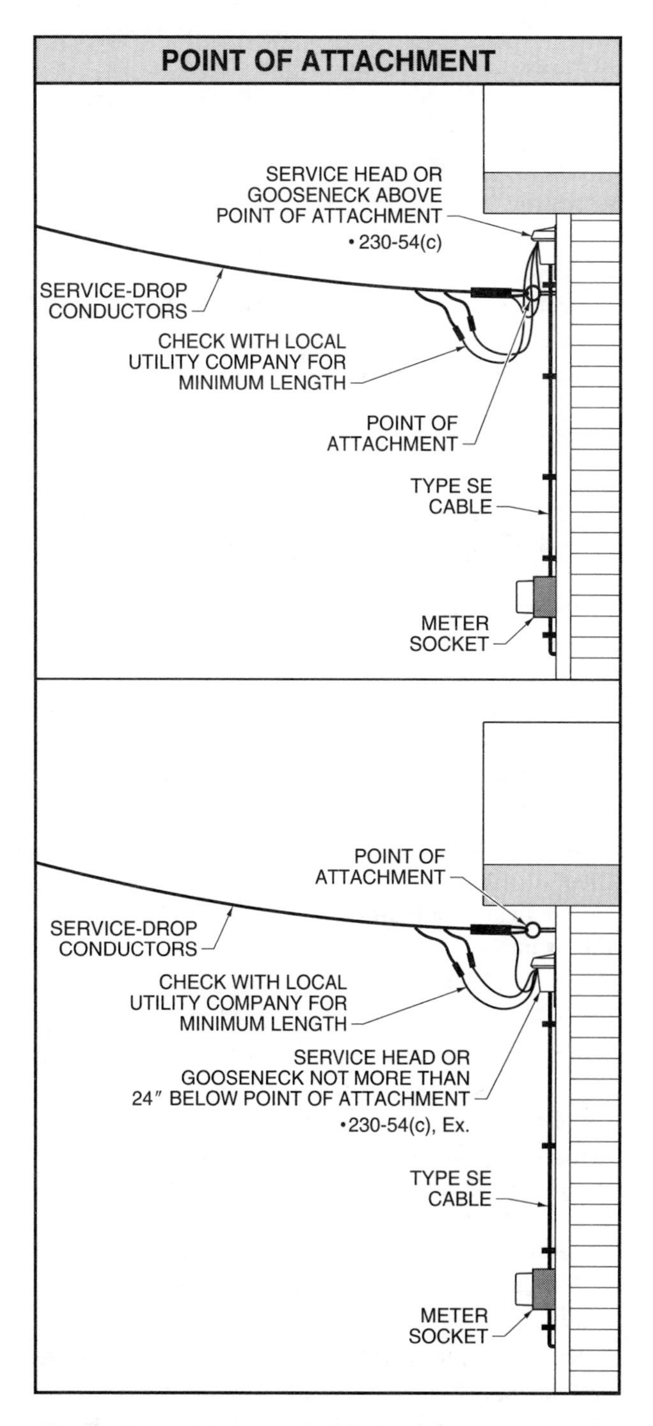

Figure 9-28. Where practicable, the service head or gooseneck shall be installed above the point of attachment of the service-drop conductors.

SERVICE EQUIPMENT – 230, PARTS E AND F

Designers and installers of electrical systems are required to ensure that the service equipment selected for the electrical system is suitable for the intended use. Article 100 defines service equipment as all of the necessary equipment for the control of electrical power to a building or a structure. Service equipment typically includes the main CB, or switch and fuses, and any accessories.

No matter what equipment is selected, 230-62 requires that the energized parts shall be either enclosed or guarded. Enclosing or guarding the energized parts protects them from accidental contact by personnel and from physical damage. Proper work clearances shall be provided about service equipment per 230-64. Work space clearances shall be in accordance with those specified in Section 110-16.

AIR Ratings – 230-65

The service equipment shall be suitable for the available system short-circuit current. Section 110-9 requires that all equipment intended to break current at fault levels shall have an ampere interrupting rating (AIR) sufficient for the available short-circuit available. Section 230-65 reiterates this requirement for service equipment.

Any service equipment selected for use in an electrical distribution system shall have an AIR sufficient for the available fault current at its supply terminals. For example, a 100 A main CB with an AIR of 16,000 A could not be installed in service equipment where the available fault current is 23,000 A.

The UL's *Panelboard Marking Guide* requires that each panelboard be marked with the phrase "Short-Circuit-Current-Rating" and the rating in RMS symmetrical amperes. Such markings list the maximum fault current to which the panelboard could be subjected and still withstand the magnetic forces generated by the fault current and ensure that the OCPDs are capable of clearing the circuit under fault conditions.

The available short-circuit current depends on several factors. Typically, as the size of the electrical system increases, the available fault current also increases. Conversely, the further away from the main power supply the equipment is located, the lower the available fault current. The effect of the requirement of 230-65 is that designers and installers of electrical services should contact their local utility company to determine how much available fault current is present at the location in which the service equipment is being installed.

SERVICE-ENTRANCE CONDUCTORS

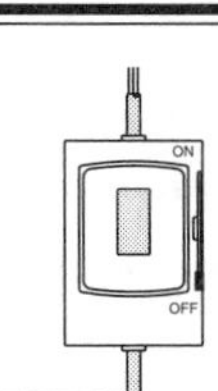

For several NEC® cycles, attempts have been made to place a limitation in 230-70(a) on the distance which service-entrance conductors can run in a building or structure before entering the service disconnecting means. CMP-4 has continually rejected such proposals on the basis that each installation presents unique requirements. For example, if a large oil heating tank is installed directly inside the point of entrance of the service conductors, enough of the service-entrance conductors could be installed to place the service disconnecting means at a readily accessible location. This does not, however, permit installers to run the service-entrance conductors to the other side of the basement, merely for convenience. If there are any doubts about the location of the service-disconnecting means, check with the AHJ before beginning the installation.

Identification – 230-66

In addition to the AIR rating, installers and designers of electrical distribution systems shall ensure that all service equipment is identified as being suitable for use as service equipment. All service equipment rated at 600 V or less shall be marked to show such suitability per 230-66. Marking is typically in the form of a label which is attached to the equipment.

Some equipment may also be marked to indicate that the equipment is "suitable only for use as service equipment." Equipment with this type of marking usually has a factory connection or bonding between the neutral terminal bar and the enclosure. This equipment is not suitable for use in sub-panel applications. Section 230-66 also states that individual meter socket enclosures are not considered service equipment and therefore need not be marked as suitable for service equipment.

Disconnecting Means – 230-70

A means shall be provided to disconnect all service-entrance conductors from all of the other conductors in the building. This isolates the electrical power from the building in the event of a fire or life-safety hazard. The service disconnect is used to accomplish this objective.

Location – 230-70(a). Service disconnecting means are not permitted to be installed in bathrooms. In addition, the service disconnecting means shall be installed in a readily accessible location. Article 100 defines a readily accessible location as one that is not obstructed and is therefore capable of being reached quickly, without requiring the use of ladders or other portable means.

The service disconnecting means is permitted to be installed either outside or inside of the building or structure. When the service disconnecting means is installed inside of the building or structure, it shall be installed nearest the point of entrance of the service conductors. There is no maximum distance established that the unprotected service conductors are permitted inside the building or structure. The location shall be at the first readily accessible point nearest the entrance of the service conductors.

Marking – 230-70(b). The service disconnecting means shall be clearly and permanently marked to distinguish its purpose. As with other identification requirements in the NEC®, such as 110-22, this is designed to aid any personnel who might be working on the service equipment or emergency operations personnel who may need to quickly identify the service disconnecting means and disconnect power from the building.

Suitable for Use – 230-70(c). Service equipment shall be suitable for the location in which it is installed. For example, if the service disconnect is located outside the building, as permitted by 230-70(a), then it shall be suitable or identified for wet locations. Likewise, if service equipment is installed in a hazardous location, then it shall be suitable for the particular Class and Division for which it is installed.

Maximum Number of Disconnects – 230-71. The service disconnecting means shall consist of not more than six switches or CBs. The six devices are permitted to be mounted in a single enclosure, installed in separate enclosures provided they are grouped, or installed in or on a switchboard. The two to six service disconnects are required to be grouped and clearly marked to indicate the load which they serve per 230-72(a).

The purpose of limiting the maximum number of service disconnects and requiring that they be grouped together is to ensure that the entire service, or services, as permitted by 230-2, can be shut down at a single location with no more than six operations of the hand. Permission to install the two to six service disconnects gives latitude to the electrical designer, yet provides for the quick interruption of power in the event of an emergency. See Figure 9-29.

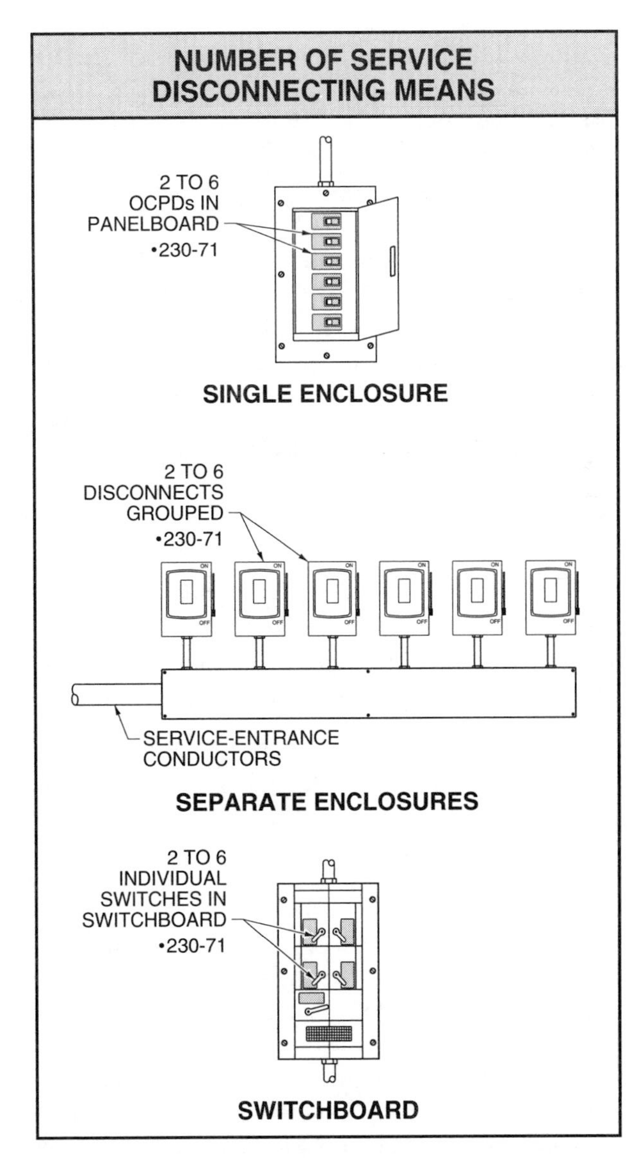

Figure 9-29. The service disconnecting means shall consist of not more than six switches or CBs.

Grouping – 230-72. The two to six disconnects shall be grouped per 230-72(a). Each disconnect shall be marked to indicate its load. Grouping facilitates installation and maintenance of the disconnects in addition to providing a central location to turn the disconnects OFF.

Section 230-72(b) permits the one or more service disconnects installed for a fire pump, per 230-2, Ex. 1, or for emergency systems per 230-2, Ex. 2, to be installed at a location which is sufficiently remote from the other service disconnecting means. This helps to ensure that the emergency power or fire pump service is not inadvertently disconnected by emergency personnel trying to interrupt normal power to the building.

Rating – 230-79. As a general rule, the service disconnecting means is required to have a rating which is equal to or greater than the load to be carried. The service load shall be calculated per 220. Subparts (a), (b), (c), and (d) to 230-79 establish minimum service disconnecting ratings.

For limited service loads that consist of a single branch circuit, 230-79(a) requires a minimum service disconnect rating of 15 A. For service loads that consist of not more than two branch circuits, 230-79(b) requires a minimum service disconnect rating of 30 A. For service loads to a one-family dwelling with either an initial connected load of 10 kVA or more, or six or more 2-wire branch circuits, 230-79(c) requires a minimum service disconnect rating of 100 A.

For all other service loads, 230-79(d) requires a minimum service disconnect rating of 60 A. If two to six disconnects are installed per 230-71, then 230-80 requires that the total of all the service disconnecting switches or CBs be equal to or greater than the ratings required by 230-79. For example, if the initial computed load of a one-family dwelling is 10 kVA, and two or more service disconnects are installed, the combined rating of the disconnects is required to be at least 100 A.

Line-Side Connections – 230-82. In general, there shall be no connections on the line or supply side of the service disconnecting means. Conductors connected to the line side of the service disconnect are not provided with any overcurrent protection and are not permitted for most installations. Often, however, line-side taps or connections are necessary for some applications. Section 230-82 lists eight specific applications where such connections are permitted.

The most common application on the line side of the service disconnecting means is for the connection of metering equipment. Meters are permitted to be connected to the line side of the service disconnecting means per 230-82, Ex. 3. The meters shall be rated above 600 V, nominal and the meter sockets or enclosures shall be grounded per 250.

Another application in which line-side taps are permitted is for cable limiters or other current-limiting devices. These devices are permitted to be connected to the line side of the service disconnecting means per 230-82, Ex. 1. Such devices are intended to protect downstream conductors and equipment from dangerous let-through currents that might result from short circuits or ground faults.

Taps on the line side are also commonly used to supply power for fire pumps. Article 695 contains the provisions under which this connection can be made and 230-82, Ex. 5 permits such a connection to be made.

Overcurrent Protection – 230-90

In general, service-entrance conductors shall be provided with overcurrent protection. Overcurrent protection for service equipment consists mainly of overload protection for the conductors. Service overcurrent protection, in most cases, does not provide protection against short circuits or ground faults that may occur on the line side of the service overcurrent protection. Section 230-90(a) specifies that each ungrounded conductor shall be provided with this protection by placing an OCPD in series with the conductor. The maximum rating or setting of the OCPD shall not exceed the allowable ampacity of the conductors.

There are particular installations where application of this rule can be difficult. Motors, for example, can pull up to six times their full-load running current at start-up. In these cases, the OCPD shall be sized to permit the motor to start. A rating higher than the allowable ampacity is permitted by 230-90(a), Ex. 1 when necessary to handle motor-starting currents. See Figure 9-30.

The requirement for overload protection does not include the grounded conductor. Opening the grounded conductor without opening the ungrounded conductors could result in dangerous voltages at the equipment. Section 230-90(b) prohibits OCPDs from being placed in series with a grounded conductor unless the device is a CB, which simultaneously opens all conductors of the circuit.

In order to protect the service equipment, 230-91(a) requires that the location of the service OCPD be within the service disconnecting means or immediately adjacent to it. In addition, each occupant of a multi-occupancy building is required by 230-91(b) to have access to the service overcurrent devices. Note that such access is not required where the building management has continuous building supervision. In these occupancies, the service OCPDs need only be accessible to authorized management personnel.

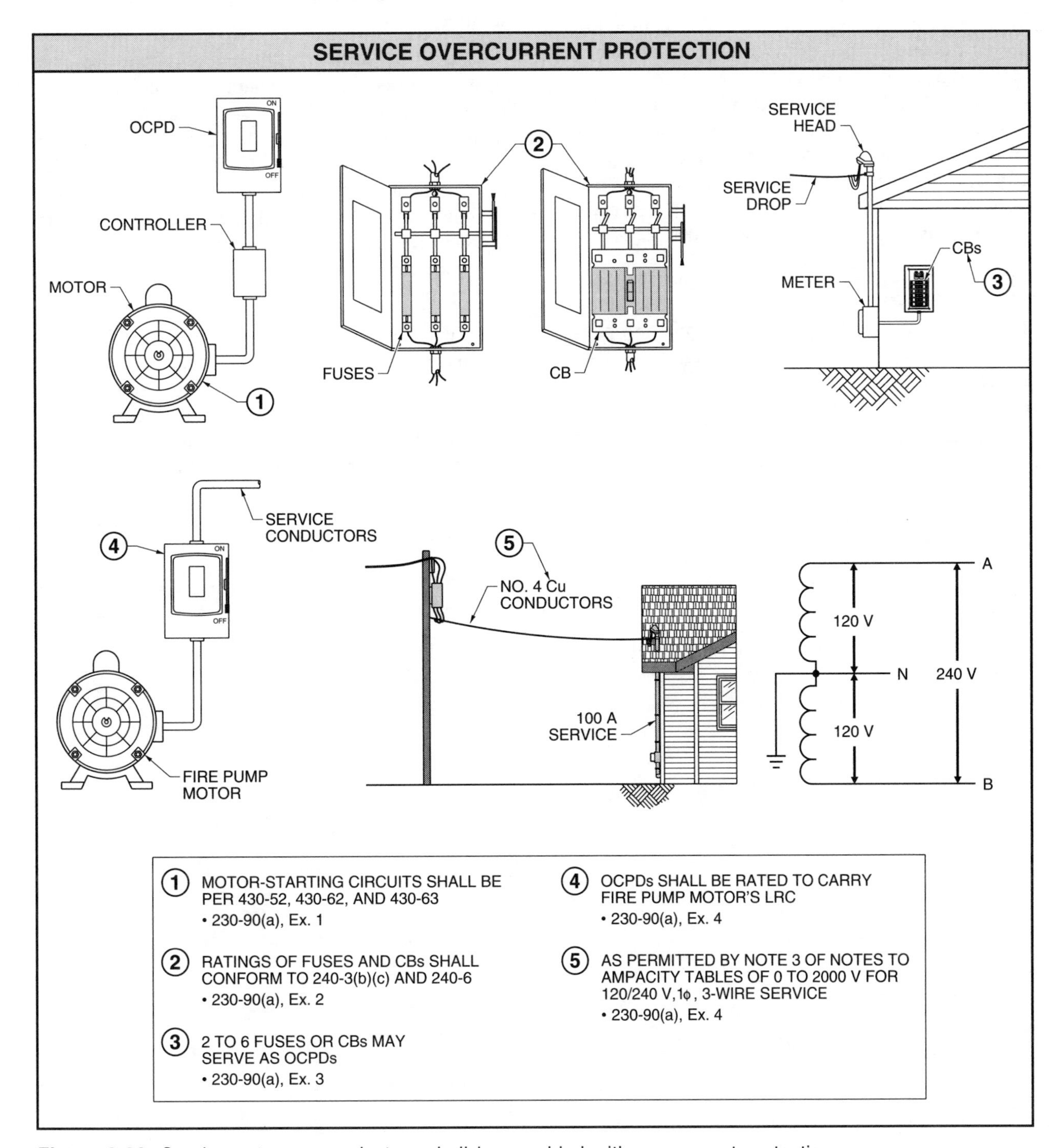

Figure 9-30. Service-entrance conductors shall be provided with overcurrent protection.

Ground-Fault Protection of Equipment (GFPE) – 230-95

There is another type of protection required by the NEC® which is sometimes confused with, but is in fact very different from, GFCI protection. GFCI protection provides protection for personnel by disconnecting the circuit in about 1⁄40 of a second in the event of a 5 mA discrepancy between the current flowing on the ungrounded conductor and the current returning on the grounded conductor. GFCI protection is most commonly found on 125 V, 15 A and 20 A receptacles. Ground-fault protection of equipment (GFPE), on the other hand, provides protection for equipment. GFPE is designed to operate at settings not greater than 1200 A and with a maximum delay of 1 second for fault currents equal to or greater than 3000 A.

GFPE protection is designed to ensure that arcing faults of wye-connected systems operating at over 150 V to ground, but less than 600 V phase-to-phase, do not occur on service disconnects rated 1000 A or more. Studies have shown that these systems, particularly the 277/480 V, 3ϕ, 4-wire, solidly-grounded, wye-connected systems have a tendency to permit arcing ground faults to occur which can severely damage the service equipment.

The GFPE requirements are tightly drawn to cover these types of systems when the service disconnect is rated at 1000 A or more. For example, if a 1000 A rated service disconnecting means is installed with an 800 A OCPD and conductors are rated at 800 A, GFPE is still required. It is the rating of the service disconnect, not the ampacity of the conductors or the rating or setting of the OCPD, that is the determining factor.

The setting of the GFPE shall allow the device to open the system in one second for ground faults equal to or greater than 3000 A per 230-95(a). The maximum setting permitted for the GFPE is 1200 A. Section 230-95(c) is one of the few instances in the NEC® where a specific performance test of a system is required. Upon completion of the GFPE system, 230-95(c) requires that a test of the system be done and a written record of the test be made and be available to the AHJ.

In the 1990 NEC®, the provisions for GFPE protection were extended to feeder disconnects as well as service disconnects because the same arcing ground faults can occur regardless of whether the circuit is a feeder or a service. Section 215-10 refers to 230-95 and incorporates all of the provisions for service disconnects to feeders.

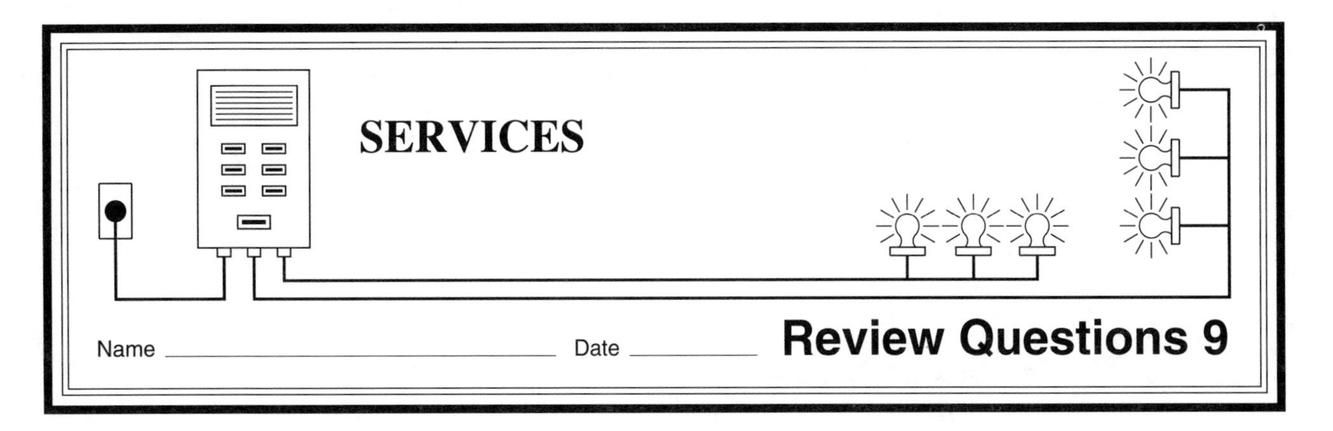

Review Questions

__________ **1.** The ________ is the electrical supply, in the form of conductors and equipment, that provides electrical power to the building or structure.

__________ **2.** The ________ is the conductors that extend from the overhead utility supply system to the service-entrance conductors at the building or structure.

__________ **3.** The ________ is the underground service conductors that connect the utility's electrical distribution system with the service-entrance conductors.

__________ **4.** The ________ is the point of connection between the local electrical utility company and the premises wiring of the building or structure.

T F **5.** Occupancies which utilize a fire pump for life safety protection are permitted to have a separate service provided solely for the fire pump.

T F **6.** Fittings for use with service masts shall be identified for the use.

T F **7.** GFPE is designed to operate at settings not greater than 200 A.

__________ **8.** ________ conductors are the conductors that extend from the service disconnecting means to the service point.

__________ **9.** ________ is the written approval of the AHJ.

__________ **10.** As a general rule, there shall be a(n) ________′ minimum clearance from final grade for all service-drop conductors.

T F **11.** As a general rule, only service conductors shall be installed in service raceways.

T F **12.** A panelboard is suitable for use as service equipment when not more than six main disconnecting means are provided.

__________ **13.** The service head or gooseneck shall not be located more than ________″ below the point of attachment.

A. 6
B. 12
C. 18
D. 24

________ **14.** Individual meter socket enclosures ________ service equipment.

A. are not considered
B. need not be marked as suitable for
C. both A and B
D. neither A nor B

________ **15.** The service disconnecting means shall consist of not more than ________ switches or CBs.

A. two
B. four
C. six
D. neither A, B, nor C

________ **16.** As a general rule, the service disconnecting means is required to have a rating which is ________ the load to be carried.

A. 75% of
B. equal to or less than
C. equal to or greater than
D. 125% of

T F **17.** Conductors located on the line or supply side of the service point are covered by the NEC®.

T F **18.** Service conductors are provided overcurrent protection by overcurrent protection devices.

T F **19.** Service-drop conductors are required to be insulated or covered.

________ **20.** The 8′ minimum vertical clearance for service-drop conductors shall be maintained in all directions for a minimum distance of ________′.

________ **21.** A(n) ________ is an assembly consisting of a service raceway, guy wires or braces, service head, and any fittings necessary for the support of service-drop conductors.

________ **22.** Direct-buried conductors emerging from the ground shall be protected from physical damage up to a point which is at least ________′ above finished grade.

________ **23.** A 100 A service is required for any one-family dwelling with an initial net computed load of ________ kVA or more.

________ **24.** Type ________ cable is the most commonly used wiring method for 120/240 V, 1φ, 3-wire services in one-family dwellings.

T F **25.** Service disconnecting means are permitted to be installed in bathrooms.

T F **26.** A minimum service disconnect rating of 30 A is required for service loads that consist of not more than two branch circuits.

________ **27.** ________ is all of the necessary equipment to control the supply of electrical power to a building or a structure.

________ **28.** Service conductors are permitted to be routed within a building if they are installed under not less than ________″ of concrete beneath the building or structure.

________ **29.** The minimum size for most service-drop conductors is No. ________.

A. 8 Cu
B. 6 Al
C. both A and B
D. neither A nor B

______ **30.** SE cable shall be supported at intervals not exceeding ______″ and within ______″ of every service head, gooseneck, or point of connection to a raceway or enclosure.

A. 12; 12
B. 12; 30
C. 30; 12
D. 30; 30

Clearance from Ground

______ **1.** The clearance from final grade at A is ______′.

A. 10
B. 12
C. 15
D. 18

______ **2.** The clearance from final grade at B is ______′.

A. 10
B. 12
C. 15
D. 18

______ **3.** The clearance from final grade at C is ______′.

A. 10
B. 12
C. 15
D. 18

______ **4.** The clearance from final grade at D is ______′.

A. 10
B. 12
C. 15
D. 18

______ **5.** The clearance from final grade at E is ______′.

A. 10
B. 12
C. 15
D. 18

______ **6.** The minimum clearance at F is ______″.

A. 10
B. 12
C. 15
D. 18

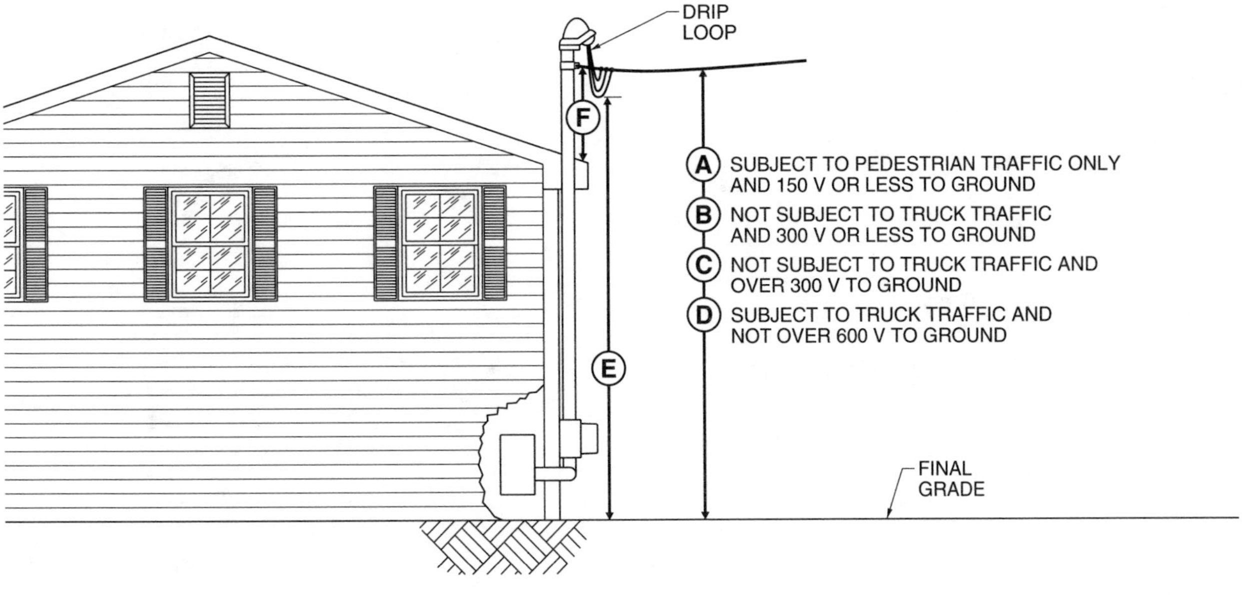

Connectors

_______ **1.** Compression

_______ **2.** Mechanical

_______ **3.** Quick-tap

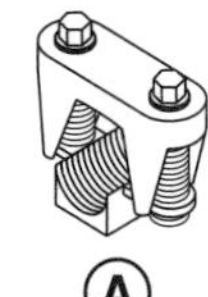
Ⓐ

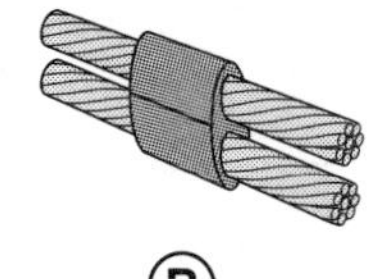
Ⓑ

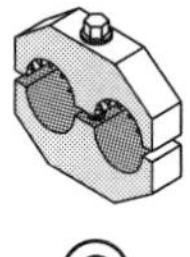
Ⓒ

Service Conductor Clearances

_______ **1.** The minimum clearance at A is ________.

A. no minimum C. 3′
B. 18′ D. 6′

_______ **2.** The minimum clearance at B is ________.

A. no minimum C. 3′
B. 18′ D. 6′

_______ **3.** The minimum clearance at C is ________.

A. no minimum C. 3′
B. 18′ D. 6′

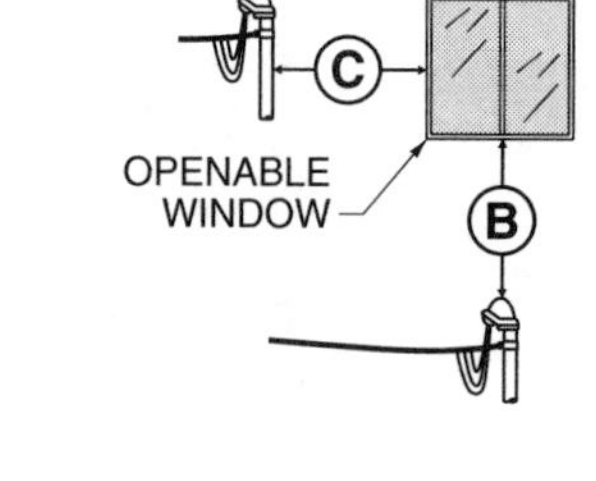

Service Equipment

_______ **1.** Loadcenter

_______ **2.** CB enclosure

_______ **3.** Panelboard

_______ **4.** Service disconnect switch

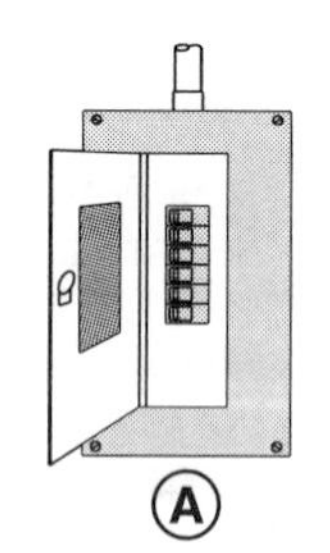
Ⓐ

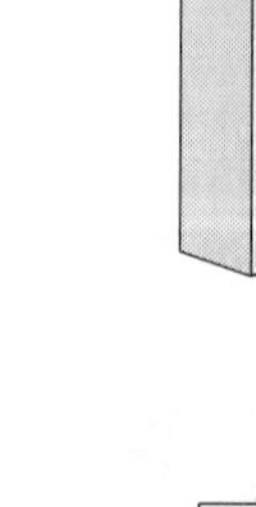
Ⓑ

Ⓒ

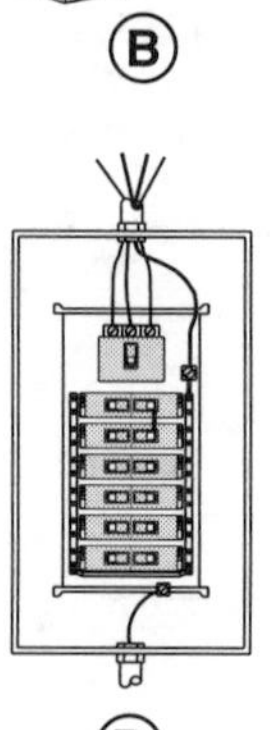
Ⓓ

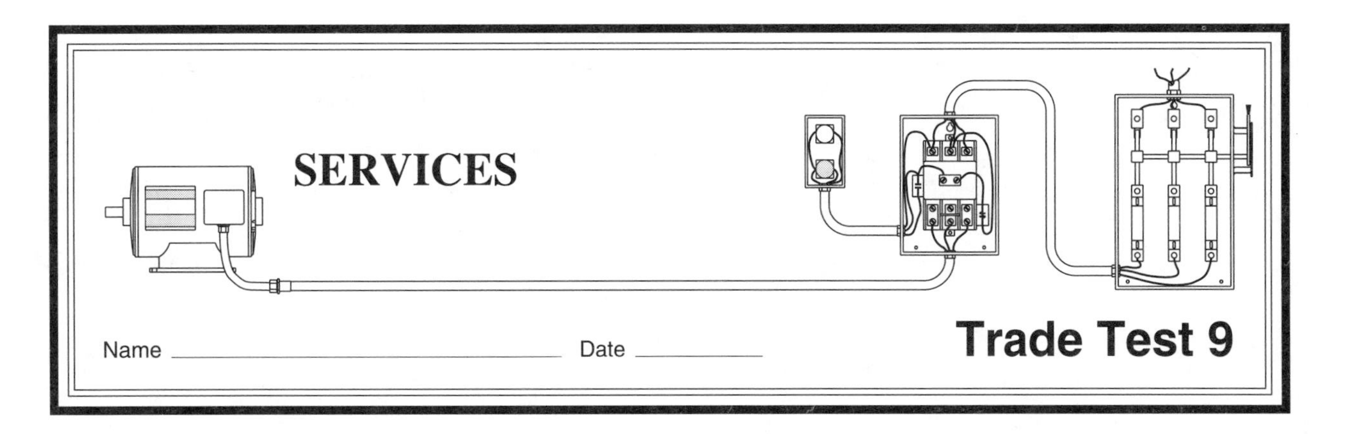

NEC®	Answer	
______	______	**1.** Determine the minimum clearance for 277/480 V service conductors which are installed above a rooftop subject to vehicular traffic.
______	______	**2.** What is the minimum clearance from ground permitted for the point of attachment of the service-drop conductors to a building or structure?
______	______	**3.** A one-family dwelling has an initial net computed load of 9.5 kVA and five branch circuits. Determine the minimum ampacity for the service ungrounded conductors.
______	______	**4.** What is the maximum number of sets of service-entrance conductors that are permitted to be connected to a single service-drop for one building with four separate occupancies?
______	______	**5.** Determine the minimum size of the Al ungrounded service-entrance conductors for a structure with a limited load of two, 2-wire branch circuits.
______	______	**6.** A 14′ vertical run of Type SE cable is installed on the outside of a building between the service head and the meter socket. What is the minimum number of cable supports needed for this installation?
______	______	**7.** Two separate service disconnecting means are grouped together to supply the service for a one-family dwelling. The ratings of the service disconnects are 30 A and 60 A. The initial computed net load of the building is 10.75 kVA. Does this installation meet the requirements of Article 230, Part F?
______	______	**8.** Determine the minimum size permitted for Al service-lateral ungrounded conductors which supply a limited load consisting of a single branch circuit.
______	______	**9.** *See Figure 1.* Determine the minimum clearance from the roof required for the service-drop conductors at A.

SERVICE-DROP CONDUCTORS
SERVICE MAST
• 230-28
A
12″
4″
ROOF SLOPE

FIGURE 1

_______ _______ **10.** Determine the minimum clearance from a building fire escape for service conductors installed as multiconductor cable without an overall outer jacket.

_______ _______ **11.** A 120/240 V, 1ϕ service drop is installed above a public street which is subject to truck traffic. Determine the minimum vertical clearance from ground for the service-drop conductors.

_______ _______ **12.** *See Figure 2.* Determine the minimum size Cu service-drop conductors required for the limited load, single branch-circuit service.

_______ _______ **13.** *See Figure 3.* Determine the minimum clearance from the roof for the service-drop conductors at A.

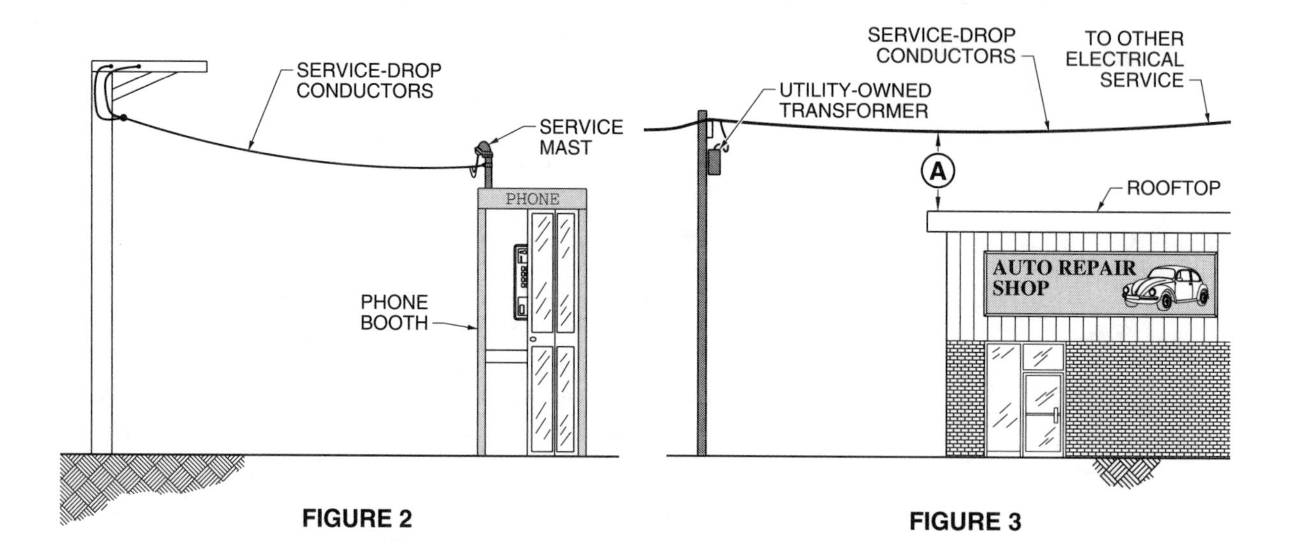

FIGURE 2

FIGURE 3

_______ _______ **14.** *See Figure 4.* Determine the minimum clearance from the openable windows required for the service-drop conductors at A.

_______ _______ **15.** *See Figure 5.* Is this installation a violation of Article 230 of the NEC®?

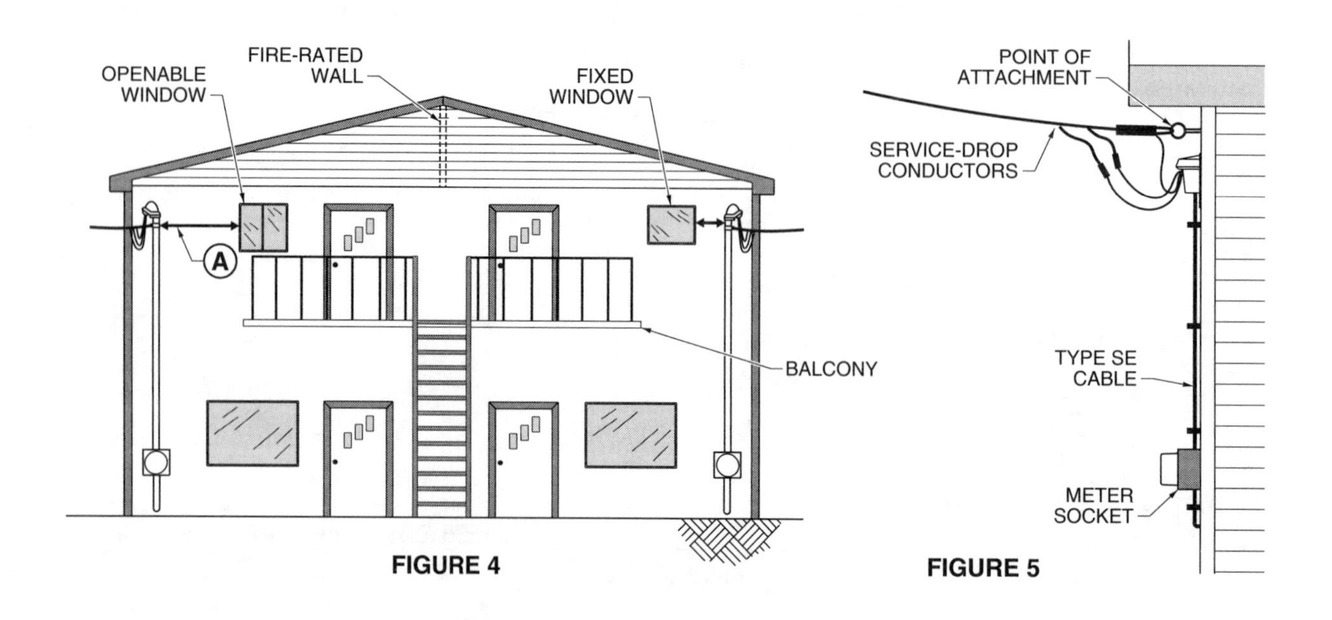

FIGURE 4

FIGURE 5

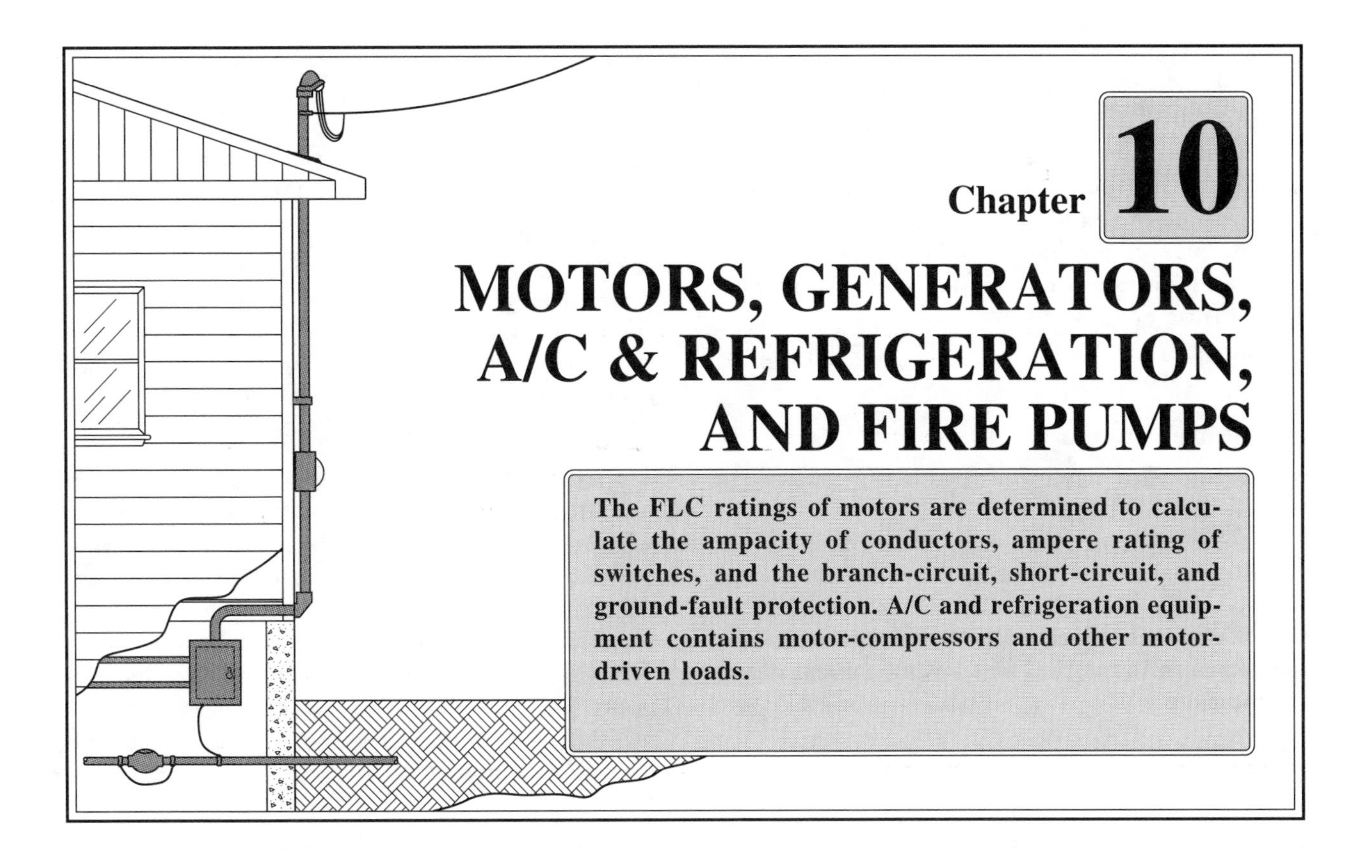

Chapter 10

MOTORS, GENERATORS, A/C & REFRIGERATION, AND FIRE PUMPS

The FLC ratings of motors are determined to calculate the ampacity of conductors, ampere rating of switches, and the branch-circuit, short-circuit, and ground-fault protection. A/C and refrigeration equipment contains motor-compressors and other motor-driven loads.

MOTORS – 430

The one line drawing shown in the beginning of 430 is a guide for the installer. This line drawing should be referenced to size the conductors and fuses or CBs for feeders or branch circuits of a motor circuit, and to size the overload protection for the motor itself. Section 430-5 requires that motors and controllers shall comply with the applicable provisions of other articles and sections of the NEC®. Three things cause problems when electricians deal with the calculations in 430:

(1) AC inductive loads are rated in W or kW. Because the windings of AC inductive loads offer inductive reactance to the circuit, watts are not the simple calculation of $W = V \times A$. The correct formula to use is $W = V \times A \times PF$. PF is the power factor, which is expressed as a percentage. If the load is resistive, and does not contain a significant amount of inductance, then $W = V \times A$ may be used.

(2) When finding the current rating of motors, 1 HP is often considered to be 746 W. However, this can only be true if the motor is 100% efficient and no motor is 100% efficient. The correct formula is 1 HP = 746 W × *Eff*. Eff is the efficiency, which is expressed as a percentage. The closer a motor runs to its HP rating, the more efficient it becomes. If efficiency is not considered when determining the current ratings of motors, a good rule of thumb is to use 1 HP = 900 W.

(3) Always read 430-6 before any work covered in 430 is started. Section 430-6 addresses the proper method of determining the current rating of a motor. Article 430 covers motors, motor branch-circuit and feeder conductors and their protection, motor overload protection, motor control circuits, motor controllers, and motor control centers.

Ampacity and Motor Ratings – 430-6

The most common mistake in applying the requirements of 430 is made when determining the motor full-load current (FLC) rating. This is an important determination since all of the calculations are based on this number. The required ampacity and motor ratings shall be determined by applying 430-6(a)(b)(c).

General Motor Applications – 430-6(a). Section 430-6(a) covers all motor installations except for torque motors and AC adjustable voltage motors. The values given in the appropriate FLC tables are used where the current rating of a motor is used to determine the ampacity of conductors supplying the motor, ampere rating of switches, and the branch-circuit, short-circuit, and ground-fault protection (fuse or CB):

Table	Motor
430-147	DC
430-148	1ϕ
430-149	2ϕ
430-150	3ϕ

The actual current rating marked on the motor nameplate shall not be used. In sizing conductors (which determines raceway sizes) and fuses or CBs, the amperage in the FLC tables is the current rating for the motor.

MOTOR FLC

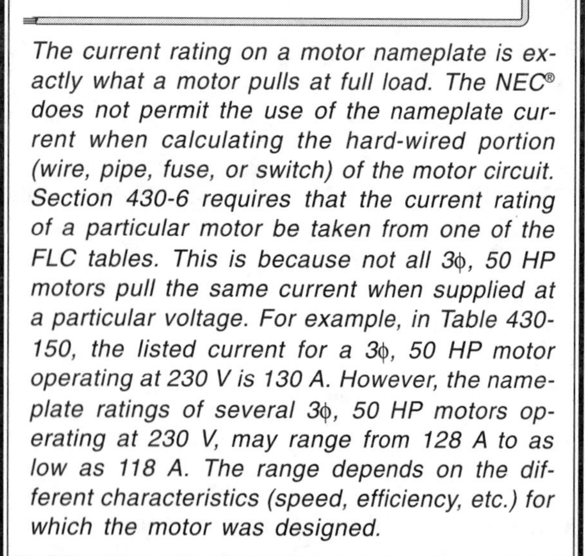

The current rating on a motor nameplate is exactly what a motor pulls at full load. The NEC® does not permit the use of the nameplate current when calculating the hard-wired portion (wire, pipe, fuse, or switch) of the motor circuit. Section 430-6 requires that the current rating of a particular motor be taken from one of the FLC tables. This is because not all 3ϕ, 50 HP motors pull the same current when supplied at a particular voltage. For example, in Table 430-150, the listed current for a 3ϕ, 50 HP motor operating at 230 V is 130 A. However, the nameplate ratings of several 3ϕ, 50 HP motors operating at 230 V, may range from 128 A to as low as 118 A. The range depends on the different characteristics (speed, efficiency, etc.) for which the motor was designed.

In the early 1920s, the NEC® required that the motor conductors and fuses be selected using the nameplate current. The problem was in the replacement of a defective motor. Unless all characteristics of the replacement motor matched the original motor, larger wire sizes, raceway sizes, OCPD sizes, etc. could be required. Therefore, the NEC® provides the FLC tables for motors. The current shown on a general-type motor nameplate never exceeds the current listed on FLA Tables 430-147 through 430-150.

Torque Motors – 430-6(b). The locked-rotor current (LRC) on the motor nameplate shall be the rated current in determining the ampacity of the branch-circuit conductors, the ampere rating of the motor overload protection, and the ampere rating of the motor branch-circuit ground-fault protection. Torque motors represent 5% to 7% of motor applications.

AC Adjustable Motors – 430-6(c). The maximum operating current on the motor or control nameplate shall be used in determining the ampacity of the conductors, ampere rating of switches, and branch-circuit, short-circuit, and ground-fault protection of AC adjustable voltage motors. If the maximum operating current does not appear on the nameplate, the ampacity shall be based on 150% of the values given in Tables 430-149 and 430-150. AC adjustable voltage motors represent 3% to 5% of motor applications.

Marking on Motors and Multimotor Equipment – 430-7

The motor nameplate shall contain the following 14 pieces of information, if applicable:

(1) Manufacturer's name.

(2) Rated volts and full-load amps.

(3) Rated frequency and number of phases.

(4) Rated full-load speed.

(5) Rated temperature rise or insulation system class and rated ambient temperature.

(6) Time rating (5, 15, 30, 60 minutes, or continuous).

(7) Rated horsepower if ⅛ HP or more.

(8) Code letter or locked-rotor amps if AC motor rated ½ HP or more.

(9) Design letter for B, C, D, or E motors.

(10) Secondary volts and full-load amps if wound-rotor induction motor.

(11) Field current and voltage for DC-excited synchronous motors.

(12) Winding type of DC motor.

(13) Motors provided with thermal protection shall be marked "Thermally Protected" or "TP."

(14) Small motors rated 100 W or less shall be marked "Impedance Protected."

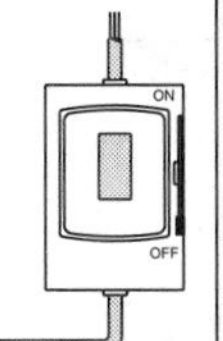

MOTOR NAMEPLATES

The amperes listed on the motor nameplate are the number of amps the motor pulls at full load. At no load, the motor pulls from 50% to 65% of the nameplate rating.

The rated full-load speed is the rpm the motor produces at full load. The rpm is higher when the motor is underloaded and lower when the motor is overloaded.

The frame (FR) number refers to the physical size of the motor. Shaft size, mounting hole dimensions, etc. can be obtained from manufacturer's literature and charts.

Multimotor and Combination-Load Equipment – 430-7(d). The nameplate on multimotor and combination-load equipment shall contain the following six items:

(1) Manufacturer's name.

(2) Voltage rating.

(3) Frequency.

(4) Number of phases.

(5) Minimum ampacity for supply conductors.

(6) Maximum amps rating of the OCPD that provides short-circuit and ground-fault protection.

Marking on Controllers – 430-8

The controller in a motor circuit brings the motor on line and drops it out upon command. A controller may be a relay, contactor, starter, or combination starter. Controllers shall be marked with the following information:

(1) Manufacturer's name.

(2) Rated voltage.

(3) Rated current or HP.

(4) Any other information to indicate suitability.

Branch Circuit – Single Motor – 430-22

The *motor branch circuit* is that point from the last fuse or CB in the motor circuit out to the motor. The minimum elements within a motor branch circuit are the motor branch-circuit conductors, the motor OL protective device, and the motor branch-circuit, short-circuit, and ground-fault protective device (fuse or CB).

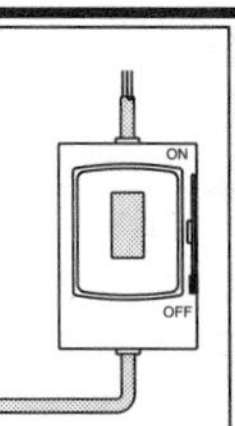

RELAYS, CONTACTORS, AND STARTERS

The difference between a relay and a contactor is the size of the load. Relays are rated for small loads (up to 15 A or 20 A) while contactors are rated for 40 A or more. The difference between a contactor and a starter is that contactors do not provide overload (OL) protection. Starters provide OL protection within the same enclosure as the contactor. A combination starter contains a disconnect switch, fuse or CB, contactor, and OL protective device within the same enclosure.

Sizing Conductors. Section 430-22(a) requires that the branch-circuit conductors supplying a single motor have an ampacity of at least 125% of the FLC rating as indicated on the appropriate FLC table. For example, Table 430-150 indicates the appropriate FLC for a 3ϕ motor. The appropriate FLC multiplied by 1.25 gives the ampacity required for the branch-circuit conductors per 430-22. Table 310-16 is used to select the proper size conductor. See Figure 10-1.

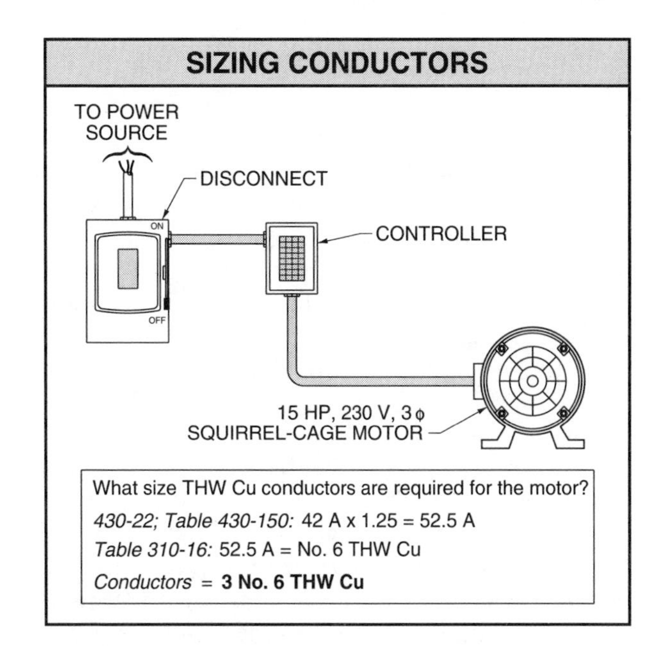

Figure 10-1. The appropriate FLC multiplied by 1.25 gives the ampacity required for branch-circuit conductors.

Sizing Raceways. Once the conductors are properly sized, the appropriate size raceway can be selected. For a metallic raceway that is an approved EGC, all of the conductors are of the same size. The tables in Chapter 9, Appendix C are used to select the proper size, depending on which raceway is being installed. See Figure 10-2.

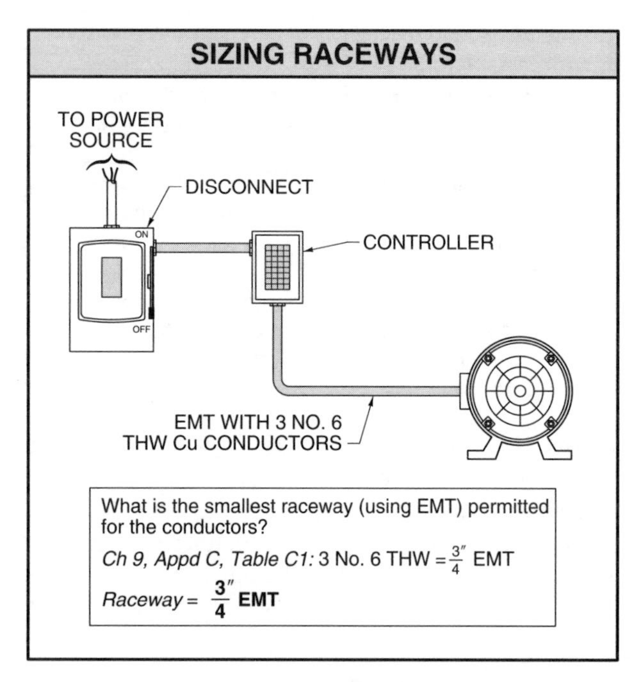

Figure 10-2. Raceways for motor conductors are sized by the tables in Chapter 9, Appendix C.

Sizing Overload Protection. The three forms of protection that all electrical circuits require are overload protection, short-circuit protection, and ground-fault protection. In most circuits, all three types of protection are provided by the fuse or CB. In the motor circuit, the fuse or CB has to be large enough to allow the starting current to flow to the motor. As a result, it is not capable of protecting the motor when an overload occurs. In most motor circuits, the fuse or CB is referred to as the short-circuit, ground-fault protector. It only opens if there is a short or a ground fault in the motor circuit. Therefore, there is a need for another device in a motor circuit to provide overload protection. The purpose of the motor overload protector is to provide the motor, and the motor circuit with overload protection. See Figure 10-3.

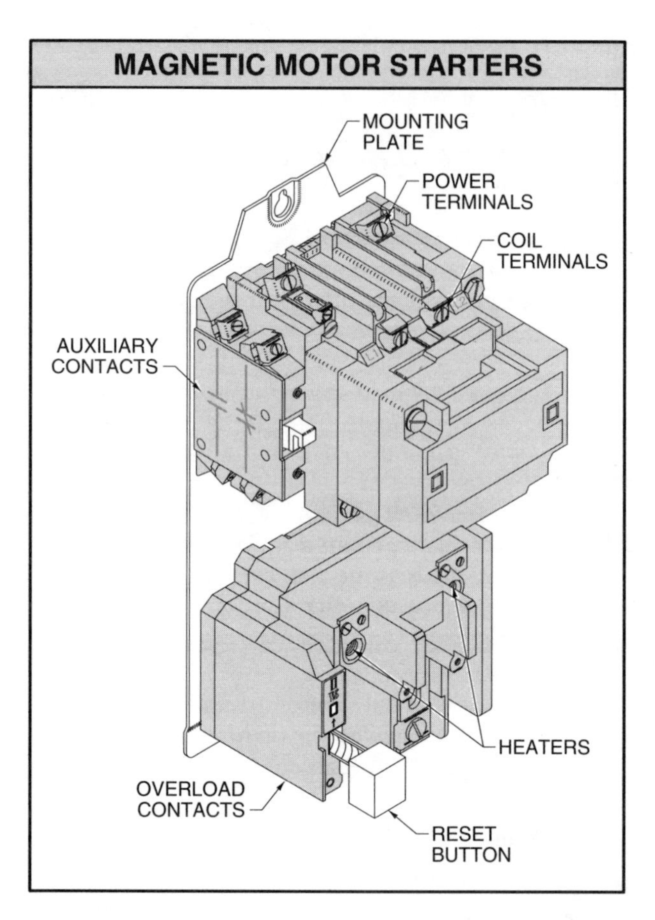

Figure 10-3. The motor overload protector provides the motor and the motor circuit with overload protection.

Overload currents are any currents exceeding the design of the motor. If an overload persists for a long enough period of time, it can cause damage to the insulation on the motor windings and the conductors supplying the motor. Section 430-6 requires that the nameplate current be used to determine the motor overload protection.

Section 430-32(a)(1) requires that the overload protective device for a motor more than 1 HP which is rated for continuous-duty be selected by multiplying the nameplate current by the following percentages:

- Nameplate service factor not less than 1.15 – 125%
- Nameplate temperature rise not more than 40°C – 125%
- All other motors – 115%

The percentage used is dependent upon the motor nameplate rating for service factor and temperature rise. Over 90% of the motors installed today use the 125% factor. If the service factor or temperature rise is not marked on the nameplate, multiply the nameplate current by 115%. See Figure 10-4.

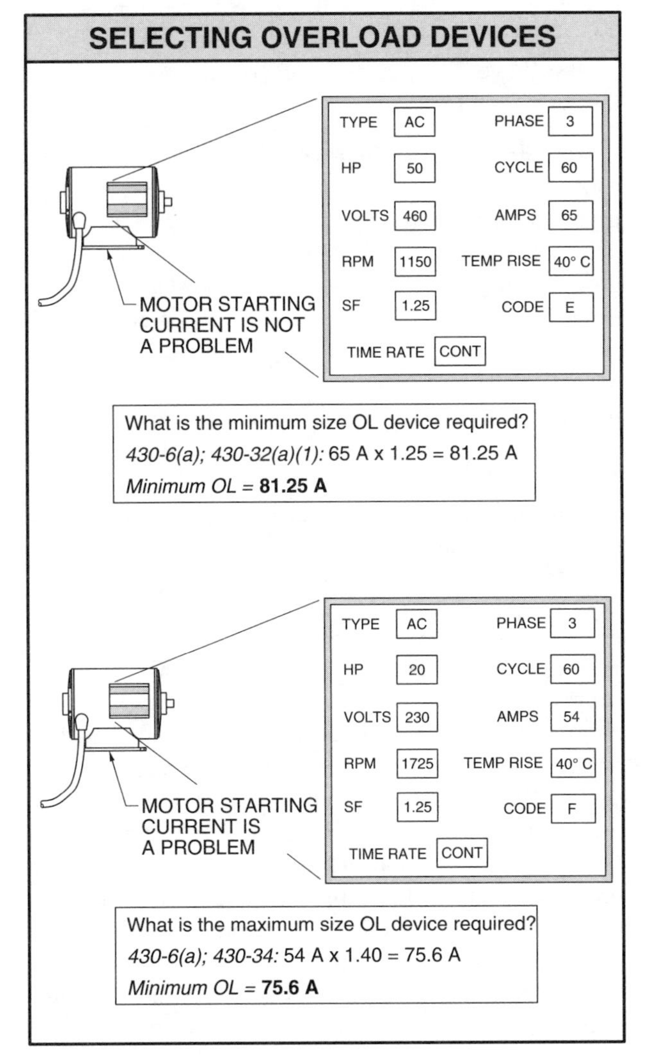

Figure 10-4. The motor nameplate current shall be used to determine the overload protection.

Section 430-32(a)(2) permits a thermal protector that is an integral part of the motor (the thermal sensor is inside the motor) to protect continuous-duty motors rated more than 1 HP. The nameplate shall indicate that this is a thermally-protected motor. The motor FLC, in this example, shall come from the FLC tables and not the nameplate. The table current is increased by one of the following percentages:

Table FLC Rating	% of Increase
Not exceeding 9 A	170
9.1 A through 20 A	156
More than 20 A	140

OVERLOAD PROTECTION WITH EUTECTIC SOLDER

The most common device used to provide overload protection in a motor circuit is the solder pot overload relay. This is one device made of two parts, the heater (which is in the power supply circuit) and the overload contacts (which are wired in the motor control circuit). All current that goes to the motor has to go through the resistive heater. The heater is in close proximity to the solder pot. This pot contains eutectic solder. The unique quality of eutectic solder is that it remains a solid until it reaches 202°F ± 2°, then it instantly turns to a liquid.

A wheel with serrated teeth on the outer edge has a shaft that goes into the solder pot. The wheel cannot turn if the pot contains solid solder. Attached to the serrated teeth on the wheel is a spring-loaded reset arm. When there is an overload and excessive current goes through the heater, the eutectic solder melts into a liquid state, allowing the spring-loaded reset arm to turn the wheel and open the OL contacts. The OL contacts are wired in series with the coil of the magnetic starter. When the OL contacts open, the coil drops out and the starter disconnects the motor from its supply. After the solder has had a chance to cool and returns to a solid, the reset button may be pushed in and the OL contacts are reclosed.

The larger a motor's full-load table current, the less the percentage of the increase in sizing the thermal protector that is an integral part of the motor. Section 430-34 allows the next higher size overload relay to be used if the one selected per 430-32(a)(1) does not permit the motor to start or carry its load. In going to the next larger size, the following percentage of motor nameplate FLC ratings shall not be exceeded:

- Motors with service factor not less than 1.15% – 140%
- Motors with temperature rise not more than 40°C – 140%
- All other motors – 130%

The class ratings of overload relays are Classes 10, 20, and 30 per 430-34, FPN. An overload of 40% requires a certain period of time to generate the necessary heat to melt the eutectic solder and open the OL contacts. An overload of 90% requires less time to generate the same amount of heat, so it responds faster. The more severe the overload, the less time it takes for the OL contacts to open. A Class 20 OL relay takes longer for its solder to melt than a Class 10 OL relay for the same amount of current. A Class 30 OL relay carries the same current for a longer period of time than a Class 20 OL relay.

Sizing Fuses and CBs. Section 430-52(b) requires that the motor branch-circuit, short-circuit, and ground-fault protection (fuse or CB) be large enough to carry the motor starting current. See Figure 10-5. Section 430-52(c)(1) requires this protective device have a rating or setting not exceeding the value calculated when using Table 430-152. Table 430-152 indicates the "Maximum Rating or Setting." Section 430-52(c)(1) requires a protective device with a rating or setting not exceeding the value calculated per the values in Table 430-152 to be used. If, when selecting the motor branch-circuit fuse or CB, the calculation made per Table 430-152 does not result in a standard size, the next lower size shall be selected. All numbers shall be rounded down in size, not up. Per 430-52(c)(1), Ex., if the values of the fuse or CB determined by using Table 430-152 do not correspond to standard sizes listed in 240-6, the next higher standard size shall be permitted. *Note:* The NEC® suggests rounding down, but permits rounding up.

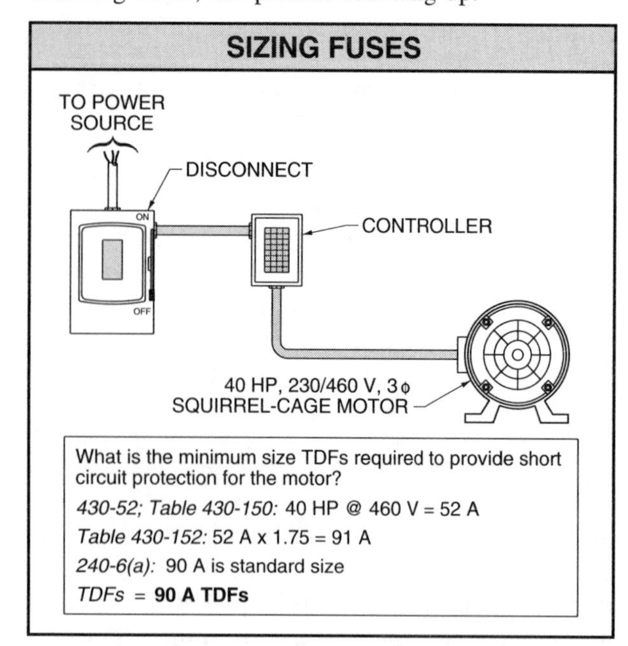

Figure 10-5. The motor protection shall be large enough to carry the motor starting current.

Sizing Disconnects. The fusible disconnect is determined by the size of fuse being used in the motor branch circuit. Section 430-110 requires that motor disconnects installed in motor circuits rated 600 V or less have an ampere rating of at least 115% or the motor's FLC.

Sizing Feeder Conductors with More than One Motor – 430-24. When there is a common feeder supplying several motor branch circuits, the feeder conductor shall be sized per 430-24, and the short-circuit, ground-fault protection of the feeder shall be sized according to 430-62. Per 430-24, feeder conductors supplying several motors shall have an ampacity equal to or larger than the sum of the FLC ratings of all motors being supplied by the feeder, plus 25% of the highest rated motor in the group. See Figure 10-6. The FLC ratings shall come from the appropriate FLC table. Similar to the branch circuit for a metallic raceway that is an approved EGC, all circuit conductors are the same size. Thus, Chapter 9, Appendix C is used to determine the appropriate size raceway.

Sizing Feeder Fuses or CBs – 430-62(a). A feeder supplying motors which are bolted in place (not portable) and sized per 430-24 shall be provided with a protective device sized per the following: If the individual motor branch circuit fuses or CBs were selected using Table 430-152, take the size of the largest fuse or CB protecting the different motor branch circuits and add the FLCs (as determined by the appropriate FLC table) of the other motors in the group. See Figure 10-7. If there are two sets of fuses that are both the largest size of the fuses in the group, either set of fuses may be used as the largest fuse in the group.

Motor Control Circuits – 430, Part F

A *motor control circuit* is the circuit of a control apparatus or system which carries electric signals directing the performance of the controller, but does not carry the main power current. See 430-71. The motor control circuit is the circuit that controls the operation of the magnetic coil within the controller itself. Within this circuit are pilot devices and indicating devices. A *pilot device* is a sensing device that controls the motor controller. An *indicating device* is a pilot light, buzzer, horn, or other type of alarm. Often, the wiring of a motor control circuit is very elaborate and requires 10 times the amount of wiring as the motor power circuit.

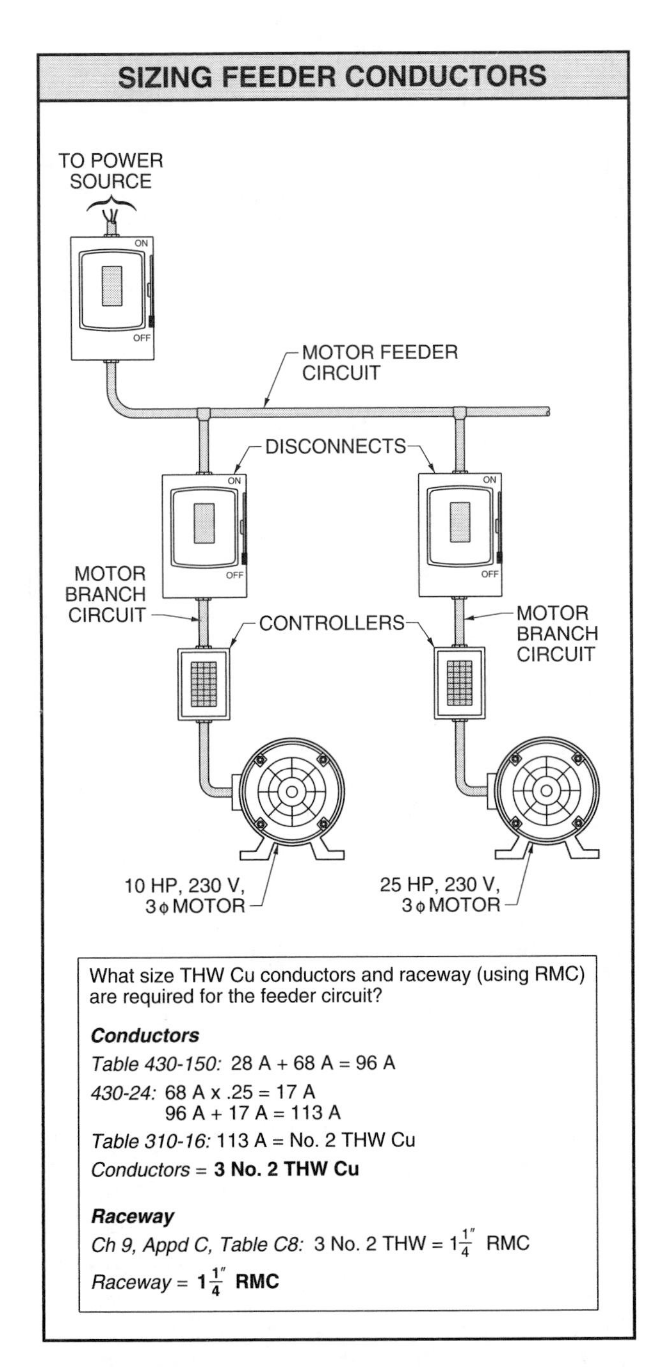

Figure 10-6. Feeder conductors supplying several motors shall have an ampacity equal to or larger than the sum of all FLC ratings plus 25% of the highest rated motor in the group.

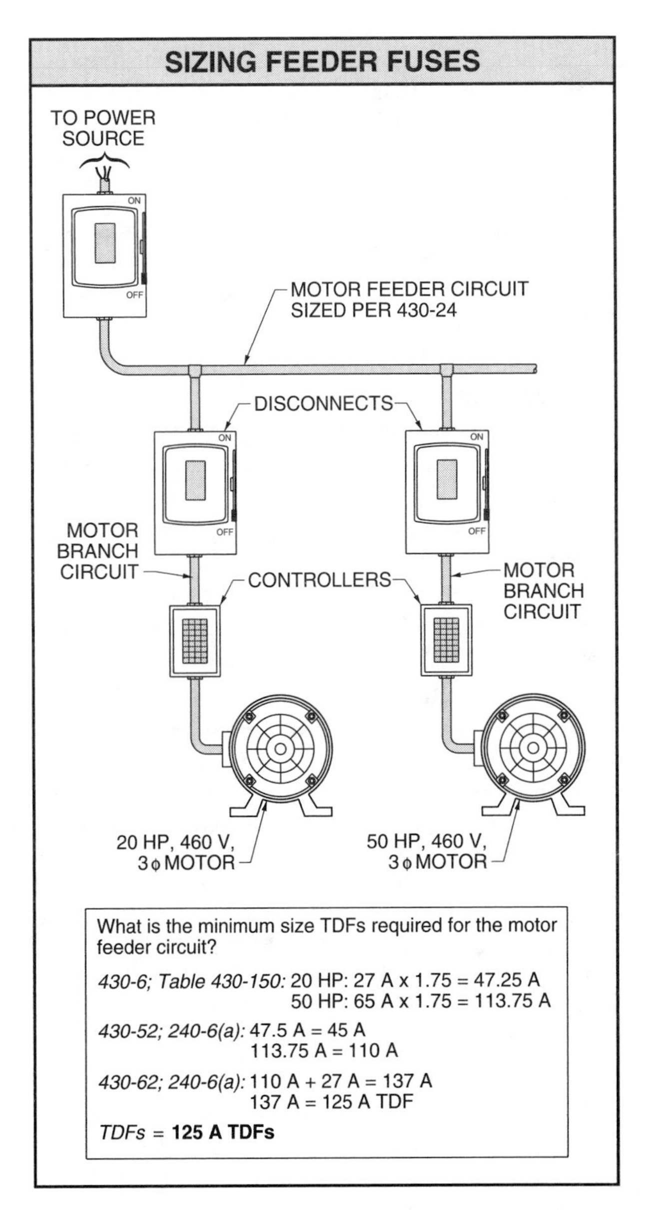

Figure 10-7. Fuses for feeder conductors supplying several motors are sized by adding the largest fuse of the branch circuit and the FLC of the other motors.

Overcurrent Protection – 430-72. The motor control circuit includes all of the pilot devices, indicating equipment, and magnetic coils within the circuit. The main concern of 430 regarding the control circuit is the protection of these items, the conductors, and the control circuit transformer.

Motor control circuits are permitted to receive their power from one of two sources: (1) the control circuit may be tapped from the load side of the motor branch-circuit fuse or CB or (2) the control circuit may be supplied from a separate source. When the control circuit is supplied from a separate source, the circuits shall be protected against overcurrents per 725-23. Section 430-72(b) requires that the overcurrent device for conductors shall not exceed the values shown in Table 430-72(b), Column A.

The NEC® does not require that the control circuit conductors have an ampacity large enough to carry the current required of a control circuit. This current is too small to be significant as most magnetic starter coils pull less than ½ A. Section 430-72 does, however, require that the control circuit conductors be provided with protection in accordance with their ampacity. In some cases, the overcurrent device may be higher than the ampacity of the conductor.

Section 430-72(a) permits a motor control circuit to originate in one of two ways: (1) It may be tapped from the load side of a motor branch-circuit fuse or CB. A control circuit tapped from the load side is not considered a branch circuit and is considered protected by either a supplementary or branch-circuit overcurrent device. (2) Other motor control circuits may be derived from a panelboard or a control transformer. In this case, the control conductors shall be protected per 725-23.

The general rule for protecting control circuit conductors is that the overcurrent protection for conductors shall not exceed the values shown in Table 430-72(b), Column A. Four exceptions to this rule permit other types of protection depending on where the control conductors are located.

Section 430-72(b), Ex. 1 permits control circuit conductors to have short-circuit, ground-fault protection only according to the amperages shown in Table 430-72(b), Column B provided these conductors do not extend beyond the motor control equipment enclosure.

Section 430-72(b), Ex. 2 permits the control circuit conductors to have short-circuit, ground fault protection only according to the amperages shown in Table 430-72(b), Column C provided these conductors extend beyond the motor control equipment enclosure.

Section 430-72(b), Ex. 3 permits control circuit conductors supplied from a 2-wire, single-voltage transformer secondary to have overcurrent protection provided by the transformer's primary overcurrent protection device shall not exceed the ampere rating in Table 430-72(b) times the secondary-to-primary transformer ratio. See Figure 10-8.

Section 430-72(b), Ex. 4 requires short-circuit, ground-fault protection for control circuit conductors that regulate circuits which would cause a hazard if they were to go out on overload, i.e., the control circuit for a fire pump.

Section 430-72(c) requires that a motor control circuit transformer be protected per 450-3. This is often accomplished by a primary-side protective device rated not more than 125% or 167% of the primary current.

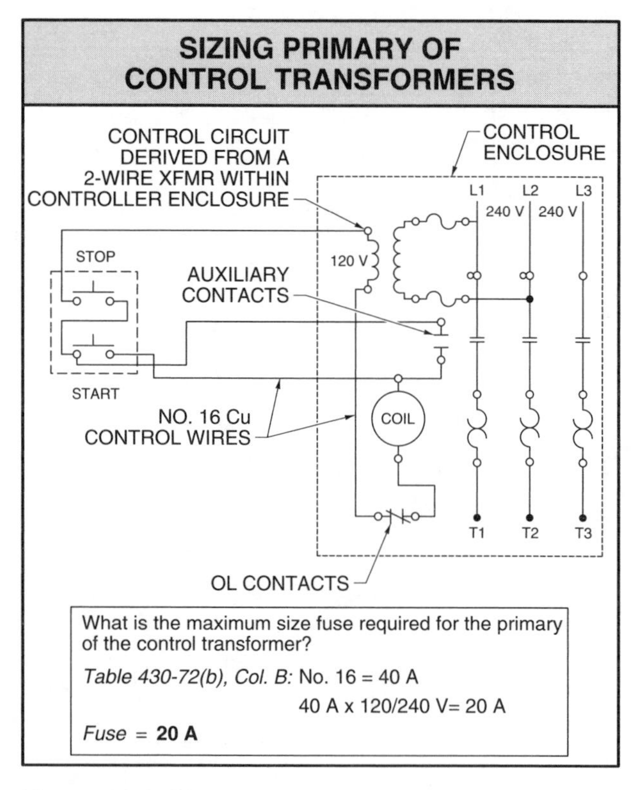

Figure 10-8. The primary of control transformers is sized per Table 430-72(b), Column A.

Section 430-72(c), Ex. 1 eliminates the requirement of protecting the control transformer if it is rated 50 VA or less, provided it is part of the motor controller and located within the controller enclosure. The intent here is that the protection is provided by the primary overcurrent devices.

Section 430-72(c), Ex. 2 permits a control transformer with a rated primary current of 2 A or less to be considered protected at up to 500% of the rated primary current, provided the overcurrent device is in each of the ungrounded conductors of the supply circuit.

Control Circuit Disconnects – 430-74. When the disconnect to the motor supply is opened, the supply to the motor control circuit shall also be opened. This disconnecting means is permitted to have two or more separate devices to accomplish the disconnecting of the motor power supply and the supply to the control circuit provided where separate devices are installed, they shall be located immediately adjacent to one another. There are two exceptions to this general rule.

Section 430-74(a), Ex. 1 pertains to motor control circuits that contain more than 12 motor control circuit conductors which are required to be disconnected. This is usually found in industrial locations with complex control circuits of multimotor machinery. The control circuit disconnect means are permitted at locations that are not immediately adjacent to each other provided the following two conditions are met: (a) Access to the energized parts is limited to qualified persons only. (b) A warning sign is permanently located on each door or cover of equipment that permits access to the live parts of the control circuit. This sign shall specify the location and identification of each of the disconnects.

Section 430-74(a), Ex. 2 states that if the opening of the control disconnect means could result in an unsafe condition for personnel or property, then the following is permitted: the control circuit disconnect means may be located where they are not immediately adjacent to each other provided (a) and (b) of Ex. 1 are complied with. When the control circuit transformer is mounted within the controller enclosure, it shall be connected to the load side of the disconnect means for the motor control circuit.

Motor Controllers – 430, Part G

The *controller* is the device in a motor circuit which turns the motor ON or OFF. Controllers are also known as starters. See 100 and 430-81(a). The two basic types of controllers are manual and magnetic. A manual controller requires that someone physically opens the contacts by pressing a pushbutton or throwing a switch. A magnetic controller operates when the magnetic coil is energized. The sensing devices placed into the control circuit that control the magnetic coil circuits are the pilot devices. Some sensing devices, such as a float switch, can act as a pilot device or a controller, depending on their function in the circuit. If the contacts within the float switch control the magnetic coil of the controller, the float switch is considered a pilot device. See Figure 10-9. If the contacts of the float switch bring the motor in and out, the float switch is considered a controller.

Section 430-81 requires a suitable controller for all motors. In the case of a stationary motor that is ⅛ HP or less and its winding impedance is high enough to prevent damage to the motor when the rotor is in a locked position, 430-81(b) permits the branch-circuit protective device to act as the controller. Section 430-81(c) permits the attachment plug and receptacle to act as the controller for portable motors of ⅓ HP or less. See Figure 10-10.

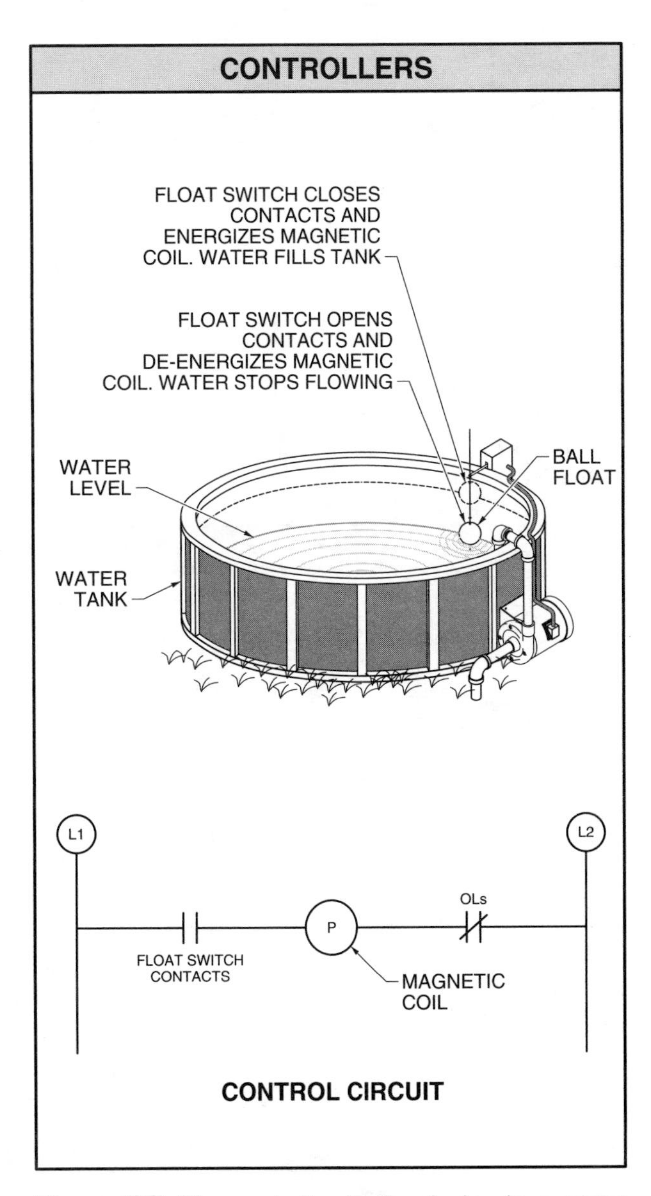

Figure 10-9. The controller is the device in a motor circuit which turns the motor ON or OFF.

Ratings – 430-83. The motor controller shall be rated at the same HP rating, or greater, than the motor itself. See Figure 10-11. The HP rating shall be selected at the motor's application voltage. There are five exceptions to this general rule.

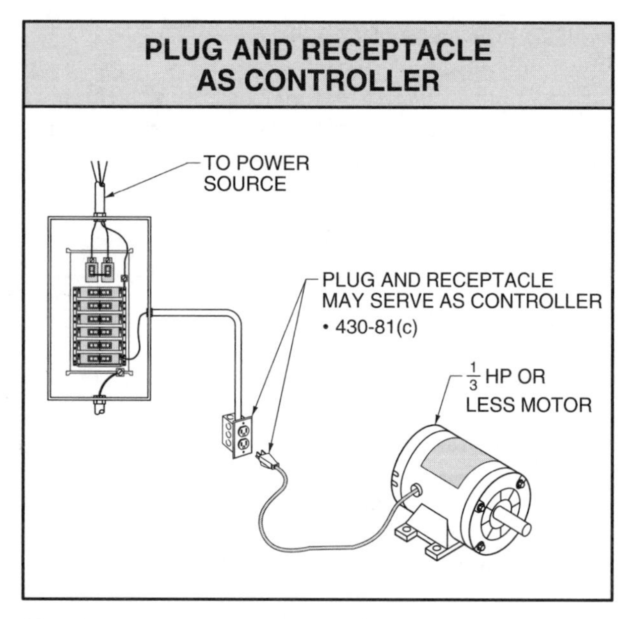

Figure 10-10. A plug and receptacle may serve as a controller for portable motors up to ⅓ HP.

Section 430-83(a), Ex. 1 requires that a Design E motor rated more than 2 HP shall have a controller which complies with the following two items: (1) the controller shall be identified as rated for use with a Design E motor. (2) The controller HP rating shall be 1.4 times the rating of the motor for motors rated 3 HP through 100 HP. For motors rated more than 100 HP, the controller shall be rated not less than 1.3 times the HP rating of the motor.

Section 430-83(a), Ex. 2 allows a general-use switch with an ampere rating twice that of the motor FLC rating to be used as a controller provided it is a stationary motor rated 2 HP or less and 300 V or less.

On AC circuits, a general-use snap switch suitable for AC only shall be permitted to control a motor whose rating is 2 HP or less and 300 V or less, provided the current rating of the motor is not more than 80% of the ampere rating of the switch.

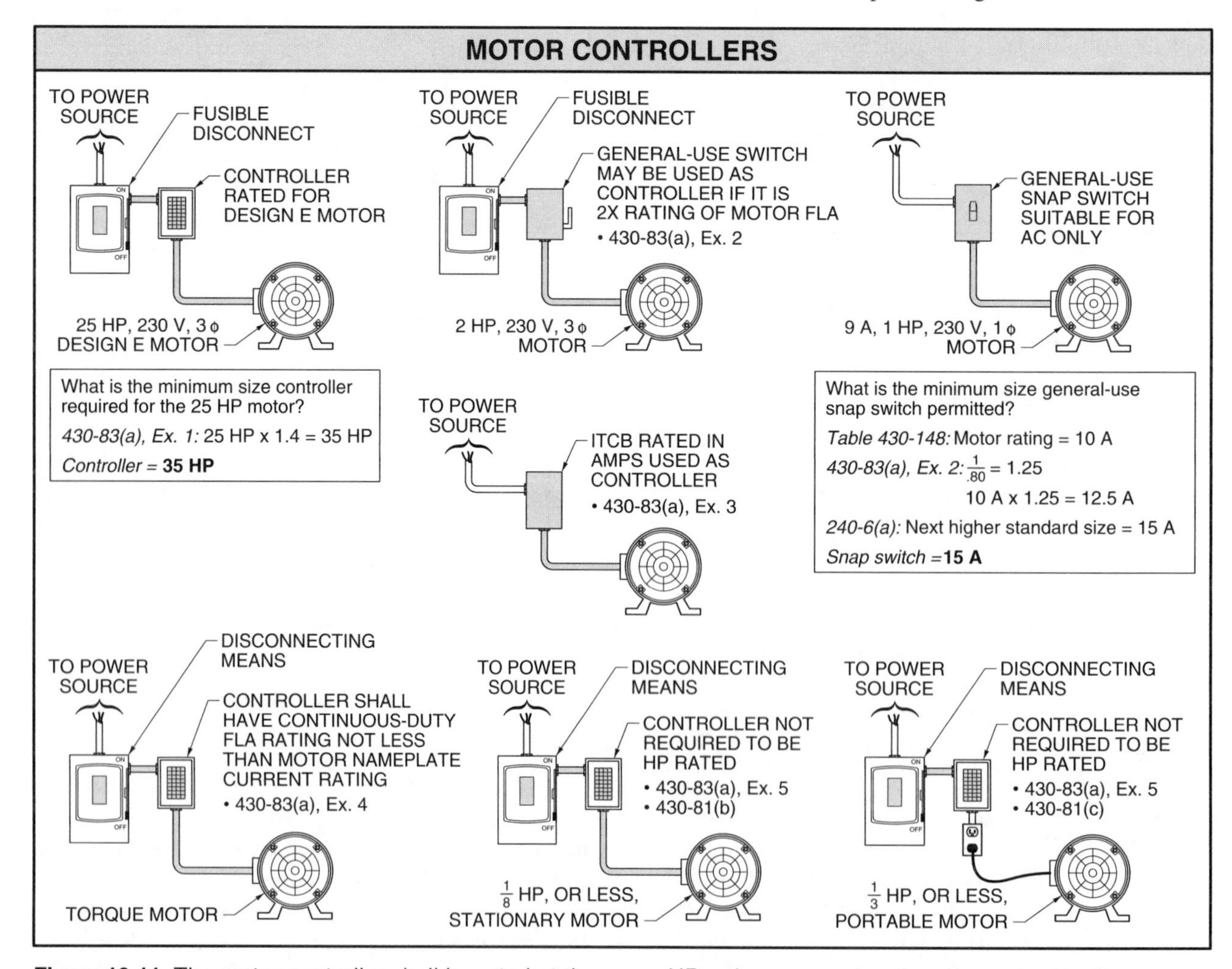

Figure 10-11. The motor controller shall be rated at the same HP rating, or greater, than the motor itself.

Section 430-83(a), Ex. 3 permits a branch-circuit ITCB rated in amperes only to be used as the motor controller. Section 430-83(a), Ex. 4 requires that a torque motor controller shall have a continuous-duty, FLC rating the same as or more than the nameplate current rating on the motor.

Section 430-83(a), Ex. 5 allows devices permitted by 430-81(b), stationary motors of ⅛ HP or less, and 430-81(c), portable motors of ⅓ HP or less, to not be rated in HP.

Section 430-83(b) requires that controllers identified with a slash voltage rating only be used on electrical systems in which the voltage-to-ground does not exceed the lower voltage rating marked on the controller and that the line voltage (phase-to-phase voltage) of the system is not greater than the higher voltage rating marked on the controller.

CONTROLLER RATINGS

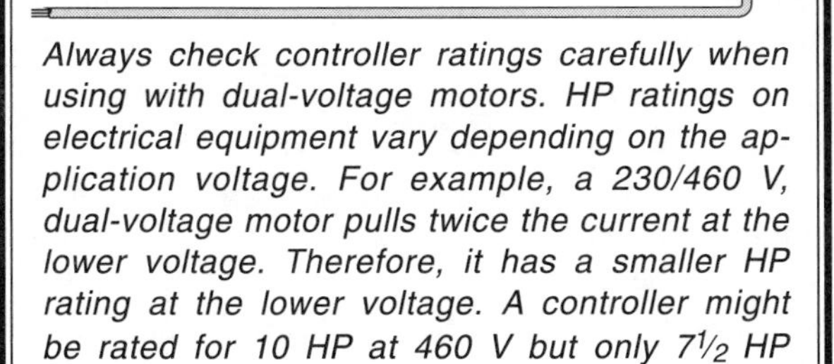

Always check controller ratings carefully when using with dual-voltage motors. HP ratings on electrical equipment vary depending on the application voltage. For example, a 230/460 V, dual-voltage motor pulls twice the current at the lower voltage. Therefore, it has a smaller HP rating at the lower voltage. A controller might be rated for 10 HP at 460 V but only 7½ HP at 230 V.

Opening of Conductors – 430-84. The controller is not required to open all conductors supplying power to the motor, provided the motor is stopped. If the controller is also serving as the disconnecting means, it shall open all of the ungrounded conductors to the motor.

Number of Motors Served by Each Controller – 430-87. Each motor shall be provided with its own controller. The exception to this rule permits a single controller to serve several motors provided the motors are rated 600 V or less and the controller has an HP rating equal to or greater than the sum of the motor's HP ratings controlled by this one controller. This exception is permitted for any one of the following three conditions: (1) where all of the motors are part of a single machine or piece of apparatus, (2) where all of the motors are protected by one overcurrent device, and (3) where the groups of motors are located in a single room within sight from the controller location.

Motor Control Centers – 430, Part H

A *motor control center (MCC)* is an assembly of one or more enclosed sections with a common power bus and primarily containing motor control units. See 430-92. An MCC is a large enclosure containing many sections or compartments. Within these enclosures are disconnect switches, starters, relays, contactors, fuses, CBs, and any other devices found in a motor or motor control circuit. Section 430-1, Ex. 1 requires that MCCs are to be installed per 384-4.

MCC Overcurrent Protection – 430-94. An MCC shall have its overcurrent protection based on the rating of the common power bus within the MCC. This overcurrent protection shall be provided in one of the following two locations: (1) ahead of the MCC, (2) a main overcurrent device mounted within the MCC.

MCC as Service Entrance Equipment – 430-95. An MCC is permitted to be used as service equipment provided it is equipped with a single main disconnecting means. The exception to this rule permits a second service disconnect to supply additional loads. Where a grounded conductor is provided, the MCC shall be equipped with a main bonding jumper within one of the sections. The purpose of this bonding jumper is to connect the grounded conductor on its supply side to the control center equipment ground bus. This main bonding jumper shall be sized per 250-79(d).

Grounding of MCC – 430-96. MCCs consisting of more than one section shall be bonded together with an EGC sized per Table 250-95. The EGC shall terminate on this grounding bus or, in the case of a single-section MCC, to the grounding termination it provides.

Busbars and Conductors within MCC – 430-97. The busbars within an MCC shall be protected from physical damage and be secured firmly in place. Except for control wiring and any required interconnections, the only conductors permitted for termination in a vertical section are those conductors intended for use in that section. The exception to this rule permits conductors to travel horizontally through vertical sections provided they are isolated from the busbars by a suitable barrier. When viewing the busbars from the front of the MCC, 430-97(b) requires the bus bars to be arranged A, B, C, from front to back, top to bottom, and left to right.

Marking of MCC – 430-98. MCCs shall be marked with the manufacturer's name, trademark, or other identification which indicates the organization responsible for the product. The markings shall include voltage rating, common power bus current rating, and MCC short-circuit rating.

Disconnecting Means – 430, Part I

The general intent of 430 regarding the motor disconnecting means is that it be rated in HP which corresponds to the motor it serves. The ideal location for the disconnecting means is readily accessible and in sight from the controller, the motor, and the driven machinery location.

Location of Disconnecting Means – 430-102. The disconnecting means shall be installed within sight from the controller location. The specified equipment is to be visible and not more than 50′ apart.

Section 430-102(a), Ex. 1 permits the disconnecting means for a controller of a motor circuit over 600 V to be located out-of-sight of the controller. The disconnecting means shall be capable of being locked in the open position. The controller shall be marked with a warning label indicating the location of the disconnecting means.

Per 430-102(a), Ex. 2, a single disconnecting means is permitted to be located adjacent to a group of controllers that are coordinated and mounted in proximity to one another provided the installation is a multimotor continuous process machine.

The disconnecting means shall be installed within sight from the motor and the driven machinery per 430-102(b). Section 430-102(b), Ex. is the same as 430-102(a), Ex. 1. The disconnect may be installed out-of-sight of the motor and driven machinery provided the disconnect is capable of being locked in the open position.

If the disconnecting means is not capable of being locked in the open position, and is located out-of-sight of the controller, an additional disconnecting means shall be located within sight of the motor.

Operation of Disconnecting Means – 430-103. The disconnecting means shall open all of the ungrounded conductors supplying the motor. No pole on the disconnecting means shall operate independently.

Type of Disconnecting Means – 430-109. The motor disconnecting means shall be a listed product. See Figure 10-12. It shall be one of the following three types: (1) a switch rated in HP, (2) a CB (provided the handle is operable from outside of the enclosure), or (3) a molded-case switch. There are eight exceptions to this rule.

A motor disconnect switch for a Design E motor rated over 2 HP shall comply with the following per 430-109, Ex. 1: (1) be marked for use with a Design E motor, (2) have a HP rating equal to or more than 1.4 times the rating of the motor for motors rated 3 HP up to and including 100 HP, or have a HP rating equal to or more than 1.3 times the rating of the motor for motors rated over 100 HP. Section 430-109, Ex. 2 permits the branch circuit fuse or CB to serve as the disconnect provided it is a stationary motor of ⅛ HP or less.

Section 430-109, Ex. 3 permits a general-use switch to act as the motor disconnect provided it has an ampere rating not less than twice the FLC rating of the motor and that the stationary (not portable) motor is rated 2 HP or less and 300 V or less. On AC circuits, general-use snap switches suitable only for use on AC circuits shall be permitted to be used as a motor disconnect switch for motors rated 2 HP or less and 300 V or less provided the switch has an ampere rating of at least 125% of the motor FLC rating.

Section 430-109, Ex. 4 states that for motors over 2 HP up to and including 100 HP, the disconnecting means required for a motor with an autotransformer-type controller shall be permitted to be a general-use switch if all of the following three provisions are met:

(1) The motor drives a generator which is provided with overload protection.

(2) The controller is capable of interrupting the LRC of the motor and is provided with no-voltage release (known in the field as 2-wire control) and has running overload protection not exceeding 125% of the motor FLC rating.

(3) Separate fuses or an ITCB rated or set at not more than 150% of the motor FLC rating are provided in the motor branch circuit.

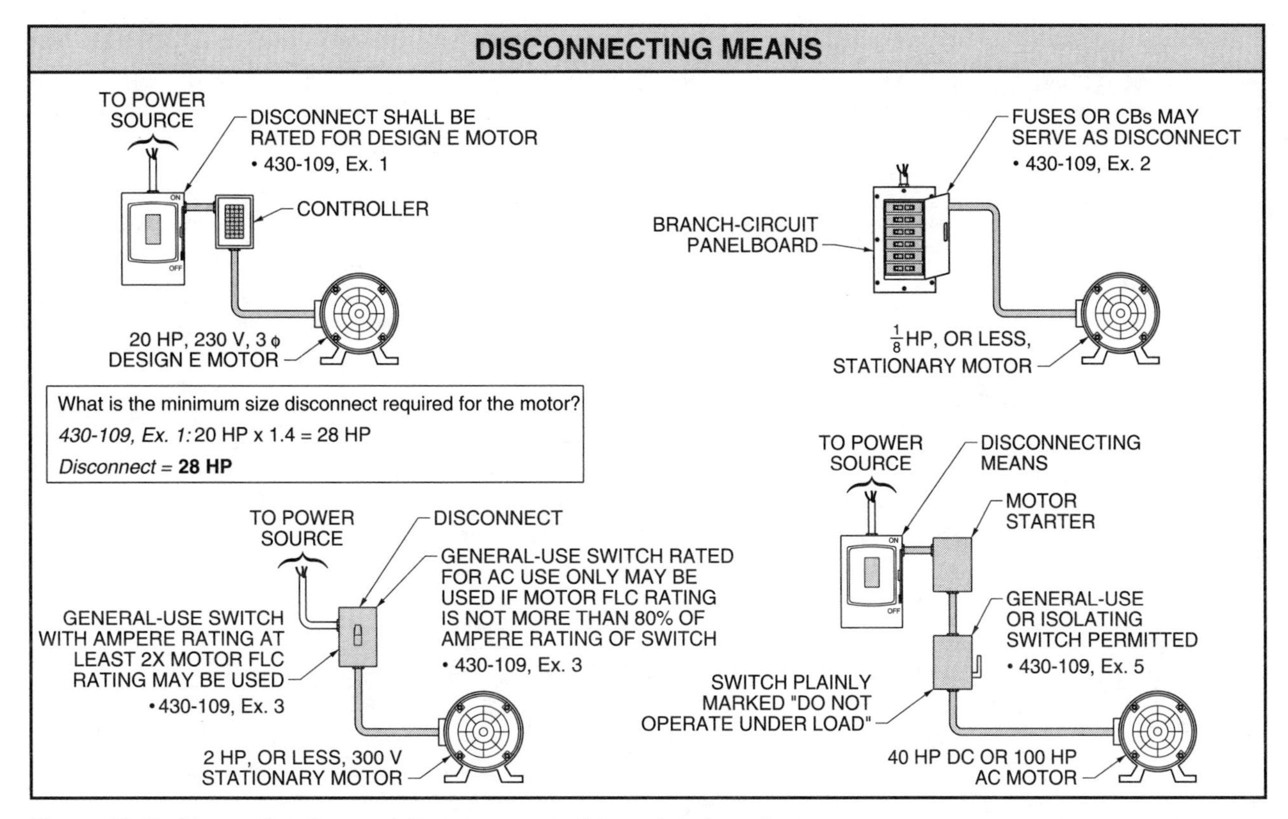

Figure 10-12. The motor disconnecting means shall be a listed product.

Section 430-109, Ex. 5 permits DC stationary motors rated more than 40 HP or AC motors rated more than 100 HP to use a general-use or isolating switch where the switch is plainly marked, "Do not operate under load." An isolating switch has no interrupting rating and is not intended to be operated while the circuit is live.

For cord- and plug-connected motors, 430-109, Ex. 6 permits a HP-rated plug and receptacle with a rating equal to or more than the motor rating to serve as the motor disconnect provided the motor is other than a Design E motor or is a Design E motor rated 2 HP or less. For a Design E motor rated more than 2 HP, a plug and receptacle used as the disconnecting means shall have a HP rating not less than 1.4 times the motor rating.

A general-use switch is permitted as a disconnecting means for torque motors per 430-109, Ex. 7. An ITB, which is part of a listed combination starter, is permitted to serve as the disconnecting means per 430-109, Ex. 8.

Ampere Rating and Interrupting Capacity of Disconnecting Means – 430-110. If the disconnect is rated in HP, the size of the fuse determines the size disconnect to be used. Section 430-110(a) is the general rule which requires that the disconnecting means for motor circuits have an ampere rating of at least 115% of the FLC rating of the motor for circuits rated 600 V or less. See Figure 10-13.

The disconnecting means for torque motors shall have a rating of at least 115% of the motor nameplate current per 430-110(b). Section 430-110(c) requires that where two or more motors are used together, or one or more motors and other loads are served by the same disconnect, the ampere and HP ratings of the combined load shall be determined by one of the following three methods:

(1) The rating of the disconnect is determined from the sum of all currents at the full-load and the locked rotor conditions.

(2) The ampere rating of the disconnect shall not be less than 115% of all currents at the full-load condition.

(3) For motors not shown in Tables 430-147 through 430-150, the LRC shall be assumed to be six times the nameplate FLC.

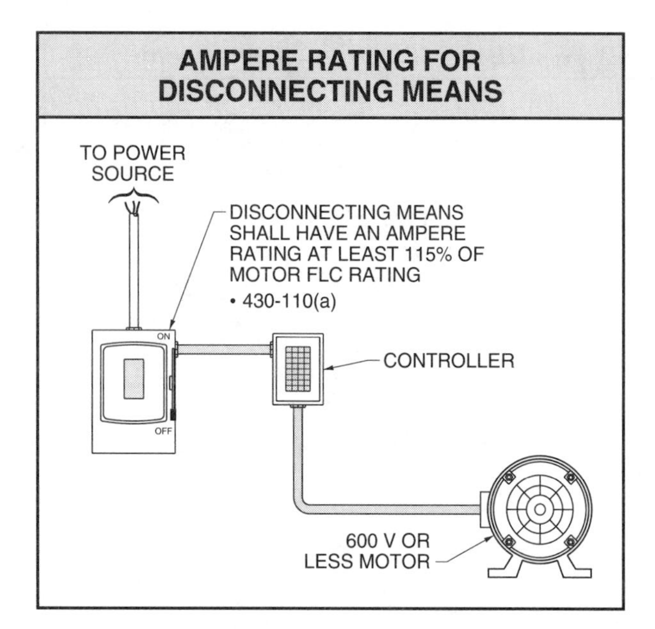

Figure 10-13. The disconnecting means for motor circuits shall have an ampere rating of at least 115% of the motor FLC rating.

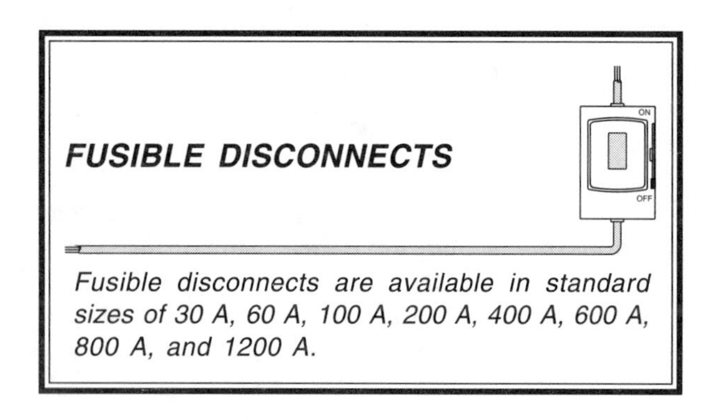

Switch or CB as Controller and Disconnecting Means – 430-111. The controller is permitted to be the disconnect means where it consists of a manually-operable switch, ITCB, or an oil switch with a rating which does not exceed 100 A or 600 V. The switch or CB used as a controller shall be protected by branch-circuit, short-circuit, and ground-fault protective devices which open all ungrounded conductors.

Motors Served by a Single Disconnecting Means – 430-112. Each motor shall have its own disconnecting means. The exception to this general rule permits one disconnecting means to serve more than one motor. See Figure 10-14. The exception has three conditions:

(1) A number of motors are used on a single machine or piece of apparatus.

(2) More than one motor is protected by one set of branch-circuit protective devices.

(3) More than one motor is in the same room within sight from the disconnecting means.

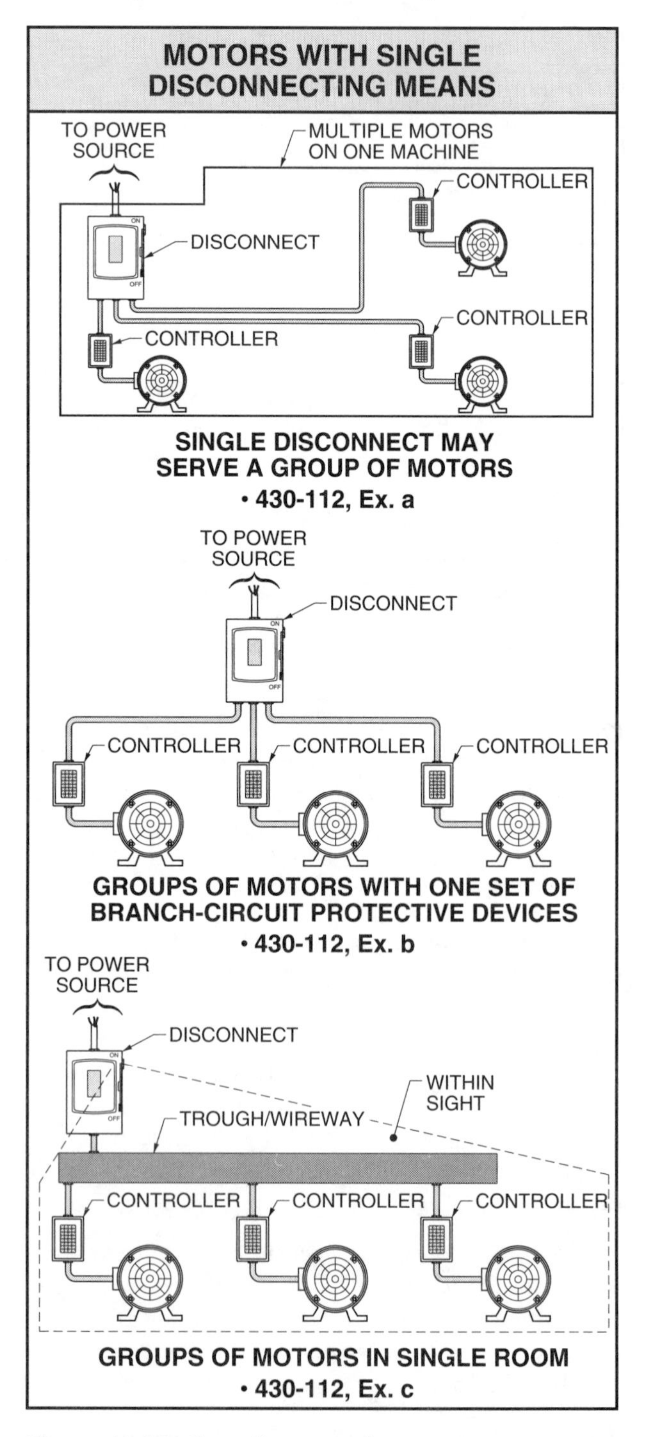

Figure 10-14. One disconnecting means may serve more than one motor.

Disconnecting Means for Energy from More than One Source – 430-113. A motor and motor-operated equipment with more than one source for receiving electrical power shall have a disconnecting means for each source of power adjacent to the equipment being served. There are two exceptions to this general rule.

Per 430-113, Ex. 1, when a motor receives power from more than one source, the disconnecting means for the main power supply to the motor is not required to be immediately adjacent to the motor, provided the disconnecting means for the controller is capable of being locked in the open position. Per 430-113, Ex. 2, a separate disconnecting means is not required for remote control thermostat wiring, rated not more than 30 V, which is isolated and ungrounded.

Other Motors

Over 90% of the motors in use today are single speed, AC induction motors. Multispeed motors, duty-cycle motors, wound rotor motors, and synchronous motors are commonly used in industry. The Design E motor is slowly finding its way into commercial and industrial applications. It does, however, draw an extremely high inrush current.

MULTISPEED MOTORS

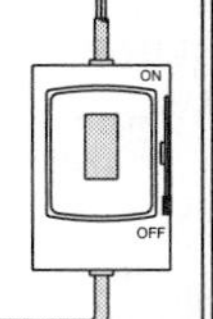

Multispeed motors are controlled by multispeed controllers. The most popular form of multispeed motors are two-speed and three-speed motors. A two-speed, multispeed motor is wound as two different motors. Each speed has its own set of "T" leads. It may be wound as a six-pole motor for low speed or as a four-pole motor for high speed. The two-speed controller for this motor has two controllers inside the one enclosure. A single set of supply conductors on the line side of the controller has the ampacity to safely supply the motor for each speed. On the load side of the controller, two sets of conductors go to the motor, one set for each speed.

Multispeed Motors – 430-22(a). The branch-circuit conductors on the line side of the controller shall be sized at 125% of the highest of the FLC ratings shown on the motor nameplate. The branch-circuit conductors on the load side of the controller shall be sized at 125% of the current rating of the winding that these conductors supply. See Figure 10-15.

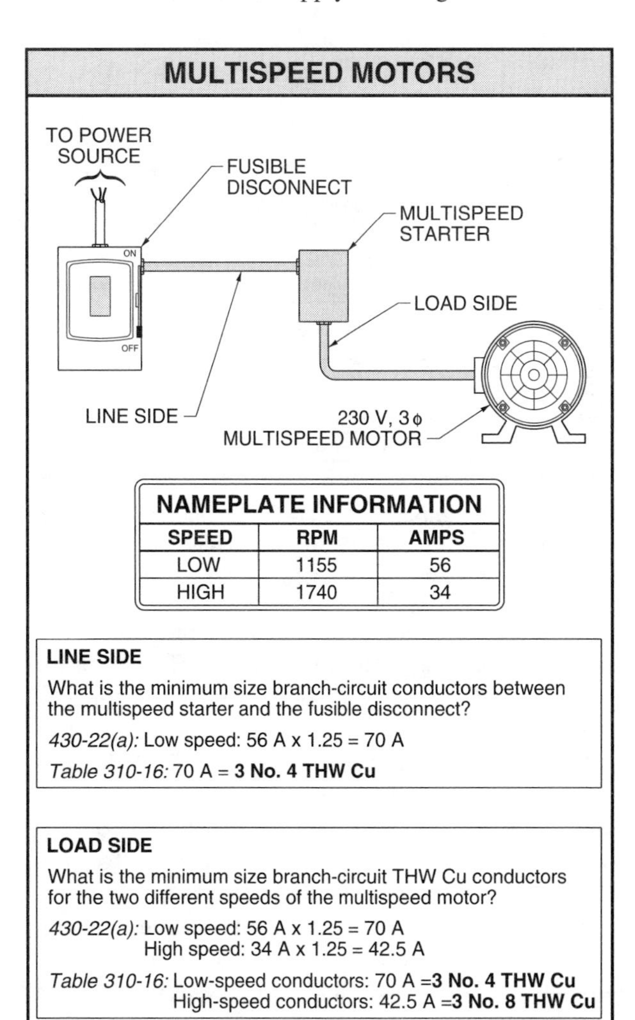

Figure 10-15. Line side branch-circuit conductors are sized at 125% of the FLC rating and load side branch-circuit conductors are sized at 125% of the current rating of the winding.

Duty-Cycle Motors – 430-22(a), Ex. 1. Table 430-22(a), Ex., is used to size branch-circuit conductors for duty-cycle motors. In applying this table, select the classification of service and find the percentage of nameplate current according to the motor's duty-cycle rating which is on the nameplate. See Figure 10-16.

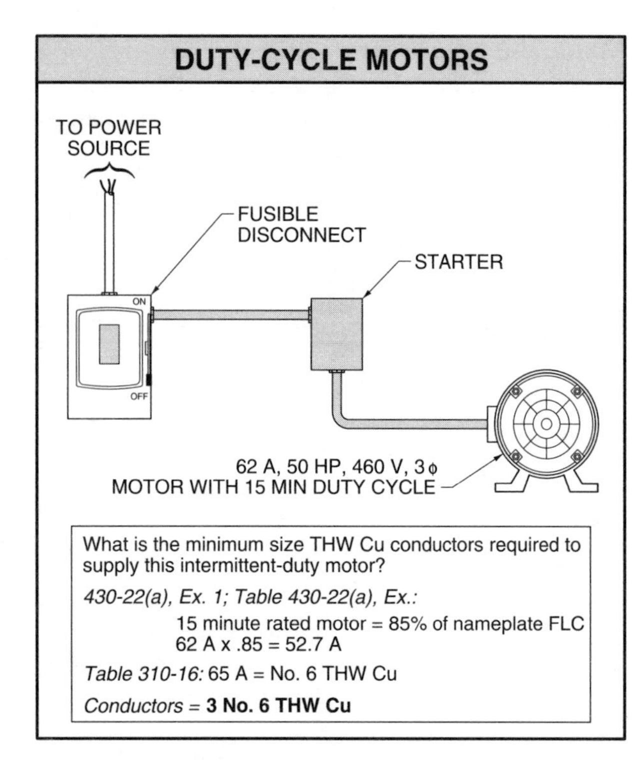

Figure 10-16. Table 430-22(a), Ex. is used to size branch-circuit conductors for duty-cycle motors.

DUTY-CYCLE RATINGS

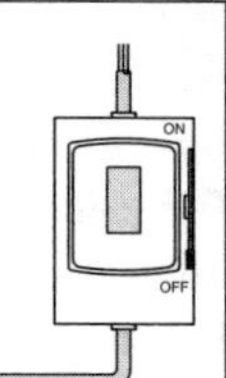

Motors are rated with 5 minute, 15 minute, 30 minute, 60 minute, and continuous duty-cycle ratings. A motor with a 5 minute rating is designed to operate at its rated HP for periods not to exceed 5 minutes. After this, a cooling or nonoperating period is required before the motor is operated again. The branch-circuit conductors also take advantage of this cooling off period and, therefore, do not have to be as large as a motor with a larger duty cycle.

Wound-Rotor Motors – 430-23. A wound-rotor motor is similar to a squirrel cage motor. The main difference is that there is a bank of resistors connected to the rotor through slip rings. Upon starting, all of the resistance is connected to the rotor for high-starting torque. Once the motor comes up to speed, the amount of resistance is controlled, which varies the torque and speed. These motors are seldom found in commercial installations. They are used in sewage treatment plants as lift pumps and in some crane applications. The full-load secondary current of a wound-rotor motor is found on the motor nameplate.

Per 430-23(a), for continuous-duty operation, the secondary leads between the motor and the controller shall be sized at least 125% of the secondary FLC of the motor. This rating is found on the motor nameplate. Per 430-23(b), for other than continuous duty operation, the secondary conductors may be sized less than 125% of the secondary current. These conductors shall have an ampacity, in percent of full-load secondary current, not less than that specified in Table 430-22(a). Section 430-23(c) requires the ampacity of the conductors between the controller and the resistor banks to be the allowable percentages of Table 430-23(c), depending on the resistor duty classification.

Synchronous Motors. Large, 3ϕ synchronous motors start like a squirrel-cage motor. The magnetic field on the rotor is a result of the induced current flow in the rotor caused by the cutting of the rotor bars by the rotating magnetic field in the stator. When the rotor reaches 90–95% of its rated speed, there is little slip. As a result, the magnetic field on the rotor is too weak to overcome the bearing and windage losses to catch up with the rotating magnetic field on the stator. At this point an outside source of DC current, which is independent of induction, is sent through the rotor bars. The rotor current increases, thus strengthening its magnetic field, and the rotor locks in step with the rotating magnetic field on the stator and spins at synchronous speed. In small 1ϕ synchronous motors, the rotor is a permanent magnet.

A/C AND REFRIGERATION EQUIPMENT – 440

Article 440 applies to electric motor-driven air-conditioning and refrigeration equipment. It also includes the branch circuits and control equipment of these circuits. There is a great deal of similarity between 440 and 430. The main difference is the special considerations that 440 has for circuits supplying hermetic refrigerant motor-compressors. A *hermetic refrigerant motor-compressor* is a combination of a compressor and motor enclosed in the same housing, having no external shaft or shaft seals, with the motor operating in the refrigerant. See 440-2.

The branch-circuit selection current is marked on the equipment nameplate. This is the current used in determining the ampacity of the branch-circuit conductors, disconnecting means, and size of controller. Air-conditioning and refrigeration equipment not only contain a motor-compressor, but often a fan motor is also included. The branch-circuit selection current is always equal to or greater than the load current.

The calculations performed in 430 always use motor FLC. In 440, the calculations are based on load current or branch-circuit selection current. When the branch-circuit selection current is marked on the nameplate of a piece of A/C or refrigeration equipment, it shall be used in selecting the size of the disconnecting means, controller, branch-circuit conductors, and overcurrent protection devices.

Section 440-3 requires that the provisions of 422 (appliances) apply to such equipment as room air conditioners, household refrigerators and freezers, water coolers, and beverage dispensers. If air-conditioning and refrigeration equipment is installed in hazardous locations, motion picture and TV studios, or other similar areas, these articles shall also be followed.

Marking on Hermetic Refrigerant Motor-Compressor Nameplate – 440-4

The nameplate of a hermetic refrigerant motor-compressor shall be marked with the manufacturer's name, phase, voltage, and frequency. The rated load current of the motor-compressor shall be marked on the motor nameplate or the nameplate of the equipment in which the compressor is used. Where the motor-compressor is internally protected with a thermal device, the equipment nameplate shall be marked with the words "Thermally-Protected System."

For multimotor and combination-load equipment, 440-4(b) requires that the equipment nameplate provide the manufacturer's name, voltage, frequency, number of phases, minimum supply-circuit conductor ampacity, and the maximum rating of the branch-circuit, short-circuit, and ground-fault protection device.

Marking on Controllers – 440-5

A controller suitable for a motor-compressor shall be marked with the manufacturer's name, voltage, phase, FLC and LRC or HP rating, and other data to indicate the motor-compressor for which it is suited.

Ampacity and Rating of Equipment and Compressor Motor – 440-6

The FLC rating on the motor-compressor is to be used in determining the size of the branch-circuit conductor ampacity, disconnecting means, short-circuit, ground-fault protection, controller, and the motor overload protector.

If the branch-circuit selection current is available on the equipment nameplate, it shall be used in lieu of the FLC. The overload protector shall be sized from the FLC rating. If the equipment nameplate does not indicate the FLC rating, it shall be taken from the motor-compressor nameplate.

Highest-Rated Motor – 440-7

Section 440-7 references 430-24 to size the conductors for two or more motors supplied by the same circuit. The conductors shall have an ampacity equal to 125% of the largest motor in the group plus the sum of the FLC ratings of all of the other motors in the group. See Figure 10-17.

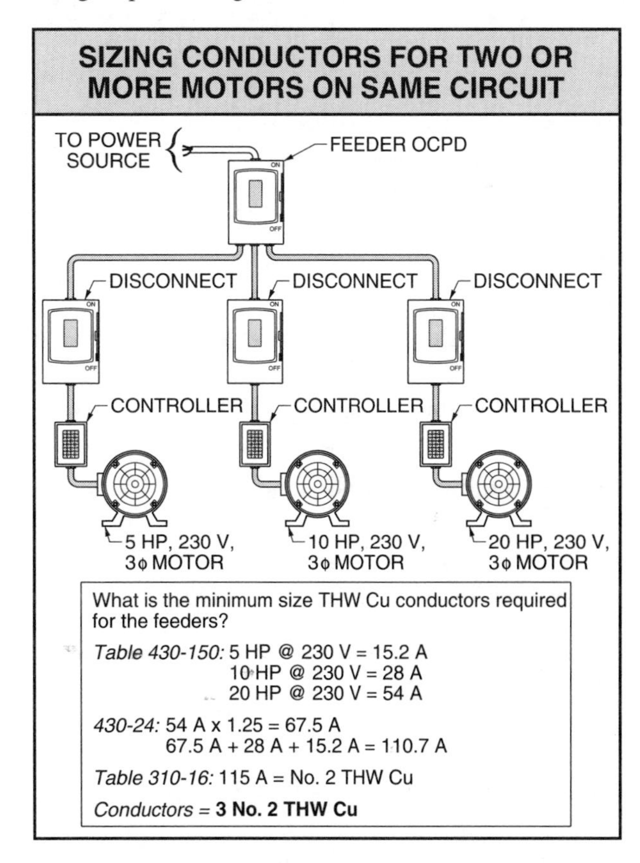

Figure 10-17. Conductors for two or more motors on the same circuit shall have an ampacity equal to 125% of the largest motor plus the sum of the FLC rating of the other motors in the group.

For determining the minimum size short-circuit, ground-fault protection when two or more motors are supplied by the same circuit, refer to 430-62(a). The highest branch-circuit overcurrent device shall be listed and then added to the sum of the FLC ratings of the remaining motors in the group. See Figure 10-18.

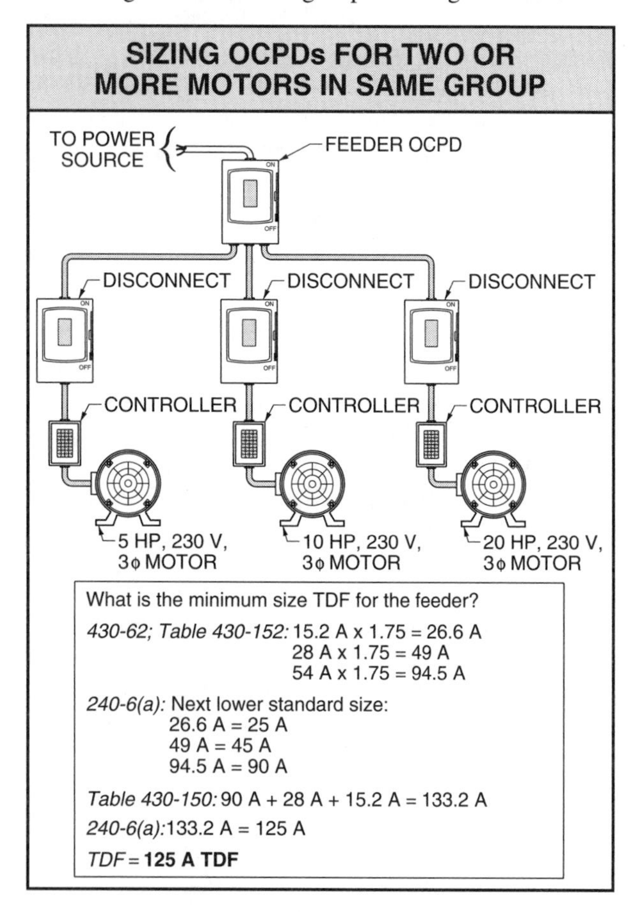

Figure 10-18. OCPDs for two or more motors on the same circuit are calculated per 430-62(a).

Single Machine – 440-8

Air-conditioning and refrigeration equipment shall be considered as a single machine, even though the system may contain several motors. Section 430-112, Ex. may be applied to determine how many disconnecting means are required.

Disconnecting Means – 440-11

The disconnecting means for air-conditioning and refrigeration equipment shall be capable of disconnecting all of the equipment including the motor-compressor from the supply circuit.

Rating and Interrupting Capacity – 440-12(a)(1). The disconnecting means for a refrigerant motor-compressor shall be at least 115% of the nameplate rated FLC or branch-circuit selection current, whichever is greater. The disconnecting means may be a HP-rated switch, a CB, or other type switch per 430-109.

Hermetic Refrigerant Motor-Compressor – 440-12(a)(2). If the nameplate rating on the motor-compressor or the equipment nameplate rating, is expressed in amperage and not HP, the following method may be used to determine the HP rating of the disconnect:

(1) Refer to the appropriate FLA Tables 430-148 through 430-150 and select the HP that corresponds to this current rating.

(2) If the LRC is marked on the nameplate, refer to Table 430-151A for 1ϕ motors or Table 430-151B for polyphase motors, and select the HP that corresponds to this LRC.

(3) Use the larger of the two HP ratings. See Figure 10-19.

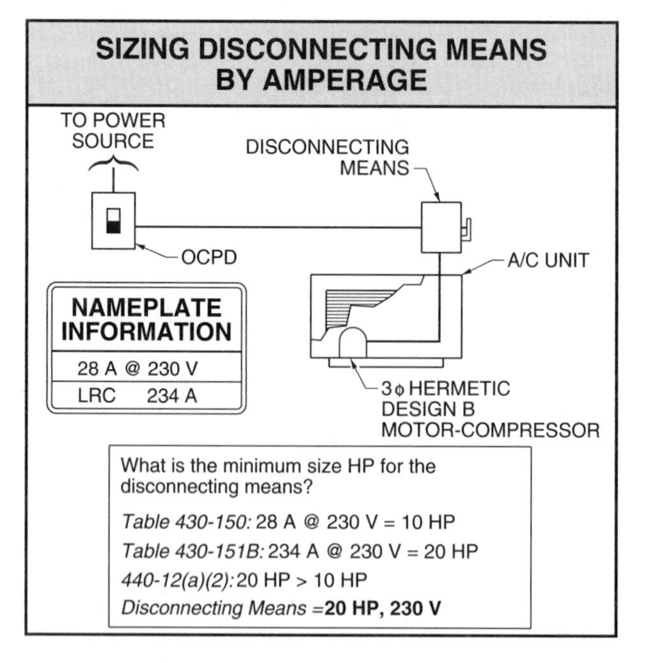

Figure 10-19. The disconnecting means is selected by the amperage if the nameplate rating is in amperage and not in HP.

If a CB is used as the disconnect for a hermetic motor-compressor, it shall be rated at least 115% of the nameplate FLA or the branch-circuit selection current, whichever is greater. See Figure 10-20.

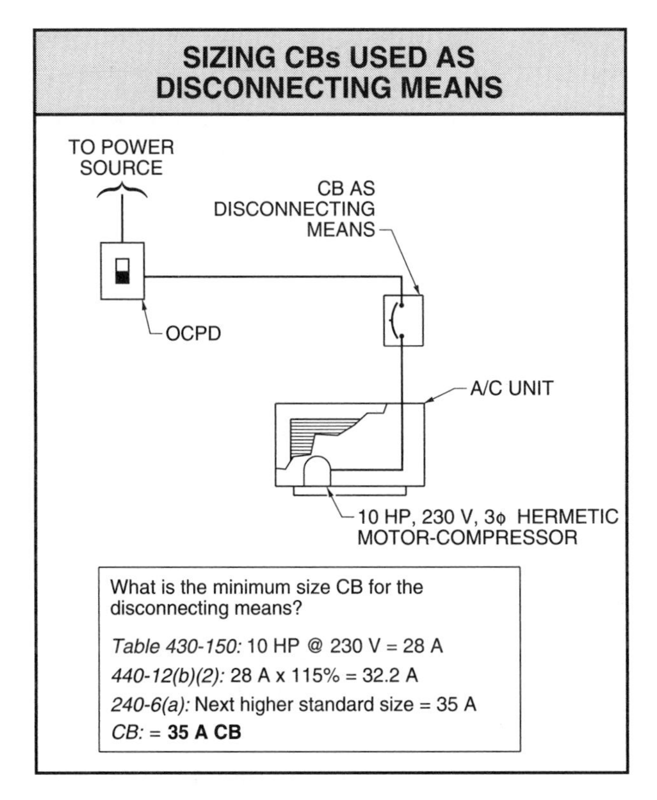

Figure 10-20. A CB used as a disconnecting means for a motor-compressor shall be rated at least 115% of the nameplate FLA or branch-circuit selection current, whichever is greater.

If a disconnecting means serves two or more hermetic motor-compressor or is used in combination with other motors or other loads, and the combined load may be on at the same time, the disconnect shall be sized as follows:

(1) Add each hermetic compressor load.

(2) Add each standard motor load.

(3) Add all other loads (if present).

The disconnecting means is then sized at 115% of the total HP.

Disconnecting Means for Cord-Connected Equipment – 440-13. Room air conditioners, household refrigerators, freezers, water coolers, and beverage dispensers that are cord-connected may use the plug and receptacle as the disconnecting means provided the installation complies with 440-63. Section 440-63 allows 1φ, 250 V or less room A/Cs to use the cord and plug as a disconnecting means provided the manual controls for the unit are readily accessible and located within 6′ of the floor, or an approved manual switch is readily accessible and within sight of the A/C unit. See Figure 10-21.

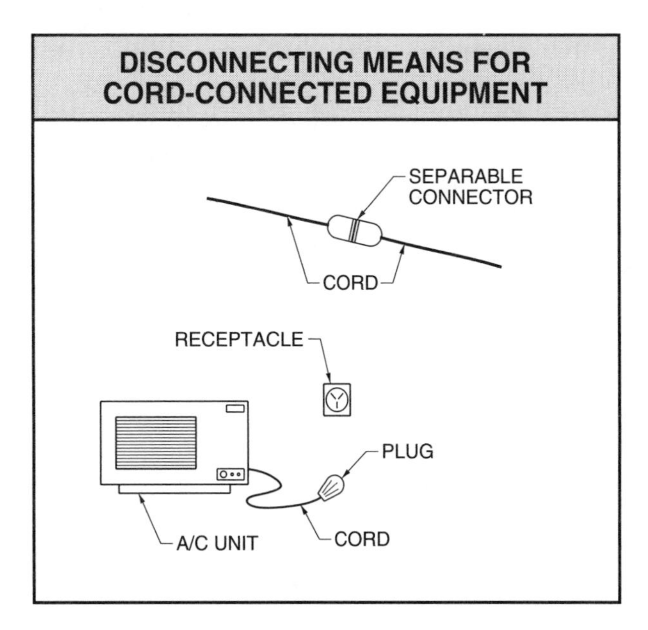

Figure 10-21. A cord and plug, or a separable connector, may serve as the disconnecting means for a room A/C unit.

Location of Disconnecting Means – 440-14. The disconnecting means for fixed-wired A/C and refrigeration equipment shall be installed within sight of the equipment and readily accessible to the user. When the equipment is not within sight of the disconnecting means, another switch shall be provided at the equipment. See Figure 10-22.

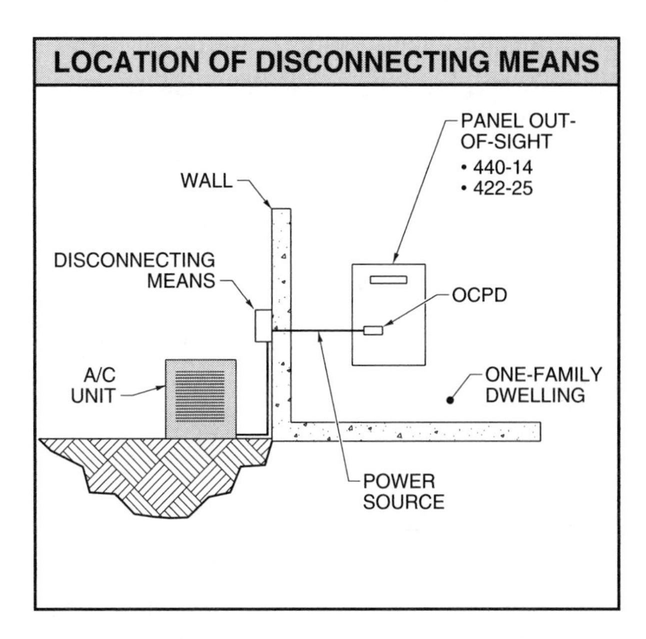

Figure 10-22. If the service panel is not within sight of the A/C unit, an additional disconnecting means shall be installed within sight.

The exception to this general rule permits the disconnect to be out-of-sight (50′ or more) of the unit provided the following conditions exist:

(1) Disconnect is capable of being locked open.

(2) Equipment is essential to an industrial process.

(3) Installation is a facility where conditions of maintenance and supervision ensure that only qualified persons service the equipment.

Branch-Circuit Fuses or CBs

When the listed A/C equipment nameplate states a certain maximum fuse size, it means the equipment was tested and listed for fuse protection only. The listing does not cover the equipment if CBs are used. If the equipment is tested and listed for fuses or CBs of the HACR type, it is indicated on the equipment nameplate. Listed CBs which have been found suitable for HACR installations are marked "Listed HACR Type."

Individual Motor-Compressor – 440-22(a). The rating or setting of the overcurrent protection shall not be greater than 175% of the nameplate FLC rating or the nameplate branch-circuit rating, whichever is greater. If the overcurrent protection does not permit the motor to start, a larger rating may be used provided it does not exceed 225% of the nameplate FLA or the nameplate branch-circuit rating, whichever is greater. See Figure 10-23.

Several Motors – 440-22(b). The rating or setting of the overcurrent protection shall be selected according to the number of hermetic motor-compressors, or the combination of hermetic motor-compressors and regular motors installed on the circuit. See Figure 10-24. Section 440-22(b)(1) requires that when two or more hermetic compressor motors are installed on the same circuit, the overcurrent protection device shall be selected according to the following:

(1) Take 175% of the largest motor's nameplate FLC or branch-circuit selection current rating, whichever is greater.

(2) Add to this the sum of the other motor's nameplate FLC ratings.

If this does not permit the compressor motors to start, increase the largest motor percentage from 175% to 225% plus the sum of the other motor nameplate FLC.

(3) Round ampacity down to next standard size OCPD.

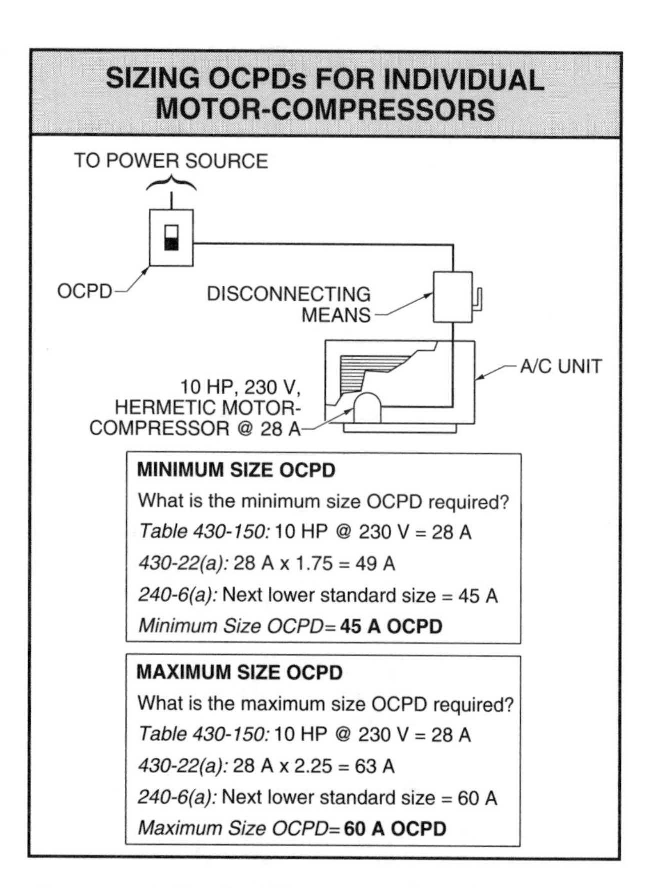

Figure 10-23. The OCPD shall not be larger than 175% of the nameplate FLC or branch-circuit rating, whichever is larger, per 440-22(a).

Section 440-22(c) requires that manufacturer's value for the overload protection device (heater) when marked on the equipment nameplate shall never be exceeded.

Branch-Circuit Conductors – 440, Part D

Section 440-32 requires that the branch-circuit conductors supplying an A/C unit containing an individual motor-compressor shall not be sized according to 430-22 as other motor loads are. On A/C units, the marked rated load current or the marked branch-circuit selection current on the equipment nameplate, if available, shall be used to determine the size of the branch-circuit conductors.

Compressor and Other Loads – 440-33. Conductors supplying one or more compressors and additional equipment such as fan motors shall have an ampacity of at least the sum of three items. See Figure 10-25. The three items are:

(1) Rated-load or branch-circuit selection current ratings, whichever is greater, of the motor-compressor(s).

(2) FLC ratings of other motors.

(3) 25% of largest motor or motor-compressor rating in the group.

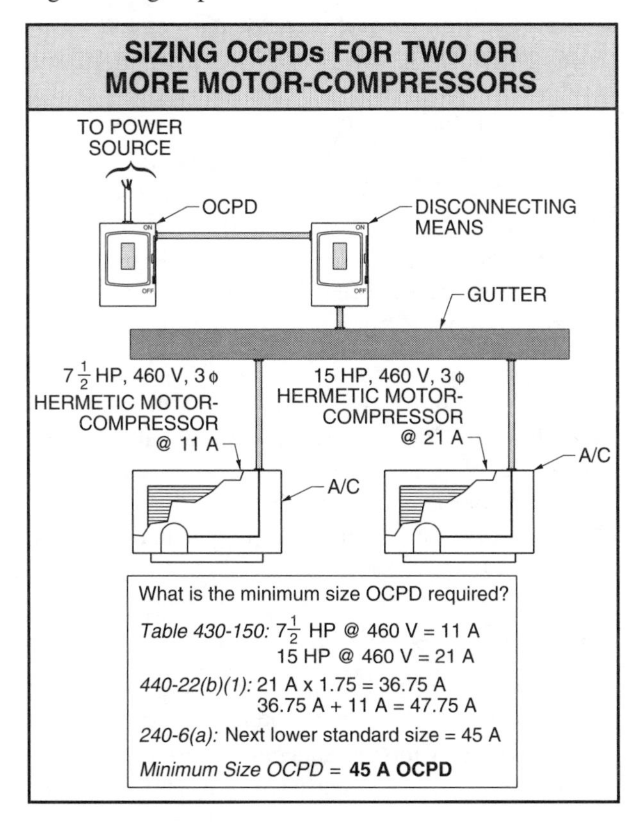

Figure 10-24. The OCPD for two or more hermetic motor-compressors on the same circuit is sized per 440-22(b).

Controllers for Compressor Motors – 440-41, Part E

Controllers which are used with motor-compressors shall be sized not less than the compressor nameplate FLA rating or branch-circuit selection current, whichever is larger, and the LRC rating of the motor-compressor.

In selecting a controller that is rated in HP only, the following shall be done to convert the compressor current ratings into HP:

(1) Refer to Tables 430-148 through 430-150 and select the corresponding HP rating as compared to the motor-compressor nameplate current at the proper voltage.

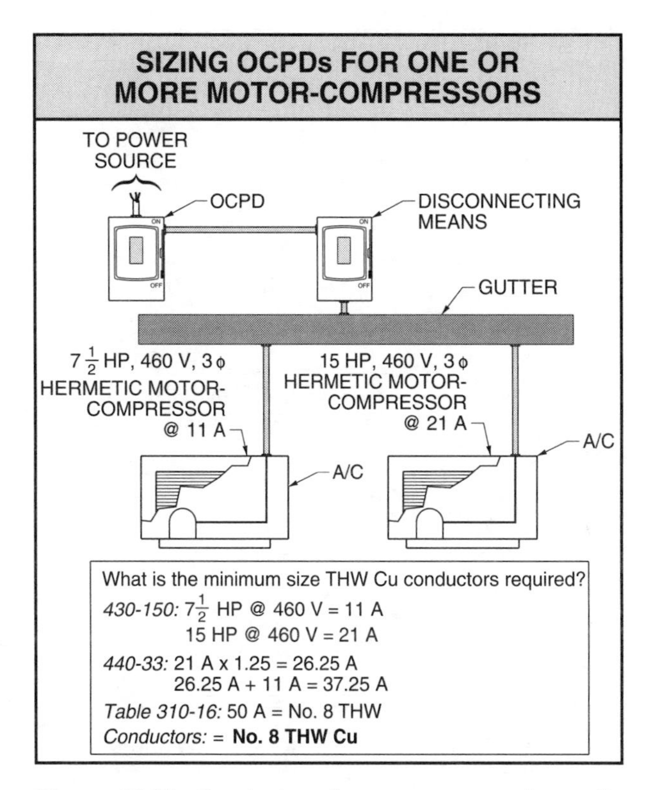

Figure 10-25. Conductors for one or more hermetic motor-compressors are sized per 440-33.

(2) Refer to Table 430-151A or 430-151B and select the corresponding HP rating as compared to the motor-compressor nameplate LRC. In both of these selections, if the nameplate current does not match the table current, the next higher value shall be used in determining the equivalent HP rating. See Figure 10-26.

Branch-Circuit Overload Protection – 440, Part F

Overload protection is required for the motor-compressor, branch-circuit conductors, and the motor control apparatus from excessive heating due to motor overloads. An *overload* is a small-magnitude overcurrent, that over a period of time, leads to an overcurrent which may operate the overcurrent protection device (fuse or CB). An overload does not include short circuits or ground faults.

The general rule of 440-52(a) is that the overload protection shall be rated at not more than 140% of the rated-load current of the motor-compressor. See Figure 10-27. Section 440-52(a)(2) permits the overload protection to be an integral part of the motor-compressor.

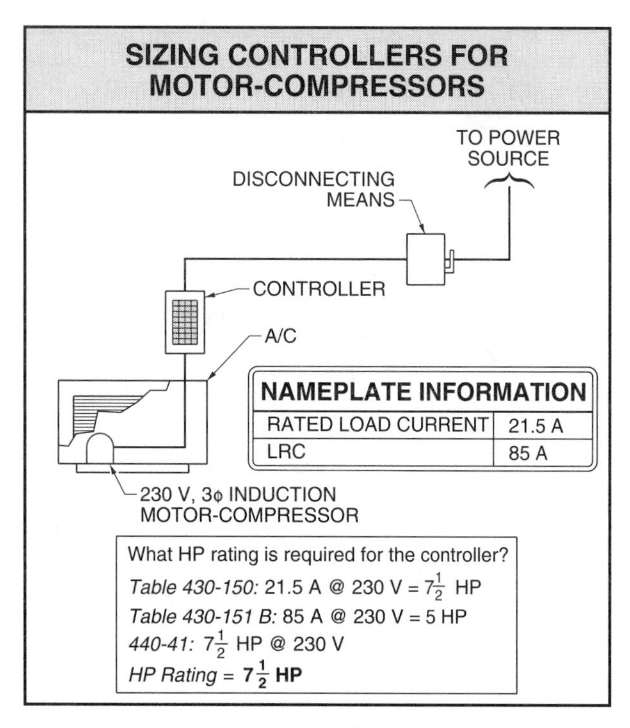

Figure 10-26. The controller for motor-compressors is sized per 440-41.

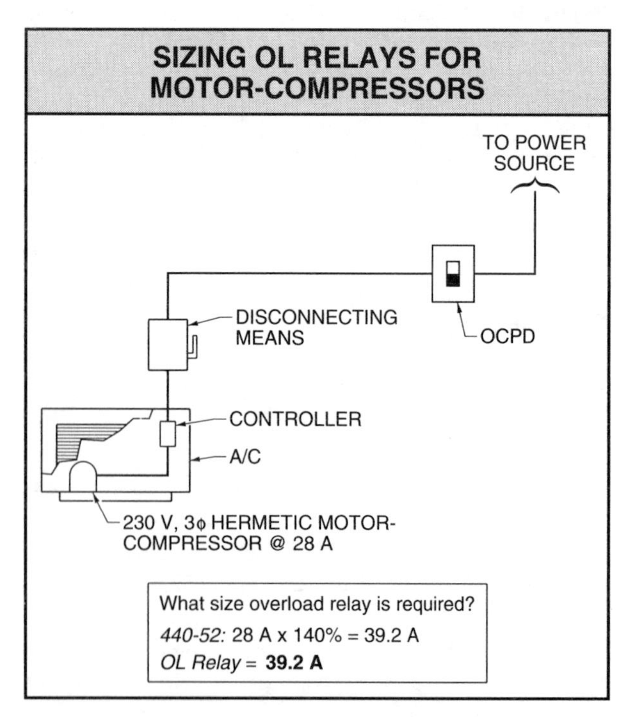

Figure 10-27. Overload protection for motor-compressors is sized per 440-52.

Overload Relays – 440-53. Overload relays do not provide short-circuit or ground-fault protection for the motor-compressor or the equipment associated with it. The overload relay requires time to open the motor circuit. The more severe the overload, the less time is required for the overload relay to open the control circuit. A short-circuit, ground-fault protector shall respond, if not instantly, within cycles of the fault.

Compressors and Equipment of 15 A or 20 A Circuits Not Cord- and Plug-Connected – 440-54. Hard-wired equipment and motor-compressors that are rated for 15 A or 20 A, 120 V, 1φ or 15 A, 240 V, 1φ branch circuits shall be provided with overload protection. A separate overload protector shall not exceed 140% of the FLC rating of the hermetic motor. The fuse or CB providing short-circuit protection shall have sufficient time delay to permit the compressor and other motors to start and bring their loads up to speed.

Compressors and Equipment of 15 A or 20 A Circuits that are Cord- and Plug-Connected – 440-55. Cord- and plug-connected compressors and equipment shall be rated no greater than 20 A on 120 V circuits. They shall be rated no higher than 15 A on 240 V circuits.

Room A/Cs – 440, Part G

Article 440-60 recognizes room A/C units which are window-mounted, console, or the in-wall type as being A/C appliances. Units that operate at 250 V, 1φ or less may be cord- and plug-connected or hard-wired. Units over 250 V, 1φ and 3φ shall be hard-wired.

Grounding – 440-61. Cord- and plug-connected room A/C units shall be grounded according to 250-45. The EGC shall be run within the supply cord of each unit. Hard-wired systems shall be grounded according to 250-42.

Branch-Circuit Requirements – 440-62. A room A/C is considered as a single motor unit when the following four condition are met:

(1) It is cord- and plug-connected.

(2) It is rated not more than 40 A, 250 V, 1φ.

(3) The total rated load current is marked on the unit nameplate.

(4) The rating of the fuse or CB does not exceed the branch-circuit conductors or the receptacle rating, whichever is less.

Section 440-62(b) requires that the rating of the A/C unit shall not exceed 80% of the branch-circuit conductor's ampacity. See Figure 10-28. Section 440-

62(c) requires that the rating of the A/C unit shall not exceed 50% of the branch-circuit conductor's ampacity where lighting or other appliances are on the same branch circuit. See Figure 10-29.

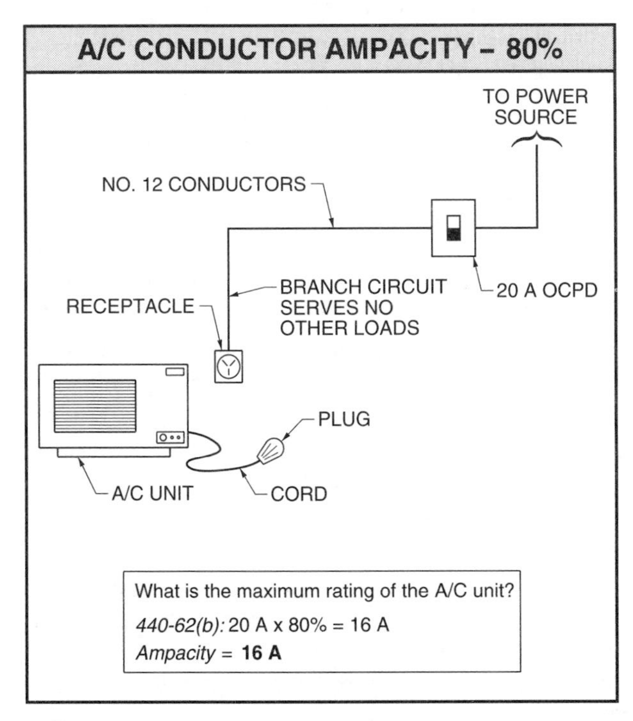

Figure 10-28. The rating of the A/C unit shall not exceed 80% of the branch-circuit conductor's ampacity.

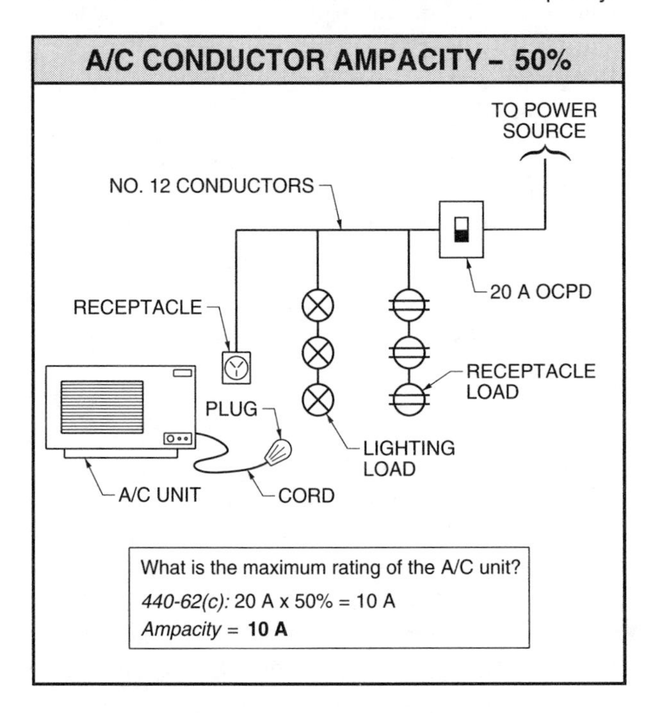

Figure 10-29. The rating of the A/C unit shall not exceed 50% of the branch-circuit conductor's ampacity with lighting or other appliances on the same circuit.

Disconnecting Means – 440-63. A separate disconnect is required for each A/C. The attachment plug and receptacle shall be permitted to serve as the disconnecting means if the unit is 1ϕ and not over 250 V, provided one of the following two provisions are met:

(1) The manual controls on the unit are readily accessible and located within 6′ of the floor.

(2) An approved manual switch is installed within sight from the unit and is readily accessible.

The length of flexible cords used to supply A/C units shall not exceed 10′ for 120 V units or 6′ for 208 V or 240 V units.

Generators – 445

Section 445-2 requires that generators shall be suitable for the location where they are installed. Section 445-3 requires that the generator nameplate shall provide the following information:

(1) Manufacturer's name.

(2) Rated frequency.

(3) Power factor.

(4) Number of phases if an AC generator.

(5) Rating in kW or kVA.

(6) Normal volts and amperes that correspond to the rating.

(7) Rated rpm.

(8) Insulation system class and rated ambient temperature or rated temperature rise.

(9) Time rating.

Overcurrent Protection – 445-4. Most AC generators are designed so that during periods of time when there is an excessive overload, the voltage diminishes enough to limit the current and power output to levels that do not damage the generator. Section 445-4 requires constant-voltage generators shall be protected from overloads. It is common practice not to protect the AC generator exciter, the reason being not to shut down the generator due to an inadvertent opening of the exciter overcurrent protection.

Section 445-4(b) permits overcurrent protection in only one conductor of a 2-wire DC generator. This overcurrent device shall be actuated by the entire generated current other than the current in the shunt field. The shunt field shall not be opened by the overcurrent device. If the shunt field were to open, a high voltage could be induced which would damage the field winding and the generator.

Section 445-4(c) does not require overcurrent protection for generators operating at 65 V or less that are part of a motor-generator set provided the motor protection device opens when the generator delivers 150% of its rated FLC.

Section 445-4, Ex. advises that the AHJ may prefer to have the generator operate until it fails, rather than have it automatically shut down and permit a greater hazard to personnel. An overload-sensing device shall be permitted to be connected to an annunciator or other indicating device which would permit authorized personnel an opportunity to have an orderly shut down of load-side equipment.

Ampacity of Conductors – 445-5. The phase conductors between the generator terminals and the first overcurrent device shall have an ampacity not less than 115% of the generator nameplate current rating. Neutral conductors are permitted to be sized according to 220-22. Conductors that carry ground-fault currents shall be sized according to 250-23(b).

Protection of Live Parts – 445-6. All live parts of generators that operate at over 50 V to ground shall not be exposed to unqualified persons.

FIRE PUMPS – 695

Article 695 is a new article in the 1996 NEC®. Section 695-1(a) covers the installation of electrical power sources, interconnecting circuits, and the switching and control equipment used only for fire pump drivers. Article 695 does not cover any built-in wiring, quality control of the system (performance, maintenance, and testing), or pressure maintenance pumps. These pumps are covered under the general requirements of 430.

Section 695-2 requires that the installation of conductors and equipment shall comply with Chapters 1 through 4 of the NEC® except as modified in this article. The material denoted by ^{X} is extracted from NFPA 20, *Fire Pump Standard*, which still has the primary jurisdiction for performance, maintenance, and testing while NFPA 70 has primary responsibility for the electrical construction and installation standards.

Power Source to Electric-Motor Driven Fire Pumps – 695-3

Power shall be supplied to an electrical motor-driven fire pump by either a service, on-site generator, or both. The power source shall be located and installed in such a fashion to minimize the possibility of damage by fire.

Section 695-3(b) requires that where power is obtained from a separate service, or a tap ahead of a service disconnecting means, it shall comply with 230-2, Ex. 1, 230-72(b), and 230-82, Ex. 5. A tap ahead of the service disconnecting means shall not be made within the disconnecting means enclosure.

Section 695-3(c) requires that the supply conductors shall directly connect the fire pump controller to the power source. Per 695-3(c), Ex. 1 the disconnecting means and overcurrent device between the power supply and the fire pump controller are permitted provided the following five items are complied with:

(a) The OCPD shall be able to carry indefinitely the sum of the LRC of the fire pump motor(s) and the pressure maintenance pump motor(s) and the FLA of any associated fire pump accessory equipment.

(b) The disconnecting means shall be lockable in the ON position and shall be marked "Suitable For Use As Service Equipment."

(c) A placard with letters no less than 1″ in height stating "Fire Pump Disconnecting Means" shall be displayed externally on the disconnecting means.

(d) A placard shall be placed next to the fire pump controller stating the location of the disconnecting means and the location of the key (if locked).

(e) The disconnecting means shall be in the closed position and supervised by either a lock or a seal or a remote or local signaling device.

Power Wiring – 695-8

The supply conductors shall be physically routed outside the building and installed as service conductors. If it is not possible to route these conductors outside the building, the conductors shall comply with 230-6, which requires the conductors to be installed under or enclosed within no less than 2″ of concrete. Per 695-8(c), fire pump conductors shall be protected against short-circuit currents only.

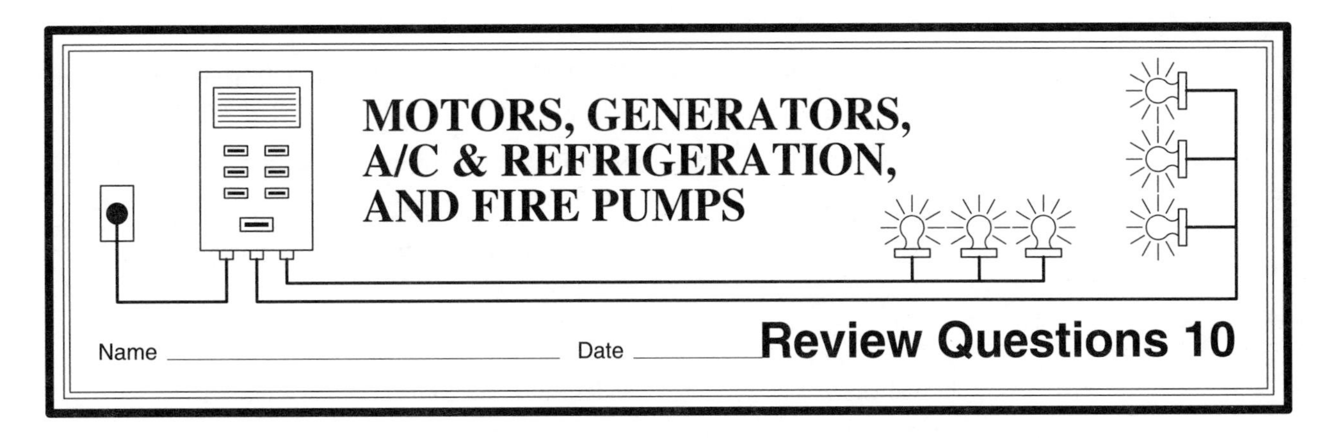

Review Questions

_______ **1.** Motor nameplates shall include the rated horsepower if _______ HP or more.

_______ **2.** The _______ branch circuit is that point from the last fuse or CB in the motor circuit out to the motor.

_______ **3.** Branch-circuit conductors supplying a single motor shall have an ampacity of _______% of the FLC rating as indicated on the appropriate FLC table.

_______ **4.** The _______ is the device in a motor circuit which turns the motor ON or OFF.

_______ **5.** A(n) _______ is an assembly of one or more sections with a common power bus and primarily containing motor control units.

A. ACM
B. CCM
C. MCC
D. MCA

_______ **6.** A(n) _______ refrigerant motor-compressor is a combination compressor and motor in the same housing, having no external shaft or shaft seals, with the motor operating in the refrigerant.

_______ **7.** A(n) _______ is a small-magnitude overcurrent which, over a period of time, leads to an overcurrent that may operate the fuse or CB.

_______ **8.** Cord- and plug-connected compressors and equipment shall be rated no greater than _______ A on 120 V circuits.

_______ **9.** The length of flexible cords used to supply A/C units shall not exceed _______′ for 120 V units.

_______ **10.** Table _______ is used to determine the current ratings of 3ϕ motors.

_______ **11.** The code letter or locked-rotor amps shall be listed on the nameplates of motors rated _______ HP or more.

T F **12.** Small motors rated 150 W or less shall be marked "Impedance Protected."

T F **13.** The larger a motor's full-load table current, the less the percentage of the increase in sizing the thermal protector that is an integral part of the motor.

T F **14.** A plug and receptacle may serve as a controller for portable motors up to ⅓ HP.

__________ **15.** Feeder conductors supplying several 3ϕ motors shall have an ampacity equal to or larger than the sum of all FLC ratings plus ________% of the highest rated motor in the group.

__________ **16.** Over ________% of the motors in use today are single speed, AC induction motors.

__________ **17.** A CB used as a disconnecting means for a motor-compressor shall be rated at least ________% of the nameplate FLA or branch-circuit selection current, whichever is greater.

A. 100
B. 115
C. 125
D. 140

__________ **18.** A/C units that operate at ________ or less may be cord- and plug-connected.

A. 480 V, 3ϕ
B. 208 V, 3ϕ
C. 250 V, 1ϕ
D. neither A, B, nor C

T F **19.** The winding type shall be given on the nameplate of a DC motor.

T F **20.** The rated current or HP shall be given on the nameplate of a motor controller.

__________ **21.** A(n) ________ device is a pilot light, buzzer, horn, or other type of alarm.

T F **22.** The overcurrent protection for an MCC may be located ahead or within the MCC.

T F **23.** A 50 A fusible disconnect is a standard size.

T F **24.** A/C and refrigeration equipment with several motors is considered as a single machine.

__________ **25.** Equipment shall be within ________′ to be considered within sight.

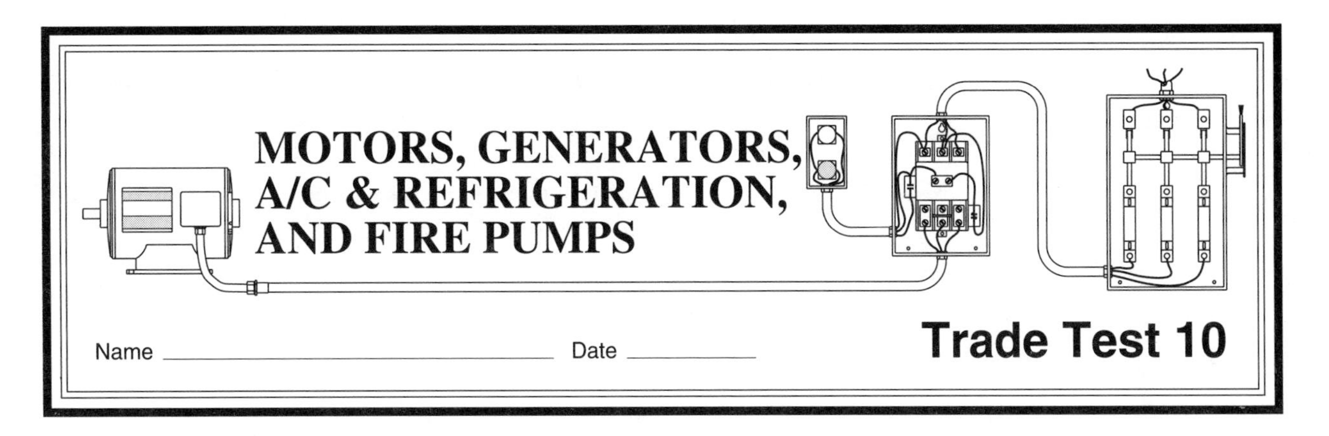

NEC®	Answer	
________	________	**1.** *See Figure 1.* What size THW Cu conductors are required for the motor?
________	________	**2.** *See Figure 2.* What is the smallest raceway (using EMT) permitted for the conductors?

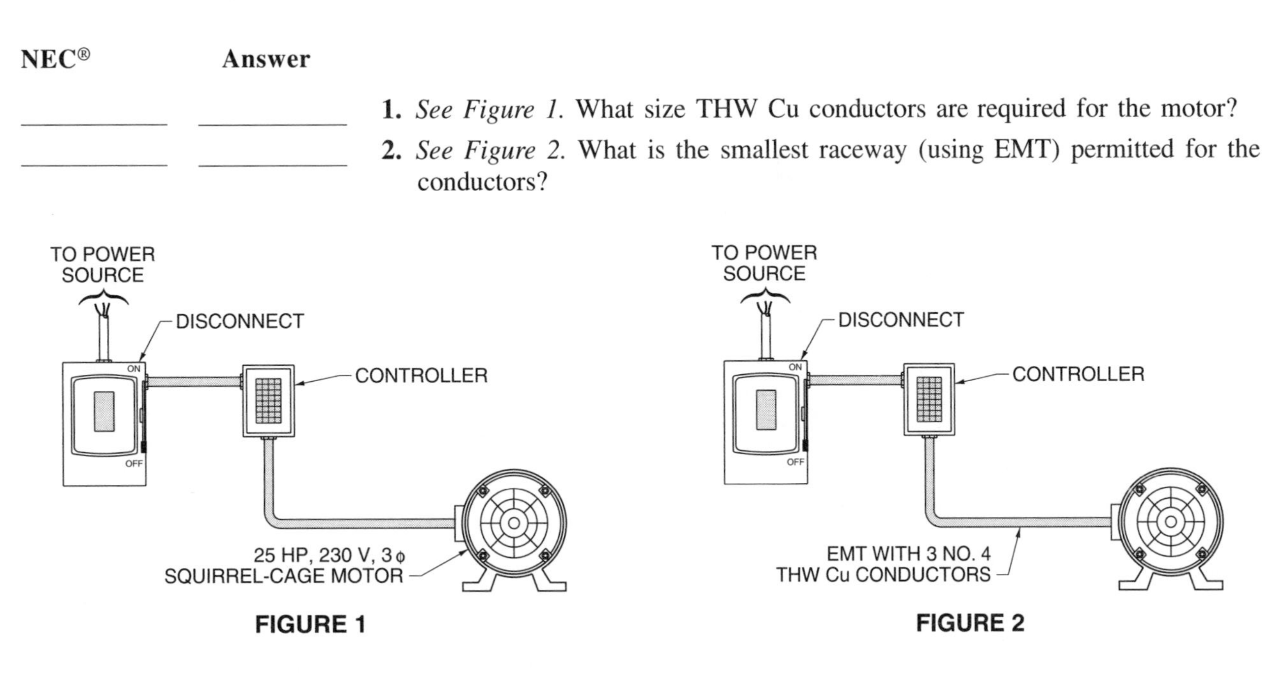

FIGURE 1

FIGURE 2

________	________	**3.** *See Figure 3.* What is the minimum size OL device required?
________	________	**4.** *See Figure 4.* What is the minimum size OL device required?
________	________	**5.** *See Figure 5.* What is the minimum size TDFs required to provide short circuit protection for the motor?

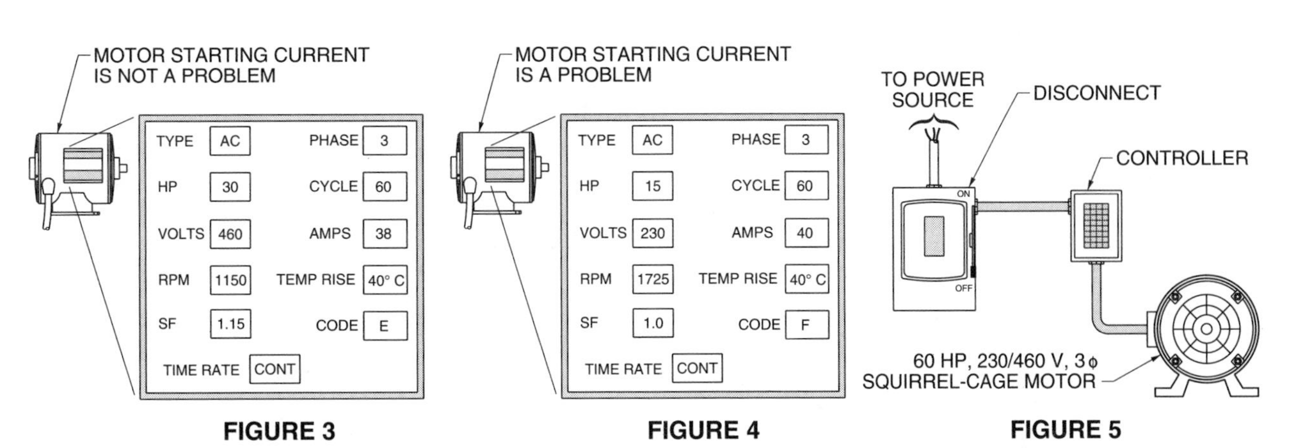

FIGURE 3

FIGURE 4

FIGURE 5

__________ __________
__________ __________

6. *See Figure 6.* What size THW conductors and raceway (using EMT) are required for the feeder circuit?

__________ __________

7. *See Figure 7.* What is the minimum size TDFs required for the motor feeder circuit?

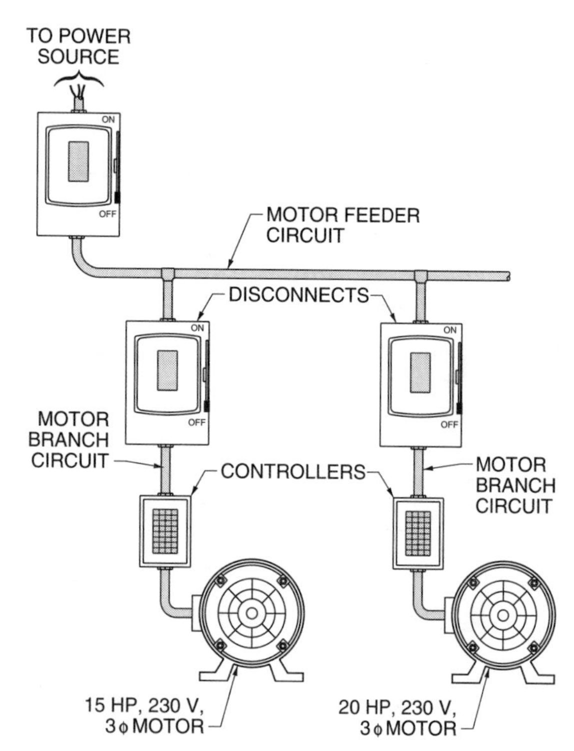

FIGURE 6

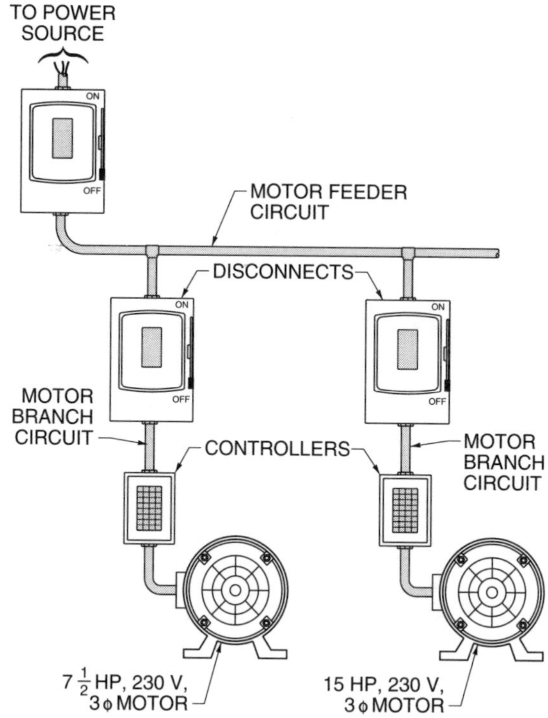

FIGURE 7

__________ __________
__________ __________

8. *See Figure 8.* What is the minimum size disconnect required for the motor?

9. *See Figure 9.* What is the minimum size branch-circuit conductors between the multispeed starter and the fusible disconnect?

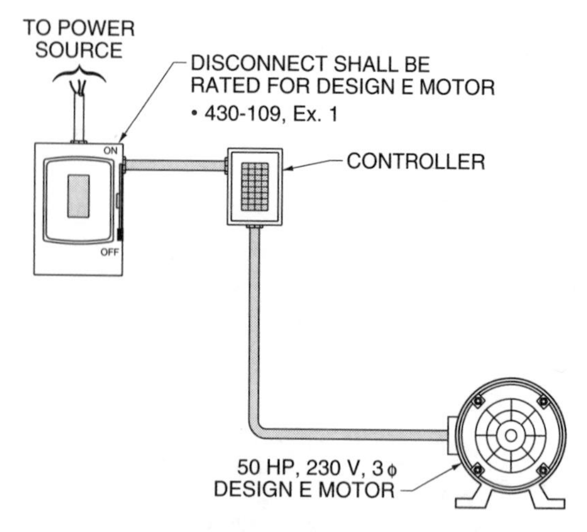

FIGURE 8

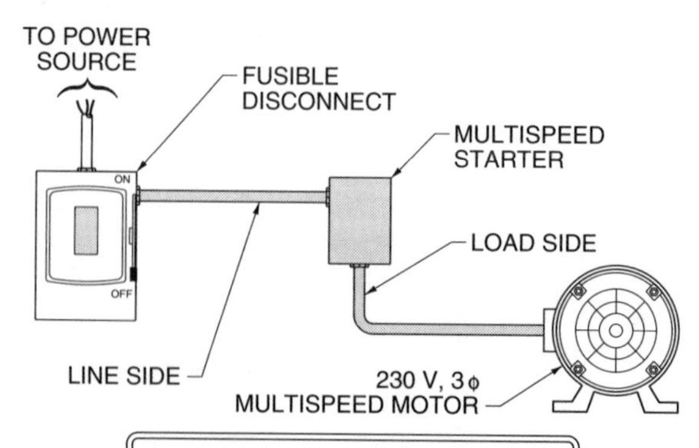

NAMEPLATE INFORMATION		
SPEED	**RPM**	**AMPS**
LOW	1155	52
HIGH	1740	29

FIGURE 9

__________ __________ **10.** *See Figure 10.* What is the minimum size THW Cu conductors required to supply this intermittent-duty motor?

__________ __________ **11.** *See Figure 11.* What is the minimum size THW Cu conductors required for the feeders?

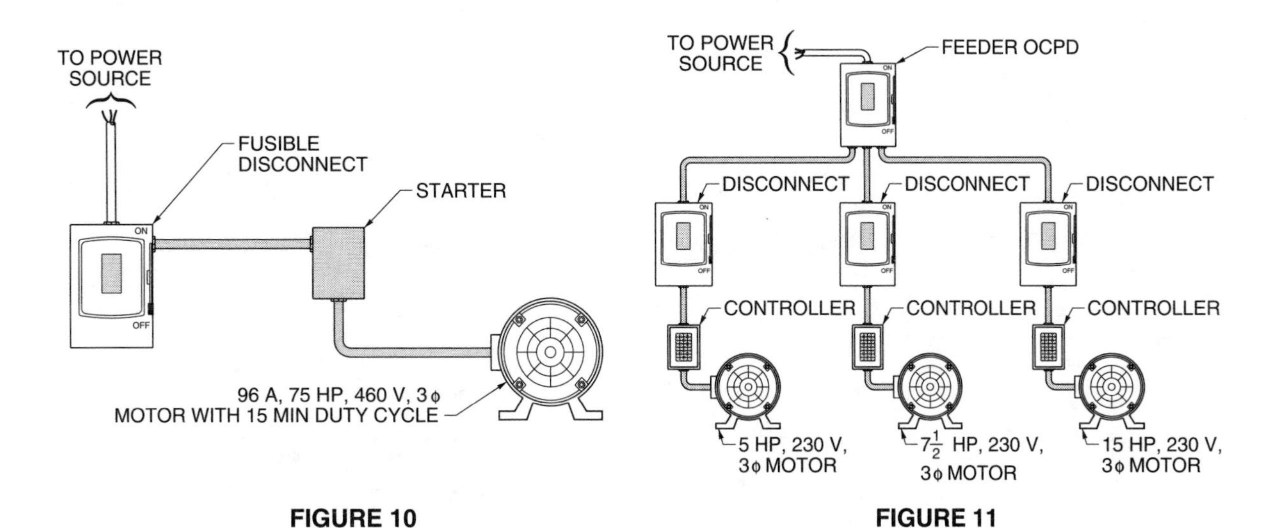

FIGURE 10 **FIGURE 11**

__________ __________ **12.** *See Figure 12.* What is the minimum size HP for the disconnecting means?

__________ __________ **13.** *See Figure 13.* What is the minimum size CB for the disconnecting means?

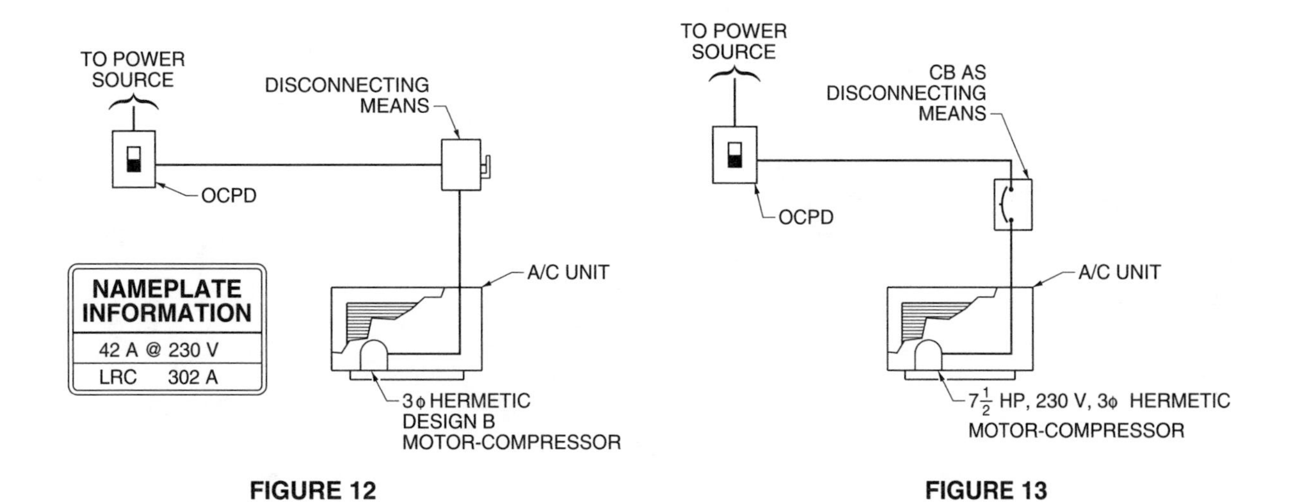

FIGURE 12 **FIGURE 13**

_____ _____ **14.** *See Figure 14.* What is the minimum size OCPD required?

_____ _____ **15.** *See Figure 15.* What is the minimum size THW Cu conductors required?

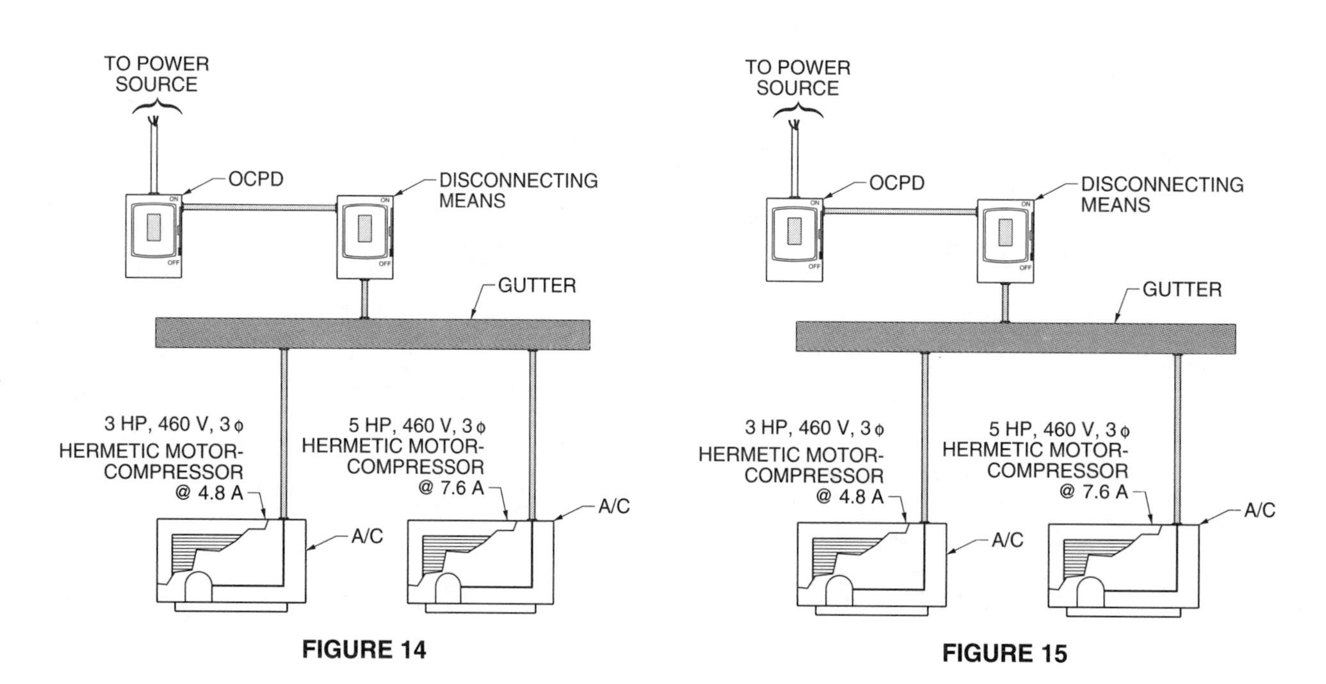

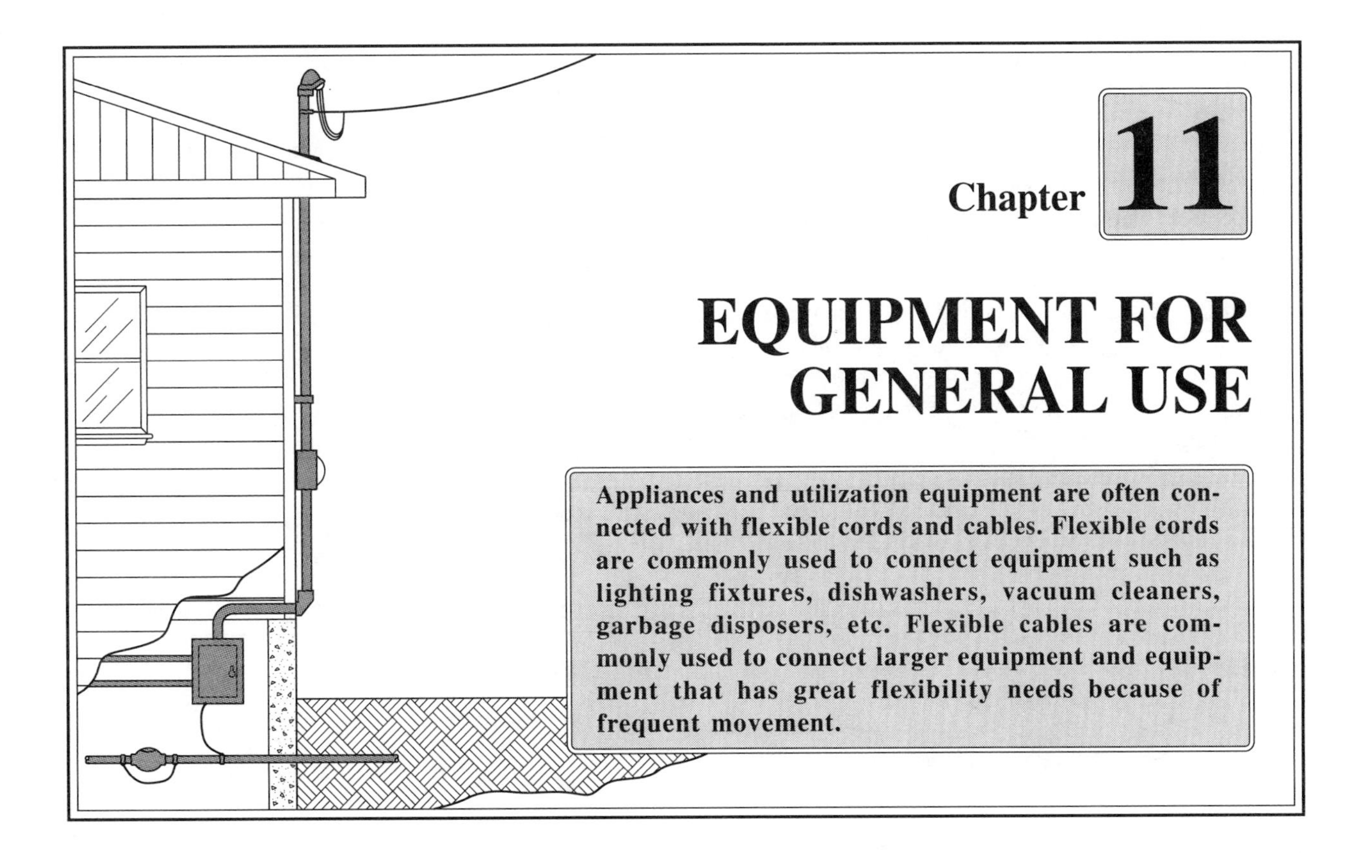

Chapter 11

EQUIPMENT FOR GENERAL USE

Appliances and utilization equipment are often connected with flexible cords and cables. Flexible cords are commonly used to connect equipment such as lighting fixtures, dishwashers, vacuum cleaners, garbage disposers, etc. Flexible cables are commonly used to connect larger equipment and equipment that has great flexibility needs because of frequent movement.

FLEXIBLE CORDS AND FLEXIBLE CABLES – 400

An *appliance* is any utilization equipment which performs one or more functions, such as clothes washing, air conditioning, cooking, etc. Appliances are generally constructed in standard sizes and types and are used in residential or commercial occupancies as opposed to industrial occupancies. *Utilization equipment* is any electrical equipment which uses electrical energy for electronic, electromechanical, chemical, heating, lighting, etc. Appliances and utilization equipment may be connected with flexible cords and cables.

A *flexible cord* is an assembly of two or more insulated conductors, with or without braids, contained within an overall outer covering and used for the connection of equipment to a power source. A *flexible cable* is an assembly of one or more insulated conductors, with or without braids, contained within an overall outer covering and used for the connection of equipment to a power source.

The primary difference between flexible cords and flexible cables is in their designated use. Flexible cords are commonly used to supply equipment such as lighting fixtures, dishwashers, vacuum cleaners, garbage disposers, etc. Flexible cables are commonly used for the connection of larger equipment and equipment that has great flexibility needs because of frequent movement. Typical equipment connected with flexible cables includes elevators, stage and lighting equipment, and electric vehicle charging equipment.

Table 400-4 contains a list of flexible cords and cables. Information on the construction of the cord or cable, type of insulation and covering, and permitted uses for each cord or cable is provided. Cords or cables, other than those listed in Table 400-4, shall be subjected to a special investigation to determine their suitability for the intended use.

Markings – 400-3; 400-6

Each flexible cord or cable and its associated fittings shall be suitable for the intended use and location. For example, cords intended for use in a wet location shall be suitable for wet locations. The Use column in Table 400-4 indicates cords that are suitable for use in wet locations. Likewise, if the cord is required to be of hard or extra-hard use, as is required in 305-4(c) for temporary branch-circuit wiring, the cord shall be identified in Table 400-4 as suitable for such use.

Marking of flexible cords and cables may be accomplished by means of a printed tag attached to the coil or reel of the cord or cable per 400-6(a). See Figure 11-1. The information required to be included on the printed tag is:

- Maximum rated voltage for which the cord or cable is listed.
- Proper letters for identifying the cord or cable.
- Type of cord or cable.
- Manufacturer's name, trademark, or other distinctive markings.
- AWG size or circular mil area of the conductors. (Some specific cords and cables are required to be surface-marked at intervals not exceeding 24″ with the type, size, and number of conductors.)

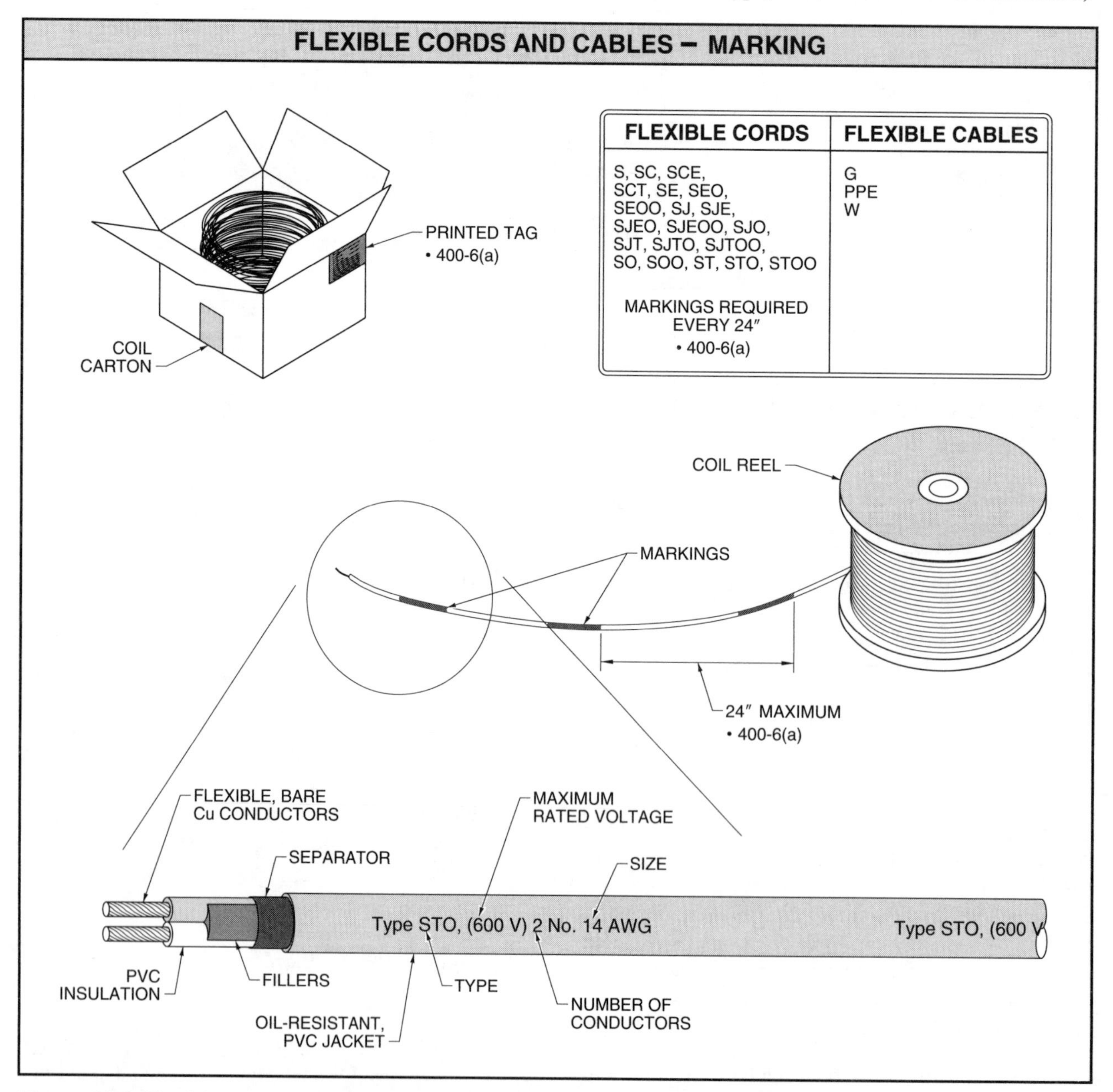

Figure 11-1. Flexible cords and cables shall be marked.

Other optional cord or cable markings may be surface marked per 400-6(b). These markings can be used to indicate special cable or cord characteristics such as LS for limited smoke characteristics or sunlight resistant for cords or cables that can be exposed to the direct rays of the sun.

Ampacities – 400-5

Flexible cords and cables, like any other conductor, have a maximum allowable ampacity which shall not be exceeded. As with other conductors, the allowable ampacity depends upon the type of insulation used and the number of conductors contained within the flexible cord or cable. Table 400-5(A) provides the allowable ampacities for many types of flexible cords and cables. Two allowable ampacities are listed for many of the cord and cables.

The allowable ampacities under Column A apply to multiconductor cords connected to utilization equipment in which only three of the conductors are current-carrying. Column B applies to multiconductor cords connected to utilization equipment in which only two of the conductors are current-carrying. Because the amount of current-carrying conductors is limited to two, Column B permits higher allowable ampacities.

The same rules that apply to other conductors for determining when they are current-carrying also apply to flexible cords and cables. If a neutral conductor only carries unbalanced current from conductors of the same circuit, it need not be considered current-carrying. If the circuit is a 3-wire circuit consisting of two phase wires and the neutral of a 4-wire, wye connected, 3ϕ system, the neutral shall be considered a current-carrying conductor. In addition, if a major portion of the load on a 3ϕ, 4-wire wye circuit consists of nonlinear loads, such as electric-discharge lighting and computer equipment, the neutral shall also be considered a current-carrying conductor.

Section 400-5 includes the applicable derating values for flexible cords and cables depending upon the number of current-carrying conductors. As with other conductors, the EGC is not considered a current-carrying conductor for the purposes of determining the number of conductors in the flexible cord or cable.

Table 400-5(B) can be used to determine the ampacity of flexible cable types SC, SCE, SCT, PPE, G, and W. Ampacities shall be selected from Columns D, E, and F depending upon the number of current-carrying conductors within the flexible cable. The same derating provisions which apply for Table 400-5(A) apply for Table 400-5(B).

Overcurrent Protection – 400-13

Overcurrent protection for flexible cords and cables is treated separately from other conductors. Per 400-13, flexible cords not smaller than No. 18 AWG shall be considered protected by the overcurrent devices specified by 240-4. Tinsel cords or other flexible cords of sizes smaller than No. 18 AWG, with similar construction characteristics, can also be protected against overcurrents by overcurrent protection devices specified in 240-4.

In general, flexible cords and cables shall be protected against overcurrents per ampacities listed in Tables 400-5(A) and 400-5(B). If a flexible cord or tinsel cord, approved for use with a specific listed appliance or portable lamp, is connected to a branch circuit installed per 210, the minimum cord size is determined by the specific branch-circuit ratings. See Figure 11-2.

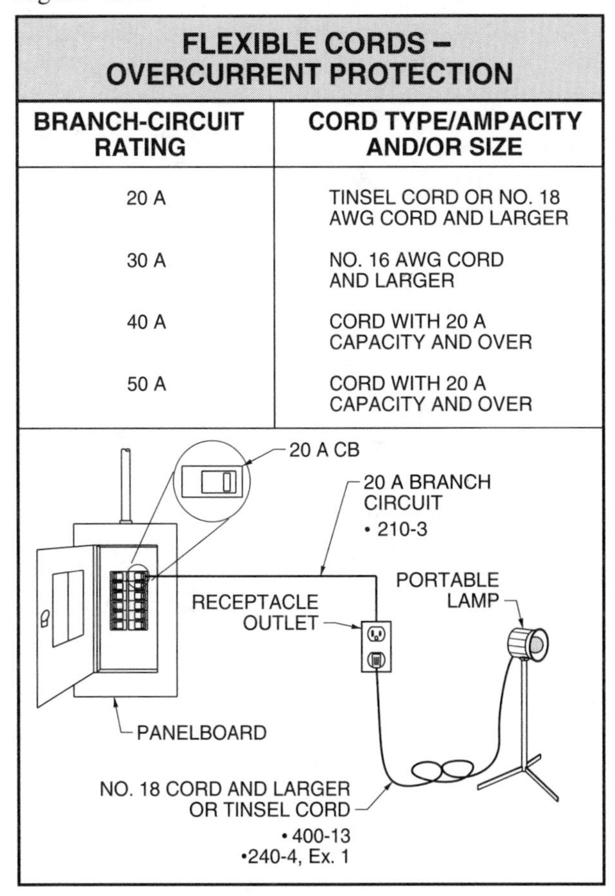

FLEXIBLE CORDS – OVERCURRENT PROTECTION

BRANCH-CIRCUIT RATING	CORD TYPE/AMPACITY AND/OR SIZE
20 A	TINSEL CORD OR NO. 18 AWG CORD AND LARGER
30 A	NO. 16 AWG CORD AND LARGER
40 A	CORD WITH 20 A CAPACITY AND OVER
50 A	CORD WITH 20 A CAPACITY AND OVER

Figure 11-2. Branch-circuit OCPDs are permitted to provide overcurrent protection for flexible cords and cables if of sufficient size or ampacity.

If flexible cords are used in listed extension cords, 240-4, Ex. 3, permits a 20 A overcurrent device to protect extension cords constructed with No. 16 AWG and larger conductors. The use of extension cords is covered by 305, Temporary Wiring.

Installation – 400-7 through 400-14

As with any wiring method, the most important step, prior to the installation of the flexible cords and cables, is to review the permitted uses. For flexible cords and cables, 400-7 contains the permitted uses and 400-8 contains the uses not permitted. Electrical installers and design personnel should review these sections to determine if the use of flexible cords and cables is permitted for the specific installation.

Another installation practice which requires review is splices. In general, 400-9 requires that flexible cords shall not contain splices. Likewise, 400-10 requires that flexible cords shall be installed in a manner that provides strain-relief for the cord or the cable at joints and terminals.

Uses Permitted – 400-7. Eleven uses are permitted for flexible cords and cables per 400-7(a). See Figure 11-3. Uses permitted are:

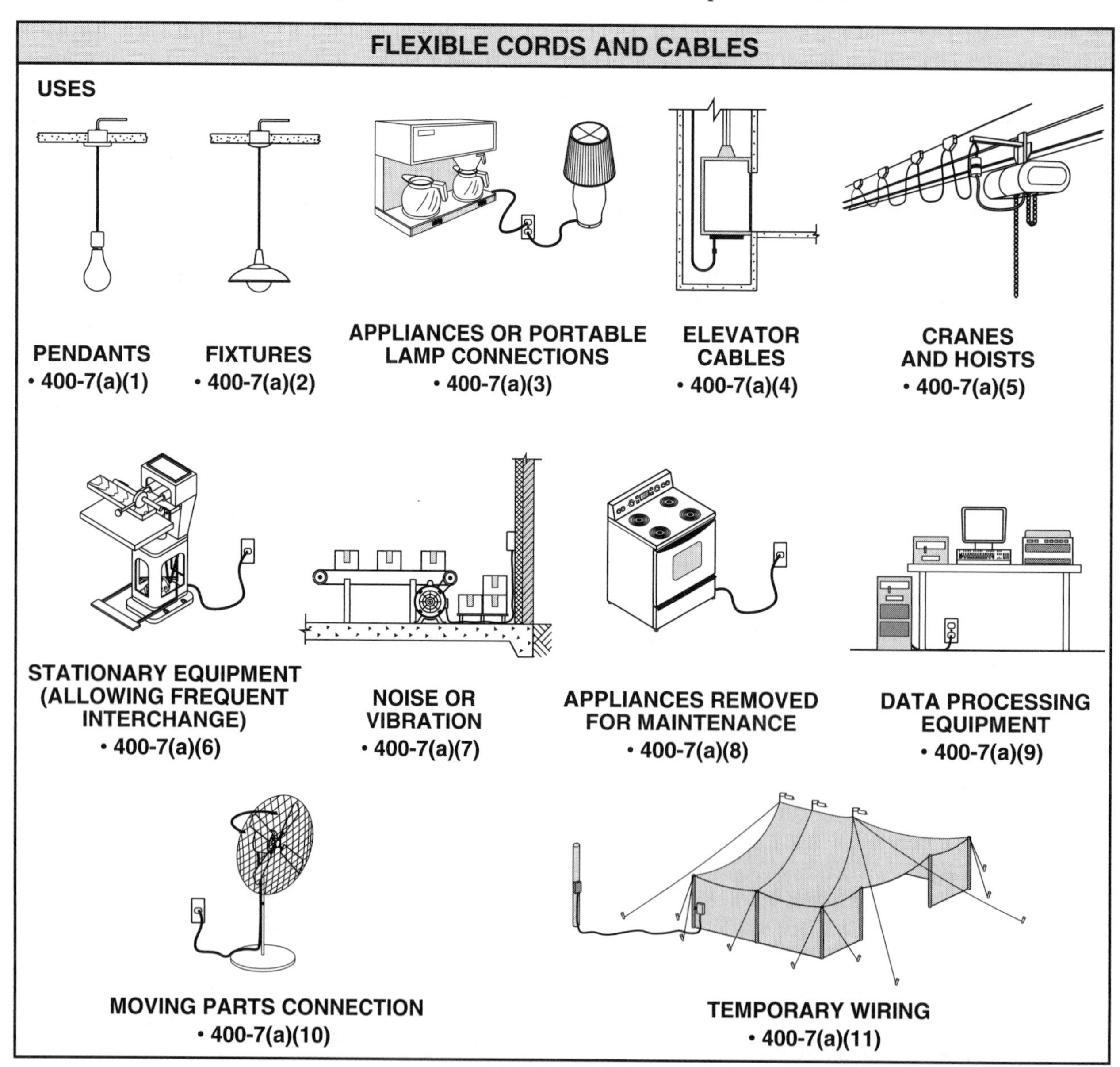

Figure 11-3. Eleven uses are permitted for flexible cords and cables.

(1) Pendants – *Pendants* are hanging light fixtures that use flexible cords to support the lampholder.

(2) Fixtures – Flexible cables are used to wire hanging fixtures.

(3) Appliances or portable lamps – Flexible cords are also permitted to supply lighting fixtures and other portable lamps or appliances. An attachment plug is required and shall be supplied from a receptacle outlet.

(4) Elevator cables – Flexible cables are required to follow elevator cars.

(5) Cranes and hoists – Flexible cables follow the moving parts of cranes and hoists.

(6) Stationary equipment with frequent interchange – Flexible cables provide flexibility in equipment requiring frequent interchange. An attachment plug is required and shall be supplied from a receptacle outlet.

(7) Noise or vibration transmission prevention – Machines subject to vibration are commonly wired with flexible cables.

(8) Appliances removed for maintenance – One of the most common uses for flexible cords and cables is for the supply of appliances that need to be removed for repair or periodic maintenance. An attachment plug is required and shall be supplied from a receptacle outlet.

(9) Data processing equipment – If the data processing equipment is installed in information technology equipment rooms, the provisions of 645-5 shall be met.

(10) Moving parts – The electrical connection of moving parts require great flexibility in the wiring method and flexible cords and cables meet the requirements.

(11) Temporary wiring – Flexible cords and cables provide wiring for temporary hookups.

Sections 305-4(b) and 305-4(c) permit flexible cords and cables to be used for feeder and branch-circuit wiring where temporary power is required. Such flexible cords and cables shall be identified in Table 400-4 for hard usage or extra-hard usage.

Uses Not Permitted – 400-8. Six uses for flexible cords and cables are specifically prohibited per 400-8. See Figure 11-4. Uses prohibited are:

(1) Substitute for building wiring – If an extension to an existing branch circuit is required to supply a new appliance, the wiring method for the extension of the branch circuit could not be flexible cord or cable as this would be a substitute for fixed wiring.

(2) Passed through walls, ceilings, or floors – To ensure that flexible cords and cables are not subject to physical damage, they are not permitted to be passed through walls, ceilings, or floors.

(3) Passed through doorways, windows, or similar openings – To ensure that flexible cords and cables are not subject to physical damage, they are not permitted to be run through doorways, windows, or similar openings.

(4) Attached to building surfaces – To ensure that flexible cords and cables are not subject to physical damage, they are not permitted to be attached to building floors, walls, ceilings, etc.

(5) Concealed behind building walls, ceilings, or floors – Flexible cords and cables are not permitted to be concealed behind building walls, ceilings, or floors.

(6) Installed in raceways unless permitted elsewhere in NEC® – For example, 680-20(b) permits the flexible cord supplying a wet-niche fixture to be installed within a raceway which extends from the forming shell to the deck box or other suitable junction box.

FLEXIBLE CORDS AND CABLES

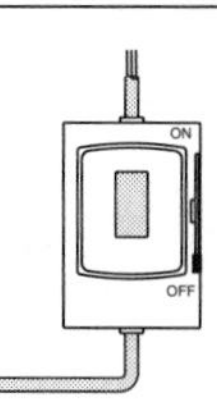

Section 400-7 lists 11 specific uses permitted for flexible cords and cables. Despite this, designers and installers of electrical systems should be very cautious when deciding to use flexible cord or cable as a wiring method. Part of the problem stems from the fact that the first item listed under uses not permitted is "as a substitute for the fixed wiring of the structure." Some electrical inspectors give very little latitude for the use of flexible cords and cables. These inspectors maintain that other wiring methods, such as FMC or LTFMC, can accomplish everything that flexible cord or cable can and the use of cable constitutes a "substitution" for fixed wiring. Always check with the AHJ before choosing flexible cords or cable as a wiring method.

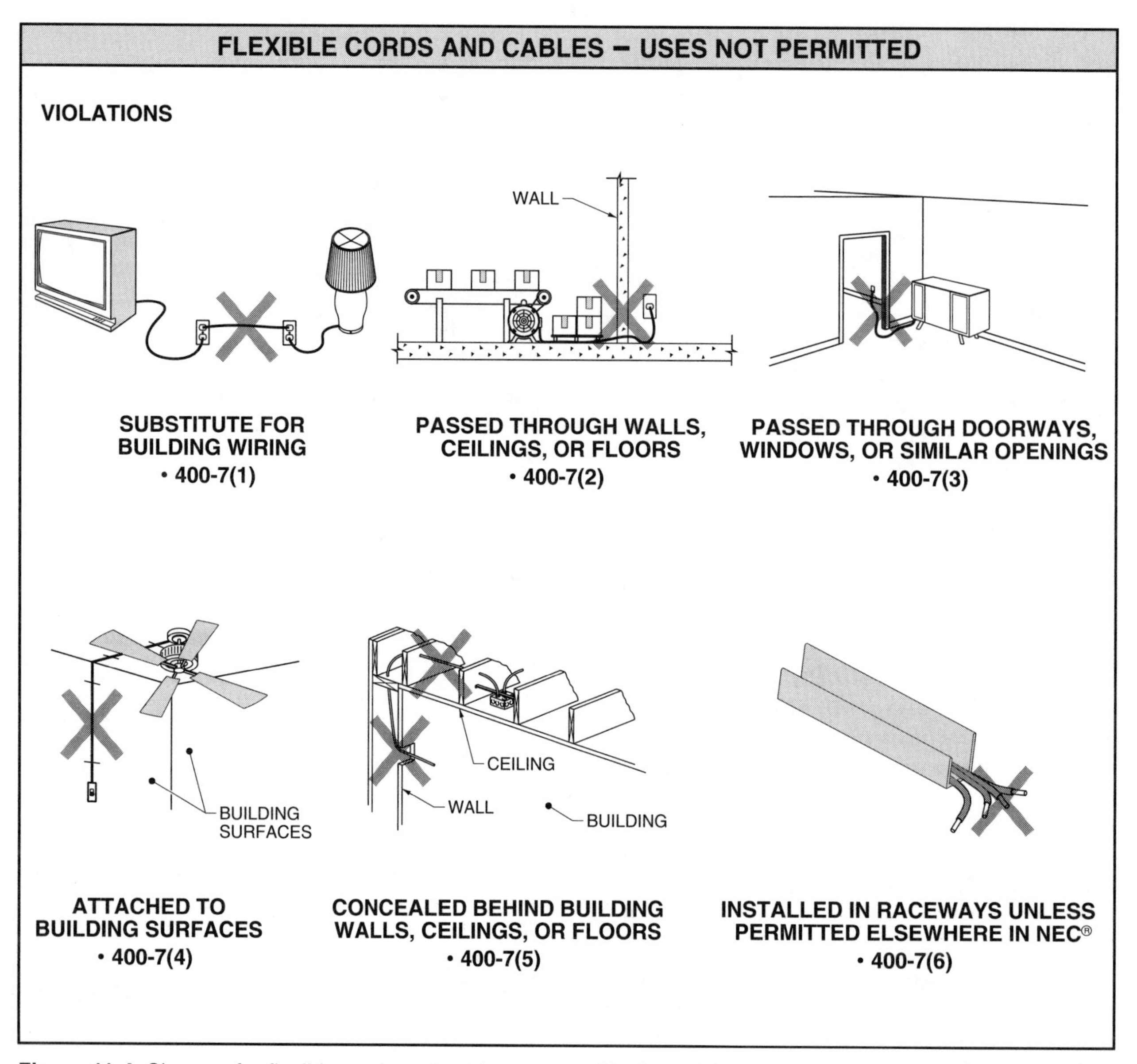

Figure 11-4. Six uses for flexible cords and cables are specifically prohibited.

Section 400-8, Ex. permits one connection to the building surface when a suitable strain-relief device or tension take-up device is needed for flexible cord or cable which supplies a piece of equipment which is required to travel or have some movement. In this case, the maximum length of the cord or cable is 6′ measured from the supply termination to the take-up device.

Splices – 400-9. Generally, flexible cords and cables shall be installed in continuous lengths without splices or taps. Requiring that the cord or cable be continuous ensures that the integrity of the conductors is not affected. If the cord types are either hard-service or junior hard-service cords, 400-9 permits splices for repairs. Frequently, in the course or use, these types of cables are damaged. This permits the cable to be repaired instead of replaced, provided the conductors are No. 14 and larger and the completed splice retains the insulation, outer sheath properties, and usage characteristics of the original cord. All splices shall be per 110-14(b).

Terminations – 400-10. Suitable devices or installation methods shall be used to ensure that tension is not transmitted from flexible cords or cables to the termination point. See Figure 11-5. Strain-relief fittings are an example of the type of fitting designed for this purpose. Installation methods that meet the requirements of this section are knotting the cord or winding with tape. Additionally, 400-14 requires that where flexible cords or cable pass through holes in covers of junction boxes or other similar enclosures, a suitable bushing or fitting shall be installed to protect the cord or cable.

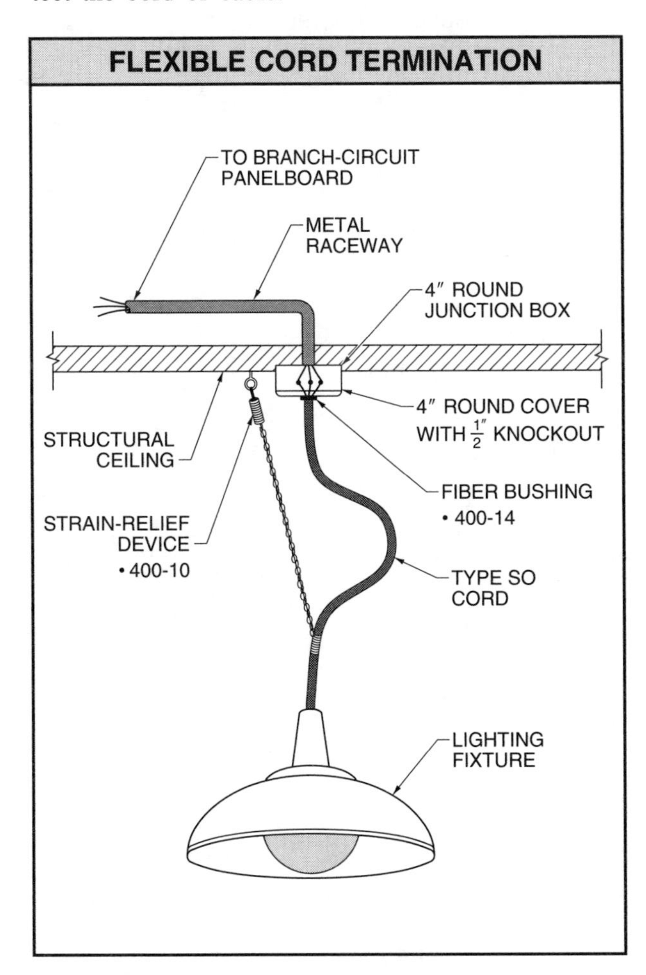

Figure 11-5. Flexible cords shall be provided with strain-relief devices or other suitable means to ensure that the weight is not transmitted to the termination or splice.

LIGHTING FIXTURES – 410

Article 410, Lighting Fixtures, Lampholders, Lamps and Receptacles contains the requirements for all types of lighting fixtures including incandescent filament lamps, arc lamps, electric-discharge lamps, and pendants. The international term for lighting fixture is luminaire. A *luminaire* is a complete lighting fixture consisting of the lamp or lamps, reflector or other parts to distribute the light, lamp guards, and lamp power supply per 410-1, FPN. See Figure 11-6.

The NEC® does have other provisions for special lighting systems. Article 411 covers lighting systems which operate at 30 V or less. Arc lamps, when used in theaters and similar locations, are also covered in 520. Article 680 contains the special provisions for lighting fixtures installed in or near swimming pools, fountains, spas, and similar locations.

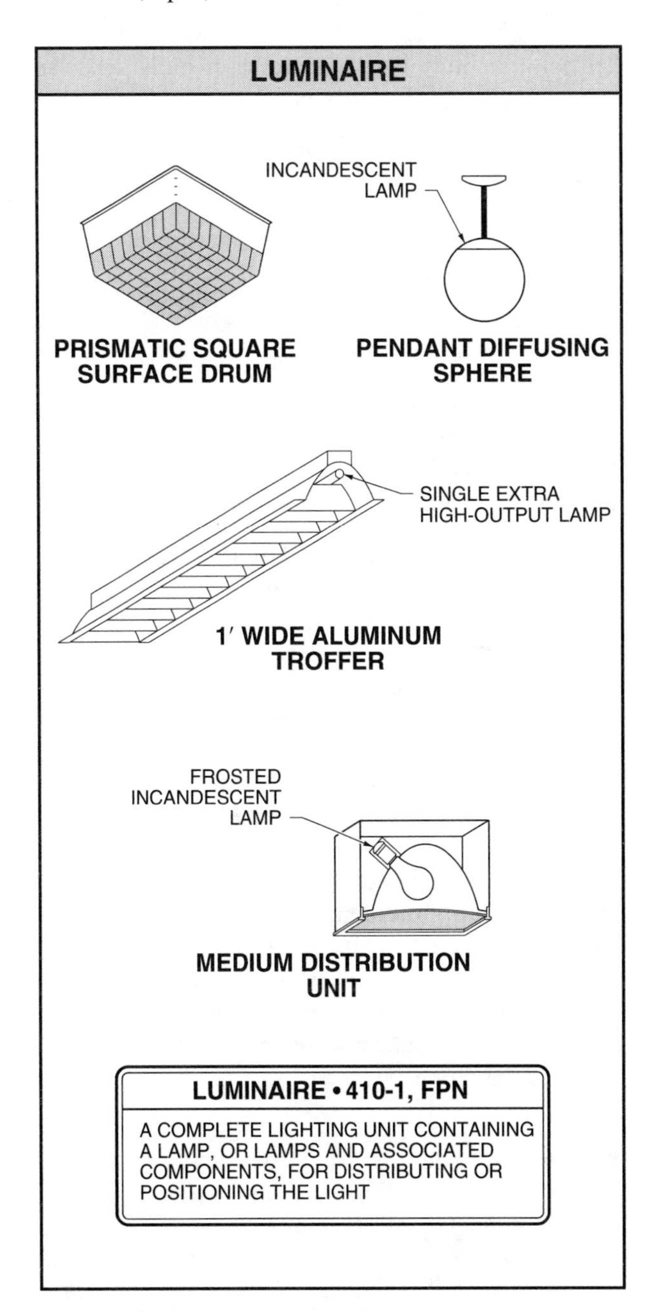

Figure 11-6. The international term for lighting fixture is luminaire.

Fixture Locations – 410, Part B

Fixtures, lampholders, and lamps are required to be constructed with no live parts normally exposed to contact. However, as with most electrical equipment, special precautions are required when the electrical equipment is installed in some types of locations. For example, precautions should be taken when lighting fixtures are installed in wet or damp locations. These locations present additional life-safety concerns for persons who may be required to operate the fixtures. Likewise, lighting fixtures installed in bathrooms can create additional hazards because of the constant presence of moisture.

Another location that has continually resulted in problems with lighting fixtures has been clothes closets. Although some light source is often needed in these spaces, great care is required in the selection of the lighting fixture and its placement because of the presence of easily-ignitable materials.

Precaution should be taken when installing lighting fixtures in hazardous locations. All electrical equipment installed in hazardous locations shall conform to the applicable provisions of 500 through 517.

Wet and Damp Locations – 410-4(a). The presence of moisture in wet and damp locations increases the potential for hazard from use of electrical equipment. Fixtures installed in either wet or damp locations shall be identified for such use. Fixtures identified for use in wet locations shall be marked *Suitable for Wet Locations*. Fixtures identified for use in damp locations shall be marked *Suitable for Wet Locations* or *Suitable for Damp Locations*. Fixtures identified for wet locations can be used in damp locations, but fixtures identified for damp locations cannot be used in wet locations. These markings ensure that the fixture is constructed for the type of location. For example, fixtures marked for wet or damp location use are constructed so that water does not enter or accumulate in wiring compartments or other places where electrical parts are included.

Bathrooms – 410-4(d). Nowhere is the potential for electric shock from lighting fixtures greater than in the bathroom. The presence of moisture and the fact that persons may be immersed in water present a great risk to personnel. Therefore, 410-4(d) establishes a zone above bathtubs from which cord-connected fixtures, hanging fixtures, lighting track, pendants, and ceiling fans shall not be located. The zone extends 8′ vertically from the top of the bathtub rim and 3′ horizontally. The zone includes the space directly over the tub. See Figure 11-7.

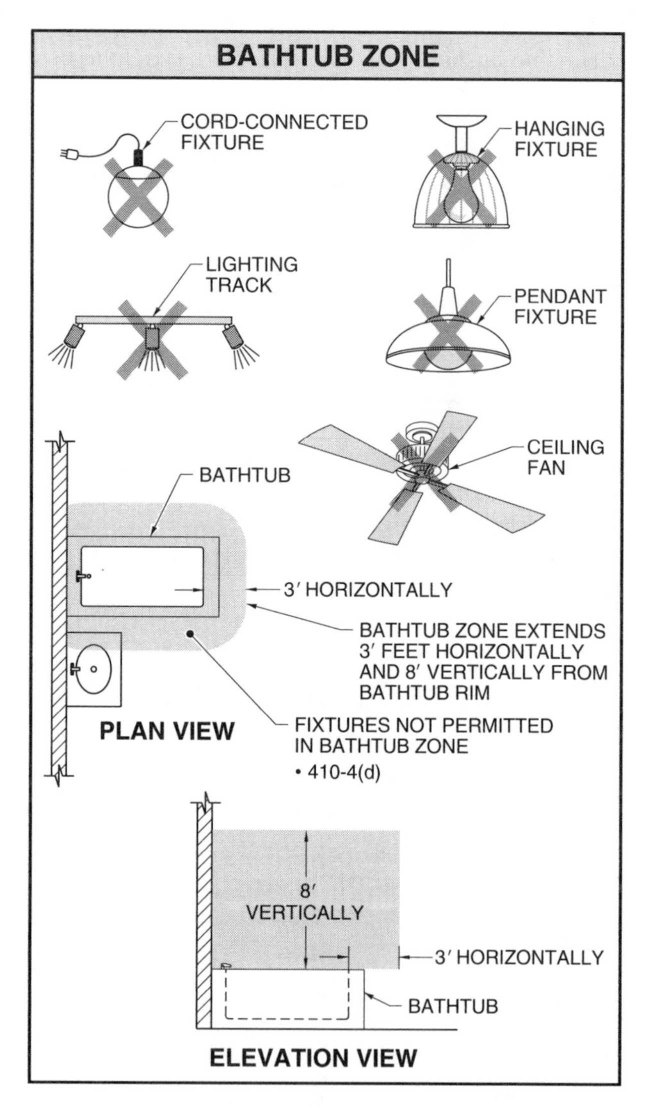

Figure 11-7. Cord-connected fixtures, hanging fixtures, lighting track, pendants, and ceiling fans are not permitted to be installed in the bathtub zone.

Clothes Closets – 410-8. The use of lighting fixtures in clothes closets has been a problem for many years. This is largely because the space used for storage is often overfilled, resulting in easily-ignitible materials coming in direct contact with the fixture lamp. Pendants and incandescent lighting fixtures with open or partially-enclosed lamps are not permitted to be installed in clothes closets. See Figure 11-8.

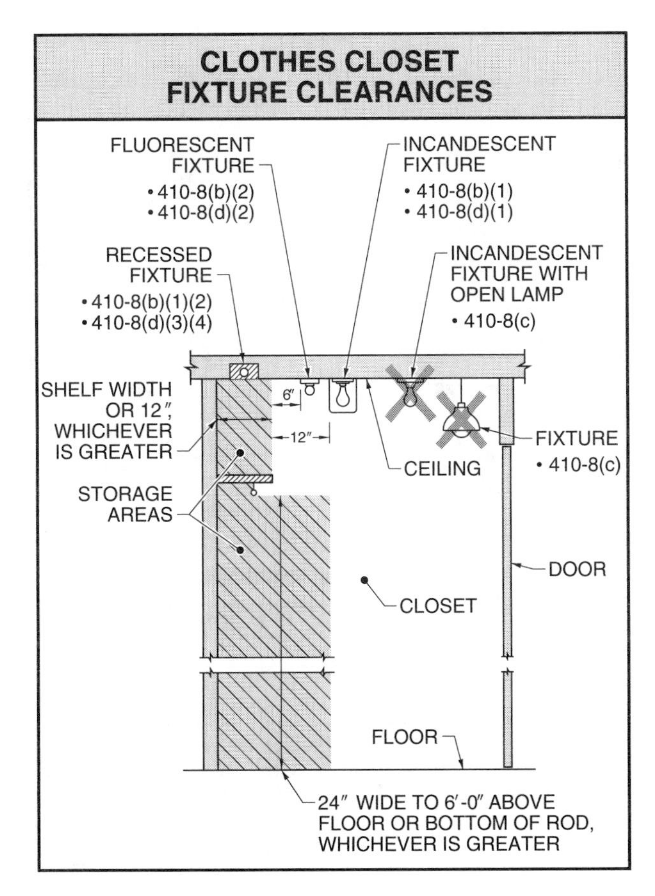

Figure 11-8. Pendants and incandescent lighting fixtures with open or partially-enclosed lamps are not permitted to be installed in clothes closets.

Fully-enclosed incandescent fixtures are permitted provided they are surface-mounted on the wall above the door or on the ceiling and have a minimum of 12" free space from the nearest storage space point or they are recessed in the wall or on the ceiling with a minimum clearance of 6" free space from the nearest storage space point. Fluorescent lighting fixtures, because of their cooler operating temperatures, are a better choice for installation in clothes closets. Additionally, because they operate at a lower temperature there is greater latitude in their placement within closets.

Surface-mounted fluorescent fixtures can be installed above the door or on the ceiling with a minimum clearance of only 6" free space from the nearest storage space point. Recessed fluorescent fixtures can be installed in the wall or on the ceiling provided there is a minimum clearance of 6" free space from the nearest storage space point. Section 410-8 includes a figure which can be used in helping to determine the closet storage space.

Fixture Supports – 410, Part D

Lighting fixtures shall be adequately supported. Support requirements vary depending upon the type of fixture selected for the installation. In some cases fixtures can be supported directly from the outlet box. If the fixture exceeds a certain weight, it shall be supported independently of the outlet box. Fixtures installed in suspended ceilings also shall be adequately supported. Generally, this is accomplished by the use of fixture clips which attach the fixture to the ceiling assembly. See 410-16(c).

Lampholders – 410-15(a). *Lampholders* are devices designed to accommodate a lamp for the purpose of illumination. Lampholders consist of a lamp socket, lamp support mechanism, and a protective covering. See Figure 11-9. Section 410-15(a) prohibits fixtures that weigh more than 6 lb, or exceed 16" in any direction, from being supported by the screw shell of the lampholder.

Metal Poles – 410-15(b). Metal poles are commonly used in outdoor locations to support lighting fixtures. Section 410-15(b) permits metal poles to be used provided four conditions are met. First, the metal poles shall be provided with an accessible handhole. In addition to providing support for the fixture, the metal pole also serves as an enclosure for the supply conductors. An accessible handhole of at least 2" × 4", with a raintight cover, ensures that access to the supply conductors or cable terminations. If a hinged base is provided on the metal pole, handholes can be omitted if the metal pole is less than 20′ in height, a grounding terminal is accessible within the hinged base, and both parts of the hinged base are bonded. See Figure 11-10.

The second condition requires that a terminal for grounding the pole be provided in a location accessible from the handhole. In the event that the metal pole does not exceed 8′ in height, the handhole and grounding terminal can be omitted if the supply wiring method continues, without splice, to a fixture mounted on the metal pole.

The third condition requires that metal raceways or other EGCs be bonded to the metal pole with an EGC recognized by 250-91(b). The fourth condition requires that the conductors within the vertical metal pole be adequately supported. Support requirements for vertical conductors are provided in 300-19.

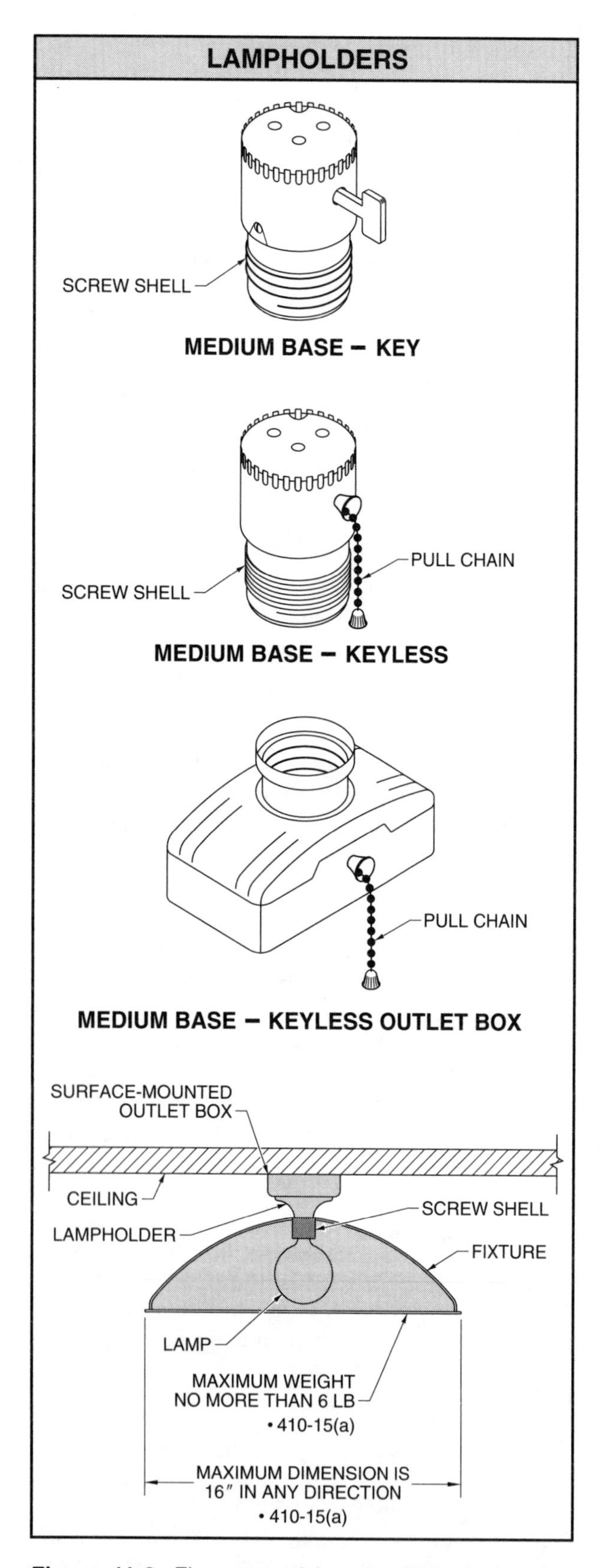

Figure 11-9. Fixtures and lampholders shall be securely supported.

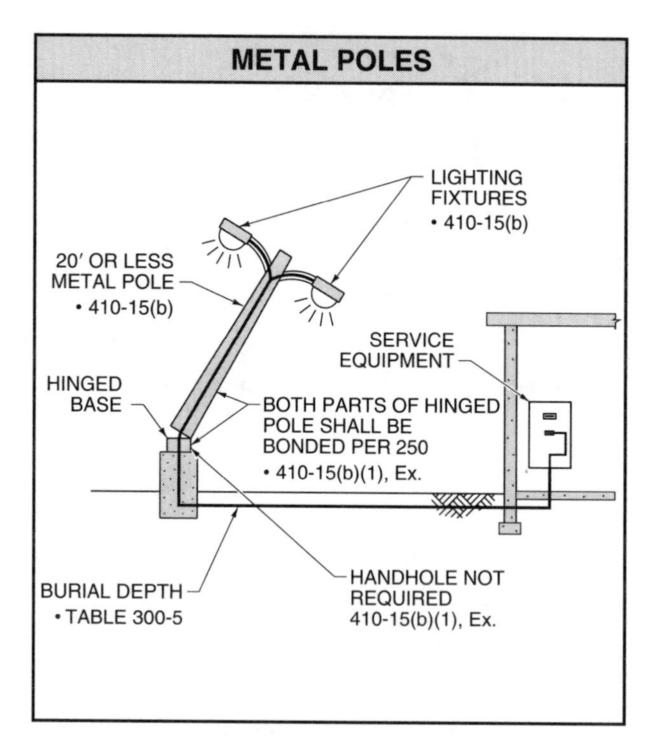

Figure 11-10. Handholes are not required in 20′ or less metal poles which support lighting fixtures if the base is hinged.

Outlet Boxes – 410-16(a). The vast majority of lighting fixtures installed are supported directly from the outlet box from which they are supplied. Section 410-16(a) permits fixtures up to 50 lb to be supported by the outlet box. Outlet boxes are covered by 370. Support requirements for boxes are contained in 370-23, and the general requirements for outlet boxes are contained in 370-27. Outlet boxes used for lighting fixtures shall be designed for the purpose. The outlet box shall be designed or installed in a manner that permits the lighting fixture to be attached to the box.

An independent means of support shall be provided for lighting fixtures that weigh over 50 lb. This can be accomplished by directly mounting the fixture assembly to the building structure or by use of rods, pipe, or straps which attach to the building structure.

Suspended Ceilings – 410-16(c). Another common means for supporting lighting fixtures is by the use of the framing members of suspended ceilings. Lay-in type fixtures are available which take the place of the standard 2′ × 2′ or 2′ × 4′ ceiling tiles. Section 410-16(c) requires that fixtures installed in suspended ceilings shall be securely fastened to the ceiling framing members. Fastening shall be by mechanical means such as screws, bolts, or rivets. Clips are also permitted. Hurricane clips were, at one time, the most common means used to secure fixtures to the ceiling framing members. New fixtures have been designed with the supporting mechanism built into the fixture. See Figure 11-11.

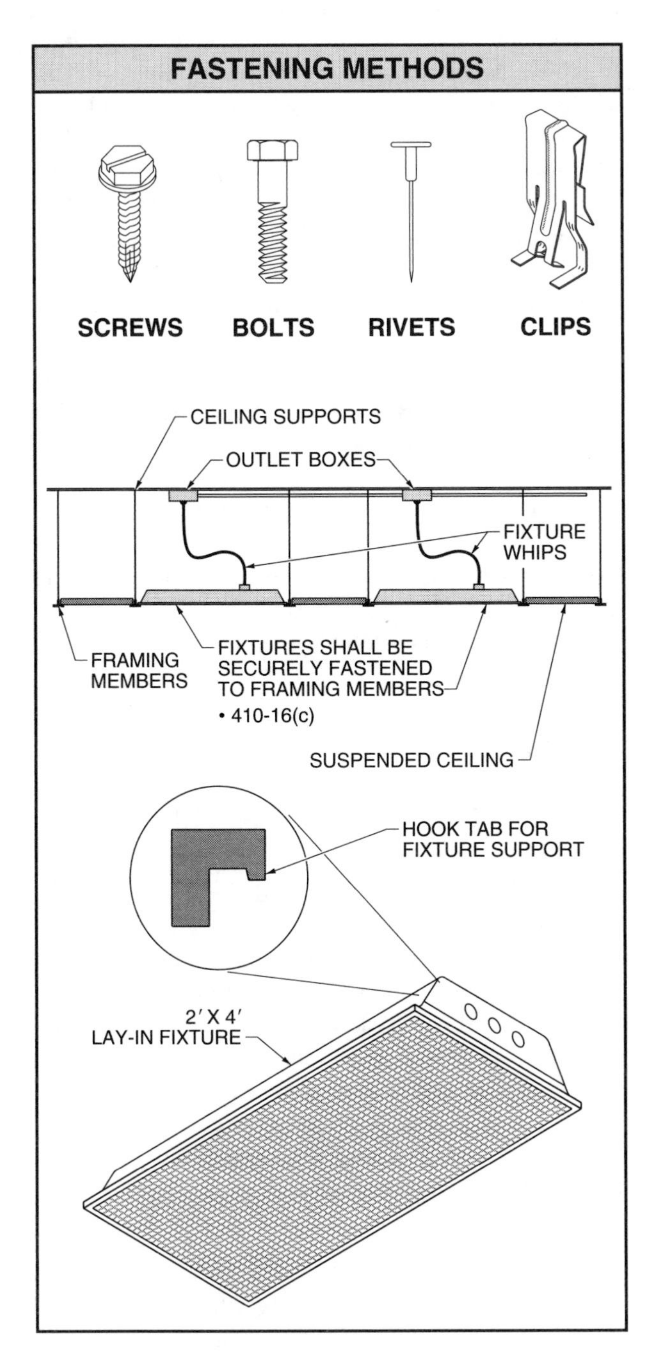

Figure 11-11. Lay-in type fixtures shall be securely fastened to the ceiling framing members by suitable mechanical means.

LAY-IN LIGHTING FIXTURES

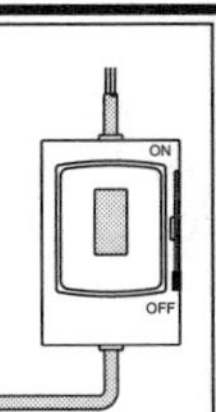

Section 410-16(c) has led to much confusion in the field regarding the support of lay-in lighting fixtures. Note that the requirement for attachment to the building structure only applies to the framing members of the suspended ceiling system. Many job specifications require, however, that the fixtures be independently supported by the building structure. In these cases, separate support wires are typically installed from the corners of the fixture to the actual building structure. While this practice is not required by the NEC®, it does provide a better electrical installation. Always check the job specifications before deciding how individual fixtures are to be supported.

Trees – 410-16(h). In some cases, installing metal poles for the support of outside lighting fixtures is impractical. Section 410-16(h) permits trees to support outdoor lighting fixtures and their associated equipment such as photovoltaic control devices. See Figure 11-12. However, several other NEC® provisions shall be considered when using trees for the support of lighting fixtures.

Section 225-26 prohibits trees from being used to support overhead conductor spans. This is because the trees are growing and moving continually. Connecting overhead spans directly to the trees would put a tremendous amount of stress on the conductors. Another option is to supply the lighting fixtures from underground raceways or direct-buried cables. Section 300-5(d) requires that direct-buried conductors and cables which emerge from the ground shall be protected from the required burial depth to a point at least 8′ above finished grade.

Fixture Wiring – 410, Part F

Fixture wiring shall be adequately protected from physical damage. The conductors selected for fixture wiring shall be carefully chosen to ensure that their insulation type, size, and ampacity are suitable for the specific installation. When conductors are installed within the fixture assembly, fixtures are not permitted to be used as raceways for circuit conductors. There are, however, exceptions to this general rule which are helpful when designing or installing conductors in lighting fixtures. Article 402 contains the provisions for the construction specifications of fixture wires. In general, fixture wires are not permitted to be used as branch-circuit conductors. Fixture wires are for use within lighting fixtures and for connecting lighting fixtures to the branch-circuit supply conductors. See 402-10 and 402-11.

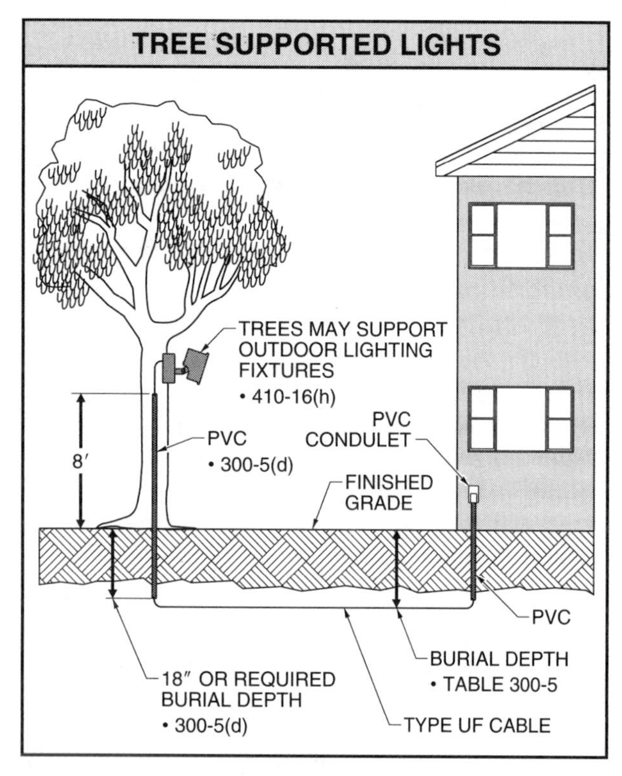

Figure 11-12. Trees are permitted to support outdoor lighting fixtures and their associated equipment.

Physical Damage – 410-22. Section 410-22 requires that fixtures shall be wired in a neat manner. The accumulation of extra conductors and excess wiring shall be avoided. Installing fixture wires in a neat manner ensures that the conductors are not subject to physical damage. For example, if excess wiring is not removed, a conductor could be pinched when placing the cover on the lighting fixture. In addition, the conductors shall be installed within the fixture in a manner that does not subject them to excessive temperatures which could lead to damage of the conductors. Many fixtures contain ballasts or other electronic components that can generate substantial heat. Conductors shall be selected and arranged so that their temperature rating is not exceeded.

Insulation – 410-24(a). Ensuring that fixture wiring be done with conductors suitable for the environmental conditions which are present is closely related to protecting conductors from physical damage. Section 410-24(a) requires that the conductor insulation be suitable for the current, voltage, and temperature to which the conductors are subjected. Section 410-25 lists the permitted fixture wires for mogul-base lampholders and other than mogul-base, screw-shell lampholders. Section 410-31 requires that any branch-circuit conductor located within 3″ of a ballast, within a ballast compartment, shall have an insulation rating not lower than 90°C (194°F). Section 400-31 lists eight specific insulation types suitable for use within 3″ of a ballast. *Note:* One of the insulation types, THW, is normally a 75°C conductor. A review of the THW insulation properties in Table 310-13 does, however, reveal that THW insulation is permitted for special applications within electric-discharge lighting equipment at the 90°C rating.

Size and Ampacity – 410-24(b). In general, fixture conductors shall not be smaller than No. 18 AWG. The branch-circuit conductors supplying the fixtures are sized per 220. For example, 220-3(a) requires that the branch-circuit rating and the branch-circuit conductor size for a continuous lighting load shall have an allowable ampacity equal to or greater than the noncontinuous load plus 125% of the continuous load.

Fixture wires are sized per 402. Fixture wires shall be of a type listed in Table 402-3, which provides information on insulation type, maximum operating temperature, and applicable application provisions. In general, fixture wires shall not be installed in a manner which could cause the operating temperature to exceed the maximum temperature rating listed in Table 402-3. The ampacity of fixture wires is determined per 402-5. Table 402-5 list the following ampacities for fixture wires:

- No. 18 – 6 A
- No. 16 – 8 A
- No. 14 – 17 A
- No. 12 – 23 A
- No. 10 – 28 A

Cord-Connected – 410-30. Cord-connected lampholders and fixtures are covered in 410-30. See Figure 11-13. Metal lampholders, when supplied by a flexible cord, shall be equipped with an insulating bushing at the inlet per 410-30(a). If the inlet is threaded, the bushing shall provide at least a 3/8″ nominal pipe size opening. Otherwise, the opening shall be sized appropriately for the cord.

Fixtures that adjust or can be aimed are not required per 410-30(b) to be equipped with an attachment plug or cord connector if the portion of the cord which is exposed is of the hard or extra-hard usage type. The installation of the cord should not be subject to physical damage. Only enough cord to permit adjusting or aiming the fixture shall be installed.

A listed fixture or listed fixture assembly may be supplied by a cord-connected wiring method per 410-30(c) provided the listed fixture or fixture assembly is located directly below the outlet box which supplies the fixture. In addition, the flexible cord shall be continually visible for its entire length, shall not be subject to physical damage or strain, and shall terminate at the outer end of the cord in a grounding-type attachment plug. A listed fixture or listed fixture assembly is permitted to be installed without an attachment plug per 410-30(c)(1) provided the fixture or assembly incorporates a cord and canopy.

Fixtures as Raceways – 410-31. The general rule of 410-31 does not permit fixtures to be used as raceways for circuit conductors. These conductors are subject to excessive heat which may be generated within the fixture or they may be subject to physical damage. There are, however, applications in which the fixtures can be used as raceways. The first is when the fixtures have been tested and are listed for use as raceways per 410-31, Ex. 1. Usually, the fixture includes a label which states that the fixture can be used as a raceway.

The second application per 410-31, Ex. 2, permits fixtures to be used as raceways when the fixtures are designed to be connected in an end-to-end fashion or the fixtures are connected together with a recognized Chapter 3 wiring method. In these cases, the fixtures can be used as raceways only for the through conductors of a 2-wire or multiwire branch circuit supplying the fixtures. This exception is commonly used in department store lighting applications where many rows of fixtures are connected together. Typically, in these applications, a multiwire branch circuit consisting of three phase conductors and one grounded conductor are pulled through a row of fixtures. Fixtures are alternately connected to either the A, B, or C phase and the grounded conductor.

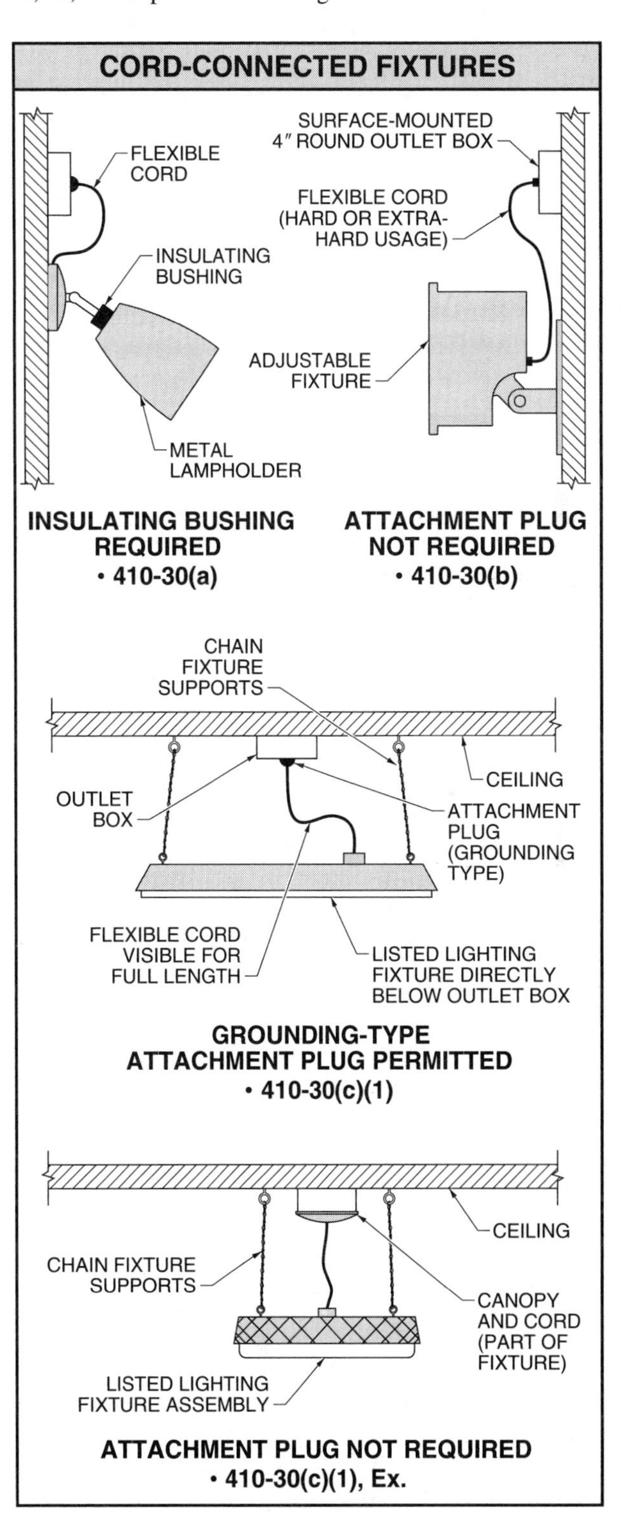

Figure 11-13. Lampholders and fixtures may be connected with flexible cord.

Fixtures are permitted by 410-31, Ex. 3 to be used as a raceway for an additional 2-wire branch circuit used to supply a fixture or fixtures installed per 410-31, Ex. 2. For example, an application of this exception is when nightlight circuits are installed for selected fixtures in the row. Such fixtures are not used for emergency purposes, but to permit individual control of some fixtures for purposes such as energy conservation.

Flush and Recessed Fixtures – 410, Part M

Flush or recessed fixtures are fixtures which are placed in recessed cavities of walls or ceilings. Because the fixture assembly is located above the ceiling or within the wall, special measures are required to ensure that the fixtures operate safely. Temperature is the greatest concern with these types of fixtures. Heat, associated with the supply conductors and the operation of the fixture, can pose a real threat to combustible materials which may be closely associated with the fixtures. The supply wiring to these fixtures shall be suitable for the temperatures that are likely to be encountered.

Incandescent Fixtures – 410-65(c). Typically, incandescent fixtures operate at a much higher temperature than fluorescent fixtures. Therefore, recessed incandescent fixtures are required by 410-65(c) to be thermally protected. *Thermally protected* is designed with an internal thermal protective device which senses excessive operating temperatures and opens the supply circuit to the fixtures. Excessive temperatures can result from overlamping. *Overlamping* is installing a lamp of a higher wattage than for which the fixture is designed. Poor heat dissipation from the fixture, such as when insulation surrounds the fixture assembly, can also cause excessive temperatures.

Recessed incandescent fixtures are permitted to be installed without thermal protection by 410-65(c), Ex. 1 when they are identified for and used in poured concrete. This is permitted because the poured concrete surrounding the fixture is not combustible.

Listed recessed incandescent fixtures which are constructed and designed with equivalent temperature characteristics of thermally protected fixtures are permitted to be installed per 410-65(c), Ex. 2 without thermal protection. An example of one type of fixture that meets the requirements of this exception is a fixture designed to prevent overlamping. These fixtures do not accept lamps of a higher wattage than the fixture is designed for, thereby providing a level of protection equivalent to thermal protection.

RECESSED LIGHTING

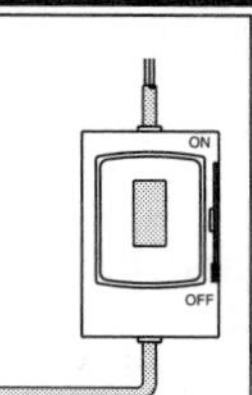

The installation of thermal insulation around recessed lighting fixtures is a concern to electricians and electrical contractors. Quite often, the electrician installs the recessed fixture and wires it long before insulation is installed. It can be difficult, under these circumstances, to ensure that the 3″ clearance is maintained. Additionally, spray insulation is becoming more popular, and maintaining clearances can be more difficult. Therefore, it is good design and installation practice to specify and use fixtures identified for direct thermal insulation contact whenever recessed fixtures are installed in a space which is to be insulated. Although these fixtures are more expensive than conventional fixtures, they do provide protection against excessive temperatures that could result in a fire.

Combustible Materials – 410-65(a). Fixtures shall be installed in a manner that does not subject adjacent combustible materials to temperatures in excess of 90°C (194°F) per 410-65(a). Often, recessed fixtures are installed in direct contact with combustible building materials such as wood framing members. If the fixture type is incandescent, significant heat can be generated and pose a potential for fire if not properly dissipated. If the fixture type is fluorescent, the lamp operates at a lower temperature, but the fixture ballast can generate significant heat. If a recessed fixture is installed in buildings of fire-resistant construction, 410-65(b) permits a temperature higher than 90°C (194°F) but not in excess of 150°C (302°F) if the fixture is of fire-resistant construction and so identified.

Clearance and Installation – 410-66. One of the greatest concerns when installing recessed fixtures is ensuring that thermal insulation is not installed too close around the fixture. Often, recessed fixtures are installed in spaces that are provided with thermal insulation. If the insulation is installed too close to the fixture, heat can be trapped and not properly dissipated. These high heat spots are a potential for fire if adjacent combustible materials reach their fire point.

Section 410-66 provides strict requirements for locating recessed fixtures near combustible materials and thermal insulation. A minimum of ½″ clearance shall be provided for recessed portions of lighting fixtures from all combustible materials per 410-66(a). This ½″ clearance does not, however, apply to the support points of the fixture.

A minimum clearance of 3″ shall be provided for all recessed fixtures from thermal insulation per 410-66(b). This clearance requirement applies to the fixture enclosure, wiring compartment, and ballast. Fixtures are permitted to be installed, per 410-66(b), Ex., in direct contact with thermal insulation if they are identified for such use. See Figure 11-14.

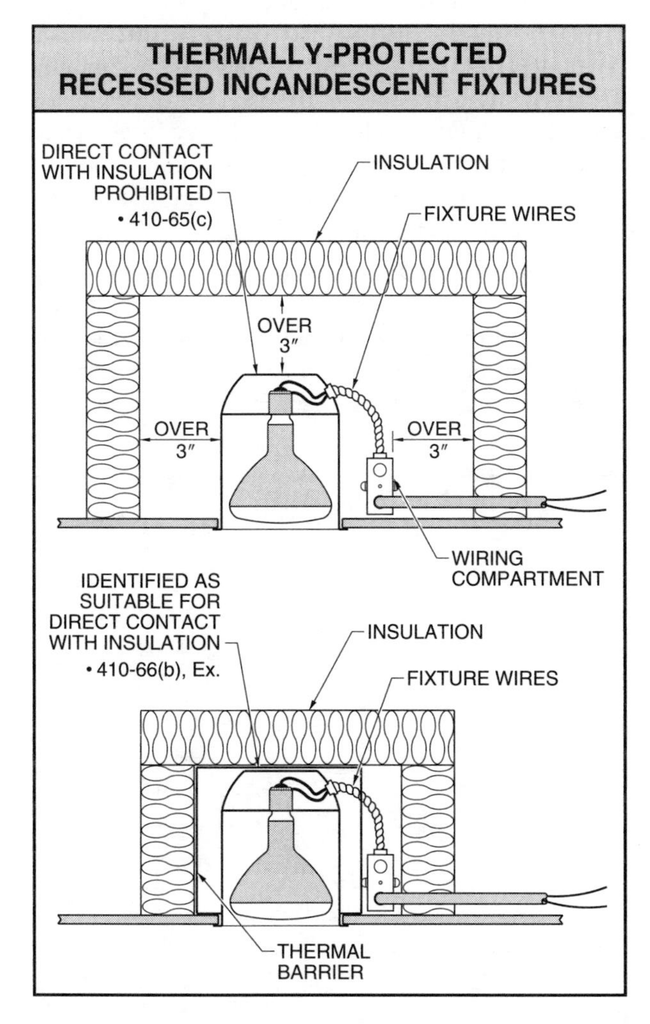

Figure 11-14. Thermally-protected fixtures provide additional safety measures by ensuring that overheated fixtures do not provide a source of ignition for combustible materials.

Wiring – 410-67. All wiring for recessed fixtures shall be done with conductors that have an insulation type suitable for the temperature that may be encountered per 410-67(a). If heat is not properly dissipated from the wiring compartment, the insulation rating of the conductors could be exceeded leading to conductor insulation degradation and possible failure. Branch-circuit wiring that supplies the fixture is permitted per 410-67(b) to terminate within the fixture if the conductors are provided with an insulation type which is suitable for temperature encountered.

Tap conductors which are run to recessed fixtures are covered in 410-67(c). If the branch-circuit conductors are not suitable for the temperature encountered in a fixture wiring compartment, an outlet box can be installed at a point which is at least 1′ from the fixture. Tap conductors are permitted to be run from the outlet box to the fixture. The tap conductors shall be suitable for the temperature encountered and at least 4′ and not more than 6′ in length. Tap conductors are permitted in a suitable raceway or they can be Type AC or MC cable. The 1′ spacing of the outlet box ensures that heat generated in the fixture does not affect the wiring in the outlet box. The tap conductor length requirement permits movement of the fixture for maintenance or repair purposes. See Figure 11-15.

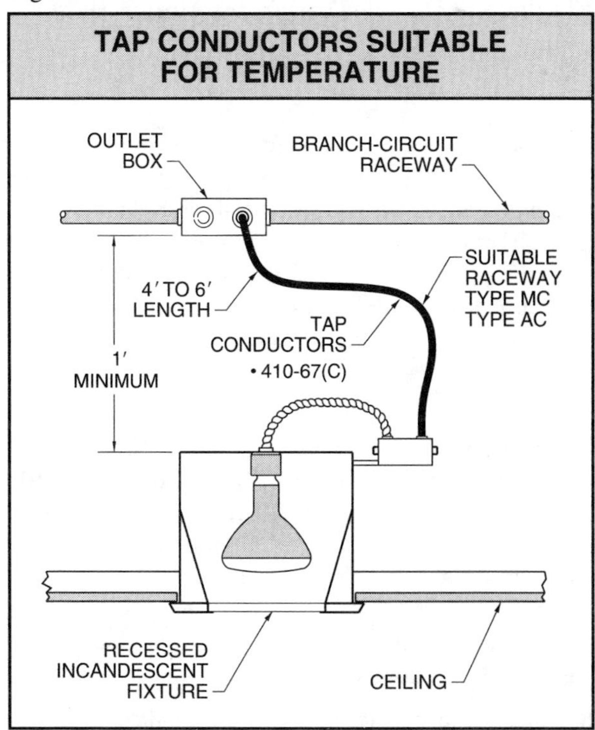

Figure 11-15. Tap conductors are used to connect branch-circuit wiring to the fixture wiring.

Markings – 410-70. Recessed incandescent lighting fixtures shall be marked with the maximum allowable wattage of lamps for use in the fixture. The marking shall be completed with permanent letters at least ¼″ high in a location that is visible during relamping. The *Underwriters Laboratories Fixture Marking Guide* identifies some other common markings found on recessed incandescent lighting fixtures.

Fixtures which are suitable for use in direct contact with thermal insulation shall be marked "Type IC," or "Inherently Protected." If a fixture is not marked "Type IC," it shall be installed with a minimum of 3″ clearance from thermal insulation on all sides and the top. Such fixtures are typically marked "WARNING – RISK OF FIRE. Do not install insulation within 3″ of fixture sides or wiring compartment nor above fixture in such a manner as to entrap heat."

Fixtures suitable for installation only in poured concrete shall be marked "For Installation Only In Poured Concrete." Another type of marking that is common on recessed lighting fixtures is suitability for use only in environmental air handling spaces. Fixtures that are identified for such use are marked "Suitable Only for Installation in Environmental Air Handling Spaces."

Electric-Discharge Equipment Under 1000 Volts – 410, Part P

An *electric-discharge lighting fixture* is a lighting fixture that utilizes a ballast for the operation of the lamp. The ballast provides a high starting voltage which is induced for striking the arc to light the lamp. Electric-discharge lighting fixtures are commonly used throughout the electrical industry for all types of applications. Provisions for electric-discharge lighting systems which operate at 1000 V or less are given in 410, Part P.

Electric-discharge lamps operate at a much cooler temperature than incandescent lamps. This permits their use in locations such as clothes closets per 410-8. Installers of electric-discharge lighting fixtures should, however, be aware of temperatures around the ballast compartment. Some fixture assemblies incorporate remote ballasts to give the designer and installer greater latitude in specifying and selecting the ballast location.

Ballasts – 410-73(e). All ballasts used in fluorescent lighting fixtures installed indoors shall be thermally protected per 410-73(e). This thermal protection is integral to the ballast and operates much like thermally-protected, recessed lighting fixtures. The ballast thermal protection device opens when the operating temperature of the ballast rises too high, interrupting the supply current to the lighting fixture.

Excessive operating temperatures can be caused by improper dissipation of heat away from the ballast, burned-out fixture lamps, or the wrong ballast for the application. Replacement ballasts shall be the thermally-protected type. Ballasts that meet the requirements for thermal protection in 410-73(e) are listed and identified by UL as Class P ballasts.

High-Intensity Discharge – 410-73(f). A *high-intensity discharge (HID) lighting fixture* is a lighting fixture that generates light from an arc lamp contained within an outer tube. The sustained arc produces a constant light source. The three most common HID lamps in use are mercury-vapor, sodium-vapor, and metal-halide. HID fixtures are used throughout the electrical industry in all types of applications. Some of the more common applications are for outdoor lighting and high bay areas in industrial establishments.

All recessed HID fixtures shall be thermally protected per 410-73(f). If an HID fixture is installed in any location so that the ballast is remote from the fixture, the ballast is also required to be thermally protected. Thermally-protected HID fixtures shall be identified. See Figure 11-16.

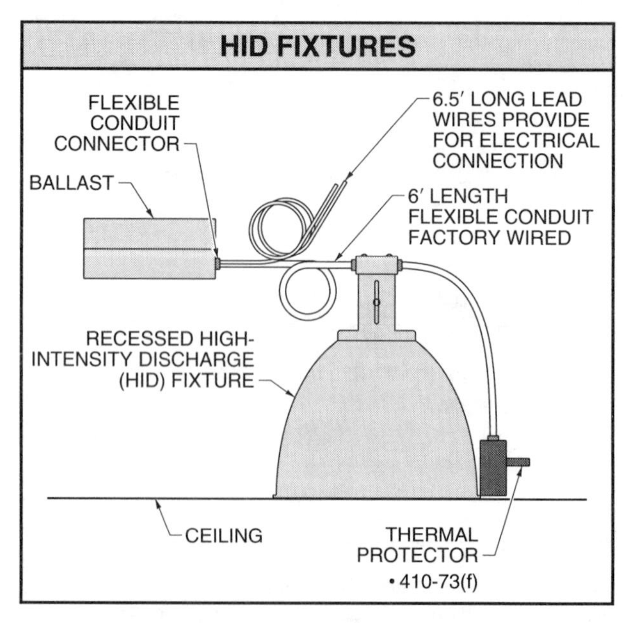

Figure 11-16. All recessed HID fixtures shall be thermally protected.

Lighting Track – 410, Part R

Lighting track is an assembly consisting of an energized track and lighting fixture heads which can be positioned in any location along the track. The track also supports the lighting fixture heads. The NEC® does not set maximum or minimum lengths for lighting track, but 410-102 requires that each 2′ of lighting track be considered 150 VA for the purposes of determining load calculations. Per 410-102, FPN, the 150 VA calculation is not intended to limit the length of the lighting track or the number of lighting fixtures that can be installed. See Figure 11-17.

Installation – 410-101. All lighting track shall be permanently installed and permanently connected to a branch circuit per 410-101. The use of lighting track with temporary wiring, or in an otherwise nonpermanent manner, is prohibited. In addition, the components for use on lighting track, such as the fittings, shall be identified for such use.

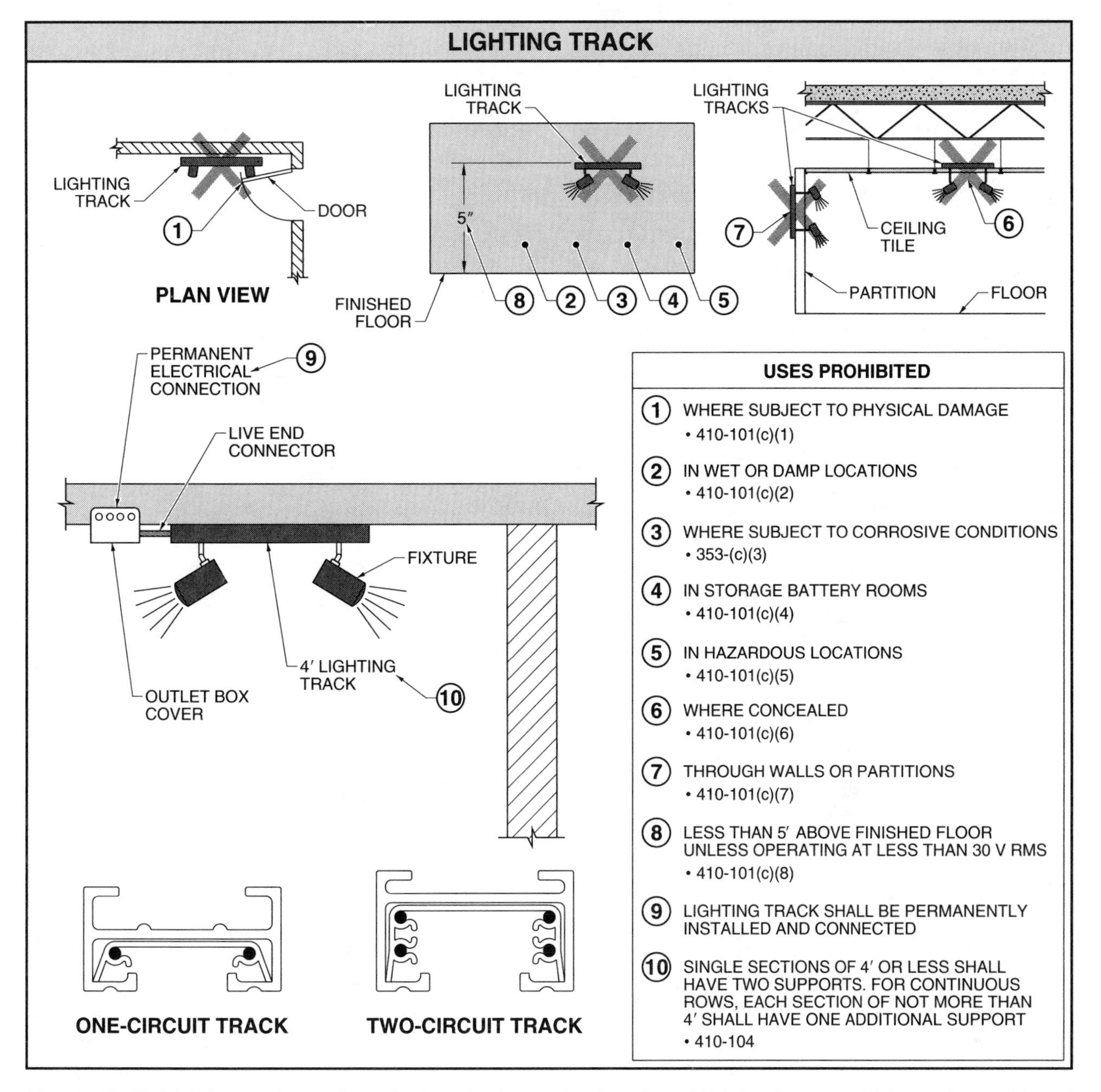

Figure 11-17. Lighting track consists of a length of energized track and lighting fixtures which can be positioned anywhere on the track.

The load on the lighting track shall be limited to that for which the lighting track is rated per 410-101(b). Installing excess lighting fixtures on the lighting track can subject the track to a larger load than for which it is designed or rated. Some lighting track systems are designed to be cut in the field to a specific length. Lighting track that is identified for field cutting includes instructions on the proper methods of cutting.

Prohibited Uses – 410-101(c). Eight limitations are placed on the use of lighting track by 410-101(c). As with most electrical equipment, lighting track shall not be used where it is subject to physical damage or in wet or damp locations. Lighting track is not permitted per 410-4(d) above bathtubs in a zone which extends 3′ horizontally and 8′ vertically from the bathtub rim. Lighting track is also not permitted to be installed in storage battery rooms, hazardous locations, or other areas subject to corrosive vapors. Lighting track is not permitted to be installed in concealed locations and is not permitted to extend through walls or partitions. It is not permitted for use at less than 5′ above the finish floor level unless it operates at less than 30 V RMS and is protected from physical damage.

Support Requirements – 410-101(d). The lighting fixtures installed on lighting track shall be designed specifically for the track on which they are to be installed. The lighting track provides the support for the lighting fixtures and the fixtures shall be securely attached to the lighting track. Fixtures which are not designed for use on lighting track cannot be modified for use on lighting track. The *UL Fixture Marking Guide* indicates that listed lighting track will be marked "CAUTION – To reduce the risk of fire and electrical shock, use only fixture assembles marked for use with __ track." The name of the manufacturer is placed in the blank space. The lighting fixtures are marked with a similar label to ensure that the lighting fixtures are used with the correct lighting track.

Mounting – 410-104. Because the lighting track supports the weight of the lighting fixtures, minimum fastening requirements are necessary to ensure a safe lighting track installation. A minimum of two lighting track supports shall be installed for individual pieces of lighting track 4′ or less in length. If the lighting track is installed in a continuous row, each individual section which is 4′ or less in length shall be provided with one additional support per section. The *UL Fixture Marking Guide* requires that lighting track systems designed to have field-drilled mounting holes shall be provided with installation instructions detailing the location of the mounting holes.

RECEPTACLES – 410, PART L

Receptacles are contact devices installed at outlets for the connection of cord-connected electrical equipment. *Devices* are any unit of an electrical system that carries, but does not use electricity. The receptacle does not use electricity, but merely provides a point on the electrical system from which electricity can be obtained. The NEC® contains installation provisions for several types of receptacles used throughout the electrical industry. Article 410 is concerned primarily with receptacle types, ratings, and installation requirements. Other NEC® articles, such as 210 and 220, also contain specific requirements for installing and using receptacles.

Receptacles are available in many voltage and current ratings and in many configurations. All receptacles, however, can be classified as either standard receptacles or special-use receptacles. See Figure 11-18. Standard receptacles consist of the grounding and nongrounding types. Grounding receptacles are used for new installations and nongrounding receptacles are limited for use in replacement applications. Specialty receptacles are receptacles designed for a specific electrical application. Isolated-ground receptacles, hospital-grade receptacles, and ground-fault circuit interrupter receptacles are examples of specialty receptacles.

Grounding Receptacles

Since the 1968 NEC®, 210-7(a) has required that all receptacles installed on 15 A or 20 A branch circuits be the grounding type. *Grounding receptacles* are receptacles which include a grounding terminal connected to a grounding slot in the receptacle configuration. The grounding terminal is connected to an equipment grounding conductor which connects noncurrent-carrying metal parts of equipment to the system grounded conductor or grounding electrode conductor at the service.

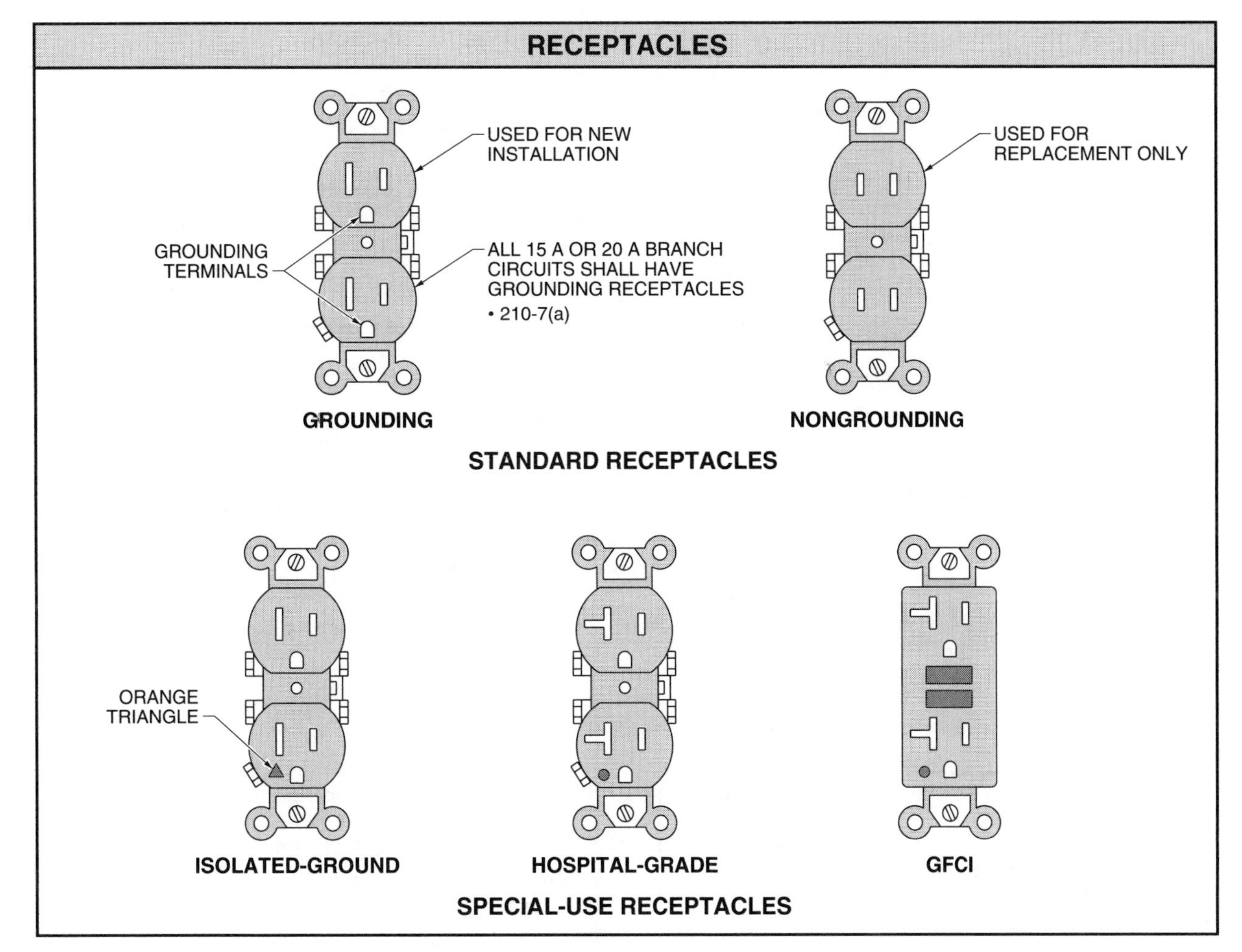

Figure 11-18. All receptacles are classified as standard receptacles or special-use receptacles.

Another type of grounding receptacle is the self-grounding receptacle. *Self-grounding receptacles* are grounding-type receptacles which utilize a pressure clip around the 6 × 32 mounting screw to ensure good electrical contact between the receptacle yoke and the outlet box. Self-grounding receptacles are permitted to be installed without a bonding jumper between the outlet box and the receptacle per 250-74, Ex. 2, if the contact devices or yokes are listed for the purpose.

Nongrounding Receptacles

Nongrounding receptacles are receptacles with two wiring slots for branch-circuit wiring systems that do not provide an equipment grounding conductor. These electrical systems were commonly installed until the late 1960s. The NEC® permits nongrounding receptacles to be used for replacement purposes only. Existing nongrounding receptacle outlets that do not contain an equipment grounding conductor can be replaced with nongrounding receptacles. See 210-7(d).

Isolated-Ground Receptacles

Isolated-ground receptacles are receptacles in which the grounding terminal is isolated from the device yoke or strap. See Figure 11-19. The isolation ensures a clean equipment ground for electronic equipment which may be adversely effected by noise in the equipment grounding path. Isolated-ground receptacles installed for the purposes of reducing electromagnetic interference, shall be identified with an orange triangle on the face of the receptacle per 410-56(c). In addition, isolated-ground receptacles shall be used only with isolated equipment-grounding conductors.

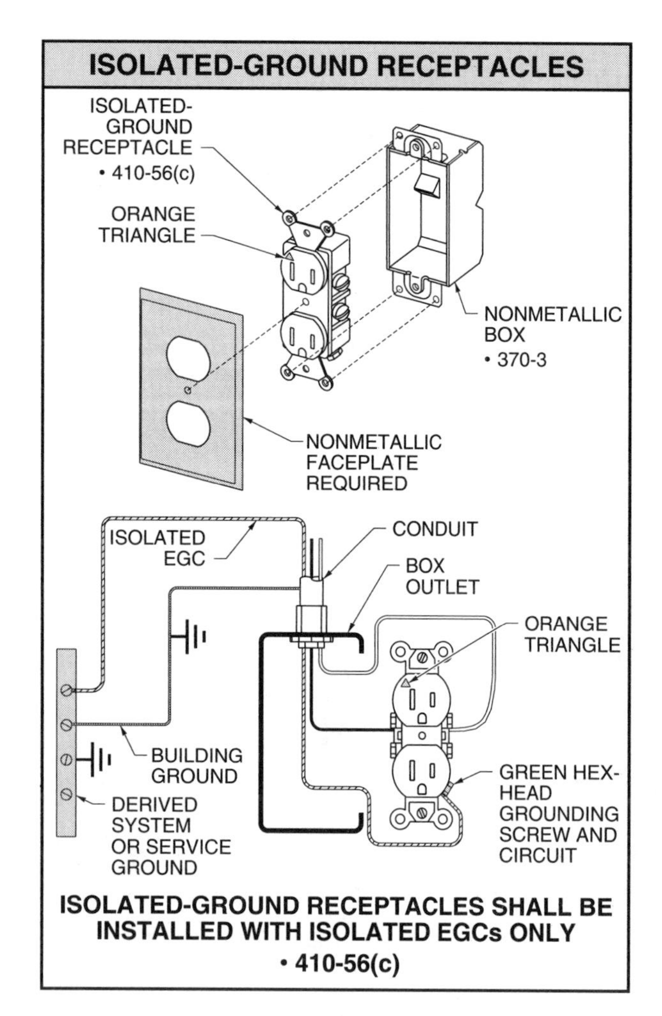

Figure 11-19. Isolated-ground receptacles shall be installed with isolated equipment grounding conductors only.

Isolated-ground receptacles are not permitted to be installed with regular equipment grounding conductors. Isolated-ground receptacles installed in nonmetallic boxes require nonmetallic faceplates.

Hospital-Grade Receptacles

Hospital-grade receptacles are the highest grade receptacles manufactured for the electrical industry. These receptacles undergo the most stringent testing and quality control of any receptacle. They are designed for use in health care facilities and other areas, such as schools and plants where they are subject to excessive abuse.

The NEC® requires that all patient bed location receptacles be listed and identified as "hospital grade" per 517-18(b). Hospital grade receptacles are identified by a green dot on the face of the receptacle.

GFCI-Type Receptacles

Ground-fault circuit interrupter receptacles are devices which interrupt the flow of current to the load when a ground fault occurs which exceeds a predetermined value of current. Class A GFCIs are commonly used for personnel protection and are designed to operate when the ground fault current flow has a value between 4 mA and 6 mA. See 210-8.

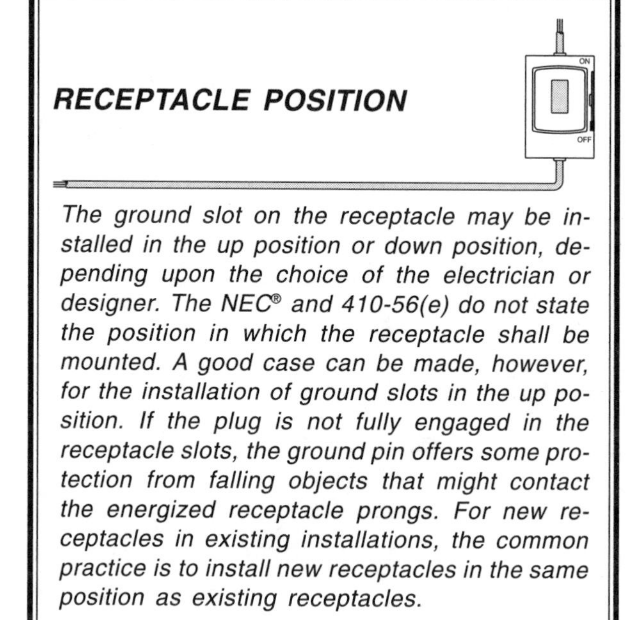

The ground slot on the receptacle may be installed in the up position or down position, depending upon the choice of the electrician or designer. The NEC® and 410-56(e) do not state the position in which the receptacle shall be mounted. A good case can be made, however, for the installation of ground slots in the up position. If the plug is not fully engaged in the receptacle slots, the ground pin offers some protection from falling objects that might contact the energized receptacle prongs. For new receptacles in existing installations, the common practice is to install new receptacles in the same position as existing receptacles.

Installation – 410-56; 410-57

Because receptacle outlets are the most common point of connection to the electrical system, several safety measures concerning their installation and use are included in the NEC®. The NEC® contains provisions regarding the installation of the receptacle outlet and faceplates and the use of receptacles in damp and wet locations. In addition, 410-56(f) requires that 15 A and 20 A attachment plugs and connectors be constructed in a manner that does not permit any exposed current-carrying parts, with the exception of the prongs, blades, or pins. Such attachment plug or connectors shall be constructed with dead fronts. A *dead front* is a cover required for the operation of a plug or connector.

Position – 410-56(e). Section 410-56(e) contains provisions for the installation of receptacles mounted in boxes. If the box is set back from the wall opening, the receptacle mounting yoke or strap shall be firmly seated against the wall surface. If the box is mounted flush with the wall or projects beyond the wall surface, the receptacle yoke or strap shall be firmly seated against the box. These provisions ensure good electrical contact between the box and the receptacle for maintaining grounding continuity. The faceplates shall be installed to completely cover the entire wall opening.

Metal faceplates are required by 410-56(d) to be grounded. This can be accomplished by use of the No. 6 × 32 metal screws which attach the faceplate to the receptacle yoke.

This GFCI receptacle is installed in a horizontal position.

Interchangeability – 410-56(h). Receptacles, attachment plugs, and cord connectors shall be designed and constructed so that they cannot be interchanged with other receptacles, attachment plugs, and cord connectors of a different voltage or current rating. This ensures that the equipment will be properly protected and the personnel who might be operating the equipment can do so in a safe manner. For example, nongrounding receptacles shall be constructed so that grounding-type attachment plugs cannot be inserted into them. However, 410-56(h), Ex. permits T-slot 20 A receptacles to be constructed so that 15 A attachment plugs can be inserted.

Raised Covers – 410-56(i). Receptacles which are installed in raised covers shall be attached to the raised cover by more than a single screw. See Figure 11-20. For many years the receptacle was attached to the raised cover by means of a single No. 6 × 32 screw which was provided to attach the receptacle faceplate. If the receptacle screw loosened, the grounding continuity between the receptacle and the raised cover could be interrupted. Raised covers are now designed to provide at least two points of connection between the receptacle and the raised cover. However, 410-56(i), Ex. permits listed devices, assemblies, or box covers to be installed with a single screw if they are identified for such use.

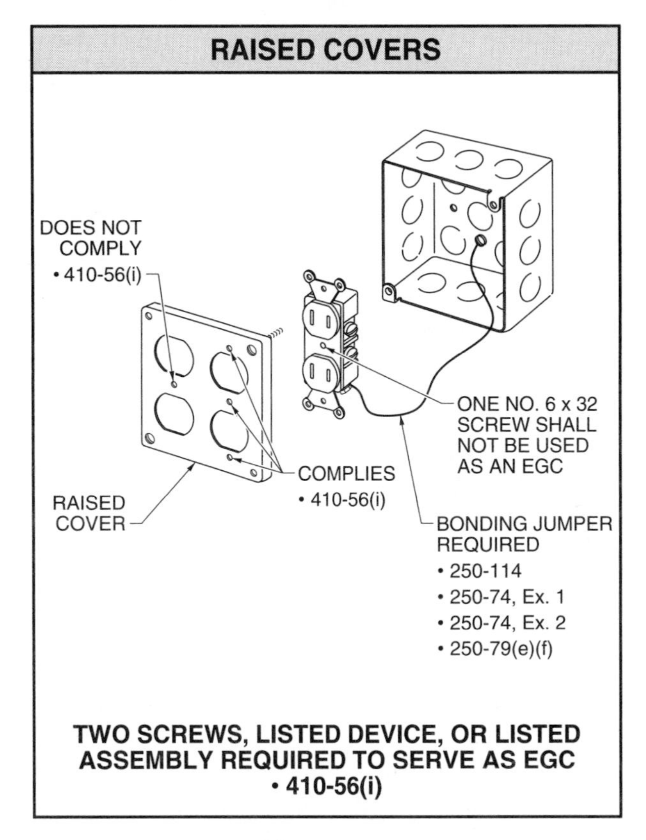

Figure 11-20. Single screws are not permitted to secure receptacles to raised covers.

Wet and Damp Locations – 410-57(a). Receptacles installed in damp or wet locations present additional hazards to those who might be using cord- and plug-connected equipment. Because of the presence of moisture, the NEC® requires additional safety provisions for these receptacles. See Figure 11-21. Any receptacle installed in a damp location, either indoors or outdoors, shall be provided with a weatherproof (WP) enclosure. The weatherproof requirement does not apply when an attachment plug cap is inserted into the receptacle.

Receptacles in wet locations, however, are required to be installed in weatherproof enclosures without the integrity being affected when the attachment plug cap is inserted. Per 410-57(b), Ex., enclosures that are not weatherproof when the attachment plug is inserted are permitted to be used if the receptacle is installed in a wet location for use with portable tools or equipment normally connected to the outlet only when it is attended.

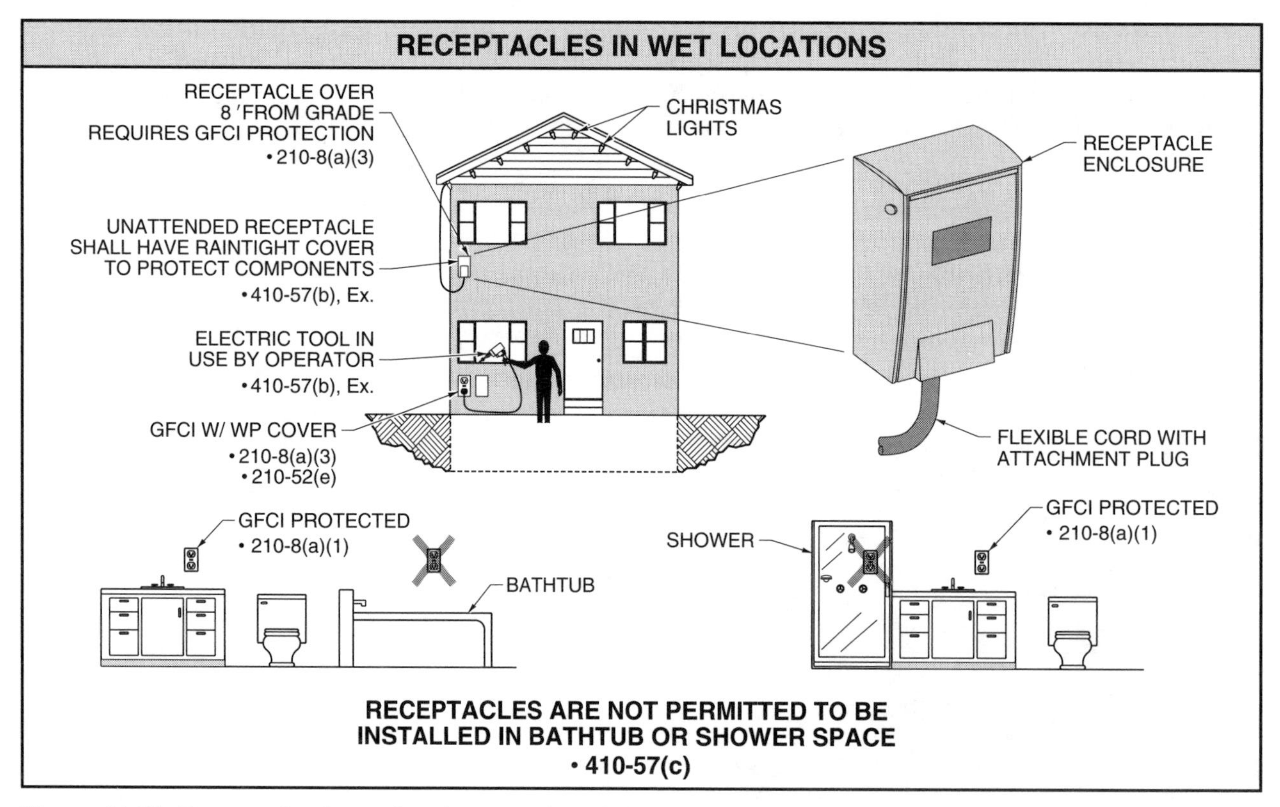

Figure 11-21. Receptacles in wet locations require additional installation provisions to protect personnel.

Another wet location which presents additional hazards to personnel operating equipment is bathrooms and shower spaces. Therefore, 410-57(c) does not permit a receptacle outlet to be installed within a bathtub or shower space.

APPLIANCES – 422

Article 422 of the NEC® contains installation provisions for appliances, including their control, protection, and disconnecting means.

An *appliance* is any utilization equipment which performs one or more functions, such as clothes washing, air conditioning, cooking, etc. Utilization equipment uses electrical energy to do some useful work, such as lighting, heating, chemical, etc. While 422 states that the article covers appliances in all types of occupancies, typically the term appliance does not include industrial equipment, but only equipment that is used in residential or commercial occupancies.

Installation Requirements – 422-6

All appliances shall be installed in an approved manner (acceptable to the AHJ). Section 110-3(b) also requires that listed or labeled equipment shall be installed, used, or both in accordance with any instructions included in the listing or labeling. This requirement is particularly important with appliances. The listing or labeling instructions for some appliances, such as ceiling fans, may include additional installation requirements that the NEC® does not address.

Per 422-2, all appliances shall be installed so that they have no live parts normally exposed to contact. However, exposed live parts necessitated by the design of the appliance are permitted by 422-2, Ex. Examples of these types of appliances are toasters and grills.

Appliances are permitted to be connected by flexible cords if such a connection is needed for frequent interchange of the appliance per 422-8(c)(1). If frequent maintenance and/or repair is needed, appliances are permitted to be connected by flexible cords per 422-8(c)(2), provided the appliances are intended for flexible cord connection.

RECEPTACLE OUTLET LOCATIONS

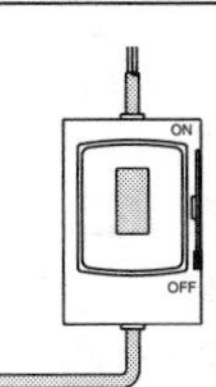

The location in which receptacle outlets are installed is significant to electrical designers and installers. The type of receptacle cover required depends upon whether the location is classified as a damp or wet location. See 100 for definitions under the term location.

All outdoor locations are not necessarily wet locations. Locations outdoors under open porches or other-partially protected areas can be classified as damp locations. The significance of 410-57(a)(b) is that spring-type receptacle covers are permitted in damp locations but not in wet locations, unless the receptacle is intended for use with portable tools or portable equipment while attended by personnel.

Branch-Circuit Ratings – 422, Part B. Part B contains the requirements for sizing branch circuits used to supply appliances. Part B does not apply to the conductors which are an integral part of the appliance.

The rating of an individual branch circuit which supplies an appliance shall be not less than the marked rating of the appliance per 422-4(a). If the appliance has combination marked loads, the branch-circuit rating shall be suitable for the marked appliance rating and the combination load. Article 100 defines an individual branch circuit as a circuit which supplies only one utilization equipment. If the appliance is the motor-operated type and markings are not provided, 422-4(a), Ex. 1 permits the branch circuit to be sized per 430, Part B.

For a continuously-loaded, nonmotor-operated appliance, 422-4(a), Ex. 2 requires the branch-circuit rating to be not less than 125% of the marked rating, unless the branch-circuit device and its assembly are listed for continuous loading at 100%. For household cooking appliances, 422-4(a), Ex. 3 permits the branch-circuit rating to be sized per Table 220-19. Per 422-4(b), the rating for branch circuits supplying combination loads consisting of appliances and other loads shall be determined by 210-23.

Overcurrent Protection – 422-5. Branch circuits supplying appliances are required to be protected against overcurrents by 240-3. In general, the conductors used to supply appliances shall be protected against overcurrents in accordance with their ampacities as listed in Table 310-16. Section 240-3 does, however, contain 13 subparts which are, in essence, exceptions to this general rule.

If a protective device rating is marked on an appliance, the branch-circuit device rating shall not exceed that marking per 422-5. Some appliances, such as air conditioners, are typically marked to include maximum branch-circuit rating information on their nameplates.

Central Heating – 422-7. Since the 1990 NEC®, 422-7 has required that an individual branch circuit supply all central heating equipment. This does not apply to fixed electric space-heating equipment, which is covered by 424. This rule removes the possibility that an overcurrent on another piece of equipment, connected to the same branch circuit as the central heating equipment, could result in the loss of the heating equipment when the OCPD opens. However, 422-7, Ex. does permit auxiliary and associated equipment, such as power-actuated valves, pumps, humidifiers, etc. to be connected to the same branch circuit that supplies the central heater.

Specific Appliances – 422-8(d). Specific appliances, such as kitchen waste disposers, built-in dishwashers, and trash compactors can be cord- and plug-connected. See Figure 11-22. Kitchen waste disposers are permitted to be hard-wired or cord- and plug-connected per 422-8(d)(1). All kitchen waste disposers that are cord- and plug-connected shall be provided with flexible cord which is identified for the purpose and they shall be terminated with a grounding-type attachment plug. In addition, the following conditions shall be met:

(a) The cord shall be at least 18″ but no longer than 36″.

(b) The receptacle shall be located so that it does not subject the flexible cord to physical damage.

(c) The receptacle shall be installed in an accessible location.

If the kitchen waste disposer is hard-wired, the installation shall comply with the type of wiring method selected. If a kitchen waste disposer is protected by a system of double insulation, and is distinctly marked to indicate such a system, the kitchen waste disposer is not required to be grounded per 422-8(d), Ex.

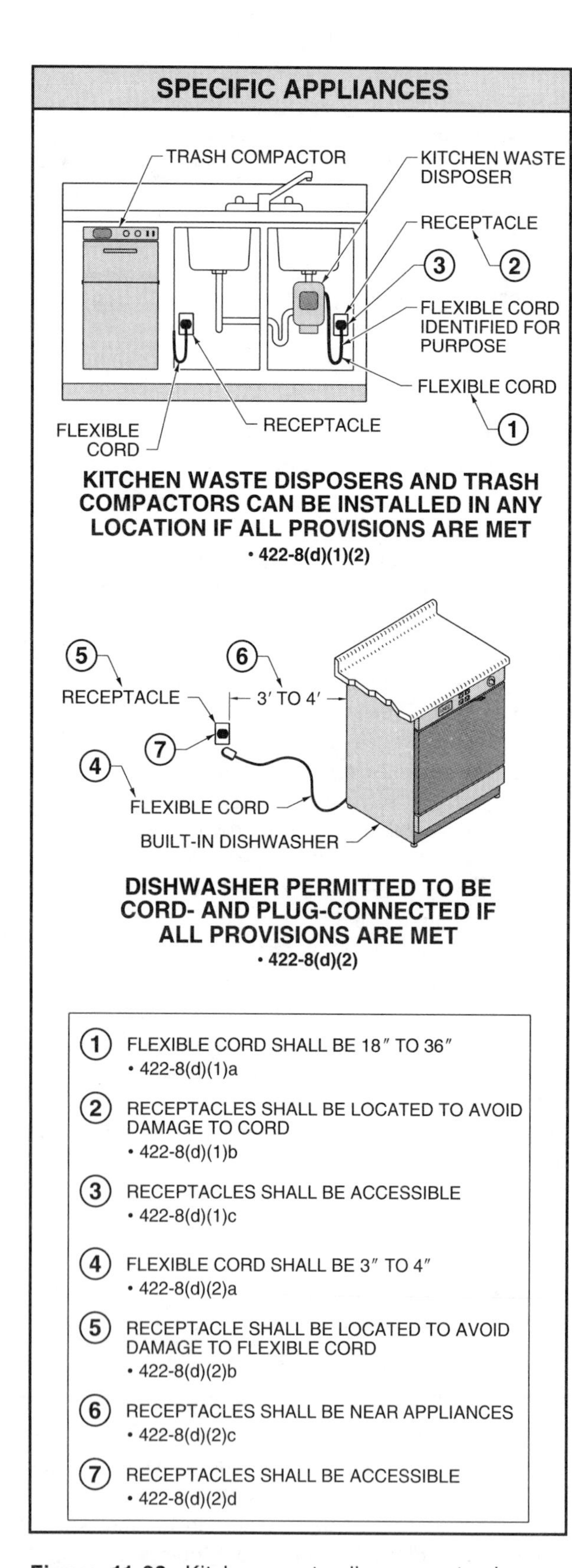

Figure 11-22. Kitchen waste disposers, trash compactors, and dishwashers are permitted to be cord- and plug-connected.

Trash compactors and built-in dishwashers are permitted be hard-wired or cord- and plug-connected per 422-8(d)(2). Trash compactors and built-in dishwashers that are cord- and plug-connected shall be provided with flexible cord which is identified for the purpose and they shall be terminated with a grounding-type attachment plug. In addition, the following conditions shall be met:

(a) The cord shall be 3′ to 4′ in length.

(b) The receptacle shall be located so that it does not subject the flexible cord to physical damage.

(c) The receptacle shall be located adjacent to the appliance.

(d) The receptacle shall be installed in an accessible location.

If the trash compactor or built-in dishwasher is hard-wired, the installation shall comply with the type of wiring method selected. If the trash compactor or built-in dishwasher is protected by a system of double insulation, and is distinctly marked to indicate such a system, the trash compactor or dishwasher is not required to be grounded per 422-8(d), Ex.

Water Heaters – 422-14. Each storage- or instantaneous-type water heater shall be provided with a temperature-limiting device to disconnect all ungrounded supply conductors per 422-14. This high-limit device is in addition to any thermostat that may be used for control purposes. The high-limit device shall be capable of sensing water temperature and be either trip-free or contain a replacement element.

The branch circuit supplying the water heater is sized by 422-14(b). For storage-type water heaters of not more than 120 gal., the branch-circuit rating shall be at least 125% of the nameplate rating of the water heater. The FPN references 422-4(a), Ex. 2, which contains requirements for sizing continuously loaded branch circuits. In essence, the water heater is treated like a continuously-operated appliance.

Wall and Counter-Mounted Ovens – 422-17. Wall and counter-mounted ovens can also be hard-wired or cord- and plug-connected for ease of servicing. If a separable plug or connector and receptacle combination is used for wall or counter-mounted ovens, two general provisions must be met. The cord and plug connection shall not be installed as the disconnecting means required by 422-20 and the installation shall be approved for the temperature to which it is subjected.

Ceiling Fans – 422-18. Ceiling fans, because of their location and the dynamics of their load, can pose a great danger to personnel if they are not properly installed. Section 422-18 divides ceiling fans into two parts. Ceiling fans which weigh 35 lb or less are covered in 422-18(a). Ceiling fans which weigh more than 35 lb are covered in 422-18(b).

Ceiling fans which weigh 35 lb or less can be supported by outlet boxes if the box is listed for such use. Ceiling fans which weigh more than 35 lb, including any accessories which may be attached to the fan, shall be supported in a manner which is independent of the outlet box. See Figure 11-23. Typically, bolt rods or other support means are directly connected to the building structure for fans weighing over 35 lb. See Section 370-23(a)(b) and 370-27(c).

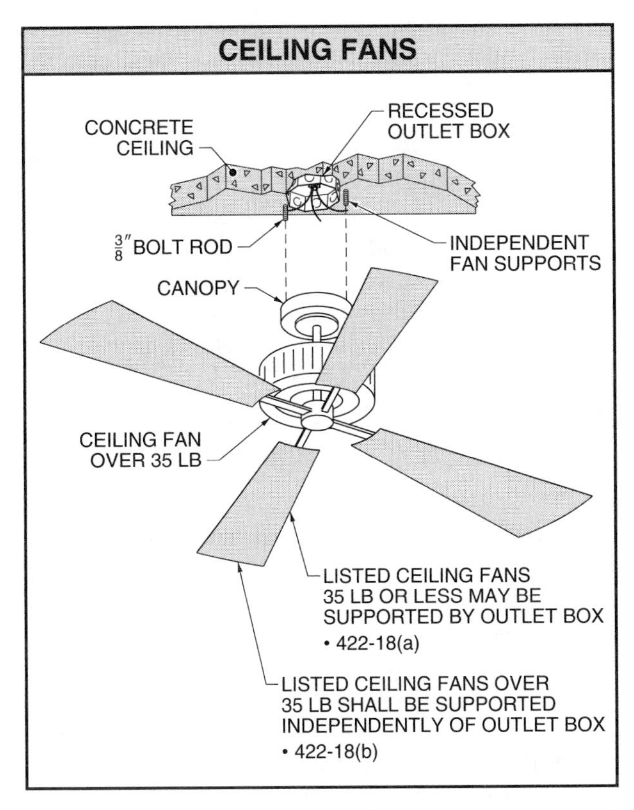

Figure 11-23. Ceiling fans weighing over 35 lb shall not be supported by the outlet box.

Disconnecting Means – 422, Part D

The disconnecting means provides a way to remove the supply of electricity from the appliance. This is a critical safety requirement for workers who may be attempting to service or repair the appliance.

Article 422, Part D contains the requirements for providing a means of disconnecting each appliance from the source of supply. In general, every appliance shall be provided with a disconnecting means that disconnects each ungrounded conductor from the supply source. In the event that an appliance is supplied by more than one source, such as where power conductors and control conductors are both used, each source shall have a separate disconnecting means and the disconnecting means shall be grouped and properly identified.

Permanently-Connected Appliances – 422-21. A *permanently-connected appliance* is a hard-wired appliance that is not cord- and plug- connected. Section 422-21 divides permanently-connected appliances into two groups. Appliances rated at not over 300 VA or 1/8 HP are covered in 422-21(a). Appliances rated over 300 VA or 1/8 HP are covered in 422-21(b).

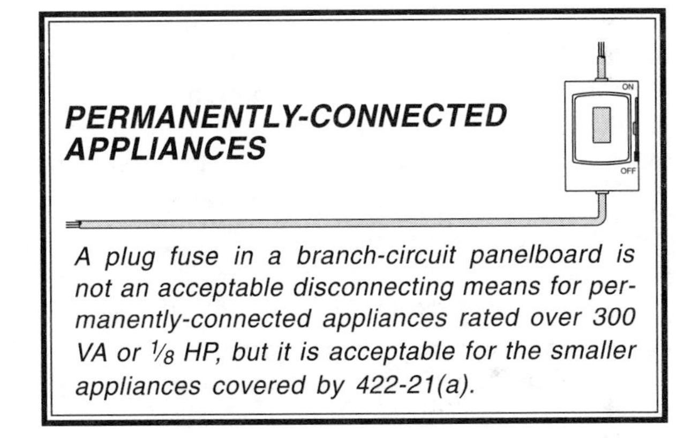

For smaller appliances, 422-21(a) permits the disconnecting means for the appliance to be the branch-circuit overcurrent device. For example, the disconnecting means for a 250 VA, permanently-installed appliance could be the branch-circuit fuse or CB.

For larger appliances, 422-21(b) permits the disconnecting means to be a branch-circuit switch or CB. The switch or CB shall be within sight from the appliance or it shall be capable of being locked in the open position. *Within sight* is visible and no more than 50′ from the object.

Cord- and Plug-Connected Appliances – 422-22. The provisions for disconnecting cord- and plug-connected appliances are listed in 422-22. The general rule of 422-22(a) permits a separate attachment plug or connector and a receptacle to serve as the disconnecting means for the appliance if the plug and receptacle are accessible. If the plug and receptacle are not accessible, as is the case if they are located behind a fixed appliance, another disconnecting means meeting the requirements of 422-21 shall be provided.

If the cord and plug is provided at the rear of the range, and is accessible from the front of the range by removal of the oven drawer, the cord and plug is permitted to serve as the disconnecting means per 422-22(b). Attachment plugs and receptacles installed to serve as the disconnecting means shall have a rating which is at least equal to the rating of the connected appliance per 422-22(c). Attachment plugs or connectors shall be designed to protect against contact with live parts, have a sufficient interrupting capacity to handle available current, and avoid interchangeability with receptacles of lesser ratings per 422-22(d).

Unit Switches – 422-25. A *unit switch* is a switch designed to control a specific unit load. Section 422-25 permits unit switches to serve as the disconnecting means if the unit switch opens all ungrounded conductors and has a marked OFF position designated on the switch. In addition to these two requirements, other means for disconnection shall also be provided depending upon the type of occupancy in which the appliance is installed. See Figure 11-24.

For multifamily dwellings, the other disconnecting means shall be located within the dwelling unit or on the same floor as the dwelling unit per 422-25(a). The disconnecting means can control lamps and other appliances such as would be the case if a branch-circuit overcurrent device in a panelboard were used.

For two-family dwellings, 422-25(b) permits the other disconnecting means to be located either inside or outside the dwelling unit that contains the connected appliance. The disconnecting means can control lamps and other appliances such as would be the case if a branch-circuit overcurrent device in a panelboard were used.

For one-family dwellings, 422-25(c) permits the other disconnecting means to be the service disconnecting means. The service disconnecting means can be located anywhere in the dwelling.

For all other occupancies, 422-25(d) permits the branch-circuit switch or CB to serve as the other disconnecting means. The branch-circuit switch or CB shall be readily accessible.

Motor-Driven Appliances – 422-27. For motor-driven appliances of more than ⅛ HP, the provisions of 430, not 422, shall be applied. Section 422-27 requires that the disconnecting means for a permanently-connected, motor-driven appliance of more than ⅛ HP shall be located in sight of the motor controller and shall comply with the 430, Part I. However, 422-27, Ex. does not require that switches or CBs used to meet the other disconnecting means required by 422-25 to be located within sight of the motor controller.

Safety Provisions – 422-23; 422-24

Several safety provisions are designed to protect personnel who use the appliances covered by 422. The first safety provision is contained in 422-23, which covers proper polarity for cord- and plug-connected appliances. The second provision is contained in 422-24, which covers immersion protection for personnel. While both of these sections deal with design considerations, they are important safety requirements for the installation of cord- and plug-connected appliances.

Cord- and Plug-Polarity – 422-23. Section 422-23 requires that appliances with a manually-operated, line-connected, single-pole control switch, an Edison-base lampholder, or a 15 A or 20 A receptacle shall be equipped with an attachment plug that is polarized or of the grounding type. A polarized attachment plug is designed so that the ungrounded and grounded prongs cannot be inserted into a receptacle in a manner that causes an opposite or reverse polarity connection. This provision is designed to protect unqualified persons who might attempt to service or relamp an appliance equipped with an edison-base lampholder. An attachment plug that is polarized or of the grounding type ensures that a reverse polarity condition cannot occur which could increase the risk of electrical shock.

IDCI – 422-24. IDCIs are a relatively new form of personnel protection. An *immersion detection circuit interrupter (IDCI)* is a circuit interrupter designed to provide protection against shock when appliances fall into a sink or bathtub. An IDCI disconnects the source of supply regardless of whether the appliance is in the ON or OFF position when it is immersed in water.

The NEC® requires that portable, freestanding hydromassage units and hand-held hair dryers be equipped with IDCIs when they are manufactured. IDCIs offers protection for persons who are using these types of appliances in occupancies without ground-fault circuit protection.

Markings – 422, Part E

The general rule of 422-30(a) requires each electric appliance to be supplied with a nameplate that contains the identifying name and rating in volts and amperes, or volts and watts, for the appliance. The appliance shall be marked if it is designed for use on a specific frequency or if it requires external overload protection.

All markings shall be in a location that is visible or at least easily accessible after the installation of the appliance per 422-30(b). If the appliance contains field-replaceable heating elements rated over 1 A, they shall be legibly marked with either the manufacturer's part number or the rating in volts and amperes or volts and watts.

When an appliance consists of a motor and other loads, the marking requirements of 422-32 must be followed. Two marking options are permissible. Per 422-32(a), in addition to the markings required by 422-30, the appliance shall be marked with the minimum supply circuit conductor ampacity and the maximum rating of the circuit overcurrent protective device. Per 422-32(a), Ex. 1 and 2, the provisions of this section do not apply if the appliance is factory-equipped with cords and attachment plugs that comply with 422-30 or the appliance's minimum supply conductor ampacity and maximum OCPD rating are both less than 15 A and the provisions of 422-30 are also met.

An alternate marking method is provided in 422-32(b). This method consists of the rating of the largest motor in volts and amperes and the other loads rating in volts and amperes or volts and watts, plus the required markings listed in 430-30. The alternate marking method does not apply if the appliance is factory-equipped with cords and attachment plugs that comply with 422-30. The ampere rating shall be permitted to be omitted from the required markings for appliances which contain a motor with a 1/8 HP or less rating or a nonmotor load of 1 A or less.

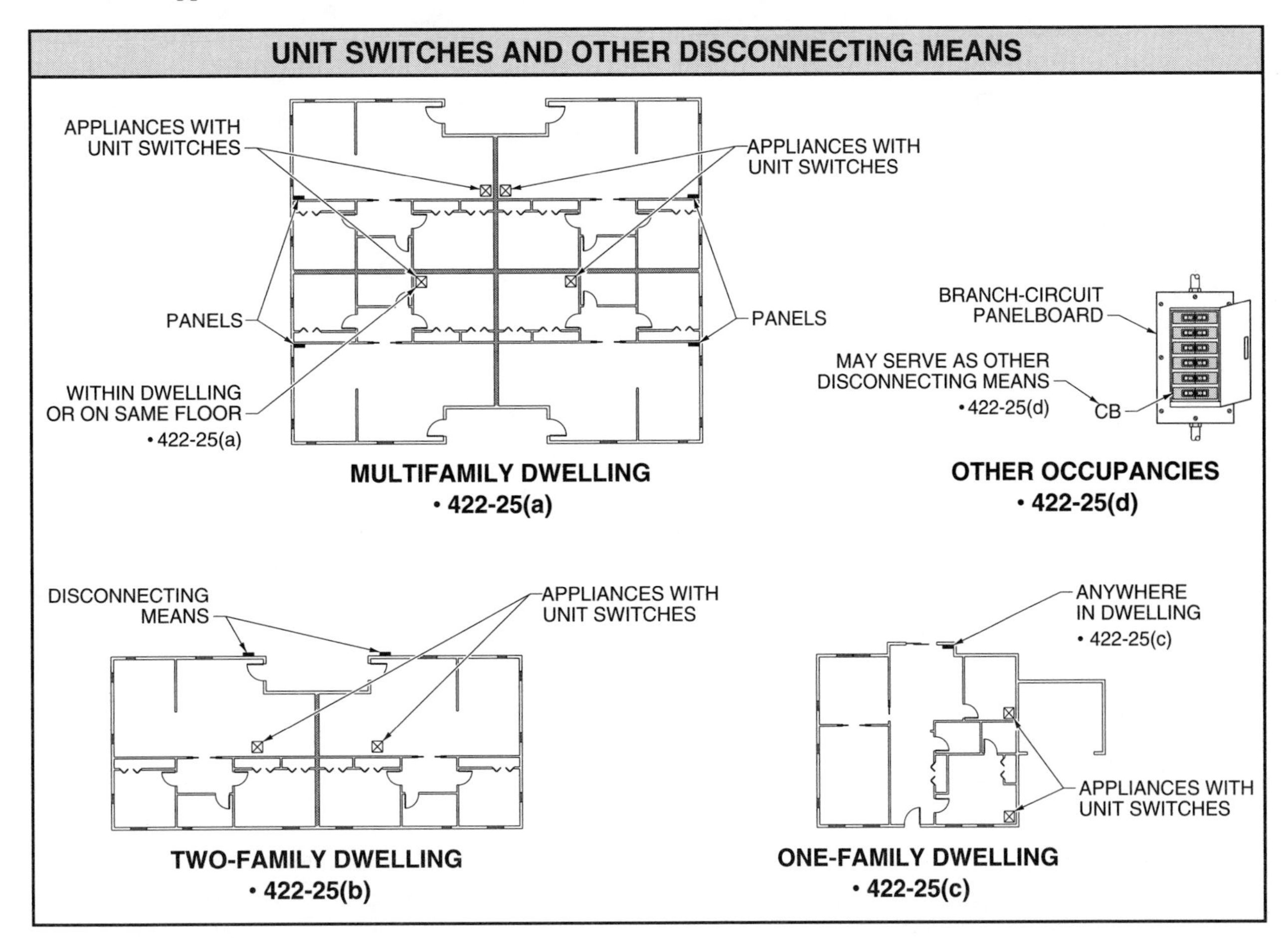

Figure 11-24. Unit switches shall have an additional or other disconnecting means which acts a backup when the unit switch is used as a disconnecting means.

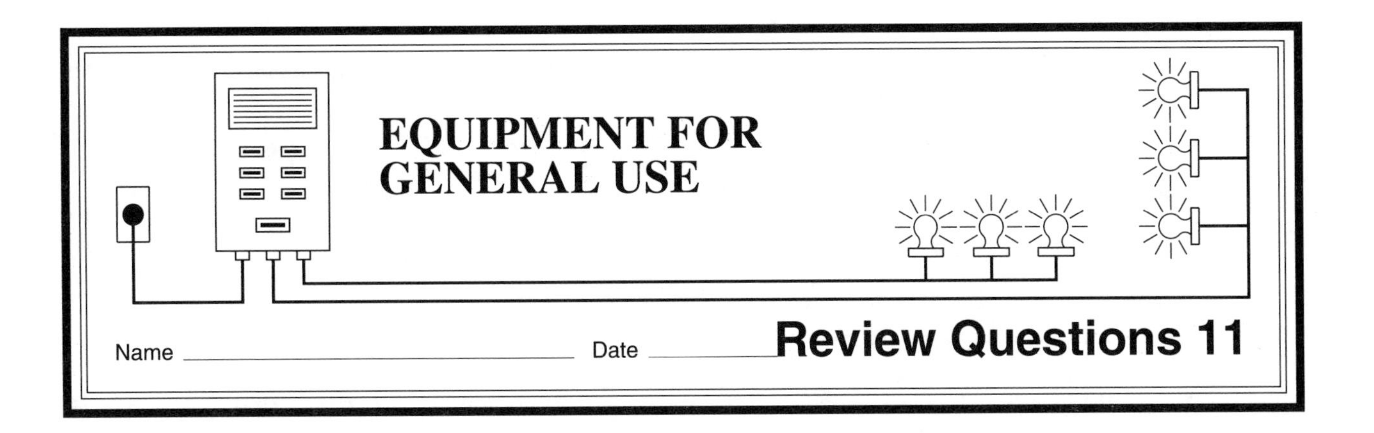

Review Questions

______ **1.** The major difference between flexible cords and flexible cables is the number of insulated ______.

______ **2.** A(n) ______ is a complete lighting fixture consisting of the lamp or lamps, reflector, lamp guards, and lamp power supply.

T F **3.** Generally, flexible cords and cables shall be installed in continuous lengths without splices or taps.

T F **4.** Lampholders shall not support lamps weighing over 6 lb.

T F **5.** Trees may be used to support outdoor lighting fixtures.

______ **6.** An outlet box for a recessed incandescent fixture shall be at least ______′ from the fixture if the branch-circuit conductors are not suitable for the temperature.

A. ½
B. 1
C. 2
D. 4

______ **7.** An electric-discharge lighting fixture utilizes a(n) ______ for the operation of the lamp.

______ **8.** Each 2′ of lighting track is considered ______ VA for the purposes of determining load calculations.

______ **9.** ______-grade receptacles are the highest grade receptacles manufactured by the electrical industry.

______ **10.** The flexible cord for a kitchen waste disposer shall be from ______″ to ______″ in length.

A. 12; 24
B. 18; 30
C. 18; 36
D. 24; 36

______ **11.** Ceiling fans which weigh ______ lb or less can be supported by outlet boxes if the box is listed for such use.

______ **12.** A(n) ______ switch is a switch designed to control a specific unit load.

______ **13.** Flexible cords not smaller than No. ______ AWG are considered protected by the overcurrent device.

__________ **14.** Handholes are not required in ________′ or less metal poles which support lighting fixtures if the base is hinged.

__________ **15.** Any branch-circuit conductor located within ________″ of a ballast, within a ballast compartment, shall have an insulation rating not lower than 90°C.

__________ **16.** A minimum of ________″ clearance shall be provided for recessed portions of lighting fixtures from all combustible materials.

T F **17.** All recessed HID fixtures shall be thermally protected.

T F **18.** Class A GFCIs are designed to operate when the ground fault current flow is between 10 mA and 15 mA.

__________ **19.** A(n) ________ is any piece of utilization equipment which performs one or more functions.

__________ **20.** Within sight is visible and no more than ________′ from the object.

__________ **21.** The horizontal bathtub zone in which electric fixtures are not permitted is ________′ from the bathtub rim.

__________ **22.** The vertical bathtub zone in which electric fixtures are not permitted is ________′ from the bathtub rim.

__________ **23.** An independent means of support shall be provided for lighting fixtures that weigh over ________ lb.

T F **24.** All lighting track shall be permanently installed and permanently connected to a branch circuit.

T F **25.** Metal faceplates are required to be grounded.

Clothes Closets

__________ **1.** The minimum distance at A is ________″.

__________ **2.** The minimum distance at B is ________″.

__________ **3.** The minimum depth at C is ________′.

__________ **4.** The minimum width at D is ________″.

__________ **5.** The minimum width at E is ________″.

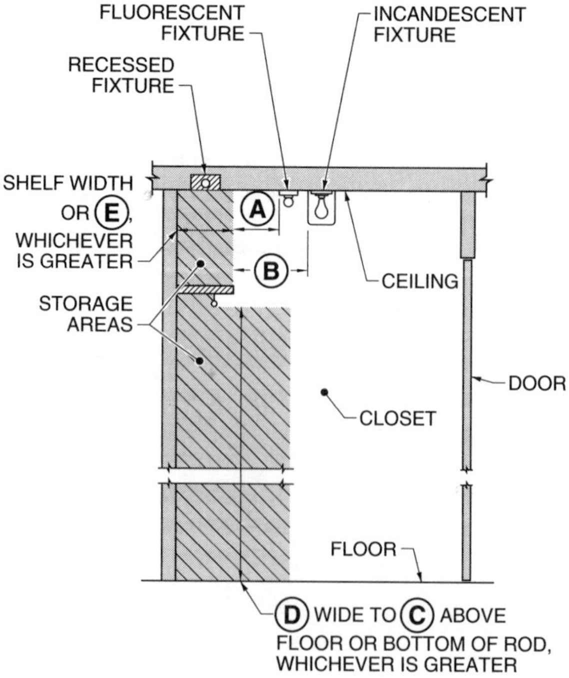

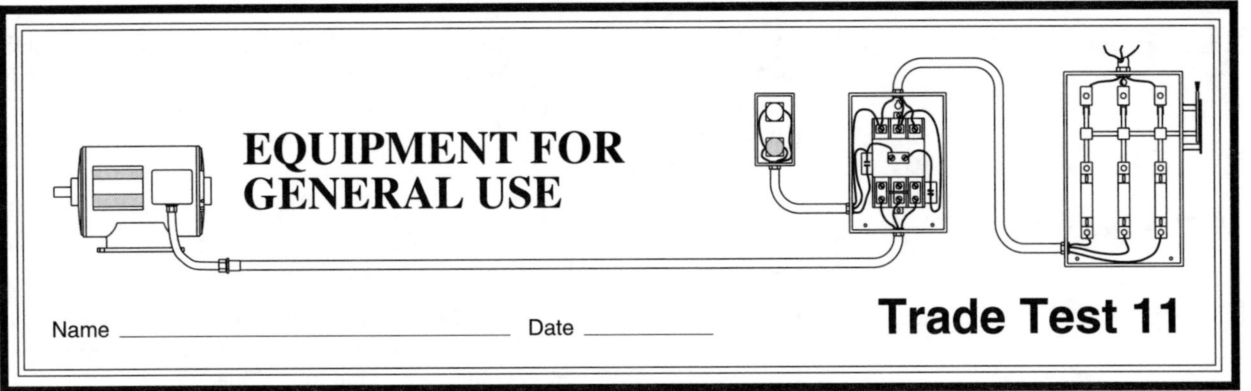

NEC® **Answer**

1. Determine the ampacity of each conductor of a No. 16/3 multiconductor Type SPT-2 flexible cord used to supply utilization equipment which is fastened-in-place. Two of the conductors are current-carrying conductors.
2. Determine the ampacity of each conductor of a No. 18/4 multiconductor Type SRDT flexible cord used to supply fastened-in-place utilization equipment. All of the conductors are current-carrying conductors.
3. Determine the ampacity of each THW conductor of a No. 8/3 multiconductor Type W flexible cable used to supply fastened-in-place utilization equipment. Two of the conductors are current-carrying conductors.
4. Determine the ampacity of each TW conductor of a No. 4/4 multiconductor Type G flexible cable used to supply fastened-in-place utilization equipment. All of the conductors are current-carrying conductors.
5. Lighting fixtures are installed outdoors under an open porch of a one-family dwelling. The fixtures are marked "Suitable for Damp Locations." Does this installation violate the provisions of Article 410 of the NEC®?
6. *See Figure 1.* Does this installation violate Article 410 of the NEC®?

SHELF
14″
CEILING
PARTIALLY-ENCLOSED INCANDESCENT FIXTURE
CLOTHES ROD
CLOSET

FIGURE 1

7. An electrical installation requires that recessed HID fixtures are to be installed in a ceiling of poured concrete decks. The HID fixtures are not provided with thermal protection. Does this installation violate Article 410 of the NEC®?
8. A 12′ length of lighting track is installed in the master bedroom of a one-family dwelling. How much VA shall be allocated for the lighting track when determining the calculated load?

______ ______ **9.** A 15 A GFCI general-use receptacle is installed outdoors under an open porch of a one-family dwelling. The receptacle is protected from the direct effects of the weather. The enclosure for the receptacle is weatherproof only when the receptacle is covered. Does this installation meet the requirements of Article 410 of the NEC®?

______ ______ **10.** Determine the minimum branch-circuit rating for an individual branch circuit which is used to supply a fastened-in-place appliance. The marked rating on the appliance is 35 A. The appliance is not continuously loaded.

______ ______ **11.** Determine the minimum branch-circuit rating for an individual branch circuit which is used to supply a fastened-in-place appliance. The marked rating on the appliance is 24 A. The appliance is continuously loaded.

______ ______ **12.** *See Figure 2.* Does this installation violate the provisions of Article 422 of the NEC®?

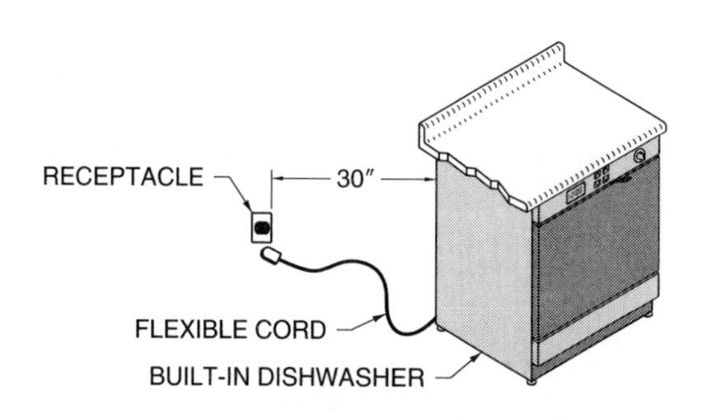

FIGURE 2

______ ______ **13.** *See Figure 3.* Does this installation violate the provisions of Article 422 of the NEC®?

______ ______ **14.** *See Figure 4.* Does this installation violate the provisions of Article 410 of the NEC®?

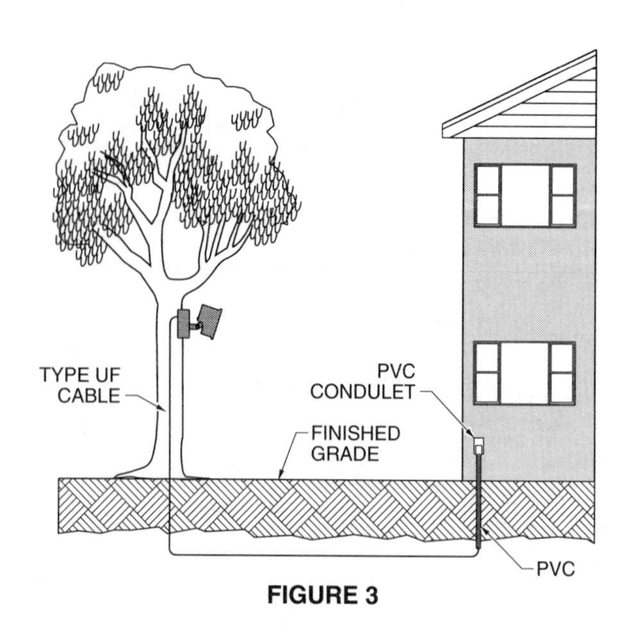

FIGURE 3

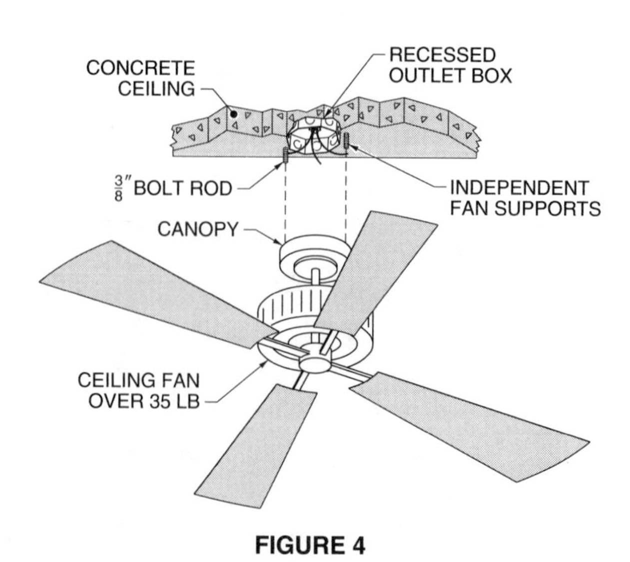

FIGURE 4

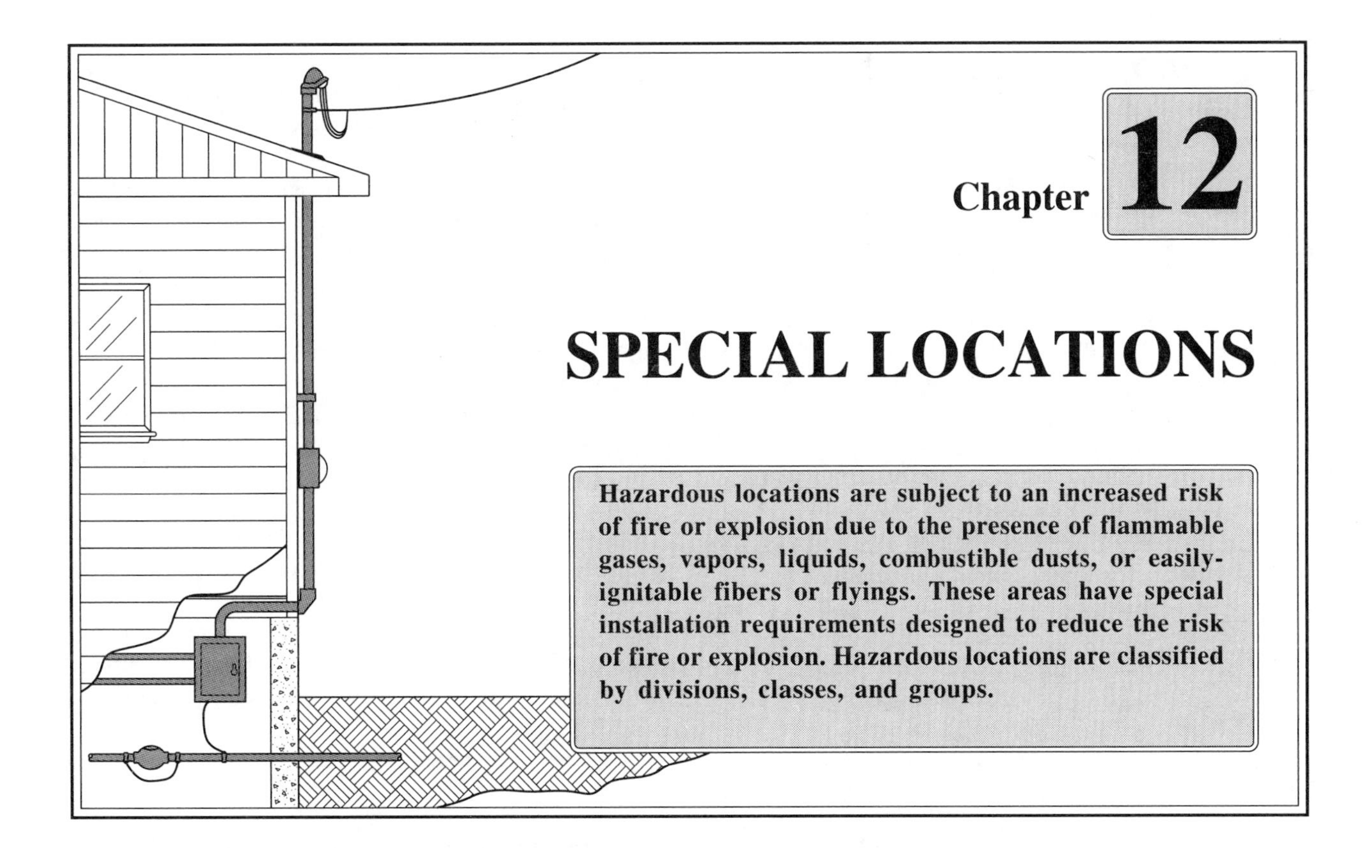

Chapter 12

SPECIAL LOCATIONS

Hazardous locations are subject to an increased risk of fire or explosion due to the presence of flammable gases, vapors, liquids, combustible dusts, or easily-ignitable fibers or flyings. These areas have special installation requirements designed to reduce the risk of fire or explosion. Hazardous locations are classified by divisions, classes, and groups.

HAZARDOUS LOCATIONS – 500

Special locations are grouped in Chapter 5 because they present unique installation and design characteristics which increase the risk of hazards due to the work which is performed within the occupancy or from materials which are present within the occupancy. A *hazardous location* is a location where there is an increased risk of fire or explosion due to the presence of flammable gases, vapors, liquids, combustible dusts, or easily-ignitable fibers or flyings.

If any of these are present in the atmosphere or there is the likelihood that they could become present, special electrical installation requirements apply. This is because electrical energy can provide the source of ignition for these hazardous substances and cause an explosion or fire if sufficient quantities of the hazardous substance and oxygen are present. Not all hazardous substances, however, pose the same risk for fire or hazard. Each of the hazardous substances has different characteristics, such as flash point or combustion point, which can affect the likelihood of a fire or explosion occurring.

The NEC® contains a classification system for hazardous substances which groups them according to their type and potential for hazard. This classification system is the basis for determining the installation requirements when hazardous substances are present.

The standard classification system used by the NEC® incorporates the use of a Class, Division, and Group system to determine the type of hazardous location. See Figure 12-1. The three classes which are used, depending upon the properties of the hazardous material which may be present are Class I, Class II, and Class III locations.

A *Class I location* is a hazardous location in which sufficient quantities of flammable gases and vapors are present in the air to cause an explosion or ignite the hazardous materials. A *Class II location* is a hazardous location in which sufficient quantities of combustible dust are present in the air to cause an explosion or ignite the hazardous materials. A *Class III location* is a hazardous location in which easily-ignitible fibers or flyings are present in the air, but not in a sufficient quantity to cause an explosion or ignite the hazardous materials.

HAZARDOUS LOCATIONS	
Classes	Likelihood that a flammable or combustible concentration is present.
I	Sufficient quantities of flammable gases and vapors present in air to cause an explosion or ignite hazardous materials.
II	Sufficient quantities of combustible dust are present in air to cause an explosion or ignite hazardous materials.
III	Easily-ignitable fibers or flyings are present in air, but not in a sufficient quantity to cause an explosion or ignite hazardous materials.
Divisions	Location containing hazardous substances.
1	Hazardous location in which hazardous substance is normally present in air in sufficient quantities to cause an explosion or ignite hazardous materials.
2	Hazardous location in which hazardous substance is not normally present in air in sufficient quantities to cause an explosion or ignite hazardous materials.
Groups	Atmosphere containing flammable gases or vapors or combustible dust.

Class I	Class II	Class III
A B C D	E F G	none

Figure 12-1. The standard classification system used by the NEC® incorporates the use of a Class, Division, and Group system to determine hazardous locations.

Each of the three classes also contains two divisions. A *division* is the classification assigned to each class based upon the likelihood of the presence of the hazardous substance in the atmosphere. A *Division 1 location* is a hazardous location in which the hazardous substance is normally present in the air in sufficient quantities to cause an explosion or ignite the hazardous materials. A *Division 2 location* is a hazardous location in which the hazardous substance is not normally present in the air in sufficient quantities to cause an explosion or ignite the hazardous materials.

A Division 2 location can also occur when ignitible concentrations of the hazardous substance exist because of frequent maintenance operations, leakage, or in cases in which a failure of the equipment or process could cause a release of the hazardous substance. Division 2 locations can also occur when ignitible concentrations of hazardous substances are prevented by the use of positive mechanical ventilation. Division 2 locations are also frequently assigned to those areas which are adjacent to Division 1 locations. The Division 2 classification is required because the hazardous substances can migrate or move into the location from the adjacent Division 1 location.

Per 500-2, each room or area within an occupancy shall be considered separately for the purposes of determining area classification. Whenever possible, design personnel and installers of electrical systems should select locations that result in the least amount of equipment to be located in hazardous locations.

DETERMINING CLASSIFICATIONS OF HAZARDOUS LOCATIONS

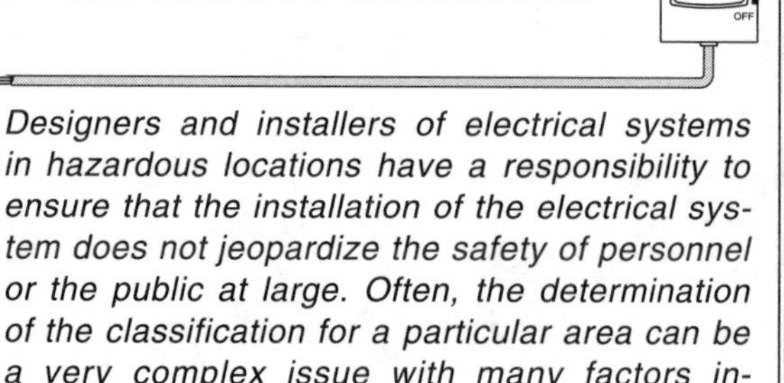

Designers and installers of electrical systems in hazardous locations have a responsibility to ensure that the installation of the electrical system does not jeopardize the safety of personnel or the public at large. Often, the determination of the classification for a particular area can be a very complex issue with many factors involved. Designers and installers should determine the correct classification of the area. If sufficient information is not available to determine the correct classification of the area, the owners of the building or structure where the classified area is located, their engineering support staff, and the AHJ should be consulted before designing or installing electrical systems for these areas.

Class I, Division 1 Locations – 500-5(a). A Class I, Division 1 location is any location in which sufficient quantities of flammable gases or vapors are present in the air, under normal conditions, to pose the threat of explosion or ignition of the hazardous substances. If sufficient quantities do not exist during normal operations, but service or maintenance operations are performed which could cause ignitable concentrations of the flammable gases or vapors to exist frequently, the area is also classified as Class I, Division 1. Additionally, an area is classified as Class I, Division 1 if a process or equipment failure could cause a release of the hazardous substances in sufficient quantities for an ignition or explosion to occur.

Locations where liquified flammable gases or liquified flammable liquids are transferred from one container to another are classified as Class I, Division 1 locations. Additionally, per 500-5(a), FPN 1, the following occupancies are also classified as Class I, Division 1 locations:

- Spray booth interiors.
- Areas adjacent to spraying or painting operations using volatile flammable solvents.
- Open tanks or vats of volatile flammable liquids.
- Drying or evaporation rooms for flammable solvents.
- Areas where fats and oil extraction equipment using flammable solvents are operated.
- Cleaning and dyeing plant rooms which use flammable liquids.
- Gas generator rooms.
- Pump rooms for flammable gases or volatile flammable liquids which do not contain adequate ventilation.
- Refrigeration or freezer interiors which store flammable materials.
- All other locations where sufficient ignitable quantities of flammable gases or vapors are likely to occur during routine operations.

Some of these Class I, Division 1 locations have sufficient quantities of hazardous substances present continuously or for long periods of time per 500-5(a), FPN 2. Designers and installers of electrical systems should avoid placing electrical equipment in these locations. In areas where electrical equipment is essential to the process, only equipment that is identified for the specific location and intrinsically safe systems per 504 should be used.

Class I, Division 2 Locations – 500-5(b). A Class I, Division 2 location is any location in which flammable gases are handled, processed, or used, but which pose no threat of explosion or ignition of the hazardous substances under normal conditions. The flammable gases or vapors are confined to containers or closed systems. The only way that the hazardous substances could be introduced into the atmosphere is through accidental rupture or breakdown of the containers.

An area is also classified as Class I, Division 2 if ignitible concentrations of flammable gases and vapors are prevented by the use of positive mechanical ventilation. In these locations, the area could become hazardous if the ventilation system failed to operate. An area could also be classified as Class I, Division 2 if it is adjacent to a Class I location and the possibility of ignitible concentrations migrating from the Class I location exists.

Per 500-5(b), FPN 1, if flammable gases or vapors are used and, in the judgment of the AHJ, their use does not pose a threat of explosion or hazard unless an accident or failure of a ventilating system occurred, the area is classified as Class I, Division 2. Similarly, if flammable liquids are stored in sealed containers that are unlikely to be opened, unless by accident, the area is classified as Class I, Division 2. Additionally, areas with piping which contain flammable gases or vapor are classified as Class I, Division 2 if the piping does not contain any valves, checks, meters, or similar devices. If the piping contains these types of devices, the likelihood of leaks is increased to a point at which the area would probably have to be classified as Class I, Division 1.

Class II, Division 1 Locations – 500-6(a). A Class II, Division 1 location is any area in which sufficient combustible dust is present in the area, under normal operating conditions, to produce explosive or ignitable mixtures. Areas are also classified as Class II, Division 1 if mechanical failure of equipment or processes could cause explosive mixtures of combustible dusts to be produced.

Additionally, if any of the electrically-conductive dusts of Group E are present in sufficient quantities to present a hazard, the area is classified as Class II, Division 1. Electrically-conductive dusts, such as aluminum and magnesium, pose a greater threat for ignition or explosion because these dusts can conduct electricity if they gather on live parts of electrical equipment. Per 500-3(b)(1), FPN, some electrically-conductive dusts have additional hazards associated with them because of their low-ignition temperatures.

Per 500-6(a), FPN, some of the combustible dusts which are nonconductive include:

- Grain and grain products.
- Pulverized sugar and cocoa.

- Dried egg and milk powders.
- Pulverized spices.
- Starch and pastes.
- Potato and woodflour.
- Oil meal from beans and seeds.
- Dried hay.
- Any other organic materials that may produce combustible dusts during their use or handling.

Class II, Division 2 Locations – 500-6(b). A Class II, Division 2 location is any location where combustible dusts may be present, but not in quantities sufficient to produce explosive mixtures or interfere with the successful operation of electrical equipment or apparatus. Class II, Division 2 locations occur where combustible dust is suspended in the air as a result of occasional malfunctioning of equipment which handles or processes the material which produces the dust. Per 500-6(b), FPN 1, a key factor in the classification of these locations is the quantity of dusts which may be present and the adequacy of the dust removal systems in these locations. Per 500-6(b), FPN 2, if the quantity of dust is limited by the manner in which it is handled, the area may not require classification.

Class III, Division 1 Locations – 500-7(a). A Class III, Division 1 location is any location in which easily ignitible fibers or flyings are handled, manufactured, or used. The fibers or flyings may be present in the air, but not in quantities which would be sufficient to produce ignitible mixtures. Per 500-7(a), FPN 1, some locations that are classified as Class III, Division 1 include:

- Portions of rayon, cotton, or other textile mills.
- Manufacturing and processing plants for combustible fibers, cotton gins, and cotton seed mills.
- Flax-processing plants.
- Clothing manufacturing plants.
- Woodworking plants.
- Other establishments involving similar hazardous processes or conditions.

Per 500-7(a), FPN 2, some of the types of materials that are defined as easily ignitable fibers or flyings include:

- Rayon
- Cotton
- Sisal or henequen
- Istle
- Jute
- Hemp
- Tow
- Cocoa fiber
- Oakum
- Baled waste kapok
- Spanish moss
- Excelsior
- Other similar materials

Class III, Division 2 Locations – 500-7(b). A Class III, Division 2 location is any location where easily ignitable fibers or flyings are stored or handled. Per 500-7(b), Ex., if easily ignitable fibers or flyings are in the process of being manufactured, the location is not a Class III, Division 2 location, but is a Class III, Division 1 location. Classification for Class III materials depend on the use of the material rather than the quantity which is present.

Group Classifications – 500-3(a)(b)

Different flammable gases and vapors have different degrees of risk associated with them. Since the 1937 NEC®, flammable gases and vapors have been grouped according to their level of hazard. A *group* is an atmosphere containing flammable gases or vapors or combustible dust. Different gases in the various groups have different ignition temperatures.

Class I locations contain four groups (A, B, C, and D) of flammable gases or vapors, each with similar explosion and ignition characteristics. Class II locations contain three groups of combustible dusts which are grouped on similar electrical resistivity. There are no groups for Class III locations.

The use of the group system primarily applies to the manufacturers of equipment which is to be installed in hazardous locations. The group system ensures that the electrical equipment installed in a particular hazardous location is suitable for the hazardous substance that may be present in the air. Section 500-3(a)(1 – 4) contains the specific flammable gases and vapors for Class I Groups A – D and 500-3(b)(1 – 3) contains the listing of Class II Groups E – G.

Wiring Methods

Chapter 5 of the NEC® has very strict requirements for wiring methods in hazardous locations. The installation of the electrical system in hazardous locations shall not create additional hazards. This is a major concern for designers and installers alike because electrical energy has the potential to serve as a source of ignition for fires and explosions in these locations.

Wiring method selection in hazardous locations closely follows the classification system for these locations. Wiring methods are selected based on the class and the division in which they are to be installed. In general, the wiring methods for Division 1 locations are more strict than the wiring methods for Division 2. Unlike area classifications, which are all contained in 500, wiring method selections are contained in separate articles as follows:

- Class I locations – 501
- Class II locations – 502
- Class III locations – 503

Class I, Division 1 Wiring Methods – 501-4(a). Because of the nature of the hazardous substances, the selection of wiring methods is strictly limited. In general, only RMC, steel IMC, and Type MI cable are permitted as wiring methods in Class I, Division 1 locations. For RMC and IMC, all threaded joints shall have at least five threads fully engaged. Type MI cable shall be terminated with fittings approved for the location. The weight of the cable shall not be transferred to the terminations.

INSTALLING TYPE MI CABLE IN HAZARDOUS LOCATIONS

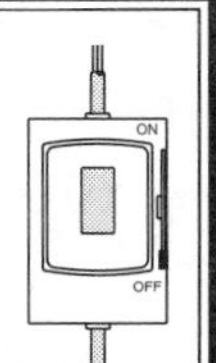

Article 330 contains the requirements for installing Type MI cable. Type MI cable is a mineral-insulated cable with a wide variety of applications through the electrical industry. When Type MI cable is listed as an acceptable wiring method for Class I, Division 1 locations, special explosionproof fittings are required. Standard MI cable termination fittings are not acceptable for use in Class I, Division 1 locations because of the lack of a threaded joint connection between the cable and the connector.

INSTALLING IMC IN HAZARDOUS LOCATIONS

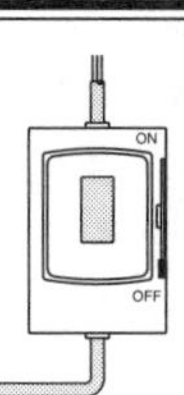

Article 345 contains the requirements for IMC. It is available constructed of steel and of other nonferrous metals such as aluminum. When IMC is listed as an acceptable wiring method, as in 230-43(4), either steel or aluminum is permitted. When the wiring method, however, specifically states "steel IMC," as in 501-4(a), aluminum IMC is not a permitted wiring method. The UL's Electrical Construction Materials Directory (White Book) requires that if IMC is constructed of other than steel, a designation such as "aluminum" or "wrought iron" shall be included in the product name marking on the conduit.

All boxes and fittings used in Class I, Division 1 locations shall be designed for threaded connections and be explosionproof. An *explosionproof apparatus* is equipment which is enclosed in a case that is capable of withstanding any explosion that may occur within it, without permitting the ignition of flammable gases or vapors on the outside of the enclosure. See 100. The explosionproof enclosure shall prevent arcs, flashes, and sparks from igniting the flammable substances by adequately containing them within the enclosure.

The explosionproof apparatus shall keep the temperature on the outside of the enclosure to a low enough temperature to prevent the ignition of the hazardous substances. When equipment requires a flexible connection, such as with motors or equipment which is subject to vibration, flexible fittings listed for Class I locations are permitted to be used.

In general, RNMC, like PVC, is not permitted in Class I, Division 1 locations. However, 501-4, Ex. 1, does permit RNMC to be used where encased in a concrete envelope at least 2″ thick. The RNMC shall be buried at least 2′ under the earth and either RMC or steel IMC shall be used for the last 2′ of the underground run to the point of emergence or to the point of connection to the aboveground raceway. A separate EGC shall be included in the RNMC to ground any noncurrent-carrying parts of metal equipment.

In addition to Type MI cable, 501-4, Ex. 2 permits Type MC cable with a gas/vaportight continuous corrugated aluminum sheath to be installed in Class I, Division 1 locations in industrial establishments with restricted public access. The conditions of maintenance shall ensure that only qualified persons service the installation. See Figure 12-2.

Class I, Division 2 Wiring Methods – 501-4(b). In addition to threaded RMC and threaded steel IMC, which are permitted in Class I, Division 1 locations, enclosed gasketed busways and enclosed gasketed wireways are also permitted to be used. Several different cable assemblies are permitted to be installed in Class I, Division 2 locations. These include Types MI, MC, MV, and TC cables with approved terminations. Additionally, Type PLTC, per 725, and Type ITC in cable trays, raceways, supported by messenger wire, or directly buried if listed for such use are also permitted. Of these cable assemblies Types MI, MC, PLTC, and ITC shall be permitted to be installed in cable trays provided they are installed in a manner which does not impose stress or strain on the terminations. Unlike Class I, Division 1 locations, boxes, fittings, and joints are not required to be explosionproof, unless specifically required, as in 501-3(b)(1), 501-6(b)(1), and 501-14(b)(1).

Several different wiring methods, for equipment requiring flexibility, are permitted in Class I, Division 2 locations. FMC, LTFMC, and LTFNMC are permitted with approved fittings. Additionally, extra-hard usage flexible cord with approved fittings can be used.

Class II, Division 1 Wiring Methods – 502-4(a). The wiring methods in a Class II, Division 1 location closely follow the requirements for Class I, Division 1 locations. In general, only RMC, steel IMC, and Type MI cable are permitted as wiring methods in Class II, Division 1 locations. Fittings for Type MI cable shall be approved and the cable shall be installed in a manner which does not transmit stress and strain to the terminations. Section 502-4(a), Ex. contains the same provisions for using Type MC cable in Class II, Division 1 locations. Listed Type MC cable with a gas/vaportight continuous corrugated aluminum sheath is permitted to be installed in Class II, Division 1 locations in industrial establishments with restricted public access, provided that the conditions of maintenance ensure that only qualified persons service the installation.

Section 502-4(a)(1) requires that all boxes and fittings used in Class II, Division 1 locations have threaded bosses and be equipped with close-fitting covers which contain no openings through which dust can enter. All fittings and boxes which contain splices, taps, terminations, etc., and which are installed in areas where conductive combustible dusts are present, shall be approved for Class II locations. Where equipment which requires flexible connections is installed in Class II, Division 1 locations, 502-4(a)(2) permits dusttight flexible connectors, LTFMC and LTFNMC with approved fittings, or flexible cord of the extra-hard usage type to be used.

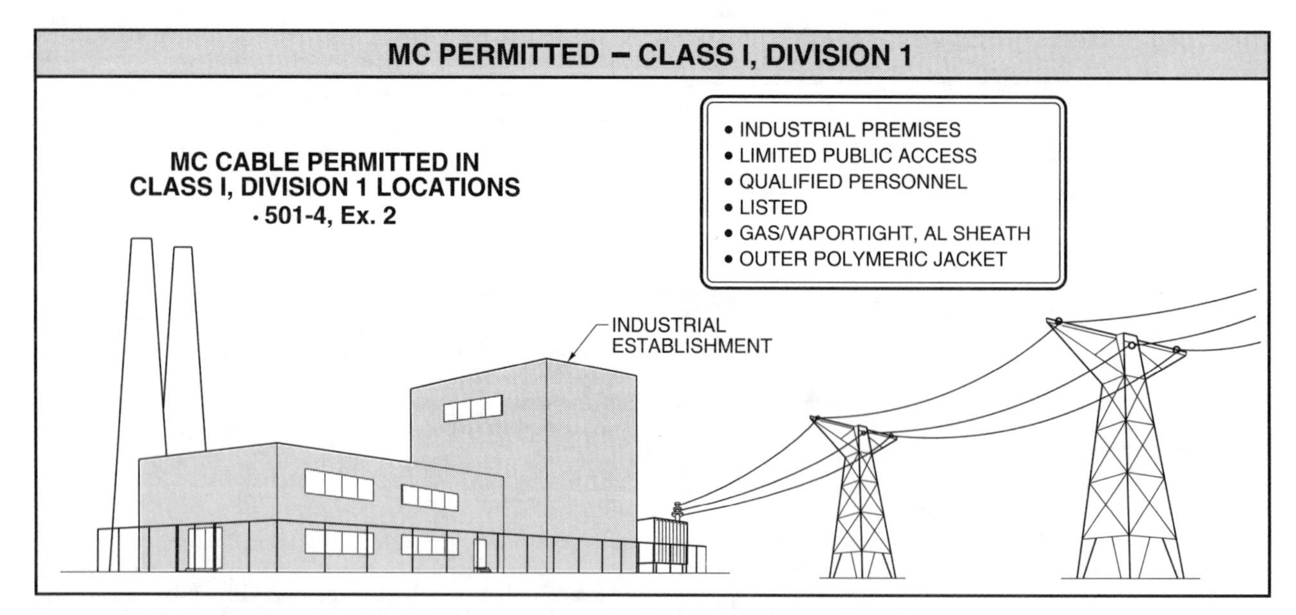

Figure 12-2. Type MC cable is permitted to be used in Class I, Division 1 locations.

Class II, Division 2 Wiring Methods – 502-4(b). The permitted wiring methods for Class II, Division 2 locations are RMC, IMC, EMT, and dusttight wireways. In addition, Type MC and MI cable with approved fittings are also permitted. Type PLTC, ITC, MC, and TC are also permitted where installed in cable trays. Type MC and TC cables are permitted in ladder, ventilated trough, or ventilated channel cable trays provided they are installed in single layers, with a space not less than the larger cable diameter between the two adjacent cables.

Section 502-4(b)(2) permits the same wiring methods used for flexibility for Class II, Division 1 locations to be used for Class II, Division 2 locations. Wireways, fittings, and boxes that contain splices, terminations, taps, etc., and that are installed in Class II, Division 2 locations shall be designed to minimize the entrance of dust and shall be equipped with telescoping or close-fitting covers or other means to prevent the escape of arcs or sparks. They shall contain no openings through which combustible materials might be ignited by any sparks or arcs within the enclosure.

Class III, Division 1 Wiring Methods – 503-3(a). Five raceways permitted to be used in Class III, Division 1 locations are RMC, IMC, RNMC, EMT, and wireways. In addition, Type MI and MC cable are permitted provided they are used with approved termination fittings. All boxes and fittings used in Class III, Division 1 locations are required to be dusttight per 503-3(a)(1). *Dusttight* is construction that does not permit dust to enter the enclosing case under specified test conditions. See 100. *Dustproof* is construction in which dust does not interfere with the successful operation of the equipment.

Per 500-3(a)(2), flexible wiring methods which are permitted to be used in Class III, Division 1 locations are dusttight flexible connectors, LTFMC, LTFNMC with approved fittings, and flexible cord.

Flexible cord used in Class III, Division 1 locations shall be approved for extra-hard usage, contain a separate EGC, be terminated in an approved manner, be supported in a manner that does not put strain on the terminations, and be equipped with a suitable means to prevent the entrance of easily ignitible fibers or flyings into the box or enclosure.

Class III, Division 2 Wiring Methods – 503-3(b). Any of the wiring methods used for Class III, Division 1 locations are permitted to be used for Class III, Division 2 locations. In addition, 503-3(b), Ex. permits open wiring installed per 320 to be used in sections, compartments, or areas used solely for storage provided physical protection of the conductors is used where the conductors are not run in roof spaces and are out of reach of sources of potential physical damage. Physical protection of the conductors is required per 320-14 when they are run within 7′ of the floor. Physical protection can be accomplished by the use of guard strips, running boards, boxing, and insertion in other raceways such as RMC, IMC, RNMC, and EMT.

Conduit Seals

A *conduit seal* is a fitting which is inserted into runs of conduit to isolate certain electrical apparatus from atmospheric hazards. Conduit seals contain provisions for pouring a sealing compound into the fitting to accomplish the isolation. See Figure 12-3.

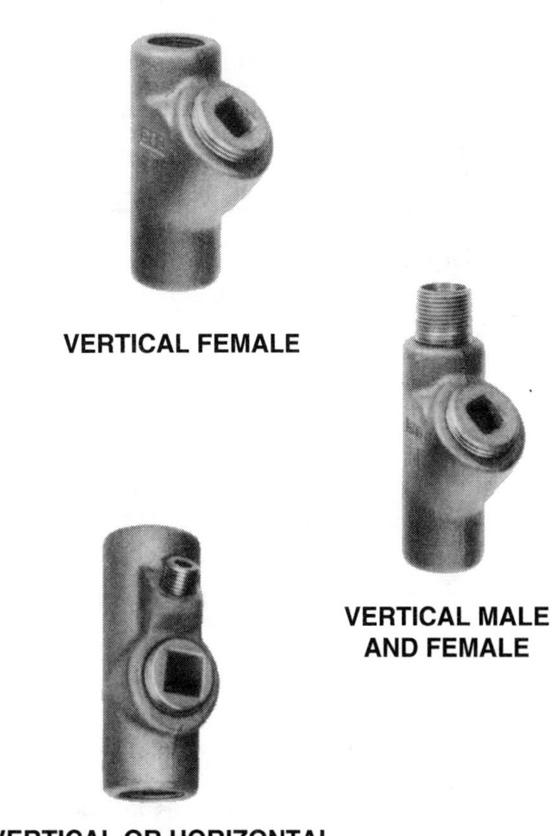

Crouse-Hinds Division, Cooper Industries, Inc.

Figure 12-3. A conduit seal is a fitting which is inserted into runs of conduit to isolate certain electrical apparatus from atmospheric hazards.

CONDUIT SEALS

Conduit seals shall be installed properly to ensure that the hazardous substance does not migrate through the raceway system. Several factors are involved in selecting the proper seal fitting for the application. Seal fittings can be designed for either vertical or horizontal mounting in the conduit system and in male and female configurations. In some cases, the seal fittings may be mounted in any position.

Conduit seals shall be installed in accordance with the manufacturer's instructions. These instructions may contain additional installation and location specifications. For example, the preferred side of a classified area boundary where the seal shall be installed may be given. Always read the manufacturer's instructions. In addition, the sealing compound used to construct the seal also contains a set of installation instructions and directions for making the seal. Always read these instructions to make a good conduit seal.

The purpose of the conduit seal is to minimize the passage of gases and vapors and to prevent the passage of flames from one area of an electrical installation to another per 501-5, FPN 1. For example, if a run of conduit leaving a class location and entering an unclassified area is installed without conduit seals, the flammable gases or vapors could migrate through the conduit system from the hazardous location to the unclassified area. Requirements for conduit seals are contained in the individual article which contains the requirements for the particular class and division. Installers of conduit seals should ensure that the correct type of seal is installed. See Figure 12-4.

Class I, Division 1 Conduit Seals – 501-5(a). Section 501-5(a) contains four subparts which contain separate provisions for conduit seals in Class I, Division 1 locations. Per 501-5(a)(1), each conduit run shall be provided with a sealing fitting within 18″ of any enclosure that contains electrical devices which may produce arcs, sparks, or excessively high temperatures during their operation. Examples of such electrical devices are switches, CBs, relays, fuses, resistors, contactors, etc. With the exception of explosionproof unions, fittings, and conduit bodies no other apparatus shall be installed between the seal fitting and the enclosure. See Figure 12-5.

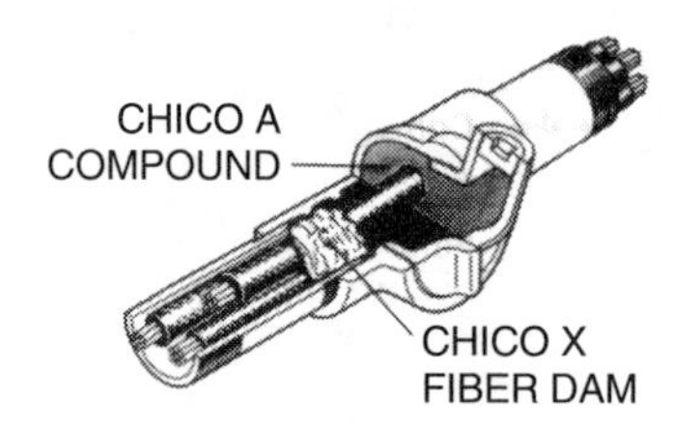

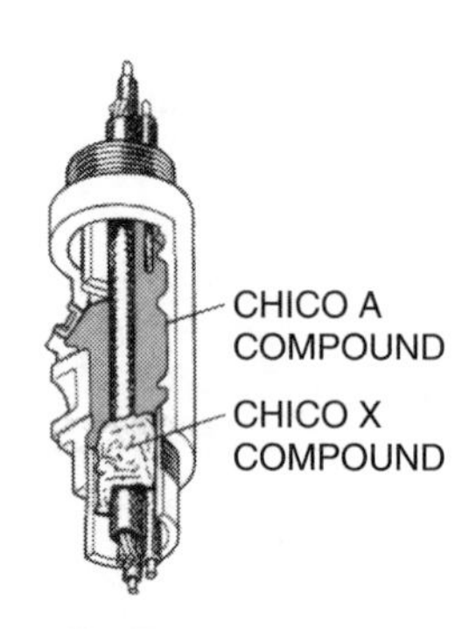

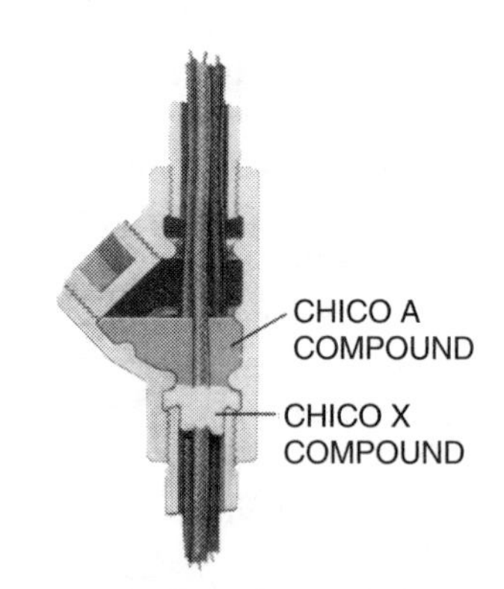

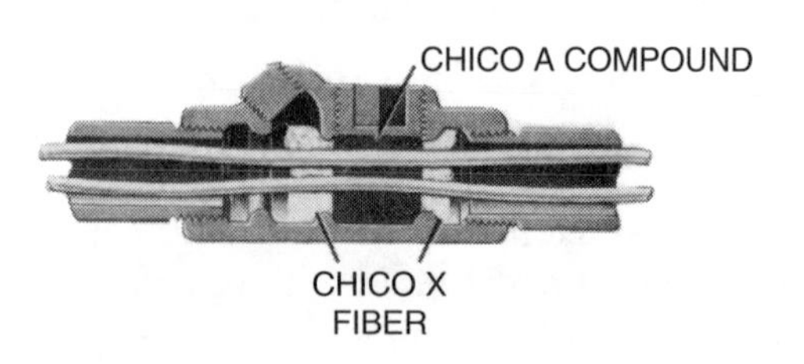

Crouse-Hinds Division, Cooper Industries, Inc.

Figure 12-4. Conduit seals are constructed using sealing compound (Chico A), fiber (Chico X), and approved conduit seal fittings.

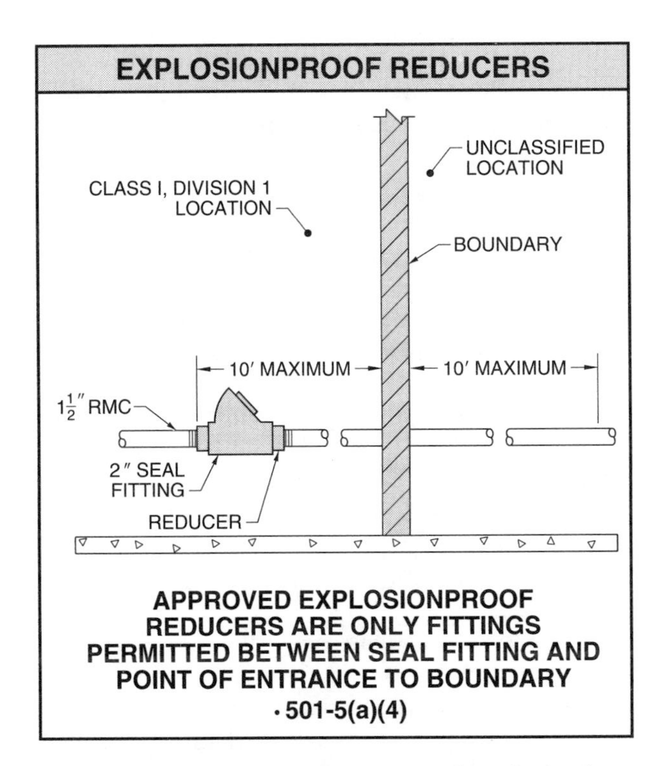

Figure 12-5. Conduit seal fittings are installed at hazardous location boundaries to prevent communication of the hazardous substance between the classified and unclassified areas.

Section 501-5(a)(1), Ex. permits the conduit seal to be omitted provided the size of the conduit does not exceed 1½″ and the current-interrupting contacts of any electrical devices contained in the enclosure are:

(1) Enclosed in a hermetically sealed chamber which prevents the entrance of gases or vapors.

(2) Immersed in oil per 501-6(b)(1)(2).

(3) Enclosed within a factory-sealed, explosionproof chamber in an enclosure approved for the location and marked "factory sealed" or the equivalent.

Section 501-5(a)(2) requires that each 2″ or larger conduit run shall be provided with a conduit seal fitting within 18″ of an enclosure or fitting which contains splices, taps, or wire terminals.

Section 501-5(a)(3) permits a single seal to be used for two or more enclosures, which are connected together by nipples or runs of conduit not exceeding 36″ in length, provided the location of the seal is not more than 18″ from either enclosure. Proper location of the conduit seal ensures that only a minimum amount of the hazardous substance can enter the conduit system. See Figure 12-6.

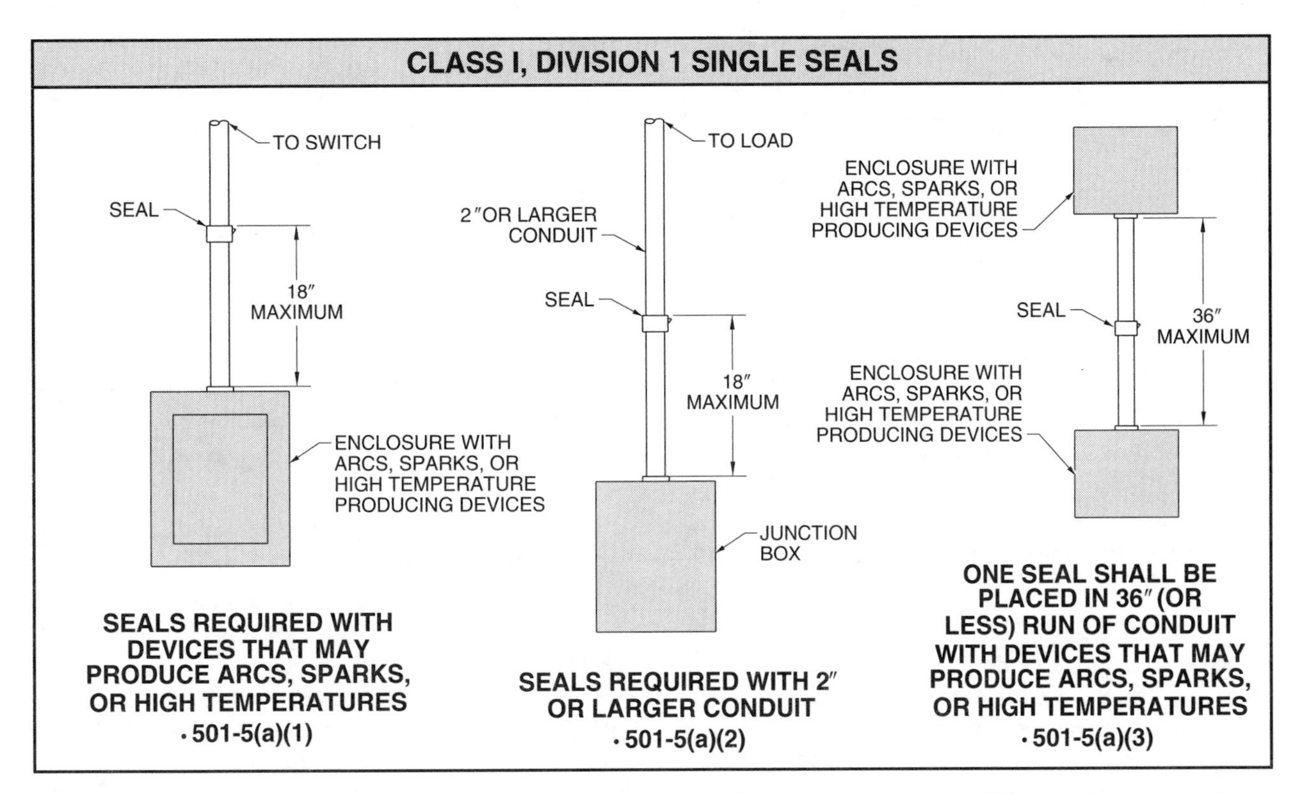

Figure 12-6. Proper location of the conduit seal ensures that only a minimum amount of the hazardous substance can enter the conduit system.

Section 501-5(a)(4) requires seals for conduit runs which leave Class I, Division 1 locations. The location of the seal is permitted on either side of the boundary provided it is within 10′ of the boundary. See Figure 12-7. With the exception of explosionproof reducers, no boxes or fittings are permitted to be installed between the boundary and the conduit seal. If, however, an unbroken run of metal conduit passes completely through a Class I, Division 1 location and terminates in an unclassified area, and there are no fittings less than 12″ from each boundary, seals are not required.

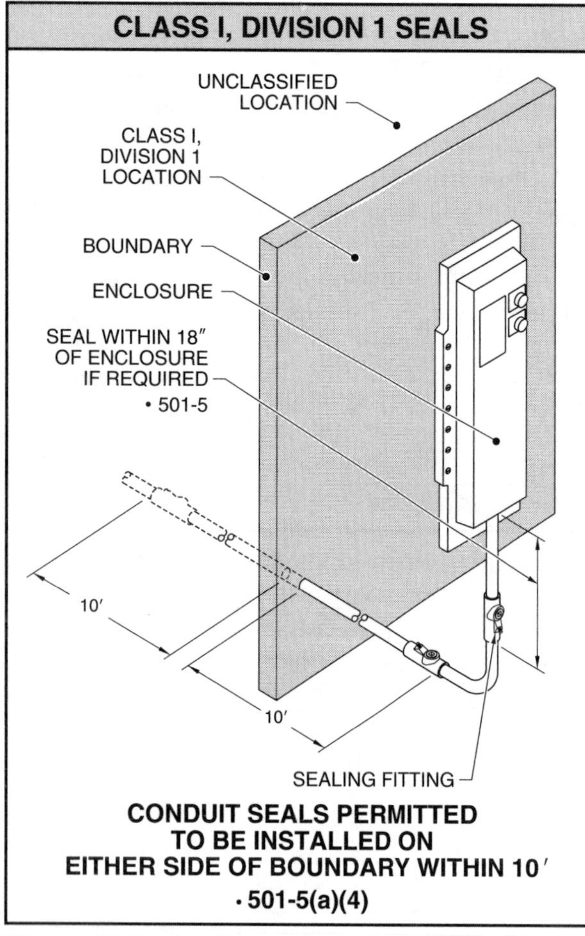

Figure 12-7. Conduit seals at hazardous locations boundaries are permitted to be installed on either side of the boundary provided they are located within 10′ of the boundary.

Class I, Division 2 Conduit Seals – 501-5(b). Conduit seals shall be installed for all connections to explosionproof enclosures which are required to be approved for Class I locations per 501-5(b)(1). The installation of the seals shall meet the provisions of 501-5(a)(1)(2)(3).

Section 501-5(b)(2) contains provisions for conduit runs which leave Class I, Division 2 locations. Conduit seals are required for these runs. The location of the seal is permitted on either side of the boundary. It should be placed in a location which minimizes the amount of gas or vapor that may have entered the conduit system within the Division 2 location. With the exception of explosionproof reducers, no boxes or fittings are permitted to be installed between the boundary and the conduit seal.

As with Class I, Division 1 locations, conduit seals are not required for an unbroken run of metal conduit which passes completely through a Class I, Division 2 location if there are no fittings less than 12″ from each boundary and if the conduits terminate in an unclassified area. See 501-5(b), Ex. 1.

For outdoor locations, or locations in which the conduit system is contained within a single room, conduit seals shall not be required where a transition is made to cable tray, cablebus, ventilated busway, Type MI cable, or open wiring. The conduit system shall terminate at an unclassified location. It is not permitted to terminate at any enclosure which contains a source of ignition during normal operation. See 501-5(b), Ex. 2.

Conduit seals can also be omitted for installations in which the conduit system enters a Class I, Division 2 location from an unclassified area that is unclassified because of Type Z pressurization. Type Z pressurization requirements are contained in NFPA 496-1993, *Standard for Purged and Pressurized Enclosures for Electrical Equipment.*

Section 501-5(b), Ex. 4 contains five conditions, which if present, do not warrant the inclusion of conduit seals where segments of aboveground conduit systems pass from a Class I, Division 2 location into an unclassified location.

Class I, Divisions 1 and 2 General Seal Requirements – 501-5(c). Whether installed in a Division 1 or Division 2 location, conduit seals installed in Class I locations shall meet the provisions of 501-5(c). All seal fittings shall be accessible and contain provisions for integral sealing. The compound used for accomplishing the seal shall be approved and shall prevent the passage of gas or vapors. The compound shall have a melting point of at least 200°F and shall not be affected by the surrounding atmosphere or liquids.

When the seal is completed, it shall have a minimum thickness equivalent to the trade size of the conduit. In no case shall this be less than ⅝″ thick. Fittings used only for sealing purposes shall not contain splices or taps. Likewise, other fittings containing splices and taps shall not be filled with seal compound.

An excessive number of conductors in a conduit seal could result in an ineffective seal and permit transmission of the hazardous substance through the conduit system. The cross-sectional area of the conductors contained within a seal fitting are not permitted to exceed 25% of the cross-sectional area of a conduit of the same trade size unless it is approved for a higher percentage fill. See Figure 12-8. This provision ensures that the sealing compound can surround and fill the seal in a manner which effectively prevents the passage of gases and vapors.

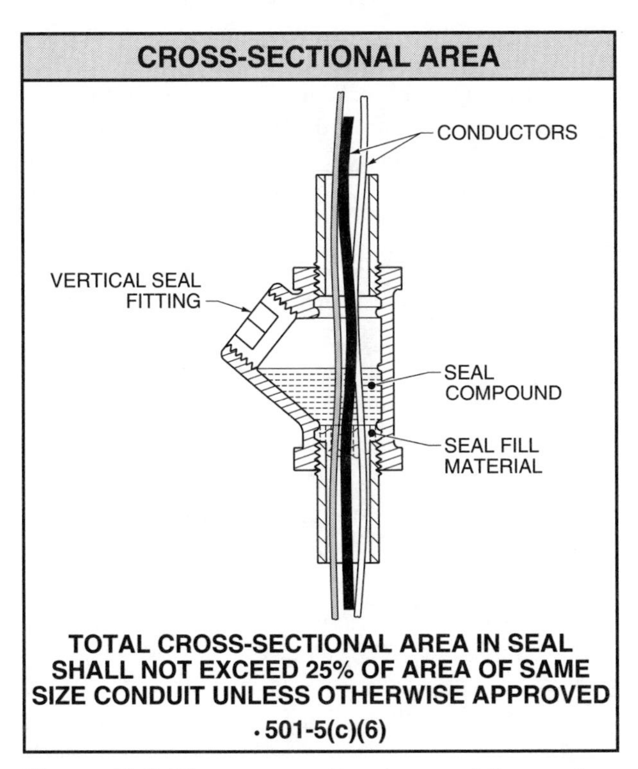

Figure 12-8. The cross-sectional area of the conductors contained within a seal fitting are not permitted to exceed 25% of the cross-sectional area of a conduit of the same trade size unless it is approved for a higher percentage fill.

Class II, Divisions 1 and 2 Conduit Seals – 502-5. Because of the nature of Class II locations, conduit seals are used to prevent the entrance of dust into a dust-ignitionproof enclosure through the raceway. *Dust-ignitionproof* is enclosed in a manner which prevents the entrance of dusts and does not permit arcs, sparks, or excessive temperature to cause ignition of exterior accumulations of specified dust. See 502-1. Three methods are listed to accomplish the conduit seal:

(1) A permanent and effective seal may be installed. The seals are not required to be explosionproof. Other sealing putties, such as electrical sealing putty, is an acceptable method of providing the seal. All sealing fittings, where used, shall be accessible.

(2) A horizontal raceway of not less than 10′ in length may be installed. Seals are not required for raceways which permit the communication between a dust-ignitionproof enclosure and an enclosure in an unclassified location.

(3) A vertical raceway of not less than 5′ in length may be installed. It shall extend downward from the dust-ignitionproof enclosure.

CABLE SEALS

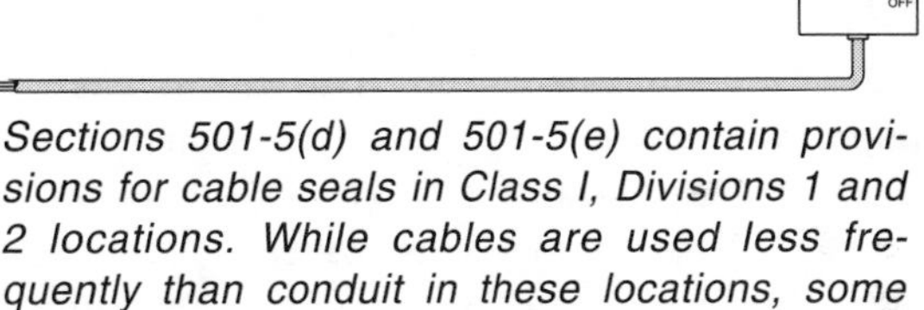

Sections 501-5(d) and 501-5(e) contain provisions for cable seals in Class I, Divisions 1 and 2 locations. While cables are used less frequently than conduit in these locations, some cables are permitted to be used under certain conditions.

Cables used in Class I locations shall be sealed for the same reasons that conduits are sealed. The provisions for installing cable seals for multiconductor cables, twisted pair, and shielded cables are included in this section. Note that each multiconductor cable which is installed in a conduit shall be considered a single cable if the cable is not capable of transmitting gases or vapors. These cables in the conduit are sealed as if they were individual conductors within the conduit.

Class III, Divisions 1 and 2 Conduit Seals. Class III locations contain easily ignitible fibers and flyings. Contamination of these fibers and flyings is not likely to occur through conduit systems, therefore there are no conduit seal requirements for Class III, Divisions 1 and 2 locations.

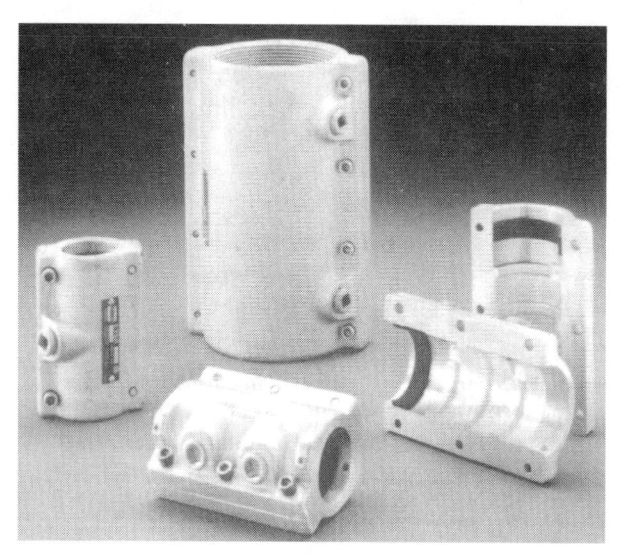

Crouse-Hinds Division, Cooper Industries, Inc.

Sealing fittings for Class II, Division 2 locations are available in hub sizes from 3/4″ to 4″.

Switches, CBs, Motor Controllers, and Fuses

When electrical equipment is installed in hazardous locations, care shall be given to ensure that such equipment does not provide an ignition source for the hazardous substances. Electrical equipment such as switches, CBs, motor controllers, and fuses are especially problematic because their operation frequently involves sparking and arcing which can provide the necessary ignition source. When such electrical equipment is installed in explosionproof enclosures, ignition is prevented by isolating the spark or arc from the hazardous atmospheres. Additionally, the explosionproof enclosure is designed to withstand an internal explosion without permitting ignition of the external hazardous atmosphere.

Class I locations involve flammable gases and vapors, therefore the most stringent requirements for electrical equipment installed in hazardous locations are for Class I locations. The requirements for switches, CBs, motor controllers, and fuses installed in Class II and III locations can be relaxed somewhat because the nature of the hazardous substance is not as volatile.

Equipment identified as being suitable for Division 1 locations is also permitted to be used in Division 2 locations. Equipment that is marked Division 2, however, is only permitted to be installed in Division 2 locations. Equipment which is not marked to show a division is suitable for both Division 1 and 2 locations.

Class I, Division 1 Electrical Equipment – 501-6(a). When switches, CBs, motor controllers, or fuses are installed in Class I, Division 1 locations, 501-6(a) requires that they be installed in enclosures which shall be approved as a complete assembly for use in Class I, Division 1 locations. Additionally, all electrical apparatus installed in Class I, Division 1 locations shall be approved for use in Class I, Division 1 locations. See Figure 12-9.

Class I, Division 2 Electrical Equipment – 501-6(b). The general rule for switches, CBs, motor controllers, or fuses located in Class I, Division 2 locations requires that the electrical equipment be installed in enclosures approved for Class I, Division 2 locations. There are, however, certain installation provisions which permit the electrical equipment to be installed in general-purpose enclosures. General-purpose enclosures are not approved for Class I locations but switches, CBs, motor controllers, or fuses can be contained in them if one of the following four conditions is met:

(1) The current-interrupting components of the equipment are contained within a hermetically-sealed chamber.

(2) The current make-and-break components of the equipment are oil-immersed, of the general-purpose type, and the immersion is a minimum of 2″ for power contacts and 1″ for control contacts.

(3) The current interruption occurs within a factory-sealed explosionproof chamber approved for the location.

(4) All switching is done with solid-state components, without contacts, and in which the surface temperature of the enclosure does not exceed 80% of the ignition temperature, in °C, of the gas or vapor involved.

When fused or unfused isolating switches are installed for transformer or capacitor banks which are not required to open circuits in their normal performance, general-purpose enclosures are permitted in place of enclosures approved for Class I, Division 1 locations. When fuses are installed for the protection of motors, appliances, or lamps but not for overcurrent protection per 501-6(b)(4), the fuses are permitted to be either the standard plug or cartridge type.

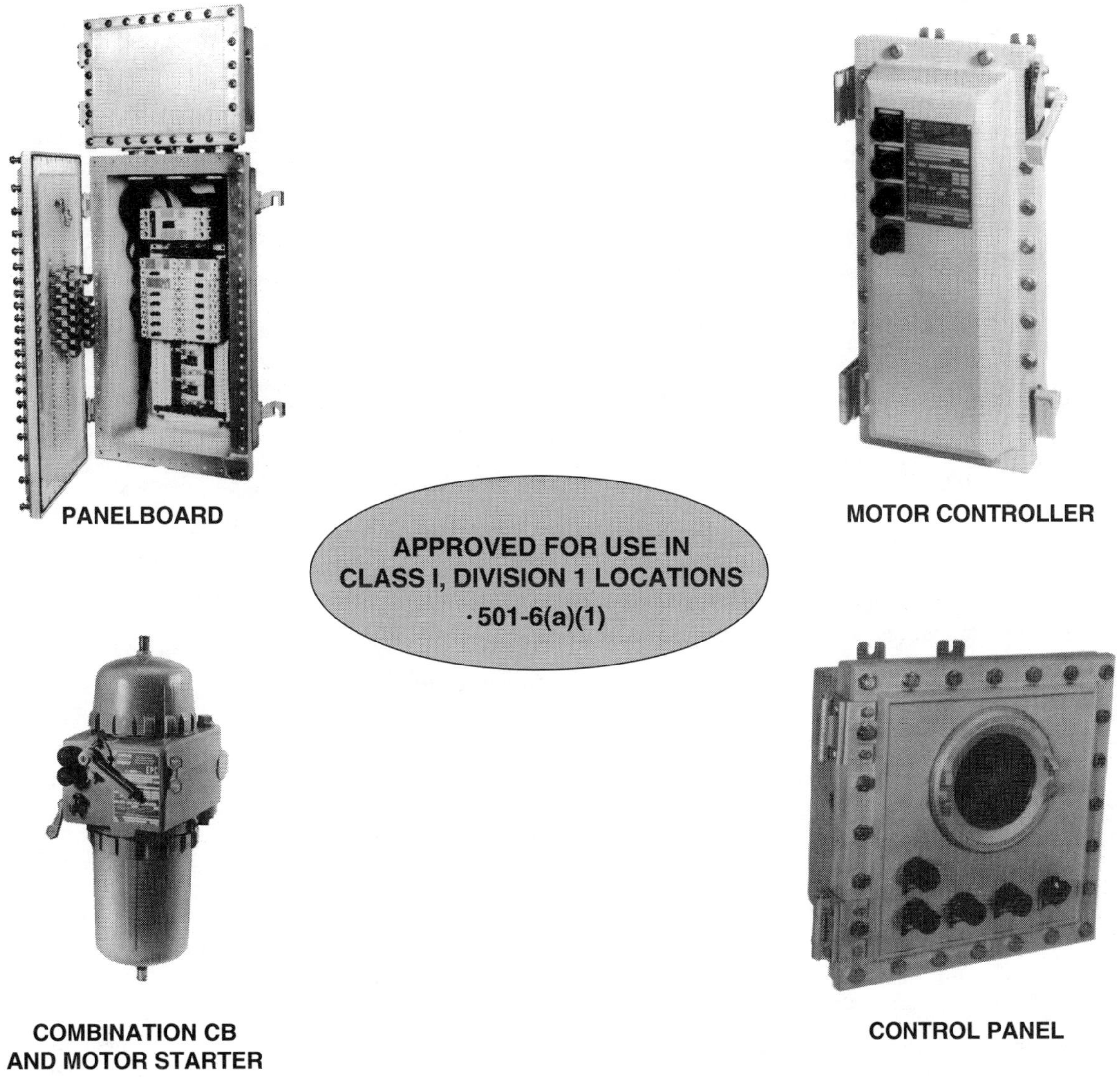

PANELBOARD

MOTOR CONTROLLER

APPROVED FOR USE IN
CLASS I, DIVISION 1 LOCATIONS
• 501-6(a)(1)

COMBINATION CB
AND MOTOR STARTER

CONTROL PANEL

Crouse-Hinds Division, Cooper Industries, Inc.

Figure 12-9. All electrical apparatus installed in Class I, Division 1 locations shall be approved for use in Class I, Division 1 locations.

Where used, fuses shall be installed within enclosures approved for the location. General-purpose enclosures are permitted for fuses in which the operating element is immersed in oil or other approved liquid, or installations in which the operating element of the fuse is enclosed in a hermetically-sealed chamber. Additionally, nonindicating, silver-sand, current-limiting fuses can be installed in general-purpose enclosures.

Section 501-6(b)(4) permits not more than 10 sets of fuses or 10 CBs to be installed in general-purpose enclosures provided they are not intended to be used as switches for the interruption of current and they are used to protect circuits supplying lamps in fixed positions only. Approved cartridge fuses, which are part of a lighting fixture and used for supplementary protection, are permitted to be installed in Class I, Division 2 locations.

Class II, Division 1 Electrical Equipment – 502-6(a). If switches, CBs, motor controllers, or fuses are installed in Class II, Division 1 locations, and their purpose is to interrupt current during normal operation, they shall be installed in dust-ignitionproof enclosures which, in addition to the electrical apparatus, shall be approved as a complete assembly for Class II locations.

Tight-fitting metal enclosures are permitted for disconnecting or isolating switches that are not intended to interrupt current provided they are not installed where dusts may be of the conductive nature, such as:

(1) Enclosures are provided with close-fitting or telescoping covers or other effective means to prevent the escape of possible ignition sources.

(2) Enclosures contain no openings from which sparks or arcs or other burning material could possibly ignite exterior accumulations of dust or other combustible materials.

If metal dusts, such as magnesium, aluminum, aluminum bronze, etc. are present, enclosures for switches, CBs, motor controllers, and fuses shall be specifically approved for the location.

Class II, Division 2 Electrical Equipment – 502-6(b). When switches, CBs, motor controllers, or fuses are installed in Class II, Division 2 locations, they shall be installed in enclosures which are identified as dusttight.

Class III, Divisions 1 and 2 Electrical Equipment – 503-4. When switches, CBs, motor controllers, or fuses are installed in Class III, Division 1 or 2 locations, they shall be installed in enclosures which are identified as dusttight.

Motors and Generators

The decision to install motors in hazardous locations should be weighed very carefully. By their nature, motors and generators tend to operate at very high temperatures which could possibly provide a source of ignition in Class I and some Class II locations. Additionally, arcing and sparking can be a normal occurrence in the operation of some motors and generators. For this reason, motors and generators should be installed outside of hazardous locations, particularly Class I and II locations, whenever practicable.

Class I, Division 1 Motors and Generators – 501-8(a). Section 501-8(a) lists four options that are permitted when motors, generators, or other electrical rotating machinery are to be installed in a Class I, Division 1 location. The first option is to use motors or generators that are approved for Class I, Division 1 locations. Motors and generators approved for Class I, Division 1 are typically available and widely used for Group D and some Group C locations. Motors and generators for Group A (acetylene) and Group B (hydrogen, fuel, and combustible gases with at least 30% hydrogen or equivalent gases or vapors) locations are not typically available and one of the other three options should be used.

The second option is to use a totally-enclosed motor or generator. These motors and generators are supplied with positive-pressure ventilation from a source of clean air. The discharge of the motor or generator is to an unclassified or otherwise safe area. The control of the motor or the generator shall be such that they cannot be energized until ventilation has been established and the enclosure has been purged with at least 10 volumes of air. The control circuit shall also de-energize the motor or generator whenever the air supply is lost.

The third option is to select a totally-enclosed, inert gas-filled motor or generator. This is the same design as the second option, except an inert gas is used instead of clean air to purge the enclosure. The same control requirements for de-energizing the motor or generator shall be met for this option as well. Totally-enclosed motors and generators are designed to prevent external surface temperatures which exceed 80% of the ignition temperature of the gas or vapor which is involved. They include safety devices that de-energize the equipment if the 80% temperature limitation is exceeded.

The fourth option relies on the use of a motor or generator of a type that is submerged in a liquid which is only flammable when it is vaporized and mixed with air. The motor or generator could also be "submerged" in a gas or vapor which has a pressure greater than atmospheric pressure and is only flammable when mixed with air. Either method shall include a control measure which prevents the motor or generator from being energized until the enclosure has been adequately purged and de-energizes the motor or generator upon a loss of the liquid, gas, or vapor which surrounds the motor or generator.

Class I, Division 2 Motors and Generators – 501-8(b). The requirements for motors, generators and other electrical rotating machinery which are installed in Class I, Division 2 locations depend on the type of motor or generator selected for the location. If the motor or generator has sliding contacts, centrifugal or other types of switches, or other integral resistance devices it shall be approved for Class I, Division 1 locations. If any of these devices, however, are contained in separate enclosures approved for Class I, Division 2 locations, then the motors or generators need not be approved for Class I, Division 1 locations.

Motors or generators which do not contain any of these devices, such as squirrel-cage induction motors, can be installed in Class I, Division 2 locations in non-explosionproof enclosures. Designers and installers should be aware, however, as 501-8(b), FPN 1 indicates, that the internal and external temperatures of the motors or generators shall be considered, even though the enclosures are not required to be explosionproof. This is because if the surface motor temperature is raised excessively during motor operation, the surface temperature could become a possible source of ignition for some hazardous atmospheres.

Class II, Division 1 Motors and Generators – 502-8(a). Two options are permitted when motors, generators, or other electrical rotating machinery are to be installed in a Class II, Division 1 location. The first option is to install motors or generators which are approved for use in Class II, Division 1 locations. Motors or generators approved for Class II, Division 1 locations are identified as dust-ignitionproof.

The second option is to use totally-enclosed motors or generators. These motors or generators shall be pipe-ventilated and designed so that dust does not enter the piping. Additionally, motors or generators installed in Class II, Division 1 locations shall be designed so that their maximum surface temperatures do not cause the dust to carbonize or become so dry that spontaneous ignition could occur.

Class II, Division 2 Motors and Generators – 502-8(b). Motors or generators used for Class II, Division 2 locations shall be one of the following totally-enclosed types:

- Nonventilated
- Pipe-ventilated
- Water-air cooled
- Fan-cooled

Dust-ignitionproof motors or generators are also permitted provided they have no external openings and their maximum full-load external surface temperature is in accordance with 500-3(f). Table 500-3(f) establishes ignition temperatures for which the equipment was approved based on the specific Class II, Group which is present.

Section 502-8(b), Ex. permits additional installation options if the AHJ determines that the likely accumulations of nonconductive, nonabrasive dusts is moderate and the motors or generators are easily accessible for maintenance and cleaning. The first option permits standard open or nonenclosed motors or generators to be installed in Class II, Division 2 locations if the motors or generators do not contain integral resistance devices, sliding contacts, centrifugal switches, or other switching mechanisms.

The second option permits standard open or nonenclosed motors, with integral resistance devices, sliding contacts, centrifugal switches, or other switching mechanisms to be installed in Class II, Division 2 locations if the sliding contacts, centrifugal switches, or other switching mechanisms are enclosed within a dusttight housing which does not contain any openings or means of ventilation. The last option permits self-cleaning, textile motors to be used provided the motors are of the squirrel-cage type.

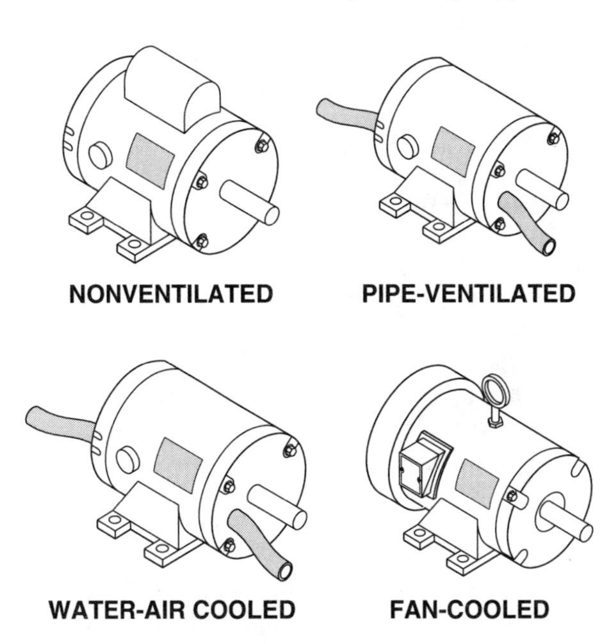

Motors or generators used for Class II, Division 2 locations shall be nonventilated, pipe-ventilated, water-air cooled, or fan-cooled.

Class III, Divisions 1 and 2 Motors and Generators – 503-6. Motors or generators used for Class III, Divisions 1 and 2 locations shall be one of the following totally-enclosed types:

- Nonventilated
- Pipe-ventilated
- Fan-cooled

Section 503-6, Ex. permits three additional options when the AHJ determines that the likely accumulations of nonconductive, nonabrasive dusts will be moderate and the motors or generators are easily accessible for maintenance and cleaning. The three options include:

(a) Self-cleaning textile motors are permitted to be used provided the motors are of the squirrel-cage type.

(b) Standard open or nonenclosed motors or generators are permitted to be installed in Class II, Division 2 locations if the motors or generators do not contain integral resistance devices, sliding contacts, centrifugal switches, or other switching mechanisms.

(c) Standard open or nonenclosed motors with integral resistance devices, sliding contacts, centrifugal switches, or other switching mechanisms are permitted to be installed in Class II, Division 2 locations if the sliding contacts, centrifugal switches, or other switching mechanisms are enclosed within a dusttight housing which does not contain any openings or means of ventilation.

Lighting Fixtures

The need for lighting fixtures is apparent in all types of hazardous locations. Even if it is desirable from a design point of view to keep lighting fixtures out of hazardous locations, local building codes often require that lighting be provided for the area. There are, however, several concerns when installing lighting fixtures in Class I, II, or III locations. The primary concern is to ensure that the lighting fixture does not provide the source of ignition for the hazardous materials which may be present in the atmosphere. While this is more of a concern for Class I and II locations, excessive temperature of the lighting fixture should be considered before selecting a lighting fixture for a particular class and division.

Another consideration is that the lighting fixtures are installed in areas where they could be subjected to physical damage. This adds the possibility that a breakage of a lamp could provide the necessary source of ignition for the hazardous substance.

Portable lighting equipment is another concern when selecting lighting fixtures for hazardous locations. Personnel working in classified areas shall ensure that any portable lighting equipment which they are using in the area is approved for the class and division in which it is being used.

Class I, Division 1 Lighting Fixtures – 501-9(a). Section 501-9(a)(1) requires that all lighting fixtures used in Class I, Division 1 locations shall be approved as an assembly for use in Class I, Division 1 locations. Fixtures, which are approved, shall be marked to indicate the maximum wattage of the lamps for which they are approved. If the fixture is designed for portable use, it shall be specifically approved, as a complete assembly, for that use. Temporary lighting fixtures constructed of individual lighting components assembled on the job, are not permitted to be used in Class I, Division 1 locations.

Section 501-9(a)(2) requires that all fixed and portable lighting fixtures used in Class I, Division 1 locations shall be protected against physical damage. This can be accomplished by the use of a fixture or lamp guard or by the specific location of the fixture. See Figure 12-10.

Section 501-9(a)(3) covers the use of pendant or hanging fixtures in Class I, Division 1 locations. Pendant fixtures are required to be supported by, and supplied through, either threaded RMC stems or threaded steel IMC stems. Loosening of any threaded joints shall be guarded against by providing a setscrew or other equally effective method. If conduit stems are required to be longer than 12″, the stem shall be provided with effective bracing to guard against excessive lateral movement. The bracing shall be installed at a level which is not more than 12″ above the lower end of the stem.

In lieu of bracing, a flexible fitting or connector which is approved for Class I, Division 1 locations may be used on conduit stems longer than 12″. The location of the flexible fitting or connector shall not be more than 12″ from the point of attachment to the box or fitting which supports the lighting fixture.

Section 501-9(a)(4) requires that all boxes, fittings, or box assemblies used to support lighting fixtures in Class I, Division 1 locations shall be approved for Class I locations.

Class I, Division 2 Lighting Fixtures – 501-9(b). Portable lighting equipment is required by 501-9(b)(1) to be approved as a complete assembly for Class I, Division 1 locations. Section 501-9(b)(2) requires all lighting fixtures installed in Class I, Division 2 locations to be protected against physical damage by fixture guards or by location. If the fixture utilizes lamps or bulbs that could cause the surface temperature of the fixture to exceed 80% of the ignition temperature of the gas or vapor involved, the fixtures shall be approved for Class I, Division 1 locations. See Figure 12-11.

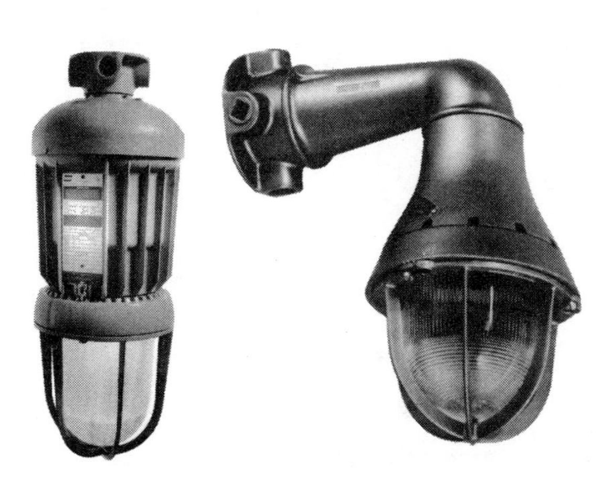

FIXED LIGHTING
CLASS I, DIVISION 1 LOCATIONS
· 501-9(a)(1)(2)

PORTABLE LIGHTING
CLASS I, DIVISION 1 LOCATIONS
· 501-9(a)(1)

Crouse-Hinds Division, Cooper Industries, Inc.

Figure 12-10. All fixed and portable lighting fixtures used in Class I, Division 1 locations shall be protected against physical damage.

FIXED LIGHTING
CLASS I, DIVISION 1 LOCATIONS
· 501-9(a)(2)(3)

PORTABLE LIGHTING
CLASS I, DIVISION 1 LOCATIONS
· 501-9(a)(1)

Crouse-Hinds Division, Cooper Industries, Inc.

Figure 12-11. Fixtures for hazardous locations shall be constructed so that their external surface temperatures cannot ignite the surrounding gas or vapor which may be present in the atmosphere.

Per 501-9(b)(3), pendant fixtures shall be suspended by either threaded RMC stems, threaded steel IMC stems, or by other approved means. If conduit stems are required to be longer than 12″, the stem shall be provided with effective bracing to guard against excessive lateral movement. The bracing shall be installed at a level which is not more than 12″ above the lower end of the stem.

In lieu of bracing, an approved flexible fitting or connector may be used on conduit stems longer than 12″. The location of the flexible fitting or connector shall not be more than 12″ from the point of attachment to the box or fitting which supports the lighting fixture.

Per 501-9(b)(4), switches, which are part of a fixture assembly or part of an individual lampholder, in Class I, Division 2 locations shall be installed per 501-6(b)(1).

The last requirement for lighting fixtures in Class I, Division 2 locations covers starting and control equipment for electric-discharge lamps. Per 501-9(b)(5), in general, such equipment shall comply with 501-7(b).

Class II, Division 1 Lighting Fixtures – 502-11(a). Section 502-11(a)(1) requires that all lighting fixtures used in Class II, Division 1 locations shall be approved for use in Class II locations. Fixtures, which are approved, shall be marked to indicate the maximum wattage of the lamps for which they are approved. Fixtures installed in locations where electrically conductive dusts such as magnesium, aluminum, aluminum bronze powders, etc. may be present shall be approved for the specific location.

Per 502-11(a)(2), all fixtures used in Class II, Division 1 locations shall be protected against physical damage. This can be accomplished by the use of a fixture or lamp guard or by the specific location of the fixture.

Per 502-11(a)(3), pendants or hanging fixtures in Class II, Division 1 locations are required to be suspended by threaded RMC stems, threaded steel IMC stems, by chains with approved fittings, or by other approved means. Loosening of any threaded joints shall be guarded against by providing a set-screw or other equally effective method. If conduit stems are required to be longer than 12″, the stem shall be provided with effective bracing to guard against excessive lateral movement. The bracing shall be installed at a level which is not more than 12″ above the lower end of the stem.

In lieu of bracing, a flexible fitting or connector which is approved for the location may be used on conduit stems longer than 12″. The location of the flexible fitting or connector shall not be more than 12″ from the point of attachment to the box or fitting which supports the lighting fixture. See Figure 12-12. Flexible cord is permitted between an outlet box or fitting and a pendant fixture provided it is of the hard-usage type and the cord is not used to support the lighting fixture.

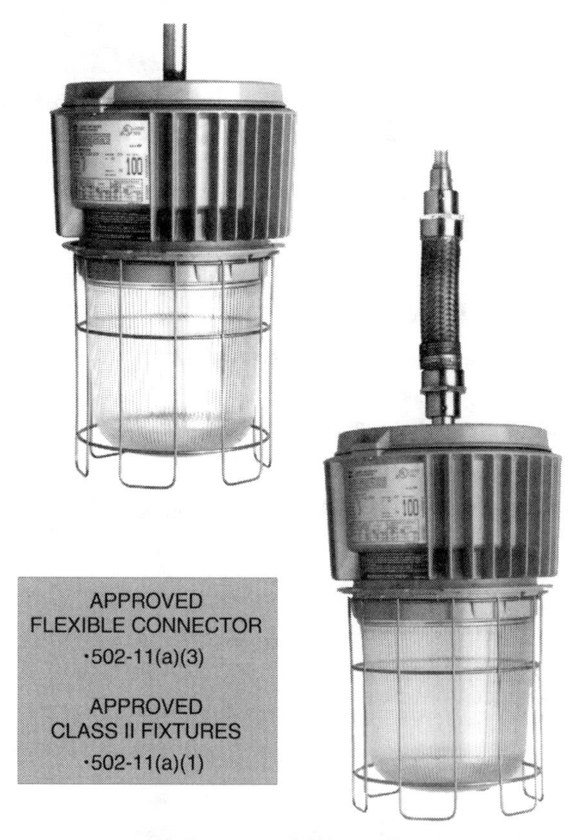

Crouse-Hinds Division, Cooper Industries, Inc.

Figure 12-12. Pendant fixtures in Class II, Division 1 locations shall be suspended by threaded RMC or steel IMC or an approved flexible connector.

Per 502-11(a)(4), all boxes, fittings, or box assemblies that are used to support lighting fixtures in Class II, Division 1 locations shall be approved for Class II locations.

Class II, Division 2 Lighting Fixtures – 502-11(b). Per 502-11(b)(1), portable lighting equipment is required to be approved for Class II locations. Fixtures for Class II, Division 2 locations shall be marked to clearly indicate the maximum wattage of lamps for which they are approved.

Per 502-11(b)(2), if fixed lighting fixtures are not approved for Class II, Division 2 locations, then they shall provide enclosures for the lamps and lampholders that are designed to minimize dust accumulations and prevent sparks, arcs, or other burning material from escaping the enclosure. In addition, the fixtures shall be marked to indicate the maximum wattage permitted for the lamps in order to limit the surface temperature which the fixture may reach.

Per 502-11(b)(3), all lighting fixtures installed in Class II, Division 2 locations shall be protected against physical damage by fixture guards or by location. Per 502-11(b)(4), pendant fixtures shall be suspended by either threaded RMC stems, threaded steel IMC stems, chains with approved fittings, or by other approved means. If conduit stems are required to be longer than 12″, the stem shall be provided with effective bracing to guard against excessive lateral movement. The bracing shall be installed at a level which is not more than 12″ above the lower end of the stem.

In lieu of bracing, an approved flexible fitting or connector may be used on conduit stems longer than 12″. The location of the flexible fitting or connector shall not be more than 12″ from the point of attachment to the box or fitting which supports the lighting fixture.

The last requirement for lighting fixtures in Class II, Division 2 locations covers starting and control equipment for electric-discharge lamps. Per 502-11(b)(5), in general, such equipment shall comply with the provisions of 502-7(b).

Class III, Divisions 1 and 2 Lighting Fixtures – 503-9. Per 503-9(a), all lighting fixtures used in Class III, Division 1 or 2 locations for fixed lighting shall include enclosures for lamps and lampholders which are designed to minimize the entrance of fibers or flyings and to prevent sparks, arcs, or other burning material from escaping the enclosure. In addition, the fixtures shall be marked to indicate the maximum wattage permitted for the lamps in order to limit the surface temperature which the fixture may reach.

Per 503-9(b), all lighting fixtures installed in Class III, Division 1 or 2 locations which may be subject to physical damage shall be protected by suitable fixture guards. Per 503-9(c), pendant fixtures shall be suspended by either threaded RMC stems, threaded IMC stems, threaded metal tubing of equivalent thickness, or by chains with approved fittings. If conduit stems are required to be longer than 12″, the stem shall be provided with effective bracing to guard against excessive lateral movement. The bracing shall be installed at a level which is not more than 12″ above the lower end of the stem.

In lieu of bracing, an approved flexible fitting or connector may be used on conduit stems longer than 12″. The location of the flexible fitting or connector shall not be more than 12″ from the point of attachment to the box or fitting which supports the lighting fixture.

Federal Signal Corporation

These explosionproof strobe lights are listed for use in Class I, Division 1 and 2, Groups C and D; Class II, Division 1 and 2, Groups E, F, and G; and Class III, Divisions 1 and 2 locations.

Section 503-9(d) contains the last requirement for lighting fixtures in Class III, Division 1 or 2 locations. In general, portable lighting fixtures shall meet the same requirements of those for fixed lighting contained in 503-9(a). Portable lighting equipment shall be equipped with handles and shall be protected with substantial guards. Exposed, noncurrent-carrying metal parts shall be grounded and exposed current-carrying parts are not permitted in the design of the fixtures. Lampholders are not permitted to include provisions for receiving attachment plugs and the lampholders may not be switched.

Grounding

Grounding in hazardous locations takes on added importance because of the presence of hazardous atmospheres that could be ignited by arcing ground faults. In addition to ensuring the operation of the overcurrent device, low-impedance ground paths ensure that arcing faults do not occur and that the surface temperatures of equipment are not raised excessively by the ground faults. If metal conduits are used as the EGC, care shall be taken to ensure that all threaded connections are tight to minimize the possibility of arcing or sparking across the joints.

Class I, Divisions 1 and 2 Grounding – 501-16. In general, all grounding shall be done in accordance with the provisions of 250. Section 501-16(a) covers bonding requirements and requires that bonding jumpers with proper fittings or other approved means be used where bonding is required in a Class I location. Locknut-bushings or double locknuts cannot be used for bonding purposes. Bonding is required around all intervening raceways, fittings, boxes, enclosures, etc. and between Class I locations and the point of grounding for separately derived systems.

Section 501-16(b) contains special requirements for the use of FMC and LTFMC when installed per 501-4(b). If either of these conduits are installed in a manner in which they are the sole equipment grounding path, they shall be installed with bonding jumpers in parallel with the conduit. The bonding jumper may be installed either inside or outside of the conduit and in accordance with 250-79.

Section 501-16(b), Ex. permits listed LTFMC to be installed without a bonding jumper in Class I, Division 2 locations, provided the total length of the conduit does not exceed 6′. In addition, the fittings used are listed for grounding, overcurrent protection for the circuit conductors is limited to 10 A or less, and the load served by the conductors is not a power utilization load.

Class II, Divisions 1 and 2 Grounding – 502-16. In general, all grounding shall be done in accordance with the provisions of 250. Section 502-16(a) covers bonding requirements and requires that bonding jumpers with proper fittings or other approved means be used where bonding is required in a Class II location. Locknut-bushings or double locknuts cannot be used for bonding purposes. Bonding is required around all intervening raceways, fittings, boxes, enclosures, etc. and between Class I locations and the point of grounding for separately derived systems.

Section 502-16(b) contains special requirements for the use of FMC and LTFMC when installed in accordance with 502-4. If either of these conduits are installed, they shall be installed with bonding jumpers in parallel with the conduit. The bonding jumper may be installed either inside or outside of the conduit and in accordance with 250-79.

Section 502-16(b), Ex. permits listed LTFMC to be installed without a bonding jumper in Class I, Division 2 locations, provided the total length of the conduit does not exceed 6′. In addition, the fittings used are listed for grounding, overcurrent protection for the circuit conductors is limited to 10 A or less, and the load served by the conductors is not a power utilization load.

Class III, Divisions 1 and 2 Grounding – 503-16. In general, all grounding shall be done in accordance with the provisions of 250. Section 503-16(a) covers bonding requirements and requires that bonding jumpers with proper fittings or other approved means be used where bonding is required in a Class I location. Locknut-bushings or double locknuts cannot be used for bonding purposes. Bonding is required around all intervening raceways, fittings, boxes, enclosures, etc. and between Class I locations and the point of grounding for separately derived systems.

Section 503-16(b) contains special requirements for the use of FMC and LTFMC when installed in accordance with 503-3. If either of these conduits are installed in a manner in which they are the sole equipment grounding path, they shall be installed with bonding jumpers in parallel with the conduit. The bonding jumper may be installed either inside or outside of the conduit and in accordance with 250-79.

Section 503-16(b), Ex. permits listed LTFMC to be installed without a bonding jumper in Class I, Division 2 locations, provided the total length of the conduit does not exceed 6′. In addition, the fittings used are listed for grounding, overcurrent protection for the circuit conductors is limited to 10 A or less, and the load served by the conductors is not a power utilization load.

INTRINSICALLY SAFE SYSTEMS – 504

Article 504, Intrinsically Safe Systems, contains an alternate method for installation of electrical systems in Class I, II, or III locations. An *intrinsically safe system* is a system with an assembly of intrinsically safe apparatus and associated apparatus which is interconnected and used in hazardous locations to supply equipment. See 504-2. The concept behind intrinsic safety is that the energy level of the circuit or equipment, which may be generated by sparking, arcs, or other short circuits or ground faults is insufficient to cause ignition of a flammable or otherwise hazardous substance. Some of the factors involved in determining the intrinsic safety of a circuit or equipment include the voltage rating of the circuit, the current levels present, the surface temperature of the equipment, and the associated apparatus.

In general, 504-4 requires that all intrinsically safe apparatus and associated apparatus be approved. Installers of intrinsically safe systems should take care to ensure that the installation of the system does not jeopardize the integrity of the intrinsically safe system. Per 504-20, intrinsically safe systems are permitted to be installed with any of the wiring methods suitable for unclassified locations.

Seal requirements to prevent the passage of flammable gases and vapors from classified to unclassified areas are also required for intrinsically safe systems. Section 504-70 requires that, in general, conduits or cables which are required to be sealed in Class I and II locations shall be sealed.

Section 504-30(a) contains the provisions for maintaining proper separation (1) from open wiring, (2) in raceways, cable trays, and cables, and (3) within enclosures. Section 504-30(b) also requires that intrinsically safe circuits be properly separated from other intrinsically safe circuits by either (1) installation in a grounded metal shield or (2) by minimum insulation thickness. Such severe separation requirements ensures that intrinsically safe systems stay isolated from circuits and equipment which could, under fault conditions or otherwise abnormal conditions, introduce energy levels into the intrinsically safe system and ignite flammable or other hazardous materials which may be present. Installation of intrinsically safe systems should only be done under the direct supervision of qualified personnel.

ZONE CLASSIFICATION SYSTEM

In 1996, an alternate method for classifying hazardous locations was included in the NEC®. Article 505, Class I, Zone 0, 1, and 2 Locations incorporates a classification system which has been widely used in Europe. The zone classification system is based on the International Electrotechnical Commission (IEC) standards for area classification. Since the inclusion of this system is new, there are no NFPA or ANSI standards available for guidance in using the zone clarification system. In addition, there are many unanswered questions as to how the two different classification systems can coexist. At this time, designers and installers of electrical systems should know that this optional method is in the NEC® and they should follow closely the development of this system in future editions of the NEC®. If 505 is used for area classification wiring selection or equipment selection, it shall be done under the supervision of a qualified Registered Professional Engineer. The standard and optional hazardous location systems shall not be commingled. See Figure 12-13.

SERVICE AND REPAIR GARAGES - 511

Article 511 contains provisions for installing electrical systems in commercial garages used for repair of self-propelled vehicles and storage of volatile flammable liquids used for fuel or power. The inclusion of volatile flammable liquids requires that areas of these garages be classified areas and the methods for installation of electrical systems be closely controlled. In addition to 511, other articles may also need to be considered when designing or installing electrical systems in commercial garages. If, for example, the commercial garage includes an area in which flammable fuel is transferred to vehicle fuel tanks, then the requirements of 514 shall also be followed.

Classifications

Most classifications in commercial garages are either Class I, Division 1 or Class I, Division 2 because vehicle fuels consist of volatile, flammable liquids. Several additional factors should be considered when classifying commercial garage areas. Often these garages include pits or other areas below the floor surface. These pits are used to access the vehicle from underneath. Such below-the-floor areas can introduce additional hazards as they provide an area in which flammable gases or vapors may collect, especially if gasoline were to leak from a vehicle. Additional areas which should be considered are the areas above Class I, Division 1 locations, the area around fuel dispensing pumps, the area around dispensing drums for flammable liquids or solvents, and in some garages, such as auto body shops or paint-spraying areas.

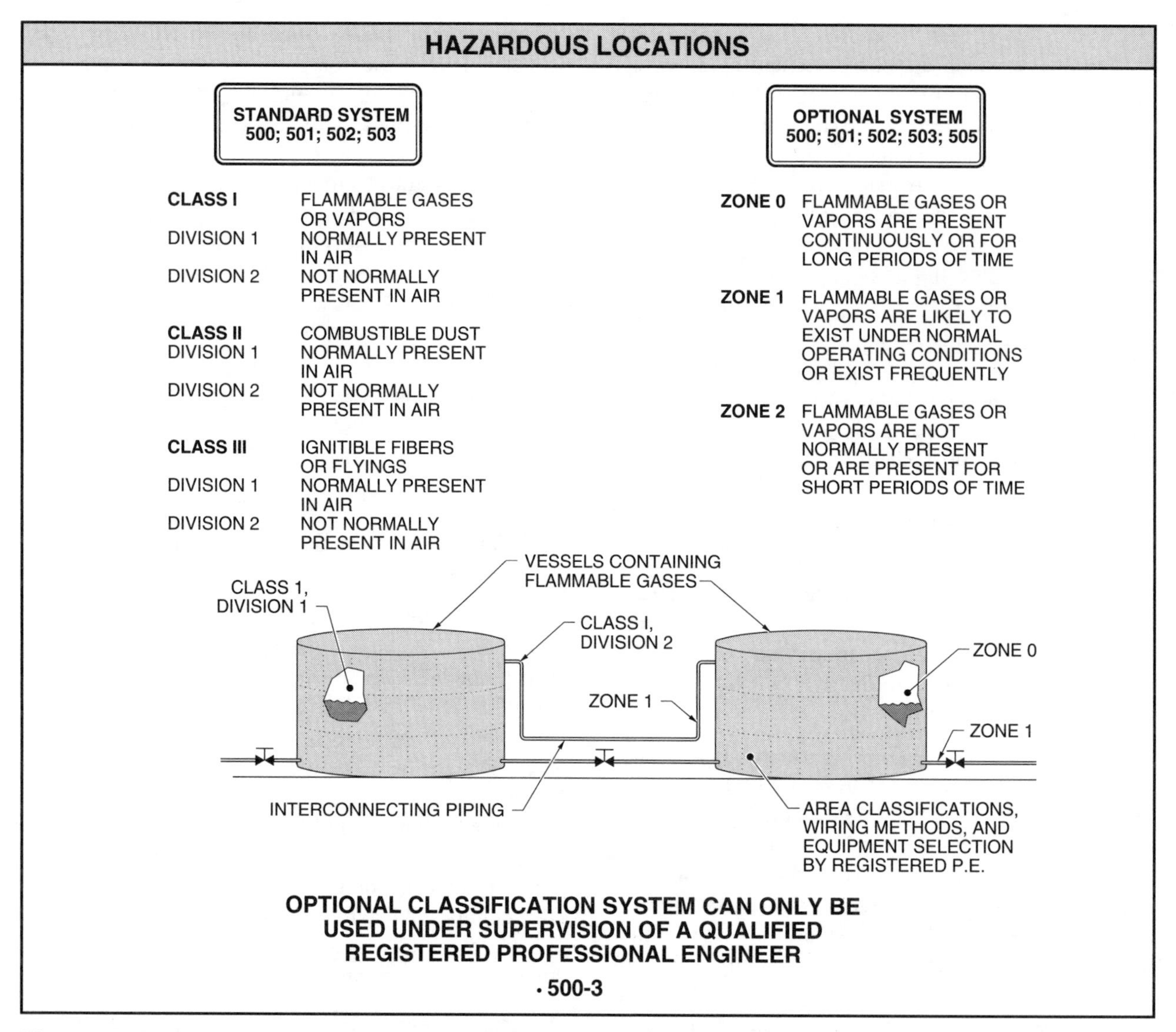

Figure 12-13. The standard and optional hazardous location systems must not be commingled.

Locations – 511-2. Locations, such as parking garages, where no repair work is performed which requires the use of electrical equipment, open flame, welding, etc. are not classified areas. However, these areas shall be adequately ventilated to remove exhaust fumes from the engines. Automobile showrooms are also not classified as hazardous locations.

Floor Space – 511-3(a). Pits or other depressions in the floor of commercial garages are potentially dangerous collection points for flammable gases or vapors. The most common area classification is for the immediate service area in commercial garages. For floor spaces, the entire area up to a level of 18″ above the floor shall be classified as a Class I, Division 2 location. However, 511-3(a), Ex. permits the space to be unclassified if the AHJ determines that there is adequate mechanical ventilation which ensures a minimum of four air changes per hour. By providing a minimum of four air changes per hour, the hazardous atmosphere can be controlled, removing the potential for hazard. See Figure 12-14.

Pits or Depressions – 511-3(b). Any pit or depression below the surface of a floor in a commercial garage shall be classified as a Class I, Division 1 location. The Class I, Division 1 classification shall extend from the pit or depression up to the floor level. If the AHJ, however, determines that there is a minimum of six air changes per hour exhausted at the floor level of the pit or depression, the area may be classified as a Class I, Division 2 location.

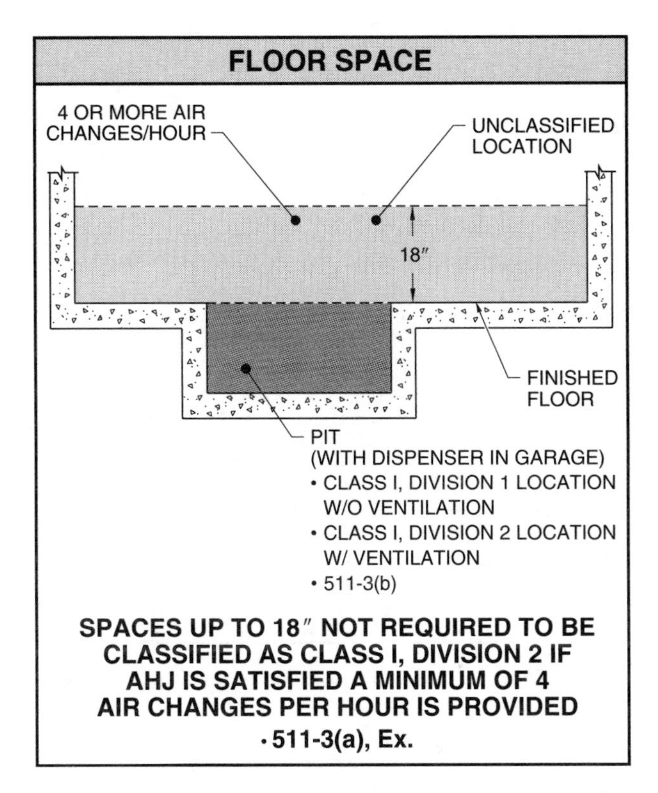

Figure 12-14. Pits or other depressions in the floor of commercial garages are potentially dangerous collection points for flammable gases or vapors.

Per 511-3(b), Ex., the clarification of lubrication and service rooms, which do not have provisions for dispensing, shall be classified per Table 514-2. This exception considers the recent development of lubritoriums which have no provisions for fuel dispensing and are designed for quick lubrication and oil changes.

Other Locations – 511-3(c)(d). Section 511-3(c) permits the areas adjacent to defined locations in which flammable vapors are not likely to be released to be unclassified, provided that mechanical ventilation is used to ensure a minimum of four air changes per hour. This includes areas such as stock rooms, switchboard rooms, and other similar locations. These rooms may also be unclassified areas if they are effectively isolated from the classified areas by walls or partitions.

Section 511-3(d) permits other adjacent areas to be unclassified if, in the opinion of the AHJ, adequate ventilation, air pressure differential, or physical spacing has removed the possibility of ignition from occurring.

Wiring Methods – 511

Specific wiring methods are required to minimize the potential hazards of installing electrical systems in commercial garages. The selection of the wiring method depends upon the classification of the area in which the wiring is to be installed. As with all classifications, the more likely the presence of the hazardous substance, the stricter the wiring method.

Class I Locations – 511-4. All wiring installed in any area within a commercial garage, which is classified as Class I, shall be installed in a wiring method which meets the applicable provisions of 501. Wiring methods in Class I, Division 1 locations shall be selected from 501-4(a) and wiring methods for Class I, Division 2 locations shall be selected from 501-4(b). Section 511-4, Ex. permits RNMC (PVC) to be installed in Class I locations, provided it is buried under not less than 2′ of cover and either threaded RMC or threaded steel IMC is used for the last 2′ of the underground run up to the point of emergence or point of connection to the aboveground raceway. See Figure 12-15.

Above Class I Locations – 511-6. The area above Class I locations often contains electrical equipment which is used in commercial garages. The equipment and wiring to such equipment shall not provide a source of ignition for the hazardous substances which may be present below. Section 511-6(a) requires that all fixed wiring in these locations shall be installed in one of the following raceways: RMC, EMT, FMC, RNMC, LTFMC, and LTFNMC. In addition, Type MI, MC, PLTC, and TC cables and manufactured wiring systems are permitted. Ceiling outlets or extensions to the floor below are permitted to be installed with cellular metal or cellular concrete floor raceways, provided such raceways have no connections leading through or into any Class I locations above the floor.

Equipment

Article 511 has several specific requirements for the various types of equipment used in commercial garages. The purpose of the requirements for portable lighting, fixed lighting, battery and vehicle charging equipment, and GFCI protection is two-fold.

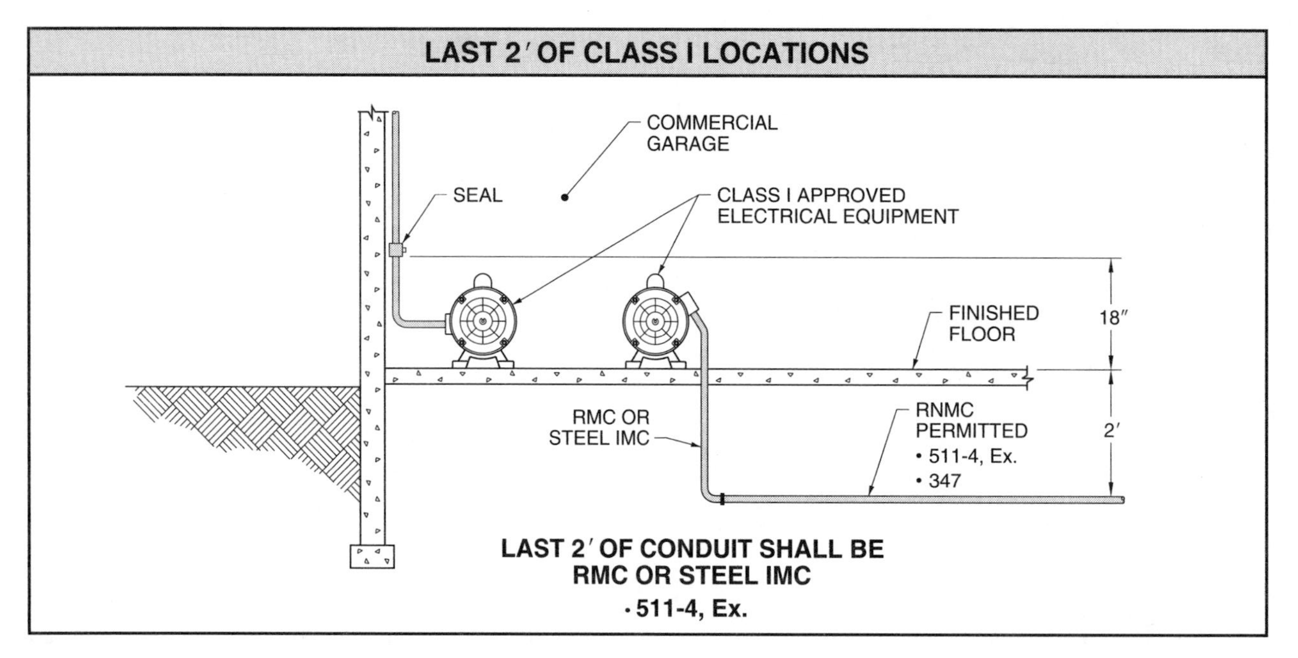

Figure 12-15. Where suitably protected and installed per 347, PVC is a permitted wiring method for Class I, Division 1 locations in commercial garages.

First, the requirements ensure that the equipment which may be used in classified areas is designed and installed in a manner which does not increase the hazards associated with the flammable gases or vapors which may be present. Secondly, the requirements help to ensure that all workers within commercial garages are protected from the hazards associated with work in these types of spaces.

Portable Lighting – 511-3(f). Portable lighting equipment shall be equipped with a handle, hook, and lampholder. It shall be protected with substantial guards. Exposed surfaces shall be constructed with nonconducting material or shall be covered with insulation.

Lampholders are not permitted to include provisions for receiving attachment plugs and the lampholders shall not be switched. The outer shell of the lighting fixture is required to be constructed of molded composition or other suitable materials. All portable lighting fixtures shall be approved for Class I, Division 1 locations unless they are installed in a manner which prevents their usage in the locations that are classified as Class I, Division 1 or 2. See Figure 12-16.

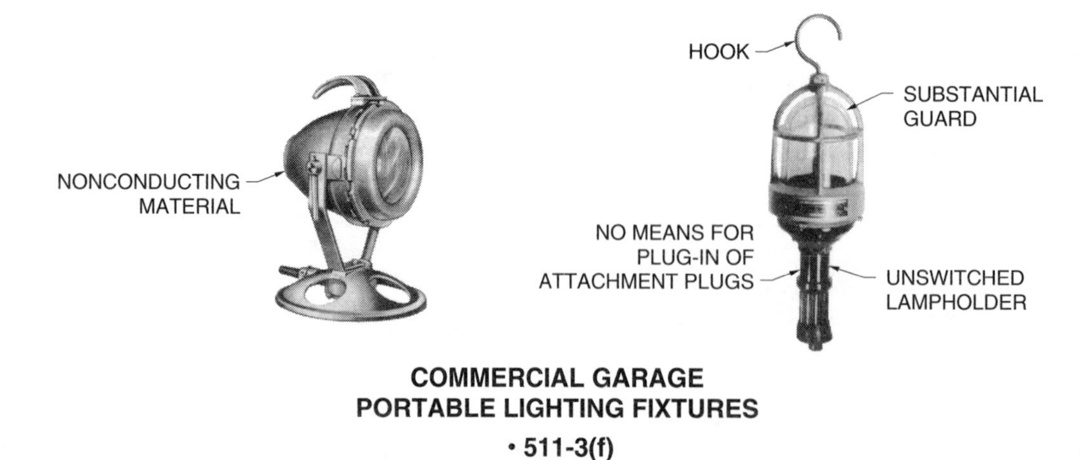

Crouse-Hinds Division, Cooper Industries, Inc.

Figure 12-16. Portable lighting equipment used in Class I, Division 1 locations in commercial garages shall be approved for the location.

Arcing Equipment – 511-7. If electrical equipment which may be installed above the floor level in commercial garages is capable of producing arcs, sparks, or releasing particles of hot metal, and is located less than 12′ above the floor, it shall be totally enclosed or designed so that it prevents the release of arcs, sparks, or particles of hot metal. Examples of the types of equipment to which 511-7 applies includes cutouts, switches, charging panels, motors, generators, and other equipment which incorporates make-and-break or sliding contacts. This section does not apply to equipment such as receptacles, lamps, or lampholders.

Fixed Lighting – 511-7(b). Unless the lighting fixtures are totally-enclosed or constructed so that the escape of arcs, sparks, or hot metal particles is prevented, all lamps and lampholders for fixed lighting shall be installed at a location which is not less than 12′ above the floor level if the fixtures are subject to physical damage. This includes lanes which commonly contain vehicle traffic.

Battery Charging Equipment – 511-8. Battery charging equipment is frequently found in commercial garages. Such equipment usually consists of the battery charger, control equipment, and the batteries themselves. The installation of any of this battery charging equipment is prohibited in any of the Class I, Division 1 or 2 locations covered in 511-3.

Vehicle Charging Equipment – 511-9. Electrical vehicle charging equipment is covered by 625. Such equipment which is installed in commercial garages shall be installed per 625. Connectors and the point of connection between chargers and vehicles shall not occur within the Class I, Division 1 or 2 locations listed in 511-3.

GFCI Protection – 511-10. To provide protection to personnel who may be working in commercial garages, 511-10 requires that all 15 A or 20 A, 125 V, 1ϕ receptacles installed for the connection of electrical diagnostic equipment, electrical hand tools, or portable lighting equipment shall be provided with GFCI protection. This provision ensures that personnel who might be working with equipment when water or other liquids are present are protected against the hazards of electrical shock. See Figure 12-17.

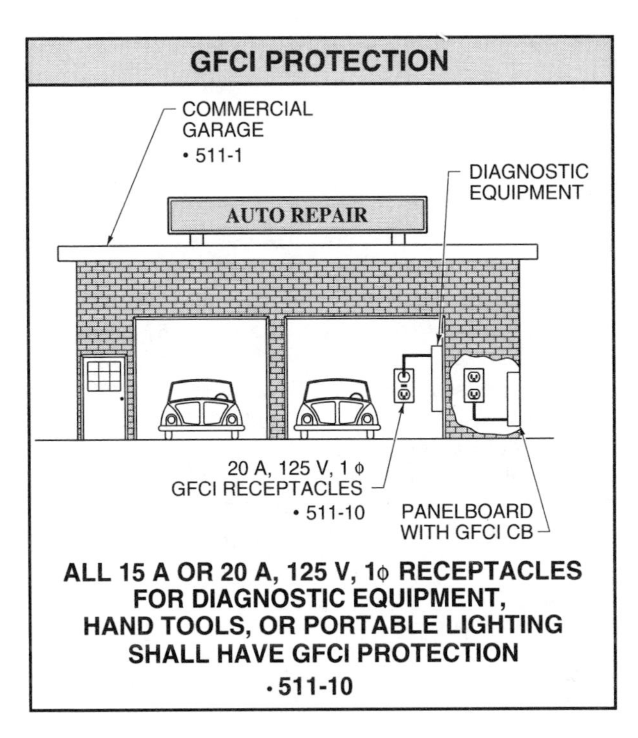

Figure 12-17. Where GFCI protection is required for personnel in commercial garages, it can be provided by either GFCI receptacles or from receptacles protected by a GFCI CB.

GASOLINE AND SERVICE STATIONS – 514

Gasoline and service stations present many of the same problems for designers and installers of electrical systems that are encountered with commercial garages. Volatile flammable liquids and liquefied flammable gases may be present and are frequently transferred to the fuel tanks of many different kinds of self-propelled vehicles or other approved containers. Table 514-2 is used when Class I liquids are stored, handled, or dispensed to determine proper area classification and delineation.

Gasoline and service stations are locations where gasoline, other volatile flammable liquids, or liquefied flammable gases are transferred to vehicle fuel tanks or other approved storage containers. Section 514 applies only to these specific areas of the service station. Other areas in service stations include service rooms, repair rooms, offices, sales rooms, etc. These areas are not covered by 514. They are subject to the provisions of 510 or 511. If the AHJ determines that flammable liquids, such as gasoline, which have a flash point below 38°C (100°F) will not be handled, the area is permitted to be unclassified.

QUICK-LUBE FACILITIES

The recent development of "quick-lube" facilities has led to some interesting classifications of these types of service stations. Table 514-2 was revised during the 1996 NEC® cycle to allow for these types of service stations which do not contain provisions for dispensing fuel, but have the potential for the presence of flammable gases or vapors. Table 514-2 now permits pits or depressions in quick-lube facilities to be unclassified if exhaust ventilation is provided at a rate of not less than 1 cfm/sq ft of floor area at all times that the building is occupied. This provision also extends to periods when vehicles may be parked over or near the pit or depression. The exhausted air shall be taken from a point which is within 12″ of the floor of the pit or depression. Designers and installers should check with the AHJ for proper area classification whenever electrical work is to be performed in these types of facilities.

Installation Requirements

Because volatile flammable liquids and gases are present in gasoline and service stations, several installation requirements shall be considered before designing or installing electrical systems in these occupancies. These considerations include:

- The areas of the occupancies shall be carefully classified to ensure that the proper wiring method and installation requirement for the particular class and division can be followed. Table 514-2 is designed to assist in making that determination based upon the location and extent of the classified area.
- Special disconnecting means are required for circuits supplying dispensing equipment. Additionally, other dispensing equipment controls may be required depending on whether the service station is attended or unattended.
- Seal requirements may be required to minimize the passage or communication of hazardous substances from classified to unclassified areas.
- Special grounding provisions may be required for classified areas within the gasoline or service station.
- Wiring methods are based upon the location within the service station and the classification of the area.

Class I Locations – Table 514-2. Table 514-2 lists the Class I and Division locations which may be found within gasoline and service stations. Some of the locations listed in the Table are areas around underground tanks, dispensing devices, remote pumps, lubrication rooms with and without dispensing, sales and storage rooms, equipment enclosures, and other vapor-processing equipment locations. Class I locations shall not extend beyond solid partitions, walls, or roofs. Any equipment installed within the Class I locations listed in Table 514-2 shall be wired per the Class I, Division 1 or 2 provisions contained in 501.

Wiring Above Class I Locations – Table 514-2. All installations of equipment and wiring in areas above locations which are Class I locations, as per Table 514-2, shall comply with the provisions of 511-6 and 511-7.

Disconnecting Means – 514-5. For safety reasons, it is important to be able to quickly disconnect all circuit conductors which supply or pass through gasoline dispensing equipment. Section 514-5 requires that such disconnection be provided in the form of clearly identifiable and readily accessible switches or other acceptable means. The switch shall be located away from the dispensing equipment and shall simultaneously open all of the circuit conductors, including the grounded conductor from the supply source. Single-pole CBs which use handle ties to accomplish the simultaneous opening of the circuit conductors are not permitted. See Figure 12-18.

For attended service stations, 514-5(b) requires that the disconnecting means be installed not more than 100′ from the dispensers in a location which is acceptable to the AHJ. For unattended service stations, 514-5(c) requires that the disconnecting means shall also be installed in a location acceptable to the AHJ and more than 20′ but less than 100′ from the dispensers. In addition, additional disconnecting means shall be provided for each group of dispensers or the outdoor equipment which is used to control the dispensers.

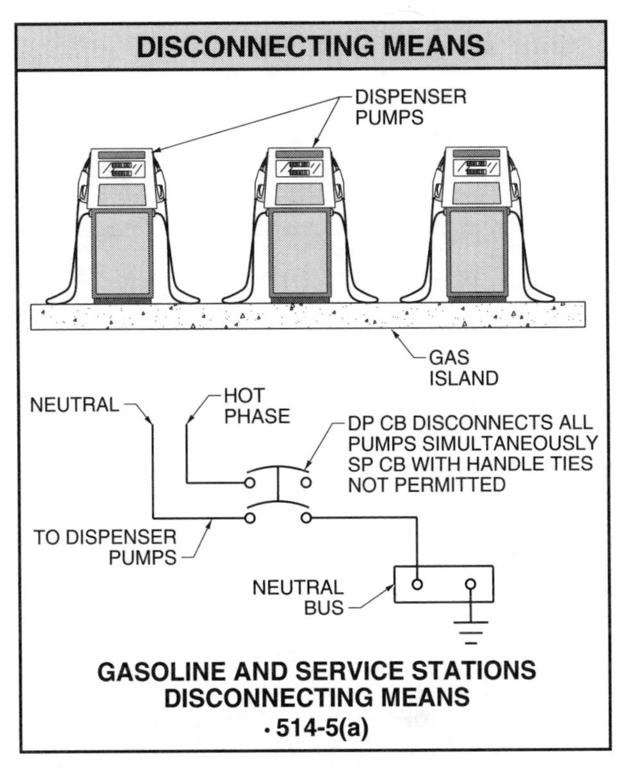

Figure 12-18. In addition to the circuit ungrounded conductors, the grounded conductor (neutral) shall be disconnected from the source of supply if it supplies or passes through dispensing equipment in service stations.

Conduit Seals – 514-6. Two specific areas within gasoline and service stations are required to have conduit seals installed to prevent the communication of hazardous substances from a classified area to an unclassified one. Per 514-6(a), an approved seal shall be installed for each conduit entering or leaving a fuel dispenser. The seal fitting shall be installed so that it is the first fitting after the conduit emerges from the earth or conduit. See Figure 12-19.

Per 514-6(b), all Class I, Division 1 or 2 conduit seal requirements listed in 501-5 shall be followed in gasoline and service stations. In addition, the requirements in 501-5(a)(4) and 501-5(b)(2) are amended to include vertical as well as horizontal runs of conduit. In any Class I, Division 1 or 2 location in a gasoline or service station, both vertical and horizontal runs of conduit which leave a classified area shall have a conduit seal installed within 10′ of the boundary.

Grounding – 514-16. All metal raceways, metal-jacketed cables, and all other metal, noncurrent-carrying parts of equipment shall be grounded per 250. Grounding in Class I locations shall also meet the requirements of 501-16. See Figure 12-20.

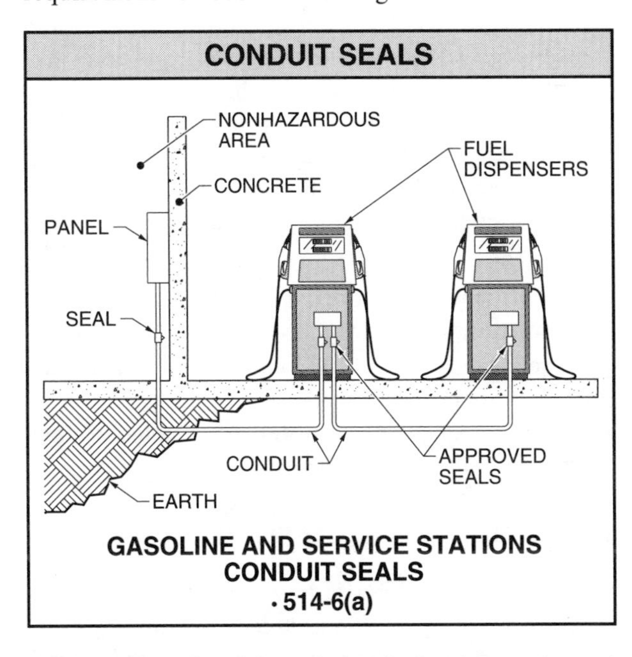

Figure 12-19. Conduit seals for dispensing equipment shall be the first fitting installed after the conduit leaves the concrete or the earth.

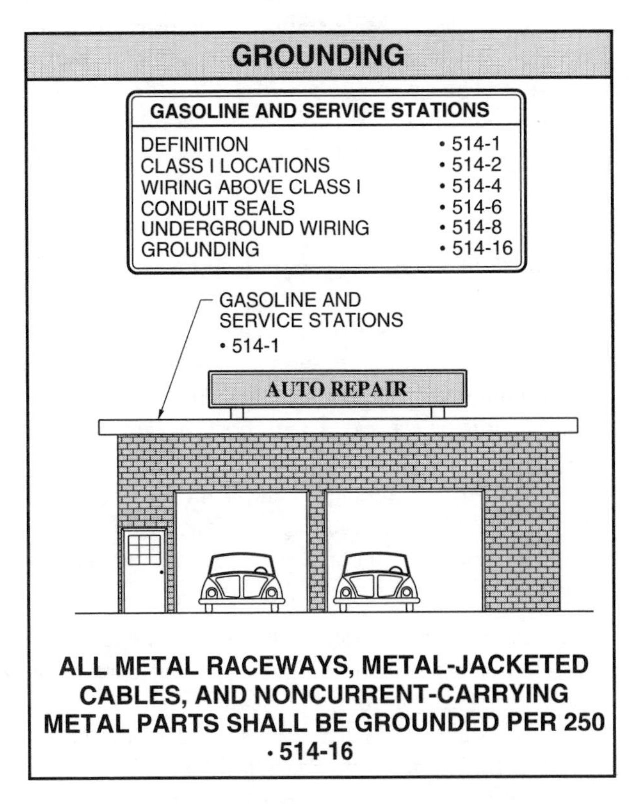

Figure 12-20. Grounding in Class I, Division 1 locations of gasoline and service stations shall be in accordance with 501-16.

Underground Wiring – 514-8. The two general wiring methods permitted for underground wiring in gasoline and service stations are threaded RMC and threaded steel IMC. In addition, two exceptions permit alternate wiring methods to be used as well. Per 514-8, Ex. 1, MI cable may be used when installed in accordance with 330. MI cable for use in underground runs shall be suitably protected against both corrosive conditions and physical damage.

Per 514-8, Ex. 2, RNMC is permitted to be installed provided it is buried under at least 2′ of cover and either threaded RMC or threaded steel IMC is used for the last 2′ of the run to emerge from the ground or connection to the aboveground raceway. A separate EGC shall be installed to maintain electrical continuity of the raceway.

HEALTH CARE FACILITIES – 517

A *health care facility* is a location, either a building or a portion of a building, which contains occupancies such as, hospitals, nursing homes, limited or supervisory care facilities, clinics, medical and dental offices, and either movable or permanent ambulatory facilities. See 517-3. *Note:* This list is not all inclusive. There may be other similar facilities that are not listed that meet the criteria for health care facilities. *Note:* The definition specifically addresses portions of buildings. Section 517-3 contains many definitions which should be reviewed prior to attempting to apply the provisions of the article.

Unlike the normal grounding requirements in 250, receptacles and equipment in patient care areas shall be grounded by an insulated copper conductor which is required to be installed in a metal raceway. This "redundant grounding" is to ensure the safety of patients who are very vulnerable to electrical shock hazards. Section 517-13(a), Ex. 1 does permit the use of Type MI, MC, or AC cables to be used in these areas, but only the outer jackets of these cables provides an acceptable grounding return path. MC cable with interlocking metal tape armor does not meet this requirement. See Figure 12-21.

Section 517-18(c) requires that all 15 A and 20 A, 125 V receptacles installed in the patient care areas of pediatric wards, rooms, or other areas shall be tamper resistant. A tamper-resistant receptacle is designed to limit improper access to its energized contacts. A receptacle cover is permitted to be used in place of a tamper-resistant receptacle per 517-18(c), Ex. See Figure 12-22.

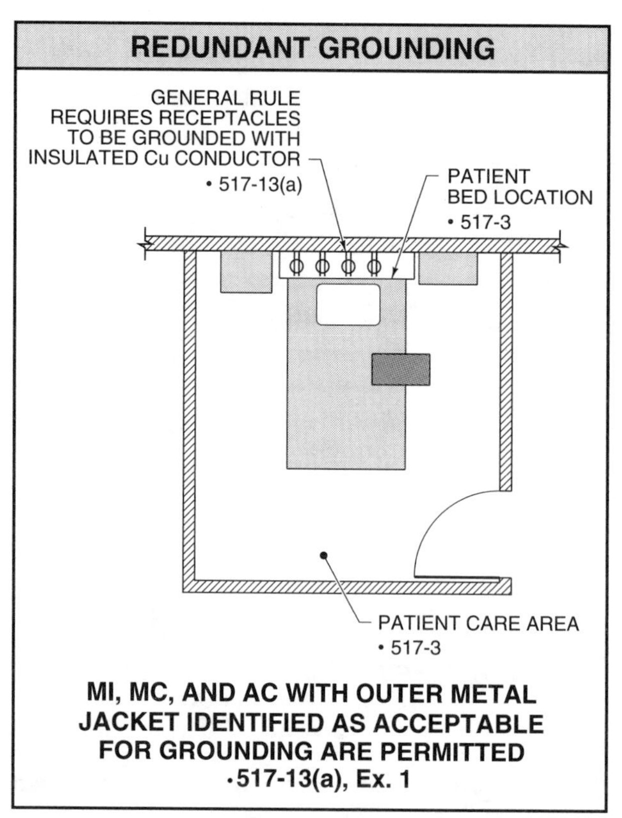

Figure 12-21. Type MI, MC, or AC cables installed in patient care areas of health care facilities shall be listed and shall be constructed with an outer metal sheath which provides an acceptable grounding return path.

Wet locations in health care facilities pose additional hazards for patients. Section 517-20 requires that all receptacles and fixed equipment installed in these types of locations shall be provided with GFCI protection if the interruption of power can be tolerated. If the interruption of power cannot be tolerated, an isolated power system shall be installed. Bathroom receptacles, in other than critical care areas where the toilet and basin are installed within the patient room, shall be provided with GFCI protection for personnel. See Figure 12-23.

PLACES OF ASSEMBLY – 518

A *place of assembly* is a building, structure, or portion of a building designed or intended for use by 100 or more persons. While the requirements contained in 518 apply to buildings or portions of buildings which are intended for use by 100 or more persons, they do not contain information on how to determine the maximum number of persons for which the occupancy is designed. Per 518-2, FPN, this determination may be made by the local building codes or the NFPA Life Safety Code in the jurisdiction in which the occupancies are located.

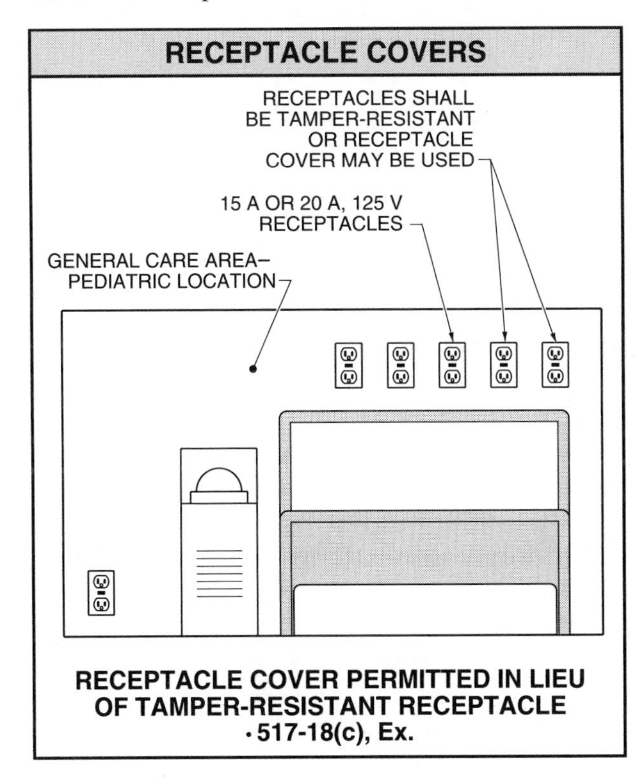

Figure 12-22. Receptacle covers, which limit access to the energized parts of a receptacle, are permitted in place of tamper-resistant receptacles in pediatric locations of health care facilities.

Because of the frequent presence of large groups of people, special installation requirements apply to places of assembly. The intent of these provisions is to ensure that the installation of the electrical system does not jeopardize the safety of the inhabitants.

Wiring Methods – 518-4. In general, all permanent wiring shall be installed in metal raceways. Nonmetallic raceways are permitted if they are encased in not less than 2″ of concrete. Type MI and MC cables are also permitted to be used in places of assembly. For those buildings or portions of the building that are not required by local building codes to be of fire-rated construction, NM and AC cables, ENT, and RNMC are also permitted wiring methods.

Per 518-3(b), Ex. 1, temporary wiring is permitted to be laid directly on the floor if it is located so that it is not accessible to the general public. The flexible cords or cables of temporary wiring laid directly on the floor shall be the hard or extra-hard usage type. See Figure 12-24.

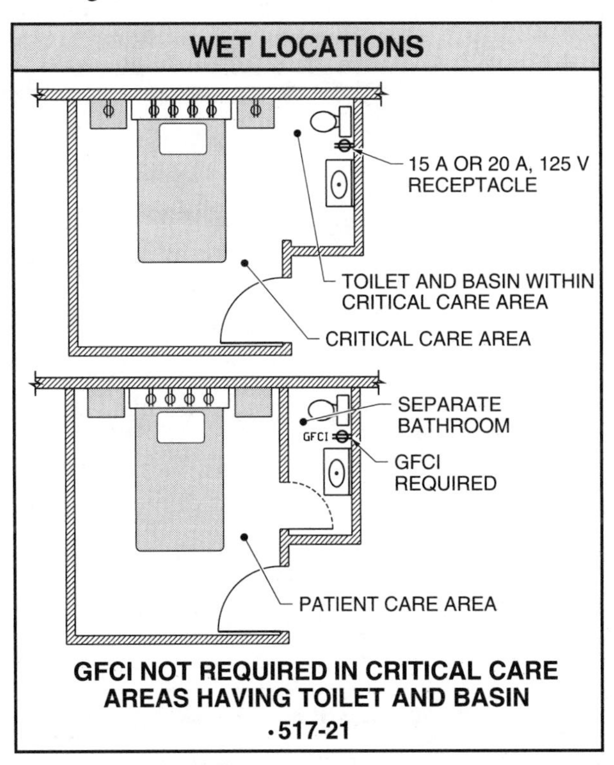

Figure 12-23. Bathroom receptacles in general patient care areas shall be provided with GFCI protection.

Per 518-3(b), Ex. 2, the GFCI requirements for temporary wiring in 305-6 do not apply to places of assembly. See Figure 12-25. In addition, 518-4, Ex. 3 permits ENT and RNMC to be used in restaurants, conference and meeting rooms in hotels or motels, dining facilities, and church chapels, provided that either the raceways are concealed within walls, floors, or ceilings which have at least a 15-minute finish rating or the raceways are installed above suspended ceilings which provide a thermal barrier constructed of a material which also has a 15-minute finish rating.

Power Distribution – 518-5. Portable power distribution equipment such as panelboards and switchboards used in places of assembly shall be supplied only from listed power outlets. The power outlets shall have sufficient voltage and current ratings for the load supplied and they shall be protected by overcurrent devices. In addition, the overcurrent devices shall be installed in a location which is not accessible to the general public.

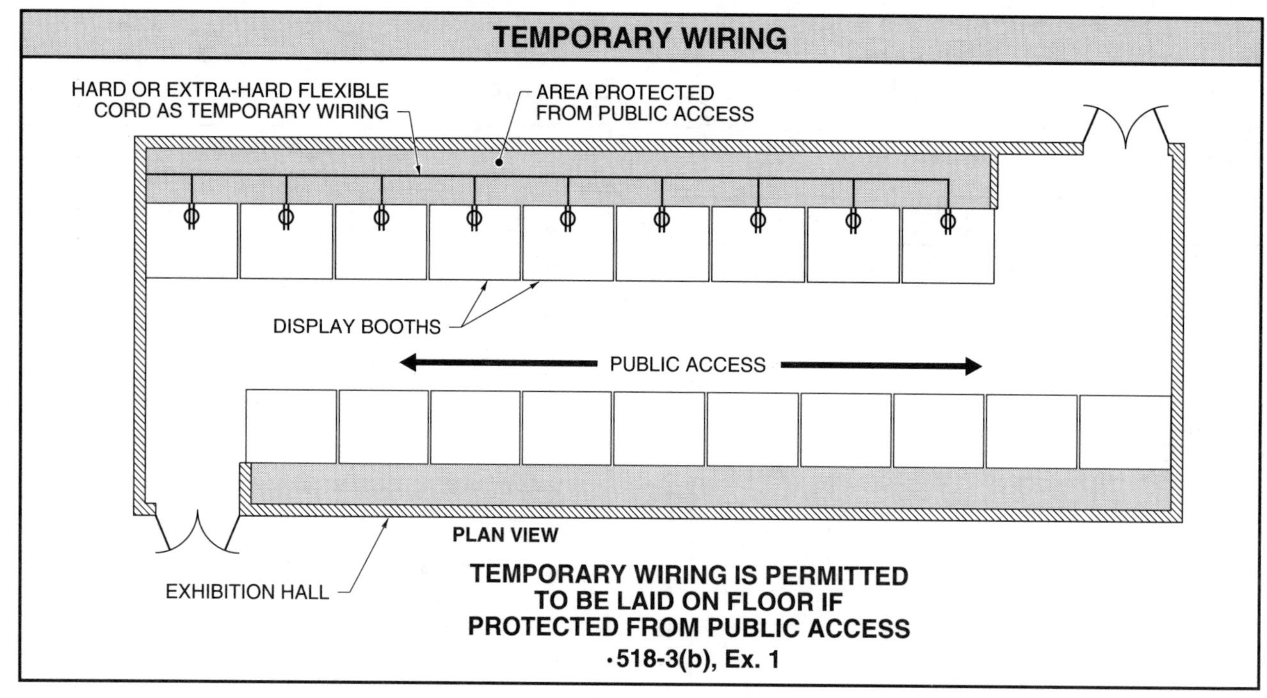

Figure 12-24. Temporary wiring is permitted to be laid directly on the floor if it is protected from public access.

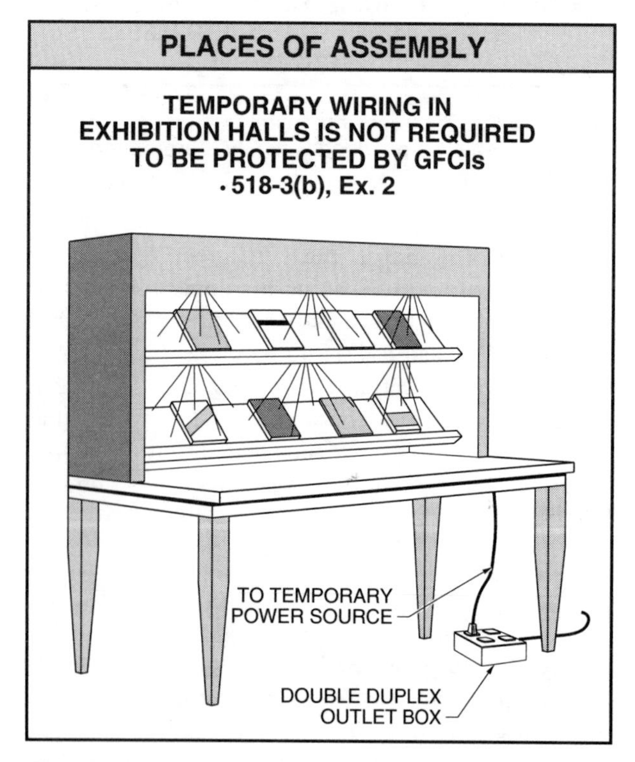

Figure 12-25. The GFCI requirements for temporary wiring in 305-6 do not apply to places of assembly.

Review Questions

__________ **1.** A(n) ________ is a location where there is an increased risk of fire or explosion due to the presence of flammable gases, vapors, liquids, combustible dusts, or easily-ignitable fibers or flyings.

__________ **2.** The standard classification system used by the NEC® to determine the type of hazardous location is the ________ system.

A. Zone
B. Class, Division, and Group
C. Zone, Class, and Division
D. neither A, B, nor C

__________ **3.** A(n) ________ is a fitting which is inserted into runs of conduit to isolate certain electrical apparatus from atmospheric hazards.

T F **4.** Intrinsically safe systems may be used in Class I, II, or III locations to supply equipment.

__________ **5.** A place of assembly is a building, structure, or portion of a building designed or intended for use by ________ or more persons.

A. 12
B. 48
C. 50
D. 100

__________ **6.** The optional classification system contains ________ zones.

T F **7.** Each room or area within an occupancy shall be considered separately for the purpose of determining area classification.

T F **8.** A Class I, Division 2 location is any location in which flammable gases pose a threat of explosion or ignition under normal conditions.

__________ **9.** A Class III, Division ________ location is any location where easily ignitable fibers or flyings are stored or handled.

__________ **10.** Class I locations contain ________ groups of flammable gases or vapors.

__________ **11.** All boxes and fittings used in Class I, Division 1 locations shall be ________.

A. explosionproof
B. designed for threaded connections
C. A and B
D. neither A nor B

T F **12.** Any wiring method permitted for use in Class III, Division 1 locations is also permitted for use in Class III, Division 2 locations.

__________ **13.** The most stringent requirements for electrical equipment installed in hazardous locations are for Class ________ locations.

__________ **14.** Motors or generators used in Class ________, Division 2 locations shall be nonventilated, pipe-ventilated, water-air cooled, or fan-cooled.

__________ **15.** Electrical equipment which may produce arcs or sparks shall be totally enclosed if located less than ________′ above the floor in commercial garages.

__________ **16.** The disconnecting means for gasoline dispensers in attended service stations shall be no more than ________′ from the dispensers.

T F **17.** All 15 A and 20 A, 125 V receptacles installed in patient care areas of pediatric wards, rooms, or other areas shall be tamper resistant.

__________ **18.** A spray booth interior is a Class I, Division ________ location.

__________ **19.** A(n) ________ is an atmosphere containing flammable gases, vapors, or combustible dust.

T F **20.** Portable lighting equipment used in Class I, Division 1 locations is permitted to include provisions for receiving attachment plugs.

Hazardous Locations

__________ **1.** Class I, Division 1

__________ **2.** Class I, Division 2

__________ **3.** Class II, Division 1

__________ **4.** Class II, Division 2

__________ **5.** Class III, Division 1

__________ **6.** Class III, Division 2

A. Combustible dust normally present in air

B. Combustible dust not normally present in air

C. Flammable gases or vapors normally present in air

D. Flammable gases or vapors not normally present in air

E. Ignitible fibers or flyings normally present in air

F. Ignitible fibers or flyings not normally present in air

Class I, Division 1 Seals

__________ **1.** If a seal is required at A, it shall be within ________″ of the enclosure.

__________ **2.** A seal is required at B if within ________′ of the boundary.

__________ **3.** A seal is required at C if within ________′ of the boundary.

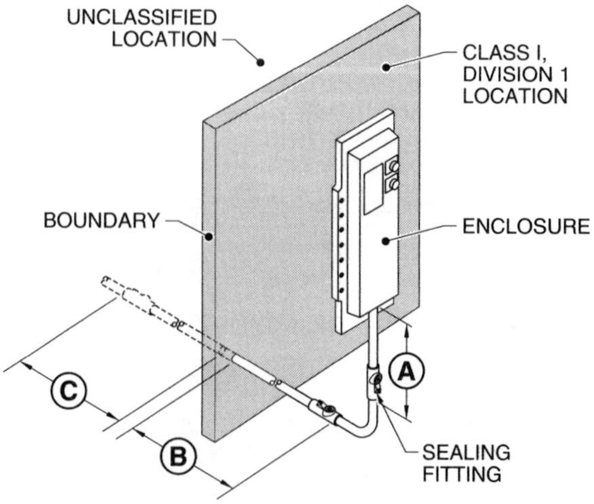

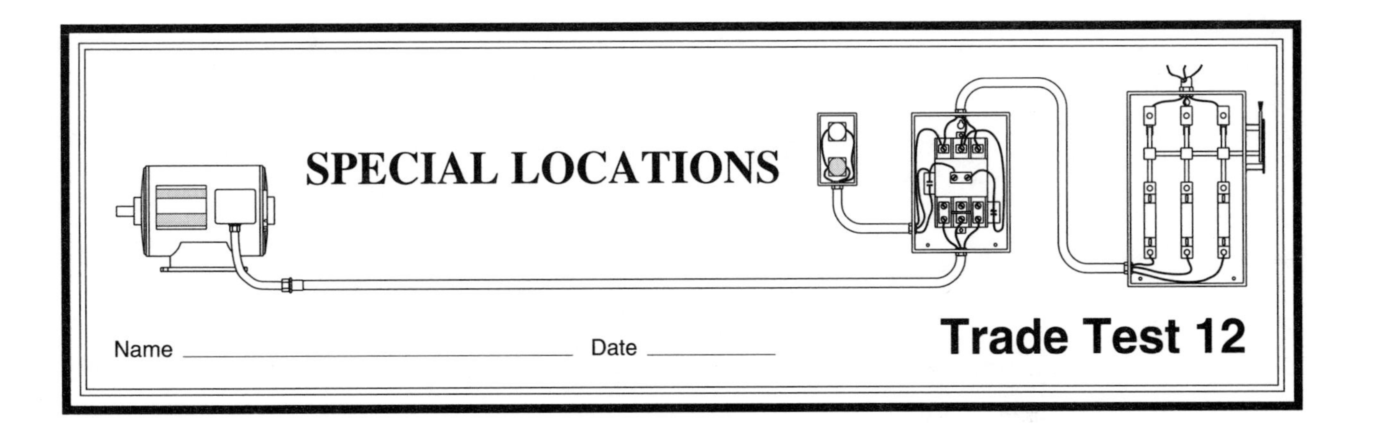

NEC®	Answers	
______	______	**1.** Determine the Class and Division for a location in which sufficient quantities of combustible dust are normally present in the air to cause an explosion or ignite hazardous materials.
______	______	**2.** Determine the Class and Division for a location in which sufficient quantities of easily-ignitable flyings are normally present in the air during the manufacturing process, but not in sufficient quantities to cause an explosion or ignite hazardous materials.
______	______	**3.** *See Figure 1.* Determine the Class and Division for the location.

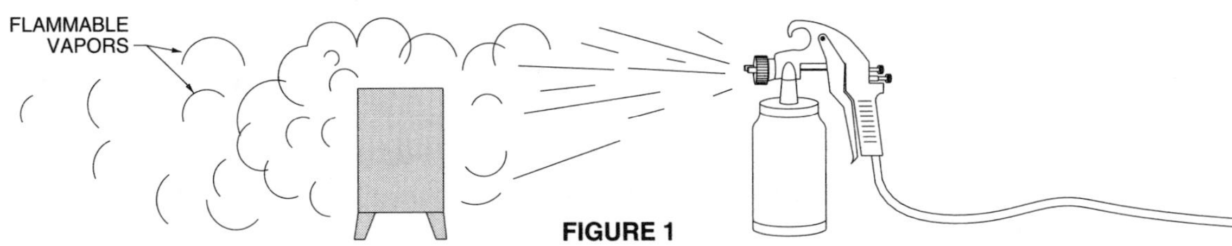

FIGURE 1

______	______	**4.** A motor is installed in a Class III, Division 1 location. The supply conductors to the motor are pulled in RMC for the entire length up to the last 3′. The last 3′ consists of FMC, which is used to provide flexibility for the motor. Does this installation violate Article 503 of the NEC®?
______	______	**5.** *See Figure 2.* Determine the maximum distance from the Class I, Division 1 boundary that the conduit seal is permitted to be installed.
______	______	**6.** *See Figure 3.* Determine the maximum distance from the electrical panelboard that the seal is permitted to be installed.

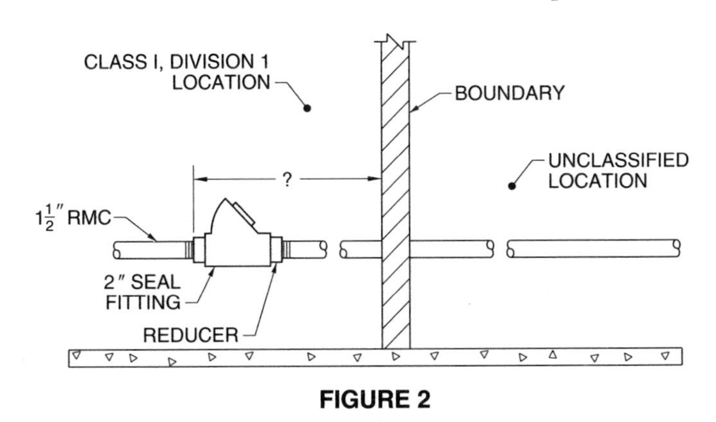

FIGURE 2

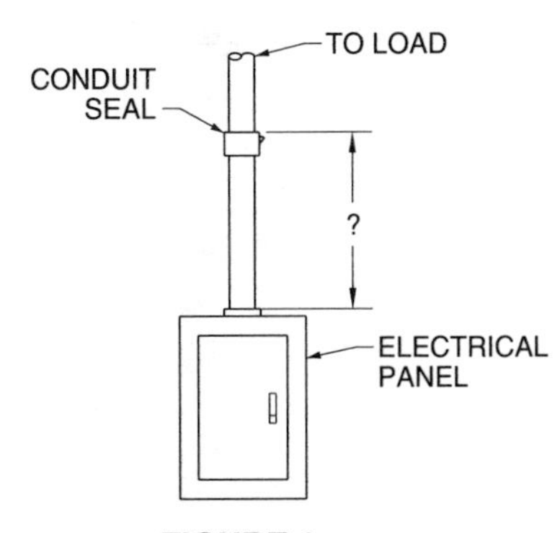

FIGURE 3

_____________ _____________ **7.** *See Figure 4.* Does this installation violate Article 501 of the NEC®?

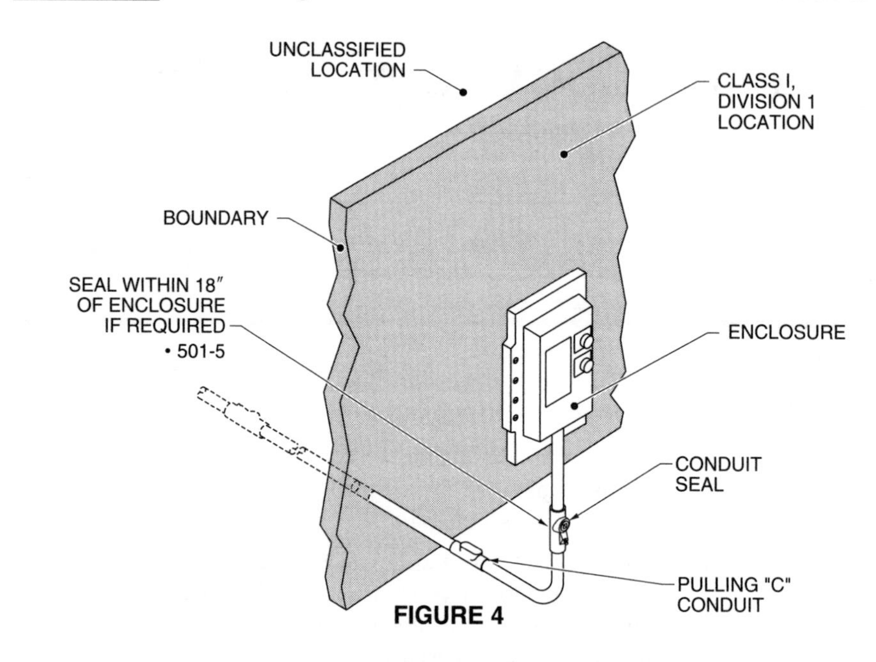

FIGURE 4

_____________ _____________ **8.** *See Figure 5.* The total area of the conductors is .1311 sq in. The total area of the sealing fitting is listed at .400 sq in. Does the installation of this conduit seal violate the provisions of Article 501 of the NEC®?

_____________ _____________ **9.** *See Figure 6.* Does the installation of the underground raceway violate Article 511 of the NEC®?

_____________ _____________ **10.** Determine the Class and Division for a space up to 18″ above the floor and 3′ horizontally from a lubrication pit in a gasoline-dispensing service station.

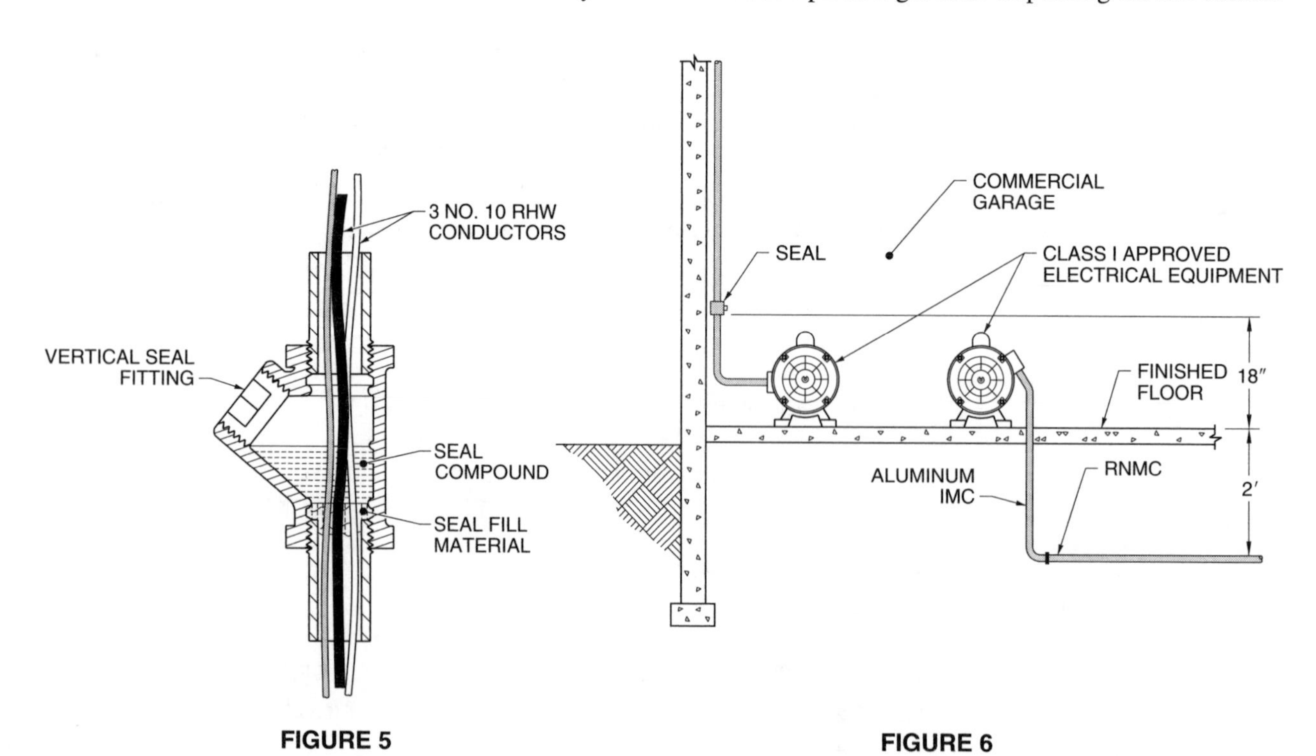

FIGURE 5

FIGURE 6

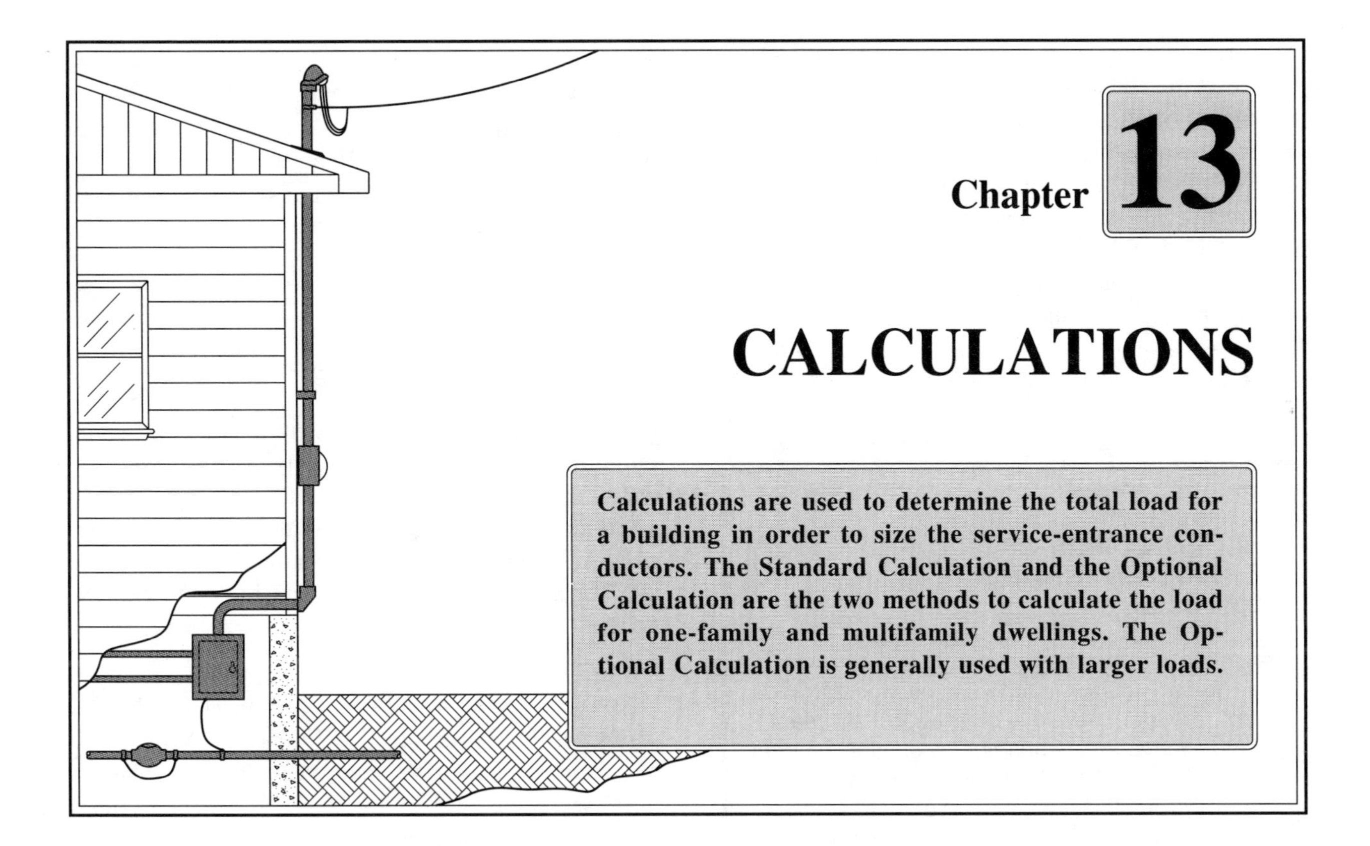

CALCULATIONS

"Calculation" is a widely used but undefined term in the NEC®. Typically, the term is used to refer to a set of procedures which are followed to arrive at a given value. Often, there are several variables which have an effect on the calculation. For example, the allowable ampacity of a given conductor and the size of an EGC for a branch circuit are variables in electrical calculations. The term is also used in a much broader sense to represent the necessary steps to determine the size of service-entrance conductors for a building or structure. This procedure involves a series of related steps which lead to a "calculated" load for the building or structure.

The Standard Calculation and the Optional Calculation are the two methods to calculate the load for one-family and multifamily dwellings per 220. In general, the Optional Calculation is less complex and easier to use than the Standard Calculation, and is used with larger loads. In either case, the calculation can be used to determine feeder or service loads.

Demand is the amount of electricity required at a given time. The concept behind the use of the term is that the total calculated load is rarely placed on the electrical system. This is because all of the electrical loads are not on at the same time. For example, a household electric range may have a nameplate rating of 14 kW. The electric range has four burners, one broiler, one heating element, and accessories such as fans, lights, and timers. Rarely, if ever, would all of these be used at the same time. Therefore, the concept of demand permits the 14 kW range to be considered as a smaller load for the purpose of determining how much load to allow for the range when calculating the total feeder or service load for the building.

BRANCH-CIRCUIT CALCULATIONS – 220-3

Section 220-3 is used to determine branch-circuit loads for various types of occupancies. Branch-circuit ratings for continuous and noncontinuous loads are covered by 220-3(a). The *branch-circuit rating* is the ampere rating or setting of the overcurrent device protecting the conductors. Continuous loads are loads that are expected to continue for three hours or more. The minimum rating for a branch circuit which supplies continuous and noncontinuous loads shall be the noncontinuous load plus 125% of the continuous load. Additionally, per 220-3(a), the minimum branch-circuit conductor size, without applying any derating provisions, shall have an ampacity which is at least equal to or greater than the noncontinuous load plus 125% of the continuous load.

FEEDER LOADS AND SERVICE-ENTRANCE LOADS

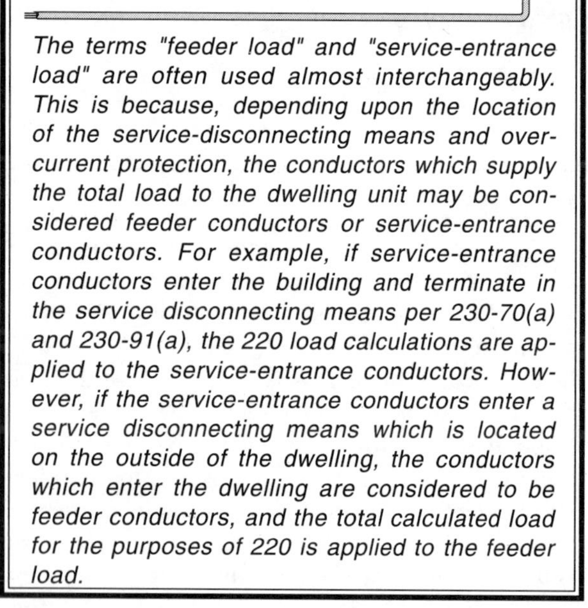

The terms "feeder load" and "service-entrance load" are often used almost interchangeably. This is because, depending upon the location of the service-disconnecting means and overcurrent protection, the conductors which supply the total load to the dwelling unit may be considered feeder conductors or service-entrance conductors. For example, if service-entrance conductors enter the building and terminate in the service disconnecting means per 230-70(a) and 230-91(a), the 220 load calculations are applied to the service-entrance conductors. However, if the service-entrance conductors enter a service disconnecting means which is located on the outside of the dwelling, the conductors which enter the dwelling are considered to be feeder conductors, and the total calculated load for the purposes of 220 is applied to the feeder load.

General Lighting Loads – Table 220-3(b)

The general lighting load for an occupancy is calculated by multiplying the area (in square feet) times the unit load per square foot (in VA). The unit load is taken from Table 220-3(b). Dwelling units, for example, require a minimum of 3 VA per square foot for general lighting load calculations while office buildings require 3½ VA per square foot.

The total area of an occupancy is determined from the outside dimensions of the building or structure. To find the area, multiply the overall outside dimensions. For example, a building with overall outside dimensions of 32′ × 40′ contains 1280 sq ft (32′ × 40′ = 1280 sq ft). When determining the total area for dwelling units, only habitable space is considered. Open porches, garages, or unused or unfinished spaces not suitable for future use are not included in the computation.

Receptacle and Other Loads – 220-3(c)

For general-use receptacle outlets in dwelling units, including one-family, two-family, and multifamily dwellings and in guest rooms of hotels and motels, a footnote to Table 220-3(b) states that no other calculations are required for such outlets which are rated 20 A or less. These outlets are considered part of the general lighting load and are included in the 3 VA per square foot which is listed in Table 220-3(b).

Another footnote to Table 220-3(b) states that for office buildings and banks, 1 VA per square foot shall be added to the 3½ VA per square foot for general-purpose receptacles in these types of occupancies when the actual number of receptacle outlets is unknown. Specific outlets which are not used for general illumination are listed in 220-3(c)(1 – 7).

For example, a sign and lighting outlet is calculated at 1200 VA for each outlet which is required to be installed by 600-5(a). Other receptacle outlets are calculated at 180 VA per outlet. Each single or multiple receptacle on a single strap is calculated at 180 VA. See Figure 13-1.

A *multioutlet assembly* is a metal raceway with factory-installed conductors and attachment plug receptacles. These assemblies can be constructed in the field or they can be factory-assembled. Article 353 contains the provisions for installing these assemblies, but 220 determines how to calculate the load for the assemblies. For other than dwelling units or the guest rooms of motels and hotels, 220-3(c), Ex. 1 permits each 5′, or fraction thereof, of fixed multioutlet assembly of each separate and continuous length, to be considered as one outlet with a calculated load of only 180 VA unless appliances are likely to be used simultaneously. If the likelihood exists that appliances will be used simultaneously, each 1′, or fraction thereof, shall be considered at 180 VA.

For calculations involving show windows, 220-3(c), Ex. 3 permits a calculated load of at least 200 VA to be used for each linear foot of show window, measured horizontally along the window base, to be used instead of the specified unit load per outlet.

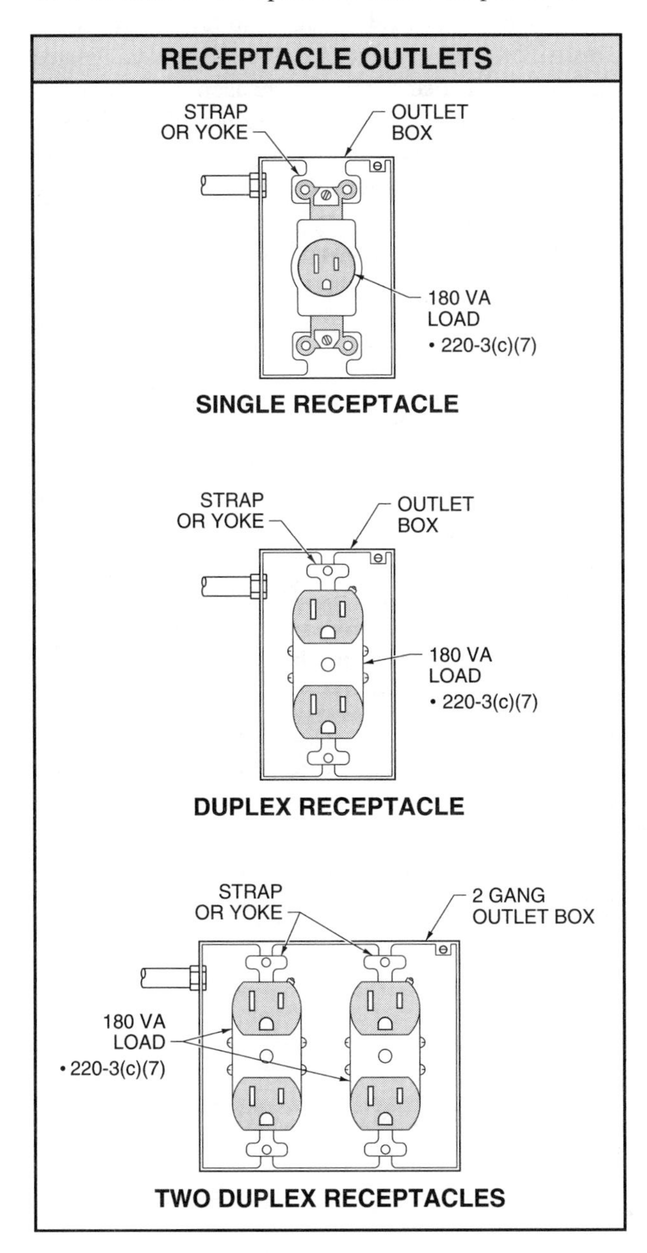

Figure 13-1. Each single or multiple receptacle on a single strap is calculated at 180 VA.

Branch Circuits Required

The minimum number of branch circuits required for lighting and appliances shall be determined from the values obtained from the 220-3 calculations. For specific loads not covered by 220-3, additional branch circuits shall also be provided. In addition to these branch-circuit requirements, 220-4(b) and (c) require separate branch-circuits for small appliances and the laundry circuit. The minimum number of branch circuits is determined from the total computed load. The minimum number changes, depending upon the size or rating of the circuits used.

Whenever possible, the required branch circuits shall be evenly divided across the branch-circuit panelboard. Section 220-4(d) requires that branch-circuit overcurrent devices and circuits need only be provided for connected loads. Future loads need not be considered.

RECEPTACLE LOADS

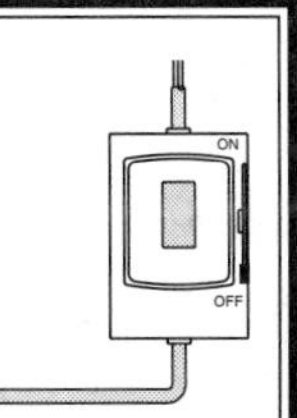

Receptacle loads are treated differently in dwelling units than in other occupancies because of the nature of the connected loads and the use of the receptacles. In dwelling units, 210-52(a) requires that specific rooms have receptacle outlets so that no point on the wall space is more than 6′ from an outlet. This applies regardless of whether there is a load for the receptacle or not. Essentially, these outlets are for general- or convenience-use purposes.

In commercial buildings, however, no such receptacle placement rule exists. The theory for placement of receptacles in commercial buildings is if there is a load to be connected, a receptacle outlet shall be installed. Therefore, each of these receptacle outlets counts as 180 VA. Note that the footnote to Table 220-3(b) does permit a calculation of 1 VA per square foot for office buildings where the actual number of receptacle outlets is unknown. In general, for commercial or industrial installations, a maximum of ten 15 A receptacles may be placed on the same branch circuit ($15\ A \times 120\ V = \frac{1800\ VA}{180\ VA}$ *per receptacle* $= 10$ *receptacles). Likewise, a maximum of thirteen 20 A receptacles may be placed on the same branch circuit* ($20\ A \times 120\ V = \frac{2400\ VA}{180\ VA}$ *per receptacle* $= 13.33 = 13$ *receptacles). For dwelling unit branch circuits, there is no limitation to the number of permitted receptacles.*

ONE-FAMILY DWELLINGS - STANDARD CALCULATION

The Standard Calculation for One-Family Dwellings contains six individual calculations that are performed before the minimum size service or feeder conductors required for the calculated load can be determined. Demand factors may be applied to four of these loads. These six loads, their NEC® references, and those to which demand factors may be applied are:

1. General Lighting – *Table 220-3(b)* (Demand Factors)
2. Fixed Appliances – *220-17* (Demand Factors)
3. Dryer – *220-18; Table 220-18* (Demand Factors)
4. Cooking Equipment – *Table 220-19* (Demand Factors)
5. Heating or A/C – *220-21*
6. Largest Motor – *220-14*

General Lighting – 220-3(b)

The general lighting load for one-family dwellings consists of three separate loads which are calculated individually, added together, and then a demand factor may be applied. The first portion of the general lighting load is calculated from the total square footage of the dwelling. See Figure 13-2.

For multiple-story dwelling units, the total square footage is determined by multiplying the area of one floor by the number of floors. The footnote to Table 220-3(b) indicates that general-use receptacles in one-family dwellings are considered outlets for general illumination and are included in the 3 VA per square foot calculation.

Small Appliances – 220-16(a). The second portion of the general lighting load consists of the calculation for small appliance branch circuits. Small appliance branch circuits supply all receptacle outlets located in the kitchen, pantry, breakfast room, dining room, or similar areas of dwelling units. A minimum of two small appliance branch circuits is required per 210-52 for each dwelling unit.

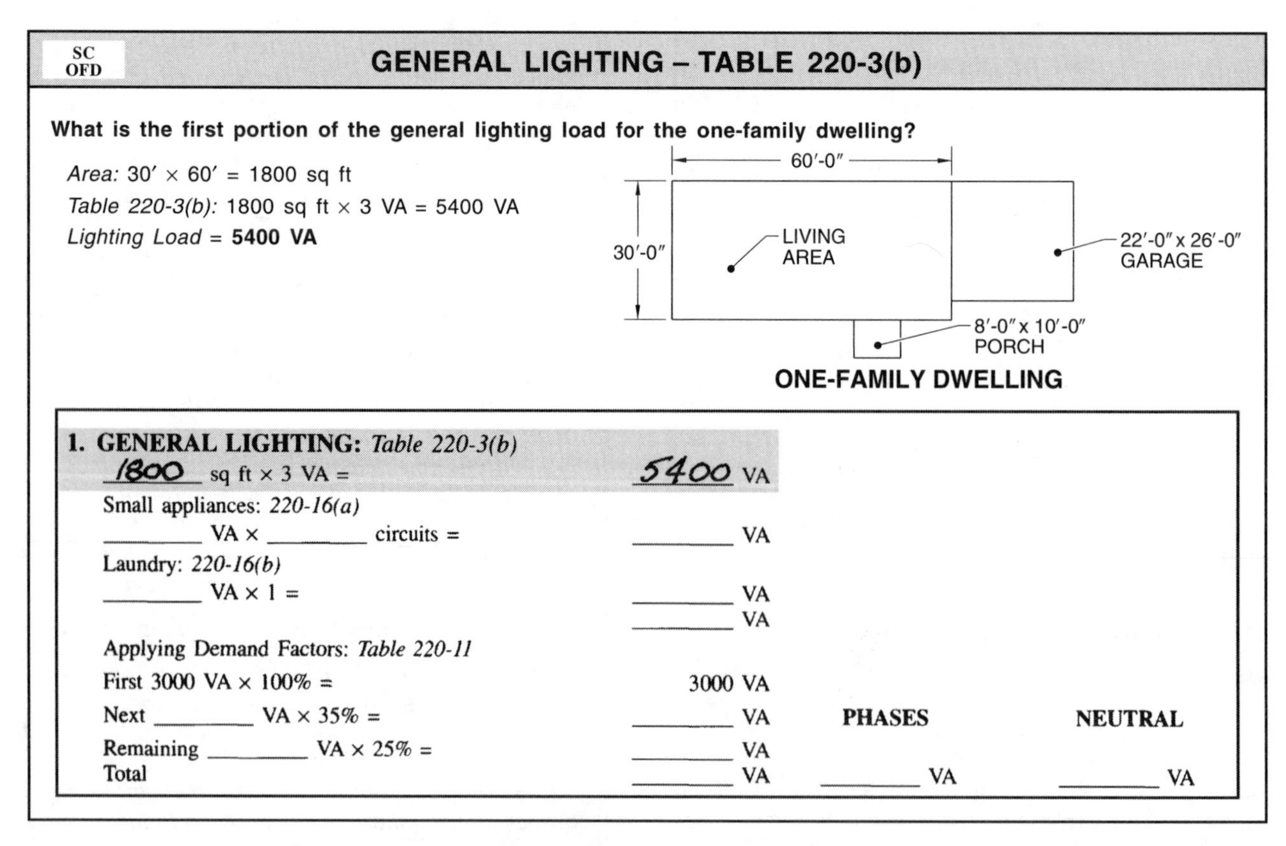

Figure 13-2. The general lighting load for an occupancy is calculated by multiplying the area (in square feet) by the unit load per square foot.

Per 220-16(a), each small appliance branch circuit shall be calculated at 1500 VA when determining the feeder load. If the dwelling has more than two small appliance branch circuits, each circuit shall be computed at 1500 VA. The small appliance load is included with the general lighting load and is permitted to be derated by the demand factors of Table 220-11. See Figure 13-3.

Laundry – 220-16(b). The last portion of the general lighting load consists of the laundry load. Sections 210-52(f) and 220-4(c) require that each dwelling be provided with a 20 A branch circuit to supply laundry receptacle outlet(s). Section 220-16(b) requires that, for the purposes of calculating feeder load, each laundry circuit shall be computed at 1500 VA. A laundry circuit shall always be provided for one-family dwellings. The laundry load is included with the general lighting load and is permitted to be derated by the demand factors of Table 220-11. See Figure 13-4.

General Lighting Load – Demand Factors – Table 220-11. Section 220-11 permits the demand factors of Table 220-11 to be applied to the general lighting load of dwelling units, hospitals, hotels and motels, storage warehouses, and other occupancies. For dwelling units, the first 3000 VA of the general lighting load is computed at 100%. The portion of the total general lighting load from 3001 VA to 120,000 VA is computed at 35%, and the remainder over 120,000 VA is computed at 25%. These demand factors cannot be applied when determining the number of branch circuits for general illumination per 220-4. See Figure 13-5.

Fixed Appliances – 220-17

Four or more appliances which are fastened-in-place in a one-family dwelling are computed at 75% of the total load for all four appliances. This demand factor applies to the nameplate rating of the appliance but does not include electric ranges, clothes dryers, space-heating equipment, or A/C equipment which are served by the same feeder in a one-family, two-family, or multifamily dwelling. See Figure 13-6.

Dryer – 220-18

The total load for household clothes dryers is computed at 5000 VA or the nameplate rating, whichever results in the greater value. Table 220-18 lists demand factors which are applied to household clothes dryers. The first four clothes dryers are computed at 100%. Subsequent dryers are computed per the demand factor percentages listed. See Figure 13-7.

SC OFD

SMALL APPLIANCES – 220-16(a)

What is the small appliance load for a one-family dwelling with three 20 A, 120 V small appliance branch circuits?

220-16(a): 1500 VA × 3 = 4500 VA
Small Appliance Load = **4500 VA**

1. GENERAL LIGHTING: *Table 220-3(b)*
________ sq ft × 3 VA = ________ VA
Small appliances: *220-16(a)*
1500 VA × 3 circuits = 4500 VA
Laundry: *220-16(b)*
________ VA × 1 = ________ VA
________ VA

Applying Demand Factors: *Table 220-11*
First 3000 VA × 100% = 3000 VA
Next ________ VA × 35% = ________ VA — **PHASES** — **NEUTRAL**
Remaining ________ VA × 25% = ________ VA
Total ________ VA — ________ VA — ________ VA

Figure 13-3. Each small appliance branch circuit is calculated at 1500 VA.

SC OFD

LAUNDRY – 220-16(b)

What is the laundry load for a one-family dwelling with one 20 A, 120 V laundry branch circuit?

220-16(b): 1500 VA × 1 = 1500 VA
Laundry Load = **1500 VA**

1. GENERAL LIGHTING: *Table 220-3(b)*

________ sq ft × 3 VA =	________ VA		
Small appliances: *220-16(a)*			
________ VA × ________ circuits =	________ VA		
Laundry: *220-16(b)*			
1500 VA × 1 =	1500 VA		
	________ VA		
Applying Demand Factors: *Table 220-11*			
First 3000 VA × 100% =	3000 VA		
Next ________ VA × 35% =	________ VA	**PHASES**	**NEUTRAL**
Remaining ________ VA × 25% =	________ VA		
Total	________ VA	________ VA	________ VA

Figure 13-4. Each one-family dwelling shall have a 1500 VA laundry branch circuit.

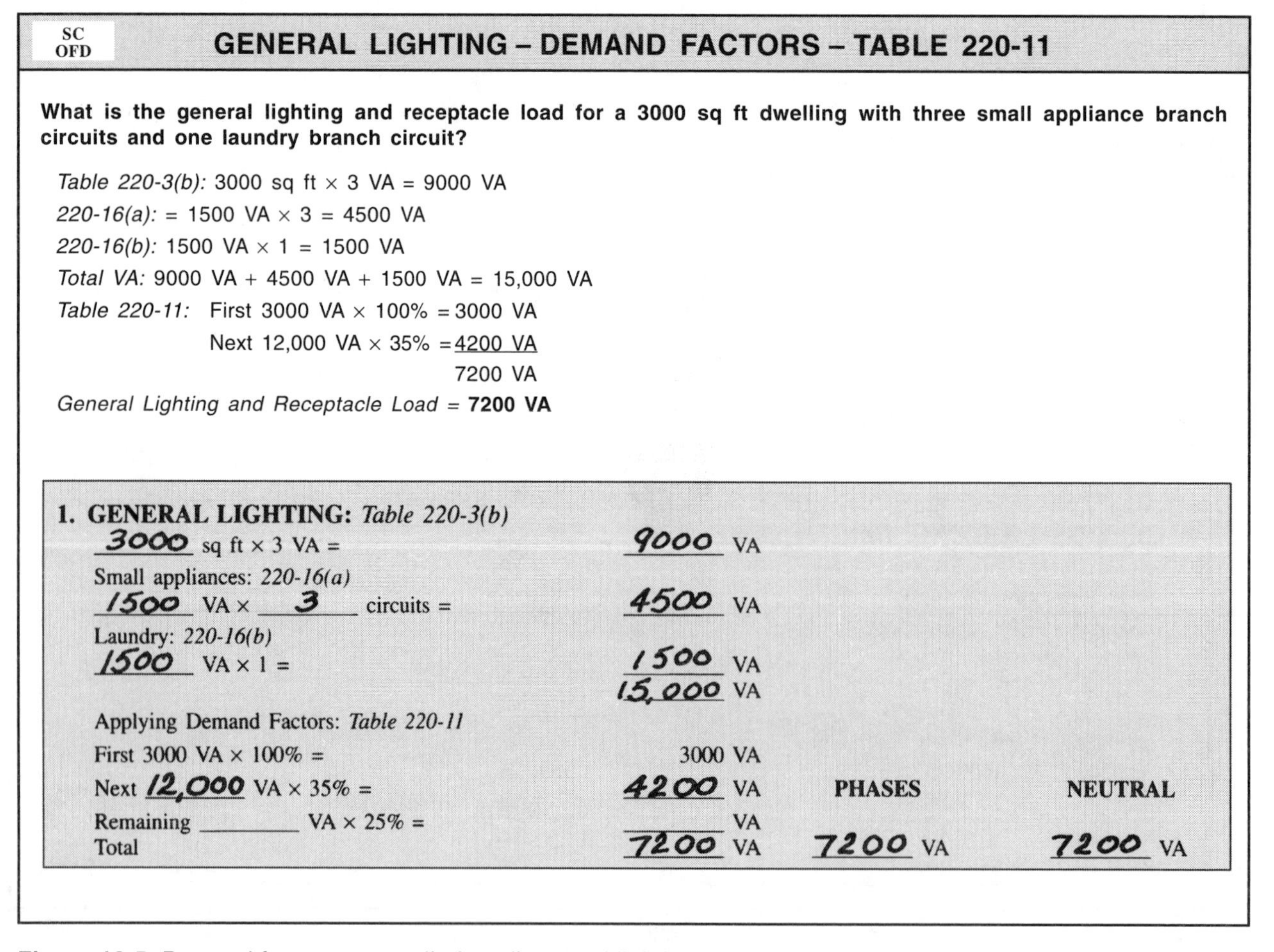

SC OFD

GENERAL LIGHTING – DEMAND FACTORS – TABLE 220-11

What is the general lighting and receptacle load for a 3000 sq ft dwelling with three small appliance branch circuits and one laundry branch circuit?

Table 220-3(b): 3000 sq ft × 3 VA = 9000 VA
220-16(a): = 1500 VA × 3 = 4500 VA
220-16(b): 1500 VA × 1 = 1500 VA
Total VA: 9000 VA + 4500 VA + 1500 VA = 15,000 VA
Table 220-11: First 3000 VA × 100% = 3000 VA
Next 12,000 VA × 35% = 4200 VA
7200 VA
General Lighting and Receptacle Load = **7200 VA**

1. GENERAL LIGHTING: *Table 220-3(b)*

3000 sq ft × 3 VA =	9000 VA		
Small appliances: *220-16(a)*			
1500 VA × 3 circuits =	4500 VA		
Laundry: *220-16(b)*			
1500 VA × 1 =	1500 VA		
	15,000 VA		
Applying Demand Factors: *Table 220-11*			
First 3000 VA × 100% =	3000 VA		
Next 12,000 VA × 35% =	4200 VA	**PHASES**	**NEUTRAL**
Remaining ________ VA × 25% =	________ VA		
Total	7200 VA	7200 VA	7200 VA

Figure 13-5. Demand factors are applied to all general lighting loads of 3001 VA and greater for dwelling units.

SC
OFD

FIXED APPLIANCES – 220-17

What is the fixed appliance load (line and neutral) for a one-family dwelling with the following 120 V appliances?

Water heater – 7500 VA
Dishwasher – 1650 VA
Garbage disposer – 950 VA
Trash compactor – 700 VA
Attic fan – 1400 VA

220-17: 7500 VA + 1650 VA + 950 VA + 700 VA + 1400 VA = 12,200 VA

12,200 VA × 75% = 9150 VA

Fixed Appliance Load: Line = **9150 VA**

Neutral = **9150 VA**

2. FIXED APPLIANCES: *220-17*

Dishwasher =	1650 VA			
Disposer =	950 VA			
Compactor =	700 VA			
Water heater =	7500 VA			
Attic fan =	1400 VA			
=	VA			
=	VA			(120 V Loads × 75%)
Total	12,200 VA	× 75% = 9150 VA	9150 VA	9150 VA

Figure 13-6. Four or more fixed appliances in a one-family dwelling are computed at 75% of the total load for the four appliances.

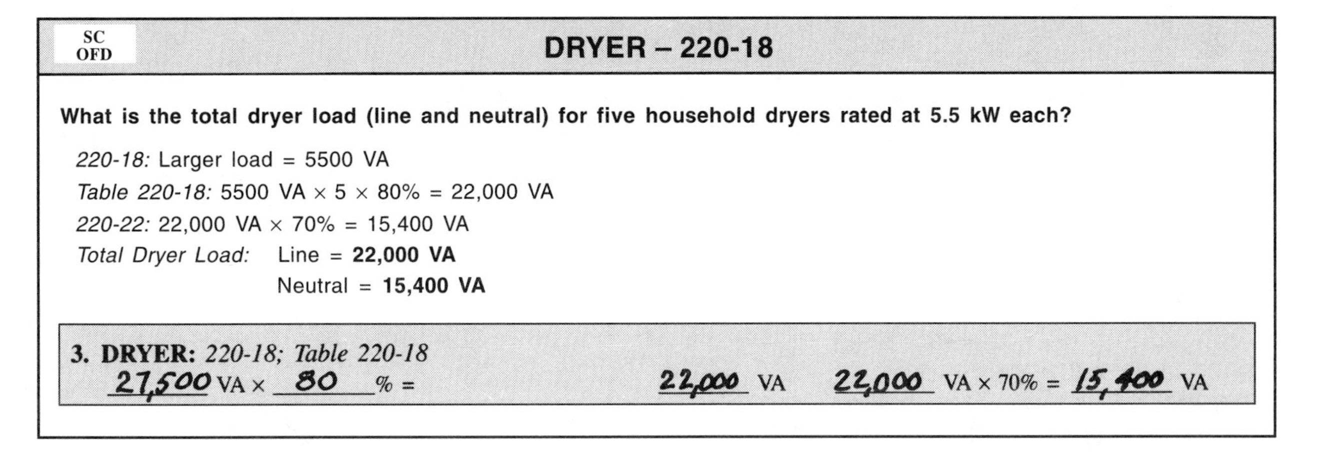

Figure 13-7. For household dryers, the total load is computed at 5000 VA or the nameplate rating, whichever is greater.

Cooking Equipment – Table 220-19; Notes

Household electric ranges rated in excess of 1¾ kW are computed per Table 220-19. This table consists of three demand columns and five footnotes. Column A is used for household electric ranges greater than 8¾ kW, but less than 12 kW. Single ranges up to 12 kW are permitted by Column A to be computed at 8 kW. Column A values are provided in kW while the other columns provide a demand factor percentage. See Figure 13-8.

SC OFD

COOKING EQUIPMENT – TABLE 220-19, COL A

What is the range demand load for three household electric ranges rated at 9.5 kW each?

Table 220-19; Col A, Notes: 28,000 VA demand = 14,000 VA
220-22: 14,000 VA × 70% = 9800 VA
Range Demand Load: Line = **14,000 VA**
Neutral = **9,800 VA**

4. COOKING EQUIPMENT: *Table 220-19; Notes*

Col A 28,500 VA × ______ % = 14,000 VA
Col B ______ VA × ______ % = ______ VA
Col C ______ VA × ______ % = ______ VA
Total 14,000 VA 14,000 VA × 70% = 9800 VA

Figure 13-8. Household ranges are computed per the values of Table 220-19.

Table 220-19, Note 1 applies to individual ranges rated over 12 kW, but not more than 27 kW which are all of the same rating. When more than one of these ranges is present in a one-family dwelling, the maximum demand in Column A shall be increased by 5% for each additional kW, or major fraction thereof, by which the rating of the individual range exceeds 12 kW. For the purposes of these calculations, a major fraction is considered to be .5 or greater. See Figure 13-9.

Note 2 also applies to Column A, but is used for applications in which individual ranges are rated over 8¾ kW through 27 kW and are of unequal ratings. In this case, an average value for all of the ranges of unequal values is determined by adding all of the individual ranges, using 12 kW for any range rated at less than 12 kW, and dividing by the number of ranges. The Column A demand is then increased 5% for each kW, or major fraction thereof, by which the average value exceeds 12 kW. See Figure 13-10.

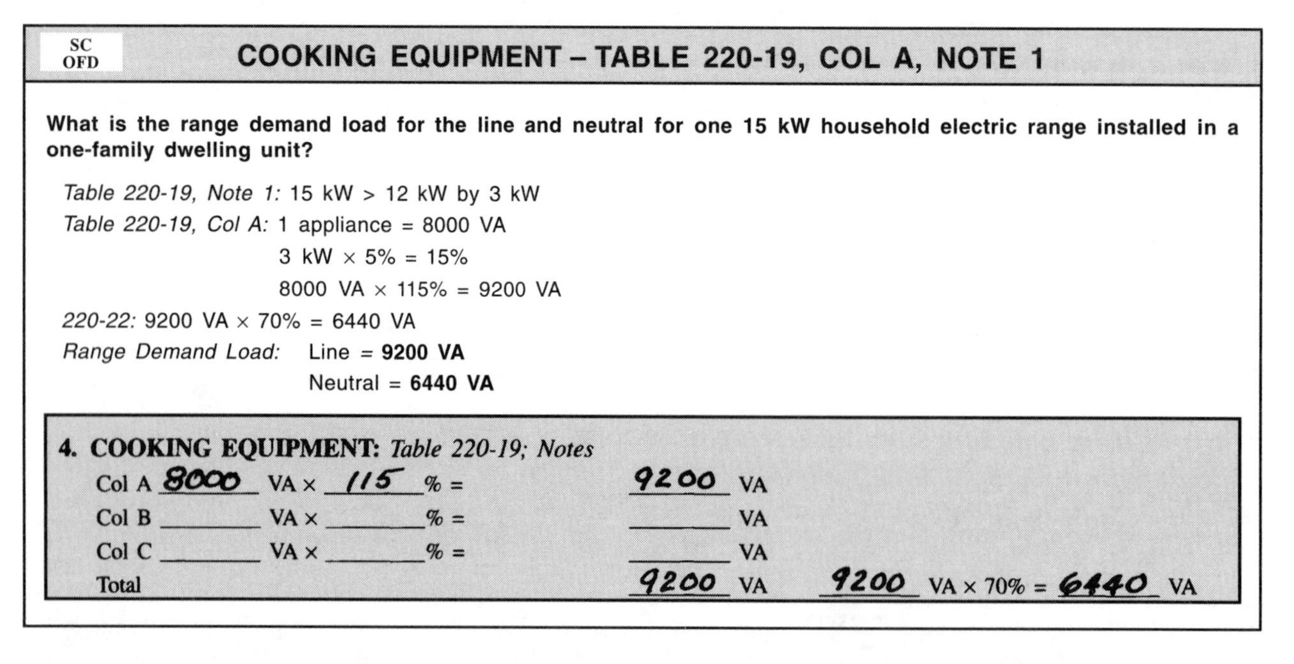

SC OFD

COOKING EQUIPMENT – TABLE 220-19, COL A, NOTE 1

What is the range demand load for the line and neutral for one 15 kW household electric range installed in a one-family dwelling unit?

Table 220-19, Note 1: 15 kW > 12 kW by 3 kW
Table 220-19, Col A: 1 appliance = 8000 VA
3 kW × 5% = 15%
8000 VA × 115% = 9200 VA
220-22: 9200 VA × 70% = 6440 VA
Range Demand Load: Line = **9200 VA**
Neutral = **6440 VA**

4. COOKING EQUIPMENT: *Table 220-19; Notes*

Col A 8000 VA × 115 % = 9200 VA
Col B ______ VA × ______ % = ______ VA
Col C ______ VA × ______ % = ______ VA
Total 9200 VA 9200 VA × 70% = 6440 VA

Figure 13-9. Per Note 1, the maximum demand for individual ranges in Table 220-19, Column A is increased by 5% for each additional kW, or major fraction thereof, by which 12 kW is exceeded through 27 kW.

SC OFD

COOKING EQUIPMENT – TABLE 220-19, COL A, NOTE 2

What is the range demand load for the line and neutral for three household electric ranges with nameplate ratings of 12 kW, 16 kW, and 20 kW?

Table 220-19, Note 2: 12 kW + 16 kW + 20 kW = $\frac{48\text{ kW}}{3}$ = 16 kW

16 kW > 12 kW by 4 kW

Table 220-19, Col A: 4 kW × 5% = 20%

Demand = 14 kW

14,000 VA × 120% = 16,800 VA

220-22: 16,800 VA × 70% = 11,760 VA

Range Demand Load: Line = **16,800 VA**

Neutral = **11,760 VA**

4. COOKING EQUIPMENT: *Table 220-19; Notes*

Col A	14,000 VA ×	120 % =	16,800 VA		
Col B	______ VA ×	______ % =	______ VA		
Col C	______ VA ×	______ % =	______ VA		
Total			16,800 VA	16,800 VA × 70% =	11,760 VA

Figure 13-10. Per Note 2, the maximum demand for ranges of unequal values is increased by 5% of the average kW exceeding 12 kW through 27 kW, or major fraction thereof.

Note 3 applies to Column A when individual ranges rated over 1¾ kW through 8¾ kW are used. While Column A could be used for these ranges, Columns B or C provides smaller kW demands. Note 3 permits the adding together of the nameplate ratings of all household cooking appliances rated between 1¾ kW and 8¾ kW and multiplying the sum of these cooking appliances by the appropriate demand from either Column B or Column C. If the cooking appliances are of different ratings that fall under both Columns B and C, the demand factors for each column shall be applied to the appliance for that column and then all of the results added together. See Figure 13-11.

Note 4 applies to branch-circuit loads and permits the load to be calculated per Table 220-19. For single wall-mounted ovens or single counter-mounted cooking units, the branch-circuit load shall be the nameplate rating of the appliance. For installations from a common branch-circuit in which a single counter-mounted cooking unit is installed with not more than two wall-mounted ovens, Note 4 permits all of the cooking appliances to be added together and treated like a single range, provided all of the cooking appliances are located in the same room. See Figure 13-12.

Note 5 is an informational note. It permits the use of Table 220-19 for calculating the total demand for household ranges rated over 1¾ kW which are used in instructional programs. Such installations are typically found in schools where cooking skills are taught.

Heating or A/C – 220-21

Heating or A/C loads often represent the largest loads in one-family dwellings. Fortunately, these are noncoincidental loads. *Noncoincidental loads* are loads that are not on at the same time. Section 220-21 permits the smaller of the two noncoincidental loads to be omitted from the load calculation, provided it is unlikely that the loads will be in use simultaneously. Heating and A/C loads are both calculated at 100% of the nameplate rating, and the one with the largest load is used in determining the total feeder or service demand. See Figure 13-13.

SC OFD

COOKING EQUIPMENT – TABLE 220-19, COL B AND C, NOTE 3

What is the range demand load for the line and neutral for four household electric ranges with nameplate ratings of 4 kW, 4 kW, 6 kW, and 6 kW?

Table 220-19, Note 3: 4 kW + 4 kW + 6 kW + 6 kW = 20 kW
Table 220-19, Col C: 4 appliances = 50%
20 kW × 50% = 10 kW = 10,000 VA
220-22: 10,000 kW × 70% = 7000 VA
Range Demand Load: Line = **10,000 VA**
Neutral = **7000 VA**

4. COOKING EQUIPMENT: *Table 220-19; Notes*

Col A	____ VA ×	____ % =	____ VA		
Col B	____ VA ×	____ % =	____ VA		
Col C	20,000 VA ×	50 % =	10,000 VA		
Total			10,000 VA	10,000 VA × 70% =	7000 VA

Figure 13-11. Per Note 3, nameplate ratings between 1¾ kW and 8¾ kW are added and then multiplied by the appropriate demand from Columns B or C.

SC OFD

COOKING EQUIPMENT – TABLE 220-19, NOTE 4

What is the branch circuit line and neutral load for one 6 kW oven and one 8 kW cooktop installed on the same branch circuit? The oven and cooktop are located in the same room.

Table 220-19, Note 4: 6 kW + 8 kW = 14 kW
Treat as one range
Table 220-19, Note 1: 14 kW − 12 kW = 2 kW
2 kW × 5% = 10%
Table 220-19, Col A: One range = 8 kW
8 kW × 110% = 8800 VA
220-22: 8800 VA × 70% = 6160 VA
Branch-Circuit Load: Line = **8800 VA**
Neutral = **6160 VA**

4. COOKING EQUIPMENT: *Table 220-19; Notes*

Col A	8000 VA ×	110 % =	8800 VA		
Col B	____ VA ×	____ % =	____ VA		
Col C	____ VA ×	____ % =	____ VA		
Total			8800 VA	8800 VA × 70% =	6160 VA

Figure 13-12. Per Note 4, cooking appliance ratings are added together and treated like a single range when located in the same room.

SC OFD	HEATING OR A/C – 220-21

What is the total heating and A/C load (line and neutral) for a one-family dwelling unit with a 240 V central heating load of 16,000 VA and an A/C load of 12,000 VA?

220-21: Largest load = 16,000 VA

Heating and A/C Load: Line = **16,000 VA**

Neutral = **0 VA**

5. HEATING or A/C: *220-21*

Heating unit = 16,000 VA × 100% =	16,000 VA		
A/C unit = 12,000 VA × 100% =	— VA		
Heat pump = ______ VA × 100% =	______ VA		
Largest Load	16,000 VA	16,000 VA	______ VA

Figure 13-13. The largest of the heating or A/C load is used.

Largest Motor – 220-14

All household motor loads are calculated per 430. Feeders for several motors (430-24) are calculated using the sum of the full-load table currents of all motors plus 25% of the largest motor. The motor full-load current ratings are calculated under the fixed appliance portion of the calculation, per 220-17, at 100%. The largest motor portion of the calculation allows for the 25% that is not covered under 220-17. See Figure 13-14.

Demand Factors – Neutral – 220-22

The *feeder neutral load* is the maximum unbalance between any of the ungrounded conductors and the grounded conductor. Section 220-22, however, permits several appliances to have the neutral demand calculated at 70% of the ungrounded conductors. In addition to the neutral demand for household cooking appliances and clothes dryers, an additional deduction of 70% is permitted for that portion of the neutral conductor demand in excess of 200 A. This reduction only applies to 3-wire DC, 1φ AC, 3φ 4-wire, and 2φ systems. This 70% reduction does not apply to portions of the load that consist of nonlinear loads when the supply system is 3φ, 4-wire, wye-connected or 3φ, 4-wire, wye-connected utilizing two phase wires and the neutral. A nonlinear load is a load where the current wave shape does not follow the applied voltage wave shape. See Figure 13-15.

DETERMINING LARGEST MOTOR LOAD

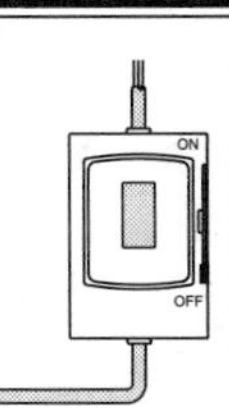

It is good practice to check with the AHJ prior to sizing service-entrance or feeder conductors for a new dwelling. Different areas may interpret differently or require calculations to be performed in a specific manner. This is often the case in determining the largest motor load. Some areas use all of the connected loads in determining the largest motor load. In these cases, the A/C compressor motor is frequently the largest motor load. In other areas, the A/C load is not considered a motor load. And still in other areas, the A/C load only counts if it is not used when determining the largest of the heating or A/C loads per 220-21. In any event, it is good practice to know how the AHJ determines the largest motor load.

Conductor Ampacity

After the six loads (General Lighting, Fixed Appliances, Dryer, Cooking Equipment, Heating or A/C, and Largest Motor) are individually calculated, the results of the individual calculations are added to determine the maximum load, in VA, for both the ungrounded and grounded conductors. This total VA value is used to determine the minimum ampacity for the ungrounded and grounded conductors.

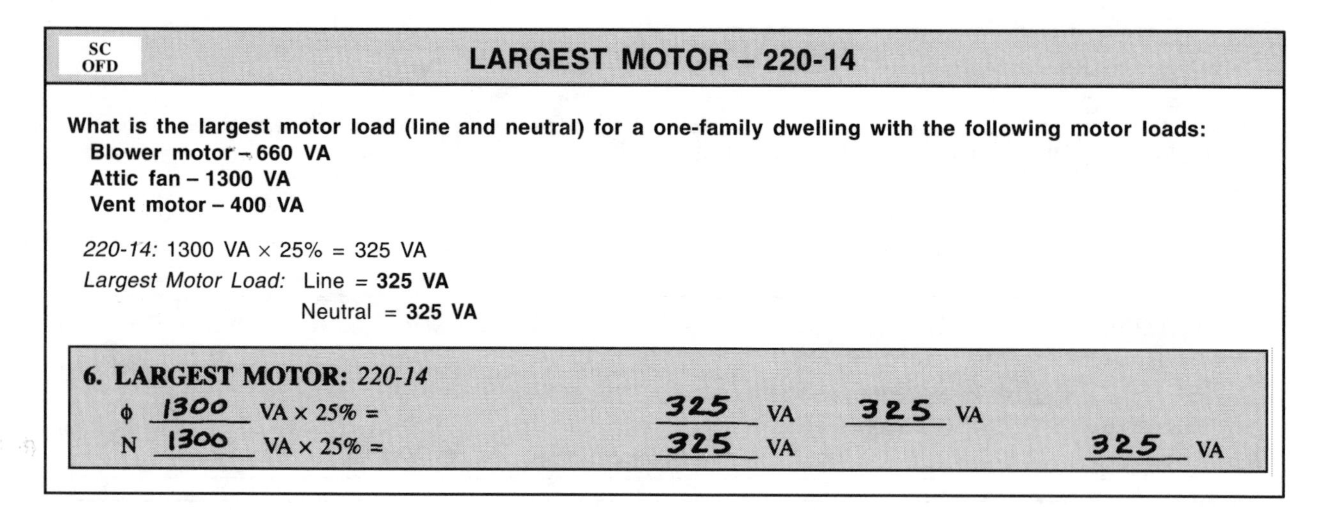

SC OFD

LARGEST MOTOR – 220-14

What is the largest motor load (line and neutral) for a one-family dwelling with the following motor loads:
Blower motor – 660 VA
Attic fan – 1300 VA
Vent motor – 400 VA

220-14: 1300 VA × 25% = 325 VA
Largest Motor Load: Line = **325 VA**
Neutral = **325 VA**

6. LARGEST MOTOR: *220-14*
ϕ 1300 VA × 25% = 325 VA 325 VA
N 1300 VA × 25% = 325 VA 325 VA

Figure 13-14. The largest motor load is multiplied by 25%.

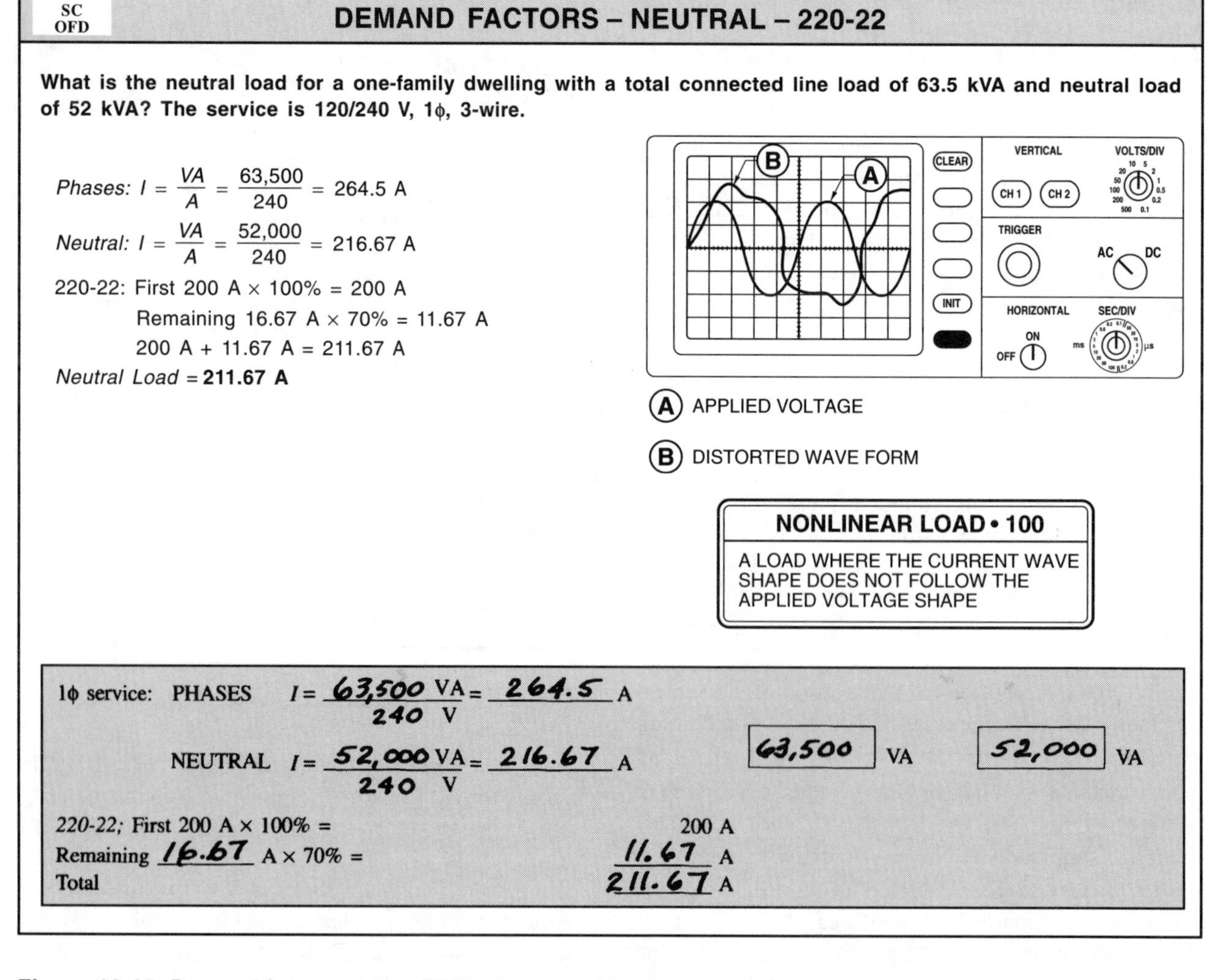

SC OFD

DEMAND FACTORS – NEUTRAL – 220-22

What is the neutral load for a one-family dwelling with a total connected line load of 63.5 kVA and neutral load of 52 kVA? The service is 120/240 V, 1ϕ, 3-wire.

Phases: $I = \frac{VA}{A} = \frac{63,500}{240} = 264.5$ A

Neutral: $I = \frac{VA}{A} = \frac{52,000}{240} = 216.67$ A

220-22: First 200 A × 100% = 200 A
Remaining 16.67 A × 70% = 11.67 A
200 A + 11.67 A = 211.67 A
Neutral Load = **211.67 A**

(A) APPLIED VOLTAGE

(B) DISTORTED WAVE FORM

NONLINEAR LOAD • 100
A LOAD WHERE THE CURRENT WAVE SHAPE DOES NOT FOLLOW THE APPLIED VOLTAGE SHAPE

1ϕ service: PHASES $I = \frac{63,500 \text{ VA}}{240 \text{ V}} = 264.5$ A

NEUTRAL $I = \frac{52,000 \text{ VA}}{240 \text{ V}} = 216.67$ A

63,500 VA 52,000 VA

220-22; First 200 A × 100% = 200 A
Remaining 16.67 A × 70% = 11.67 A
Total 211.67 A

Figure 13-15. Demand factors are applied to the neutral load.

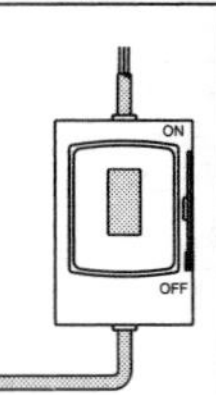

SIZING GROUNDED CONDUCTORS

The standard neutral calculation in 220 shall be performed to determine the minimum size of the grounded conductor for a one-family dwelling. Once the total load on the neutral is calculated, the ampacity of the grounded conductor is computed. The minimum size of the grounded conductor is selected using the ampacity value and Table 310-16.

It is interesting to note that during the 1996 NEC®, Note 3 was revised as it relates to sizing of the grounded conductor. Prior to the 1996 NEC®, the minimum size of the grounded conductor could be no more than two AWG sizes smaller than the ungrounded conductors. This limitation is removed in the 1996 NEC®. The grounded conductor can be any number of sizes smaller than the ungrounded conductor provided the requirements of 215-2, 220-22, and 230-42 are met.

For 1ϕ systems, the formula $I = \frac{VA}{V}$ is used to determine minimum ampacity. In this formula, I = amperes (A), VA = volt amperes, and V = voltage. After the minimum ampacity has been determined, Table 310-16 and the applicable notes are used to calculate the minimum conductor size for the feeder or service-entrance conductors.

Ungrounded Conductors. The ungrounded phase conductors are not subject to any additional demand factors. Once the phase amperes are calculated by using the formula, the size of the phase conductors is determined. Table 310-16 is used to size the ungrounded conductors. In addition, Note 3 of Notes to Ampacity Tables of 0 to 2000 Volts is applicable to 1ϕ service conductors and feeders for dwelling units that are supplied from 120/240 V, 1ϕ, 3-wire systems. Note 3 permits reduced conductor sizes for service or feeder ratings from 100 A to 400 A. See Figure 13-16.

Grounded Conductors – 220-22. An additional demand reduction of 70% is applied to the feeder or service-entrance grounded conductors. This only applies to 3-wire DC, 1ϕ AC, 3ϕ 4-wire, and 2ϕ systems that supply other than nonlinear loads. Once the total neutral load is calculated using $I = \frac{VA}{V}$, the demand factor is applied. The first 200 A is taken at 100% and the remainder is taken at 70%.

The minimum size of the neutral conductor is now selected from Table 310-16. Note 3 of Notes to Ampacity Tables of 0 to 2000 Volts permits the grounded conductor (neutral) to be smaller than the ungrounded conductors provided the requirements of 215-2, 220-22, and 230-42 are met. See Figure 13-17.

ONE-FAMILY DWELLING – OPTIONAL CALCULATION – 220-30

The Optional Calculation is designed for dwellings with large electric loads. It is an easier method than the Standard Calculation for calculating feeder and service loads for dwellings. The Optional Calculation is permitted for the ungrounded conductors of 120/240 V, 3-wire or 120/208 V service-entrance or feeder conductors with an ampacity of at least 100 A.

SIZING UNGROUNDED CONDUCTORS – NOTE 3 OF NOTES TO AMPACITY TABLES OF 0 TO 2000 VOLTS

What is the minimum size THW, Cu conductors required for a one-family dwelling with a calculated load of 200 A? The service is 120/240 V, 1ϕ, 3-wire.

Table 310-16: 200 A = No. 3/0
Note 3: 200 A = No. 2/0
Conductors: **No. 2/0 THW Cu**

Figure 13-16. Note 3 permits reduced conductor sizes for service or feeder ratings from 100 A to 400 A.

SIZING GROUNDED CONDUCTOR – 220-22; TABLE 310-16

What is the minimum size THW, Cu conductor required for the grounded conductor for a one-family dwelling with a total calculated neutral load of 225 A? The service is 120/240 V, 1φ, 3-wire.

220-22: 200 A × 100% = 200 A
25 A × 70% = 17.5 A
200 A + 17.5 A = 217.5 A
Table 310-16: 217.5 A = No. 4/0
Grounded Conductor: **No. 4/0 THW Cu**

Figure 13-17. An additional demand factor of 70% is applied to the grounded conductor.

The grounded conductor calculations for the Optional Calculation is determined per 220-22. The Optional Calculation for dwelling units consists of two groups of loads. These two groups of loads and their NEC® references are:

1. Heating or A/C – *Table 220-30(1 – 5)*
2. Other Loads – *220-30(b)*

The Optional Calculation can be used for large-capacity dwelling units and permits designers of electrical systems to take advantage of the increased diversity and demand factors with these types of dwellings and utilize smaller service-entrance or feeder conductors.

Heating or A/C – Table 220-30(1 – 5)

Table 220-30 lists five heating and A/C load selections which are considered in the Optional Calculation for one-family dwellings. The largest of these selections shall be used in determining the total feeder or service load. See Figure 13-18. These five loads are:

(1) A/C equipment including heat pump compressor (at 100% of nameplate ratings).

(2) Electric thermal storage loads (at 100% of nameplate ratings) where the usual load is expected to be continuous at the full nameplate value.

(3) Central electric space-heating equipment, including integral supplemental heat for heat pumps (at 65% of nameplate ratings).

(4) Less than four separately-controlled electric space-heating units (at 65% of nameplate ratings).

(5) Four or more separately-controlled electric space-heating units (at 40% of nameplate ratings) considered.

OC
OFD

HEATING OR A/C – TABLE 220-30

What is the heating or A/C load for a one-family dwelling with a 14 kW central electric space-heating load and a 10 kW A/C load? The service is 120/240 V, 1φ, 3-wire with a calculated ampacity of 150 A.

Table 220-30: 10,000 VA × 100% = 10,000 VA
14,000 VA × 65% = 9100 VA
Heating or A/C Load = **10,000 VA**

1. HEATING or A/C: *Table 220-30(1–5)*

Heating units (3 or less) = 14,000 VA × 65% =	9100 VA		
Heating units (4 or more) = _____ VA × 40% =	_____ VA		
A/C unit = 10,000 VA × 100% =	10,000 VA		
Heat pump = _____ VA × 100% =	_____ VA	**PHASES**	
Largest Load	10,000 VA		
Total	10,000 VA	10,000 VA	

Figure 13-18. The largest of five load selections is the "Heating or A/C Load" for the Optional Calculation.

Other Loads – 220-30(b)

This portion of the Optional Calculation is also included in Table 220-30. In addition to the largest of the heating or A/C load, 100% of the first 10 kVA of all other loads and 40% of the remainder of all other loads shall be included. Four loads shall be added together to constitute the "other loads" specified by Table 220-30. The other loads include: (1) 1500 VA for each 2-wire, 20 A small appliance branch circuit, (2) 3 VA per square foot for the general lighting and general-use receptacle load, (3) the nameplate rating of all fastened-in-place, permanently-connected appliances, and (4) the nameplate rating, in either amperes or kVA, of all motor and other low power factor loads. See Figure 13-19.

OC
OFD

OTHER LOADS – TABLE 220-30; 220-30(b)

What is the total "other loads" for a 2500 sq ft one-family dwelling with the following loads:

2 – Small appliance 20 A branch circuits
1 – 8 kW electric range
1 – 2 kW dishwasher
1 – 5 kW clothes dryer
1 – 1500 VA blower motor

220-30(b): (1) Small appliance load = 3000 VA
(2) General lighting load
2500 sq ft × 3 VA = 7500 VA
(3) Fixed appliance load = 15,000 VA
(4) Motor load = 1500 VA
Total load = 27,000 VA

Table 220-30: 10,000 VA × 100% = 10,000 VA
17,000 VA × 40% = 6800 VA
Total = 16,800 VA

Other Loads = **16,800 VA**

2. OTHER LOADS: *220-30(b)*

General lighting: *220-30(b)(2)*
2500 sq ft × 3 VA — 7500 VA

Small appliance and laundry loads: *220-30(b)(1)*
1500 VA × 2 circuits = — 3000 VA

Special loads: *220-30(b)(3)(4)*
Dishwasher = 2000 VA
Disposer = ______ VA
Compactor = ______ VA
Water heater = ______ VA
Range = 8000 VA
Dryer = 5000 VA
Blower motor = 1500 VA
______ = ______ VA
______ = ______ VA
16,500 VA — 16,500 VA

Total 27,000 VA

Applying Demand Factors: *Table 220-30*
First 10,000 VA × 100% = 10,000 VA
Remaining 17,000 VA × 40% = 6800 VA
Total 16,800 VA — 16,800 VA

Figure 13-19. Four loads constitute the "Other Loads" for the Optional Calculation.

MULTIFAMILY DWELLINGS – STANDARD CALCULATION

The Standard Calculation for multifamily dwellings uses the same six loads as used by the Standard Calculation for one-family dwellings. The only difference is that for each of the individual loads, the computed load is multiplied by the total number of units in the multifamily dwelling to arrive at the total connected load. As with the one-family dwelling calculation, demand factors are applied.

General Lighting Load – Table 220-3(b)

The general lighting load for multifamily dwellings consists of the same three loads which were used in the one-family dwelling standard calculation. The first portion of the general lighting load is calculated from the total square footage of the dwelling. Table 220-3(b) lists the unit load per square foot for dwelling units at 3 VA. For this portion of the lighting load, the total square footage of the dwelling, excluding open porches, garages, and unused or unfinished spaces is multiplied by 3 VA to determine the calculated load for one dwelling. This value is then multiplied by the number of dwelling units in the multifamily dwelling to determine the first portion of the general lighting load.

For multifamily dwellings in which the square footage of the individual units is different, the square footage of each unit is calculated separately and added together to get the total square footage of the multifamily dwelling. This value, multiplied by 3 VA, provides the first portion of the total general lighting load. The footnote to Table 220-3(b) indicates that general-use receptacles in multifamily as well as one-family dwellings are considered outlets for general illumination and are included in the 3 VA per square foot calculation. See Figure 13-20.

SC MFD

GENERAL LIGHTING – TABLE 220-3(b)

What is the first portion of the general lighting load for a multifamily dwelling consisting of three individual dwelling units with the following total areas:

Unit No. 1 – 1050 sq ft
Unit No. 2 – 950 sq ft
Unit No. 3 – 1075 sq ft

Table 220-3(b):	1050 sq ft × 3 VA	=	3150 VA
	950 sq ft × 3 VA	=	2850 VA
	1075 sq ft × 3 VA	=	3225 VA
	Total	=	9225 VA

Lighting Load = **9225 VA**

1. GENERAL LIGHTING: *Table 220-3(b)*

1050 sq ft × 3 VA × 1 units = 3150 VA
950 sq ft × 3 VA × 1 units = 2850 VA
1075 sq ft × 3 VA × 1 units = 3225 VA

Small appliances: *220-16(a)*
____ VA × ____ circuits × ____ units = ____ VA

Laundry: *220-16(b)*
____ VA × 1 × ____ units = ____ VA
____ VA

Applying Demand Factors: *Table 220-11*
First 3000 VA × 100% = 3000 VA
Next ____ VA × 35% = ____ VA
Remaining ____ VA × 25% = ____ VA
Total ____ VA

	PHASES	NEUTRAL
Total	____ VA	____ VA

Figure 13-20. The general lighting load for an occupancy is calculated by multiplying the area (in square feet) by the unit load per square foot.

Small Appliances – 220-16(a). As with one-family dwellings, a minimum of two small appliance branch circuits are required by 210-52 for each multifamily dwelling unit. Per 220-16(a), each small appliance branch circuit shall be calculated at 1500 VA when determining the feeder load. If the dwelling has more than two small appliance branch circuits, each circuit is computed at 1500 VA.

For multifamily dwellings, the small appliance load is calculated by multiplying 1500 VA times the number of small appliance branch circuits times the number of dwelling units in the multifamily dwelling. The small appliance load is included with the general lighting load and is subject to the demand factors of Table 220-11.

Laundry Load – 220-16(b). The last portion of the general lighting load for multifamily dwellings consists of the laundry circuit load. Per 210-52(f) and 220-4(c), each dwelling shall be provided with a 20 A branch circuit to supply laundry receptacle outlet(s). There are, however, two exceptions which are applicable to multifamily dwellings. Per 210-52(f), Ex. 1, a laundry receptacle may be omitted from dwellings which are part of a multifamily dwelling with separate laundry facilities accessible to all building occupants. Per 210-52(f), Ex. 2, the laundry receptacle may be omitted in other than one-family dwellings, provided laundry facilities are not installed or permitted. In either of these cases, the 1500 VA for the laundry circuit need not be included in the multifamily standard calculation. If, however, each dwelling unit is to have laundry facility provisions, the load shall be added to the total connected load of the multifamily dwelling. This is accomplished by computing 1500 VA for each dwelling unit multiplied by the number of dwelling units in the multifamily dwelling. As with the one-family Standard Calculation, the laundry load is included with the general lighting load and is subject to the demand factors of Table 220-11.

General Lighting Load – Demand Factors – Table 220-11. The demand factors of Table 220-11 are applied to the general lighting load of multifamily dwelling units. The first 3000 VA of the lighting load is computed at 100%. The portion of the general lighting load from 3001 VA to 120,000 VA is computed at 35%, and the remainder over 120,000 VA is computed at 25%. These demand factors cannot be applied when determining the number of branch circuits for general illumination per 220-4. See Figure 13-21.

Fixed Appliances – 220-17

Each fixed appliance in multifamily dwellings is calculated using the nameplate rating. For four or more appliances, the nameplate ratings are added together and multiplied by the 75% demand factor specified in 220-17. As with one-family dwellings, this demand factor applies to the nameplate rating of the appliance but does not include electric ranges, clothes dryers, space-heating equipment, or A/C equipment served by the same feeder. See Figure 13-22.

Dryer – 220-18; Table 220-18

The total load for household electric clothes dryers is computed at 5000 VA or the nameplate rating, whichever results in the greater value. Table 220-18 lists demand factors which are applied to clothes dryers. The first four clothes dryers are computed at 100%. Subsequent numbers of dryers are computed with decreasing demand factor percentages. See Figure 13-23.

Cooking Equipment – Table 220-19, Notes

Household electric ranges rated in excess of 1¾ kW shall be computed per Table 220-19. Column A values are used less frequently for multifamily dwellings since the size of the individual ranges are generally smaller in these occupancies than in one-family dwellings.

Note 3 applies when individual ranges rated over 1¾ kW through 8¾ kW are used. While Column A could be used for these ranges, Columns B or C provide smaller kW demands. Note 3 permits the nameplate ratings of all household cooking appliances rated between 1¾ kW and 8¾ kW to be added together and multiplied by the appropriate demand from either Column B or C. Note 3 is commonly used in multifamily dwellings with the smaller individual cooking units. As with the one-family calculations, if the cooking appliances are of different ratings that fall under both Columns B and C, the demand factors for each column shall be applied to the appliance for that column and then all of the results added together. See Figure 13-24.

Note 4, which applies to branch-circuit loads, can also be used frequently with multifamily dwelling units. Note 4 permits the load to be calculated per Table 220-19. For single wall-mounted ovens or single counter-mounted cooking units, the branch-circuit load shall be the nameplate rating of the appliance.

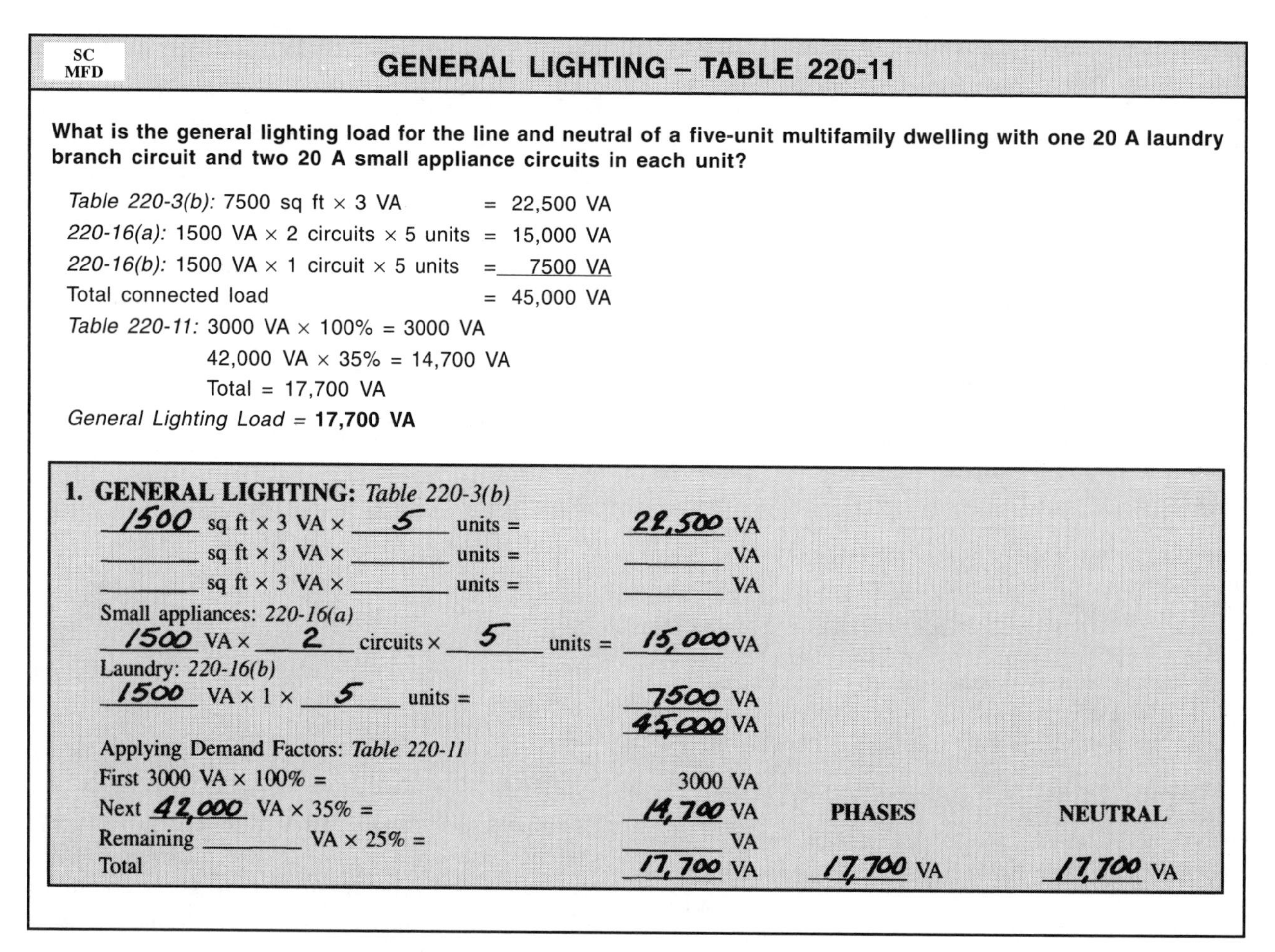

SC MFD

GENERAL LIGHTING – TABLE 220-11

What is the general lighting load for the line and neutral of a five-unit multifamily dwelling with one 20 A laundry branch circuit and two 20 A small appliance circuits in each unit?

Table 220-3(b): 7500 sq ft × 3 VA = 22,500 VA
220-16(a): 1500 VA × 2 circuits × 5 units = 15,000 VA
220-16(b): 1500 VA × 1 circuit × 5 units = 7500 VA
Total connected load = 45,000 VA
Table 220-11: 3000 VA × 100% = 3000 VA
42,000 VA × 35% = 14,700 VA
Total = 17,700 VA
General Lighting Load = **17,700 VA**

1. GENERAL LIGHTING: *Table 220-3(b)*
1500 sq ft × 3 VA × 5 units = 22,500 VA
___ sq ft × 3 VA × ___ units = ___ VA
___ sq ft × 3 VA × ___ units = ___ VA
Small appliances: *220-16(a)*
1500 VA × 2 circuits × 5 units = 15,000 VA
Laundry: *220-16(b)*
1500 VA × 1 × 5 units = 7500 VA
45,000 VA
Applying Demand Factors: *Table 220-11*
First 3000 VA × 100% = 3000 VA
Next 42,000 VA × 35% = 14,700 VA
Remaining ___ VA × 25% = ___ VA
Total 17,700 VA

PHASES 17,700 VA
NEUTRAL 17,700 VA

Figure 13-21. Demand factors of Table 220-11 are applied to the general lighting load of multifamily dwellings.

For installations from a common branch circuit in which a single counter-mounted cooking unit is installed with not more than two wall-mounted ovens, Note 4 permits all of the cooking appliances to be added together and treated like a single range, provided all of the cooking appliances are located in the same room.

Heating or A/C – 220-21

As with one-family dwellings, the heating or A/C load for multifamily dwellings is often the largest load in the multifamily dwelling calculation. Section 220-21 permits the smaller of the two noncoincidental load calculations to be omitted from the multifamily dwelling load calculation, provided it is unlikely that the loads will be in use simultaneously.

Heating and A/C loads are both calculated at 100% of the nameplate rating. These loads are multiplied by the number of units which contain the loads. The calculation with the largest load is used in determining the total feeder or service demand for the multifamily dwelling. If the individual units contain heating or A/C loads of different values, they are added together and the total loads are compared. The largest load is used in the multifamily dwelling calculation. See Figure 13-25.

Largest Motor – 220-14

Despite the fact that each unit in a multifamily dwelling may have the same motor load, 220-14 requires that only 25% of the largest motor in the entire multifamily dwelling be added to the standard calculation for multifamily dwellings. Feeders for several motors are calculated per 430-24 using the sum of the full-load table currents of all motors plus 25% of the largest motor. A motor's full-load current rating is calculated under the fixed appliance portion of the calculation at 100% per 220-17. The largest motor portion of the calculation allows for the 25% that is not covered under 220-17.

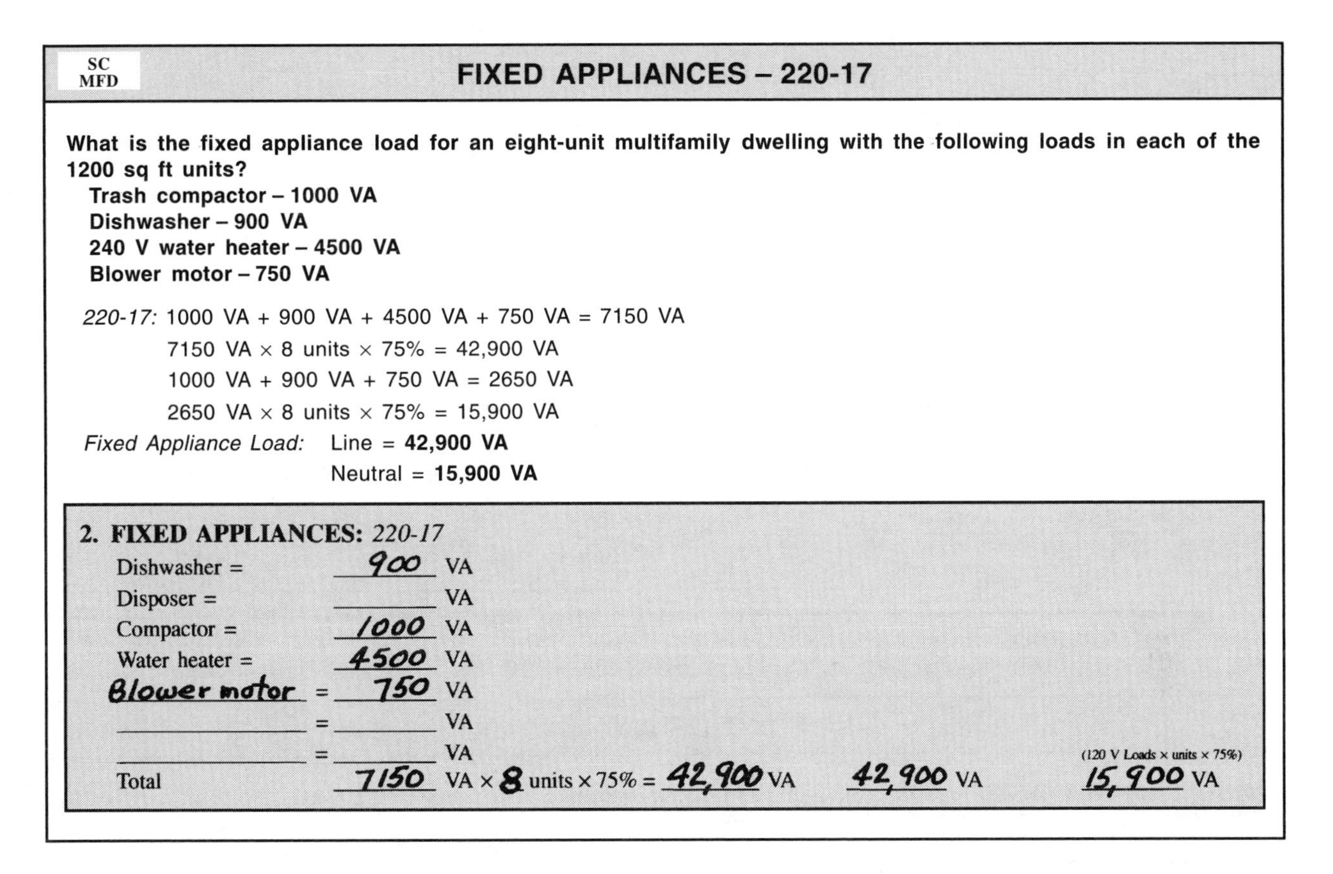

Figure 13-22. A 75% demand factor is applied to four or more fixed appliance loads in multifamily dwellings.

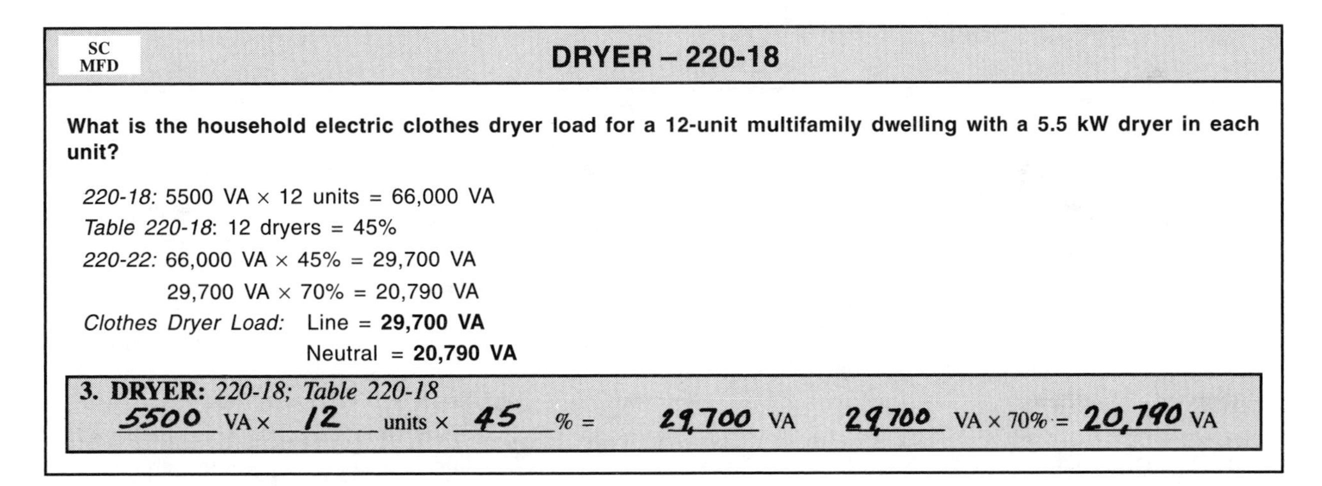

Figure 13-23. Demand factors of Table 220-18 are applied to clothes dryers in multifamily dwellings.

Demand Factors – Neutral – 220-22

Section 220-22, which permits several appliances to have the neutral demand calculated at 70% of the ungrounded conductors, is applicable to multifamily dwellings as well as to one-family dwellings. Likewise, for multifamily dwellings, the deduction of 70% is permitted for that portion of the neutral conductor demand in excess of 200 A. This reduction only applies to 3-wire DC, 1φ AC, 3φ 4-wire, and 2φ systems. The 70% reduction does not apply to portions of the load that consist of nonlinear loads when the supply system is 3φ, 4-wire, wye-connected or 3φ, 4-wire, wye-connected utilizing two phase wires and the neutral.

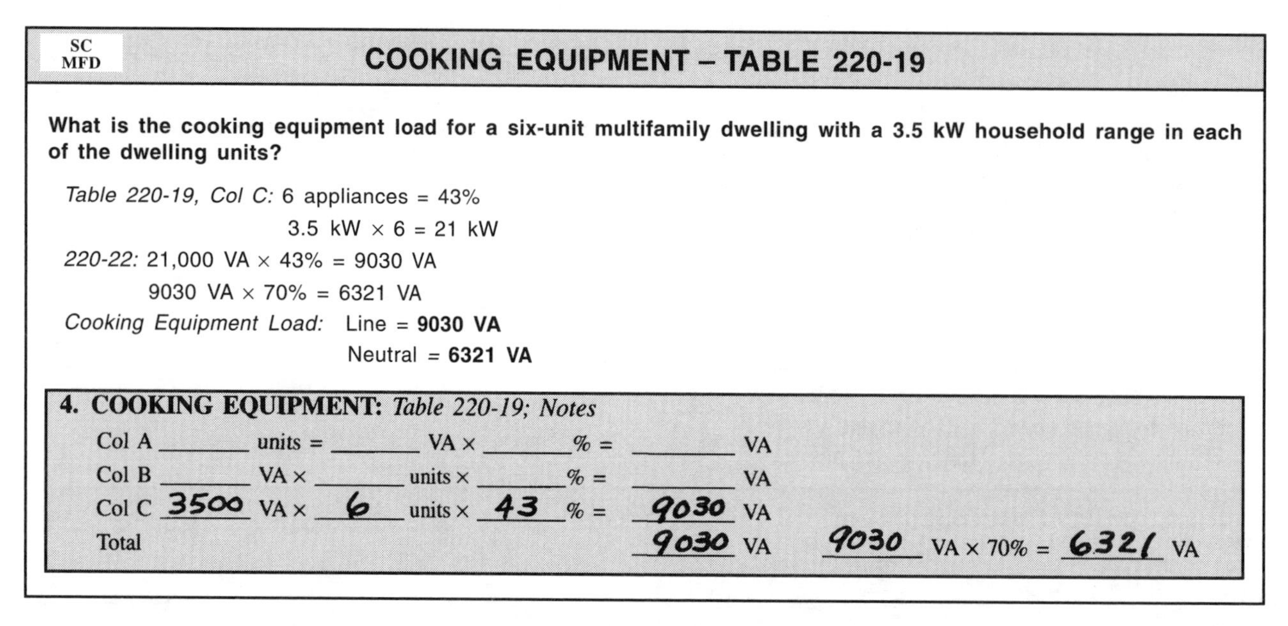

SC MFD

COOKING EQUIPMENT – TABLE 220-19

What is the cooking equipment load for a six-unit multifamily dwelling with a 3.5 kW household range in each of the dwelling units?

Table 220-19, Col C: 6 appliances = 43%

3.5 kW × 6 = 21 kW

220-22: 21,000 VA × 43% = 9030 VA

9030 VA × 70% = 6321 VA

Cooking Equipment Load: Line = **9030 VA**

Neutral = **6321 VA**

4. COOKING EQUIPMENT: *Table 220-19; Notes*

Col A ______ units = ______ VA × ______ % = ______ VA

Col B ______ VA × ______ units × ______ % = ______ VA

Col C 3500 VA × 6 units × 43 % = 9030 VA

Total 9030 VA 9030 VA × 70% = 6321 VA

Figure 13-24. Household ranges over 1¾ kW are computed per Table 220-19.

SC MFD

HEATING OR A/C – 220-21

What is the heating or A/C load for a nine-unit multifamily dwelling which contains a 6 kW, 240 V A/C and an 8 kW, 240 V heating unit in each dwelling?

Total Heating Load = 8 kW × 9 units = 72 kW

Total A/C Load = 6 kW × 9 units = 54 kW

220-21: Largest load = 72 kW

Heating or A/C Load = **72,000 VA**

5. HEATING or A/C: *220-21*

Heating unit = 8000 VA × 100% × 9 units = 72,000 VA

A/C unit = 6000 VA × 100% × 9 units = 54,000 VA

Heat pump = ______ VA × 100% × ______ units = ______ VA

Largest Load 72,000 VA 72,000 VA — VA

Figure 13-25. The largest of the heating or A/C load is used.

Conductor Ampacity

After the six loads (General Lighting, Cooking Equipment, Dryer, Fixed Appliances, Heating or A/C, and Largest Motor) are individually calculated, the results of the individual calculations are added to determine the maximum load, in VA, for both the ungrounded and grounded conductors. This total VA value is used to determine the minimum ampacity for the ungrounded and grounded conductors.

For 1φ systems, the formula $I = \frac{VA}{V}$ is used to determine minimum ampacity. In this formula, I = amperes (A), VA = volt amperes and V = voltage. After the minimum ampacity has been determined, Table 310-16 and the applicable notes are used to calculate the minimum conductor size for the feeder or service-entrance conductors.

The ungrounded phase conductors are not subject to any additional demand factors. Once the phase amperes are calculated by using the formula, the size of the phase conductors is determined. Table 310-16 can always be used to size the ungrounded conductors. In addition, Note 3 of Notes to Ampacity Tables of 0 to 2000 Volts is applicable to the feeders or service-entrance conductors for the individual dwelling units of the multifamily dwelling, but not to the service-entrance conductors which supply the entire building. If the individual feeders or service-entrance conductors are 120/240 V, 1ϕ, 3-wire, they may be sized per Note 3. The service-entrance conductors for the entire building are sized per Table 310-16 because Note 3 only applies to dwelling units.

Grounded Conductors. Section 220-22 permits an additional demand factor of 70% to be applied to the feeder or service-entrance grounded conductor for that portion in excess of 200 A for multifamily dwellings. This only applies to 3-wire DC, 1ϕ AC, 3ϕ 4-wire, and two-phase systems that supply other than nonlinear loads. Once the total neutral load is calculated using $I = \frac{VA}{V}$, the demand factor can be applied. The first 200 A is taken at 100% and the remainder is taken at 70%. The minimum size of the neutral conductor is selected from Table 310-16. Note 3 of Notes to Ampacity Tables of 0 to 2000 Volts permits the grounded conductor (neutral) to be smaller than the ungrounded conductors provided the requirements of 215-2, 220-22, and 230-42 are met.

MULTIFAMILY DWELLING – OPTIONAL CALCULATION – 220-32

In addition to the Standard Calculation, there is a multifamily dwelling Optional Calculation which can be performed if three conditions are present. First, 220-32(a)(1) requires that no dwelling within the multifamily structure be supplied by more than one feeder. The second condition is that each dwelling shall be equipped with electric cooking equipment. If the individual dwelling units have natural gas cooking equipment, the Optional Calculation could not be used. The last condition is that each of the individual dwelling units shall have either electric space-heating equipment, A/C, or both loads present.

If all three of these conditions are present, the Optional Calculation for multifamily dwellings permits the demand factors of Table 220-32 to be applied to the total computed load. The neutral load for these dwelling unit feeders or service-entrance conductors shall be computed per 220-22.

For dwelling units that qualify for the optional calculation, two loads are considered before applying the demand factors of Table 220-32. These are the house loads and the connected loads. A *house load* is an electrical load which is metered separately and supplies common usage areas. For example, typical house loads are hallway and perimeter lighting. In addition, 210-25 requires that branch circuits for central alarm, signal, communications, or other needs for public or common areas of multifamily types of occupancies be supplied from equipment that does not supply individual dwelling units. This is to ensure that a single tenant of a multifamily occupancy does not disconnect or de-energize one of these common area branch circuits and create a life safety concern for all of the occupants.

The second consideration is connected loads. Section 220-32(c) lists the five connected loads which are added together for calculating the total connected load to which the demand factors of Table 220-32 are applied.

The other loads include: (1) 1500 VA for each 2-wire, 20 A small appliance branch circuit, (2) 3 VA per square foot for the general lighting and general-use receptacle load, (3) the nameplate rating of all fastened-in-place, permanently-connected appliances, (4) the nameplate rating, in either amperes or kVA, of all motor and other low power factor loads. If the permanently connected appliance is a water heater in which the elements are interlocked such that all elements cannot be energized at the same time, the maximum load which can be connected at one time shall be used for the purposes of the calculation, and (5) the larger of the electric space-heating load or the A/C load.

Once the house load is added to the connected loads from 220-32(c), the total connected load can be calculated. This is determined by multiplying the sum of these two loads by the demand factors listed in Table 220-32. If, for example, the total connected load for the multifamily dwelling is 200,000 kVA and there are 16 units in the multifamily dwelling, the demand factor is 39%. Therefore, the total connected load is 78,000 kVA (200,000 kVA × 39% = 78,000 kVA). This value is then used to calculate the size of the ungrounded conductors to the multifamily dwelling. The size of the grounded conductor is determined using the Standard Calculation for the total neutral load. See Figure 13-26.

OC MFD

OPTIONAL CALCULATION: MULTIFAMILY DWELLING

What is the line and neutral demand load for a seven-unit multifamily dwelling with the following loads? Each unit is 975 sq ft in area and contains:

- **2 – Small appliance 20 A branch circuits**
- **1 – 12 kW electric range**
- **1 – 950 VA dishwasher**
- **1 – 6.5 kW A/C**
- **1 – Laundry branch circuit**
- **1 – 8 kW central heating unit**
- **1 – 5000 VA water heater**

1. HEATING or A/C: *220-32(c)(5)*

Heating unit = 8000 VA × 100% × 7 units = 56,000 VA
A/C unit = 6500 VA × 100% × 7 units = 45,500 VA
Heat pump = ____ VA × 100% × ____ units = ____ VA
Largest Load 56,000 VA
Total — PHASES 56,000 VA

2. OTHER LOADS: *220-32*

General lighting: *220-32(c)(2)*
975 sq ft × 3 VA × 7 units = 20,475 VA — 20,475 VA
____ sq ft × 3 VA × ____ units = ____ VA
____ sq ft × 3 VA × ____ units = ____ VA
____ sq ft × 3 VA × ____ units = ____ VA

Small appliance and laundry loads: *220-32(c)(1)*
1500 VA × 3 circuits × 7 units = 31,500 VA — 31,500 VA

Special loads: *220-32(c)(3)*
Dishwasher = 950 VA
Disposer = ____ VA
Compactor = ____ VA
Water heater = 5000 VA
Range = 12,000 VA
____ = ____ VA
____ = ____ VA
____ = ____ VA
____ = ____ VA
Total 17,950 VA × 7 units = 125,650 VA — 125,650 VA

Total Connected Load 233,625 VA

Applying Demand Factors: *Table 220-32*
233,625 VA × 44 % = 102,795 VA — 102,795 VA

NEUTRAL (Loads from Standard Calculation)

1. General lighting = 20,141 VA
2. Fixed appliances = 6650 VA
3. Dryer = – VA
4. Cooking equipment = 15,400 VA
5. Heating or A/C = – VA
6. Largest motor = – VA

Total 42,191 VA

1φ service: PHASES $I = \frac{VA}{V} =$ ____ A
3φ service: $I = \frac{VA}{V \times \sqrt{3}} =$ ____ A

NEUTRAL $I = \frac{VA}{V} =$ ____ A
$I = \frac{VA}{V \times \sqrt{3}} =$ ____ A

220-22; First 200 A × 100% = 200 A
Remaining ____ A × 70% = ____ A
____ A

Figure 13-26. Demand factors of Table 220-32 are used for multifamily dwellings by the Optional Calculation.

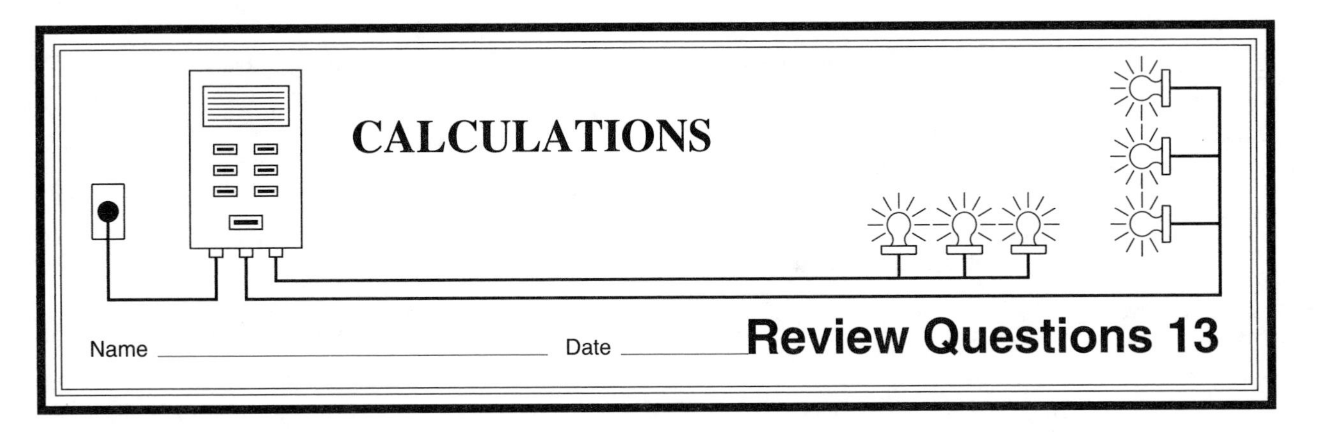

Review Questions

_________ **1.** _________ is the amount of electricity required at a given time.

_________ **2.** General lighting loads for dwelling units are calculated at _________ VA per square foot.

A. 3
B. $3\frac{1}{2}$
C. 4
D. neither A, B, nor C

_________ **3.** Each single or multiple receptacle on a single strap is calculated at _________ VA.

_________ **4.** Demand factors may be applied to _________ of the six loads of the Standard Calculation.

_________ **5.** Four or more appliances which are fastened-in-place in a one-family dwelling are computed at _________% of the total load for all four appliances (Standard Calculation).

_________ **6.** The total load for household clothes dryers is computed at _________ VA or the nameplate rating, whichever is higher (Standard Calculation).

_________ **7.** _________ loads are loads that are not on at the same time.

_________ **8.** The _________ load is the maximum unbalance between any of the ungrounded conductors and the grounded conductor.

_________ **9.** For 1ϕ systems, the formula _________ is used to determine minimum ampacity.

A. $I = \frac{V}{VA}$
B. $I = V \times VA$
C. $I = \frac{VA}{V}$
D. neither A, B, nor C

_________ **10.** Each dwelling shall be supplied with a(n) _________ A branch circuit to supply laundry receptacle(s).

_________ **11.** Heating or A/C loads are calculated at _________% of the nameplate rating and the largest load is used (Standard Calculation).

_________ **12.** After the minimum ampacity of the service has been calculated, Table _________ and applicable notes are used to determine the size of feeder or service-entrance conductors.

_________ **13.** The Optional Calculation for dwelling units consists of heating or A/C loads and ________ loads.

A. general lighting
B. small appliance
C. largest motor
D. other

_________ **14.** Household electric ranges rated in excess of ________ kW are computed per Table 220-19 (Standard Calculation).

_________ **15.** The ________ rating is the ampere rating or setting of the overcurrent device protecting the conductors.

_________ **16.** The total area of an occupancy is determined from the ________ dimensions of the building or structure.

A. inside
B. outside
C. either A or B
D. neither A nor B

_________ **17.** ________ branch circuits supply all receptacle outlets located in the kitchen, pantry, breakfast room, dining room, or similar areas of dwelling units (Standard Calculation).

_________ **18.** Feeders for several motors are calculated using the sum of the FLC tables of all motors plus ________% of the largest motor (Standard Calculation).

A. 3
B. 15
C. 25
D. neither A, B, nor C

_________ **19.** Less than four separately-controlled electric space-heating units are calculated at ________% of their nameplate ratings (Optional Calculation).

T F **20.** The Optional Calculation can be used to find the total ampacity of a multifamily dwelling with natural gas cooking equipment.

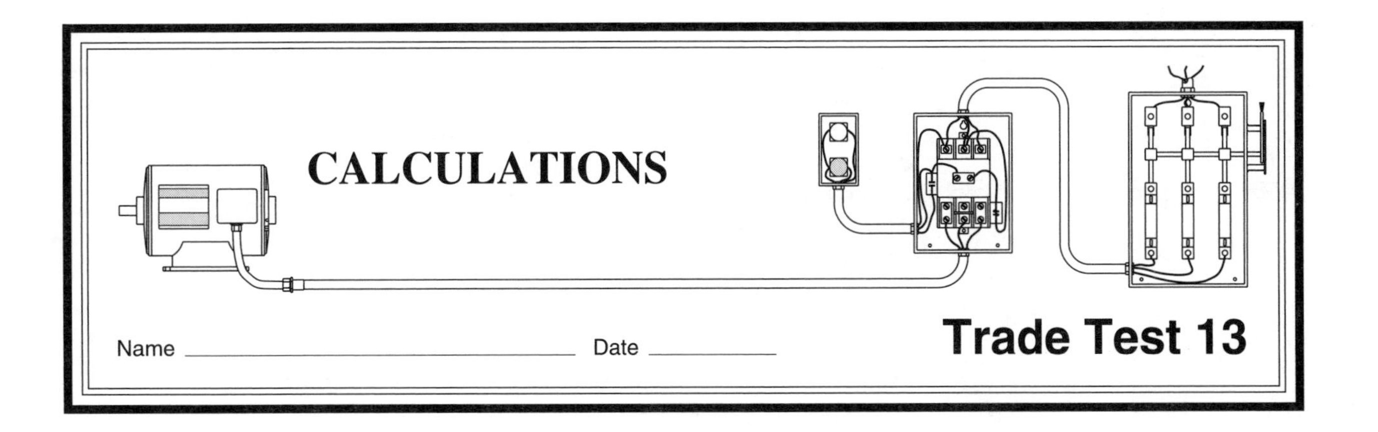

CALCULATIONS

Name ______________________ Date __________

Trade Test 13

Calculation forms in the Appendix may be copied and used to solve the following problems.

NEC®	Answer	
______	______	**1.** *Standard Calculation: One-Family Dwelling.* What is the general lighting and receptacle load (line and neutral) for a 2800 sq ft dwelling with two small appliance branch circuits and one laundry branch circuit?
______	______	**2.** *Standard Calculation: One-Family Dwelling.* What is the fixed appliance load (line and neutral) for a dwelling with the following 120 V loads? Water heater – 5000 VA Dishwater – 1200 VA Garbage disposer – 1000 VA Trash compactor – 900 VA Attic fan – 1200 VA
______	______	**3.** *Standard Calculation: Multifamily Dwelling.* What is the total dryer load (line and neutral) for seven household dryers rated at 6.0 kW each?
______	______	**4.** *Standard Calculation: One-Family Dwelling.* What is the range demand load (line and neutral) for one 17 kW household electric range?
______	______	**5.** *Standard Calculation: One-Family Dwelling.* What is the range demand load (line and neutral) for three household electric ranges with nameplate ratings of 12 kW, 19 kW, and 20 kW?
______	______	**6.** *Optional Calculation: One-Family Dwelling.* What is the total "other loads" for a 3000 sq ft dwelling with the following loads? 3 – Small appliance 20 A branch circuits 1 – 8 kW electric range 1 – 1.2 kW dishwasher 1 – 5 kW clothes dryer 1 – 1800 VA blower motor
______	______	**7.** *Standard Calculation: Multifamily Dwelling.* What is the general lighting load (line and neutral) for an eight-unit multifamily dwelling with one 20 A laundry branch circuit and three 20 A small appliance circuits in each unit?
______	______	**8.** *Standard Calculation: Multifamily Dwelling.* What is the household electric clothes dryer load (line and neutral) for a 20-unit multifamily dwelling with a 6.0 kW dryer in each unit?

______ ______ **9.** *Standard Calculation: One-Family Dwelling.* What is the total connected load (line and neutral), in VA, and the minimum service-entrance conductor ampacity (line and neutral) of a 120/240 V, 1ϕ, 3-wire service for a 2900 sq ft dwelling with the following loads?
General lighting and receptacle loads
Three small appliance loads
One laundry load
Range – 10,500 VA
Dryer – 4800 VA
A/C – 9500 VA
Central heating – 12,000 VA
Water heater (240 V) – 5000 VA
Dishwasher – 1200 VA
Trash compactor – 1000 VA
Garbage disposer – 900 VA
Blower motor – 1600 VA

______ ______ **10.** *Optional Calculation: One-Family Dwelling.* Determine the total connected load (line and neutral), in VA, and the minimum service-entrance conductor ampacity (line and neutral) for the dwelling in Problem 9.

______ ______ **11.** *Standard Calculation: Multifamily Dwelling.* What is the total connected load (line and neutral), in VA, and the minimum service-entrance conductor ampacity (line and neutral) of a 120/208 V, 3ϕ, 4-wire service which supplies six 1200 sq ft dwellings with the following loads?
General lighting and receptacle loads
Two small appliance loads
One laundry load
Range – 8000 VA
Dryer – 5000 VA
A/C – 6000 VA
Electric space heating – 8000 VA
Water heater (120 V) – 4000 VA
Dishwasher – 800 VA
Garbage disposer – 900 VA
Blower motor – 1000 VA

______ ______ **12.** *Optional Calculation: Multifamily Dwelling.* Determine the total connected load (line and neutral), in VA, and the minimum service-entrance conductor ampacity (line and neutral) for the dwelling in Problem 11.

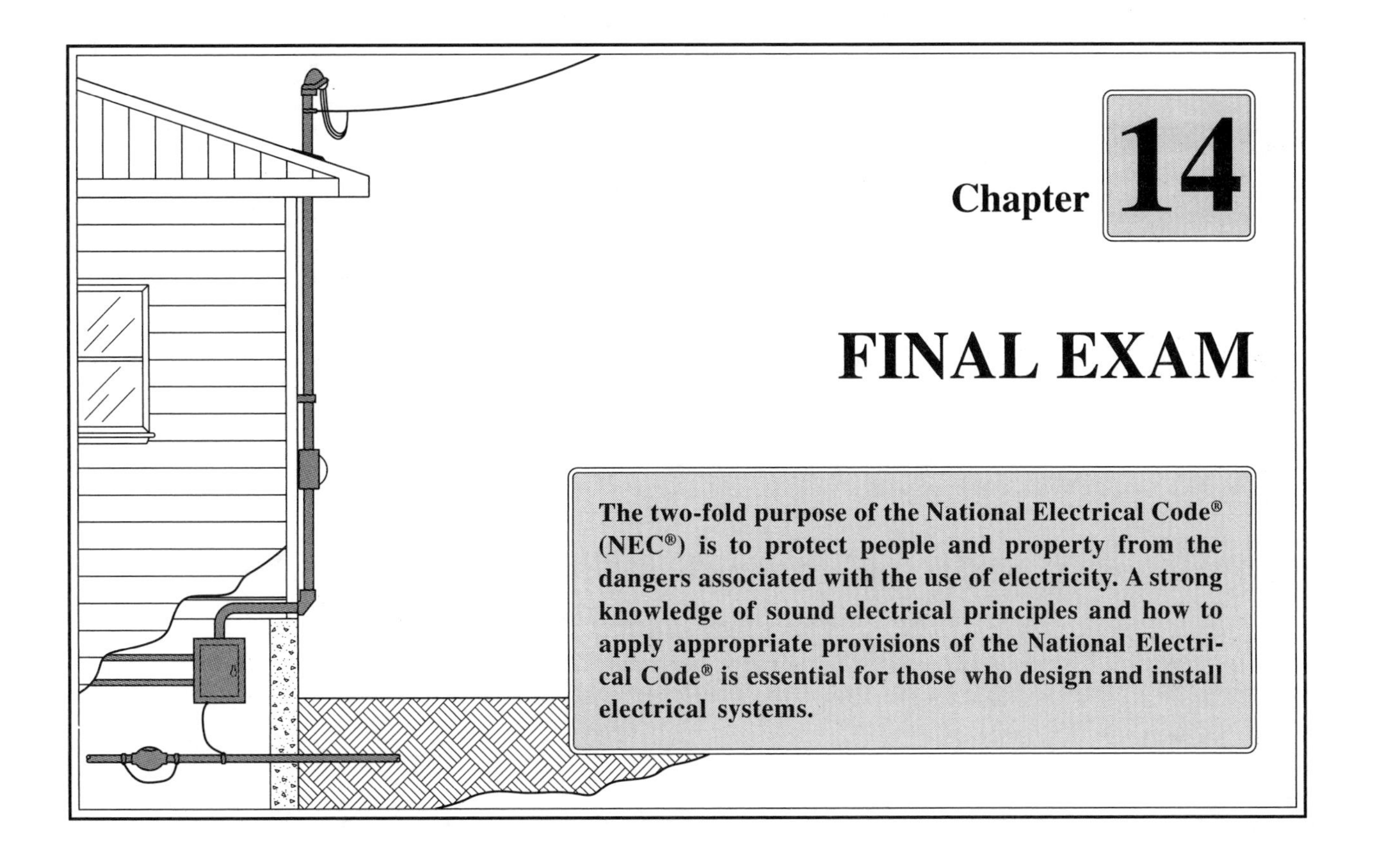

FINAL EXAM

The problems in the Final Exam of *Electrical Systems* are based on Chapters 1 – 13 and the current edition of the National Electrical Code®. All problems are in the multiple-choice format because many testing agencies use this format for their test.

Read the problems very carefully and select the response that most clearly answers the question. Record the letter representing that response in the blank space to the immediate left of the problem number. Record the NEC® reference that substantiates the answer in the blank space to the far left of the problem number. See Figure 14-1.

The option of either using or not using the NEC® to verify the answers is at the discretion of your instructor. Some instructors prefer an "open book" test while other instructors prefer a "closed book" test.

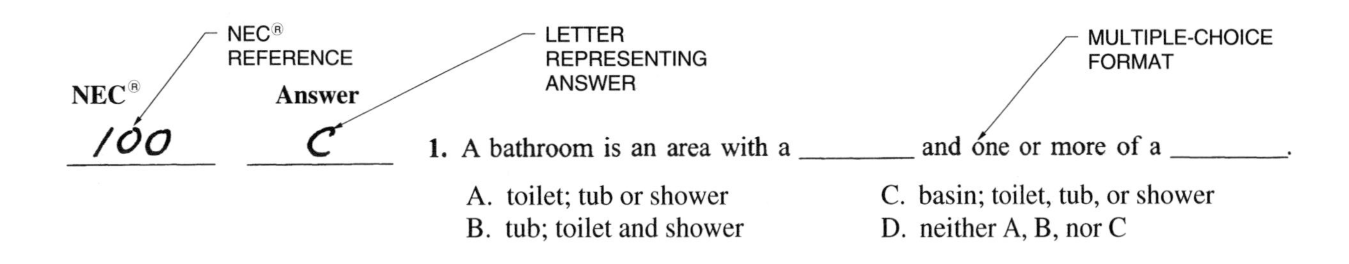

Figure 14-1. Record answers and NEC® references for problems in the Final Exam.

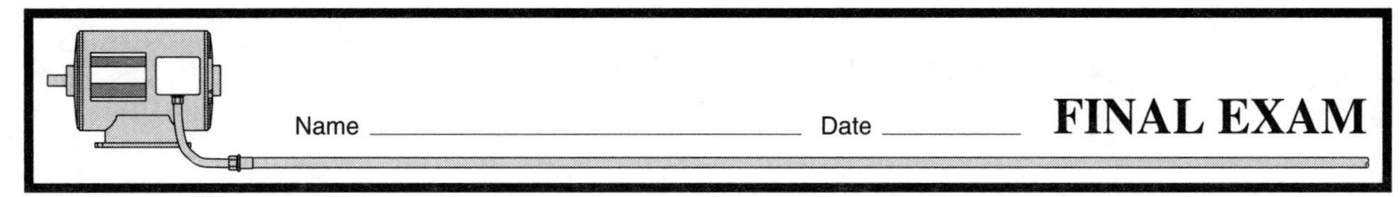

FINAL EXAM

NEC® **Answer**

______ ______ **1.** A ________ is an intentional or accidental grounding connection between an electrical circuit or equipment and the earth, or to some conducting body that serves in place of the earth.

A. short
B. ground
C. bonding jumper
D. grounded conductor

______ ______ **2.** All branch circuits used for temporary wiring shall originate in a(n) ________ power outlet or panelboard.

A. listed
B. identified
C. approved
D. weatherproof

______ ______ **3.** ________ are not covered by Article 517, Health Care Facilities.

A. Nursing homes
B. Dental offices
C. Hospitals
D. neither A, B, nor C

______ ______ **4.** Type ________ cable is not a permitted nonmetallic-sheathed cable for use in one-family dwellings.

A. NM
B. NMC
C. NMS
D. NMX

______ ______ **5.** The allowable ampacity for three No. 1/0, THW Cu conductors installed in the same raceway in an ambient temperature of 86°F is ________ A.

A. 120
B. 125
C. 150
D. 170

______ ______ **6.** Service conductors installed as open conductors without an overall outer jacket shall have a clearance of not less than ________′ from windows that are designed to be opened.

A. 1
B. 2
C. 3
D. 5

______ ______ **7.** A(n) ________ is not a permissible grounding electrode for use on a separately derived system.

A. No. 2 Cu ground ring
B. effectively-grounded building steel
C. effectively-grounded water pipe
D. $\frac{5}{8}$″ × 8″ Cu ground rod

______ ______ **8.** ________ is classified as a Group E dust.

A. Grain
B. Coal
C. Wood
D. Aluminum

______ ______ **9.** Branch-circuit conductors within ________″ of a ballast inside a ballast compartment shall have an insulation rating not lower than 90°C.

A. 3
B. 6
C. 9
D. 12

______ ______ **10.** ________ is not a permitted wiring method for wiring from the controller of a fire pump to the pump motor.

A. RMC
B. RNMC
C. IMC
D. Type MI cable

__________ __________ **11.** The maximum size of LTFMC permitted to be used is the ________″ trade size.

A. 3
B. 4
C. 5
D. 6

__________ __________ **12.** Mandatory rules of the NEC® are characterized by the use of the word ________.

A. must
B. should
C. shall
D. permitted

__________ __________ **13.** Primary transformer protection, for transformers rated 600 V or less, shall be provided by an individual overcurrent device on the primary side rated or set at not more than ________% of the rated primary current of the transformer.

A. 125
B. 150
C. 175
D. 250

__________ __________ **14.** The minimum capacity required for a standard metal box that contains four No. 12 conductors and four No. 8 conductors is ________ cu in.

A. 16
B. 21
C. 25
D. 28

__________ __________ **15.** Edison-base plug fuses shall be classified at not over 125 V and ________ A and below.

A. 10
B. 15
C. 20
D. 30

__________ __________ **16.** At least ________″ of free conductor shall be left at each outlet, junction, and switch point for splices or the connection of fixtures or devices.

A. 4
B. 6
C. 8
D. 12

__________ __________ **17.** A 15 A, 125 V, 1ϕ receptacle installed ________ of a dwelling unit does not require GFCI protection.

A. in a bathroom
B. below grade in a crawl space
C. in a finished basement
D. at a kitchen countertop

__________ __________ **18.** A(n) ________ is not classified as a place of assembly.

A. assembly hall
B. bowling lane
C. shopping center
D. skating rink

__________ __________ **19.** The full-load table current for a 10 HP, 480 V, 3ϕ squirrel-cage motor is ________ A.

A. 14
B. 28
C. 30.8
D. 32.2

__________ __________ **20.** The minimum wire-bending space required for an enclosure with No. 3/0 conductors, one wire per terminal, where the conductors do not enter or leave the enclosure through the wall opposite the terminal is ________″.

A. 3
B. 4
C. 6
D. 8

__________ __________ **21.** The minimum size THW, Cu service-entrance conductor permitted to supply a 200 A service to a one-family dwelling with a 120/240 V, 1ϕ, 3-wire service is No. ________.

A. 1
B. 1/0
C. 2/0
D. 4/0

______ ______ **22.** A lighting and appliance branch-circuit panelboard is a panelboard with more than ________% of its overcurrent devices rated at 30 A or less, for which neutral connections are provided.

A. 10 C. 20
B. 15 D. 25

______ ______ **23.** The calculated line load for six 5.5 kW household clothes dryers installed in a multifamily dwelling is ________ VA (Standard Calculation).

A. 21,300 C. 25,100
B. 23,100 D. 33,000

______ ______ **24.** The minimum front work space width required for a 100 A panelboard installed in a dwelling is ________″.

A. 24 C. 32
B. 30 D. 36

______ ______ **25.** Branch-circuit conductors supplying a single motor shall have an ampacity not less than ________ % of the motor FLC rating.

A. 100 C. 125
B. 115 D. 140

______ ______ **26.** For commercial garages, each unventilated floor space up to ________″ above the floor shall be considered a Class I, Division 2 location.

A. 6 C. 18
B. 12 D. 24

______ ______ **27.** The minimum size Cu GEC permitted for a 120/208 V, 3ϕ service which consists of two 500 kcmil Cu conductors paralleled per phase is No. ________.

A. 2 C. 2/0
B. 1/0 D. 3/0

______ ______ **28.** The household electric range demand for the line and neutral of a 15 kW, 240 V range installed in a one-family dwelling is ________ VA.

A. 8000 line; 8000 neutral C. 9200 line; 9200 neutral
B. 8000 line; 5600 neutral D. 9200 line; 6440 neutral

______ ______ **29.** The minimum burial depth for 600 V conductors installed in RMC, in a trench below 2″ thick concrete is ________″.

A. 4 C. 12
B. 6 D. 18

______ ______ **30.** EMT shall be supported at least every ________′ and within ________″ of each outlet box, junction box, device box, conduit body, or other tubing termination.

A. 5; 12 C. 10; 12
B. 5; 36 D. 10; 36

______ ______ **31.** A separate overload device is installed for a continuous-duty, 2 HP, 12 FLA motor with a marked service factor of 1.15. The motor does not have starting current problems. The maximum rating of the overload device required to protect the motor is ________ A.

A. 13.8 C. 15.6
B. 15.0 D. 16.8

________ ________ **32.** The maximum length which Type AC cable, installed from an outlet box to a lighting fixture in an accessible ceiling, shall be installed without being supported is ________′.

A. 2
B. 4
C. 6
D. 10

________ ________ **33.** The minimum ampacity for a 25′ tap conductor which is tapped from a conductor protected by a 300 A overcurrent protective device is ________ A.

A. 30
B. 100
C. 150
D. 300

________ ________ **34.** A 20 A, 40 A, and 70 A branch circuit are installed in the same 2″ RNMC. No. ________ is the minimum size Cu EGC required for this installation.

A. 12
B. 10
C. 8
D. 6

________ ________ **35.** Equipment which is required to be in sight from other equipment shall be visible and not more than ________′ away.

A. 10
B. 25
C. 50
D. 100

________ ________ **36.** A conduit seal shall be installed within ________″ of the enclosure or fitting in each 2″ or larger conduit which enters an enclosure or fitting housing terminals, splices, or taps in a Class I, Division 1 location.

A. 6
B. 12
C. 18
D. 24

________ ________ **37.** The minimum ungrounded conductor ampacity for service-entrance conductors which supply a one-family dwelling with an initial net computed load of 12 kVA shall be ________ A.

A. 60
B. 100
C. 150
D. 200

________ ________ **38.** Receptacle outlets in dwelling unit floors shall not be counted as part of the required number of receptacle outlets unless located within ________″ of the wall.

A. 6
B. 12
C. 18
D. 24

________ ________ **39.** GFPE, as specified in 230-95, shall be provided for feeder disconnects rated ________ A or more in a solidly-grounded wye system with greater than 150 V to ground, but not exceeding 600 V phase-to-phase.

A. 300
B. 600
C. 750
D. 1000

________ ________ **40.** FMC shall be securely fastened-in-place by an approved means within ________″ of each box, cabinet, conduit body, or other conduit termination and shall be supported and secured at intervals not to exceed ________′.

A. 6; 3
B. 6; 4½
C. 12; 3
D. 12; 4½

APPENDIX

Text Abbreviations 398
Residential Electrical Symbols 399–400
Architectural Symbols 401–402
Plot Plan Symbols 402
Alphabet of Lines 403
Three-Phase Voltage Values 404
Ohm's Law 404
Power Formulas–1φ, 3φ 404
AC/DC Formulas 405
Horsepower Formulas 405
Voltage Drop Formulas – 1φ, 3φ 405
Locked Rotor Current 406
Maximum OCPD 406
Efficiency 406
Voltage Unbalance 406
Power 406
Capacitors 407
Gear Reducer 407
Horsepower 407
Temperature Conversions 407
Motor Torque 407
AC Motor Characteristics 408
DC and Universal Motor Characteristics 408
Overcurrent Protection Devices 408
Enclosure Types 409
Raceways 410
Cables 410
Fuses and ITCBs 410
Common Electrical Insulations 410
Standard Calculation: One-Family Dwelling 411
Standard Calculation: Multifamily Dwelling 412
Optional Calculation: One-Family Dwelling 413
Optional Calculation: Multifamily Dwelling 414

TEXT ABBREVIATIONS			
A	Amps	k	Kilo (1000)
A/C	Air Conditioner	kcmil	1000 Circular Mils
AC	Alternating Current	kFT	1000′
AEGCP	Assured Equipment Grounding Program	kVA	Kilovolt Amps
A/H	Air Handler	kW	Kilowatt
AHCC	Ambulatory Health Care Center	kWh	Kilowatt-Hour
AHJ	Authority Having Jurisdiction	L	Line
AIR	Ampere Interrupting Rating	LFLI	Less-Flammable, Liquid-Insulated
Al	Aluminum	LP	Lighting Panel
ANSI	American National Standards Institute	LRA	Locked-Rotor Ampacity
ATCB	Adjustable-Trip Circuit Breaker	LRC	Locked-Rotor Current
AWG	American Wire Gauge	MBCSCGF	Motor Branch-Circuit, Short-Circuit, Ground-Fault
BJ	Bonding Jumper	MBJ	Main Bonding Jumper
C	Celsius	MCC	Main Control Center
CATV	Cable Antenna Television	N	Neutral
CB	Circuit Breaker	NATCB	Nonadjustable-Trip Circuit Breaker
CH	Chapter	NEC®	National Electrical Code®
CM	Circular Mils	NEMA	National Electrical Manufacturers Association
CMP	Code-making Panel	NESC	National Electrical Safety Code
CU	Copper	NFPA	National Fire Protection Association
DC	Direct Current	NO.	Number
DIA	Diameter	NTDF	Non-Time Delay Fuse
DP	Double Pole	OCPD	Overcurrent Protection Device
E	Voltage	OL	Overload(s)
EBJ	Equipment Bonding Jumper	OSHA	Occupational Safety and Health Administration
E_{ff}	Efficiency	P	Power
EGC	Equipment Grounding Conductor	PC	Personal Computer
EX.	Exception	PF	Power Factor
F	Fahrenheit	R	Resistance; Resistor
FLA	Full-Load Amps	RMS	Root Mean Square
FLC	Full-Load Current	ROC	Receipt of Comments
FPN	Fine Print Note	ROP	Receipt of Proposals
FR	Frame	SF	Service Factor
G	Ground	SP	Single-Pole
GEC	Grounding Electrode Conductor	SPCB	Single-Pole Circuit Breaker
GES	Grounding Electrode System	SQ FT	Square Foot (Feet)
GFCI	Ground Fault Circuit Interrupter	SWD	Switched Disconnect
GFPE	Ground Fault Protection of Equipment	T	Time
GR	Green	TDF	Time-Delay Fuse
HACR	Heating, Air-Conditioning, Refrigeration	TP	Thermally Protected
HP	Horsepower	UF	Underground Feeder
HRS	Hours	UL	Underwriter's Laboratory
I	Current	V	Volts
IDCI	Immersion Detection Circuit Interrupter	VA	Volt Amps
IG	Isolated Ground	VAC	Volts Alternating Current
IN.	Inch	VD	Voltage Drop
ITB	Instantaneous-Trip Circuit Breaker	W	Watts
ITCB	Inverse-Time Circuit Breaker	W	White
K	Conductor Resistivity	WP	Weatherproof

RESIDENTIAL ELECTRICAL SYMBOLS...

LIGHTING OUTLETS

- OUTLET BOX AND INCANDESCENT LIGHTING FIXTURE (CEILING, WALL)
- INCANDESCENT TRACK LIGHTING
- BLANKED OUTLET
- DROP CORD
- EXIT LIGHT AND OUTLET BOX. SHADED AREAS DENOTE FACES.
- OUTDOOR POLE-MOUNTED FIXTURES
- JUNCTION BOX
- LAMPHOLDER WITH PULL SWITCH
- MULTIPLE FLOODLIGHT ASSEMBLY
- EMERGENCY BATTERY PACK WITH CHARGER
- INDIVIDUAL FLUORESCENT FIXTURE
- OUTLET BOX AND FLUORESCENT LIGHTING TRACK FIXTURE
- CONTINUOUS FLUORESCENT FIXTURE
- SURFACE-MOUNTED FLUORESCENT FIXTURE

PANELBOARDS

- FLUSH-MOUNTED PANELBOARD AND CABINET
- SURFACE-MOUNTED PANELBOARD AND CABINET

CONVENIENCE OUTLETS

- SINGLE RECEPTACLE OUTLET
- DUPLEX RECEPTACLE OUTLET
- TRIPLEX RECEPTACLE OUTLET
- SPLIT-WIRED DUPLEX RECEPTACLE OUTLET
- SPLIT-WIRED TRIPLEX RECEPTACLE OUTLET
- SINGLE SPECIAL-PURPOSE RECEPTACLE OUTLET
- DUPLEX SPECIAL-PURPOSE RECEPTACLE OUTLET
- RANGE OUTLET (R)
- SPECIAL-PURPOSE CONNECTION (DW)
- CLOSED-CIRCUIT TELEVISION CAMERA
- CLOCK HANGER RECEPTACLE
- FAN HANGER RECEPTACLE
- FLOOR SINGLE RECEPTACLE OUTLET
- FLOOR DUPLEX RECEPTACLE OUTLET
- FLOOR SPECIAL-PURPOSE OUTLET
- UNDERFLOOR DUCT AND JUNCTION BOX FOR TRIPLE, DOUBLE, OR SINGLE DUCT SYSTEM AS INDICATED BY NUMBER OF PARALLEL LINES

BUSDUCTS AND WIREWAYS

- SERVICE, FEEDER, OR PLUG-IN BUSWAY
- CABLE THROUGH LADDER OR CHANNEL
- WIREWAY

SWITCH OUTLETS

- SINGLE-POLE SWITCH
- DOUBLE-POLE SWITCH
- THREE-WAY SWITCH
- FOUR-WAY SWITCH
- AUTOMATIC DOOR SWITCH
- KEY-OPERATED SWITCH
- CIRCUIT BREAKER
- WEATHERPROOF CIRCUIT BREAKER
- DIMMER
- REMOTE CONTROL SWITCH
- WEATHERPROOF SWITCH
- FUSED SWITCH
- WEATHERPROOF FUSED SWITCH
- TIME SWITCH
- CEILING PULL SWITCH
- SWITCH AND SINGLE RECEPTACLE
- SWITCH AND DOUBLE RECEPTACLE
- A STANDARD SYMBOL WITH AN ADDED LOWERCASE SUBSCRIPT LETTER IS USED TO DESIGNATE A VARIATION IN STANDARD EQUIPMENT (a.b)

...RESIDENTIAL ELECTRICAL SYMBOLS

COMMERCIAL AND INDUSTRIAL SYSTEMS

PAGING SYSTEM DEVICE

FIRE ALARM SYSTEM DEVICE

COMPUTER DATA SYSTEM DEVICE

PRIVATE TELEPHONE SYSTEM DEVICE

SOUND SYSTEM

FIRE ALARM CONTROL PANEL — FACP

SIGNALING SYSTEM OUTLETS FOR RESIDENTIAL SYSTEMS

PUSHBUTTON

BUZZER

BELL

BELL AND BUZZER COMBINATION

COMPUTER DATA OUTLET

BELL RINGING TRANSFORMER — BT

ELECTRIC DOOR OPENER — D

CHIME — CH

TELEVISION OUTLET — TV

THERMOSTAT — T

UNDERGROUND ELECTRICAL DISTRIBUTION OR ELECTRICAL LIGHTING SYSTEMS

MANHOLE — M

HANDHOLE — H

TRANSFORMER-MANHOLE OR VAULT — TM

TRANSFORMER PAD — TP

UNDERGROUND DIRECT BURIAL CABLE

UNDERGROUND DUCT LINE

STREET LIGHT STANDARD FED FROM UNDERGROUND CIRCUIT

ABOVE-GROUND ELECTRICAL DISTRIBUTION OR LIGHTING SYSTEMS

POLE

STREET LIGHT AND BRACKET

PRIMARY CIRCUIT

SECONDARY CIRCUIT

DOWN GUY

HEAD GUY

SIDEWALK GUY

SERVICE WEATHERHEAD

PANEL CIRCUITS AND MISCELLANEOUS

LIGHTING PANEL

POWER PANEL

WIRING – CONCEALED IN CEILING OR WALL

WIRING – CONCEALED IN FLOOR

WIRING EXPOSED

HOME RUN TO PANEL BOARD
Indicate number of circuits by number of arrows. Any circuit without such designation indicates a two-wire circuit. For a greater number of wires indicate as follows: (3 wires)
(4 wires), etc.

FEEDERS
Use heavy lines and designate by number corresponding to listing in feeder schedule

WIRING TURNED UP

WIRING TURNED DOWN

GENERATOR — G

MOTOR — M

INSTRUMENT (SPECIFY) — I

TRANSFORMER — T

CONTROLLER

EXTERNALLY-OPERATED DISCONNECT SWITCH

PULL BOX

ARCHITECTURAL SYMBOLS. . .

Material	Elevation	Plan	Section
EARTH			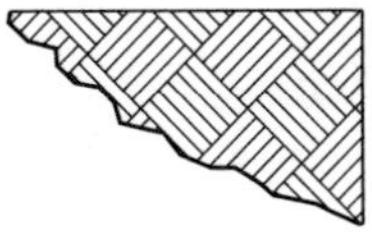
BRICK	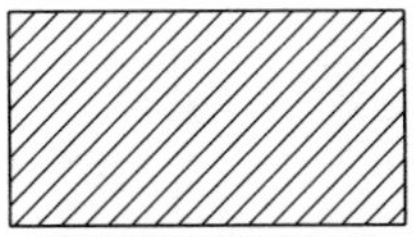 WITH NOTE INDICATING TYPE OF BRICK (COMMON, FACE, ETC.)	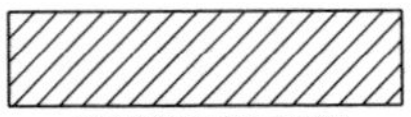 COMMON OR FACE 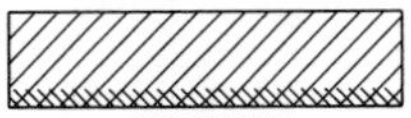 FIREBRICK	SAME AS PLAN VIEWS
CONCRETE	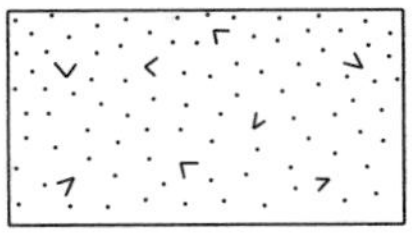	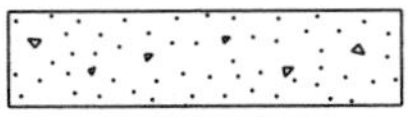 LIGHTWEIGHT 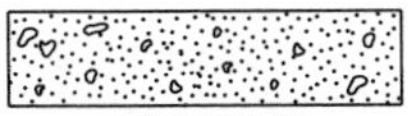 STRUCTURAL	SAME AS PLAN VIEWS
CONCRETE BLOCK	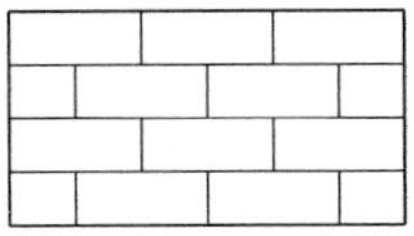	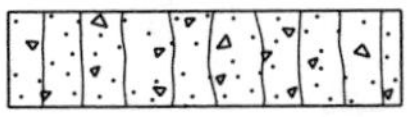 OR 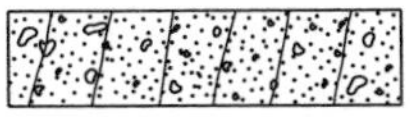	 OR 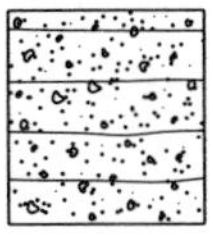
STONE	 CUT STONE 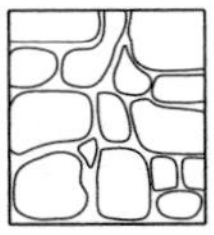 RUBBLE	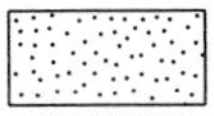 CUT STONE 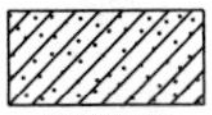 RUBBLE 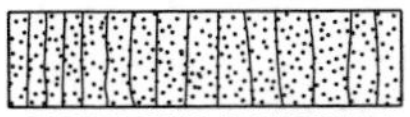 CAST STONE (CONCRETE)	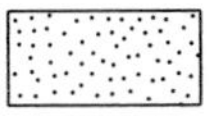 CUT STONE 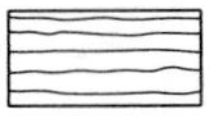 CAST STONE (CONCRETE) 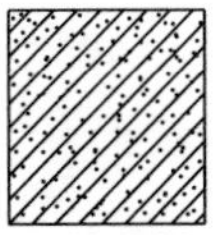 RUBBLE OR CUT STONE
WOOD	 SIDING 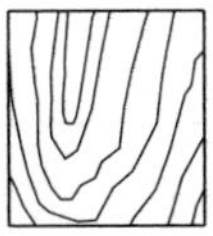 PANEL	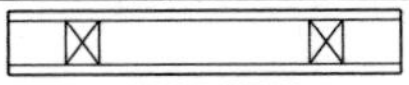 WOOD STUD 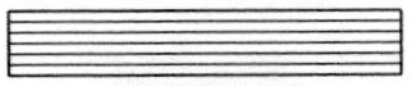 REMODELING  DISPLAY	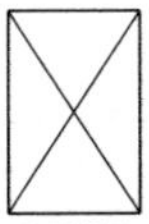 ROUGH MEMBERS  FINISHED MEMBERS
PLASTER	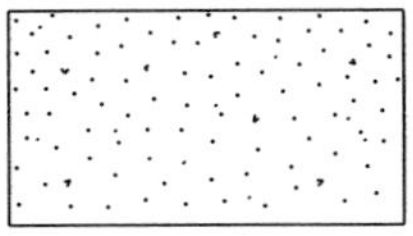	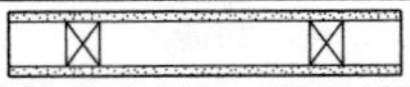 WOOD STUD, LATH, AND PLASTER 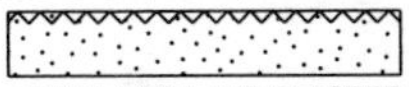 METAL LATH AND PLASTER 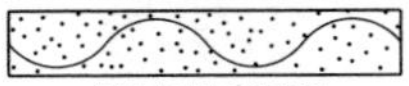 SOLID PLASTER	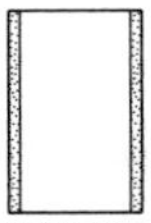 LATH AND PLASTER
ROOFING	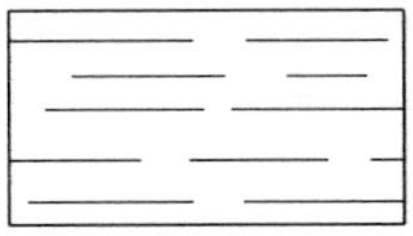 SHINGLES	SAME AS ELEVATION VIEW	
GLASS	 OR  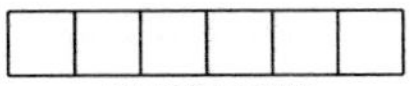 GLASS BLOCK	 GLASS 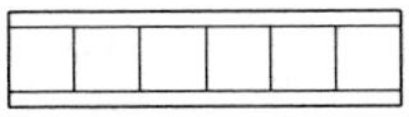 GLASS BLOCK	 SMALL SCALE 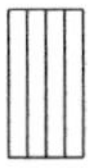 LARGE SCALE

. . .ARCHITECTURAL SYMBOLS

Material	Elevation	Plan	Section
FACING TILE	CERAMIC TILE	FLOOR TILE	CERAMIC TILE LARGE SCALE; CERAMIC TILE SMALL SCALE
STRUCTURAL CLAY TILE			SAME AS PLAN VIEW
INSULATION		LOOSE FILL OR BATTS; RIGID; SPRAY FOAM	SAME AS PLAN VIEWS
SHEET METAL FLASHING		OCCASIONALLY INDICATED BY NOTE	
METALS OTHER THAN FLASHING	INDICATED BY NOTE OR DRAWN TO SCALE	SAME AS ELEVATION	SMALL SCALE; STEEL; CAST IRON; ALUMINUM; BRONZE OR BRASS
STRUCTURAL STEEL	INDICATED BY NOTE OR DRAWN TO SCALE	OR	REBARS; SMALL SCALE; LARGE SCALE; L-ANGLES, S-BEAMS, ETC.

PLOT PLAN SYMBOLS

N — NORTH	FIRE HYDRANT	WALK	E OR — ELECTRIC SERVICE
POINT OF BEGINNING (POB)	MAILBOX	IMPROVED ROAD	G OR — NATURAL GAS LINE
UTILITY METER OR VALVE	MANHOLE	UNIMPROVED ROAD	W OR — WATER LINE
POWER POLE AND GUY	TREE	BUILDING LINE	T OR — TELEPHONE LINE
LIGHT STANDARD	BUSH	PROPERTY LINE	NATURAL GRADE
TRAFFIC SIGNAL	HEDGE ROW	PROPERTY LINE	FINISH GRADE
STREET SIGN	FENCE	TOWNSHIP LINE	+ XX.00′ — EXISTING ELEVATION

ALPHABET OF LINES

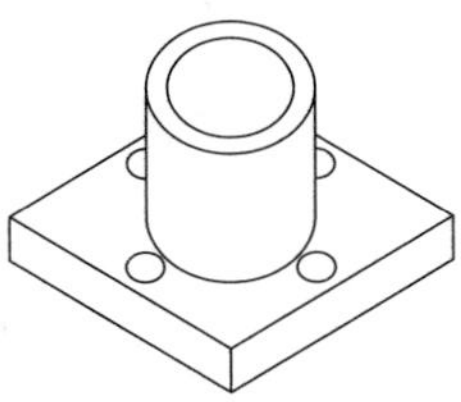

NAME AND USE	CONVENTIONAL REPRESENTATION	EXAMPLE
OBJECT LINE Define shape. Outline and detail objects.	THICK	OBJECT LINE 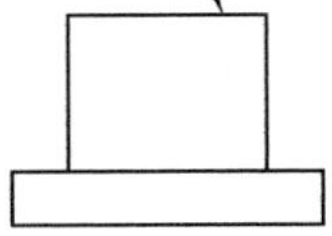
HIDDEN LINE Show hidden features.	$\frac{1}{8}''$ (3 mm) THIN $\frac{1}{32}''$ (0.75 mm)	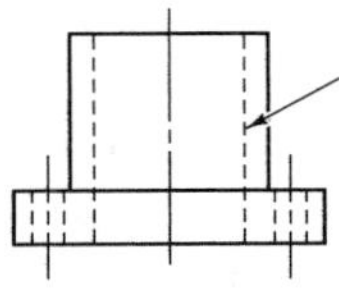 HIDDEN LINE
CENTER LINE Locate centerpoints of arcs and circles.	$\frac{1}{16}''$ (1.5 mm) THIN $\frac{1}{8}''$ (3 mm) $\frac{3}{4}''$ (18 mm) TO $1\frac{1}{2}''$ (36 mm)	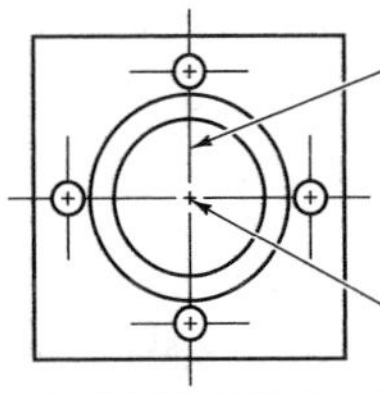 CENTER LINE CENTERPOINT
DIMENSION LINE Show size or location. EXTENSION LINE Define size or location.	DIMENSION LINE DIMENSION THIN 2′-6″ EXTENSION LINE	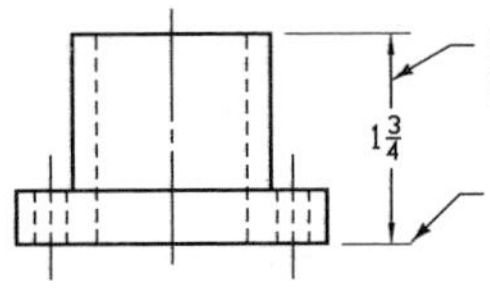 DIMENSION LINE $1\frac{3}{4}$ EXTENSION LINE
LEADER Call out specific features.	OPEN ARROWHEAD THIN X CLOSED ARROWHEAD 3X	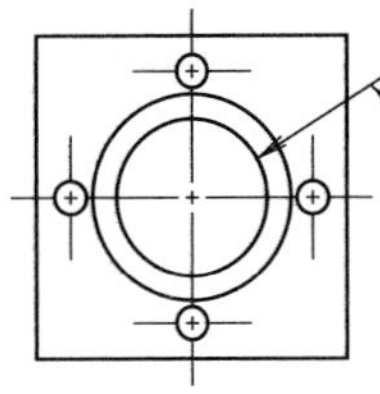 $1\frac{1}{2}$ DRILL LEADER
CUTTING PLANE Show internal features.	THICK $\frac{1}{8}''$ (3 mm) $\frac{1}{16}''$ (1.5 mm) A A $\frac{3}{4}''$ (18 mm) TO $1\frac{1}{2}''$ (36 mm)	A 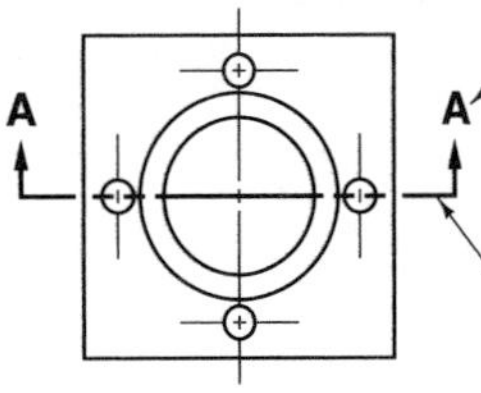 A LETTER IDENTIFIES SECTION VIEW CUTTING PLANE LINE
SECTION LINE Identify internal features.	$\frac{1}{16}''$ (1.5 mm) THIN	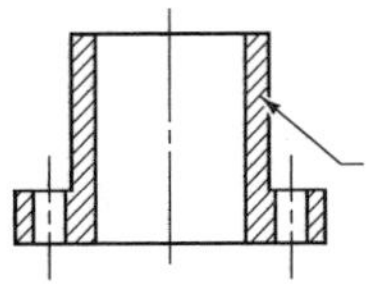 SECTION LINES
BREAK LINE Show long breaks. BREAK LINE Show short breaks.	$\frac{3}{4}''$ (18 mm) TO $1\frac{1}{2}''$ (36 mm) THIN FREEHAND THICK	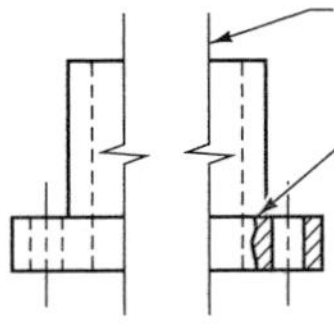 LONG BREAK LINE SHORT BREAK LINE

THREE-PHASE VOLTAGE VALUES
For 208 V × 1.732, use 360
For 230 V × 1.732, use 398
For 240 V × 1.732, use 416
For 440 V × 1.732, use 762
For 460 V × 1.732, use 797
For 480 V × 1.732, use 831
For 2400 V × 1.732, use 4157
For 4160 V × 1.732, use 7205

Ohm's Law

Ohm's law is the relationship between the voltage, current, and resistance in an electrical circuit. Ohm's law states that current in a circuit is proportional to the voltage and inversely proportional to the resistance. It is written $I = E/R$, $R = E/I$, and $E = R \times I$.

Power Formula

The *power formula* is the relationship between the voltage, current, and power in an electrical circuit. The power formula states that the power in a circuit is equal to the voltage times the current. It is written $P = E \times I$, $E = P/I$, and $I = P/E$. Any value in these relationships is found using Ohm's Law and Power Formula.

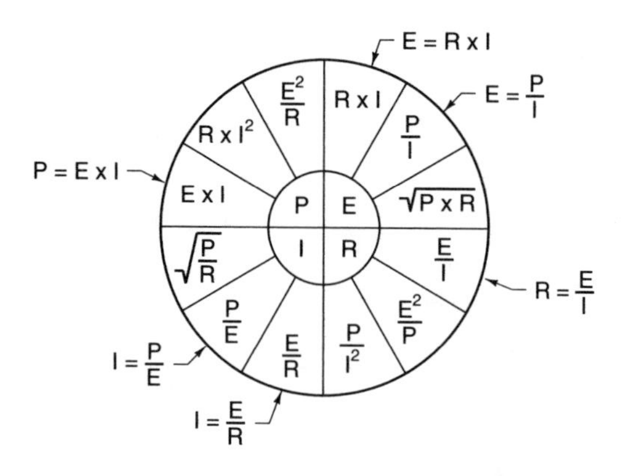

OHM'S LAW AND POWER FORMULA

Phase	To Find	Use Formula	Example: Given	Example: Find	Example: Solution
1φ	I	$I = \frac{VA}{V}$	32,000 VA, 240 V	I	$I = \frac{VA}{V}$ $I = \frac{32{,}000\ VA}{240\ V}$ I = **133 A**
1φ	VA	$VA = I \times V$	100 A, 240 V	VA	$VA = I \times V$ $VA = 100\ A \times 240\ V$ VA = **24,000 VA**
1φ	V	$V = \frac{VA}{I}$	42,000 VA, 350 A	V	$V = \frac{VA}{I}$ $V = \frac{42{,}000\ VA}{350\ A}$ V = **120 V**
3φ	I	$I = \frac{VA}{V \times \sqrt{3}}$	72,000 VA, 208 V	I	$I = \frac{VA}{V \times \sqrt{3}}$ $I = \frac{72{,}000\ VA}{360\ V}$ I = **200 A**
3φ	VA	$VA = I \times V \times \sqrt{3}$	2 A, 240 V	VA	$VA = I \times V \times \sqrt{3}$ $VA = 2 \times 416$ VA = **832 VA**

Table title: POWER FORMULAS—1φ, 3φ

AC/DC FORMULAS

To Find	DC	AC		
		1ϕ, 115 or 220 V	1ϕ, 208, 230, or 240 V	3ϕ—All Voltages
I, HP known	$\frac{HP \times 746}{E \times E_{ff}}$	$\frac{HP \times 746}{E \times E_{ff} \times PF}$	$\frac{HP \times 746}{E \times E_{ff} \times PF}$	$\frac{HP \times 746}{1.73 \times E \times E_{ff} \times PF}$
I, kW known	$\frac{kW \times 1000}{E}$	$\frac{kW \times 1000}{E \times PF}$	$\frac{kW \times 1000}{E \times PF}$	$\frac{kW \times 1000}{1.73 \times E \times PF}$
I, kVA known		$\frac{kVA \times 1000}{E}$	$\frac{kVA \times 1000}{E}$	$\frac{kVA \times 1000}{1.763 \times E}$
kW	$\frac{I \times E}{1000}$	$\frac{I \times E \times PF}{1000}$	$\frac{I \times E \times PF}{1000}$	$\frac{I \times E \times 1.73 \times PF}{1000}$
kVA		$\frac{I \times E}{1000}$	$\frac{I \times E}{1000}$	$\frac{I \times E \times 1.73}{1000}$
HP (output)	$\frac{I \times E \times E_{ff}}{746}$	$\frac{I \times E \times E_{ff} \times PF}{746}$	$\frac{I \times E \times E_{ff} \times PF}{746}$	$\frac{I \times E \times 1.73 \times E_{ff} \times PF}{746}$

E_{ff} = efficiency

HORSEPOWER FORMULAS

To Find	Use Formula	Example		
		Given	Find	Solution
HP	$HP = \frac{I \times E \times E_{ff}}{746}$	240 V, 20 A, 85% E_{ff}	HP	$HP = \frac{I \times E \times E_{ff}}{746}$ $HP = \frac{20\ I \times 240\ E \times 85\%}{746}$ $HP = \mathbf{5.5}$
I	$I = \frac{HP \times 746}{E \times E_{ff} \times PF}$	10 HP, 240 V, 90% E_{ff}, 88% PF	I	$I = \frac{HP \times 746}{E \times E_{ff} \times PF}$ $I = \frac{10\ HP \times 746}{240\ V \times 90\% \times 88\%}$ $I = \mathbf{39\ A}$

VOLTAGE DROP FORMULAS – 1ϕ, 3ϕ

Phase	To Find	Use Formula	Example		
			Given	Find	Solution
1ϕ	VD	$VD = \frac{2 \times R \times L \times I}{CM}$	240 V, 40 A, 60′ L, 16,510 CM, 12 R	VD	$VD = \frac{2 \times R \times L \times I}{CM}$ $VD = \frac{2 \times 12 \times 40 \times 60}{16{,}510}$ $VD = \mathbf{3.5}$
3ϕ	VD	$VD = \frac{2 \times R \times L \times I}{CM}$	208 V, 110 A, 75′ L, 66,360 CM, 12 R, .866 multiplier	VD	$VD = \frac{2 \times R \times L \times I}{CM}$ $VD = \frac{2 \times 12 \times 75 \times 110}{66{,}360}$ $VD = 2.98 \times .866$ $VD = \mathbf{2.58}$

* $\frac{\sqrt{3}}{2} = .866$

LOCKED ROTOR CURRENT

Apparent, 1ϕ	Apparent, 3ϕ	True, 1ϕ	True, 3ϕ
$LRC = \frac{1000 \times HP \times kVA/HP}{V}$ where LRC = locked rotor current (in amps) 1000 = multiplier for kilo HP = horsepower kVA/HP = kilovolt amps per horsepower V = volts	$LRC = \frac{1000 \times HP \times kVA/HP}{V \times \sqrt{3}}$ where LRC = locked rotor current (in amps) 1000 = multiplier for kilo HP = horsepower kVA/HP = kilovolt amps per horsepower V = volts $\sqrt{3}$ = 1.73	$LRC = \frac{1000 \times HP \times kVA/HP}{V \times PF \times E_{ff}}$ where LRC = locked rotor current (in amps) 1000 = multiplier for kilo HP = horsepower kVA/HP = kilovolt amps per horsepower V = volts PF = power factor E_{ff} = motor efficiency	$LRC = \frac{1000 \times HP \times kVA/HP}{V \times \sqrt{3} \times PF \times E_{ff}}$ where LRC = locked rotor current (in amps) 1000 = multiplier for kilo HP = horsepower kVA/HP = kilovolt amps per horsepower V = volts $\sqrt{3}$ = 1.73

MAXIMUM OCPD

$OCPD = FLC \times R_M$

where

FLC = full-load current (from motor nameplate or NEC® Table 430-150)

R_M = maximum rating of OCPD

Motor Type	Code Letter	FLC (%)				
		Motor Size	TDF	NTDF	ITB	ITCB
AC*	—	—	175	300	150	700
AC*	A	—	150	150	150	700
AC*	B–E	—	175	250	200	700
AC*	F–V	—	175	300	250	700
DC	—	⅛ to 50 HP	150	150	150	250
DC	—	Over 50 HP	150	150	150	175

EFFICIENCY

Input and Output Power Known	Horsepower and Power Loss Known
$E_{ff} = \frac{P_{out}}{P_{in}}$ where E_{ff} = efficiency (%) P_{out} = output power (W) P_{in} = input power (W)	$E_{ff} = \frac{746 \times HP}{746 \times HP + W_l}$ where E_{ff} = efficiency (%) 746 = constant HP = horsepower W_l = watts lost

VOLTAGE UNBALANCE

$$V_u = \frac{V_d}{V_a} \times 100$$

where

V_u = voltage unbalance (%)

V_d = voltage deviation (V)

V_a = voltage average (V)

100 = constant

POWER

Power Consumed	Operating Cost	Annual Savings
$P = \frac{HP \times 746}{E_{ff}}$ where P = power consumed (W) HP = horsepower 746 = constant E_{ff} = efficiency (%)	$C_{/hr} = \frac{P_{/hr} \times C_{/kWh}}{1000}$ where $C_{/hr}$ = operating cost per hour $P_{/hr}$ = power consumed per hour $C_{/kWh}$ = cost per kilowatt hour 1000 = constant to remove kilo	$S_{Ann} = C_{Ann\ Std} - C_{Ann\ Eff}$ S_{Ann} = annual cost savings $C_{Ann\ Std}$ = annual operating cost for standard motor $C_{Ann\ Eff}$ = annual operating cost for energy-efficient motor

CAPACITORS

Connected in Series		Connected in Parallel	Connected in Series/Parallel
Two Capacitors	Three or More Capacitors		
$C_T = \frac{C_1 \times C_2}{C_1 + C_2}$ where C_T = total capacitance (in µF) C_1 = capacitance of capacitor 1 (in µF) C_2 = capacitance of capacitor 2 (in µF)	$\frac{1}{C_T} = \frac{1}{C_1} + \frac{1}{C_2} + \ldots$	$C_T = C_1 + C_2 + \ldots$	1. Calculate the capacitance of the parallel branch. $C_T = C_1 + C_2 + \ldots$ 2. Calculate the capacitance of the series combination. $C_T = \frac{C_1 \times C_2}{C_1 + C_2}$

GEAR REDUCER

Output Torque	Output Speed	Output Horsepower
$O_T = I_T \times R_R \times R_E$ where O_T = output torque (in lb-ft) I_T = input torque (in lb-ft) R_R = gear reducer ratio R_E = reducer efficiency (percentage)	$O_S = \frac{I_S}{R_R} \times R_E$ where O_S = output speed (in rpm) I_S = input speed (in rpm) R_R = gear reducer ratio R_E = reducer efficiency (percentage)	$O_{HP} = I_{HP} \times R_E$ where O_{HP} = output horsepower I_{HP} = input horsepower R_E = reducer efficiency (percentage)

HORSEPOWER

Current and Voltage Known	Speed and Torque Known
$HP = \frac{E \times I \times E_{ff}}{746}$ where HP = horsepower E = voltage (volts) I = current (amps) E_{ff} = efficiency	$HP = \frac{rpm \times T}{5252}$ where HP = horsepower rpm = revolutions per minute T = torque (lb-ft)

TEMPERATURE CONVERSIONS

Convert °C to °F	Convert °F to °C
$°F = (1.8 \times °C) + 32$	$°C = \frac{(°F - 32)}{1.8}$

MOTOR TORQUE

Torque	Starting Torque	Nominal Torque Rating
$T = \frac{HP \times 5252}{rpm}$ where T = torque HP = horsepower 5252 = constant $\left(\frac{33{,}000 \text{ lb-ft}}{\pi \times 2} = 5252\right)$ rpm = revolutions per minute	$T = \frac{HP \times 5252}{rpm} \times \%$ where HP = horsepower 5252 = constant $\left(\frac{33{,}000 \text{ lb-ft}}{\pi \times 2} = 5252\right)$ rpm = revolutions per minute % = motor class percentage	$T = \frac{HP \times 63{,}000}{rpm}$ where T = nominal torque rating (in lb-in) 63,000 = constant HP = horsepower rpm = revolutions per minute

AC MOTOR CHARACTERISTICS						
Motor Type	Typical Voltage	Starting Ability (Torque)	Size (HP)	Speed Range (rpm)	Cost*	Typical Uses
1φ						
Shaded-pole	115 V, 230 V	Very low 50% to 100% of full load	Fractional ½ HP to ⅓ HP	Fixed 900, 1200, 1800, 3600	Very low 75% to 85%	Light-duty applications such as small fans, hair dryers, blowers, and computers
Split-phase	115 V, 230 V	Low 75% to 200% of full load	Fractional ⅓ HP or less	Fixed 900, 1200, 1800, 3600	Low 85% to 95%	Low-torque applications such as pumps, blowers, fans, and machine tools
Capacitor-start	115 V, 230 V	High 200% to 350% of full load	Fractional to 3 HP	Fixed 900, 1200, 1800	Low 90% to 110%	Hard-to-start loads such as refrigerators, air compressors, and power tools
Capacitor-run	115 V, 230 V	Very low 50% to 100% of full load	Fractional to 5 HP	Fixed 900, 1200, 1800	Low 90% to 110%	Applications that require a high running torque such as pumps and conveyors
Capacitor-start-and-run	115 V, 230 V	Very high 350% to 450% of full load	Fractional to 10 HP	Fixed 900, 1200, 1800	Low 100% to 115%	Applications that require both a high starting and running torque such as loaded conveyors
3φ						
Induction	230 V, 460 V	Low 100% to 175% of full load	Fractional to over 500 HP	Fixed 900, 1200, 3600	Low 100%	Most industrial applications
Wound rotor	230 V, 460 V	High 200% to 300% of full load	½ HP to 200 HP	Varies by changing resistance in rotor	Very high 250% to 350%	Applications that require high torque at different speeds such as cranes and elevators
Synchronous	230 V, 460 V	Very low 40% to 100% of full load	Fractional to 250 HP	Exact constant speed	High 200% to 250%	Applications that require very slow speeds and correct power factors

* based on standard 3φ induction motor

DC AND UNIVERSAL MOTOR CHARACTERISTICS						
Motor Type	Typical Voltage	Starting Ability (Torque)	Size (HP)	Speed Range (rpm)	Cost*	Typical Uses
DC Series	12 V, 90 V, 120 V, 180 V	Very high 400% to 450% of full load	Fractional to 100 HP	Varies 0 to full speed	High 175% to 225%	Applications that require very high torque such as hoists and bridges
Shunt	12 V, 90 V, 120 V, 180 V	Low 125% to 250% of full load	Fractional to 100 HP	Fixed or adjustable below full speed	High 175% to 225%	Applications that require better speed control than a series motor such as woodworking machines
Compound	12 V, 90 V, 120 V, 180 V	High 300% to 400% of full load	Fractional to 100 HP	Fixed or adjustable	High 175% to 225%	Applications that require high torque and speed control such as printing presses, conveyors, and hoists
Permanent-magnet	12 V, 24 V, 36 V, 120 V	Low 100% to 200% of full load	Fractional	Varies from 0 to full speed	High 150% to 200%	Applications that require small DC-operated equipment such as automobile power windows, seats, and sun roofs
Stepping	5 V, 12 V, 24 V	Very low** .5 to 5000 oz/in.	Size rating is given as holding torque and number of steps	Rated in number of steps per sec (maximum)	Varies based on number of steps and rated torque	Applications that require low torque and precise control such as indexing tables and printers
AC/DC Universal	115 VAC, 230 VAC, 12 VDC, 24 VDC, 36 VDC, 120 VDC	High 300% to 400% of full load	Fractional	Varies 0 to full speed	High 175% to 225%	Most portable tools such as drills, routers, mixers, and vacuum cleaners

* based on standard 3φ induction motor

** torque is rated as holding torque

OVERCURRENT PROTECTION DEVICES						
Motor Type	Code Letter	FLC (%)				
		Motor Size	TDF	NTDF	ITB	ITCB
AC*	—	—	175	300	150	700
AC*	A	—	150	150	150	700
AC*	B–E	—	175	250	200	700
AC*	F–V	—	175	300	250	700
DC	—	⅛ to 50 HP	150	150	150	150
DC	—	Over 50 HP	150	150	150	175

* full-voltage and resistor starting

ENCLOSURE TYPES				
Type	**Use**	**Service conditions**	**Tests**	**Comments**
1	Indoor	No unusual	Rod entry, rust resistance	
3	Outdoor	Windblown dust, rain, sleet, and ice on enclosure	Rain, external icing, dust, and rust resistance	Do not provide protection against internal condensation or internal icing
3R	Outdoor	Falling rain and ice on enclosure	Rod entry, rain, external icing, and rust resistance	Do not provide protection against dust, internal condensation, or internal icing
4	Indoor/outdoor	Windblown dust and rain, splashing water, hose-directed water, and ice on enclosure	Hosedown, external icing, and rust resistance	Do not provide protection against internal condensation or internal icing
4X	Indoor/outdoor	Corrosion, windblown dust and rain, splashing water, hose-directed water, and ice on enclosure	Hosedown, external icing, and corrosion resistance	Do not provide protection against internal condensation or internal icing
6	Indoor/outdoor	Occasional temporary submersion at a limited depth		
6P	Indoor/outdoor	Prolonged submersion at a limited depth		
7	Indoor locations classified as Class I, Groups A, B, C, or D, as defined in the NEC®	Withstand and contain an internal explosion of specified gases, contain an explosion sufficiently so an explosive gas-air mixture in the atmosphere is not ignited	Explosion, hydrostatic, and temperature	Enclosed heat-generating devices shall not cause external surfaces to reach temperatures capable of igniting explosive gas-air mixtures in the atmosphere.
9	Indoor locations classified as Class II, Groups E or G, as defined in the NEC®	Dust	Dust penetration, temperature, and gasket aging	Enclosed heat-generating devices shall not cause external surfaces to reach temperatures capable of igniting explosive gas-air mixtures in the atmosphere
12	Indoor	Dust, falling dirt, and dripping noncorrosive liquids	Drip, dust, and rust resistance	Do not provide protection against internal condensation
13	Indoor	Dust, spraying water, oil, and noncorrosive coolant	Oil explosion and rust resistance	do not provide protection against internal condensation

RACEWAYS	
EMT	Electrical Metallic Tubing
ENT	Electrical Nonmetallic Tubing
FMC	Flexible Metal Conduit
FMT	Flexible Metallic Tubing
IMC	Intermediate Metal Conduit
LTFMC	Liquidtight Flexible Metal Conduit
LTFNMC	Liquidtight Flexible Nonmetallic Conduit
RMC	Rigid Metal Conduit
RNMC	Rigid Nonmetallic Conduit

FUSES AND ITCBs	
Increase	**Standard Ampere Ratings**
5	15, 20, 25, 30, 35, 40, 45
10	50, 60, 70, 80, 90, 100, 110
25	125, 150, 175, 200, 225
50	250, 300, 350, 400, 450
100	500, 600, 700, 800
200	1000, 1200
400	1600, 2000
500	2500
1000	3000, 4000, 5000, 6000

1 A, 3 A, 6 A, 10 A, and 601 A are additional standard ratings for fuses.

CABLES	
AC	Armored Cable
BX	Tradename for AC
FCC	Flat Conductor Cable
IGS	Integrated Gas Spacer Cable
MC	Metal-Clad Cable
MI	Mineral-Insulated, Metal Sheathed Cable
MV	Medium Voltage
NM	Nonmetallic-Sheathed Cable (dry)
NMC	Nonmetallic-Sheathed Cable (dry or damp)
NMS	Nometallic-Sheathed Cable (dry)
SE	Service-Entrance Cable
TC	Tray Cable
UF	Underground Feeder Cbale
USE	Underground Service-Entrance Cable

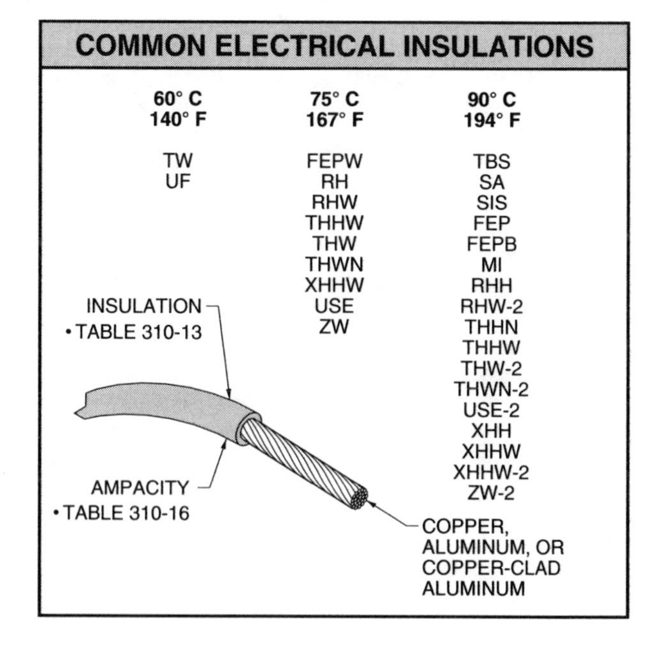

STANDARD CALCULATION: ONE-FAMILY DWELLING

1. GENERAL LIGHTING: *Table 220-3(b)*

_________ sq ft × 3 VA = _________ VA

Small appliances: *220-16(a)*

_________ VA × _________ circuits = _________ VA

Laundry: *220-16(b)*

_________ VA × 1 = _________ VA

_________ VA

Applying Demand Factors: *Table 220-11*

First 3000 VA × 100% = 3000 VA

Next _________ VA × 35% = _________ VA

Remaining _________ VA × 25% = _________ VA

		PHASES	NEUTRAL
Total	_________ VA	_________ VA	_________ VA

2. FIXED APPLIANCES: *220-17*

Dishwasher = _________ VA

Disposer = _________ VA

Compactor = _________ VA

Water heater = _________ VA

_________________ = _________ VA

_________________ = _________ VA

_________________ = _________ VA

(120 V Loads × 75%)

Total _________ VA × 75% = _________ VA _________ VA _________ VA

3. DRYER: *220-18; Table 220-18*

_________ VA × _________% = _________ VA _________ VA × 70% = _________ VA

4. COOKING EQUIPMENT: *Table 220-19; Notes*

Col A _________ VA × _________% = _________ VA

Col B _________ VA × _________% = _________ VA

Col C _________ VA × _________% = _________ VA

Total _________ VA _________ VA × 70% = _________ VA

5. HEATING or A/C: *220-21*

Heating unit = _________ VA × 100% = _________ VA

A/C unit = _________ VA × 100% = _________ VA

Heat pump = _________ VA × 100% = _________ VA

Largest Load _________ VA _________ VA _________ VA

6. LARGEST MOTOR: *220-14*

ϕ _________ VA × 25% = _________ VA _________ VA

N _________ VA × 25% = _________ VA _________ VA

1ϕ service: PHASES $I = \frac{\text{VA}}{\text{V}}$ = _________ A

NEUTRAL $I = \frac{\text{VA}}{\text{V}}$ = _________ A

[ ] VA [ ] VA

220-22; First 200 A × 100% = 200 A

Remaining _________ A × 70% = _________ A

Total _________ A

STANDARD CALCULATION: MULTIFAMILY DWELLING

1. GENERAL LIGHTING: *Table 220-3(b)*

_________ sq ft × 3 VA × _________ units = _________ VA
_________ sq ft × 3 VA × _________ units = _________ VA
_________ sq ft × 3 VA × _________ units = _________ VA

Small appliances: *220-16(a)*

_________ VA × _________ circuits × _________ units = _________ VA

Laundry: *220-16(b)*

_________ VA × 1 × _________ units = _________ VA
_________ VA

Applying Demand Factors: *Table 220-11*

First 3000 VA × 100% = 3000 VA
Next _________ VA × 35% = _________ VA
Remaining _________ VA × 25% = _________ VA
Total _________ VA

	PHASES	NEUTRAL
Total	_________ VA	_________ VA

2. FIXED APPLIANCES: *220-17*

Dishwasher = _________ VA
Disposer = _________ VA
Compactor = _________ VA
Water heater = _________ VA
_______________ = _________ VA
_______________ = _________ VA
_______________ = _________ VA

Total _________ VA × __ units × 75% = _________ VA | _________ VA | (120 V Loads × units × 75%) _________ VA

3. DRYER: *220-18; Table 220-18*

_________ VA × _________ units × _________ % = _________ VA | _________ VA × 70% = _________ VA

4. COOKING EQUIPMENT: *Table 220-19; Notes*

Col A _________ units = _________ VA × _________ % = _________ VA
Col B _________ VA × _________ units × _________ % = _________ VA
Col C _________ VA × _________ units × _________ % = _________ VA
Total _________ VA | _________ VA × 70% = _________ VA

5. HEATING or A/C: *220-21*

Heating unit = _________ VA × 100% × _________ units = _________ VA
A/C unit = _________ VA × 100% × _________ units = _________ VA
Heat pump = _________ VA × 100% × _________ units = _________ VA
Largest Load _________ VA | _________ VA | _________ VA

6. LARGEST MOTOR: *220-14*

ϕ _________ VA × 25% = _________ VA | _________ VA
N _________ VA × 25% = _________ VA | _________ VA

1ϕ service: PHASES $I = \frac{______ VA}{V} = $ _________ A

NEUTRAL $I = \frac{______ VA}{V} = $ _________ A

[ ] VA [ ] VA

3ϕ service: $I = \frac{______ VA}{V \times \sqrt{3}} = $ _________ A

$I = \frac{______ VA}{V \times \sqrt{3}} = $ _________ A

220-22; First 200 A × 100% = 200 A
Remaining _________ A × 70% = _________ A
_________ A

OPTIONAL CALCULATION: ONE-FAMILY DWELLING

1. HEATING or A/C: *Table 220-30(1–5)*

Heating units (3 or less) = ________ VA × 65% = ________ VA

Heating units (4 or more) = ________ VA × 40% = ________ VA

A/C unit = ________ VA × 100% = ________ VA

Heat pump = ________ VA × 100% = ________ VA

Largest Load ________ VA

Total ________ VA

PHASES

________ VA

2. OTHER LOADS: *220-30(b)*

General lighting: *220-30(b)(2)*

________ sq ft × 3 VA ________ VA

Small appliance and laundry loads: *220-30(b)(1)*

________ VA × ________ circuits = ________ VA

Special loads: *220-30(b)(3)(4)*

Dishwasher = ________ VA

Disposer = ________ VA

Compactor = ________ VA

Water heater = ________ VA

________ = ________ VA

________ = ________ VA

________ = ________ VA

________ = ________ VA

________ = ________ VA

________ VA ________ VA

Total ________ VA

Applying Demand Factors: *Table 220-30*

First 10,000 VA × 100% = 10,000 VA

Remaining ________ VA × 40% = ________ VA

Total ________ VA ________ VA

NEUTRAL (Loads from Standard Calculation)

1. General lighting = ________ VA
2. Fixed appliances = ________ VA
3. Dryer = ________ VA
4. Cooking equipment = ________ VA
5. Heating or A/C = ________ VA
6. Largest motor = ________ VA

Total [ ] VA

1ϕ service: PHASES $I = \frac{______ VA}{V} = $ ________ A

NEUTRAL $I = \frac{______ VA}{V} = $ ________ A

[ ] VA

OPTIONAL CALCULATION: MULTIFAMILY DWELLING

1. HEATING or A/C: *220-32(c)(5)*

Heating unit = ________ VA × 100% × ________ units = ________ VA
A/C unit = ________ VA × 100% × ________ units = ________ VA
Heat pump = ________ VA × 100% × ________ units = ________ VA
Largest Load ________ VA
Total

PHASES

________ VA

2. OTHER LOADS: *220-32*

General lighting: *220-32(c)(2)*
________ sq ft × 3 VA × ________ units = ________ VA ________ VA
________ sq ft × 3 VA × ________ units = ________ VA
________ sq ft × 3 VA × ________ units = ________ VA
________ sq ft × 3 VA × ________ units = ________ VA

Small appliance and laundry loads: *220-32(c)(1)*
________ VA × ________ circuits × ________ units = ________ VA ________ VA

Special loads: *220-32(c)(3)*
Dishwasher = ________ VA
Disposer = ________ VA
Compactor = ________ VA
Water heater = ________ VA
________________ = ________ VA
________________ = ________ VA
________________ = ________ VA
________________ = ________ VA
________________ = ________ VA

Total ________ VA × ________ units = ________ VA ________ VA

Total Connected Load ________ VA

Applying Demand Factors: *Table 220-32*
________ VA × ________% = ________ VA [ ] VA

NEUTRAL (Loads from Standard Calculation)

1. General lighting = ________ VA
2. Fixed appliances = ________ VA
3. Dryer = ________ VA
4. Cooking equipment = ________ VA
5. Heating or A/C = ________ VA
6. Largest motor = ________ VA

Total [ ] VA

1ϕ service: PHASES $I = \frac{\text{VA}}{\text{V}}$ = ________ A

NEUTRAL $I = \frac{\text{VA}}{\text{V}}$ = ________ A

3ϕ service: $I = \frac{\text{VA}}{\text{V} \times \sqrt{3}}$ = ________ A

$I = \frac{\text{VA}}{\text{V} \times \sqrt{3}}$ = ________ A

220-22; First 200 A × 100% = 200 A
Remaining ________ A × 70% = ________ A
________ A

GLOSSARY

A

accessible: Equipment admits close approach and is not guarded by locked doors, elevation, etc.

adjustable-trip CBs (ATCBs): CBs whose trip setting can be changed by adjusting the ampere setpoint, trip time characteristics, or both, within a particular range.

ambient temperature: The temperature of air around a piece of equipment.

ampacity: The current that a conductor can carry continuously, under the conditions of use.

appliance: Any utilization equipment which performs one or more functions, such as clothes washing, air conditioning, cooking, etc.

appliance branch circuit: A branch circuit that supplies energy to one or more outlets to which appliances are to be connected.

approved: Acceptable to the AHJ.

armored cable (AC): A factory assembly that contains the conductors within a jacket made of a spiral wrap of steel.

askarel: A group of nonflammable synthetic chlorinated hydrocarbons that were once used where nonflammable insulating oils were required.

autotransformers: Single-winding transformers that share a common winding between the primary and secondary circuits.

auxiliary gutter: A sheet-metal enclosure equipped with hinged or removable covers that is used to supplement wiring space.

B

bare conductor: A conductor with no insulation or covering of any type.

bathroom: An area with a basin and one or more of a toilet, tub, or shower.

bend: Any change in direction of a raceway.

bonding: Joining metal parts to form a continuous path to conduct safely any current that is commonly imposed.

box: A metallic or nonmetallic electrical enclosure used for equipment, devices, and pulling or terminating conductors.

branch circuit: The portion of the electrical circuit between the last overcurrent device (fuse or CB) and the outlets or utilization equipment.

branch-circuit rating: The ampere rating or setting of the overcurrent device protecting the conductors.

building: A stand-alone structure or a structure which is separated from adjoining structures by fire walls.

bushing: A fitting placed on the end of a conduit to protect the conductor's insulation from abrasion.

busway: A sheet metal enclosure that contains factory-assembled aluminum or copper busbars which are supported on insulators.

C

cable: A factory assembly with two or more conductors and an overall covering.

cable assembly: A flexible assembly containing multiconductors with a protective outer sheath.

cable tray system (CTS): An assembly of sections and associated fittings which form a rigid structural system used to support cables and raceways.

cadwelding: A welding process used to make electrical connections of copper to copper or copper to steel in which no outside source of heat or power is required.

cartridge fuse: A fuse constructed of a metallic link(s) which is designed to open at predetermined current levels to protect circuit conductors and equipment.

circuit breaker (CB): A device which opens and closes circuits by nonautomatic means and opens circuits automatically when a predetermined overcurrent exists.

circular mil: A measurement used to determine the cross-sectional area of a conductor.

Class I location: A hazardous location in which sufficient quantities of flammable gases and vapors are present in the air to cause an explosion or ignite the hazardous materials.

Class II location: A hazardous location in which sufficient quantities of combustible dust are present in the air to cause an explosion or ignite the hazardous materials.

Class III location: A hazardous location in which easily-ignitible fibers or flyings are present in the air but not in a sufficient quantity to cause an explosion or ignite the hazardous materials.

conductor: A slender rod or wire that is used to control the flow of electrons in an electrical circuit.

conduit body: A conduit fitting that provides access to the raceway system through a removable cover at a junction or termination point.

conduit seal: A fitting which is inserted into runs of conduit to isolate certain electrical apparatus from atmospheric hazards.

continuous load: A load in which the maximum current may continue for three hours or more.

controller: The device in a motor circuit which turns the motor ON or OFF.

cover: The shortest distance measured between a point on the top surface of any direct-buried conductor, cable, conduit, or other raceway and the top surface of finished grade, concrete, or similar cover.

covered conductor: A conductor not encased in a material recognized by the NEC®.

current-limiting fuses: Fuses that open a circuit in less than ½ of a cycle to protect the circuit components from damaging short-circuit currents.

current transformer: A transformer that creates a constant ratio of primary to secondary current instead of attempting to maintain a constant ratio of primary to secondary voltage.

D

damp location: A partially protected area subject to some moisture.

dead front: A cover required for the operation of a plug or connector.

demand factor: The ratio of the maximum demand of a system, or part of a system, to the total connected load of a system or the part of the system under consideration.

device: Any unit of an electrical system that carries, but does not use electricity.

devices: Electrical components, such as receptacles and switches, that are designed to carry, but not utilize, electricity.

device box: A box which houses an electrical device.

division: The classification assigned to each Class based upon the likelihood of the presence of the hazardous substance in the atmosphere.

Division 1 location: A hazardous location in which the hazardous substance is normally present in the air in sufficient quantities to cause an explosion or ignite the hazardous materials.

Division 2 location: A hazardous location in which the hazardous substance is not normally present in the air in sufficient quantities to cause an explosion or ignite the hazardous materials.

dry location: A location which is not normally damp or wet.

dry-type transformer: A transformer which provides air circulation based on the principle of heat transfer.

dust-ignitionproof: Enclosed in a manner which prevents the entrance of dusts and does not permit arcs, sparks, or excessive temperature to cause ignition of exterior accumulations of specified dust.

dustproof: Construction in which dust does not interfere with the successful operation of the equipment.

dusttight: Construction that does not permit dust to enter the enclosing case under specified test conditions.

dwelling: Contains eating, living, sleeping space, and permanent provisions for cooking and sanitation.

dwelling unit: A dwelling with one or more rooms used by one or more people for housekeeping.

dynamic load: A load that produces a small but constant vibration.

E

Edison-base fuse: A plug fuse that incorporates a screw configuration which is interchangeable with fuses or other ampere ratings.

effectively grounded: Grounded with sufficient low impedance and current-carrying capacity to prevent hazardous voltage buildups.

electrical metallic tubing (EMT): A lightweight tubular steel raceway without threads on the ends.

electric-discharge lighting fixture: A lighting fixture that utilizes a ballast for the operation of the lamp.

enclosure: The case or housing of equipment or other apparatus which provides protection from live or energized parts.

equipment: Any device, fixture, apparatus, appliance, etc. used in conjunction with electrical installations.

equipment bonding jumper (EBJ): A conductor that connects two or more parts of the EGC.

equipment grounding conductor (EGC): An electrical conductor that provides a low-impedance path between electrical equipment and enclosures and the system grounded conductor and GEC.

equipotential plane: An area in which all conductive elements are bonded or otherwise connected together in a manner which prevents a difference of potential from developing within the plane.

explosionproof apparatus: Equipment which is enclosed in a case that is capable of withstanding any explosion that may occur within it, without permitting the ignition of flammable gases or vapors on the outside of the enclosure.

exposed: As applied to wiring methods is on a surface or behind panels which allows access.

F

feeder: All circuit conductors between the service equipment or the source of a separately derived system and the final branch-circuit overcurrent device.

feeder neutral load: The maximum unbalance between any of the ungrounded conductors and the grounded conductor.

fitting: An electrical system accessory that performs a mechanical function.

flash point (fire point): The temperature at which liquids give off vapor sufficient to form an ignitable mixture with the air near the surface of the liquid.

flexible cable: An assembly of one or more insulated conductors, with or without braids, contained within an overall outer covering and used for the connection of equipment to a power source.

flexible cord: An assembly of two or more insulated conductors, with or without braids, contained within an overall outer covering and used for the connection of equipment to a power source.

flexible metal conduit (FMC): A raceway of metal strips which are formed into a circular cross-sectional raceway.

fuse: An overcurrent protection device with a fusible link that melts and opens the circuit when an overload condition or short circuit occurs.

G

general-purpose branch circuit: A branch circuit that supplies a number of outlets for lighting and appliances.

generator: A device that is used to convert mechanical power to electrical power.

ground: A conducting connection between electrical circuits or equipment and the earth.

grounded: Connected to the earth or a conducting body connected to the earth.

grounded conductor: A conductor that has been intentionally or grounded.

ground fault: An unintentional connection between an ungrounded conductor and any grounded raceway, box, enclosure, fitting, etc.

grounding conductor: The conductor that connects electrical equipment or the grounded conductor to the grounding electrode.

grounding electrode conductor: The conductor that connects the grounding electrode(s) to the grounded conductor and/or the EGC.

grounding receptacles: Receptacles which include a grounding terminal connected to a grounding slot in the receptacle configuration.

group: An atmosphere containing flammable gases or vapors or combustible dust.

H

hazardous location: A location where there is an increased risk of fire or explosion due to the presence of flammable gases, vapors, liquids, combustible dusts, or easily-ignitable fibers or flyings.

health care facility: A location, either a building or a portion of a building, which contains occupancies such as, hospitals, nursing homes, limited or supervisory care facilities, clinics, medical and dental offices, and either movable or permanent ambulatory facilities.

hermetic refrigerant motor-compressor: A combination of a compressor and motor enclosed in the same housing, having no external shaft or shaft seals, with the motor operating in the refrigerant.

high-intensity discharge (HID) lighting fixture: A lighting fixture that generates light from an arc lamp contained within an outer tube.

I

identified: Recognized as suitable for the use, purpose, etc.

immersion detection circuit interrupter (IDCI): Circuit interrupter designed to provide protection against shock when appliances fall into a sink or bathtub.

impedance: The total opposition to the flow of current in a circuit.

indicating device: A pilot light, buzzer, horn, or other type of alarm. Often, the wiring of a motor control circuit is very elaborate and requires ten times the amount of wiring as the motor power circuit.

individual branch circuit: A branch circuit that supplies only one piece of utilization equipment.

in sight from: Visible and not more than 50′ away.

instantaneous-trip CBs (ITBs): CBs with no delay between the fault or overload sensing element and the tripping action of the device.

instrument transformer: A transformer used to reduce higher voltage and current ratings to safer and more suitable levels for the purposes of control and measurement.

insulated conductor: A conductor covered with a material classified as electrical insulation.

intermittent load: A load in which the maximum current does not continue for three hours.

interrupting rating: The maximum amount of current that an OCPD can clear safely.

intrinsically safe system: A system with an assembly of intrinsically safe apparatus and associated apparatus which is interconnected and used in hazardous locations to supply equipment.

inverse-time CBs (ITCBs): CBs with an intentional delay between the time when the fault or overload is sensed and the time when the CB operates.

isolated-ground receptacles: Receptacles in which the grounding terminal is isolated from the device yoke or strap.

isolation transformer: A transformer that utilizes a shield between the primary and secondary windings and a transformer ratio of 1:1 to ensure that the load is separated from the power source.

J

junction box: A box in which splices, taps, or terminations are made.

K

kick: A single bend in a raceway.

L

labeled: Equipment acceptable to the AHJ to which a label has been attached.

lampholders: Devices designed to accommodate a lamp for the purpose of illumination.

less-flammable liquid: An insulating oil which is flammable but has reduced flammable characteristics and a higher fire point.

lighting and appliance branch-circuit panelboard: A panelboard with more than 10% of its branch-circuit fuses or CBs rated at 30 A or less (15 A, 20 A, 25 A, or 30 A).

lighting outlet: An outlet intended for the direct connection of a lampholder, a lighting fixture, or pendant cord terminating in a lampholder.

lighting outlets: Outlets that provide power for lighting fixtures.

lighting track: An assembly consisting of an energized track and lighting fixture heads which can be positioned in any location along the track.

line surge: A temporary increase in the circuit or system voltage or current that may occur as a result of fluctuations in the electrical distribution system.

liquid-filled transformer: A transformer that utilizes some form of insulating liquid to immerse the core and windings of the transformer to aid in the removal of heat generated by the transformer windings.

liquidtight flexible metal conduit (LTFMC): A raceway of circular cross section with an outer liquidtight, nonmetallic, sunlight-resistant jacket over an inner helically-wound metal strip.

liquidtight flexible nonmetallic conduit (LTFNMC): A raceway of circular cross section in one of three types.

listed: Equipment or material approved by the AHJ in a list.

luminaire: A complete lighting fixture consisting of the lamp or lamps, reflector or other parts to distribute the light, lamp guards, and lamp power supply.

M

main bonding jumper (MBJ): The connection at the service equipment that ties together the EGC, the grounded conductor, and the GEC.

metal-clad cable (MC): A factory assembly of one or more conductors with or without fiber-optic members.

mil: .0001″.

milliampere (mA): $\frac{1}{1000}$ of an ampere (1000 mA = 1 A).

mineral oil: A chemically untreated insulating oil that is distilled from petroleum.

motor branch circuit: The point from the last fuse or CB in the motor circuit out to the motor.

motor control center (MCC): An assembly of one or more enclosed sections with a common power bus and primarily containing motor control units.

motor control circuit: The circuit of a control apparatus or system which carries electric signals directing the performance of the controller, but does not carry the main power current.

multifamily dwelling: A dwelling with three or more dwelling units.

multiple receptacle: A single device with two or more receptacles.

multioutlet assembly: A metal raceway with factory-installed conductors and attachment plug receptacles. A flush or surface raceway which contains conductors and receptacles.

multiwire branch circuit: A branch circuit with two or more ungrounded conductors having a potential difference between them, and is connected to the neutral or grounded conductor of the system.

N

nipple: A short piece of conduit or tubing that does not exceed 24″ in length.

nominal voltage: Any voltages within an acceptable range.

nonadjustable-trip CBs (NATCBs): Fixed CBs designed without provisions for adjusting either the ampere trip setpoint or the time-trip setpoint.

noncoincidental loads: Loads that are not on at the same time.

nonflammable liquid: A liquid that is noncombustible and does not burn when exposed to air.

nongrounding receptacles: Receptacles with two wiring slots for branch-circuit wiring systems that do not provide an equipment grounding conductor.

nonlinear load: A load where the wave shape of the steady-state current does not follow the wave shape of the applied voltage.

nonmetallic-sheathed cable (NM): A factory assembly of two or more insulated conductors having an outer sheath of moisture-resistant, flame-retardant, nonmetallic material.

non-time delay fuses (NTDFs): Fuses that may detect an overcurrent and open the circuit almost instantly.

O

offset: A double bend in a raceway, each containing the same number of degrees.

one-family dwelling: A dwelling with one dwelling unit.

outlet: Any point in the electrical system where current supplies utilization equipment.

outlet box: A box which houses a piece of utilization equipment.

overcurrent: Any current in excess of that for which the conductor or equipment is rated.

overlamping: Installing a lamp of a higher wattage than for which the fixture is designed.

overload: A small-magnitude overcurrent, that over a period of time, leads to an overcurrent which may operate the overcurrent protection device (fuse or CB).

oxidation: The process by which oxygen mixes with other elements and forms a type of rust-like material.

oxide: A thin, but highly resistive coating that forms on metal when exposed to the air.

P

panelboard: A single panel or group of assembled panels with buses and overcurrent devices, which may have switches to control light, heat, or power circuits.

panic hardware: Door hardware designed to open easily in an emergency situation.

parallel conductors: Two or more conductors that are electrically connected at both ends to form a single conductor.

pendants: Hanging light fixtures that use flexible cords to support the lampholder.

permanently-connected appliance: A hard-wired appliance that is not cord- and plug-connected.

phase-to-ground voltage: The difference of potential between a phase conductor and ground.

phase-to-phase voltage: The maximum voltage between any two phases of an electrical distribution system.

pilot device: A sensing device that controls the motor controller.

place of assembly: A building, structure, or portion of a building designed or intended for use by 100 or more persons.

plug fuse: A fuse that uses a metallic strip which melts when a predetermined amount of current flows through it.

potential transformer: A transformer which steps down higher voltages while allowing the voltage of the secondary to remain fairly constant from no-load to full-load conditions.

power panelboard: A panelboard with more than 10% of its branch-circuit fuses or CBs rated over 30 A or more.

premises wiring: Basically all interior and exterior wiring installed on the load side of the service point or the source of a separately derived system.

pull box: A box used as a point to pull or feed electrical conductors into the raceway system.

R

raceway: A metal or nonmetallic enclosed channel for conductors.

raceway system: An enclosed channel of metal or nonmetallic materials used to contain the wires or cables of an electrical system.

readily accessible: Capable of being reached quickly.

receptacle outlets: Outlets that provide power for cord- and plug-connected equipment.

receptacles: Contact devices installed at outlets for the connection of cord-connected electrical equipment.

redundant grounding: Grounding with two separate grounding paths.

rigid metal conduit (RMC): A conduit made of metal. It is the universal raceway.

rigid nonmetallic conduit (RNMC): A conduit made of materials other than metal.

S

self-grounding receptacles: Grounding type receptacles which utilize a pressure clip around the 6 × 32 mounting screw to ensure good electrical contact between the receptacle yoke and the outlet box.

separately derived system: A system that supplies premises with electrical power derived or taken from storage batteries, solar photovoltaic systems, generators, transformers, or converter windings.

service: The electrical supply, in the form of conductors and equipment, that provides electrical power to the building or structure.

service conductor: A conductor that extends from the service point to the service disconnecting means.

service conductors: The conductors from the service point or other source of power to the service disconnecting means.

service drop: The conductors that extend from the overhead utility supply system to the service-entrance conductors at the building or structure.

service-entrance cable (SE): A single or multiconductor assembly with or without an overall covering.

service-entrance conductors – overhead systems: Conductors that connect the service equipment for the building or structure with the electrical utility supply conductors.

service-entrance conductors – underground systems: Conductors that connect the service equipment with the service lateral.

service equipment: All of the necessary equipment to control the supply of electrical power to a building or a structure.

service lateral: The underground service conductors that connect the utility's electrical distribution system with the service-entrance conductors.

service mast: An assembly consisting of a service raceway, guy wires or braces, service head, and any fittings necessary for the support of service-drop conductors.

service point: The point of connection between the local electrical utility company and the premises wiring of the building or structure.

short circuit: The unintentional connection of two ungrounded conductors that have a potential difference between them. The condition that occurs when two ungrounded conductors (hot wires), or an ungrounded and a grounded conductor of a 1ϕ circuit, come in contact with each other.

single receptacle: A single contact device with no other contact device on the same yoke.

special permission: The written approval of the authority having jurisdiction.

step-down transformer: A transformer with more windings in the primary winding, which results in a load voltage which is less than the applied voltage.

step-up transformer: A transformer with more windings in the secondary winding, which results in a load voltage which is greater than the applied voltage.

strut-type channel raceway: A surface raceway formed of moisture-resistant and corrosion-resistant metal.

supervised installation: An electrical installation in which the conditions of maintenance are such that only qualified persons monitor or service the electrical equipment.

surface raceway: An enclosed channel for conductors which is attached to a surface.

switch: A device, with a current and voltage rating, used to open or close an electrical circuit.

switchboard: A single panel or group of assembled panels with buses, overcurrent devices, and instruments.

T

temperature rise: The amount of heat that an electrical component produces above the ambient temperature.

thermally protected: Designed with an internal thermal protective device which senses excessive operating temperatures and opens the supply circuit to the fixtures.

time delay fuses (TDFs): Fuses that may detect and remove a short circuit almost instantly, but allow small overloads to exist for a short period of time.

torque: A turning or twisting force, typically measured in foot pounds (ft-lb).

transformer: A device that converts or "transforms" electrical power at one voltage or current to another voltage or current.

transformers: Electrical devices that contain no moving parts and are used primarily to convert electrical power at one voltage and current rating to another voltage and current rating.

two-family dwelling: A dwelling with two dwelling units.

Type I building: A building in which all structural members (walls, columns, beams, girders, trusses, arches, floors, and roofs) are constructed of approved noncombustible or limited-combustible materials.

Type II building: A building that does not qualify as Type I construction in which the structural members (walls, columns, beams, girders, trusses, arches, floors, roofs, etc.) are constructed of approved noncombustible or limited-combustible materials.

Type S fuse: A plug fuse that incorporates a screw and adapter configuration which is not interchangeable with fuses of another ampere rating.

U

unfinished basement: The portion of area of a basement which is not intended as a habitable room, but is limited to storage areas, work areas, etc.

unit switch: A switch designed to control a specific unit load.

utilization equipment: Any electrical equipment which uses electrical energy for electronic, electromechanical, chemical, heating, lighting, etc. purposes.

V

voltage-to-ground: The difference of potential between a given conductor and ground.

W

wet location: Any location in which a conductor is subjected to excessive moisture or saturation from any type of liquids or water.

wireway: A metallic or nonmetallic trough with a hinged or removable cover designed to house and protect conductors and cables.

within sight: Visible and no more than 50′ from the object.

INDEX

A

A/C and refrigeration equipment, 284–292
access, 18
accessible, 5–6
AC/DC general-use snap switch, 88–89
AC general-use snap switch, 88
adjustable trip CBs, 116, 129
aluminum, 105
ambient temperature, 7, 110
American Wire Gauge (AWG), 14
ampacity, 7, 109–112
appliance, 299, 320–325
  branch circuit, 138
    ratings, 321
  disconnecting means, 323–324
  installation requirements, 320
  markings, 325
  motor-driven, 324
  overcurrent protection, 321
  safety provisions, 324
  specific, 321–322
approved, 8
armored cable, 56–58
Askarel, 210
  -insulated transformers, 220
authority having jurisdiction (AHJ), 4–5
autotransformers, *211*
auxiliary gutters, 69–*70*

B

back-fed devices, 94, 96
backfill, 28
bare conductor, 8
bathroom, 145
bend, 48–*49*
bonding, 186–191
  over 250 Volts, 188
  services, 187
box, 70
box supports, 76–77
branch circuit, 137–141
  calculations, 365–368
  maximum loads, *151*
  permissible loads, 151–152
  ratings, 138, 148–152, 365
  voltage limitations, 141–143
building, *8*
buried splices and taps, 28–*29*
bushing, 48
busways, 67–69
  branches from, 68–69
  reduction in size of, *69*

C

cable, 102–*103*
cable assemblies, 56–63
cable tray systems (CTS), 63–65
cadwelding, 69
calculations, 365–386
  branch-circuit, 365–367
  multifamily dwellings, 380–386
  one-family dwellings, 368–379
cartridge fuses, 126–127
  markings, 127
circuit breaker (CB), 83, 122–*123*
  markings, 129–130
circular mils (CM), 14
clamp fill, 75–76
clear spaces, 17
CO/ALR snap switches, 89
conductor, 8, 14, 101–109
  bare, 101–*102*
  color code, 106–107
  construction, 104
  covered, 102
  fill, 75
  insulated, 102
  protection from physical damage, 25–*26*
  sizes, 14, 108
  solid, *106*
  stranded, *106*
  support methods, 34
  taps, 115
  temperature limitations, 103–*104*
  weight, 104
conduit body, 72–*74*
conduit seals, 337–341
continuous load, 9, 94
cooling, 15
copper, 104
copper-clad aluminum, 105–106
cover, 27
covered conductor, 9
crawl spaces, 146
current-limiting fuses, 127
current transformers, 213

D

damp location, 12, 179
dead front, 318
demand, 365
demand factor, 158
device, *9*, 316
device box, 70
device fill, 76
division, 332
Division 1 location, 332–334, 336–354
Division 2 location, 332–334, 336–337, 340–342, 344–346, 348–352, 357
dry location, 12
dry-type transformers, 205–208
duct wiring, 36
dust-ignitionproof, 341
dustproof, 337
dusttight, 337
dwelling, *10*
dwelling unit, *10*, 139, 145, 152
dynamic load, 77

E

Edison-base fuses, 125
effectively grounded, 166
EGC fill, 76
electrical
  connections, 16
  continuity of metal raceways and enclosures, 30–*31*
  installations, 14–18
    approval, 14
electrical metallic tubing (EMT), 49–*50*
electric-discharge lighting fixture, 314
  ballasts, 314
enclosures, 85, 177
entrance, 18
equipment, 14–15, 179
equipment bonding jumper (EBJ), 189
equipment grounding conductors, 107–108, 168–*169*, 196–198
equipotential plane, 186
explosionproof apparatus, 335
exposed, 10

F

feeder, 137, 158–160
  neutral load, 375
fine print notes (FPNs), 3
fire pumps, 292
fittings, *11*
fixed equipment, 179
flashpoint, 210
flexible cables and cords, 299

- ampacities, 301
- installation, 302
- markings, 299–301
- overcurrent protection, 301–302
- splices, 304
- terminations, 304–305
- uses not permitted, 303–304
- uses permitted, 302–303

flexible metal conduit (FMC), 49–*51*
floor boxes, 77
flush and recessed fixtures, 312–314
- clearance and installation, 312–313
- markings, 314
- wiring, 313

fuses, 115, 122–*123*
- voltage ratings, 127

G

garage and accessory buildings, 145–146
gasoline and service stations, 355–358
- installation requirements, 356

general purpose branch circuit, 138
generators, 176–177, 291–292
- portable, 176–177
- vehicle-mounted, 177

ground, 166
grounded, 166
grounded conductors, 83–*84*, 167–*168*
ground fault, 13, 114
ground-fault circuit-interrupter protection for personnel, 144–148
grounding, 165–198
grounding conductor, *168*
grounding electrode conductor, 170, 194–196
grounding electrode system, 191–194
grounding enclosures, 87–88
grounding systems, 170–177
- AC, 170–173
- DC, 173–174

ground movement, 28

H

hazardous locations, 331–350
- Class I location, 331–336, 338–340, 342–349, 351–354, 356–357
- Class II location, 331–337, 341, 343–345, 348–350
- Class III location, 331–332, 334, 337, 341, 344–345, 349–350
- grounding, 349–350
- group classification, 334
- lighting fixtures, 346–349
- motors and generators, 344–346
- wiring methods, 335–337

hazardous shutdown, 114
headroom, *18*
health care facilities, 358
hermetic refrigerant motor-compressor, 284
- ampacity and ratings, 285
- disconnecting means, 286
- markings on controllers, 285
- markings on nameplate, 285

high-intensity discharge (HID) lighting fixture, 314

I

identified, 11
illumination, *18*
immersion detection circuit interrupter (IDCI), 324
impedance, 166
indicating device, 274
individual branch circuit, 138
indoor wet locations, 29–*30*
in sight from, *11*
instantaneous-trip CBs (ITBs), 129
instrument transformer, 213
insulated conductor, 9
intermittent load, 9
interrupting rating, 130
intrinsically safe systems, 350–351
inverse-time CBs (ITCBs), 115, *129*
isolation transformers, 213

J

junction box, 72–*73*

K

kick, 48–*49*
kitchens, 147
knife switches, 85–86

L

labeled, 11
lamp holders, 307
length of conductors at outlet box, 32
less-flammable liquid, 210
lighting and appliance branch-circuit panelboard, 93
lighting fixtures, 305–316
- flushed and recessed, 312–314
- locations, 306
- supports, 307
- wiring, 309–312

lighting outlets, 12–*13*, 157
lighting track, 315–316
- installation, 315–316
- mounting, 316
- prohibited uses, 316
- support requirements, 316

line surge, 165
liquid-filled transformers, 209–210
liquidtight flexible metal conduit (LTFMC), 52–53
liquidtight flexible nonmetallic conduit (LTFNMC), 52–53
listed, 11–12
locations, *12*
luminaire, 305

M

main bonding jumper (MBJ), 188–*189*
mechanical and electrical continuity of conductors, 31
mechanical continuity of raceways and cables, 31
mechanical installation, 15
metal-clad cable (MC), 58–*59*
metallic wireways, 65–66
mil, 14
mineral oil, 210
motor, 269–284
- AC adjustable, 270
- branch circuit, 271
  - sizing conductors, 271
  - sizing disconnects, 274
  - sizing fuses and CBs, 274
  - sizing overload protection, 272–274
  - sizing raceways, 272
- control center (MCC), 279
  - grounding, 279
  - marking, 280
  - overcurrent protection, 279
- control circuits, 274
- controller, 277
  - ratings, 277
- disconnecting means, 280–283
- FLC, 269–270
- markings, 270–271
- multispeed, 283
- synchronous, 284
- torque, 270
- wound rotor, 284

mounting, 15
multifamily dwelling, *10*
multioutlet assembly, 55–*56*, 364
multiple receptacle, 150
multiwire branch circuit, 138–139

N

National Electrical Code (NEC®), 1
- correlating committee, 2
- definitions, 5–6

deletions, 2–3
enforcement, 4
exceptions, 3
extracted text, 2–3
future use, 5
process, 1
proposals, 1–2
purpose and intent, 4
revisions, 2–3
scope, 4
National Fire Protection Agency (NFPA), 1
annual meeting, 2
nipple, *47*
nominal voltage, 13
nonadjustable-trip CBs (NATCBs), 129
noncoincidental loads, 371
nonflammable liquid, 210
nonlinear load, *139*
nonmetallic-sheathed cable (NM), 60–62
nonmetallic wireways, 66–67
non-time delay fuses (NTDFs), 126

O

offset, 48–*49*
oil-insulated transformers, 220–221
one-family dwelling, *10*
outdoors, 146
outlet, 12–*13*, 70–72
outlet box, 70
overcurrent, 13, 112–*113*
devices, 122–130
protection, 112–116
ampere ratings, 115–116
protection device location, 116–122
grounded conductor, 116–*117*
ungrounded conductor, 116
overhead service-drop conductors, 245–250
clearances, 246–249
size and rating, 246
overhead service locations, 256–257
drip loops, 257
point of attachment, 257
overlamping, 312
overload, 13, 112, 289
oxidation, 210
oxide, 105

P

panelboard, 89–96
construction specifications, 96
grounding, 94, *96*
wire bending space, 96
panic hardware, 223
parallel conductors, 108–*109*
pendant boxes, 77
pendants, 303
permanently-connected appliance, 323
phase-to-ground voltage, 165
phase-to-phase voltage, 165
pilot device, 274
place of assembly, 358–360
plug fuses, 123–125
position of knife switch, 85–86
potential transformers, 213
power panelboard, 93
premises location, 121–122
preventing heating effect in metallic parts, 34–35
conductors grouped together, *34*
single conductors passing through metal, 35
preventing spread of toxic fumes, 36–38
protection against corrosion, 29–*30*
indoor wet locations, 29–*30*
protection of conductors, 114
protection of equipment, 114
pull box, 72–73

R

raceway, 13, 45
exposed to different temperatures, *30*
installations, 33–34
systems, 45–56
used as means of support, 31
readily accessible, 7
receptacle outlets, 12–*13*
receptacles, 316–320
GFCI-type, 318
grounding, 316–317
hospital-grade, 318
isolated-ground, 317–*318*
nongrounding, 317
self-grounding, 317
receptacles and cord connectors, 143–144
removing devices in multiwire circuits, 32
replacing receptacles, 143–144
required boxes, conduit bodies, or fittings, 32–34
required outlets, 152–157
rigid metal conduit (RMC), 45–47
rigid nonmetallic conduit (RNMC), 47–49
room A/Cs, 290–291

S

securing and supporting raceways, cable, assemblies, boxes, and cabinets, 31
securing integrity of fire-resistant-rated walls, *35*
separately derived systems, *174*
service, 237–241
service and repair garages, 351
classifications, 351–353
equipment, 353–355
wiring methods, 353
service conductors, 121, 137, 177, 238
service drop, 239
service-entrance cable (SE and USE), 62–63
service-entrance conductors, 251–257
– overhead systems, 238
protection, 254–255
size and rating, 251–252
supports, 255–256
– underground systems, 239
wiring methods, 254
service equipment, 137, 178, 241, 258–262
AIR ratings, 258
disconnecting means, 259
ground fault protection of equipment, 262
identification, 258
overcurrent protection, 260–261
service lateral, 239–*240*
service limitations, 241–245
service mast, 249–*250*
service point, *240*
short circuit, 13, 113
single receptacle, *150*
snap switches, 89
spacing conductor supports, 24
Standards Council Issuance, 2
step-down transformer, 205
step-up transformer, 205
strut-type channel raceway, 54–*55*
supervised installation, 224
support fitting fill, 76
supporting conductors in a vertical raceway, 34
surface raceway, 53–*54*
switchboard, 89–93
clearances, 92–93
grounding, 92
switch connections, 83–84
switches, 83–89

T

tap rules, 117–121
temperature rise, 214
thermally protected, 312
time delay fuses (TDFs), 126–127
torque, 105
transformer, 174, 205–229
grounding, 215
guarding, 213–*214*
installation, 213–221

location, 215–216
overcurrent protection, 223–229
vaults, 221
doorways, 223
location, 222
ventilation, 223
ventilation, 214–*215*
two-family dwelling, *10*
Type I building, 218
Type II building, 218
Type S fuses, 125

U

underground bushings, 28
underground installations, 27–*29*
minimum burial depths of cables and raceways, *27*–28
protection from damage, 28
underground service-lateral conductors, 250–251
size and rating, 251
unfinished basement, 147
unit switch, 324
usage of equipment, 14
utilization equipment, 138, 299, 320

V

voltage
nominal, 13
to ground, 13

W

wet bar sinks, 147–*148*
wet location, 12, 85, 102, 179
wireways, 65–67
wiring
in ducts for dust, loose stock, or vapor removal, 36
in ducts or plenums used for environmental air, 36–37
in other spaces used for environmental air, 37
in spaces used for data processing systems, 37–38
wiring methods, 23–38
cables and nonmetallic tubing through metal framing members, 25
cables and raceways in shallow grooves, 25
cables and raceways parallel to framing members, 25
cables through spaces behind panels designed to allow access, 25
conductors of different systems, *24*
conductors of same circuit, 23–*24*
within sight, 323
working clearances, 16–*17*
work space clearance, 16

Z

zone classification system, 351